Springer Series in Geomechanics and Geoengineering

Series editor

Wei Wu, Universität für Bodenkultur, Vienna, Austria
e-mail: wei.wu@boku.ac.at

Geomechanics deals with the application of the principle of mechanics to geomaterials including experimental, analytical and numerical investigations into the mechanical, physical, hydraulic and thermal properties of geomaterials as multiphase media. Geoengineering covers a wide range of engineering disciplines related to geomaterials from traditional to emerging areas.

The objective of the book series is to publish monographs, handbooks, workshop proceedings and textbooks. The book series is intended to cover both the state-of-the-art and the recent developments in geomechanics and geoengineering. Besides researchers, the series provides valuable references for engineering practitioners and graduate students.

More information about this series at http://www.springer.com/series/8069

Wei Wu · Hai-Sui Yu
Editors

Proceedings of China-Europe Conference on Geotechnical Engineering

Volume 2

 Springer

Editors
Wei Wu
Institut für Geotechnik
Universität für Bodenkultur
Vienna, Austria

Hai-Sui Yu
Faculty of Engineering
University of Leeds
Leeds, UK

ISSN 1866-8755   ISSN 1866-8763   (electronic)
Springer Series in Geomechanics and Geoengineering
ISBN 978-3-319-97114-8   ISBN 978-3-319-97115-5   (eBook)
https://doi.org/10.1007/978-3-319-97115-5

Library of Congress Control Number: 2018949380

This Springer imprint is published by the registered company Springer Nature Switzerland AG
The registered company address is: Gewerbestrasse 11, 6330 Cham, Switzerland

Foreword

Nigh on a century after Karl Terzaghi published his epoch-making book "Erdbaumechanik auf bodenphysikalischer Grundlage" in 1925, geotechnical engineering has developed from its infancy to a full-fledged engineering discipline. Europe is the cradle of modern soil mechanics and geotechnical engineering. The old continent still hosts the finest researchers, engineers, contractors, and manufacturers in geotechnical engineering. However, the construction activities, as the driving force for research and innovation, have subsided considerably in Europe. A major player in the construction sector is China with its huge domestic market as well as its ambitious Belt and Road Initiative abroad. For many years, China has the most construction activities of the world with such impressive infrastructure projects such as the Three Gorges Dam, South–North Water Transport, High-Speed Railway. This conference will link the birthplace of modern soil mechanics with the country with most construction and research activities in geotechnical engineering. It offers a welcoming opportunity to take stock of the state-of-the-art practice and the current research trends in China, Europe, and beyond.

The responses following our call for papers were overwhelming and echoed well beyond China and Europe. We have received about 400 papers from 35 countries, which make this conference a truly international event. The contributions in this proceedings cover virtually all areas of geotechnical engineering including constitutive model; numerical simulation; micro–macro relationship; laboratory testing; monitoring, instrumentation, and field test; foundation engineering; underground construction; innovative geomaterials; environmental geotechnics; cold regions geotechnical engineering; transportation and hydraulic engineering; unsaturated soils in waste management and CO_2 storage; geohazards–risk assessment, mitigation, and prevention. The proceedings provide an excellent overview of the current geoengineering research and practice in China, Europe, and beyond.

I am indebted to the plenary speakers, session organizers, and authors. The generous support from the City of Vienna and PORR, Austria, represented by Dr. Schön Harald (CEO), is gratefully acknowledged. My co-workers in Vienna, Dr. Wang Shun, Dr. He Xu-Zhen, and Dr. Lin Jia, spent many hours reading and correcting the papers. Lastly, I am grateful to my dear wife Ms. Wang Jing-Xiu, who made every effort to make this conference a success.

April 2018   Wei Wu

Contents

Part VII: Foundation Engineering

Finite Element Analysis of Deep Excavation: A Case Study 885
Tugce Aktas, Oguz Calisan, and Erdal Cokca

**Excavation and Peripherical Earth Retaining Solutions
for a Building at the Intendente Square, in Lisbon, Portugal** 889
Vanessa Aleixo, Rui Tomásio, and Alexandre Pinto

Bearing Capacity of Uplift Piles with End Gates 893
Ahmed S. Al-Suhaily, Ahmed S. Abood, and Mohammed Y. Fattah

**Gravity Grouting and Its Future Alternative for Soil
Reinforcement Systems** 898
Mohammad Zahidul Islam Bhuiyan, Shanyong Wang,
Scott William Sloan, Daichao Sheng, and Liang Kee Ming

On Safety Alarm in Deep Excavations 902
Xue-shan Cao, Li-su E, and Xi-yang Lai

Case Study: Moxy Hotel Bored Piles Wall 906
Cláudia Carvalho and Alexandre Pinto

**Soil Characterisation Based on Pipejacking Parameters
and Spoil Characteristics** 910
Wen-Chieh Cheng, James C. Ni, and Jack Shuilong Shen

**Vertical Response of 2 × 2 Pile Group Under Rotating Machine
Induced Vibrations** 915
Shiva Shankar Choudhary, Sanjit Biswas, and Bappaditya Manna

Precast Concrete Piles in Europe – AARSLEFF's Experience 919
Piotr Dziadziuszko and Dariusz Sobala

Case Study: BIM and Geotechnical Project in Urban Area –
Infinity Tower 923
João Gondar and Alexandre Pinto

Study on Penetration Resistance of Bucket Foundation Breakwater
by Centrifuge Model Tests 927
Yunfei Guan and Yongyong Cao

On Deformation of Bridge Pile Foundations Adjacent to Excavations . . . 932
Jiaqi Guo, Xiao Zhang, Junhua Xiao, and Nan Wu

Experimental Study on Dynamic Penetration of OMNI-Max
Anchor in Clay 937
Congcong Han, Jun Liu, and Wei Zhao

Effect of Vertical Load on Lateral Behavior of Pile in Clay 942
Ben He, Kanmin Shen, Yi Hong, and Lizhong Wang

Dynamic Responses of Mono-piles in the Presence of Scour Holes 947
Rui He, Tao Zhu, Bo Ma, and Wenyan Tao

Random Field Simulation of Foundation Settlement of Soft Soil
in Southern China 952
Linchong Huang, Shuai Huang, and Yu Liang

Unified Theoretical Approach to Ultimate Horizontal Pullout
Capacity of Vertical Strip Anchor Plate in Sand 956
Wei Hu, Chengbi Long, Wenhua Gao, and Chaofeng Zeng

Joint Research into the Behaviour of Driven Piles 961
R. J. Jardine and Z. X. Yang

3D Settlement Analysis of Underpinning Piles Under Raft
Foundation Subjected to Nonuniform Vertical Loading 973
Volkan Kalpakcı, Şevki Öztürk, H. Murat Algın, and A. Burak Ekmen

Application of Semi-active Control Strategy for the Wall
Retaining Granular Fills 978
Nisha Kumari and Ashutosh Trivedi

Ground Improvement by the In-Tube Deep Dynamic
Compaction Method 983
Ping Li, Jianhui Tang, and Chengwen Fan

Field Study on Load Transfer Mechanism of XCC Pile Raft
Foundation 987
Ping Li, Chengwen Fan, Jianying Lai, Xuanming Ding, and Ke Cheng

Stability Analysis of L-Shape Slurry Trench During Concrete
Diaphragm Wall Installation in $c - \varphi$ Soils 992
Wei Liu, Peixin Shi, Qiang Tang, and Fei Wang

Experimental Studies on Model Single Pile and Pile Groups Subjected to Torque 997
Sagar Mehra and Ashutosh Trivedi

Use of Drainage Holes on Reinforced Concrete Pipe Piles to Accelerate Soil Consolidation 1001
Guoxiong Mei, Pengpeng Ni, Meijuan Xu, and Yanlin Zhao

Ageing Effect on Pull-Out Capacity of Driven Micropiles in Dunkirk Sand 1005
Imane Salama, Christophe Dano, Matias Silva, and Miguel Benz Navarrete

The Joint Inspector - A New Method of Quality Control for Diaphragm Walls 1009
Nikolaus Schneider

Analysis of Cofferdam Stability for Foundation of a Bridge Pillar 1013
Monika Sulovska and Jakub Stacho

Dynamic Response of Embedded Block Foundation Under Vertical Vibration 1017
Kavita Tandon, Rohit Ralli, B. Manna, G. V. Ramana, and M. Datta

Free Vibration Calculations of an Euler-Bernoulli Beam on an Elastic Foundation Using He's Variational Iteration Method 1022
Alima Tazabekova, Desmond Adair, Askar Ibrayev, and Jong Kim

Vertical Capacity of Large and Deep Barrette Pile for Bangkok Subsoils 1027
Wanchai Teparaksa, Mike Sinkinson, and Jirat Teparaksa

Application Examples of Elastic Support Layers for High-Rise Buildings in China 1032
Alexander Tributsch, Silke Appel, and Bertram Grass

CPT-Based Approach to Study the Load-Displacement Behavior of Driven Piles by the New Method of Stress Characteristics 1036
Fatemeh Valikhah, Abolfazl Eslami, and Mehdi Veiskarami

On Trench Construction of Diaphragm Wall in Medium-Coarse Sand: Slurry Composition and Construction Optimization 1041
Jingyu Wang, Jianguo Liu, Longlong Fu, Weitao Ye, and Guangwei Xu

Tunneling Induced Building Settlement Based on Variational Principle 1046
Xinjiang Wei, Xiao Wang, Xingfu Yu, and Lianying Zhou

Measured and Predicted Response of a Post-grouted Pile in Cohesionless Soil 1051
Xiong Xiao, Shanyong Wang, Scott Sloan, and Daichao Sheng

Parameter Identification of the High-Fill Foundations
Using Response Surface Method 1055
Ming Xu, Dehai Jin, and Erxiang Song

Distribution Patterns of Horizontal Foundation Modulus
Under Pre-excavation Dewatering 1059
Xiu-Li Xue, Miao-Kun Li, and Chao-Feng Zeng

New Method for Determining Foundation Bearing Capacity Based
on Plate Loading Test 1064
Guanghua Yang, Yan Jiang, Chuanbao Xu, Zhiyun Li, Fuqiang Chen,
and Kai Jia

Numerical Investigation of T-Shaped Soil-Cement Column
Supported Embankment Over Soft Ground 1068
Yaolin Yi, Pengpeng Ni, and Songyu Liu

Experiment Study on Wave Induced Excessive Pore Pressure
Around Near-Sea Foundation Pit 1072
Hong-wei Ying, Ding-ye Xu, and Cheng-wei Zhu

Centrifugal Model Tests on Working Behavior of Composite
Foundation Reinforced by Rigid Piles with Caps
Under Embankment 1077
Jian-Lin Yu, Jia-Nan Zhong, Jun-Yuan Li, Ri-Qing Xu,
and Xiao-Nan Gong

On Performance of Anchored Retaining Piles and Stability
of Excavations 1081
Xiaoyi Yuan, Longzhu Chen, and Chunyu Song

Responses of Retaining Wall to Pre-excavation Dewatering
Under Staggered Layout of Dewatering Wells 1085
Chao-Feng Zeng, Zhi-Cheng Yuan, Xiu-Li Xue, and Wei Hu

Experimental Study of Temperature Effect on Binding Properties
of Resin Anchors 1090
Zha Wenhua and Liu Zaobao

An Innovative Bolt Fastener for Steel Tube Bracing in Deep
Excavations 1094
Mingju Zhang, Meng Yang, Pengfei Li, and Dechun Lu

Seepage Stability Calculation of Excavation Base due
to Groundwater Level Fluctuation 1098
Lisha Zhang, Hongwei Ying, Di Wang, Kanghe Xie, and Cheng-wei Zhu

Design Charts and Simplified Approach for the Bearing Capacity
of Strip Footings on Sand Overlying Clay 1102
Haizuo Zhou, Gang Zheng, and Jiapeng Zhao

Field Tests on Behavior of Pre-bored Grouted Planted Pile in Soft
Soil Area with Existing Pile Foundation 1106
Jiajin Zhou, Xiaonan Gong, and Rihong Zhang

Effect of Vertical Shaft Resistance on the Lateral Behavior
of Large-Diameter Pile Foundation 1111
Ming-xing Zhu, Hong-qian Lu, Guo-liang Dai, and Zhi-hui Wan

Part VIII: Underground Construction

Tunnelling in Urban Environments: Protecting Sensitive Buildings 1117
Eduardo E. Alonso

Validation of an In-Situ Stress Level Rating Method 1128
Fei Chen and Jianhui Deng

Dynamic Response of Underground Structure Under Bidirectional
Shaking in Layered Liquefiable Ground 1132
Renren Chen, Rui Wang, and JianMin Zhang

Non-destructive Technology for Underground Utility Mapping:
A Case Study 1136
Kulin Dave and Silky Agrawal

Numerical Investigation on the Influence of the Excavation Rate on
the Mechanical Response of Deep Tunnel Fronts in Cohesive Soils 1140
Luca Flessati and Claudio di Prisco

Choosing Optimal Parameters of Mine Air Conditioning Systems 1144
A. F. Galkin and I. V. Kurta

Monitoring System for Previous Convergence Measurements
in Tunnel Construction 1149
Stephan Großwig and Maria-Barbara Schaller

Numerical Investigation of Impact of Non-circular Tunneling
in Sensitive Soft Clay Layers 1152
Tadashi Hashimoto, Yujian Liu, Teruo Nakai, Hossain Md. Shahin,
Yaohong Zhu, and Zibo Dong

Numerical Analysis of Prefabricated Column-Base Connections
in Tunnels 1156
Haixi Jiang, Longjin Li, Jie Cao, and Haitao Yu

Experimental Study on the Properties of Geocell-Reinforced
Embankments 1160
Lihua Li, Feilong Cui, Zhi Hu, and Henglin Xiao

Study on the Joint Bending Stiffness of Large-Diameter Shield Tunnel:
A Case Study 1164
Yu Liang and Linchong Huang

Coal Pillar Design by Numerical Simulation to Protect
an Inclined Tunnel 1168
Baoguo Liu, Yi Qi, Xiaomeng Shi, and Yan Wang

On Stability of Shallow Tunnel by Model Test and Numerical
Simulation 1172
Nader Moussaei, Mostafa Sharifzedeh, Kourosh Sahriar,
and Mohammad Hossein Khosravi

Semmering Base Tunnel, Austria 1177
Michael Proprenter and Enrico Soranzo

Track Differential Settlement in a Culvert-Embankment
Transition Zone Due to Adjacent Shield Tunneling: A Case Study 1181
Yao Shan, Yi Lu, Li Su, and Longlong Fu

Effects of Fault Zones on Failure Mechanism of Tunnel,
an Experimental Study 1186
Ba Thao Vu, Hehua Zhu, Qianwei Xu, and Xiaoying Zhuang

Field Study on the Behavior Super-Long and Large-Diameter
Grouted Drilled Shafts 1192
Zhihui Wan, Guoliang Dai, Weiming Gong, and Mingxing Zhu

Numerical Modeling of the Effects of Ground Fissure Dislocation
on Metro Tunnel 1197
Ming Wu, Jianbing Peng, Yahong Deng, and Xin Liu

Drift Ratio Limit for the Seismic Design of Underground Structures . . . 1201
Mian Xiao, Renren Chen, Rui Wang, and Jianmin Zhang

Numerical Analysis on Interaction Between Closely Spaced
Shield Tunnels 1206
Congcong Xiong, Binglong Wang, Zhi Liu, Weitao Ye, Wei Zhao,
and Jianbing Long

Hydraulic Response on Shield Tunnel Face Subjected to Water
Level Fluctuation 1211
Hongwei Ying, Hua Wei, Jinhong Zhang, and Chengwei Zhu

Undrained Stability of Tunnel with a Longitudinal Gradient 1216
Fei Zhang, Guangyu Dai, Hu Wang, Yufeng Gao, and G. H. Lei

Shaking Table Test of Tunnel-Shaft Junction in Shield Tunnel 1221
Jinghua Zhang, Haitao Yu, and Yong Yuan

Nonlinear Analysis on Buried Pipelines Effected by Tunnelling 1226
Chenrong Zhang and Haili Li

**Adequate Numerical Simulation of Tail Void Grouting
for Tunneling in Saturated Soil** 1230
Chenyang Zhao, Arash A. Lavasan, and Tom Schanz

**Torque Fluctuation and Penetration Analysis of Shield Tunnel
in Rock–Soil Mixed Ground** 1234
Yu Zhao, Quanmei Gong, Runlai Zhang, and Jie Xia

Semi-analytical Solutions on Seepage Field of Twin Tunnels 1239
Chengwei Zhu, Hongwei Ying, Xiaonan Gong, Huawei Shen,
and Xiao Wang

Analytical Solutions for Seepage Field of Underwater Tunnel 1244
Chengwei Zhu, Hongwei Ying, Xiaonan Gong, Huawei Shen,
and Xiao Wang

Part IX: Environmental Geotechnics

**Using a Complementary Evapotranspiration Relationship to Estimate
Surface Suction for Soil-Atmosphere Interaction Analysis** 1251
Hossein Assadollahi and Hossein Nowamooz

**Numerical Modeling of Leakage, Transport and Remediation
of Mixed DNAPL and LNAPL in the Unsaturated Clayey Soil
Underlain by Saturated Sandy Soil** 1256
Yuzhang Bi, Yanjun Du, Xingyuan You, Kaixuan Yuan, and Jin Ni

**Long Root Grasses in Pyroclastic Soils: Vegetation Growth
and Effects on Induced Soil Suction** 1260
Vittoria Capobianco, L. Cascini, and V. Foresta

**A Simple Model for Estimating Shear Strength
of Root-Soil Composite** 1264
Jiulong Ding, Faning Dang, and Songhe Wang

**Field Pilot Scale Ex-Situ S/S of Electroplating Industrial
Contaminated Soil Using Two Novel Binders** 1269
Yasong Feng, Yanjun Du, Weiyi Xia, and Krishna R. Reddy

**Three-Dimensional Numerical Analysis of the Air Phase Flow
During Air Sparging in Sands** 1274
Z. B. Liu, S. Y. Liu, Z. L. Chen, Y. Wang, L. L. Lu, and G. Y. Du

**Long-Term Performance of a Three-Layer Capillary Barrier Cover
System in Humid Climates** 1278
Jian Liu, Yuedong Wu, Rui Chen, and C. W. W. Ng

NaCl Activation of Steel Slag upon Component Adjustment 1282
Li Liu, Qianwen Liu, Yongfeng Deng, and Yu Zhao

Bioengineering for Slope Stabilisation Using Plants: Hydrological
and Mechanical Effects...................................... 1287
C. W. W. Ng, A. K. Leung, and J. J. Ni

Effect of Liner Consolidation on Contaminant Transport Through
a Landfill Bottom Liner System 1304
Hefu Pu, Jinwei Qiu, Junjie Zheng, and Rongjun Zhang

New Method to Reduce Porosity of Rockfill Materials
with Composite Slurry 1308
Tao Wang, Sihong Liu, and Yan Feng

Soil Nutrient Effects on Suction and Volumetric Water Content
in Heavily Compacted Vegetated Soil.......................... 1312
R. Tasnim, J. L. Coo, C. W. W. Ng, and V. Capobianco

Heat Transfer in a Geosynthetics Composite Liner System
Containing Wrinkles 1316
Mayu Tincopa, Abdelmalek Bouazza, and R. Kerry Rowe

Root Reinforcement to Stabilize Slopes 1320
Rick Veenhof and Wei Wu

Engineering Properties and Mechanism of Solidifying Materials
Treated Sludge.. 1324
Zhenhua Wang, Wei Xiang, Qingbing Liu, and Xueting Wu

Experimental Study on Effects of NaCl Solutions on Soil-Water
Characteristic Curves of Expansive Soil........................ 1328
Xiujuan Yang, Wojciech T. Sołowski, Henghui Fan, and Jinqian Dang

Part X: Cold Regions Geotechnical Engineering

A Double-Yield-Surface Model for Frozen Saline Sandy Soil
Incorporating Particle Crushing.............................. 1335
Dan Chang and Yuanming Lai

A New Strength Criterion for Frozen Clay Considering
Temperature Effect 1340
Dun Chen, Wei Ma, Yanhu Mu, Zhiwei Zhou, Dayan Wang, and Lele Lei

A Constitutive Model for Frozen Granular Soils 1345
Roberto Cudmani, Jian Sun, and Wei Yan

The Crystallization and Salt Expansion Characteristics
of a Silty Clay 1350
Jianhong Fang, Xu Li, Jiankun Liu, and Chenyinan Liu

**Experimental Study on Temperature Threshold for Warm Frozen
Sand in Terms of Mechanical Properties** 1355
Xueluan Guo, Junlin Zhao, Yansong Wang, and Jilin Qi

Freeze-Thaw Processes of Soils in Active Layers in Northeast China ... 1359
Ruixia He, Huijun Jin, and Xiaoli Chang

**Effect of Thermokarst Lake on Foundation Under Embankment
in Permafrost Regions** 1364
Xiaoying Hu, Yu Sheng, and Erxing Peng

**Experimental Study on Anti-frost Jacking of Belled Pile
in Seasonally Frozen Ground Regions** 1368
Xubin Huang and Yu Sheng

**Experimental Study on the Interaction Between Pipe and Soil
Under Frost Heave Condition** 1372
Long Huang and Yu Sheng

**A Novel Preparation Device for Remolded Hollow Cylinder
Specimen of Frozen Soil** 1376
Lele Lei, Dayan Wang, Yongtao Wang, Dun Chen, Yan Guo, and Wei Ma

**Centrifuge Model Test on Performance of Thermosyphon Cooled
Sandbags Supporting Warm Oil Pipeline Buried in Thawing
Permafrost** 1380
Guoyu Li, Hongyuan Jing, Nikolay Volkov, Wei Ma, and Fei Wang

Experimental Study on Creep Behavior for Thawed Saturated Clay ... 1385
Bo Lin, Feng Zhang, and Decheng Feng

**Compaction Behavior of Clay-Gravel Mixtures Under Normal
and Low Temperature** 1390
Yang Lu, Sihong Liu, Meng Yang, and Yonggan Zhang

**Experimental Study on the Strength Characteristics of Frozen Clay
on π Plane** 1394
Yanhu Mu, Wei Ma, Dun Chen, Zhiwei Zhou, Dayan Wang, and Lele Lei

**Experimental Study of the Mechanical Properties of Coarse-Grained
Soils from High Altitude and Cold Areas Under Freeze-Thaw Cycle** ... 1399
Yonglong Qu, Guoliang Chen, Fujun Niu, Wankui Ni, Yanhu Mu,
and Tao Chen

**Research on Bearing Capacity of Pile Foundation with Impact
of Subpermafrost Water** 1403
Xiangyang Shi, Dongqing Li, and Ze Zhang

Experimental Study on Compressibility of Frozen Saturated
Chinese Standard Sand 1407
Xiaoyu Sun, Yansong Wang, Junlin Zhao, and Jilin Qi

A Simple Equation for Predicting Freezing Point of Saline Soft Clay ... 1412
Qinze Wang, Songhe Wang, Jilin Qi, Fengyin Liu, and Peng An

Field Investigation on Thermo-Mechanical Behaviors
of Pre-melting Lime Pile in Permafrost Regions.................. 1416
Jiliang Wang and Chenxi Zhang

Study on Changes in Integrity Decay of Sandstone Subjected
to Freeze-thaw Cycling...................................... 1420
Liping Wang, Ning Li, Jilin Qi, Yanzhe Tian, and Shuanhai Xu

Influence of Warm Oil Pipeline on Underlying Permafrost and
Cooling Effect of Thermosyphon Based on Field Observations 1424
Fei Wang, Guoyu Li, Wei Ma, Yanhu Mu, Yuncheng Mao, and Bo Wang

Study on the Stability of Spread-Footing Foundations
on Permafrost Regions 1429
Zhi Wen, Zhizhong Sun, and Shujuan Zhang

Nonlinear Numerical Analysis of Thaw Consolidation of Ice
Rich Frozen Soil... 1433
Xiaoliang Yao, Boxiang Dang, and Jilin Qi

Cooling Performance of a Composite Embankment
for High-Grade Highways in Permafrost Regions 1438
Mingyi Zhang, Yuanming Lai, Wansheng Pei, Qihao Yu,
and Zhongrui Yan

Experimental Study of Elastic Properties of Saturated Clay
Subjected to Freeze-Thaw Cycles.............................. 1442
Feng Zhang, Bo Lin, Tao Li, and Decheng Feng

Numerical Modeling of Rate-Dependence Behavior of Saturated
Frozen Soil... 1447
Qiyin Zhu, Xiangyu Shang, Lianfei Kung, and Li Gang

Part XI: Geohazards-Risk Assessment, Mitigation and Prevention

Centrifuge Modelling of Slope Instability Due to Leakage
of Buried Pipes.. 1453
Kit Chan, Limin Zhang, Hong Zhu, and Te Xiao

On Geological Hazards in Georgia............................ 1458
Diana Egiazarova and Zurab Tchkonia

Numerical Modelling of Main Shock and Aftershock Line
of Chuya Earthquake 27.09.2003, Altay, Russia 1462
Mikhail Eremin and Pavel Makarov

Geotechnical Issues of Landslides in Ukraine: Simulation,
Monitoring and Protection 1466
Iurii Kaliukh, Gennadiy Fareniuk, and Iegor Fareniuk

Back Analysis Algorithm Based on Particle Swarm Optimization 1470
Abdoulie Fatty and A. J. Li

Seismic Performance of Geosynthetic-Reinforced Earth Retaining
Walls Subjected to Strong Ground Motions 1474
Domenico Gaudio, Luca Masini, and Sebastiano Rampello

Effect of Air Compression and Counterflow on Shallow Landslides
Under Intense Rainfall 1479
Tongchun Han, Shiguo Ma, and Riqing Xu

Upper Bound Stability Analysis of MSW Slope Layered by Fill
Age Considering Particle Compressibility 1483
Maosong Huang, Xinping Fan, Haoran Wang, and Xilin Lu

Modeling of Landslides and Assessment of Their Impact on
Infrastructure Objects in Ukraine 1487
Olena Ivanik

The Ultimate Lateral Soil Pressure on Stabilizing Piles in Slopes 1491
Guoping Lei and Wei Wu

Influence of Bulk Density and Slope on Debris Flows Deposit
Morphology: Physical Modelling 1495
Shuai Li, Xiaoqing Chen, Gongdan Zhou, Dongri Song,
and Jiangang Chen

Simulation of Building Failure by Landslide Impact 1500
Hongyu Luo and Limin Zhang

Characteristics and Application of Micropiles in Slope Engineering . . . 1504
Li Ma, Yifu Hu, Desheng Gu, and Chong Jiang

Influence of Wetting-Drying Cycle in Road Cut Slope in Loess
in Northwest China 1508
Yuncheng Mao, Guoyu Li, Wei Ma, Yanhu Mu, and Fei Wang

Numerical Investigation of Drilled Shafts Near an Embankment
Slope Under Combined Torque-Lateral Load Scenario 1512
Aigul Mussabayeva, Jong Kim, Deuckhang Lee, Taeseo Ku,
and Sung-Woo Moon

Evaluation of Residual Shear Strength of Landslide Reactivated Soil . . . 1516
V. Senthilkumar and S. S. Chandrasekaran

**Investigation of Rockfall Impact Against Gravel Cushion
via a Discrete Element Approach** 1521
Weigang Shen, Tao Zhao, Feng Dai, Jiawen Zhou, and Nuwen Xu

**Impact of Precipitation on Dissipation of Pore Pressure
in Colluvium of the Carpathian Flysch Landslide** 1526
J. Stanisz, P. Krokoszyński, and R. Kaczmarczyk

Impact of Root System on Soil Strength in Shallow Landslide 1530
J. Stanisz, Ł. Kaczmarek, T. Zydroń, A. Gruchot, T. Wejrzanowski,
and P. Popielski

**A Simple Method for Evaluating Progressive Failure Process
of Rainfall-Induced Shallow Landslide** 1535
Yang Tang, KunLong Yin, and Wei Wu

**Dimensionless Stability Charts for c-φ Slopes with Tension Cracks
Subject to Seismic Action** 1539
S. Utili and A. Abd

**Introduction to the Badong Field Test Site for Landslide Research
in the Three Gorges Reservoir Area of China** 1546
Jinge Wang, Wei Xiang, Aijun Su, and Chengren Xiong

**Factors Influencing Landslide Deformation from Observations
in the Three Gorges Reservoir** 1551
Beibei Yang, Suzanne Lacasse, Kunlong Yin, and Zhongqiang Liu

Modelling of Castaño Viejo Tailings Flow Case History 1556
Francisco Zabala, Gustavo Navarta, and Luciano A. Oldecop

**Soil - Structure Interaction at the Bogatići Landslide in Bosnia
and Herzegovina** 1561
Sabid Zekan, Mato Uljarević, Majda Mešić, and Alen Baraković

**Evaluation of Stability Analysis Methods of Embankments
on Soft Clays** 1565
Akzhunis Zhamanbay, Jong Kim, and Sung-Woo Moon

**Centrifuge Model Test on Excavation-Induced Failure of Soil
Slopes Overlying Bedrock** 1570
Yiying Zhao and Ga Zhang

Discrete Element Analyses of Earthquake-Induced Landslide 1574
Tao Zhao, Giovanni B. Crosta, and Nuwen Xu

Deformation and Instability Mechanism of Reservoir Landslide:
A Case Study 1579
Changjun Zhao and Minghui Xu

Effect of Slope Angle on Stabilizing Piles in C-φ Soil 1583
Mingxing Zhu, Hongqian Lu, Weiming Gong, and Zhihui Wan

Part XII: Unsaturated Soils and Energy Geotechnics

Energy Utilisation and Ground Temperature Distribution
of a Field Scale Energy Pile Under Monotonic and Cyclic
Temperature Changes 1591
Mohammed Faizal and Abdelmalek Bouazza

Thermo-Mechanical Behavior of Reinforced Concretes
for Energy Piles 1595
Wei Huang, Wei Xiang, and Jin Luo

Flow Behaviour of Fractured Geothermal Reservoir Rocks
Under In-Situ Stress and Temperature Conditions 1600
W. G. P. Kumari and P. G. Ranjith

Calculation of Osmotic Suctions for Bentonite in Saline Solutions 1604
Xiaoyue Li, X. J. Zheng, and Yongfu Xu

Principle of Effective Stress is an Approximated Method 1609
Chenggang Zhao, Zhaoyang Song, Jian Li, Guoqing Cai, and Weihua Li

Thermal Conductivity of Unsaturated Soil: Equivalent
Microstructure Approach 1613
Dariusz Łydżba, Adrian Różański, and Damian Stefaniuk

Modelling Hydro-Chemo-Mechanical Behaviour of Active Clays
Through the Fabric Boundary Surface 1618
Mario Manassero, Andrea Dominijanni, and Nicolò Guarena

Soil Texture Based Approach for Thermal Conductivity Evaluation . . . 1627
Adrian Różański

Experimental Study on Highly Compacted Bentonite Aggregates
Subjected to Wetting and Drying 1632
Haiquan Sun, David Mašín, and Jan Najser

Interpretation of Plate Load Tests on Unsaturated Sand 1636
Yi Tang and Zhengyin Cai

Thermo-Mechanical Behaviour of Inventory Materials
in a Packed Bed Thermal Energy Storage 1640
Xuetao Wang and Christoph Niklasch

A New Setup to Measure Hydraulic Properties of Unsaturated Soils . . . 1644
Tiande Wen, Longtan Shao, and Xiaoxia Guo

Part XIII: Geotechnics in Transportation, Structural and Hydraulic Engineering

Analytical Study on the Effect of Moving Surface Load
on Underground Tunnel 1651
Zhigang Cao, Si Sun, Zonghao Yuan, and Yuanqiang Cai

Influence of Matric Suction on the Long-Term Behavior
of Fouled Road Base Materials Under Traffic Loading 1656
Jingyu Chen, Yuanqiang Cai, Zhigang Cao, and Chuan Gu

Land Subsidence in Shanghai and Its Influence
on Transportation Infrastructure 1660
Weiwei Cao, Mingguang Li, Yujin Shi, Jinjian Chen, and Jianhua Wang

Geotechnical Properties of Phosphogypsum and Its Use
in Road Engineering 1664
Barbara Cichy, Cezary Kraszewski, and Leszek Rafalski

Mechanically Stabilized Earth Walls and Uneven Reinforcement
Lengths (Trapezoidal Walls) – Design Development and Challenges . . . 1668
Ching Dai and James Livingston

Field Test Study of Long-Term Displacement of Bridge Foundation
Subjected to Lateral Loads 1672
Guangming Yu, Weiming Gong, and Guoliang Dai

Geotechnical Solutions for Linear Transport Infrastructure
in Mining Areas 1677
Jacek Kawalec

Kinematics of Piled Embankments with Defective Piles 1682
Louis King, Abdelmalek Bouazza, and Stephen Dubsky

Seismic Behavior of Pile Supported Railway Track 1687
Jinsun Lee, Mintaek Yoo, and Yunwook Choo

Initial Sinking Method for Large Open Caisson in a Highway
Bridge Project 1692
Peng Li, Erxiang Song, and Tianliang Zheng

Mechanism of Isolating Piles in Reducing Tunnel Settlement
of Hong Kong-Zhuhai-Macao Bridge Project 1697
Peng Li, Erxiang Song, Abbas Haider, and Xiaodong Liu

Lateral Decompression Behaviors of a Hard Claystone
in Excavation-Damaged Zone of Galleries 1702
Zaobao Liu, Jianfu Shao, Shouyi Xie, and Nathalie Conil

**Dynamic Shakedown of Cohesive-Frictional Materials Under
Moving Traffic Load** 1707
Yuchen Dai, Jiangu Qian, Xiaoqiang Gu, and Maosong Huang

**Influence of Rock Joint Orientation on the Natural Frequency
of Dam-Foundation System** 1711
Ajay Rampal, Prasun Halder, Bappaditya Manna, and K. G. Sharma

**Sheet Pile Bridge Abutments: Faster, Economic and Viable
Solution for Urban Transportation Needs** 1716
Abhishek Jain and Anirban Sen

**Robust Geotechnical Design of Transition Zone Material
Parameters Based on Fuzzy Set Theory** 1721
Yao Shan, Li Su, and Yi Lu

**Tentative Investigation of Structure Size Effect of High-Filled
Geotechnical Structures** 1726
Erxiang Song, Tianliang Zheng, and Yufei Kong

**Numerical Analysis of Stone Columns for Road Embankment
Construction** 1730
Jakub Stacho, Jana Frankovska, and Peter Mušec

**Using the FE-Method for Lock Repair Measures
at the Main-Danube-Canal** 1734
Oliver Stelzer and Annette Richter

**Performance Evaluation of Ballast-Subballast Interface Stabilized
with Geogrids** 1738
Kumari Sweta and Syed Khaja Karimullah Hussaini

**The Relation Between Static and Dynamic Shakedown Limits
of Slab Track Substructures Under Moving Train Loads** 1742
Juan Wang and Shu Liu

**Modelling Stress Distribution in Subgrade Due to Construction
of Enlarged Embankment** 1746
Heng Wang, Cong Mou, Jianwen Ding, and Xing Wan

**Geosynthetic Strain During Filling Stage of Geosynthetic-Reinforced
Pile-Supported (GRPS) Embankments** 1750
Haiyun Yan, Binglong Wang, and Chengyu Liu

**Design Improvement of Sandy Soil Levees in Hydro-engineering
Projects, China** 1754
Chunbao Yang, Feng Zhu, An Zhang, and Shuli Jiang

Deflection Mechanism and Stress Analysis of Large Steel Structure Derrick in Mine 1759
Zhi-shu Yao, Ming-kai Liu, Xiao-jian Wang, Bin Tang, and Wei-pei Xue

Numerical Simulation on Pore Pressure in Electro-Osmosis Combined with Vacuum Preloading 1763
Tianjiao Zhang, Fanglei Zhan, Jian Zhou, Cunyi Li, and Xiaonan Gong

Author Index... 1767

Part VII: Foundation Engineering

Finite Element Analysis of Deep Excavation: A Case Study

Tugce Aktas[1,2(✉)], Oguz Calisan[2], and Erdal Cokca[1]

[1] Middle East Technical University, Ankara, Turkey
tugceaktas@calisangeo.com
[2] Calisan Geotechnical, Ankara, Turkey

Abstract. World population increases day by day. In order to meet this increasing population's basic needs such as sheltering, transportation and social activities, necessity for underground structures gradually increase. As a result, deep excavations become more important each passing day while excavations deepen. In order to design the optimum system which ensures the safety of environment, roads, structures and the excavation, the excavation properties should be modelled so that it reflects the actual geometry and boundary conditions. The ground parameters, which have crucial effect on the solution, should also be determined precisely. In this study, a deep excavation with a depth of 33.0 m performed on over-consolidated Ankara Clay is investigated. The excavation system was designed to be consisted of double-row of piles supported by multilevel anchors. Inclinometer and load cell installations on the shoring elements have been made for the inspection of the pile wall displacements and anchor forces. Back analysis is made in order to obtain the ideal ground parameters which give the measured displacements and anchor forces. This analysis is conducted by a series of Plaxis analysis in which ground parameters are changed iteratively in order to find their values that give the measured displacements and forces. For these Plaxis analyses, three different models namely; Hardening Soil, HS small and Mohr Coulomb are used and ground parameters are determined. Finally, ground parameters obtained by back analysis for different soil models and the parameters determined from field tests are compared and a correlation is proposed to simulate the behavior of over consolidated Ankara clay.

Keywords: Deep excavation · Back analysis · Ankara Clay
Ground displacement

1 Introduction

A new residence and business center has been built in Ankara, which is the capital of Turkey. The construction of the 42 stories of this building with 8 basements and one ground floor required an excavation ranging from 20.0 m to 33.0 m in depth. Excavation is conducted mainly on Ankara clay which is an overconsolidated clay with reddish brown color. The Ankara clay is composed of clayey, sandy and gravely levels in variable thicknesses also there are layers of lime formations in many locations. The presence of fissures and lime in the soil make it most difficult to obtain undisturbed samples and

© Springer Nature Switzerland AG 2018
W. Wu and H.-S. Yu (Eds.): *Proceedings of China-Europe Conference on Geotechnical Engineering*, SSGG, pp. 885–888, 2018.
https://doi.org/10.1007/978-3-319-97115-5_1

to prepare them for laboratory testing. Ordemir et al. [1] reported that determining in-situ shear strength of Ankara clay is a difficult task because of the stiffness of clay; furthermore, standard penetration tests will not give reliable results in the estimation of in situ shear strength. In order to correlate stiffness parameters of Ankara clay, excavation with depth of 33.0 m height is monitored; then, back analysis is performed on Ankara clay by a series of plaxis analysis. Main idea of this procedure is to determine stiffness parameters of the Ankara clay corresponding to displacements obtained from the site.

2   Geological Assessment of the Study Area

SPT-N and laboratory test results obtained from undisturbed samples that are taken from appropriate depths are available for the construction site. The excavation system has been designed by design engineer to be consisted of double-row of piles supported by multilevel anchors. This system is composed of 80 cm diameter piles spaced 1.0 m center to center. In order to verify the design assumptions and to monitor the performance of the deep excavation, inclinometer and load cells were installed on the side. Lateral deformation of the pile system is measured as 1.2 cm for maximum excavation depth (33.0 m) by the inclinometer. In this paper, back analysis of this very deep excavation is performed only for the maximum excavation depth. Idealization of the soil profile was made according to SK-4 and SK-7 boreholes that are near the section under consideration. According to these boreholes, site consist of only one layer, Ankara Clay, with total depth up to 50.0 m. Figure 1 shows the idealized soil profile and average geotechnical parameters.

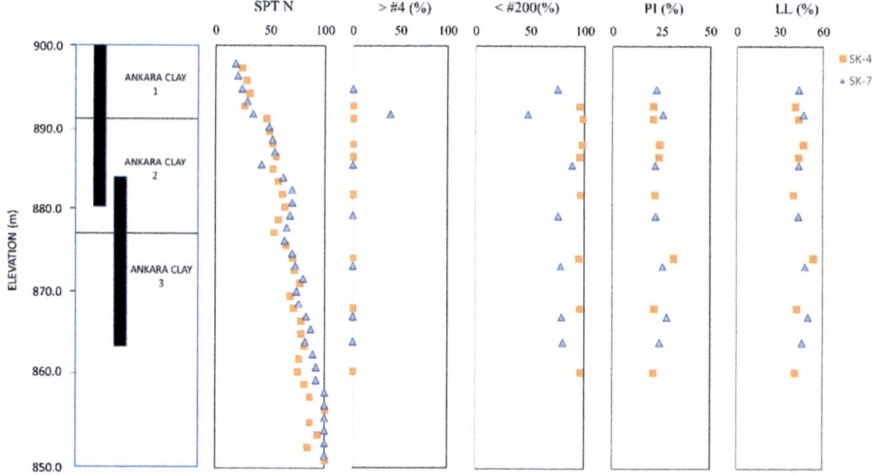

Fig. 1. Idealized Soil Profile and average geotechnical parameters (SPT N, blow numbers of standard penetration test; >#4, variation of sand percent with elevation; <#200, variation of clay percent with elevation; PI, plasticity index; LL, liquid limit).

3   Finite Element Analysis

A finite element analysis programme called Plaxis was used to undertake the analysis of the deep excavation. Three different constitutive models were employed in the analysis. Those are the simple Mohr-Coulomb failure criteria (MC), the standard Plaxis hardening soil model (HS) and the hardening soil small model (HSS). For all models, soil is assumed to be drained inside the site. This assumption is considered appropriate for intermittent bored piles where ingression of groundwater occur between the piles. MC, HS and HSS models use different type of stiffness modulus parameters; however, strength parameters are the same for all models. Strength parameters of the over consolidated Ankara clay is determined according to Sorensen and Okkels [2] in which authors correlate peak angle of shearing resistance and effective cohesion with respect to plasticity index as follows;

$$4 < PI(\%) < 50 \quad \varphi'_{oc} = 45 - 14.\log PI \text{ (deg)} \tag{1}$$

$$7 < PI(\%) < 30 \quad c'_{oc} = 30 \text{ (kPa)} \tag{2}$$

4   Discussion and Results

E' and E_{50}^{ref} parameters are obtained by the back-analysis. E' is an important input parameter of the MC model and similarly E_{50}^{ref} is an important input parameter of the HS and HSS models. A series of Plaxis analysis were conducted by changing E' and E_{50}^{ref} values while keeping input parameters of MC and HS models same; however, G_{ref}^0 parameter is also changed when altering E_{50}^{ref} value without changing other input parameters for the analysis in HSS model. Results of parametric studies are listed in Tables 1, 2 and 3 measured lateral wall displacement is obtained by the stiffness values $E' = 6200N$ for MC model, $E_{oed}^{ref} = E_{50}^{ref} = 5200N$ for HS model and $E_{oed}^{ref} = E_{50}^{ref} = 3140N$ for HSS model.

Table 1. MC model parameters

Layer	γ_u (kN/m^3)	φ' ($°$)	ν	E' (kPa)
Ankara Clay-1	19.0	25	0.3	155000
Ankara Clay-2	20.0	25	0.3	334800
Ankara Clay-3	23.0	25	0.3	545600

Table 2. HS model parameters

Layer	γ_u (kN/m^3)	φ' ($°$)	E_{50}^{ref} (kPa)	E_{oed}^{ref} (kPa)	E_{ur}^{ref} (kPa)
Ankara Clay-1	19.0	25	130000	130000	390000
Ankara Clay-2	20.0	25	280800	280800	842400
Ankara Clay-3	23.0	25	457600	457600	1372800

Table 3. HSS model parameters

Layer	$\gamma_{0.7}$	G_{ref}^0 (kPa)	E_{50}^{ref} (kPa)	E_{oed}^{ref} (kPa)	E_{ur}^{ref} (kPa)
Ankara Clay-1	0.00020	195750	78500	78500	235000
Ankara Clay-2	0.00020	423000	169000	169000	507000
Ankara Clay-3	0.00022	689000	275000	275000	826000

5   Conclusions

In this study the back analysis of 33.0 m deep excavation in Ankara clay is performed by using three constitutive soil models of Plaxis 2D and deformation modulus parameters are determined for different soil layers. Calculated stiffness parameters can be sorted (from largest to smallest) as MC, HS and HSS which means that for the same stiffness MC will give the highest displacement. Therefore, we conclude that HS and HSS provide more accurate results. Although similar results are expected for HS and HSS models, HSS model's dependency on $\gamma_{0.7}$ and G_{ref}^0 parameters makes the difference.

References

1. Ordemir, I.: Report on Ankara Clay. Faculty of Engineering, Ankara (1965)
2. Sorensen, K.K., Okkels, N.: Correlation between drained shear strength and plasticity index of undisturbed overconsolidated clays. In: Proceedings of the 18th International Conference on Soil Mechanics and Geotechnical Engineering, Paris (2013)

Excavation and Peripherical Earth Retaining Solutions for a Building at the Intendente Square, in Lisbon, Portugal

Vanessa Aleixo[✉], Rui Tomásio, and Alexandre Pinto

JetSJ-Geotecnia, Lda., Lisbon, Portugal
{valeixo,rtomasio,apinto}@jetsj.com

Abstract. In this paper are presented the main design criteria for the excavation and earth retaining wall solutions developed for the construction of the underground floors of a building located ta Rua do Benformoso, near to Intendente Square, in Lisbon down town, Portugal. Two different earth retaining wall solutions were developed considering local constraints, mainly geological features, topography, urban envelope and building architecture. The first solution is a Berlin-type wall braced by temporary ground anchors and steel struts. The second solution is a conventional retaining wall with reinforced concrete buttresses. This solution was performed at the garden back area due to the impossibility of using permanent ground anchors.

Keywords: Earth retaining wall · Urban excavation · Berlin-type walls

1 Introduction

The site construction area is about 1795 m^2 and was previously occupied by a residential building for many years. Before the excavation works started, intersecting the Lisbon Miocene soft rocks, mainly marls and weathered sandstones, the original masonry building was demolished, and only the main facades and some walls were preserved.

The new reinforced concrete building structure, without basements, has 6 upper floors and roof. The geometry of the new building has a "U" shape, with a garden area at back side.

Due to the existent topography, the site is located at the base of an urban hill, as well as the new building floor levels, the following solutions were adopted: a multi temporary ground anchored peripheral Berlin type wall and internal classical reinforced concrete wall with buttresses in order to better face the architectural demands at the garden back area.

2 Main Constraints

The main constraints related to the studied earth retaining solutions were the dense and old urban environment, where the construction area is located, the facades retention towers, as well as the geotechnical, geological and topographical scenarios.

© Springer Nature Switzerland AG 2018
W. Wu and H.-S. Yu (Eds.): *Proceedings of China-Europe Conference on Geotechnical Engineering*, SSGG, pp. 889–892, 2018.
https://doi.org/10.1007/978-3-319-97115-5_2

3   Earth Retaining Solutions

3.1   Peripheral Areas: Berlin - Type

Urban construction often involves the execution of underground floors. Due to existing surrounding structures and infrastructures, as well as the existent topography, vertical excavations, supported by earth retaining walls, had to be executed. In such scenario, Berlin-type (king post with reinforced concrete cast in situ panels) walls are one of the most suitable techniques. This technique takes advantage of the staged construction, allowing to minimize walls thickness and back structures and infrastructures displacements. Under this perspective, it is important that the design guidelines are accomplished and, mainly, the excavation stages. This technique consists in phased retaining wall execution, from top to bottom, of 0,30 m net thickness reinforced concrete panels, supported by vertical tubular micropiles with high strength steel. The panels are casted directly against the excavated soil face and braced by temporary ground anchors [1] and temporary steel struts. At the final stage, the new building slabs will assure the stability of the retaining walls, and the ground anchors and struts are deactivated.

3.2   Garden Back Area: Conventional Walls

Due to the impossibility of using permanent ground anchors, reinforced concrete traditional concrete walls with "L" shape with buttresses, founded over tubular steel micropiles, were used at the garden back area at the front of the Berlin type walls (see Fig. 1).

Fig. 1.   Retaining walls view at the Garden area (Berlin type wall finished)

The buttresses with 0,30 m thickness are 7,80 m high and executed with 3,0 m of distance. Considering the irregular geometry of the excavation area, the buttresses width is variable, and, in some cases, a cap beam connects the buttresses to the Berlin type retaining wall, improving the overall earth retaining system stiffness.

The concrete walls were executed according to the traditional method, from the bottom to the top after the excavation and execution of the concrete wall foundation (see Fig. 1).

4   Numerical Analysis

In order to study the retaining walls and the neighboring constructions behavior a numerical analysis using FEM Plaxis 2D software was performed (see Fig. 2). This analysis allows the simulation of all the construction phases and was used to calibrate the Monitoring and Survey Plan.

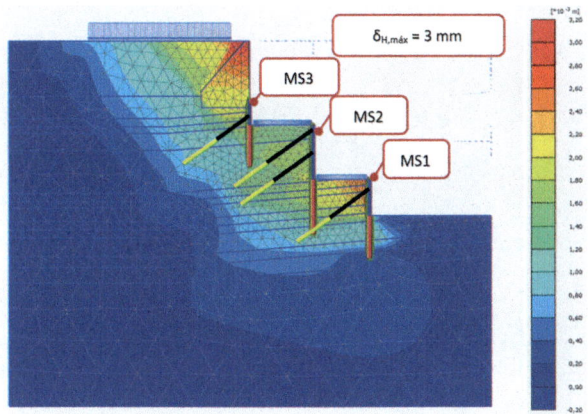

Fig. 2.   Software Plaxis output - Horizontal displacements (last excavation phase)

The analysis carried out consisted in the study of the most representative and conditioning section in order to evaluate stresses, strains, as well as the stability of the retaining walls. Through this analysis the displacements of the retaining walls were estimated, as well as the ground settlements, the tensile forces at the temporary ground anchors and the micropiles axial loads. Notwithstanding, elements such capping beams and spread foundations were analyzed considering simplified models, using concepts of the classical theory of elastic bars as well as strut and tie reinforced concrete models.

The external stability of the traditional concrete walls with "L" shape with buttresses was performed considering the classical ultimate limit states approach for this type of structures, mainly: bearing capacity, sliding and overturning.

5   Monitoring and Survey Plan

To ensure the execution of the excavation works in safe and economic conditions, a Monitoring and Survey Plan was performed. For the design of the Monitoring Plan it was considered the complexity of the works as well as the sensibility of the neighboring structures and infrastructures. The Monitoring Plan provided the following data: vertical and horizontal displacements of the retaining wall; vertical and horizontal displacements of the neighboring buildings; ground water table depth; ground anchors load; vibrations at the neighboring buildings when the excavation works intersected the Miocene bed rock (stiffer marls and sandstones).

6   Main Conclusions

The excavation work presented in this paper proves, for the existing and demanding restraints, including the local geotechnical and topographic scenarios, the efficiency and versatility of the Berlin-type wall technique. A global view after the excavation work is shown in Fig. 3.

Fig. 3. Global view after excavation work after the execution of the earth retaining walls

The adoption of this solution has many advantages, mainly: the use of equipment with reduced dimension and high versatility, the possibility to perform the final wall simultaneously with the excavation, as well as the possibility to adjust the peripherical retaining solution during the excavation works. When intersecting softer soils, the ground should be previously improved with small diameter soil cement columns.

Reference

1. Bustamante, M., Doix, B.: Une méthode pour le calcul de tirants et des micropieux injectés. Bulletin de Liaison des Laboratoires des Ponts et Chaussées, Ministère de L'Équipement, du Logement, des Transports et de la Mer, Paris, n°140, pp. 75–92 (1985)

Bearing Capacity of Uplift Piles with End Gates

Ahmed S. Al-Suhaily[1(✉)], Ahmed S. Abood[2], and Mohammed Y. Fattah[1]

[1] University of Technology, Baghdad, Iraq
Ahmed_suhaily@yahoo.com
[2] Senior Engineer (Civil), Ministry of Defense, Baghdad, Iraq

Abstract. The pile base plays an important role in increasing its bearing capacity. Several studies have been carried out to produce types of piles which are enlarged at their bases. Therefore, it was thought in this study to design a mechanism on enlarged base for the pile at a selected depth so that it does not intersect with pile driving process. The manufactured pile is square open-ended from both sides and provided with two or four gates. The gates are linked on the pile shaft at the base by steel hinges which allow the gate to rotate at an angle of (90°). The pile is driven alone while the spreader (inner Shaft) is kept inside the pile using a hammer. The driving in this stage is stopped when the penetration of the pile and spreader is equal. At the end of this stage, the spreader is extruded from the pile shaft and the pile becomes ready for loading test. This type of piles (steel piles with gates) does not demand reaching the strong soil layer or knowing its depth, because this type of piles is suitable for soft to medium soils and it provides very good results, so that the required pile length will be shorter than in ordinary piles. This means lower cost and time but with additional efforts.

Keywords: Pile · Dry sand · Tension · Enlarged base · Model

1 Introduction

Piles supporting high structures, such as tall chimneys, transmission towers, water towers, tents, electric poles, silos are required to resist uplift force due to wind. So, piles are designed to resist this tensile uplift force. Resistance to uplift is given by the friction between the pile and the surrounding soil plus weight of pile itself. The uplift resistance of vertical pile can be computed similar to friction piles. Uplift piles are invariably provided with an enlarged area at the base in the form of a bell or a bulb. Piles develop resistance to pull-out only from the skin friction developed along the embedment length. Point bearing is not included, but weight of the pile is included in uplift capacity [1].

The production of steel piles with enlarged bases is not widely used in the field of piling. Therefore, it was thought in this study to design a mechanism on enlarged base for the pile or widening at a selected depth so that it does not intersect with pile driving process or its penetration to soil layers. Subsequent paragraphs, however, are indented.

© Springer Nature Switzerland AG 2018
W. Wu and H.-S. Yu (Eds.): *Proceedings of China-Europe Conference on Geotechnical Engineering*, SSGG, pp. 893–897, 2018.
https://doi.org/10.1007/978-3-319-97115-5_3

2   Experimental Work

2.1   Preparation of Soil Bed and Pile Design Details

In the present study eight tension pile models were carried out, these tests were carried out in poorly graded sandy soil. A hole was excavated with a dimension of (1.5×1) m^2 and 1.1 m depth, the hole was braced and covered with polythene sheet in order to prevent moisture form the surround soil, the sand was placed in to the hole with layers using the raining technique in order to achieve the desired unit weight 17.11 kN/m^3 until reach 900 mm height, the surface was leveled with leveling ruler.

The manufacture pile is open-ended from both sides and provided with two gates. It consists of a pipe of square cross-section with dimensions of (30×30) mm, 1.5 mm thick and 600 mm long. The gates are linked on the pile shaft at the base by steel hinges 2 mm thick which allow the gate to rotate at an angle of 90°.

A steel shaft closed at both ends was manufactured to drive the soil inside. The open pile, it consists of a square steel section for a pile of square with dimensions of (24×24) mm with 480 mm length, and 1.5 mm.

2.2   Pile Implementation

A steel frame was designed to be used in driving the piles, the steel frame was placed over the pile. The pile was fixed in order to keep the pile vertical during driving process.

The driving process was carried out in three stages; Stage (1): the pile is driven in to the soil by the free fall of 5 kg weight from 300 mm height in addition to the weight of the shaft 5.5 kg. The driving continues until reaching the desired depth. Stage (2): an inner shaft was lubricated with oil and placed inside the pile and driven with pile in order to open the gates. Stage (3): when the desire depth is maintained, the inner shaft is removed. The piles and inner shaft are shown in Fig. 1, while Table 1 presents a summary of the testing program.

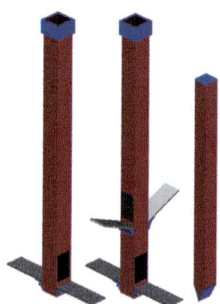

Fig. 1. The piles and inner shaft used in tests.

Table 1. Summary of tension pile models.

Case	Type	Number of gates	Length of one gate	Width of the gate	Abbreviation
1	Ordinary square pile with opened end	/	/	/	SPO1
2	Ordinary square pile with closed end	/	/	/	SPC1
3	Square pile with opened end	2	6 cm (2B)	3 cm (B)	SPG1
4	Square pile with opened end	2	9 cm (3B)	3 cm (B)	SPG2
5	Square pile with opened end	4	6 cm (2B)	3 cm (B)	SPG3
6	Square pile with opened end	4	9 cm (3B)	3 cm (B)	SPG4
7	Square pile with opened end	4	6 cm (2B)	3 cm (B)	SPG5
8	Square pile with opened end	4	9 cm (3B)	3 cm (B)	SPO6

3 Testing and Results

After finishing the driving of the pile, tension pile cap apparatus was placed on the pile and fixed with a bolt and attached to tension cable, the pulleys are placed over the steel frame and fixed with bolts in order to place the cable over the pulleys, the other end of the cable is attached to loading base which is used to carry loads. The tensile load test is carried out according to ASTM [2] quick test by applying load with increment of 5% of the anticipated failure load and using time interval of 15 min between each load increment. The results of pull out tests are shown in Fig. 2. In order to calculate the increment ratio in pull out resistance, T_{ult}, the following equation is used:

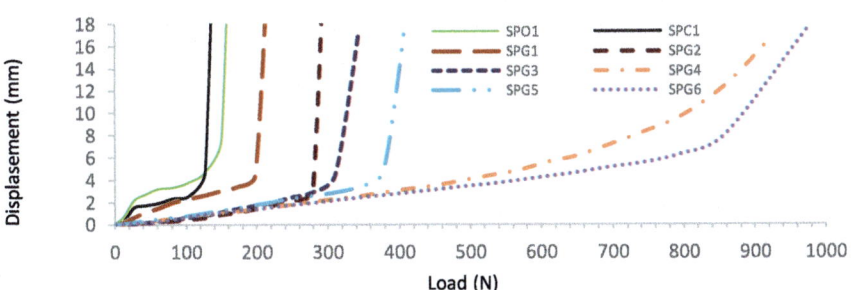

Fig. 2. Results of pull out tests.

$$\% \text{ Increasing in } T_{ult} = \frac{T_{ult} \text{ of pile \& gate} - T_{ult} \text{ of ordinary pile}}{T_{ult} \text{ of treated sand by pile}} \times 100 \qquad (1)$$

The main objective of using gate piles is to increase the surface area under the soil (A_u) in order to provide additional weight to resist the pull-out force. In the present study, the surface area under the soil is function of θ_1 (θ_1 the average angle between the pile and the gate of the lower gates), θ_2 (the average angle between the pile and the gate of the middle gates) and A (area of the gate). Table 2 illustrates the average of θ_1 and θ_2 and the surface area under the soil, the following equation shows the calculation of the surface area under the soil resulting from the opening of gates:

$$A_u = 2.A.\sin\theta_1 + 2.A.\sin\theta_2 \qquad (2)$$

Table 2. Summary of increment ratio in pull out resistance, values of θ and surface area of soil.

Case	Type of pile	Dry unit weight of soil, (kN/m^3)	$T_{ult.}$ (N)	Increase in $T_{ult.}$, %	Area of one gate	θ_1	θ_2	Surface area under soil
Case (1)	SPO1	17.11	53	/	/	/	/	B^2
Case (2)	SPC1	17.11	52	/	/	/	/	B^2
Case (3)	SPG1	17.11	115	117	$2 B^2$	50°	–	$3.1 B^2$
Case (4)	SPG2	17.11	278	425	$3 B^2$	60°	–	$5.2 B^2$
Case (5)	SPG3	17.11	282	432	$2 B^2$	60°	20°	$4.8 B^2$
Case (6)	SPG4	17.11	390	636	$3 B^2$	70°	30°	$8.6 B^2$
Case (7)	SPG5	17.11	320	504	$2 B^2$	70°	30°	$5.8 B^2$
Case (8)	SPO6	17.11	430	711	$3 B^2$	80°	40°	$9.8 B^2$

It can be noticed that as the angle θ_1 increases, the surface area under the gate increases and hence the pile base resistance increases. This component of resistance is not considered in ordinary tension piles.

4   Conclusions

1. Design and production of a type of steel pillars with gates with a mechanism for the methods of this type of substrates consists of a spreader closed ends used for the purpose of tapping the soil inside the pillars. The substrate shaft function at this stage is pushing the soil inside the substrate towards the bottom and towards the lower sides of the substrate (gate opening), thus causing the main purpose of the research work (opening the portals of the substrate) which increases the loading area and finally increases the tolerance of the substrate (Bearing Capacity). Spreader also limits the soil when hammered and works as a barrier to prevent the rise of soil in the substrate (Plugging) in all cases.

2. The main purpose of using a substrate with gates is to increase the surface area for loading and thus increase substrate tolerance. The tests and results proved the success of the work by opening the gates at an angle ranging from 20–80° and for the different types of substrates, thus increasing the loading area from (B^2 to $9.8\,B^2$) by an increase of 880%, and increasing the increment ratio in pull out resistance from (53 N to 430 N) by an increase of 711% for (Case 8). The gate opening angle depends on the opening of the gates in the substrate Spreader and the amount of energy applied in the dimming process are proportionately proportional to each other.

References

1. ASTM D 3689-07: Standard Test Methods for Deep Foundations Under Static Axial Tensile Load. American Society for Testing and Materials (ASTM) (2010)
2. Chaudhary, A., Goswami, P., Sahu, A.K.: Effect of skin resistance and enlarged base on pull out capacity of modeled piles. Electron. J. Geotech. Eng. EJGE **21**, 7513–7516 (2016). Bund. 23

Gravity Grouting and Its Future Alternative for Soil Reinforcement Systems

Mohammad Zahidul Islam Bhuiyan[1(✉)] ⓘD, Shanyong Wang[1], Scott William Sloan[1], Daichao Sheng[1], and Liang Kee Ming[2]

[1] ARC Centre of Excellence for Geotechnical Science and Engineering, Faculty of Engineering and Built Environment, University of Newcastle, Callaghan, Australia
mohammadzahidulislam.bhuiyan@uon.edu.au
[2] Liang United Engineering Studio, Kuala Lumpur, Malaysia

Abstract. Gravity grouting technique is commonly used in soil nailing and ground anchorage systems to increase pull out capacity of soil inclusions. Bond strength in between soil-grout interface estimates the pull out capacity of a grouted soil nail/ anchor. The bond strength improvement due to gravity grouting and pressure grouting is very limited and grout likely shrinks after setting, resulting in reduction of skin friction between cement grout and surrounding soil of drill hole. One of the major concerns of soil nailing techniques is excessive lateral movement or creep behaviour over the service life and a case study of instrumented ground anchor wall reported that gravity grouted soil reinforcement technique experience excessive creep behaviours. The application of fracture grouting technique in soil nailing is very new and presumably it not only provides drill hole expansion but also provides mechanical interlocking between the penetrating grout and surrounding soil, which could resist the creep behaviour of soil-nails as well as enhance the bond resistance. The application of fracture grouting in soil nailing system could also be a cost-effective method since it likely to increase the pullout resistance of soil-nails, resulting in reduction of the number of soil-nails.

Keywords: Gravity grouting · Soil nailing · Fracture grouting

1 Introduction

Soil nailing is a reinforcement technique used to reinforce in-situ soil to stabilize it more effectively and economically. Among different types of soil nailing techniques, nowadays drilled and gravity grouted soil nail system is commonly used in soil stabilization and reinforcement since it provides higher design capacity compared to driven/jacked nail systems. Bond strength in between soil-grout interface, the mobilized shear resistance, estimates the pull out capacity of a grouted soil nail as well as assesses the internal stability of soil-nailed structures, which controls the design and deformation of a soil nailing system (especially, grouted soil nailed wall for excavation). A comprehensive literature review reported that the interaction behaviour of soil-nail is complex and influenced by different uncertainties present in actual field conditions [1]. Method of installation and pre-drilling greatly influence the interface shear capacity and it damages

© Springer Nature Switzerland AG 2018
W. Wu and H.-S. Yu (Eds.): *Proceedings of China-Europe Conference on Geotechnical Engineering*, SSGG, pp. 898–901, 2018.
https://doi.org/10.1007/978-3-319-97115-5_4

the influence of overburden pressure. Therefore, pull capacity completely depends on the acting normal load around the soil nail instead of applied surcharge load. The bond strength improvement due to gravity grouting and pressure grouting is very limited and grout likely shrinks after setting, resulting in a reduction of skin friction between cement grout and surrounding soil of drill hole. Hence, the effectiveness of this type of grouting is not as satisfactory as expected.

2   A Case Study of Gravity Grouted Anchor Wall

Figure 1 illustrates the lateral displacement of the instrumented anchored wall for different depths over the period of construction in which the deflection was measured from an in-pile inclinometer installed at a specific bored pile along the retaining wall. Overall, the plots demonstrate that wall is moving towards the excavation gradually with sudden jumps in the displacement for each layer of excavation (about 3.3 m) and in some cases the jumps are much higher than expected value (5–10 mm). It is obvious that the top of CBP (contiguous bored pile) wall is moving over the time instead of being stopped (Fig. 2). This might be happened due to the creeping failure of 1st layer of installed ground anchor. However, the historical displacement behaviour (Fig. 1) shows a very steady rate of increment of displacement over the period except a sharp rise on 15

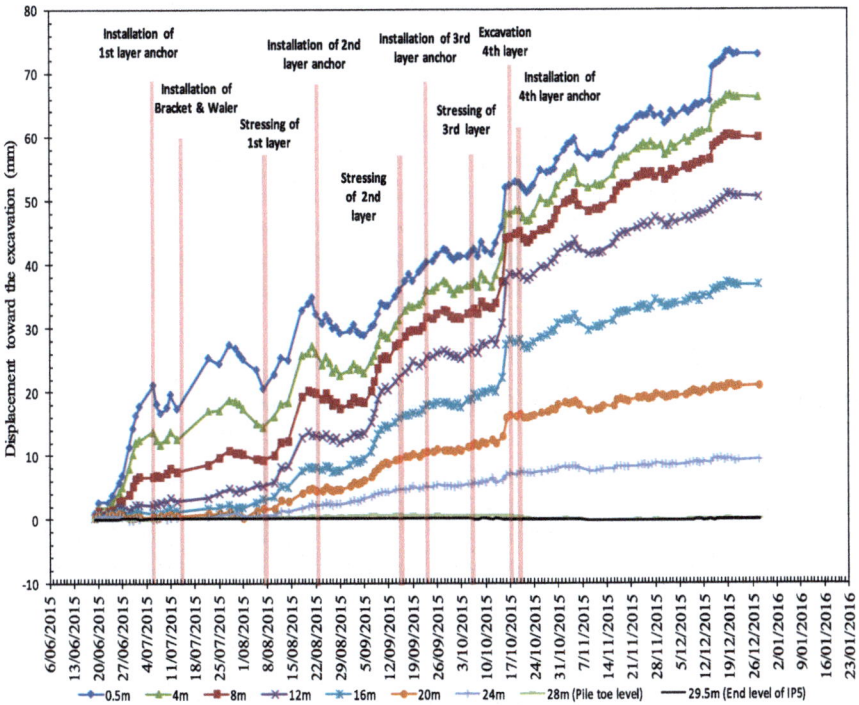

Fig. 1. History of lateral displacement at different excavation depth

October 2015 that is expected for 4th layer of excavation. Even after the installation and stressing of the final layer of anchor (4th layer), the wall illustrates a gradual increment of deflection over the post construction period instead of being stabilized/stopped. This wall deflection behaviour demonstrates the creeping of installed ground anchored and this was justified by the load cell reading of a specific anchor installed at 2nd layer.

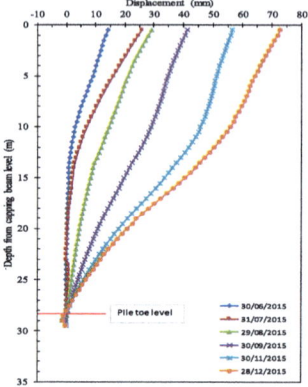
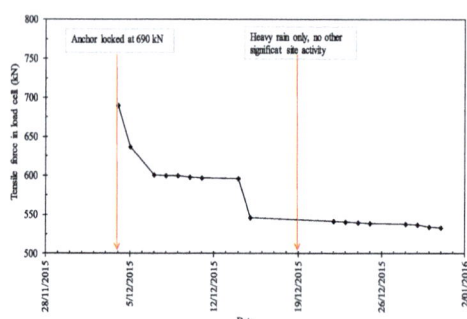

Fig. 2. Total lateral displacement for every month

Fig. 3. Lockoff load over the period

Load cells were installed along the wall to measure the condition of the lockoff load – the allowable design load applied at anchor head (Fig. 3). Figure 3 shows an interesting behaviour of changing applied tensile load at ground anchor over the time frame. It is clear that ground anchor is capable of taking the allowable design load (lockoff load) but it decreases over the time significantly due to release of gripping between soil-grout interfaces. Therefore, by considering design and remedial considerations, it might be suspected that gravity grouting system is not as effective as expected to provide sufficient bond strength since it might experience a substantial amount of deflection, which could be regarded as failure of soil retaining structures for excessive deflection.

3 Innovated Fracture Grouted Soil-Nail System

The application of fracture grouting technique in soil nailing is very new and still under development. Selection of grouting techniques is mainly dependent on the types of in-situ soils. It is seen that compaction grouting, low mobility grouting (LMG), is mainly used in granular soil especially loose sand and soft soil whereas fracture grouting can be applied for a wide range of soil fabrics [1].

Fracture grouting in drill hole may consists of three basic consecutive processes: (1) formation of grout filter-cake and expanding the cavity, (2) fracture initiation into the soil and (3) infiltration of low viscosity cement grout into surrounding soil mass and propagation with injection pressure [3]. Hypothetically, it could be expected that cavity expansion will densify the soil and provide consolidation to the boundary soil. In addition, the cementation may be caused by infiltration of injecting grout, which could enhance the shear strength of the surrounding soil mass. Fracture propagation may also create a positive mechanical interlocking into soil matrix.

The innovated soil nailing system combines compaction grouting and fracture grouting and the conceptual design is presented in Fig. 4. For the conventional soil nail (Fig. 4a), the pullout resistance is mainly from the friction between the grout and soil. For the fracture grouting soil nail (Fig. 4b), the appearance and propagation of the soil fracture network will be monitored, permitting analysis of the fracture effects on the shear resistance improvement and the development of undesired cracking. The mechanism for the interaction of soil and grout will be more complexed. However, the pullout resistance is even higher than that of compaction grouting soil nail, because the persistence pressure due to the grouts entering the fractures is much higher than the friction between the grout and soils in Fig. 4b.

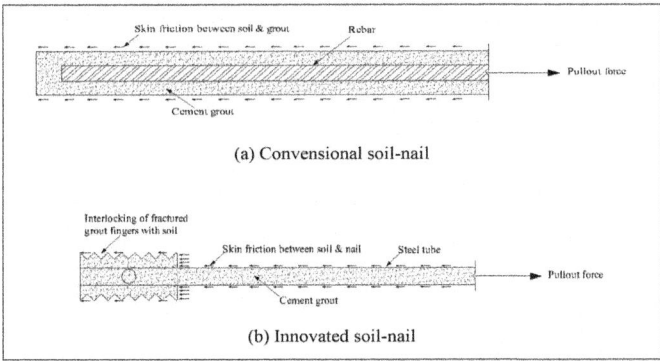

(a) Convensional soil-nail

(b) Innovated soil-nail

Fig. 4. Comparison of (a) conventional soil-nail and (b) innovated fracture grouted soil-nail

4   Conclusion

To date, the behaviour of fractured grouted soil-nail is not well understood compared to conventional grouted soil-nail. In addition, effects of injecting grout viscosity, injection rate, and the corresponding fracture propagation in heterogeneous soils are still missing. A detailed laboratory scale experimental study is going on and the findings will be reported in the future publications.

References

1. Bhuiyan, M.Z.I.: Experimental and numerical study of fracture grouted soil nail. Ph.D. confirmation report, University of Newcastle, Australia, pp. 1–144 (2017)
2. Wang, Q., Ye, X., Wang, S., Sloan, S.W., Sheng, D.: Experimental investigation of compaction-grouted soil nails. Can. Geotech. J. **54**(12), 1728–1738 (2017)
3. Wang, Q., Wang, S.Y., Sloan, S.W., Sheng, D.C., Pakzad, R.: Experimental investigation of pressure grouting in sand. Soils Found. **56**(2), 161–173 (2016)

On Safety Alarm in Deep Excavations

Xue-shan Cao[(✉)], Li-su E, and Xi-yang Lai

College of Civil and Transportation Engineering, Hohai University, Nanjing, China
x.s.cao@163.com

Abstract. Safety alarm method is the key to the deformation of deep excavation engineering control security technology. A new safety alarm system was put forward by the deformation behaviors of soil-rock mass because that the alarm value of control in the code was too preliminary, which could not effectively reflect the deformation state of deep excavation engineering. The new safety alarm method involved three kinds control indicators: the accumulative value, monitoring rate and monitoring acceleration; and four levels of alarm: blue alarm, yellow alarm, orange alarm and red alarm. Alarm levels and the safety control key indicators were corresponded to the tasks and responsibilities. The one excavation example was verified for the practical application of the alarm system.

Keywords: Deep excavation engineering · Safety monitoring · Alarm method
Control indicators · Alarm level

1 Introduction

Due to continuous growth in urbanization worldwide, a large number of metro lines are being constructed within congested urban areas of China. The excavation works in the soft ground inevitably lead to ground movements, which may cause adjacent surface buildings to deform and possibly sustain unrecoverable damages. In order to assure the safety and serviceability of nearby buildings during metro construction, it is necessary to perfect the safety alarm system of the excavation and to perform preventive measures for adjacent buildings ahead of time, which is a good way to reduce the social disputes and to optimize construction technique.

2 The Current Safety Alarm

The current safety alarm method is preliminary. First, the safety control value of excavation is not related to the excavation state, but the depth. The deeper the excavation is, the smaller the control deformation value will be. And the monitoring deformation value of excavation in the soft soil area is often many times of the alarm value. Second, although it is necessary to perform measures when the monitoring value is over the range of alarm values in most design documents, it is not sure to how to deal with the alarm and monitoring value.

© Springer Nature Switzerland AG 2018
W. Wu and H.-S. Yu (Eds.): *Proceedings of China-Europe Conference on Geotechnical Engineering*, SSGG, pp. 902–905, 2018.
https://doi.org/10.1007/978-3-319-97115-5_5

It is very difficult to achieve expected effects on the controlling deformation. The alarm means imminent danger, which is related to the destruction of engineering structure and disaster, and then the constructors often worry about the alarm triggered. However, if there are not any dangers after safety alarm triggered, the constructors will be indifferent to the alarm and considered the alarm as a mistake. Gradually the alarms will lose the function. Corresponding, the terrible accidence of "wolf is coming" is difficult to avoid.

It is impossible to prevent accidents by simple measures. When the monitoring value is over the alarm, the measures are often to strengthen monitoring to observe the alarm developing, such as increasing frequency. The method shows that the constructors are indifferent to the alarm and want to verify what and when the accidents are by monitoring. That also means waiting the accidents approaching and then entering the emergency rescue and relief situation.

It is unable to supervise the performance of the preliminary alarm method. The regulatory authorities are often unable to judge the validity of the engineering measures by construction document.

Therefore, the preliminary alarm method cannot reduce the accident occurrence possibility of engineering, which is the problem of the safety technology of the excavation engineering.

3   New Safety Alarm Method of Excavation Engineering

Soil-rock mass is a typical viscoelastic and plastic material; which deformation has time effect. The state of stress and strain of soil-rock mass varies with time under the condition of constant load. Deformation development stages of the soil-rock mass is divided into three stages. First, the deformation acceleration is negative, meaning the deformation rate decreased. Second, the deformation acceleration is zero, i.e. the deformation rate unchanged. Third, the deformation acceleration is positive and the deformation rate increased with time until the destruction. A new safety alarm system was put forward by the deformation behaviors of soil-rock mass [1].

The control key parameters system is in according with the safety state of excavation. The safety control key monitoring parameters have three kinds: the cumulative value, the rate and the acceleration. The alarm levels of excavation are: the blue, the yellow, the orange and the red alarm. The safety control key parameters and the alarm levels are all associated with engineering personnel and their responsibilities.

When the cumulative value or rate exceeds the corresponding first alarm value, the blue alarm occurs. The blue alarm indicates that the cumulative value of the excavation or the monitoring rate is larger, and reminds the engineering personnel, especially the designers that whether the deformation behaviors in the field are in the range of the design requirements. If not, design change should be carried out to ensure the structure in a safe state according to the site condition.

When the cumulative value or the rate exceeds second alarm value, or when the cumulative value and the rate all reach the first alarm value, yellow alarm is activated. The yellow alarm shows the cumulative value or the rate build up to a threat to the safety of excavation, and companying with local tiny cracks of width less than 1 mm in the structure. Those are omens of engineering accidents, and warn the engineering personnel of monitoring value beyond the range of the design requirements and accidences would happen. Therefore, the engineering personnel should carry out the remedial work to avoid accidence.

When the cumulative value or rate exceed third alarm value, or when the cumulative value and the rate all reach the second alarm value, the orange alarm is activated. The orange alarm shows the cumulative value or the rate build up to a serious threat to the safety of excavation, and companying with local significant cracks of width less than 10 mm or scarp in the building structure. Those are omens of engineering disaster and accidents. The engineering personnel should stop to continue excavation work and perform the measures to prevent deformation.

When the acceleration also exceeds the alarm value, the red alarm is started on the basic of the orange alarm. The red alarm means that the excavation disaster has occurred companying sliding failure in a big area. And the engineering personnel should stop construction, evacuate the irrelevant people, and assemble machinery, vehicles and personnel for rescue and relief work.

If measures are carried out to eliminate the possibility of deformation, and to make the situation back to the initial state, we could eliminate the alarm and renew the monitoring value from the initial. If the preventive measures reduce the occurring probability of the project disaster, the alarm threshold value should be redefined. If there is no any measure, or no effective measure which does not change the possibility of engineering disaster, the alarm should be continued and accumulate monitoring value. If the monitoring rate is reduced, it is shown that the deformation of soil-rock mass is in the first stage of stable deformation.

4   Example

Hangzhou metro pit (107.8 m long, 21 m wide, and 15.7–16.3 m deep), which was collapsed in 2008, designs diaphragm walls (31.5–34.5 m deep and 0.8 m thick) as retaining wall, and four steel struts (Φ609 mm, 16 mm thick and spacing of 3 m) [2].

The deformation of ground and diaphragm walls was displayed in the Table 1, which indicates that the blue alarm had already occurred. There was a crack in the road structure near to the construction section one month before the Hangzhou metro pit collapse [3], which indicates that the yellow alarm should have been issued. "Before the collapse, the construction workers had been found an obvious crack in the diaphragm wall, 10 meters long, width the palm of a hand" [3], that was some damage occurred, cracks appeared more than 10 mm, orange alarm was activated. At last, when the collapse occurred, the displacement rate must increase, the displacement acceleration was larger, and the red alarm occurred.

Table 1. The alarm threshold value in the design files and monitoring value [2]

Item		Cumulative value (mm)		Date	
		Design	Monitoring	Monitoring	Collapse
Settlement of ground		25	316	2008/11/15	p.m.
Deep displacement of diaphragm wall	CX49	40	43.7	2008/11/11	2008/11/15
	CX45		65	11/13	

Following the blue alarm, If the designer could carefully analyze differences between the design value and monitoring in site, and effect of construction technology on the deformation of excavation, and then could combined the actual situation, change design and the follow-up construction technology. The collapse would be avoided.

Even if the yellow alarm were motivated companying with local tiny cracks less than 1 mm, the constructors could change design according to local cracks, contractor could perform the new design and verify the effect of measures and could avoid the accidents. If the constructors could pay attention to public supervision, and the government supervision department could inspect the change design on basis of these cracks and check the effect of change design implementation, the safety of the excavation could also be ensured.

The orange alarm triggered when engineering appeared signs of instability, such as more than 10 mm crack, scarp of the slumping. If the constructors could demand that the builder should stop the construction performing local rescue efforts, and the designer should change design considering the local instability, and perform the measure to ensure the safety of the excavation, then the Engineering construction could not be out of controlling.

5   Conclusion

1. The current alarm method is a preliminary method.
2. A new safety alarm method was put forward by the deformation behaviors of soil-rock mass. The safety alarm method involves three kinds control indicators: the accumulative value, monitoring rate and monitoring acceleration; and four levels of alarm: blue alarm, yellow alarm, orange alarm and red alarm. And alarm levels and the safety control key indicators are corresponded to the tasks and responsibilities.
3. The example was verified the practical application of the safety alarm method.

References

1. Cao, X., Zhang, R., Gu, Q.: A deformation alarm method for deep foundation pit. China. 201510467880.7 [Invention Patent] (2017)
2. Ren, J.: Analysis and discussion of several excavation accidents, Changjiang University (2012)
3. Bi, S.: How About the Cause of Hangzhou Metro Disaster Is Natural, Not Human, 11,22. No. 002 (2008)

Case Study: Moxy Hotel Bored Piles Wall

Cláudia Carvalho[(✉)] and Alexandre Pinto

JETsj, Geotecnia Lda., Lisbon, Portugal
{ccarvalho,apinto}@jetsj.com

Abstract. Given the sudden rise of the construction market in Portugal, allied with the growth of tourism, a need to build more hotels has emerged. This project takes place in Lisbon, close to the Oriente Station (Metro and Railway) at Parque das Nações. The Moxy Hotel, with fifteen raised floors, will be built over a hill, thus rising from the ground. In order to respect the neighbor conditions, a bored piles earth retaining wall (BPW) was designed, reaching a maximum depth of approximately 22 m. The geological and geotechnical conditions of the site are pointed out, as well as the geometrical and topographic conditions. This BPW will be braced differently, depending on whether it is the temporary or the final stage. Therefore, one solution was designed, taking into account those two stages, aiming to find an economic-safe balance. At last, a numerical analysis is presented, where the deformations and efforts are predicted.

Keywords: Bored piles wall (BPW) · Monitoring · Deformations

1 Introduction

After a deep crisis, the construction and touristic market in Portugal has started to rise slowly, being held numerous new constructions, as well as rehabilitations, for accommodation purpose.

The Moxy Hotel will be built at Via Recíproca, at Parque das Nações, in Lisbon (see Fig. 1). The site is located near the "Gare do Oriente" train and metro station, design by Santiago Calatrava for the Universal Exhibition in 1998 (Expo 98), close to the Tagus river. It is an urban area with significant topographical level variations. The building will have a rectangular geometry, with a plan area of about 50×30 m^2. A total of fifteen stories will be built, including two semi-underground ones. The main entrance to the building will be at the South side, from the Via Recíproca. The geometry of the building, the topography of the site, the geotechnical and geological conditions, will lead to an excavation with a maximum depth of 17 m at North (see Figs. 1 and 2). As earth retaining structure, a BPW was designed, being described in Sect. 3.

© Springer Nature Switzerland AG 2018
W. Wu and H.-S. Yu (Eds.): *Proceedings of China-Europe Conference on Geotechnical Engineering*, SSGG, pp. 906–909, 2018.
https://doi.org/10.1007/978-3-319-97115-5_6

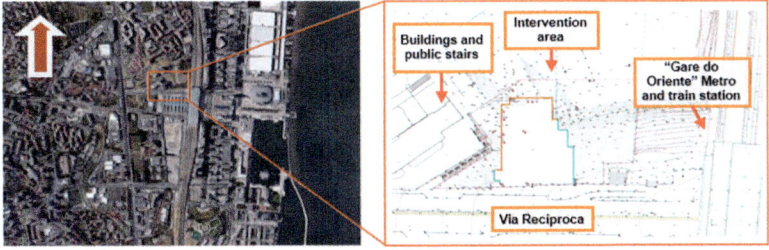

Fig. 1. Site location and conditioning

Fig. 2. Project confrontations and intervention area

2   Main Restraints

The building under study has an approximately rectangular implantation area and has 15 stories, from the Via Reciproca. Some of the most relevant restraints of this project are the Oriente Station, the surrounding structures and infrastructures, as well as the geological and topographic conditions. In this scenario, it was decided to build a solution for the earth retaining structure, using a BPW.

2.1   Geological and Geotechnical

According to the Geological-Geotechnical Study, the ground that will be intersected by the excavations works is composed by four geotechnical zones: ZG1 - highly decompressed landfills, fine silty sands and silty clays; ZG2A - soft Miocene (fine silty sands and silty clays with carbonate intercalations); ZG2B - Miocene moderately stiff (silty sands, clay sands and silts, intercalated by biocalcarenite benches); ZG3 - black and brown silty sands, thin black and brown clays, intercalated with slightly stiff biocalcarenite benches.

2.2   Topography

Regarding the topography, the building under study is located on a hill, which elevation varies from level +8 to +34 m, as shown in Fig. 2.

3 Proposed Earth Retaining Solution

Regarding the works of the earth retaining structure, considering the mentioned restraints, a BPW solution was adopted (Ø600 mm, spaced apart of 1.20 m). The pile length will vary between 11 and 22 m, to ensure an embedment length of at least 5 m. The excavation maximum depth reaches 17 m and the ground between piles is covered with reinforced and drained shotcrete. To ensure the necessary drainage conditions, geodrains are installed between piles [1, 2].

3.1 Temporary Phase

At this stage, the BPW bracing is provided by the temporary ground anchors (spaced at 2.4 m and 3.60 m) and corner shores at two or three levels. In order to ensure a better distribution of forces on the BPW and avoid stress concentrations, the anchors and shores will be installed at the distribution and capping beams.

3.2 Final Phase

At the final stage, the floor slabs will brace the BPW (see Fig. 3a and b). An exception is the west wall, where part of the structure does not lean against the BPW. There, the bracing system will comprise reinforced concrete beams, replacing the temporary ground anchors effect.

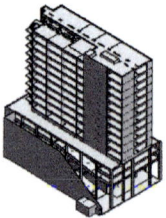 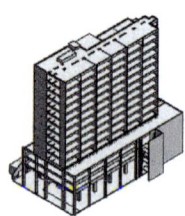 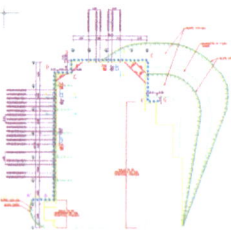

Fig. 3. (a) West BPW; (b) East BPW (both images obtained from Revit); (c) Plant solution (obtained from AutoCAD)

4 Design (Stress Strain Approach)

The BPW design was supported by a numerical analysis using a PLAXIS2D model. Geometry input, soil and interfaces properties were chosen to replicate the local conditions. Figure 4a presents the finite element mesh of one of the developed calculation model. The main scope was to predicted deformations (Fig. 4b), and efforts at the BPW for all the main construction phases.

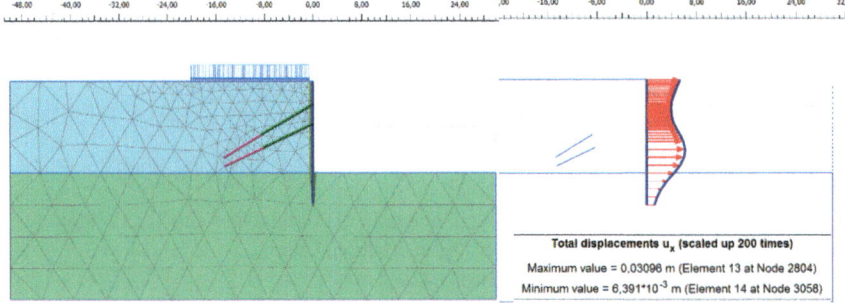

Fig. 4. (a) Numerical model deformed mesh; (b) expected horizontal displacement (u$_x$) at the end of the excavation

Regarding the horizontal displacements obtained, the maximum value registered was 3 cm at the end of the excavation, as shown in Fig. 4b. This corresponds to a relative deformation of approximately d/H = 474, whereby it is considered that the service limit state of horizontal wall deformation is verified.

A monitoring plan was designed as risk management important tool, aiming to predict the behavior of the earth retaining structure, as well as of neighboring structures and infrastructures, during and after the excavation works. The following main devices were installed: topographic targets, inclinometers and ground anchors load cells.

5   Final Remarks

Considering the restraints faced, mainly the topography and nearby constructions, it was possible to overcome them by using a BPW solution. For a maximum excavation depth of 17 m.

As usual in this kind of projects, the developed model will be calibrated and validated on time by the instrumentation and observation plan results, during the excavation works.

References

1. Pinto, A., Fartaria, C., Pita, X., Tomásio, R.: FPM41 high rise building in central Lisbon: innovative solutions for a deep and complex excavation. In: 19th International Conference on Soil Mechanics and Geotechnical Engineering, Seoul, Korea, pp 2029–2032 (2017). TC 207 (Soil Structure). ISBN 978-89-952197-5-1
2. Pinto, A., Tomásio, R., Coelho, R., Nicolas R.: Retaining structures and special foundations at the platinum tower. In: Maputo 16th African Regional Conference on Soil Mechanics and Geotechnical Engineering, Tunisia, pp. 323–329 (2015). Session 3 (Deep and Shallow Foundations). ISBN 978-9938-12-936-6

Soil Characterisation Based on Pipejacking Parameters and Spoil Characteristics

Wen-Chieh Cheng[1(✉)], James C. Ni[2], and Jack Shuilong Shen[3]

[1] School of Civil Engineering, Xi'an University of Architecture and Technology,
13, Yanta Rd, Xi'an 710055, China
w-c.cheng@xauat.edu.cn
[2] Department of Civil Engineering, National Taipei University of Technology,
1, Sec. 3, Zhongxiao E Rd, Taipei 10608, Taiwan ROC
[3] School of Naval Architecture, Ocean and Civil Engineering, Shanghai Jiao Tong University,
800, Dongchuan Rd, Shanghai 200240, China

Abstract. Possessing a few geological boreholes distributed along the tunnel alignment is likely to lead to an inability of understanding the complex geological structure of worksite and optimising the tunnelling parameters. This lack of geological borehole will result in a high potential of geo-hazards for tunnelling works. This study proposes an alternative method to dynamically determine the major and other components of ground by taking the pipejacking parameters and spoil characteristics into account. The validity of the proposed method is verified via a case study.

Keywords: Soil characterisation · Pipejacking · Torque of cutter wheel
Sieve residue

1 Introduction

Significant variation in the jacking force can not only eat up the jacking capacity, but also can cause damage to the concrete pipe itself (Zhen et al. 2014; Bergeson 2014). Mitigation of the impacts on the jacking capacity can be achieved with a comprehensive understanding of the geological structure of worksite (Cheng et al. 2017, 2018). However, it is common practice to have only a few geological boreholes deployed. An alternative method to dynamically determine the major and other components of ground is shown to be necessary in the construction stage. The objectives of this study are: (1) to propose a method to determine the major and other ground components by taking the pipejacking parameters and spoil characteristics into account and (2) to verify the validity of the proposed method through a case study.

© Springer Nature Switzerland AG 2018
W. Wu and H.-S. Yu (Eds.): *Proceedings of China-Europe Conference on Geotechnical Engineering*, SSGG, pp. 910–914, 2018.
https://doi.org/10.1007/978-3-319-97115-5_7

2   Proposed Method

Since the encountered soil type can be distinguished using the torque of cutter wheel, the associated current value was chosen as a main indicator of determining the major ground component. Tunnelling spoils retained on sieves mounted on shaker decks are used to verify the previously deduced major ground component. The more the particles retained on a sieve with the mesh opening of 2 mm, the larger the component of gravels. Otherwise, the ground is primarily consisted of sands with the particle size larger than 0.3 mm, although a small amount of sands retained on the sieve of the mesh opening of 2 mm or passed through a sieve with the mesh opening of 0.3 mm could be observed. In the case that only a few particles are retained on both the sieves, the ground mainly contains silts of smaller than 0.075 mm in diameter. Clays due to their electrical double layer structure are consisted of individual clay particles surrounded by a film of water and are to present together with gravel only on the sieve with the mesh opening of 2 mm. Manipulating the flow rate of feed line is usually adopted to scrape soil particles off the cutting face. Sands and/or silts can be easily scraped off the face by increasing the flow rate of feed line resulting in a reduction in the current of cutter wheel. However, it may not be effective in scraping off gravels and/or clays and thus the measured current value remains nearly constant while tunnelling. Additionally, the pressure in the feed and discharge lines could be significantly increased as the slurry carries an excessive amount of clays and/or silts. In short, the reduction of the current of cutter wheel in relation to an increase in the flow rate of the feed line indicates sands and/or silts. While the increase in the pressure in the feed and discharge lines corresponds to fine-grained soils. Moreover, as the slurry density can largely affect its spoil carrying ability, it should be kept at an appropriate value to ensuring an effective slurry circulation during tunnelling works. The more the clays carried, the higher the slurry density. A threshold value of 12.8 kN/m^3, which indicates some clays, was derived via a series of preliminary field measurements undertaken. Otherwise, the ground may include sands and/or silts, with the slurry density being smaller than 12.8 kN/m^3. Procedure associated with the proposed method is shown in Fig. 1.

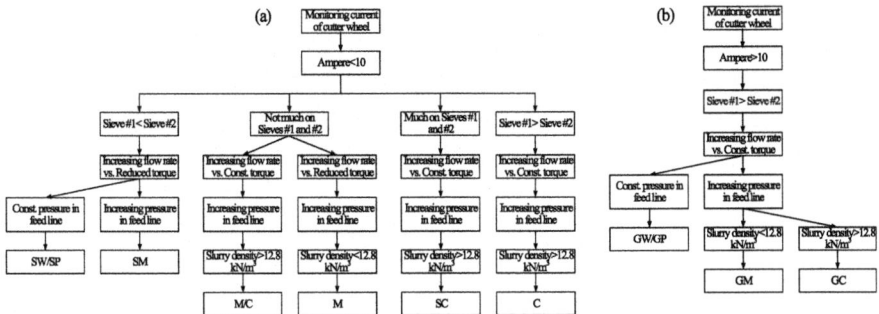

Fig. 1. Procedures associated with the proposed alternative method for soil characterisation: (a) Current of cutter wheel less than 10 A; (b) Current of cutter wheel greater than 10 A.

3   Pipejacking Project Background

The pipejacking project located in the Shulin district in Taipei, Taiwan was implemented at depths varying from 10.3 to 10.8 m below the surface. Five geological boreholes drilled from a backfill, through a silty sand layer into a dense to very dense sand and gravel layer were deployed along the design alignment. A 1500-mm diameter cutter wheel was utilised for the tunnel bore excavation. A 1-m long pipe, 1200-mm internal diameter pipe, with a thickness of 120 mm, was introduced in this project, leading to an overcut annulus of 30 mm where a special bentonite lubricant with 2% polymer was injected.

4   Analysis and Discussions

Figure 2 shows the pipejacking activities at the investigated drive. It is evident that the variation in the current of cutter wheel from the 41–53 m section was relatively small, revealing that the ground was mainly consisted of sand soils or fine-grained soils. Sieve residue present more on the 0.3-mm mesh opening sieve indicated a major sand component. It can be seen from Fig. 3 that in the 41–53 m section the current of cutter wheel was reduced with increasing the flow rate. Additionally, in this section the pressure buildup was less significant, which corresponded to sands and/or gravels. Moreover, the slurry densities either from the refreshed bentonite slurry or from the ordinary slurry were greater than 12.8 kN/m^3. It was most likely attributed to no slurry refreshing undertaken while tunnelling. Thus, based upon the above results, the ground in the 41–53 m section was mainly consisted of sands with some gravels, as labelled by "S/G" in Figs. 2 and 3.

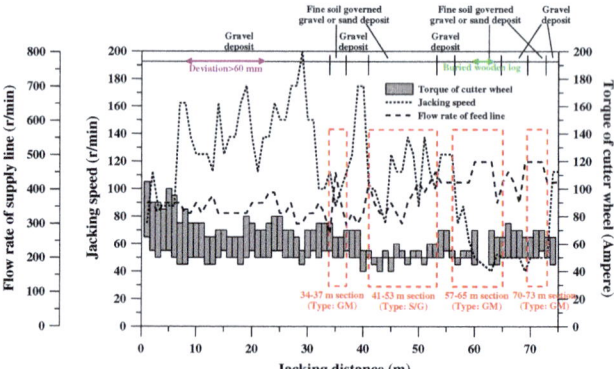

Fig. 2. Variations of torque of cutter wheel, flow rate of feed line, and jacking speed for pipe-jacking works at investigated drive.

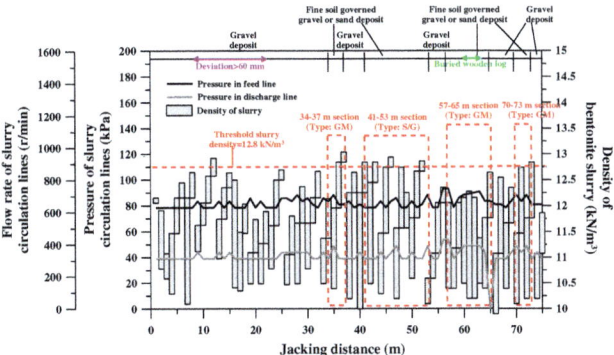

Fig. 3. Variation of pressure in feed/discharge line, and density of bentonite slurry for pipejacking works at investigated drive.

The variation in the current of cutter wheel was however greater in the rest of this drive, with more sieve residue on the 2-mm mesh opening sieve. The jacking speed was relatively low in the 34–37 m, 57–65 m, and 70–73 m sections, which indicated an ineffective slurry circulation, resulting from the reduced spoil carrying ability caused by an excessive amount of the fine-grained soils. It explains the fact that the pressure buildups present in the said three sections. The density values greater than or close to 12.8 kN/m^3 from the 34–37 m, 57–65 m, and 70–73 m sections were ascribed to no slurry refreshing undertaken and the ground in fact should include some silts. In short, for the 34–37 m, 57–65 m, and 70–73 m sections, the ground was mainly comprised of gravels with some silts, as labelled by "GM" in Figs. 2 and 3, while for the five other sections, the ground principally included gravels with some sands (their labels "G/S" are not particularly shown in Figs. 2 and 3).

5   Conclusions

The proposed method utilised five parameters to dynamically determine the major and other ground components during tunnel excavation. Since the understanding of the geological structure at the worksite due to only a few boreholes available along the tunnel alignment was limited, the application of the proposed method provided an opportunity of establishing a more comprehensive geological structure for optimising the tunnelling parameters, reducing the potential of face instability.

References

Bergeson, W.: Review of long drive microtunneling technology for use on large scale projects. Tunn. Undergr. Space Technol. **39**, 66–72 (2014)

Cheng, W.C., Ni, J.C., Arulrajah, A., Huang, H.W.: A simple approach for characterising tunnel bore conditions based upon pipe-jacking data. Tunn. Undergr. Space Technol. **71**, 494–504 (2018)

Cheng, W.C., Ni, J.C., Shen, S.L., Huang, H.W.: Investigation into factors affecting jacking force: a case study. Proc. Inst. Civ. Eng.-Geotech. Eng. **170**(4), 322–334 (2017)

Zhen, L., Chen, J.J., Qiao, P., Wang, J.H.: Analysis and remedial treatment of a steel pipe-jacking accident in complex underground environment. Eng. Struct. **59**, 210–219 (2014)

Vertical Response of 2 × 2 Pile Group Under Rotating Machine Induced Vibrations

Shiva Shankar Choudhary[1(✉)], Sanjit Biswas[2], and Bappaditya Manna[3]

[1] National Institute of Technology Patna, Patna 800005, India
shv.snkr@gmail.com
[2] National Institute of Technology Sikkim, Sikkim 737139, India
[3] Indian Institute of Technology Delhi, Delhi 110016, India

Abstract. The dynamic behavior of 2 × 2 pile group has been investigated under rotating machine induced vertical vibration using both field testing and numerical analysis. Forced vibration tests have been performed in the field on a pile group (pile length = 3 m, pile diameter = 0.114 m) embedded in silty soil. The frequency-amplitude responses of the pile group are measured for four different eccentric moments. It is found that the measured response curves are nonlinear in nature. The numerical analysis has been performed using continuum approach and superposition method to determine the nonlinear frequency-amplitude responses of the pile group. From the analysis, the boundary parameters and soil-pile separation lengths are predicted for different eccentric moments. The analytical frequency-amplitude responses of the pile foundations are compared with the dynamic field test results and it is found from the comparison curves that the nonlinear frequency-amplitude responses obtained from the analysis have a very close match with the field nonlinear response curves.

Keywords: Vertical vibration · Pile foundation · Dynamic field test

1 Introduction

Recently, the growing need of different industries such as hydraulic and nuclear power plants, petrochemical industries, oil refineries etc. has forced development agencies to explore lands with poor soil characteristics which has led to the development and construction of pile foundations. When piles are subjected to dynamic load like machine induced vibrations, special considerations have been taken for design of pile foundations because of soil nonlinearity. To study the complex nonlinear behavior of soil-pile system, tests has been conducted on piles in the field under different vibration modes. EI-Marsafawi et al. [1] conducted field tests on small-scale piles to investigate the dynamic behavior of single and group piles under harmonic vibration. Elkasbgy and Naggar [2] carried out full scale tests on driven helical steel piles to investigate the dynamic characteristics of soil-pile system under vertical vibration. Biswas and Manna [3] performed dynamic field tests and continuum approach analysis on full-scale concrete pile to investigate the response characteristics of single piles under coupled vibration. To determine the dynamic behavior of soil-pile systems, many theoretical

© Springer Nature Switzerland AG 2018
W. Wu and H.-S. Yu (Eds.): *Proceedings of China-Europe Conference on Geotechnical Engineering*, SSGG, pp. 915–918, 2018.
https://doi.org/10.1007/978-3-319-97115-5_8

methods have been developed and extensively used to design the pile foundations under machine vibration. Novak and Aboul-Ella [4] developed a theoretical solution for embedded piles considering layered soil for different modes of vibration. Novak and Sheta [5] introduced the concept of weak soil boundary zone around the pile to investigate the effect of soil nonlinearity and pile-soil separation. From the literature, it is found that the experimental verification of different theories was rarely presented, especially in the case of pile groups. Hence in this present study, dynamic field tests have been performed on a 2×2 pile group to investigate the effectiveness of a nonlinear soil model based on the continuum approach analysis.

2   Experimental Investigation

The vertical vibration pile tests are performed in front of Block III at IIT Delhi, New Delhi, India. Standard penetration test (SPT) is conducted at the site to determine the site soil characteristics. The laboratory tests are performed on soil samples collected from the SPT borehole to determine the soil properties. Based on the in-situ and laboratory tests results, the whole soil profile has been classified as inorganic silt (ML). In the study, closed ended hollow steel circular pipes having an outer diameter (d) of 114 mm, thickness of 3 mm and length (l) of 3 m are used for the dynamic pile testing. A steel plate of dimension $0.9 \text{ m} \times 0.9 \text{ m} \times 0.037 \text{ m}$ is used as pile cap. The piles are driven into undersized boreholes at $3d$ spacing by repeated blows of the SPT hammer. First, mild steel plates are placed on the pile cap to provide desired static weight. Then the mechanical oscillator is mounted on the top of the static load to generate the vibration force on the pile group. Oscillator is run by a DC motor through flexible shaft. One accelerometer is attached vertically on the center top of the loading system to measure the vertical acceleration during dynamic testing with the help of data acquisition system. The frequency-amplitude responses are measured for four varying eccentric moments ($W.e = 0.868, 1.269, 1.631$ and 1.944 Nm) under static load (W_s) of 12 kN. The complete field vibration test setup is shown in Fig. 1.

Fig. 1. Complete vertical vibration field test setup.

3 Theoretical Investigation

In this study, the continuum approach [4] and the superposition method [6] are used to predict the nonlinear dynamic responses of piles. These methods are available as a computer software named DYNA5 [7]. In this method, the complex group stiffness is determined using dynamic interaction factors. To account for the soil nonlinearity, a cylindrical boundary zone [5] is assumed around the pile which is characterized by lesser soil modulus and higher damping relative to the free-field. From the analysis, the boundary zone parameters like shear modulus reduction factor (G_m/G), weak zone soil damping (D_m), thickness ratio (t_m/R) and separation length (l_s) are predicted for different eccentric moments using trial and error technique. The variation of the boundary zone parameters with pile depth is presented in Fig. 2. Using the G_m/G ratio as zero at the upper most layer, the separation between the pile and soil is introduced in the analysis. The separation lengths are found $0.88d$ (= 0.10 m) to $1.05d$ (= 0.12 m) for lower (0.868 Nm) to higher (1.944 Nm) eccentric moments respectively.

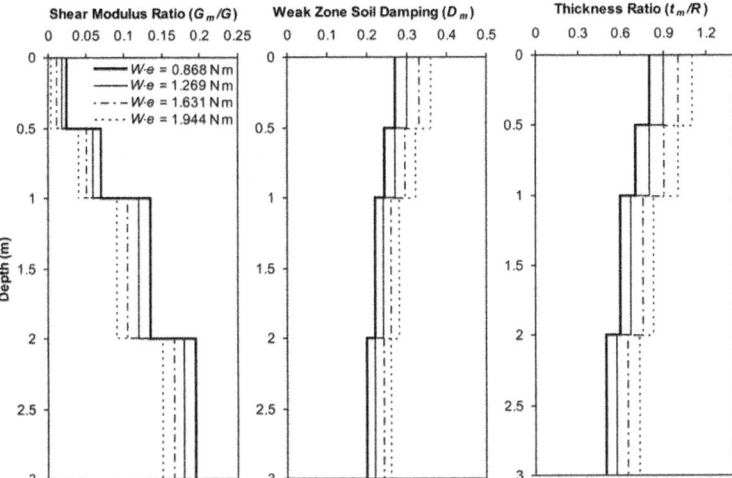

Fig. 2. Variations of boundary zone parameters with pile depth for different eccentric moments.

4 Comparison of Theoretical and Test Results

The analytical results are compared with the field dynamic test results. The comparison curve of frequency-amplitude response of pile group obtained from field testing and nonlinear analysis are presented in Fig. 3. It is found from the comparison curves that the predicted resonant frequencies and amplitude values are reasonably matching with the test results. The nonlinearity is also exhibited in the curves by both testing and nonlinear analysis as the resonant frequencies decreases and resonant amplitude values disproportionally increase with increase of eccentric moments.

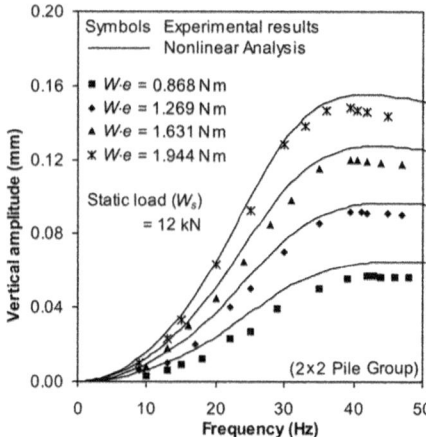

Fig. 3. Comparison of frequency-amplitude responses obtained from test and nonlinear analysis.

5   Conclusions

From the test and theoretical results, it is found the measured frequency-amplitude response curves of pile group exhibit nonlinear behavior of the soil-pile system. It is also found from the comparison curves that the predicted resonant frequencies and amplitude values are well matched with the test results. However, the accuracy of the predicted nonlinear response dependent on the choice of boundary zone parameters i.e. shear modulus reduction factor (G_m/G), weak soil damping zone (D_m), thickness ratio (t_m/R) and separation length (l_s).

References

1. El Marsafawi, H., Han, Y.C., Novak, M.: Dynamic experiments on two pile groups. J. Geotech. Eng. ASCE **118**(4), 576–592 (1992)
2. Elkasabgy, M., El Naggar, M.H.: Dynamic response of vertically loaded helical and driven steel piles. Can. Geotech. J. **50**, 521–535 (2013)
3. Biswas, S., Manna, B.: Experimental and theoretical study on the nonlinear response of full-scale single pile under coupled vibrations. Soil Dyn. Earthq. Eng. **94**, 109–115 (2017)
4. Novak, M., Aboul-Ella, F.: Impedance functions for piles embedded in layered medium. J. Eng. Mech. ASCE **104**(3), 643–661 (1978)
5. Novak, M., Sheta, M.: Approximate approach to contact problems of piles. In: Proceedings of the Dynamic Response of Pile Foundations: Analytical Aspects, New York, pp. 53–79 (1980)
6. Novak, M., Mitwally, H.: Random response of offshore towers with pile- soil-pile interaction. J. Offshore Mech. Arct. Eng. **112**, 35–41 (1990)
7. Novak, M., El Naggar, M.H., Sheta, M., El Hifnawy, L., El Marsafawi, H., Ramadan, O.: DYNA 5 - A computer program for calculation of foundation response to dynamic loads, Geotechnical Research Centre, University of Western Ontario, London, Ontario (1999)

Precast Concrete Piles in Europe – AARSLEFF's Experience

Piotr Dziadziuszko[1(✉)] and Dariusz Sobala[2]

[1] Eng., Aarsleff sp. z o.o, Aleja Wyścigowa 6, 02-681 Warsaw, Poland
pdz@aarsleff.com.pl
[2] Eng., ul. Tarnopolska 29, 35-317 Rzeszów, Poland

Abstract. This paper reviews the modern technologies of precast concrete piles applied in Europe. Supported by the long-time experience of the company in manufacturing and driving precast concrete piles on a large scale in numerous European countries, the up-to-date standards and directions of technology development have been discussed. Presented in the paper the examples of executed pile foundations for various types of structures – from single piles under the support structures of railway overhead lines, through buildings of various sizes and intended use, wind farms, harbour embankments, to large industrial plants and bridges, which verify the broad opportunities to apply precast concrete pile technology in construction engineering.

Keywords: Precast concrete piles · Pile technology · Piling works

1 Technical Requirements for Precast Piles in Europe

Technical requirements to be met to manufacture precast concrete piles and apply this technology are defined in Europe by regularly updated standards:

- EN 1992 regarding design of concrete structures [1],
- EN 12794 regarding manufacture of precast products [2],
- EN 12699 regarding execution of displacement piles [3] and
- EN 1997-1 regarding geotechnical design [4].

2 Material and Structural Solution for Precast Piles

The materials currently used in Europe to manufacture precast piles are concrete of strength class C40/50 and C50/60 with reinforcing steel ductility class "B", yield strength class fy = 500 MPa.

Commonly applied precast piles are square section between 0.2 m and 0.45 m and the length of single elements often limited to 15 m to maximize logistics at transport and installation. Yet, the installation lengths of precast piles can be significantly longer since the application of mechanical couplers is possible. The lengths of precast piles used with load bearing pile couplers varies between 15 and 80 m.

© Springer Nature Switzerland AG 2018
W. Wu and H.-S. Yu (Eds.): *Proceedings of China-Europe Conference on Geotechnical Engineering*, SSGG, pp. 919–922, 2018.
https://doi.org/10.1007/978-3-319-97115-5_9

A special type of precast piles with cast-in load bearing bolts is commonly applied under the support structures for new and modernized railway overhead lines. This precast bolt foundation element has proven to be a very cost and time effective method.

3   Manufacture

Full automation is applied in a modern manufacturing process in the field of [8]:

- manufacture of reinforcing cages,
- manufacture of mechanical couplers,
- dosing of ingredients, manufacture and application of concrete blend and
- identifying of finished prefabricated components on a storage yard (Fig. 1).

Fig. 1. Fully automated pile coupling production ensures high quality and full traceability on materials and process.

4   Pile Drivers and the Order of Work

At present precast piles are driven by means of tracked or rolling pile drivers provided with hydraulic hammers 1.5 up to 11 tones. Pile drivers are usually provided with the system of automated registration of drive parameters the analysis of which makes it possible to monitor driving process efficiency. Modern pile drivers are frequently equipped with modern hammers of high efficiency and limited sound emission level (reduced to 5 dB), which enables the execution of work in built-up areas.

The use of precast piles should be preceded by a research phase which increases the reliability of executed foundation. Test piling and the results of load-bearing capacity tests enable the optimal solution for the foundation, the choice of appropriate equipment and parameters for effective driving. Research may also be carried out during or after piling process, with the use of active design that incorporates updated piling results (observational method according to [3]).

5 Market Overview and Examples of Precast Piles Applications

Precast reinforced concrete piles are applied all over Europe to found [7] the support structures for railway overhead lines, residential buildings, other buildings of various sizes and intended use, sports facilities (e.g. stadiums), transmission towers, wind farms, chimneys, industrial buildings, road and railway embankments and bridges of various sizes [5]. The growth of precast pile production in AARSLEFF Group during last 9 years is shown in Fig. 2.

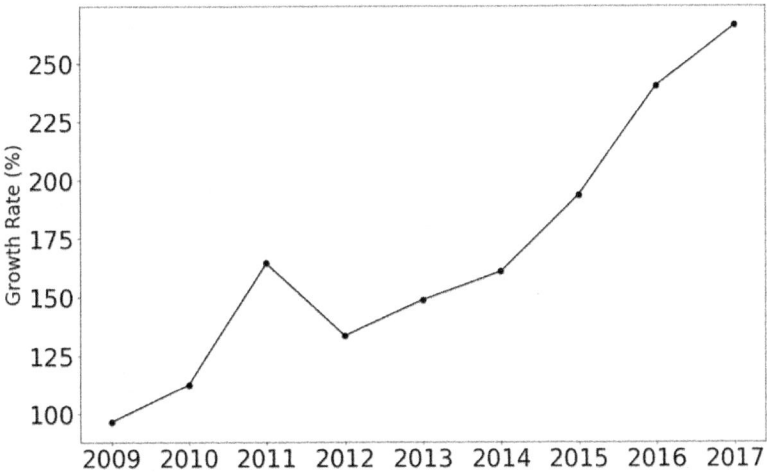

Fig. 2. Growth of precast pile production in AARSLEFF Group during the last 9 years.

Most frequently precast piles are applied in difficult terrains (e.g. organic or contaminated soil) as well as in water environment.

6 Main Advantages of Precast Piles

The most significant advantages of precast piles are:

- long-term history of application >> 100 years;
- fully-controlled manufacture process providing high-quality prefabricated pile components;
- high strength parameters, durability and resistance to aggressive environment;
- high speed and simplicity of piling work;
- possibility to incline piles significantly (especially to resist horizontal loads) and use of pile groups;
- significant simplicity of monitoring piling work (tests in load-bearing capacity) [6];
- possibility to apply in most types of soils, including organic and contaminated ones, which results in high organizational, economic and technical effectiveness of the technology.

7 Future Improvements

The precast pile technology is likely to develop in the following fields:

- **manufacture of prefabricated components** by means of making the automation of manufacture process more popular and broader in range,
- **material solutions** due to the progress in concrete technology, e.g. the use of self-repair concrete,
- **progress in piling equipment** resulting from the bigger operational capacity of pile drivers and higher effectiveness of hammers,
- **execution of piling work** by means of further automation and increase of range of monitoring along with the limitation of negative impact of piling work on the environment (e.g. by the use of silent hammers which enable driving piles in built-up areas),
- **designing** by means of further development of design methods incorporating immediate research in soils or/and data gathered during piling work,
- **organisation** by means of further, complex optimisation of all the process comprising design, manufacture, transport, driving, testing, monitoring and reporting of piling work and
- **energy piles** as a sustainable method for ground sourced heating and cooling of buildings build on piles.

References

1. EN 1992. Eurocode 2. Design of concrete structures
2. EN 12794. Precast concrete products – Foundation piles
3. EN 12699. Execution of special geotechnical works - Displacement piles
4. EN 1997-1. Eurocode 7. Geotechnical design. Part 1. General rules
5. Sobala, D., Wąchalski, K.: Modern solution for the foundation of Poland's largest Arch Bridge. In: 8th International Conference on Arch Bridges ARCH 2016, Wrocław, Poland (2016)
6. Sobala D., Tkaczyński G.: Interesting Developments in Testing Methods Applied to Foundation Piles, IOP Conference Series: Materials Science and Engineering, pp. 1–8, ISBN/ISSN: 1757-8981 (2017)
7. www.aarsleff.com, www.aarsleff.com.pl; www.aarsleff.se, www.aarsleff.co.uk
8. www.centrumpaele.dk, www.centrumpali.pl, www.centrumpile.se

Case Study: BIM and Geotechnical Project in Urban Area – Infinity Tower

João Gondar$^{(\boxtimes)}$ and Alexandre Pinto

JETsj, Geotecnia Lda., Lisbon, Portugal
{jgondar,apinto}@jetsj.com

Abstract. The Building Information Modeling (BIM) methodology is spreading across the Architecture, Engineering, and Construction (AEC) industry. The need of leaner processes from the concept to the execution of projects are opening space for this approach. One of the advantages of BIM methodology is an early coordination and communication among the different project's stakeholders, a key issue for the geotechnical project where uncertainty is high. This case study approaches the use of BIM for a geotechnical project of an excavation for a residential tower in the center of Lisbon, with a deployment area of 4600 m and a maximum excavation depth of 17.60 m. The solution proposed was a Bored Pile Wall (BPW) system with anchors and bracing slab stripes. The solution was modeled coordinated with the 3D BIM model from the architecture and was then exported to a soil analysis numerical software and a structural analysis software to improve the solution proposed.

Keywords: BIM · Collaboration · Bored Piles Wall (BPW)

1 Introduction

The recent rise in the real estate industry in Portugal associated with the urban planning decisions to release the surface area for leisure purposes are pulling some of the surface area used for the underground of buildings, creating the need of constant coordination of this kind of projects with other building project's specialties.

The rise of Building Information Modeling (BIM) concept is promoting an increase in the collaboration among the different project specialties in the construction industry. The traditional design-bid-build process where the architect handles the design to the engineers and all the documentation is produced before it is delivered to the contractor is shifting to a more design-build process, where the different project subjects must team up much earlier [1]. The process is no longer linear but collaborative.

This collaboration is especially important for projects where uncertainty is high, what is generally the case of geotechnical projects. This increased collaboration can lead to a rise in efficiency and a better time management [2].

The geotechnical construction sector is characterized by its both financial and physical risk associated with its projects [3]. The lack of early design stage integration among specialties and the accurate and timely availability of geological and geotechnical

© Springer Nature Switzerland AG 2018
W. Wu and H.-S. Yu (Eds.): *Proceedings of China-Europe Conference on Geotechnical Engineering*, SSGG, pp. 923–926, 2018.
https://doi.org/10.1007/978-3-319-97115-5_10

information are some of the challenges faced by this sector and can potentially benefit from the BIM approach.

2   Study Case

The main restrains defined were related to the topography, geological and geotechnical conditions, and the neighboring conditions.

2.1   Geological and Geotechnical Conditions

The characterization of the underground conditions was made through 9 SPT tests, done across the deployment area. The area is composed of volcanic compounds covered by a landfill deposit layer. It was divided into 4 geotechnical Zones: ZG1, regarding the landfill layer; ZG2 for pyroclastic tufts and low-quality basalts; ZG3 and ZG4 for medium to high-quality basalts.

2.2   Existing Topography

The existing topography, with the building deployment laying over a small hill, makes that the excavation depth to vary from 17.60 m to 6.25 m in opposed alignments.

2.3   Neighboring Conditions

The intervention zone is located in an urbanized area. In the west, it is limited by the Lisbon sub-urban train line, at south side by a viaduct, and North-East and East fronts are limited by road traffic and pedestrian streets (Fig. 1).

Fig. 1. Deployment area surroundings

3   Solution Proposed

The solution was proposed considering the existing restraints and with the purpose to control the soil deformation and execute the excavation with the minimum interference with the surrounding infrastructures and services, taking in account the safety, constructability, and cost associated.

The conceived solution was a Bored Pilled Wall system with a fixed 600 mm diameter pile and a displacement varying between 0.80 m and 1.20 m, according to with the geological conditions and stresses over the wall system. The total pile's length varies from 21.60 m to 10.30 m, all with a minimum embedment length of 4.00 m.

The curtain wall was braced at each floor level, compatible with the architecture, with a concrete framing of 0.35×0.60 m and the piles capped with a 0.75×0.80 concrete frame. The soil exposed between the piles will be covered by a shotcrete layer of 80 mm minimum thickness, and geodrains will be installed with a minimum 3.60 m of displacement to ensure a drained condition in the system.

In the west front, the wall will be held by one level of anchors to be installed at level −2 with 3.60 m of displacement. The remaining fronts of excavation will be stabilized with slab strips of 12.00 m width and 0.35 m minimum thickness, compatible with the architecture and the structure. The slab strips will ensure a stiff bracing to the solution [4].

These slab strips are supported by vertical steel profiles HEB260 embed in 600 mm piles, 4.00 m below the bottom level of excavation. The slab strips above level −2 are supported by slimmer slab strips of 2.50 m width that will unload in a temporary reinforced concrete structure, partially embedded in the piles at the west front.

4   Design Methodology

Among the elements received was the architecture project geometry in a 3D BIM model. The existing topography was designed in the BIM software, and then, the architecture model was linked to the file and the geographic position of the surface was coordinated with the architecture model and the existing lot boundary (Fig. 2).

Fig. 2. 3D Architecture BIM model and the Retaining BIM model (left) and the future tower (right)

The modeling of the curtain walls was done according to with the architecture 3D BIM model. The solution was then tested using a numerical analysis software (PLAXIS2D) and the displacements and forces were analyzed considering the respective soil parameters. The geometry from the BIM model and the loads from the PLAXIS2D were then exported to a structural analysis software (SAP2000) using an IFC file type. The retaining solution was then adjusted according to the results obtained (Fig. 3).

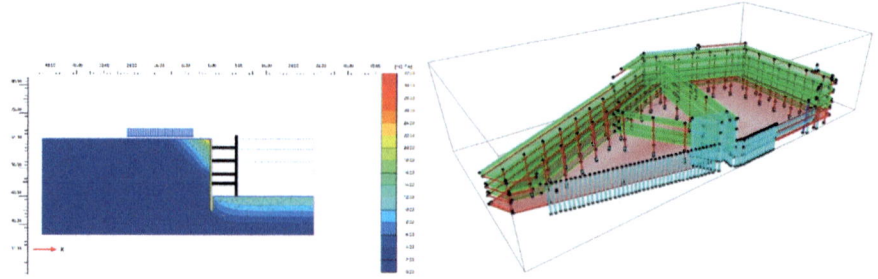

Fig. 3. Analysis in PLAXIS2D (left) and SAP2000 model imported from a BIM file

5   Final Remarks

The use of BIM methodology allowed an accurate coordination with the architecture project and promoted efficiency in terms of project documentation, especially when changes in the design were needed. The interoperability among software allowed that the geometry from the 3D BIM model could be exported, avoiding the re-modeling and possible geometry inaccuracies. The 3D visualization of the project and the restrains helped to find engineering solutions.

References

1. Azhar, S.: Building information modeling (BIM): trends, benefits, risks, and challenges for the AEC industry. Leadersh. Manage. Eng. **11**, 241–252 (2011)
2. Carmona, J., Irwin, K.: BIM: who, what, how and why - Facilities Management Software Feature.   http://www.facilitiesnet.com/software/article/BIM-Who-What-How-and-Why–7546. Accessed 29 Oct 2017
3. Pinto, A., Fartaria, C., Pita, X., Tomásio, R.: FPM41 high rise building in central Lisbon: innovative solutions for a deep and complex excavation. In: 19th International Conference on Soil Mechanics and Geotechnical Engineering Seoul, pp. 2029–2032 (2017)
4. Sterling, R.L.: Advances in underground construction help provide quality of life for modern societies. Engineering **3**(6), 780–781 (2017)

Study on Penetration Resistance of Bucket Foundation Breakwater by Centrifuge Model Tests

Yunfei Guan[✉] and Yongyong Cao

Department of Geotechnical Engineering, Nanjing Hydraulic Research Institute, Nanjing, China
gyfnhri@163.com

Abstract. This study presents the details of a newly developed bucket foundation breakwater. The bucket is sunk and penetrated into the foundation soil when it is installed. Centrifuge model tests are performed to study the penetration behaviour of the bucket foundation during the sinking process. The details of the test model and procedure are presented. It is found that the total penetration resistance increasing linearly in the muddy clay layer. An inflection point is found when the bucket reaches the silty clay layer. The penetration resistance corresponding to this inflection point can be thought to be the suction required to install the bucket foundation.

Keywords: Bucket foundation · Penetration resistance · Centrifuge model tests

1 Introduction

For the vertical breakwater built in deep clay layer under complex wave loading a bucket foundation is developed. Compared to the conventional breakwater, the breakwater with bucket foundation can be built more conveniently and economically and it requires less material (Xu et al. 2001, 2010). The breakwater consists of several bucket structures built in the clay layer. Each structure consists of a bucket at the bottom and two barrels at the top (Fig. 1). The length and width of the bottom bucket is 30 m and 20 m, respectively. The inner space is divided into 9 compartments by the clapboards. The thickness of the wall and the clapboard is 0.4 m and 0.3 m, respectively. Two upper barrels are built at the top of the bottom bucket along the shorter axis with a diameter of 8.9 m and a thickness of 0.4 m. The whole structure is transited to predetermined location by the floating dock after being built. It is sunk by draining water and applying suction to the bottom bucket. The bucket is penetrated into the foundation soil. The penetration resistance includes the frictional resistance on the bucket wall and the base resistance of the bottom of the bucket. It is important to estimate the penetration resistance to determine the suction required in the installation of the bucket. In this study centrifuge model tests are performed to investigate the penetration resistance of the bucket foundation.

© Springer Nature Switzerland AG 2018
W. Wu and H.-S. Yu (Eds.): *Proceedings of China-Europe Conference on Geotechnical Engineering*, SSGG, pp. 927–931, 2018.
https://doi.org/10.1007/978-3-319-97115-5_11

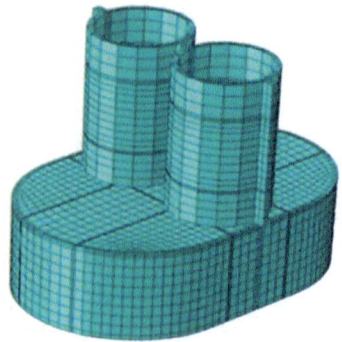

Fig. 1. Breakwater with bucket foundation

2   Penetration Behavior of Bucket Foundation into Soil in Centrifuge

2.1   Model Design and Preparation

The centrifuge model tests are performed at Nanjing Hydraulic Research Institute in China. A medium-size centrifuge with 2.25 m in rotational radium is used. It has 250 g of maximum acceleration and 50 g ton of capacity. The test box is 685 mm long, 350 mm wide and 450 mm deep. Here a scaling factor of 80 is used according to the full size of a unit of bucket foundation breakwater, and, i.e. n = 80.

2.2   Measure Sensors

The measure sensors consist of one load cell and two strain displacement transducers. The displacement transducers with a range of 150 mm are used to measure penetration displacement and are attached to with the cover plate of bucket.

It is known that stress elastic material can be estimated based on the measured strain according to the Hooke's law. 20 strain gauges are installed at 50 mm above the bottom of the bucket (4 gauges for each location) to measure the strain. The stress of this height is calculated based on the measured strain.

A driving actuator is developed to penetrate the bucket into soil during the rotation of the centrifuge. The speed of penetration is set to be 6 mm/min which is the same as the speed of sinking of the bucket in practice (Fig. 2).

2.3   Test Procedure

A sample with a silty clay layer and a muddy clay layer was prepared. After the preparation of the soil sample, the bucket is placed at a few millimeters below the soil surface to avoid tilting during sinking under its self-weight. The driving actuator is then installed along with the displacement transducers. The model is placed in the centrifuge basket which rotates with a acceleration of 80 g. During this process, the bucket

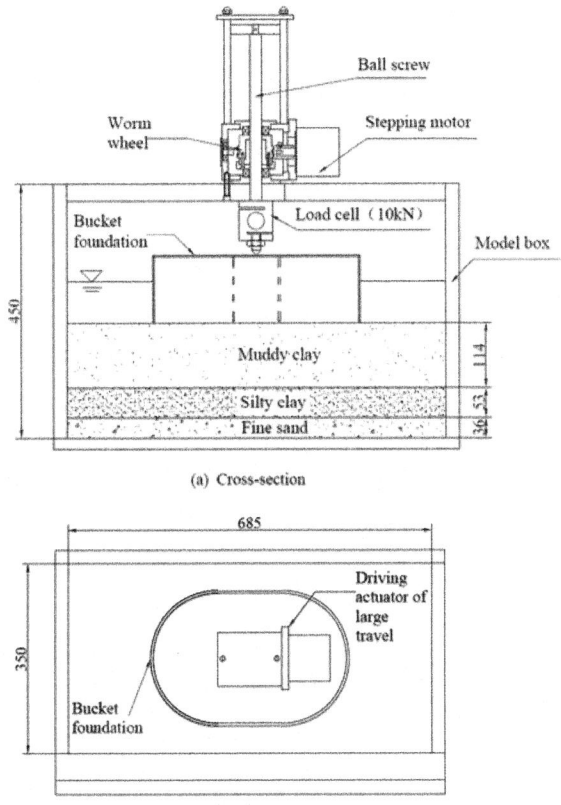

(a) Cross-section

(b) Plan view

Fig. 2. (a) Model of the bucket foundation in centrifuge; (b) Model of the bucket foundation in centrifuge

penetrates into the clay layer to reach a certain depth under its self-weight and then remain stable. The end of the sinking process can be monitored by the displacement transducers. The penetration process starts by turning on a driving actuator. The bucket is thrust into soil at a constant rate of 6 mm/min. When the base of the bucket reaches a certain position, the driving actuator is stopped and the model test ends.

3 Penetration Resistance Analysis

Five centrifuge model tests were performed. All the variables used here are in prototype scale by converting the measured values into the corresponding values in prototype scale according to the scale ratios between model and prototype.

The embedding process of the bucket includes the self-sinking stage due to its self-weight and the penetration stage under external driving force. In other words, the total penetration depth of bucket foundation is the sum of the self-sinking depth and the penetration displacement. The total resistance is the sum of the self-weight of the bucket and the external driving force. Figure 3(a) shows the development of total resistance versus penetration displacement. It can be seen that the total resistance is almost linearly increased with penetration displacement. It can be also found that the total resistance consists of the wall friction and the base resistance of that bucket. The base resistance is constant. The total resistance is proportional to the contact area between soil and the wall of the bucket which linearly increases with the penetration displacement.

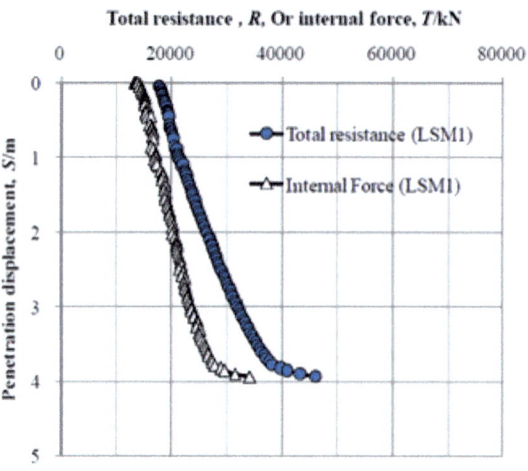

(a) Total resistance and internal force

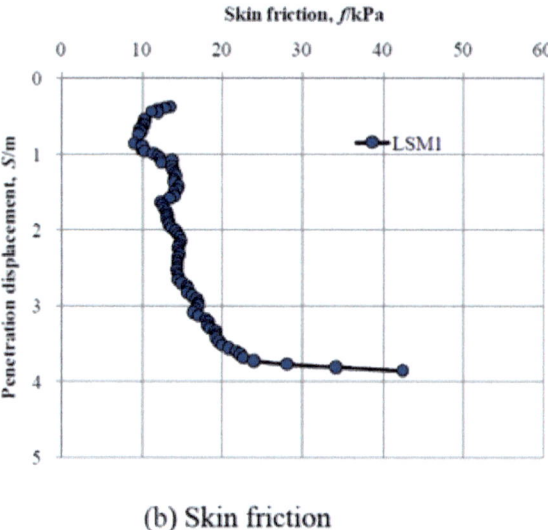

(b) Skin friction

Fig. 3. (a) Penetration behavior under external driving; (b) Penetration behavior under external driving force.

After the penetration displacement reaches a certain value, the total resistance begins to increase at an increasing rate and it leads to obvious inflection in the curve shown in Fig. 3(a). Since the wall friction and bearing stress are larger within the layer of silty clay than those within the layer of muddy clay, two resistance components will both increase significantly.

Since the layer of silty clay is served as the bearing layer of the bucket foundation, the embedding process should stop when the bucket base is rested onto the silty clay deposit. The total resistance corresponding to the inflection point of curve is regarded as the critical total resistance. For the given soil, the critical total resistance is about 40,000 kN. The value of the critical total resistance is important for the installation of the bucket because it governs the suction utilized in its embedding process. For the present prototype in this study, the required vacuum value is about 47 kPa.

References

Xu, G., Cai, Z., Gu, X., Li, Y., Wu, J., Xie, S., Wang, Y.: Centrifuge modeling for wave loading on a cylindrical breakwater. In: Proceedings of the International Symposium on Geomechanics and Geotechnics: From Micro to Macro, Shanghai, China, pp. 163–170 (2010)

Xu, G., Zhang, W., Lai, Z.: Centrifuge modeling of quay walls composed of large-diameter cylinders embedded below mud line. Ocean Eng. 19(1), 38–43 (2001). (in Chinese)

On Deformation of Bridge Pile Foundations Adjacent to Excavations

Jiaqi Guo[1(✉)], Xiao Zhang[1], Junhua Xiao[1], and Nan Wu[2]

[1] Key Laboratory of Road and Traffic Engineering of the Ministry of Education,
Tongji University, Shanghai 201804, China
476284080@qq.com
[2] School of Civil Engineering,
Beijing Jiaotong University, Beijing 100044, China

Abstract. The unloading effect of excavations can cause a change in the stress field and displacement field of surrounding strata, which may seriously influence the existing bridge pile foundations adjacent to excavations. Based on a foundation pit excavation in Kunshan, Jiangsu Province, the finite element method considering the small strain of soil is adopted to study behavior of the excavation and its influence on deformation of the nearby pile foundations. The major calculation parameters of HS-Small model are determined to well describe the deformation characteristics of the soil. Through numerical calculation and simplified analysis, the influence zones for deformation of bridge pile foundations adjacent to the excavation are divided into three types, considering various deformation controlling criteria for existing pile foundations as well as the relative spatial relationship of the pile foundation and the excavation. Dividing lines between these types of zones are the allowable displacement line and the warning displacement line. Accordingly, the adjacent piles are divided into three types. During the construction of foundation pit, it is useful to take corresponding measures depending on which influence zone that the bridge pile foundations in.

Keywords: Excavation · Pile foundation · Numerical analysis
Influence zone

1 Introduction

The unloading effect of excavations can cause a change in the stress field and displacement field of surrounding strata. How to protect the safety of the surrounding bridge pile foundations is a difficult problem in foundation pit excavations.

Based on a foundation pit excavation in Kunshan, Jiangsu Province, the finite element method considering the small strain of soil is adopted to study the behavior of the excavation and its influence on deformation of the nearby pile foundations. According to various deformation controlling criteria for existing pile foundations, the method to divide the influence zones for deformation of bridge pile foundations adjacent to excavations is proposed. Therefore, the influence of excavation on bridge pile foundations may be pre-evaluated, and the design schemes of foundation pits and the protection measures of bridge pile foundations can be made in advance.

© Springer Nature Switzerland AG 2018
W. Wu and H.-S. Yu (Eds.): *Proceedings of China-Europe Conference on Geotechnical Engineering*, SSGG, pp. 932–936, 2018.
https://doi.org/10.1007/978-3-319-97115-5_12

2 Engineering Overview

A foundation pit project in Kunshan, Jiangsu Province proposes to build three storey basements. The circumference of the 13.75 m deep square foundation pit is about 351 m. The south side of the foundation pit is adjacent to the rail transit elevated structure and an urban arterial road. The rest sides of the excavation are near different roads.

The site is mainly composed of clay, silt and sand within a depth of 70 m and has the characteristics of stratification. Soils from top to bottom can be divided into 8 layers, including ① miscellaneous fill, ②1 clay, ③ mucky silty clay, ④1 silty clay, ④3 sandy silt, ⑤3 silty clay, ⑧12 silty clay with silt and ⑧21 silty clay with sandy silt.

A large number of monitoring data showed that the soil in most geotechnical problems is in a small strain state [1]. Yin [2] verified the feasibility of Hardening-Soil Small (HSS) Constitutive Model for numerical analysis of deep foundation pit engineering. Therefore, the soil in this paper is simulated by HSS model in PLAXIS and mechanical parameters of soils are shown in Table 1.

Table 1. Mechanical parameters of soils.

Soil	T/m	$\gamma/(kN \cdot m^{-3})$	$c'/(kN \cdot m^{-2})$	$\varphi'/^\circ$	$E_{oed}^{ref}/(kN \cdot m^{-2})$	$E_{50}^{ref}/(kN \cdot m^{-2})$	$E_{ur}^{ref}/(kN \cdot m^{-2})$	$G_0^{ref}/(kN \cdot m^{-2})$	$\gamma_{0.7}/10^{-3}$
①	1.3	18.3	12	27	3.42	4.10	23.94	95.76	0.2
②1	1.1	18.7	19	28	4.40	5.28	30.80	123.20	0.2
③	4.4	17.7	11	30	2.52	3.02	17.64	70.56	0.2
④1	5.6	19.5	28	36	8.04	9.65	56.28	225.12	0.2
④3	9.9	18.7	7	34	8.60	10.32	60.20	240.8	0.2
⑤3	12.1	17.9	17	34	4.68	5.62	32.76	131.04	0.2
⑧12	31	18.4	18	34.5	5.99	7.19	41.93	167.72	0.2
⑧21	4.6	18.7	19	34	6.94	8.33	48.58	194.32	0.2

3 Control Standards of Pile Foundation Deformation

"Code for design of metro" [3] provides that the differential settlement of adjacent piers cannot exceed 20 mm. "Pile foundations handbook" [4] takes the allowable value of lateral displacement of pile top as 10 mm. Therefore, take 20 mm, 10 mm as the settlement and horizontal displacement allowable values of bridge piles respectively.

"Technical code for protection structures of urban rail transit" [5] introduces the monitoring ratio G, ratio of the measured value of the monitoring project to the value of the structural safety control index, as the basis for the classification of external project monitoring and warning levels. When G ≥ 1.0, it should start the safety emergency plan; when 0.6 ≤ G < 1.0, it should carry out monitoring and early warning; when G < 0.6, it can carry out the normal construction. 60% of the allowable displacement is set as pile warning displacement in this paper. The control standards are shown in Table 2.

Table 2. Control standards of pile foundation deformation.

Control standard	Settlement/mm	Horizontal deformation/mm
Allowable displacement	20	10
Warning displacement	12	6

4 Calculation Model

Using finite element software PLAXIS to study the influence of excavation on surrounding bridge pile foundations. According to the symmetry of the excavation and stratum, take half to calculate. The simplified model is shown in Fig. 1.

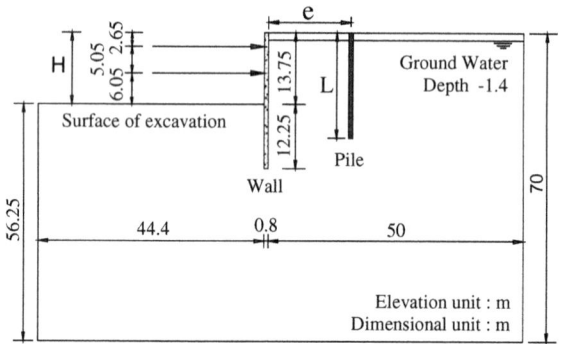

Fig. 1. Dimensions of model.

Soil parameters used in model are given in Table 1. Parameters of the diaphragm retaining wall, concrete bracing and pile foundation are set according to the design. In the process of calculation, the influence of precipitation in foundation pit is considered.

5 Establishment of Influence Zones

The deformation of the abutment pile is related to the position parameters e and h between the pile and the excavation (see Fig. 1). After calculation, extract the settlement values and horizontal deformation values at the top of the pile under different values of e and h. Import bridge pile foundation deformation data into MATLAB for interpolation. Then draw 20 mm, 12 mm contours based on settlement data and 10 mm, 6 mm contours based on horizontal deformation data. Add these four curves into one graph (see Fig. 2(a)). The shaded areas represent uncalculated areas.

Draw envelopes on the same control standard contours. As shown in Fig. 2(a), the envelopes of allowable displacement and warning displacement of pile foundation are the two contours of horizontal deformation. The envelopes of deformation controlling considers both horizontal deformation and vertical deformation of the pile. This is to say, the maximum horizontal displacement and vertical displacement of the pile foundation in the influence zones obtained by the envelopes must have at least one index beyond the control value.

Figure 2(a) shows that the range of influence zone outside the excavation gradually increases as the deformation control value decreases. However, this partition result is inconvenient in engineering practice. To this end, envelopes shown in Fig. 2(a) are reduced to polygonal lines shown in Fig. 2(b).

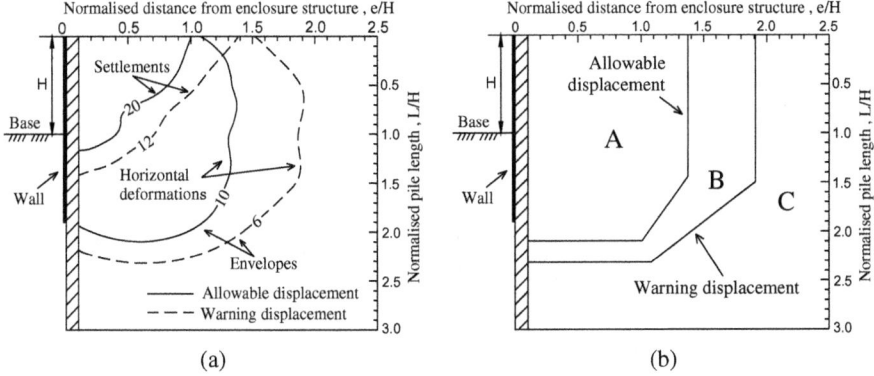

Fig. 2. Influence zones of bridge pile foundations adjacent to excavation

Demarcation lines in Fig. 2(b) divide the influence zones for deformation caused by excavation into three zones: A, B and C zones. Different engineering measures should be taken for pile foundations in different regions. The area where pile deformation is less than the warning displacement is zone C and the pile whose bottom is located in zone C is called C-type pile. C-type pile does not need to take protective measures. The area where pile deformation is greater than warning displacement but less than the allowable displacement is zone B and the pile whose bottom is located in zone B is called B-type pile. Although B-type pile is in a permissible deformation state, the deformation exceeding the allowable displacement may occur due to actual construction state changes. It is necessary to strengthen the monitoring of B-type pile. The area where pile deformation is greater than the allowable displacement is zone A and the pile whose bottom is located in zone A is called A-type pile. A-type pile is in a dangerous state and must take reinforcement protection measures.

References

1. Burland, J.B.: Ninth Laurits Bjerrum Memorial Lecture: "Small is beautiful"-the stiffness of soils at small strains. Can. Geotech. J. **26**(4), 499–516 (1989)
2. Yin, J.: Application of hardening soil model with small strain stiffness in deep foundation pits in Shanghai. Chin. J. Geotech. Eng. **32**(S1), 166–172 (2010)
3. Ministry of Housing and Urban-Rural Development of the People's Republic of China: Code for Design of Metro. China Architecture & Building Press, Beijing (2014)
4. Committee for Compiling the Pile Foundations Handbook: Pile Foundations Handbook. China Architecture and Building Press, Beijing (1995)
5. Ministry of Housing and Urban-Rural Development of the People's Republic of China: Technical Code for Protection Structures of Urban Rail Transit. China Architecture & Building Press, Beijing (2014)

Experimental Study on Dynamic Penetration of OMNI-Max Anchor in Clay

Congcong Han[✉], Jun Liu, and Wei Zhao

State Key Laboratory of Coastal and Offshore Engineering, Dalian University of Technology,
Dalian 116024, Liaoning, China
hancongcong@mail.dlut.edu.cn

Abstract. As a newly developed gravity installed anchor, the OMNI-Max anchor is an efficient anchoring foundation due to its dynamic installation and diving property during keying. However, due to the complex geometry of the anchor, it is with difficulty to predict the anchor final penetration depth embedded in the seabed. This study carried out $1g$ model tests in normally consolidated (NC) and lightly overconsolidated (LOC) clay to investigate the effects of the impact velocity, and the soil strength on the anchor final penetration depth. The anchor dynamic penetration process in the soil was recorded by a micro-electromechanical systems (MEMS) accelerometer. By integrating the anchor acceleration recorded by the MEMS accelerometer, the anchor velocity-displacement curve was obtained. Finally, an empirical formula is put forward to predict the final penetration depth of the OMNI-Max anchor. This study may be beneficial for the engineering design and application.

Keywords: Gravity installed anchor · Dynamic penetration · Clay · Model test

1 Introduction

As a gravity installed anchor, the OMNI-Max anchor provides a capacity efficient alternative for the deep-water mooring system. The OMNI-Max anchor is comprised of three discontinuous plate flukes and a mooring loading arm located towards the anchor tip (Fig. 1). Based on the previous studies, this study conducted $1g$ model tests to simulate the dynamic penetration process of the OMNI-Max anchor in normally consolidated (NC) and lightly overconsolidated (LOC) clay. The effects of the impact velocity and soil undrained shear strength on the anchor final penetration depth are thoroughly discussed. Finally, a simplified empirical formula is proposed to quickly predict the anchor final penetration depth in practical engineering.

2 Experimental Details

During the anchor dynamic penetration within the soil, the anchor total energy has been dissipated by the soil resistance acting on the anchor, i.e.

© Springer Nature Switzerland AG 2018
W. Wu and H.-S. Yu (Eds.): *Proceedings of China-Europe Conference on Geotechnical Engineering*, SSGG, pp. 937–941, 2018.
https://doi.org/10.1007/978-3-319-97115-5_13

$$E_{\text{total}} = E_k + E_p = \frac{1}{2} m_A v_i^2 + W_A z_e = \int_0^{z_e} f \, dz \qquad (1)$$

where E_{total} is the anchor total energy at the soil surface, which includes the kinetic energy, E_k, at the soil surface and the remaining potential energy, E_p, relative to the anchor final penetration depth; m_A is the anchor mass; v_i is the anchor velocity at the soil surface, which is termed as the impact velocity; W_A is the anchor effective weight; z_e is the anchor final penetration depth from the anchor tip to the soil surface; and f is the soil resistance.

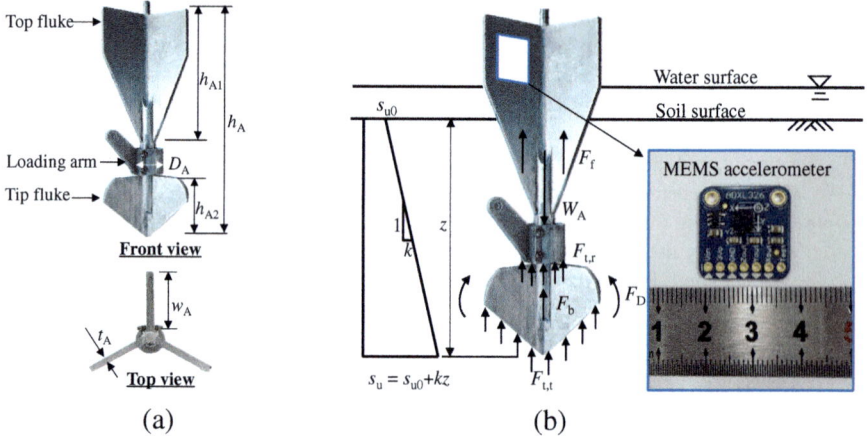

Fig. 1. OMNI-Max anchor: (a) photograph; and (b) forces acting on the anchor

The model anchor shown in Fig. 1(a), which was reduced with a scale of $\lambda_L = 50$, was a replica of the real OMNI-Max anchor in Shelton [1]. The anchor was made of stainless steel and has a prototype weight of 54.8 t. The primary dimensions of the anchor model are summarized as below: anchor length, $h_A = 181$ mm; top fluke length, $h_{A1} = 104.2$ mm; tip fluke length, $h_{A2} = 49.0$ mm; fluke width, $w_A = 37.7$ mm; fluke thickness, $t_A = 4$ mm; and ring diameter of the loading arm, $D_A = 22$ mm.

Speswhite kaolin clay (liquid limit, 65%; plastic limit, 34%; specific gravity, 2.61) was used to simulate the clayey soil in this study. Two types of soil samples, NC and LOC samples, were prepared. The soil undrained shear strengths are shown in Table 1. In the model tests, the anchor free fall process in water was ignored and the anchor was released from a predetermined height above the soil surface. The anchor was allowed to fall freely in the air before impacting within the seabed. A 10 mm-height water was maintained above the soil surface to ensure saturated conditions. A micro-electromechanical systems (MEMS) accelerometer (ADXL326 ± 16 g, see Fig. 1) was sealed in the model anchor to measure the anchor acceleration. By integrating the acceleration, the anchor velocity and the anchor fall distance can be determined.

Table 1. Testing cases details

Cases	Sample (strength in prototype)	v_i (m/s)		z_e (m)		z_e/h_A
		Model	Prototype	Model	Prototype	
1	1 (NC clay, $s_u = 2.4z$ kPa)	2.12	15	0.243	12.13	1.34
2		2.69	19	0.268	13.39	1.48
3		3.25	23	0.299	14.93	1.65
4	2 (NC clay, $s_u = 2.0z$ kPa)	2.12	15	0.273	13.67	1.51
5		2.69	19	0.308	15.39	1.70
6		3.25	23	–	–	–[§]
7	3 (LOC clay, $s_u = 10 + 2.0z$ kPa)	2.12	15	0.206	10.32	1.14
8		2.69	19	0.255	12.76	1.41
9		3.25	23	0.279	13.94	1.54

[§] The anchor dynamic penetration was failed in case 6.

3   Testing Results

3.1   Typical Result

Typical results of case 7 are plotted in Fig. 2. Based on the variation of the anchor vertical acceleration in Fig. 2(a), the anchor dynamic penetration process can be divided into four stages. In stage 1 (S1), the anchor is hanging in the air, hence the anchor acceleration $a_z = 0$. In stage 2 (S2), the anchor falls freely in the air, and the anchor acceleration $a_z = g$. In stage 3 (S3), the anchor impacts the soil. When the soil resistance is larger than the anchor self-weight, the acceleration varies from positive to negative. In stage 4 (S4), the anchor is rest within the soil, and the anchor acceleration reversely maintains zero.

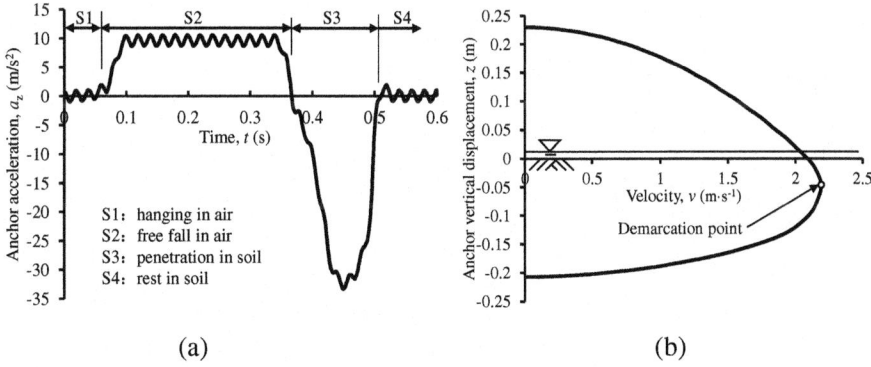

(a)   (b)

Fig. 2. Typical testing results for case 7: (a) anchor acceleration versus time; (b) anchor velocity versus anchor fall distance

The anchor velocity against the anchor displacement in Fig. 2(b) indicates that the anchor velocity first increases as the anchor falls freely in the air. The anchor velocity still exhibits an increase as the anchor impacts the soil initially, during which the soil resistance is lower than the anchor self-weight. As the anchor penetrates further into the soil, the velocity reversely decreases once the soil resistance is beyond the anchor self-weight.

3.2 Empirical Formula Based on the Anchor Total Energy

The relationship of the anchor final penetration depth and the total energy plotted in Fig. 3 can be presented as Eq. (2).

$$\frac{z_e}{D_{\text{eff}}} = \eta \left(\frac{E_{\text{total}}}{k_e A_s D_{\text{eff}}^2} \right)^r \tag{2}$$

where D_{eff} is the anchor equivalent diameter, which can be calculated by $D_{\text{eff}} = \sqrt{(4A_F/\pi)}$, A_F is the anchor frontal area; As is the anchor side surface area; and k_e is the equivalent soil strength gradient, which is calculated by $k_e = (s_{u0} + kz_e)/z_e$ (s_{u0} is the soil strength at the mudline, and k is the soil strength gradient as shown in Fig. 1); η and r are fitting parameters that can be determined via a least squares regression scheme to obtain a best-fit solution. The best-fit is obtained using $\eta = 2.02$ and $r = 0.39$. The experimental results show considerable consistency with the field tests reported by Zimmerman et al. [2].

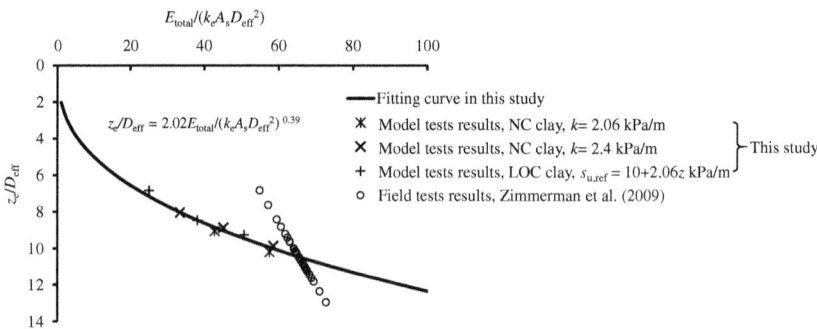

Fig. 3. Comparison of anchor final penetration depth

4 Conclusions

This study conducted $1g$ model tests to investigate the dynamic penetration of the OMNI-Max anchor in the NC and LOC clay. By using a MEMS accelerometer, the anchor history during the anchor dynamic penetration within the seabed is determined. Based on the testing results, an empirical formula based on the anchor total energy is proposed to predict the anchor final penetration for practical application.

References

1. Shelton, J.T.: OMNI-Max anchor development and technology. In: Proceedings of the Ocean Conference, Vancouver, Canada (2007)
2. Zimmerman, E.H., Smith, M.W., Shelton, J.T.: Efficient gravity installed plate anchors for deepwater mooring. In: Offshore Technology Conference, OTC20117. Houston, Texas, USA (2009)

Effect of Vertical Load on Lateral Behavior of Pile in Clay

Ben He[1,2(✉)], Kanmin Shen[1,2], Yi Hong[2], and Lizhong Wang[2]

[1] Power China Huadong Engineering Corporation Limited, Hangzhou, China
hebenzheda@126.com
[2] Zhejiang University, Hangzhou, China

Abstract. The influence of vertical load on the lateral stiffness and capacity of the pile is merely considered in the current design practice and therefore deserve a further investigation. This study presents detailed centrifuge tests results on the lateral behavior of a single pile in normal (NC) and over consolidated clay (OC), with and without application of vertical loading at the pile head. In the meantime, three-dimensional finite element analyses (FEA) were carried out to offer further insights into the effect mechanism of vertical loads on lateral pile behavior. Both physical and numerical investigation reveal that after applying the vertical load and allowing the dissipation of excess pore pressure in NC, the stress ratio (q/p') of the soil around the pile decreases while the mobilisable undrained shear strength (s_u) increases, resulting in 10% and 50% increase of the lateral initial stiffness and bearing capacity of the pile, respectively. In contrast, due to application of vertical load to a single pile in the over consolidated clay, the q/p' prior to lateral loading increases while the mobilisable s_u decreases, consequently leading to 13% and 33% reduction of the lateral initial stiffness and bearing capacity of the pile, respectively.

Keywords: Vertical load · Single pile · Lateral response · Clay

1   Introduction

Laterally loaded pile foundations are widely used for supporting structures like transmission towers, onshore and offshore wind turbines, and offshore structures. These piles are subjected to both lateral loads caused by wind, wave, and current, and the vertical load induced by self-weight. Currently in the design of pile foundations [1], its performances under lateral and vertical loading are studied separately. In this case, the lateral deflection is calculated only by the lateral load, which ignores the effect of vertical load on the lateral performance. Obviously, this simplified method is not appropriate to evaluate the performance of piled foundations under combined loads.

In view of the afore-mentioned issue, centrifuge tests were carried out to reveal the lateral behaviours of piles after applying the vertical working loads, in normally consolidated (NC) and over consolidated clay (OC). To meet the in-situ situation, re-consolidation of surrounding soil was allowed after the vertical loading. In the meantime, three-dimensional finite element analyses (FEA) were carried out to offer further insights into the effect mechanism of vertical loads on lateral pile behavior.

© Springer Nature Switzerland AG 2018
W. Wu and H.-S. Yu (Eds.): *Proceedings of China-Europe Conference on Geotechnical Engineering*, SSGG, pp. 942–946, 2018.
https://doi.org/10.1007/978-3-319-97115-5_14

2 Centrifuge Test and Preliminary Finite Element Analyses

2.1 Centrifuge Modelling

All of the centrifuge model tests reported in this study were carried out at a centrifugal acceleration of 40 g in the geotechnical centrifuge facility at the Hong Kong University of Science and Technology (HKUST). Each single model pile was made of a 420 mm long cylindrical aluminum tube (with an elastic modulus of 72 GPa and yield strength of 241 MPa). The tube had an outer and an inner diameter of 20 and 18 mm, respectively. The total length of the pile model was 43 cm, with 33 cm embedded in soil. The vertical working load, 50% of the estimated ultimate vertical capacity (Vult), was applied at the pile head by a lumped mass.

The clay adopted in the centrifuge tests was reconstituted from Speciwhite China Kaolin clay with a liquid limit (LL) of 27%, plastic limit (PI) of 61% and a specific gravity of 16.5 kN/m^3. The slurry was consolidated in two stages, i.e., preliminary consolidation at 1 g, followed by in-flight consolidation at 40 g. After the consolidation stage, micro T-bar tests were performed. The undrained shear strength and the deduced distribution of over-consolidation ratio (OCR) of the clay for NC and OC conditions are shown in Fig. 1, respectively.

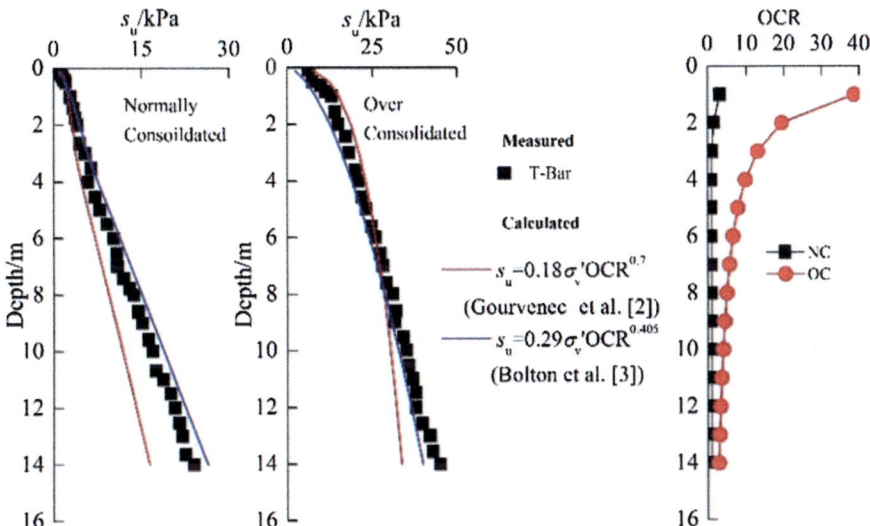

Fig. 1. Variation of OCR and undrained shear strength with depth in NC and OC clay

After installing the model pile, the excess pore pressure was fully dissipated, and then lateral load was applied at the pile head. The monotonic lateral loading period was set to 6 s, within which the soil was considered to be undrained.

2.2   Centrifuge Test Results

The measured monotonic load-deflection responses at the pile head (loading point, 2 m above the mudline in prototype) are shown in Fig. 2. It was apparent that the pile in NC clay benefited from the vertical working load (50% Vult) if the pore pressure induced by vertical load was dissipated, as both the stiffness and bearing capacity of the single pile increased. But the opposite effect was observed in OC clay. The vertical load decreased the stiffness and bearing capacity of the pile and resulted in a much softer response. In order to quantify the effect of vertical loading, the lateral capacity and initial stiffness of the pile were determined based on the method proposed by Kulhawy et al. [4]. In their method, it is suggested to fit the measured relationship between the lateral load (F) and the resulting lateral pile head displacement (δ) with the following hyperbolic curve:

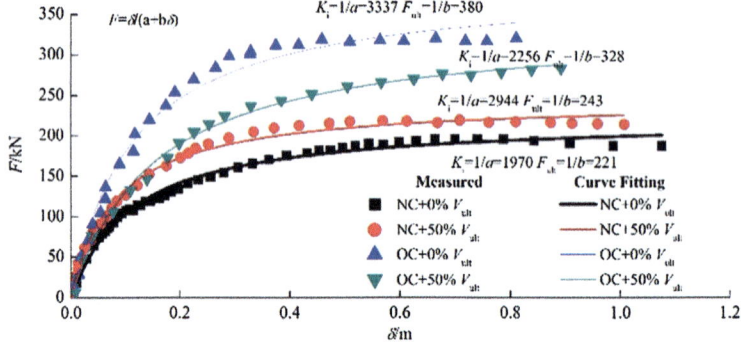

Fig. 2. Measured and fitting results of load-displacement responses at pile head

$$F = \frac{\delta}{a + b\delta} \tag{2}$$

The reciprocals of a and b represent the interpreted ultimate lateral capacity and initial pile head stiffness, as also shown in Fig. 2. Based on Kulhawy et al. [4], the ultimate lateral capacity and the initial stiffness of the pile in NC clay were increased by 10% and 50% respectively with the applied vertical load. However, in OC clay, the ultimate lateral capacity and the initial stiffness were decreased by 13% and 33%.

2.3   Preliminary Finite Element Analyses

To short this section, only the simulation results are shown as follow. Detailed introduction on finite element mesh, numerical modelling procedure and used constitutive model (hypoplastic clay model) parameters can be seen in reference [5].

In order to analyze the effect mechanism of vertical loads on lateral pile behavior, the stress path of the soil element in front of the pile (6 m below the mudline in prototype) was interpreted. As illustrated in Fig. 3, for the pile embedded in NC clay, after applying the vertical load, the consolidation process decreased the initial stress ratio q/p' and hence increased the undrained shear strength s_{u2} which can be mobilised in the following lateral loading process. The value of s_{u2} was increased by over 20% compared with the strength s_{u1} in simulation case without vertical loading.

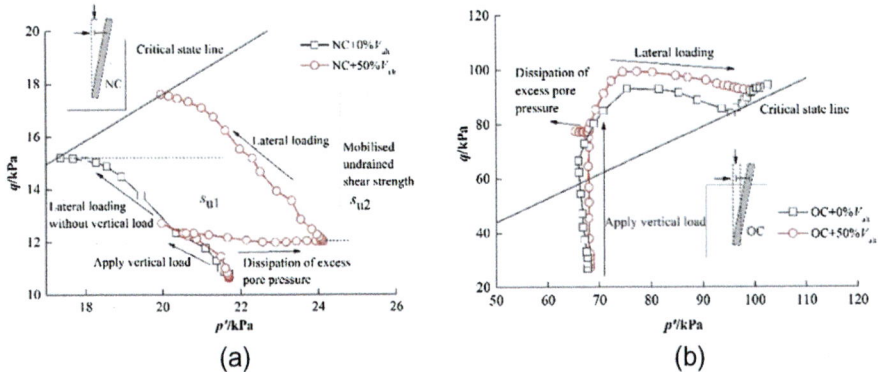

Fig. 3. Typical stress path of soil element in (a) NC and (b) OC clay

However, the difference between the stress paths in OC and NC clay was apparent. Due to the over consolidated state of the clay, only the elastic deformation was developed after applying the vertical working load, which indicated that only small pore pressure was generated in this process. The dissipation of pore pressure during the re-consolidation of the soil did not cause significant change in the stress ratio q/p', and hence undrained shear stress which can be mobilised did not change. The vertical load, however, increased the initial stress ratio q/p' in the soil and the soil shear strength had mostly been mobilized in order to resist the vertical load, which finial left only a small portion of the strength to resist the shearing by lateral loading.

3   Conclusion

In NC clay, applying the vertical load (50% Vult) and allowing the dissipation of excess pore pressure could improve the behaviour of the laterally loaded single pile. But the opposite effect was observed in OC clay. The effect mechanism of vertical loads on the lateral behaviour of the single pile could be interpreted through the initial stress ratio and mobilized shear strength of the surrounding soil: in NC clay, the vertical load and the dissipation of induced pore pressure reduced the initial stress ratio q/p' and increased undrained shear strength so that can be mobilised in the lateral loading. But in OC clay, the vertical load, on the contrary, increased the initial stress ratio q/p' and reduced the mobilisable undrained shear strength su.

References

1. American Petroleum Institute: Recommended practice for planning, designing and constructing fixed offshore platforms (2014)
2. Gourvenec, S., Acosta-Martinez, H.E., Randolph, M.F.: Experimental study of uplift resistance of shallow skirted foundations in clay under transient and sustained concentric loading. Géotechnique **59**(6), 525–537 (2009)
3. Boltton, M.D., Stewart, D.I.: The effect on propped diaphragm walls of rising groundwater in stiff clay. Géotechnique **44**(1), 111–127 (1994)
4. Kulhawy, F.H., Chen, Y.J.: A thirty-year perspective of Brom's lateral loading models, as applied to drilled shafts. In: Proceedings of the Bengt B. Broms Symposium on Geotechnical Engineering, pp. 225–240 (1992)
5. Hong, Y., He, B., Wang, L.: Cyclic lateral response and failure mechanisms of a semi-rigid pile in soft clay: centrifuge tests and numerical modelling. Can. Geotech. J. **54**(6), 806–824 (2017)

Dynamic Responses of Mono-piles in the Presence of Scour Holes

Rui He[✉], Tao Zhu, Bo Ma, and Wenyan Tao

College of Harbor, Coastal and Offshore Engineering, Hohai University, Nanjing, China
herui0827@163.com

Abstract. Mono-piles are widely used in offshore wind industry, and it will be necessary to consider the influences of scour for mono-piles under combined waves and currents. There are about two main effects on the mono-piles when scour occurs: the first is to reduce the ultimate resistances and the other is to reduce the impedances (stiffness and damping ratio). While impedances play a dominant role on the natural frequencies and dynamic responses of offshore wind turbines, the risk of resonance and fatigue will increase dramatically due to the presence of scour holes. In this paper, the evolution of dynamic characteristics of model mono-piles under different scour depths and excitation frequencies are studied by model tests. The scour depths are calculated by laboratory tests or field observations in the literature. By means of Fourier transform and modal analysis, the dynamic responses of mono-piles in time domain and frequency domain are obtained and analyzed. The important parameters for mono-piles are obtained, including the resonant frequencies and resonant amplitudes. The experimental results are compared with a dynamic numerical analysis. The results of this paper will be helpful to understand the dynamic characteristics of offshore mono-piles in complex ocean environment.

Keywords: Offshore wind turbine · Mono-pile · Scour · Resonance

1 Introduction

Mono-pile is the most popular support structure type for offshore wind turbines. Under the combined action of wave and current, the scour hole is easy to form around the pile foundation [1], and the formation of scour holes will reduce the lateral bearing capacities and change the vibration characteristics of mono-pile foundations. The type of scour can be divided into the general scour and the local scour [2]. When waves and currents are strong enough, the general scour occurs. Therefore, it is of great practical significance to study the dynamic characteristics of mono-pile foundations and ensure that these dynamic parameters are still in safe range after the formation of the scour hole. In this paper, the general scour and local scour are simulated by experiments, and the dynamic responses in time domain and frequency domain of the mono-pile foundation under lateral cyclic loading with different frequencies are obtained. Subsequently, a 3D-FEM model is established to compare with of the experimental results.

© Springer Nature Switzerland AG 2018
W. Wu and H.-S. Yu (Eds.): *Proceedings of China-Europe Conference on Geotechnical Engineering*, SSGG, pp. 947–951, 2018.
https://doi.org/10.1007/978-3-319-97115-5_15

2   Formulations

2.1   Experimental Design

A mono-pile foundation test platform is designed in this paper. The soil tank with size of $2 \times 2 \times 1.5$ m is filled with dry Nanjing quartz sand, and the mono-pile foundation is placed in the center by hammer-driven (refer to Fig. 1), and the lateral harmonic load is applied at a mass block which is welded on the top of the pile by a vibration exciter. The force is measured by a force sensor, while an accelerometer and a displacement transducer are used to measure the dynamic responses of the foundation. The lateral harmonic load is applied to the pile at a constant frequency and the time domain responses are obtained at first. With the change of the load frequency, the displacement frequency response curve can be plotted. By analyzing the frequency domain responses, the resonant frequencies and amplitudes are obtained. Through this test platform, the change of the vibration characteristics of mono-pile foundations is investigated when $0D$, $1D$ and $2D$ scour hole occur around the mono-pile [3] and the initial embedded depth is 10D (D is the diameter of the pile). The local scour holes are dug by a shovel with $45°$, while the general scour is simulated by different embedded depths of the pile.

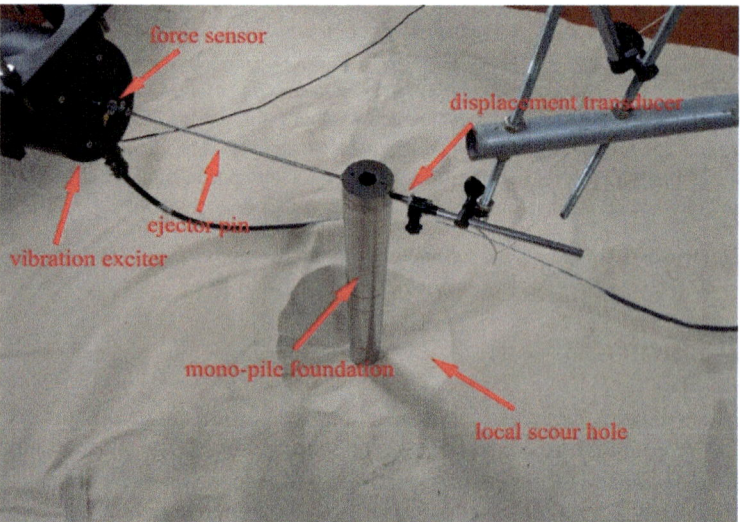

Fig. 1. Illustration of the model test.

The mono-pile model used in the test is shown in Fig. 1. It is a hollow pipe pile made of stainless steel with radius $D = 6$ cm, length $l = 90$ cm, wall thickness $h = 1$ mm. For the test sand, its particle size is 0.2–0.6 mm, with medium diameter 0.42 mm. In order to obtain uniform soil layers, the sand is prepared layer by layer. The maximum dry density is 1.57 g/cm^3, the minimum dry density is 1.35 g/cm^3, the density of the sand is 1.44 g/cm^3, the void ratio is $e = 0.84$, and the relative density is 0.45. Shear modulus of the sand at small strains are obtained by the formula given by Hardin and Drnevich [4].

After installation, a sweep frequency test is carried out, in which the steady-state sinusoidal signal is used, the frequency range is 0-200 Hz, and the sweep speed is 1 Hz/s [5]. Frequency Response Function is obtained and the resonance frequency and amplitude are generally understood, then the constant frequency test is carried out. The frequency increases from 0–180 Hz at the rate of 10 Hz and the displacement frequency response curve is obtained.

2.2   Test Data Processing

Firstly, the influence of the general scour on the vibration characteristics of the pile foundation is investigated. It can be obtained from the displacement frequency response curve in Fig. 2(a) that the resonant frequency of the mono-pile foundation is about 100 Hz, 90 Hz and 70 Hz when the general scour depth is 0D, 1D and 2D respectively. The resonant amplitude increases with the increase of the scour depth.

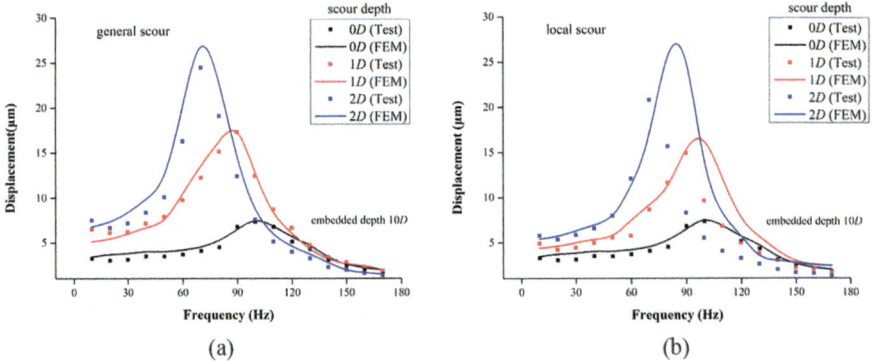

Fig. 2. Displacements of the mono-pile under different scour depths in frequency domain and the comparison of FEM and the experimental results.

Compared to general scour, the resonant frequency of local scour is very similar (Fig. 2(b)), while the resonant amplitude for local scour is relatively smaller than that of the general scour. This shows that the general scour is more dangerous than the local scour for the design of the offshore wind turbine.

2.3   Comparison of Test Results with FEM Model

In this paper, the numerical analysis is carried out using FEM in ABAQUS [7]. According to the actual model test, the symmetric boundary condition is set up. The perfect contact condition is set between the mono-pile foundation and the soil contact surface [6]. Viscoelastic material is used to depict the sand as very small forces are applied during the tests, and Rayleigh damping with alpha = 0, beta = 0.0003 is adopted in the FEM model according to the half-power bandwidth method based on frequency sweep test results. The results are shown in Fig. 2.

As shown in Fig. 2, the trend in both FEM model and model test is similar. For the general scour, the results obtained by FEM coincide with the experimental results very well, while for the local scour the comparison is not so perfect. For the local scour, the resonant amplitude obtained by FEM is much larger than that obtained in model test, which is believed to due to damping in the model test cannot be captured accurately in FEM. Damping in the model test is divided into material damping and radiative damping. In the FEM model, Rayleigh damping is used for the soil, and the displacement of the FEM model is larger than that obtained by the actual test. As to the resonant frequencies calculated by FEM, they are larger than those obtained by the test. This is because in the presence of scour hole, stress release of the surface soil causes the decay of the soil shear modulus which is of no consideration in FEM. Thus, the shear modulus in the FEM model is actually larger than that in the test.

In the actual design process of offshore wind farms, one main consideration is to avoid resonance of the whole wind turbine, and the resonant amplitudes are very difficult to predict accurately. Therefore, the FEM model established in this paper can predict the resonant frequencies of the mono-pile foundation when the damping of the system is estimated accurately, which has great significance in practical applications.

3   Conclusions

In this paper, the lateral vibration of a mono-pile in the presence of scour is studied by a 1g laboratory test. Based on the vibration test of the mono-pile foundation, the lateral resonant frequencies and amplitudes of the mono-pile foundation are obtained. The results are compared with a FEM model. From the comparison, it can be found that the resonant frequencies of mono-pile foundation can be well predicted by FEM method, while the amplitudes obtained in the FEM model is much larger than the model test results, which is believed to due to damping cannot be well captured in the FEM model.

Acknowledgements. The author would like to acknowledge the support of the Grant No. 51509082 from the National Natural Science Foundation of China, the Grant No. BK20150804 from the Natural Science Foundation of Jiangsu province.

References

1. Tseng, W.C., Kuo, Y.S., Chen, J.W.: An investigation into the effect of scour on the loading and deformation responses of monopile foundations. Energies **10**(8), 1–11 (2017)
2. Van der Tempel, J., Zaaijer, M.B., Subroto, H.: Wind turbine structural dynamics - a review of the principles for modern power generation, onshore and offshore. Wind Eng. **26**(4), 211–220 (2002)
3. Sumer, B.M., Fredsoe, J.: Scour around pile in combined waves and current. J. Hydraul. Eng. **127**(5), 403–411 (2001)
4. Hardin, B.O., Dmevich, V.P.: Shear modulus and damping in soils. J. Soil Mech. Found. Div. **98**, 667–692 (1972)

5. Lin, J.: Vertical dynamic impedance of offshore wind turbine bucket foundations. Master thesis, Hohai University (2017)
6. Kuo, Y., Achmus, M., Kao, C.: Practical design considerations of monopile foundations with respect to scour. In: Global Wind Power, pp. 29–31. Beijing (2008)
7. ABAQUS, Dassault Systèmes Simulia Corp (2012)

Random Field Simulation of Foundation Settlement of Soft Soil in Southern China

Linchong Huang, Shuai Huang, and Yu Liang[✉]

Sun Yat-sen University, Guangzhou 510275, China
liangyu25@mail.sysu.edu.cn

Abstract. In order to study the effect of spatial variability of soil parameters on foundation settlement. In this paper, the method of combining the theory of random field and numerical analysis is used to systematically analyze the settlement probability of the soft soil foundation in the south of China. The influence of spatial variability of soil parameters on probability settlement of foundation is studied. The results indicate that the settlement value of foundation increases with the increase of parameter variation coefficient, which is the most sensitive to deformation modulus. When the spatial variability of soil space is relatively large, the auto-correlation function selection has a great influence on the settlement of the foundation. The settlement value based on the exponential square is larger, and the single exponential is smaller.

Keywords: Foundation settlement · Random field · Auto-correlation functions

1 Introduction

The soft soil is widely distributed in the coastal areas of southern China, with high compressibility and low shear strength [1]. With the acceleration of infrastructure construction in the region, a large number of structures are built on soft soil foundation, so it is of great significance to study the settlement prediction of soft soil foundation. At present, the analysis methods of foundation settlement are mainly classical formula method and numerical analysis method [2, 3]. However, these two traditional methods neglect the spatial variability of soil parameters. At present, many scholars at home and abroad have done a lot of research on geotechnical engineering considering the spatial variability of soil parameters. Li et al. [4] proposed a non-invasive stochastic finite element method and applied the method to the reliability analysis of underground caverns. Amaneh [5] simulated soil mass as an anisotropic random field, study on the influence of soil spatial variability on settlement of shallow ground. Lo [6] used Latin hypercube sampling to simulate the random field, and apply it to the reliability analysis of strip foundation and slope. The theory of the random field indicates that: there is spatial correlation of soil between any two points in space. Auto-correlation functions are generally used to solve the correlation distance. Commonly used autocorrelation functions include single exponential (SNX), exponential square (SQX), cosine exponential (CSX), second-order Markov (SMK) and binary noise (BIN) [7]. This paper

© Springer Nature Switzerland AG 2018
W. Wu and H.-S. Yu (Eds.): *Proceedings of China-Europe Conference on Geotechnical Engineering*, SSGG, pp. 952–955, 2018.
https://doi.org/10.1007/978-3-319-97115-5_16

mainly studies the influence of variation of rock and soil parameters and the selection of autocorrelation function on the foundation settlement.

2 Random Field FE Simulation

In practical engineering, the soil generally obeys non-Gaussian distribution, and there is some cross-correlation between the soil parameters. This paper will simulate the related non-Gaussian random fields based on Cholesky decomposition technique with midpoint discretization [8–10]. Related non-Gaussian distribution of random field simulation needs to generate the relevant standard Gaussian random field, this paper takes the relevant standard Gaussian random field fetch index, derived correlation logarithmically random field, $S_i(x, y) = \exp(\mu_{\ln i} + \sigma_{\ln i} \cdot S_i^D(x, y))$, $(i = c, \varphi)$.

Considering the auto-correlation between any two points of the soil, which is characterized by the auto-correlation coefficient matrix K of the soil. K is solved by the theoretical autocorrelation function. The Cholesky decomposition of the autocorrelation coefficient matrix K is performed, $K = L_1 L_1^T$, and the lower triangular matrix L_1 is obtained. Considering the cross-correlation between cohesion and internal friction angle, the cross-correlation coefficient R matrix is used. Cholesky decomposition of the cross-correlation matrix, $R = L_2 L_2^T$, leads to the lower triangular matrix L_2. A set of related standard normal random sample matrices α was derived using Latin hypercube sampling, $\alpha_i = \{\alpha_i^1, \alpha_i^2, \cdots, \alpha_i^{n_e}\}$, $(i = c, \varphi)$. The relevant standard Gaussian random field, $S_i^D(x, y) = L_1 \cdot \alpha \cdot L_2^T$.

3 Example of Foundation Settlement Analysis

3.1 Deterministic Analysis

Based on ANSYS software, this paper establishes a two-dimensional foundation plane strain model. The model is shown in Fig. 1. There is a concentrated load $P = 100$ kN on the foundation. Calculated parameters are as follows: cohesion 20 kPa, internal friction angle 12°, heavy 18 kN/m³, modulus of deformation 4 MPa, Poisson's ratio 0.25.

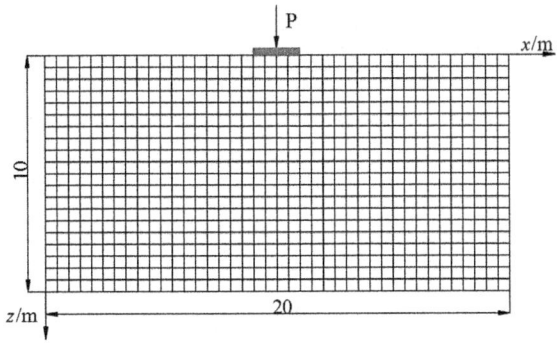

Fig. 1. Finite element model

The maximum vertical settlement is 41.18 mm by the finite element calculation, just below the foundation. In order to verify the accuracy of the model calculation results, the theoretical solution to solve the foundation settlement by the traditional hierarchical design method is 39.1 mm, the results obtained are close to each other, the error is 5.3%, which shows that the numerical simulation results are reliable.

3.2   Randomness Analysis

Figure 2 (a)–(d) are the curves of the maximum settlement average with the COV_E, COV_c, COV_φ, $\rho(c, \varphi)$, respectively. It can be seen from the figure that with the increase of coefficient of variation of soil parameters, the average value of maximum subsidence also increases, but the average value is larger than the result of deterministic analysis, which indicates that the parameter variability of soil has an important influence on settlement. From the angle of parameter sensitivity, $E > c > \varphi$. The sedimentation mean curve calculated based on the five kinds of auto-correlation functions was analyzed. Among them, the mean maximum settlement value obtained by SQX was larger and SNX was smaller.

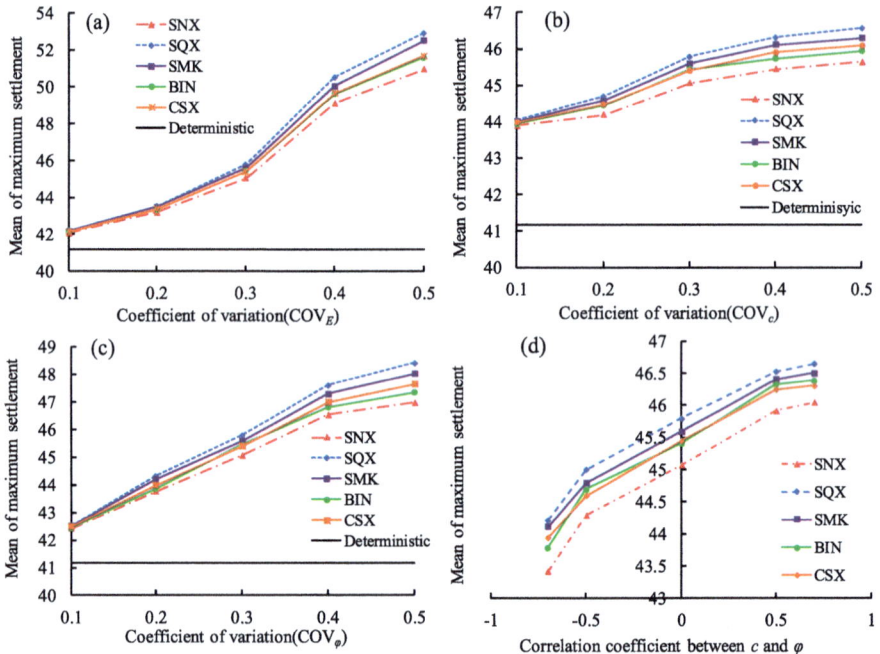

Fig. 2. Curve of foundation settlement randomness analysis

4   Summaries

This paper combines Cholesky decomposition midpoint method with Monte-Carlo method and obtains the calculation method of two-dimensional ground settlement based on random field theory, and considers the influence of the autocorrelation function selection in the random field simulation, and draws the following conclusions:

(1) The variability of soil parameters has a significant effect on the calculation results of foundation settlement, and the results of randomness analysis are greater than the results of deterministic analysis. The mean value of maximum settlement increases with the variation coefficient of the parameters, and the deformation modulus E of soil affects the calculated value of foundation settlement the most. At the same time, the foundation settlement increases with $\rho(c, \varphi)$.

(2) The selection of different auto-correlation functions has a significant effect on the calculation of foundation settlement, the settlement value based on SQX and SMK is larger, and the settlement value obtained based on SNX and BIN is smaller. The result of SNX is significantly smaller than the other types. With the increase of coefficient of variation, the influence of the selection of autocorrelation function on the settlement value also increases.

References

1. Huang, L.C., Zhou, C.Y., Li, W.H.: Modeling the Microstructure Random Fields of Soft Soil in the South of China. vol. 236, pp. 495–501. Geotechnical Special Publication (2014)
2. Yang, G.H., Li, J., Jia, K., et al.: Improved settlement calculation method for engineering practice. Chin. J. Rock Mech. Eng. **36**(S2), 4229–4234 (2017)
3. Hou, J.F., Chen, J., Kou, X.Q.: Numerical analysis of soft soil ground consolidation settlement. Appl. Mech. Mater. **638–640**, 503–506 (2014)
4. Li, D.Q., Jiang, S.H., Chen, Y.F., et al.: Reliability analysis of serviceability performance for an underground cavern using a non-intrusive stochastic method. Environ. Earth Sci. **71**(3), 1169–1182 (2014)
5. Kenarsari, A.E., Chenari, R.J.: Probabilistic settlement analysis of shallow foundations on heterogeneous soil stratum with anisotropic correlation structure. vol. 1914, pp. 1905–1914. Geotechnical Special Publication (2015)
6. Lo, M.K., Leung, Y.F.: Probabilistic analyses of slopes and footings with spatially variable soils considering cross-correlation and conditioned random field. J. Geotech. Geoenviron. Engi. **143**(9), 04017044 (2017)
7. Cao, Z., Wang, Y.: Bayesian model comparison and selection of spatial correlation functions for soil parameters. Struct. Saf. **49**, 10–17 (2014)
8. Cho, S.E., Park, H.C.: Effect of spatial variability of cross-correlated soil properties on bearing capacity of strip footing. Int. J. Numer. Anal. Methods Geomech. **34**(1), 1–26 (2010)
9. Kasama, K., Whittle, A.J., Zen, K.: Effect of spatial variability on the bearing capacity of cement-treated ground. Soils Found. **52**(4), 600–619 (2012)
10. Jiang, S.H., Dian-Qing, L.I., Zhou, C.B., et al.: Slope reliability analysis considering effect of autocorrelation functions. Chin. J. Geotech. Eng. **36**(3), 508–518 (2014)

Unified Theoretical Approach to Ultimate Horizontal Pullout Capacity of Vertical Strip Anchor Plate in Sand

Wei Hu[✉], Chengbi Long, Wenhua Gao, and Chaofeng Zeng

Hunan Province Key Laboratory of Geotechnical Engineering Stability
Control and Health Monitoring, School of Civil Engineering,
Hunan University of Science and Technology, Xiangtan 411201, Hunan, China
yilukuangben1982@163.com

Abstract. Self-developed visual horizontal drawing strip anchor plate model test and numerical simulation experiment showed that there existed a triangular soil core before anchor plate under ultimate pulling condition, whose two bottom corner's variation could reflect the symmetry of failure slip-line field indirectly. Along with the increasing of buried ratio, the upper bottom corner enlarged from φ to π/4 + φ/2, and the lower bottom corner reduced from π/2 to π/4 + φ/2 with the sum of two corners basically remained unchanged. In this course, the failure slip-line field before anchor plate evolved from asymmetric to symmetric gradually. Base on this knowledge, corresponding assumptions were put forward to construct the ultimate bearing mechanical model of vertical strip anchor plate under horizontal drawing, and ultimate mechanical equilibrium analysis method was used to derive the unified theoretical formula of ultimate bearing capacity at last. Calculation of four tests and comparison to other two theories indicated that unified theoretical solution had good applicability to vertical strip anchor plate in sand.

Keywords: Vertical strip anchor plate · Horizontal pulling · Mechanical model
Ultimate bearing capacity · Unified theoretical solution

1 Introduction

Ultimate pullout capacity of anchor plate is a key index has to be provided in the design of anchor-plate retaining structure. Current study presents the displacement and deformation laws of soil before anchor plate under ultimate pulling by integrated application of self-developed visual horizontal drawing model test and numerical simulation test. Then ultimate bearing mechanical model will be set up to derive the unified theoretical solution of ultimate bearing capacity, which will be proved by comparison to other theories and tests datum.

© Springer Nature Switzerland AG 2018
W. Wu and H.-S. Yu (Eds.): *Proceedings of China-Europe Conference
on Geotechnical Engineering*, SSGG, pp. 956–960, 2018.
https://doi.org/10.1007/978-3-319-97115-5_17

2   Bearing Mechanism Study

Self-developed visual horizontal drawing model test apparatus is shown in Fig. 1. High resolution camera was used to shoot the soil around anchor plate under ultimate pullout with buried ratio (H/h) changing from 2 to 15, where H is the buried depth and h is the height of anchor plate. Displacement field was figured by image analysis software PhotoInfor and PostViewer. In the test, friction angle of sand is 35°, dry unit weight $\gamma = 15.13\,\text{kN/m}^3$ and relative density $D_r = 50\%$. Figure 2 shows the typical displacement field vector of soil before anchor plate and its envelope angle. The result indicates that this angle enlarges from fast to slow with the increasing of buried ratio, and generally changing from $\pi/2$ to $(\pi/4 + \varphi/2) + \pi/2$.

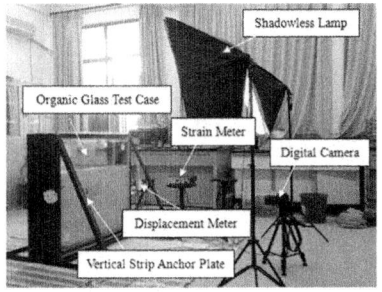

Fig. 1. Model test apparatus

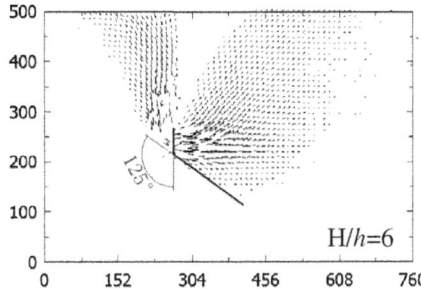

Fig. 2. Envelope angle of displacement field vector

Numerical model of vertical strip anchor plate was set up using software ABAQUS. The parameters of soil and anchor plate are the same with that of model test. The contact interface between soil and plate is supposed to be completely coarse. Typical plastic strain field of soil is shown in Fig. 3 with different buried ratios under ultimate pulling. There exists a triangular soil core moving with anchor plate and whose two bottom corner's variation can reflect the symmetry of failure slip-line field indirectly. Along with the increasing of buried ratio, the upper bottom corner enlarges from φ to $\pi/4 + \varphi/2$. Accordingly, the lower bottom corner reduces from $\pi/2$ to $\pi/4 + \varphi/2$ with the sum of two corners basically remains $\pi/2 + \varphi$, unchanged.

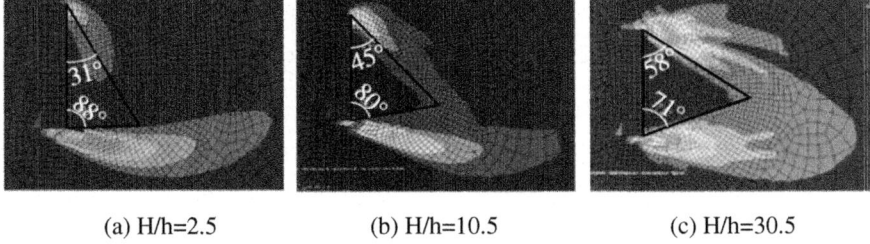

(a) H/h=2.5   (b) H/h=10.5   (c) H/h=30.5

Fig. 3. Plastic strain fields of soil before plate

3 Mechanical Model and Unified Theoretical Solution

Unified mechanical models of vertical strip anchor plate under horizontal ultimate pulling are shown in Fig. 4. γ and φ are the unit weight and friction angle of soil respectively, K_0 is the static soil pressure coefficient. $\triangle ABC$ represents the soil core before anchor plate, whose upper and lower bottom corners are α_1 and α_2, $\varphi \leq \alpha_1 \leq \pi/4 + \varphi/2$, $\pi/4 + \varphi/2 \leq \alpha_2 \leq \pi$. Log spiral CD is the slip line in downside of soil with starting radius $\overline{BC}$. Ψ is the angle between BC and BD, $\Psi = -\dfrac{2\pi}{\pi - 2\varphi}\alpha_2 + \dfrac{\pi^2}{\pi - 2\varphi}$.

Force Analysis of Soil Core. $\triangle ABC$ is shown in Fig. 5. $\triangle ABC$ satisfies the static equilibrium conditions of $\sum X = 0$, $\sum Y = 0$ and $\sum M_A = 0$.

$$T - Q_1 \cos\left(\alpha_1 - \varphi\right) - Q_2 \cos\left(\alpha_2 - \varphi\right) = 0 \tag{1}$$

$$Q_1 \sin\left(\alpha_1 - \varphi\right) - Q_2 \sin\left(\alpha_2 - \varphi\right) + W_1 + W_2 = 0 \tag{2}$$

$$\overline{AB}\left[Q_1 \cos\left(\alpha_1 - \varphi\right) + Q_2 \cos\left(\alpha_2 - \varphi\right)\right]\big/2 - \overline{BC} \cdot Q_2 \sin\alpha_2 \sin\left(\alpha_2 - \varphi\right)/2$$
$$-\overline{AC} \cdot Q_1 \cos\varphi\big/2 - Q_2 \cos\left(\alpha_2 - \varphi\right)\left(\overline{BC}\cos\alpha_2\big/2 + \overline{AC}\cos\alpha_1\right) - W_1 l_1 - W_2 l_2 = 0 \tag{3}$$

Where, W_1 and W_2 are gravities of $\triangle ABC$ and anchor plate itself respectively, Q_1 and Q_2 are resultant earth pressures of upside and downside of $\triangle ABC$ respectively, T is the ultimate pullout load, l_1 and l_2 are the arms of W_1 and W_2 to the point A respectively. If Q_1 and Q_2 are the known quantities, α_1 and α_2 can be solved by combination of Eqs. 2 and 3, substituting them into Eq. 1 will obtain T at last. Q_1 and Q_2 can be calculated by force analysis of curved line triangles ACE (or polygon ACEFG) and BCD based on static equilibrium conditions respectively.

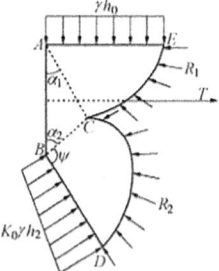

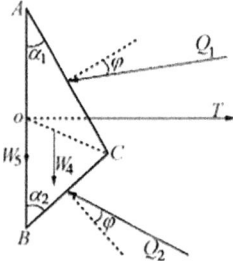

Fig. 4. Mechanical model of vertical strip anchor plate

Fig. 5. Force analysis of soil core

4   Results

New theoretical solution is used to calculate four test cases and compared with Terzaghi method and Code method. Details of four test cases about vertical strip anchor plate are shown in Table 1. Calculation in Fig. 6 indicates that Terzaghi method is generally conservative, Code method's results are very discrete, sometimes too safe and sometimes too dangerous, New method's results are closer to test values and have lesser Standard Deviation and least Variable Coefficient, which means obvious advantage over other two traditional methods.

Table 1. Test cases of vertical strip anchor

Cases	$\varphi(°)$	$\gamma(\mathrm{kN/m^3})$	h(m)	H/h
Ovesen and Strømann (1972)	30	15.5	0.075/0.0375	1.5/3.5/9.5
	38	17.1	0.015	
Neely et al. (1973)	37.5	13.82	0.0508	1/1.5/2/2.5/3/3.5/4.5
Sawwaf and Nazir (2006)	37	17.97	0.075	1.5/2/2.5
	41	18.84		
This paper	35	15.13	0.05	2/3/5/6/7/8/9/10/11/12/13/14/15

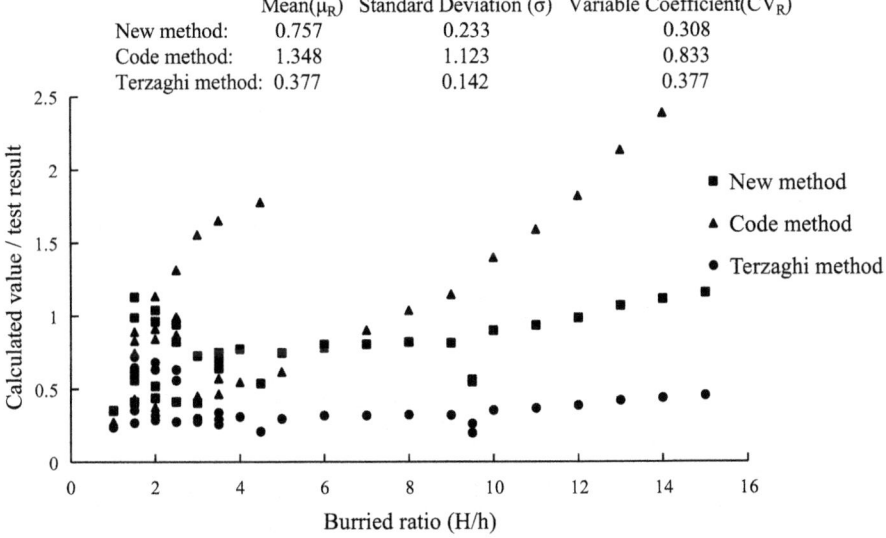

	Mean(μ_R)	Standard Deviation (σ)	Variable Coefficient(CV_R)
New method:	0.757	0.233	0.308
Code method:	1.348	1.123	0.833
Terzaghi method:	0.377	0.142	0.377

Fig. 6. Calculated value/test result versus buried ratio

5   Conclusions

Unified theoretical solution has been created for the ultimate tensile capacity of vertical strip anchor in sand. Calculation comparison shows good agreement with four test cases and advantages over other two theoretical methods.

References

Terzaghi, K.: Theoretical Soil Mechanics. John Wiley & Sons Inc., New York (1943)

CCCC FIRST HARBOR CONSULTANTS CO., LTD: Code for design and construction for quay wall of sheet pile (JTS 167-3-2009). China Communication Press, Beijing (2009)

Ovesen, N.K., Strømann, H.: Design method for vertical anchor slabs in sand. In: Performance of Earth and Earth-Supported Structures, pp. 1481–1500 (1972)

Neely, W.J., Stuart, J.G., Graham, J.: Failure loads of vertical anchor plates in sand. J. Soil Mech. Found. **SM9**, 669–685 (1973)

Sawwaf, M.E., Nazir, A.: The effect of soil reinforcement on pullout resistance of an existing vertical anchor plate in sand. Comput. Geotech. **33**, 167–176 (2006)

Joint Research into the Behaviour of Driven Piles

R. J. Jardine[1(✉)] and Z. X. Yang[2]

[1] Imperial College, London SW7 2AZ, UK
r.jardine@imperial.ac.uk
[2] Zhejiang University, Hangzhou 310058, China
zxyang@zju.edu.cn

Abstract. Large driven piles are used widely in both onshore and offshore construction. Predicting their limiting capacities and load-displacement behaviour under a range of static and cyclic, axial, lateral and moment loading conditions is critical to many engineering applications. This paper reviews relevant recent joint research by groups at Imperial College London (ICL) and Zhejiang University China (ZJU). Two tracks of enquiry are outlined: (i) assembling and analysing a major and open database of high quality load tests conducted on industrial scale piles at well characterised sites; and (ii) modelling the effective stress regime developed around piles driven in sands. Both avenues of research are vital to enabling scientifically well-founded and yet industrially credible improvements to practical pile design methods. The scope of future joint research is also outlined.

Keywords: Driven piles · Sands · Database · Stress regime · Joint research

1 Introduction

Large driven piles provide support worldwide for large bridges, harbour works and also the thousands of offshore structures that have been installed to produce oil, gas and renewable energy supplies. Designers need to ensure that their foundation piles can be driven successfully and are able to sustain the static and cyclic loads imposed safely, especially in severe marine environments. A need to reduce infrastructure costs, especially in offshore energy projects where oil prices and renewable power tariffs have reduced dramatically, has led to equal attention being given to ensuring economy in design. Up to 30% of the capital costs of offshore wind energy are associated with wind-turbine foundations. Geotechnical engineering advances have contributed to the increasing competitiveness of this important renewable energy resource.

Jardine [1] summarised findings from several recent and current research projects involving the Imperial College Geotechnics group that have contributed to advancing the design of large driven piles. These included the PISA Joint Industry study for monopiles under lateral and moment loading in sands and clays [2], work with international colleagues on a range of experimental, theoretical and database projects and large scale investigations of pile behaviour under axial loading in chalk [3–5]. The latter have involved the first large-scale offshore field tests of which we are aware where

© Springer Nature Switzerland AG 2018
W. Wu and H.-S. Yu (Eds.): *Proceedings of China-Europe Conference on Geotechnical Engineering*, SSGG, pp. 961–972, 2018.
https://doi.org/10.1007/978-3-319-97115-5_18

autonomous underwater pile tests have been conducted on the seabed. The systems shown in Fig. 1 were deployed in 40 m of water to test 1.37 m outside diameter piles driven in a succession of glacial till over low-to-medium density chalk at three locations in the Wikinger windfarm, sited in the German sector of the Baltic Sea. Other case histories that demonstrate the industrial impact of the research include those given by [6–9].

We concentrate in this paper on two strands of research that has been undertaken jointly by Imperial College London (ICL) and Zhejiang University (ZJU) with support from a Newton Advanced Fellowship awarded from the UK's Royal Society and supported by matching funding from Natural Science Foundation of China.

2   Joint ZJU-ICL Studies into the Reliability of Axial Capacity Predictions

The joint driven pile research involves first a macro-level approach that recognises the lack of international agreement on which design methods offer the most reliable predictions for axial capacity – the factor that dominates the design of multi-legged jacket structures. Here the ZJU-ICL team concentrated first on collating new databases of pile load tests, ensuring the quality of the tests and associated site investigations and adding value to the tests by conducting new experiments and analyses. Growing from earlier work summarised by [10, 11], new tests have been added while the application of stricter quality criteria have eliminated other. Table 1 summarises the 117 tests assembled for piles driven in sands that were reduced into a consistent database format and made publicly accessible by [12, 13].

The extended sand database provided the key resources that enabled the assessment of 6 different axial capacity design methods, which is summarised in Table 2. As may be seen, the physically based ICP-05 [11] and UWA-05 [17] 'CPT' methods led to the best reliability statistics when expressed as mean values and Coefficients of Variation (CoV) for the ratios of calculated Qc and measured Qm axial capacities. The Qc/Qm statistics found from the 80 'age filtered' tests in the ZJU-ICL database are summarised in Table 2. The individual Qc/Qm ratio results varied by up to ±0.1 when all 117 tests were included; this order of sensitivity to the specific dataset is typical of surveys involving around 100 piles; the statistics become far more sensitive to individual cases with smaller populations. Table 2 indicates slightly more favourable outcomes for the API [14] Main Text method than earlier studies. The Yang et al. [13] analysis also highlighted the significantly poorer statistical outcomes when the "offshore variants" of the ICP-05 and UWA-05 approaches were considered. While the latter approaches are preferred in the API [14] Commentary sections, it is better to retain the full ICP-05 procedure for use in practical design.

A parallel Joint Industry study led by NGI led to broadly compatible conclusions regarding the Q_c/Q_m ratios applying to piles driven in sand, as summarised by [2]. The ZJU-ICL joint database study is now turning to consider the behaviour of piles driven in clay.

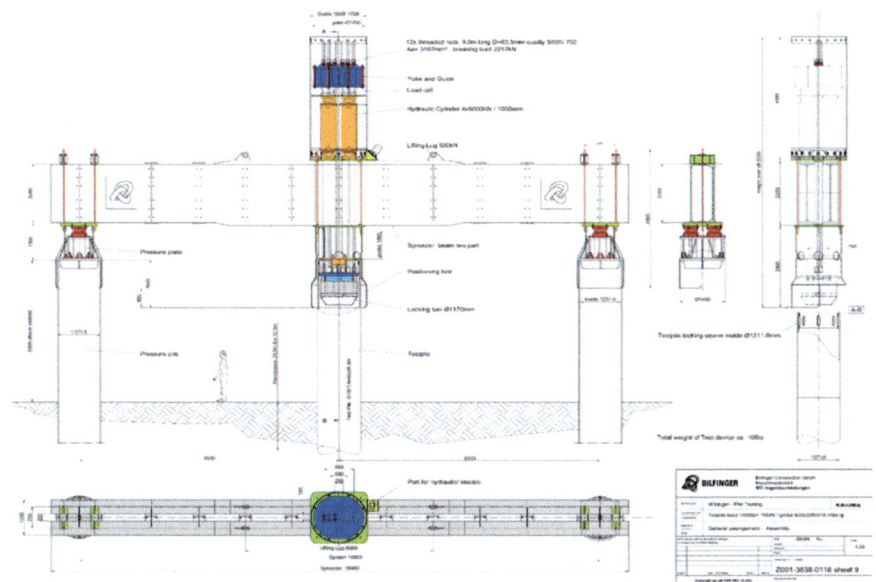

Fig. 1. Load test arrangements employed by Bilfinger for Wikinger project, German Baltic Sea, after Barbosa et al. [3]).

The NGI led Joint Industry database project identified significant shortfalls in the best current international datasets (see Lehane et al. [2]) and the ongoing ZJU-ICL study is now seeking to identify both new tests and conduct additional laboratory and field investigations to add value to existing case histories for which vital information is currently unavailable. Approaches have been made to colleagues internationally, including the ISSMGE's TC 212. The Authors would be grateful to any delegates who may have suitable test and SI data that they can contribute to the construction of an authoritative, reliable and accessible international database.

The novel features of the successful 'CPT' sand methods listed in Table 2 originated in the use of highly instrumented displacement piles to investigate the key factors that governed shaft and base failure in sands, with a focus on the local shear and radial stress distributions developed on the pile shafts during installation, equalisation and load testing to failure. Establishing reliable experimental observations of the stress regimes developed around displacement piles and understanding how these may change is vital to making any further improvements and to considering key features of field behaviour that remain poorly understood, including the marked effects of pile ageing on shaft capacity [19, 20] and the impact of cyclic loading [21].

These questions are being investigated in the second main strand of collaborative work, which focuses on the detailed processes and mechanics of the soil responses, which include consideration down to the micro-level of how individual soil grains behave around the shafts of driven piles.

Table 1. Main features of ZJU-ICL database for piles driven in sand; after Yang et al. [13]

	All entries			Filtered entries with age = 10–100 days		
	Closed	Open	All	Closed	Open	All
Number of piles	62	55	117	48	32	80
Steel	25	48	73	18	26	44
Concrete	37	7	44	30	6	36
Tension tests	10	31	41	8	16	24
Compression tests	52	24	76	40	16	56
Average length L(m)	17.6	25.2	21.2	18.9	26.0	21.8
Range of lengths L(m)	6.2–45	5.3–79.1	5.3–79.1	6.2–45	5.3–79.1	5.3–79.1
Average of diameter D(m)	0.413	0.645	0.522	0.422	0.667	0.520
Range of diameter D(m)	0.2–0.7	0.324–2.0	0.2–2.0	0.2–0.7	0.324–2.0	0.2–2.0
Average of density D_r(%)	54	60	57	54	61	57
Range of D_r(%)	28–89	30–88	28–89	31–89	30–87	30–89
Average test time after installation	35	80	61	43	28	35

Table 2. Joint ZJU-ICL database test assessment of approaches for axial capacity prediction in sands; means and coefficients of variation (CoV) for Qc/Qm ratios; summarised from Yang et al. [13]

Method	Mean	CoV
API [14], Main Text	0.88	0.55
Fugro-05 [15]	1.20	0.47
ICP-05 (full) [11]	0.94	0.30
NGI-05 [16]	1.23	0.48
UWA-05 (full) [17]	1.05	0.35
LCPC-82 [18]	1.25	0.40

3 Characterising the Effective Stress Regime Around Piles Driven in Sands

In addition to the practical macro-level studies outlined above, the joint ZJU-ICL research into driven pile behaviour has tackled the fundamental mechanics of pile driving, ageing in-situ and loading under both static and cyclic conditions. The research started with Yang's work with Imperial College and the late Professor Pierre Foray's group at the Université Grenoble Alpes' 3S-R laboratory. Their large calibration chamber was modified considerably to allow long-term, highly instrumented, model pile experiments to be conducted under closely controlled pressurized conditions. Jardine et al. [22], Zhu et al. [23], Yang et al. [24], Tsuha et al. [25], Jardine et al. [26, 27], and Rimoy et al. [20] describe multiple static and cyclic experiments with a stainless steel cyclically jacked mini-ICP pile (with 36 mm diameter, 1 m length) that could monitor shaft shear and normal stresses at three levels. As indicated in Fig. 2, dozens of soil stress sensors were also deployed within the masses of Fontainebleau sand (of both NE34 and GA39 grades) that were air-pluviated into a medium dense (75% relative density) state before being pre-loaded and aged under 150 kPa vertical stress to match the average sand state (void ratio and pressure) applying to the field and calibration chamber tests. The NE34 silica sand has both a similar grading and comparable grain shapes to the sands found at the Dunkirk site in Northern France where large scale field tests were conducted earlier, as described by Jardine et al. [19]. Parallel laboratory testing was conducted at Imperial College that employed high pressure triaxial cells to study NE34 sand's behaviour under the extreme stress conditions imposed by pile installation; [28–30].

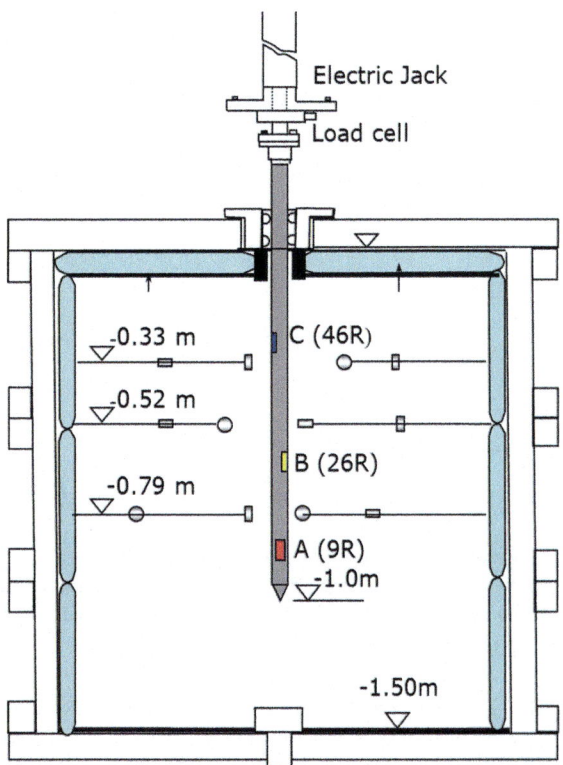

Fig. 2. Calibration chamber test arrangements at Grenoble 3S-R laboratory for mini-ICP pile experiments in NE-34 Fontainebleau sand mass equipped with multiple soil stress sensors and subjected to 150 kPa vertical stresses; after Jardine et al. [22].

Among the many findings from the research were the micro-level observations presented by Yang et al. [24] of the particle crushing that takes place beneath the pile tip and leads to a 'crust' of crushed and densified material adhering to the pile shaft, which has also been found in field tests at Dunkirk and elsewhere. This observation, along with the observations made of the soil stress system during installation that are illustrated in Fig. 3 and the pile tip loads provide benchmarks against which new theoretical predictions may be tested and, hopefully, refined.

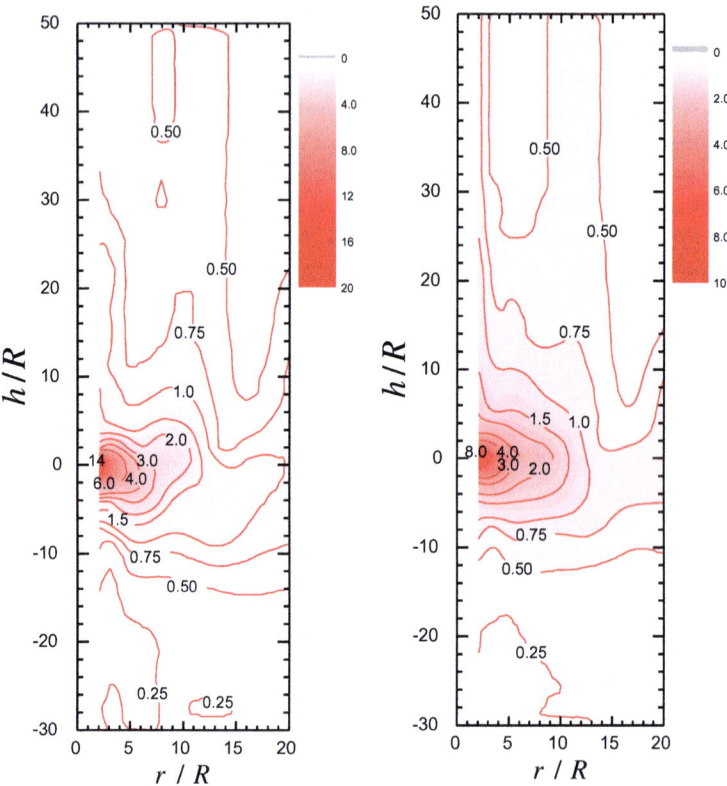

Fig. 3. Experimental observations of normalised radial effective stress regime, σ'_r/q_c developed in sand mass around the mini-ICP during (left) penetration stages and (right) paused between strokes of cyclic jacking installation process in pressurised NE34 sand; after Jardine et al. [27]. Note q_c is local CPT tip resistance.

Professor Einav's group at Sydney was the first to attempt simulations of the Calibration Chamber pile experiments. They employed a crushable soil grain model that had been calibrated to the soil element testing conducted at Imperial College and applying an Arbitrary Lagrangian Eulerian (ALE) FE technique were able to capture some key aspects of the experiments, as outlined by [31, 32]. Their simulations for the soil stress regime gave encouraging results and also indicated scope for further improvement. Particle by particle (Discrete Element Method) simulations were undertaken by [33, 34] at Imperial College in which half a million (over-sized) crushable grains were employed along with a reduced size soil volume. These analyses provided similarly good simulations of the experimentally observed pile tip resistances and grain crushing zones and gave predictions for the stress regime developed by pile installation.

However, scope exists for further improvements in the numerical analysis of displacement pile installation. In addition to seeking still better matches for the Calibration Chamber tests, major developments are required to allow analyses to be undertaken of the stress regime developed around the open-ended tubular piles employed in most large scale offshore, harbour and bridge driven pile projects.

A team led by Yang has been advancing such studies at ZJU as part of the joint work with Imperial College. The ALE technique within ABAQUS has been applied first to simulate the calibration chamber experiments described above. The ALE option is efficient as it can adopt a 2D axisymmetric mesh and employ relatively low numbers of finite elements; see Fig. 4. In the simulation, the pile is considered as a rigid Lagrangian body with a standard cone tip (D = 2R = 36 mm), a total length of 1.5 m and a 1 m final embedment (L), giving (L/R ≈ 56). The sand mass was regarded as deformable material subject to ALE adaptive mesh control. The same mesh dimensions as the calibration chamber were adopted to match any boundary effects, with width w = 0.6 m, height H = 1.5 m, w/R = 33.3. The default ALE control parameters were adopted in all the simulations attempted. The contact between the pile and soil adopted ABAQUS' built-in surface-to-surface contact laws, based on a master-slave-principle, in which the pile is treated as master surface and sand as the slave. The penalty friction formulation was adopted with a friction coefficient of 0.46, taken from the interface friction angle measurements of 25°–27° found in interface ring shear tests by Yang et al. [24].

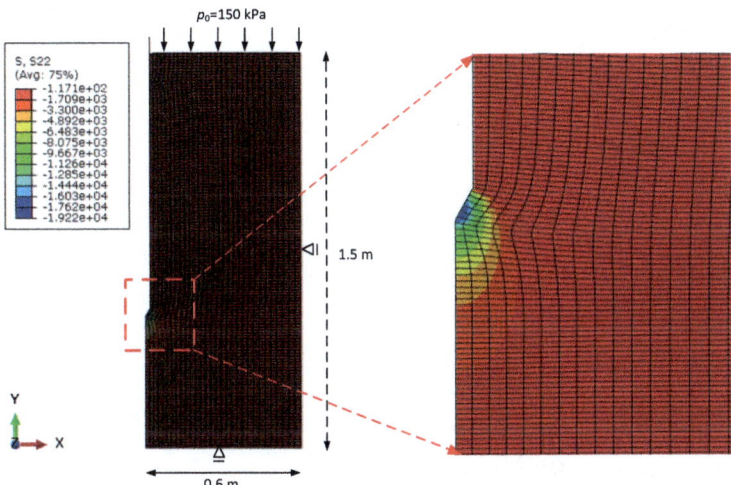

Fig. 4. Detained information of the 2D ALE model for calibration chamber experiments

A simple Mohr-Coulomb elastic-plastic soil model was adopted for the sand with the material parameters being calibrated to generate equivalent qc profiles with experiments. Based on the experimental results, the sand mass can be partitioned into different zones with the distinct mean normal stress and the plastic shear strain in each zone, as illustrated in Fig. 5. The varying dilatancy angle ψ and friction angle φ can be assigned in each zone to accommodate the state and strain-level dependency of the sand's

behaviour. Figure 6 presents the stress distributions obtained from numerical simulations. Figure 6(a) presents the normalized radial stresses measured at radial distances r/
R = 2 at three levels in sand mass, z/R = 10.6, 30.6 and 46.1, the same as those measurements made in Jardine et al. [27]. Figure 6(b) presents the normalised radial stress
contour maps. These results match well with the experimental data reported by [27].
The deformation fields developed during the pile penetration were also captured by the
numerical simulations, as illustrated in Fig. 7. Figure 7(a) presents the radial displacement contours, while Fig. 7(b) gives the radial strain contours. The numerical simulations appear to show encouraging agreement with experimental results obtained by
Arshad [35] in another calibration chamber that offered scope for Digital Image Correlation analysis of the sand movements that employed a 'half-cylindrical' arrangement
that differed from the Grenoble chamber and employed a different test sand.

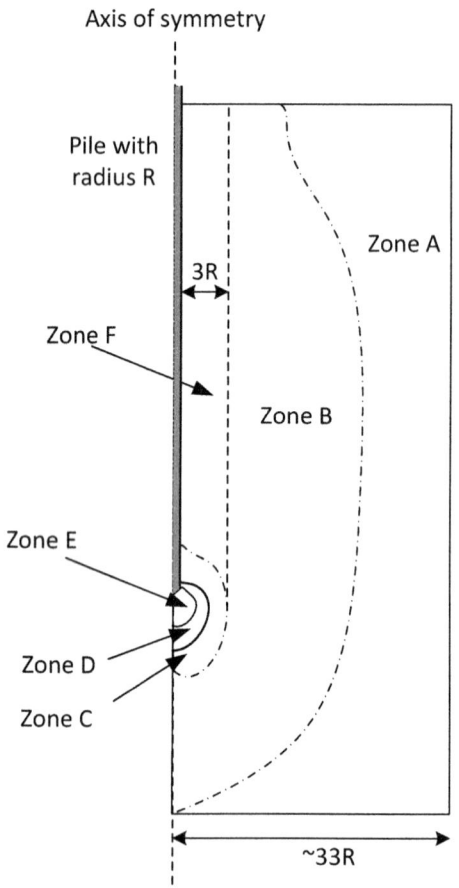

Fig. 5. Schematic diagram of zones with different dilatancy and friction angles

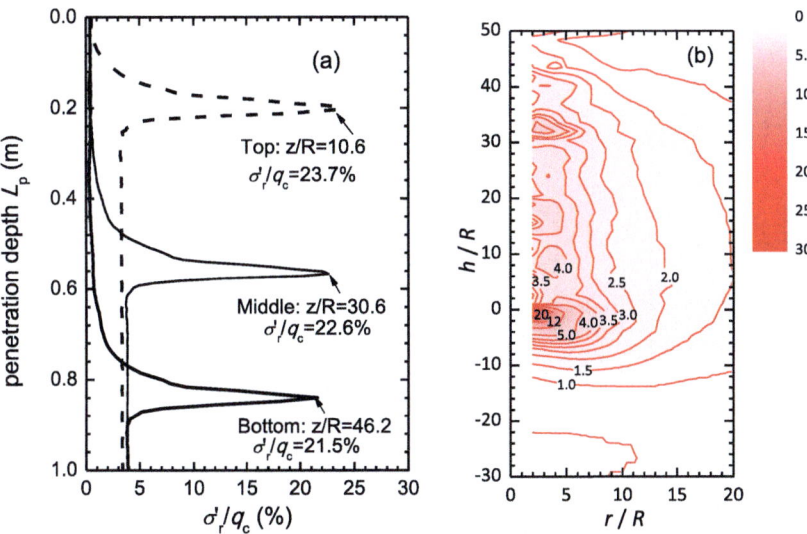

Fig. 6. ALE simulations of normalized radial stresses during penetration (a) measured at r/R = 2; (b) contours

Numerical simulations of full scale tubular piles pose more changeling tasks. Multiple elements are required across the width of the annular pile tip and these lead to very large overall elements numbers being required for precise FE analysis and render the problem beyond the capacity of conventional computation. A multiscale approach appears to be a more attractive solution, where material close to the inner and outer pile shaft could be modelled by a Discrete Element Method (DEM) approach and the far field be treated by FE. In this way, the contact between the pile surface and the sand grains, and the large deformations taking place at the soil-pile interface, can be handled more conveniently. Soil plugging phenomena can also be explored in more detail and its effect on the capacity of open-ended piles can be considered. Such work is in progress and will be reported in the near future.

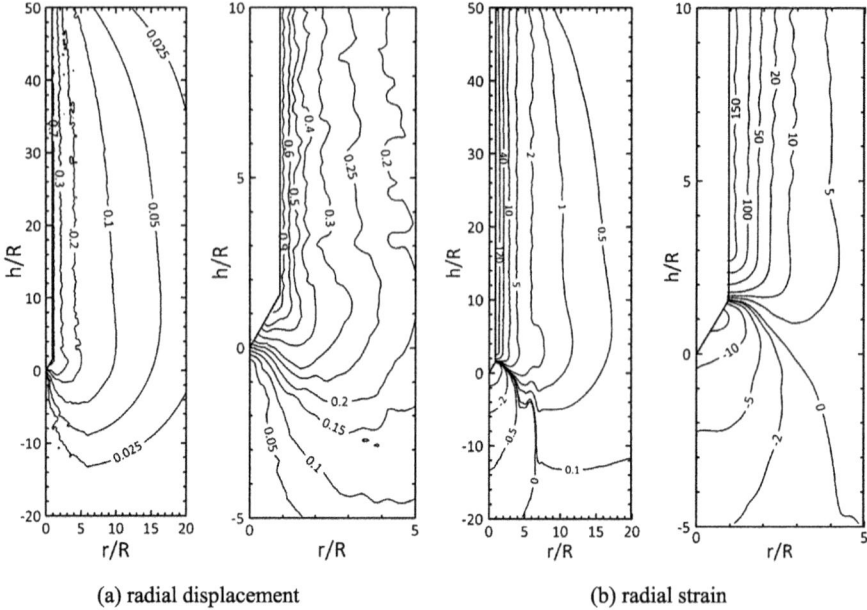

(a) radial displacement   (b) radial strain

Fig. 7. ALE simulations of radial displacement/strain during penetration (a) displacement (b) strain

4   Summary and Conclusions

A considerable demand exists, particularly from the offshore engineering sector, for improvements in the design tools available for large driven piles. Joint work between Imperial College London and Zhejiang University China is contributing to meeting this demand. Our joint paper has outlined two tracks of research that tackle the problem from (i) the macro-level, by assembling and adding value to industrial scale pile tests in a major database study and (ii) from a fundamental level that works from the grain scale upwards and includes new simulations of the detailed stress field developed around driven piles. Both avenues of research are vital to enabling scientifically well-founded and yet industrially credible improvements to practical pile design methods.

Although not described in this paper, the joint ZJU-ICL team is also conducting parallel research into other societally important geotechnical topics, which include the analysis of how the cyclic loading imposed by moving vehicles impacts on the settlements of the ground beneath road and railway pavements.

Acknowledgement. The joint research described in this study was supported by Newton Advanced Fellowship (NA160438) and Natural Science Foundation of China (51761130078) that were jointly awarded by Royal Society and NSFC.

References

1. Jardine, R.J.: Geotechnics and Energy. In: 56th Rankine Lecture (2016). To appear in Géotechnique
2. Lehane, B.M., Lim, J.K., Carotenuto, P., Nadim F., Lacasse, S., Jardine, R.J., Dijk, B.F.J.: Characteristics of unified databases for driven piles. In: Proceedings of 8th International Conference on Offshore Site Investigations and Geotechnics, SUT London, vol. 1, pp. 162–194 (2017)
3. Barbosa, P., Geduhn, M., Jardine, R., Schroeder, F., Horn, M.: Full scale offshore verification of axial pile design in chalk. In: Proceedings of International Symposium Frontiers in Offshore Geotechnics (ISFOG), vol. 1, pp. 515–520. CRC Press, London (2015)
4. Buckley, R.M., Jardine, R.J., Kontoe, S., Parker, D., Schroeder, F.: Ageing and cyclic behaviour of axially loaded piles driven in chalk. Géotechnique **68**(2), 146–162 (2018)
5. Buckley, R.M., Jardine, R.J., Kontoe, S., Lehane, B.M.: Effective stress regime around a jacked steel pile during installation ageing and load testing in chalk. Can. Geotech. J. (2018, in Press)
6. Jardine, R.J., Thomsen, N.V., Mygind, M., Liingaard, M.A., Thilsted, C.L.: Axial capacity design practice for North European Wind-turbine projects. In: Proceedings of International Symposium Frontiers in Offshore Geotechnics (ISFOG), Oslo, vol. 1, pp. 581–586. CRC Press, London (2015)
7. Argiolas, R., Jardine, R.J.: An Integrated pile foundation re-assessment to support life extension and new build activities for a mature North Sea oil field project. In: Proceedings of 8th International Conference on Offshore Site Investigations and Geotechnics, SUT London, vol. 2, pp. 695–702 (2017)
8. Hampson, K., Evans, T.G., Jardine, R.J., Moran, P., Mackenzie, B., Rattley, M.J.: Clair Ridge: Independent foundation assurance for the capacity of driven piles in very hard soils. In: Proceedings of 8th International Conference on Offshore Site Investigations and Geotechnics, SUT London, vol. 2, pp. 1299–1306 (2017)
9. Rattley, M.J., Costa, L., Jardine, R.J., Cleverly, W.: Laboratory test predictions of the cyclic axial resistance of a pile driven in North Sea soils. In: Proceedings of 8th International Conference on Offshore Site Investigations and Geotechnics, SUT London, vol. 2, pp. 636–643 (2017)
10. Chow, F.C.: Investigations into displacement pile behaviour for offshore foundations. Ph.D. thesis, Imperial College, London, UK (1997)
11. Jardine, R.J., Chow, F.C., Overy, R.F., Standing, J.R.: ICP Design Methods for Driven Piles in Sands and Clays. Thomas Telford Ltd., London (2005)
12. Yang, Z.X., Jardine, R.J., Guo, W.B., Chow, F.C.: A Comprehensive Database of Tests on Axially Loaded Piles Driven in Sands. Elsevier, Amsterdam (2015)
13. Yang, Z.X., Guo, W.B., Jardine, R.J., Chow, F.C.: Design method reliability assessment from an extended database of axial load tests on piles driven in sand. Can. Geotech. J. **54**, 59–74 (2017)
14. American Petroleum Institute: API ANSI/API recommended practice 2GEO. RP2GEO, 1st edn. API, Washington, D.C (2014)
15. Kolk, H.J., Baaijens, A.E., Sender, M.: Design criteria for pipe piles in silica sands. In: Proceedings of the International Symposium on Frontiers in Offshore Geotechnics, pp. 711–716. Taylor & Francis, London (2005)
16. Clausen, C.J.F., Aas, P.M., Karlsrud, K.: Bearing capacity of driven piles in sand, the NGI approach. In: Proceedings of the International Symposium on Frontiers in Offshore Geotechnics, pp. 677–681. Taylor & Francis, London (2005)

17. Lehane, B.M., Schneider, J.A., Xu, X.: CPT based design of driven piles in sand for offshore structures, GEO:05345. The University of Western Australia (2005)
18. Bustamante, M., Gianeselli, L.: Pile bearing capacity by means of static penetrometer CPT. In: Proceedings of the 2nd European Symposium on Penetration Testing, Amsterdam, pp. 493–500 (1982)
19. Jardine, R.J., Standing, J.R., Chow, F.C.: Some observations of the effects of time on the capacity of piles driven in sand. Géotechnique **55**(4), 227–244 (2006)
20. Rimoy, S.P., Silva, M., Jardine, R.J., Foray, P., Yang, Z.X., Zhu, B.T., Tsuha, C.H.C.: Field and model investigations into the influence of age on axial capacity of displacement piles in silica sands. Géotechnique **67**(7), 578–589 (2015)
21. Jardine, R.J., Standing, J.R.: Field axial cyclic loading experiments on piles driven in sand. Soils Found. **52**(4), 723–737 (2012)
22. Jardine, R.J., Zhu, B., Foray, P., Dalton, C.P.: Experimental arrangements for the investigation of soil stresses developed around a displacement pile. Soils Found. **49**(5), 661–673 (2009)
23. Zhu, B., Jardine, R.J., Foray, P.: The use of miniature soil stress measuring cells in laboratory applications involving stress reversals. Soils Found. **49**(5), 675–688 (2009)
24. Yang, Z.X., Jardine, R.J., Zhu, B.T., Foray, P., Tsuha, C.H.C.: Sand grain crushing and interface shearing during displacement pile installation in sand. Géotechnique **60**(6), 469–482 (2010)
25. Tsuha, C.H.C., Foray, P.Y., Jardine, R.J., Yang, Z.X., Silva, M., Rimoy, S.P.: Behaviour of displacement piles in sand under cyclic axial loading. Soils Found. **52**(3), 393–410 (2012)
26. Jardine, R.J, Zhu, B.T., Foray, P., Yang, Z.X.: Measurement of stresses around closed-ended displacement piles in sand. Géotechnique **63**(1), 1–17 (2013a)
27. Jardine, R.J, Zhu, B.T., Foray, P., Yang, Z.X.: Interpretation of stress measurements made around closed-ended displacement piles in sand. Géotechnique **63**(8), 613–628 (2013b)
28. Altuhafi, F., Jardine, R.J.: Effect of particle breakage and strain path reversal on the properties of sands located near to driven piles. In: Chung, et al. (eds.) Proceedings of Deformation Characteristics of Geomaterials, IS-Seoul, vol. 1, pp. 386–395. Hanrimwon, Seoul (2011)
29. Aghakouchak, A., Sim, W.W., Jardine, R.J.: Stress-path laboratory tests to characterise the cyclic behaviour of piles driven in sands. Soils Found. **44**(5), 917–928 (2015)
30. Altuhafi, F., Jardine, R.J., Georgiannou, V.N., Sim, W.W.: Effects of particle breakage and stress reversal on the behaviour of sand around displacement piles. Géotechnique (2018, in Press)
31. Zhang, C., Nguyen, G.D., Einav, I.: The end-bearing capacity of piles penetrating into crushable soils. Geotechnique **63**(5), 341–354 (2013)
32. Zhang, C., Yang, Z.X., Nguyen, G.D., Jardine, R.J., Einav, I.: Theoretical breakage mechanics and experimental assessment of stresses surrounding piles penetrating into dense silica sand. Géotech. Lett. **4** (January to March), pp. 11–16 (2014)
33. Ciantia, M.: Personal Communication (2016)
34. Ciantia, M.O., O'Sullivan, C., Jardine, R.J.: DEM Investigation of stress evolution around displacement piles in sand. Géotechnique (2018, Under Review)
35. Arshad, M.: Experimental study of the displacements caused by cone penetration in sand. Ph.D. thesis, Purdue University, USA (2014)

3D Settlement Analysis of Underpinning Piles Under Raft Foundation Subjected to Nonuniform Vertical Loading

Volkan Kalpakcı[1(✉)], Şevki Öztürk[2], H. Murat Algın[3], and A. Burak Ekmen[3]

[1] Hasan Kalyoncu University, Gaziantep, Turkey
volkan.kalpakci@hku.edu.tr
[2] Erzurum Technical University, Erzurum, Turkey
[3] Harran University, Sanliurfa, Turkey

Abstract. Existing rafts under the design loads sometimes experience excessive settlements or confront such a possibility in the future if the modified functionality of building is induced to increase the foundation loading. Differential settlement and deflection may also be observed in case of eccentric loading especially when these structures are built on soft soils. One of such precedents was observed in a silo structure used as a cement plant located in Douala, Cameroon. The structure that is founded on deep, soft clay with high ground water table rests on a raft supporting the storage tanks located at one side of the building conveying non-uniform loading to the existing raft. In one and a half year after completion of the construction, the silo structure had significantly settled and deflected. The under-pinning pile remediation system allowing the continuity of cement production is applied from the outside of building using the rigidly connected protruding rein-forced concrete section as a capping beam. In this study, the entire foundation system is numerically analyzed using the presented 3D finite element (FE) models. The back-analyses are used for the calibration concurring with the actual measurements of settlement and deflection at the site. The foundation systems with and without the underpinning piles are compared with each other to reveal how the remedial improvement is achieved by the presented underpinning pile system.

Keywords: Underpinning piles · Raft · Settlement · Soft clay · Finite element

1 Introduction

The utilization of underpinning piles as a remedial measure is commonly referred to get through the excessive settlement of existing raft. The experiences reveal that conventional design procedures without settlement analyses are not always correct, nor safe. Tilting and settlement problems are frequently observed at silo structures especially when they are built on soft and/or problematic soil deposits [1–3]. Such a silo structure was constructed as a part of a cement plant in Douala, Cameroon. However, the silo structure of the plant had tilted significantly in about 1.5 year after construction. After

© Springer Nature Switzerland AG 2018
W. Wu and H.-S. Yu (Eds.): *Proceedings of China-Europe Conference on Geotechnical Engineering*, SSGG, pp. 973–977, 2018.
https://doi.org/10.1007/978-3-319-97115-5_19

tilting and excessive settlements were observed, firstly a site investigation was conducted at the silo structure. Also, the silo structure was re-analyzed in detail to obtain the pressure distribution on the foundation of the silo. After identifying the problems, a piled remediation system was designed by the authors of this study to prevent further tilting of the silo structure. In this study, the silo structure is modelled by 3D finite element software Abaqus based on the technique presented previously [4]. The model parameters are calibrated according to the site measurements (tilting, settlement, etc.). After calibration of the model, the analyses are carried out for cases with and without the remediation system and the estimated further settlements for these two cases are compared with each other. Based on the analyses, it is concluded that the expected further settlements under this structure was decreased almost by 85% with the construction of the designed remediation system.

2   Investigations

The investigations may be grouped in two main categories as site investigations and analytical investigations. During the site investigations, 3 boreholes (each 35 m deep) were drilled and deep, soft clay deposits, the basic characteristics of which were formerly studied by [5, 6], were encountered in the study area. Groundwater table depth was between 0.30–0.80 m from ground surface. Pressuremeter tests were conducted at every 1 m depth along the boreholes. Pressuremeter test results are averaged to categorize the

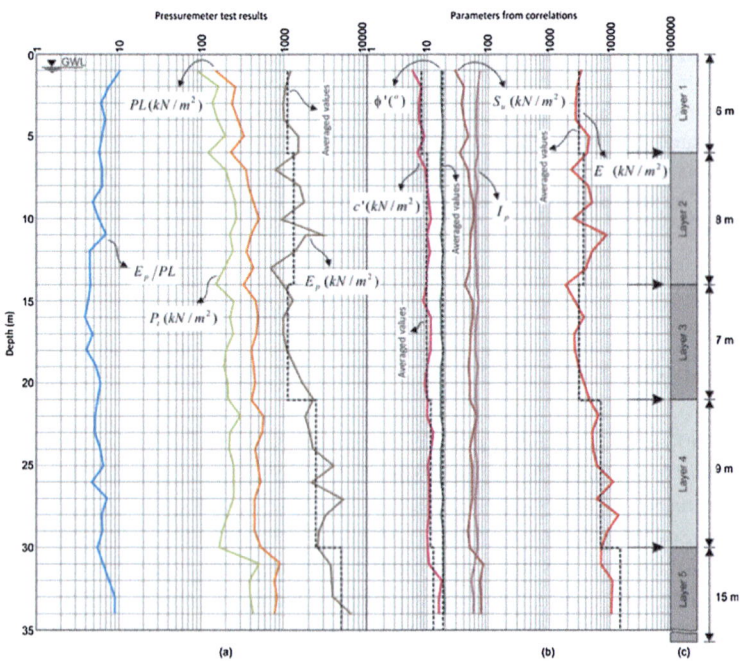

Fig. 1. Averaged pressuremeter test results and the correlated parameters.

soil layering as shown in Fig. 1(a) and the available correlations from the current literature are used to obtain the drained parameters depicted in Fig. 1(b) required for the effective stress analysis. Figure 1(c) shows the soil layers, their depths and the corresponding soil parameters utilized in the presented numerical analyses.

The structural re-analyses showed a significantly non-uniform load distribution on the raft of the structure. Moreover, the higher loads were concentrated at the tilting side of the structure due to the placement of machinery parts. Based on these findings, it was concluded that the main reasons of the tilting of the structure were the non-uniform load distribution on the foundation and the soft/deformable soil conditions.

3   Results and Discussions

The 3D FE meshes and the settlement results are shown in Figs. 2 and 3, respectively. Figure 3(a) and (b) indicate that the ultimate settlement (58.88 cm) occurs at the tapering part of the raft prior to the underpinning. The measured settlement after 1.5 year (20 cm) indicates that about one third of settlement has taken place in this period. The ultimate settlement for the raft with the underpinning piles is obtained as 10.22 cm indicating around 85% improvement in settlement (Fig. 3c and d). The analyses shown in Fig. 3 demonstrates that the remedial design is rather accurate and effective.

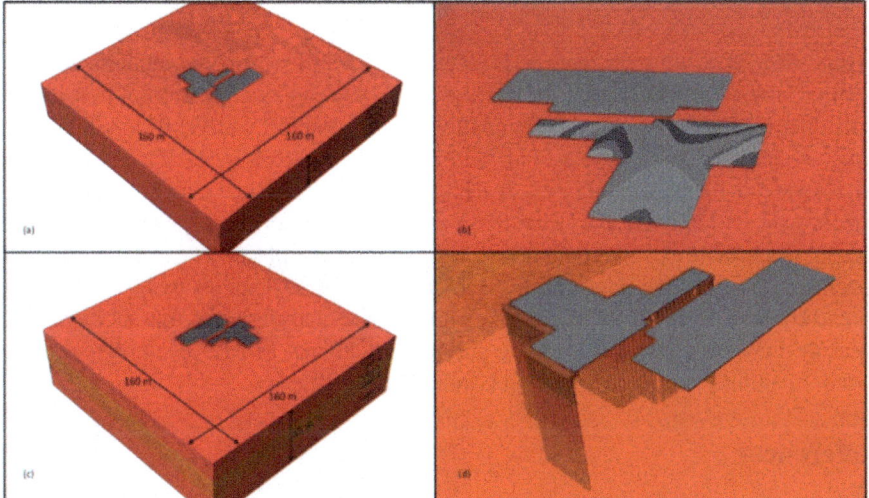

Fig. 2. 3D FE meshes. (a) the raft without underpinning piles, (b) non uniform loading pattern applied, (c) the raft with underpinning piles, (d) the close-up view.

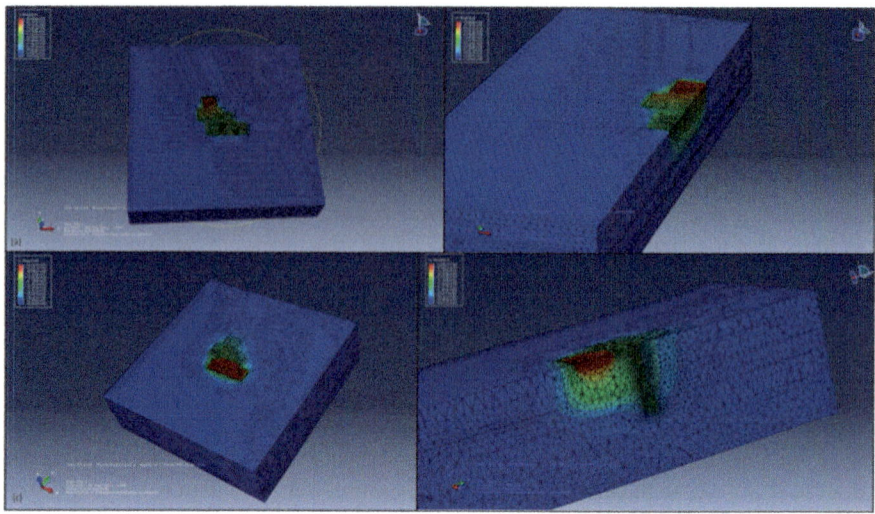

Fig. 3. 3D FE analyses results. (a and b) the raft without underpinning piles, (c and d) the raft with underpinning piles

4 Conclusions

In this study, the tilting problem of a silo structure was investigated numerically. After detailed studies, it was concluded that the problem was a combination of non-uniform load distribution acting on the foundation and soft/deformable soil conditions. The structure and the remediation system were modelled by Abaqus. Analyses were carried out to evaluate the performance of the designed system. The analyses have revealed that the ultimate settlement of the structure was 58.88 cm without underpinning piles. Reinforcing the structure by 80 cm diameter, 22 m long piles; placed at 1 m center-to-center spacing along the outer edges of the structure reduced settlements by 85% which indicates the effectiveness of such a remedial system. The analyses show that such a remedial measure may be used in preventing further settlements in case of similar problems without halting the production of such industrial facilities.

References

1. Rahal, M.A., Vuez, A.R.: Analysis of settlement and pore pressure induced by cyclic loading of silo. J. Geotech. Geoenviron. Eng. **124**(12), 1208–1210 (1998)
2. Laier, J.E., Cowles, G.D., White, M.E.: Anatomy of foundation performance involving three grain silos systematically loaded to impending failure. From Research to Practice in Geotechnical Engineering, 1st edn. ASCE Press, USA (2008)
3. Kalpakci, V.: Collapse of base soil and its consequences during a cement plant construction. Selcuk Univ. J. Eng. Sci. Technol. **5**(4), 500–510 (2017)
4. Algin, H.M.: Optimised design of jet-grouted raft using response surface method. Comput. Geotech. **74**, 56–73 (2016)

5. Wouatong, A., Kitagawa, R., Takeno, S., Felix, M.T., Daniel, N.: Morphological transformation of kaolin minerals from granite saprolite in the western part of Cameroon. Clay Science **10**(1), 67–81 (1996)
6. Ngon, G.F.N., Etame, J., Ntamak-Nida, M.J., Mbog, M.B., Mpondo, A.M.M., Gérard, M., Bilong, P.: Geological study of sedimentary clayey materials of the Bomkoul area in the Douala region (Douala sub-basin, Cameroon) for the ceramic industry. Comptes Rendus Geosci. **344**(6–7), 366–376 (2012)

Application of Semi-active Control Strategy for the Wall Retaining Granular Fills

Nisha Kumari(iD) and Ashutosh Trivedi$^{(\boxtimes)}$(iD)

Delhi Technological University, Delhi 110042, India
nishaani.ce@gmail.com, atrivedi@dce.ac.in

Abstract. The smart structures provide more efficient designs. In smart structures electronic components, sensors and actuators are well distributed and integrated, capable of controlling vibration for self-actuation and self-diagnostics. The purpose of structural control is to absorb, refract and reflect the energy induced by seismic wave propagation due to dynamic disturbances. This paper investigates semi-active control strategy for walls retaining granular material. It has two components, the seismic protection of retaining structure and the mitigation of wave induced vibration in retaining structure. An ideal semi-active motion equation of a retaining structure that consist of cantilever retaining wall with granular backfill bounded with a PZT patch was analyzed. The stress-strain analysis was performed by varying material properties and geometric configuration of the wall and granular fill. The stress- strain behavior of granular fill with PZT patch is presented in graphical form. A mathematical analysis is performed to examine the vibration induced effect of provision of PZT on in retaining structure due to the dynamic loading.

Keywords: Semi-active control · Dynamic loading · Strain analysis
Seismic waves · Cantilever retaining wall

1 Introduction

Structure vibration can be generally controlled in two ways: (1) by constructing the building using smart materials: (2) by adding controlling devices like dampers, isolators, and actuators to the building. Retaining wall stability depends on the intensity of vibration generated at the time of earthquake [1]. This vibration is analyzed in terms of acceleration and natural frequency of seismic waves. In current study used semi active control technique to control the vibration input in cantilever retaining structures to protect structures from severe dynamic loading is the use of control strategies. There are three types of control strategies namely passive control strategies, active control strategies and semi active control strategies. Which are used for controlling the vibration in smart structures [2].

© Springer Nature Switzerland AG 2018
W. Wu and H.-S. Yu (Eds.): *Proceedings of China-Europe Conference on Geotechnical Engineering*, SSGG, pp. 978–982, 2018.
https://doi.org/10.1007/978-3-319-97115-5_20

2   Semi Active Control Method

In semi active control scheme sketched in Fig. 1 the control actuators do not add mechanical energy directly to the structure, hence bounded input and bounded output stability is guaranteed.

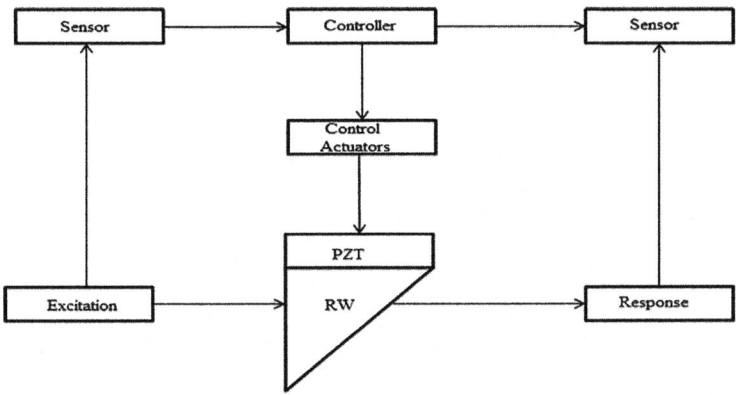

Fig. 1. Structure with semi active control strategy.

Figure 1 is a block diagram of a semi active control system with structure. Where control actuators, sensors and piezoelectric devices are highly distributed and integrated to perform the function like self-sensing, self-actuation and self-diagnosing to control the vibration [3]. Piezoelectric patches are used to counteract or cancel the dynamic vibration.

3   Average Normal Stress and Strain in Granular Backfill

In Fig. 1 Cantilever wall retaining granular backfill bounded of a PZT layer on the outer surface of the backfill. It is assumed that the PZT layer and granular backfill are firmly bounded together, PZT and retaining wall models are linear and elastic and the material properties of the PZT and backfill are homogeneous and isotropic.

The stress of individual PZT and granular fill layer is the product of strain and modulus of the material of each layer. In pure bending the normal strain $\in_n$ and normal stress σ_n in x-direction can be expressed as,

$$\in_n = \frac{Z}{R} \tag{1}$$

$$\sigma_n = E \in_n \tag{2}$$

Where 1/R is radius of curvature due to bending, E is the modulus and Z is the direction on Z-coordinate from the neutral surface. Normal strain along x-direction can be calculated as

$$\in_n = \frac{-Z}{R} = \frac{MZ}{EI} \tag{3}$$

For the PZT layer the strain in mid surface of the PZT is

$$(\in_{PZT})_{avg} = -\frac{M(Z_P - \overline{Z})}{E_P I_R + E_P I_R} = \frac{M(\overline{Z} - Z_P)}{E_R(I_R + nI_P)} \tag{4}$$

For the retaining wall shown in Fig. 2, we define a thickness ratio μ that is the ratio of the PZT thickness over the retaining wall thickness, and a modulus ratio n that is the ratio of the modulus of the PZT over the modulus of the retaining wall respectively, that is,

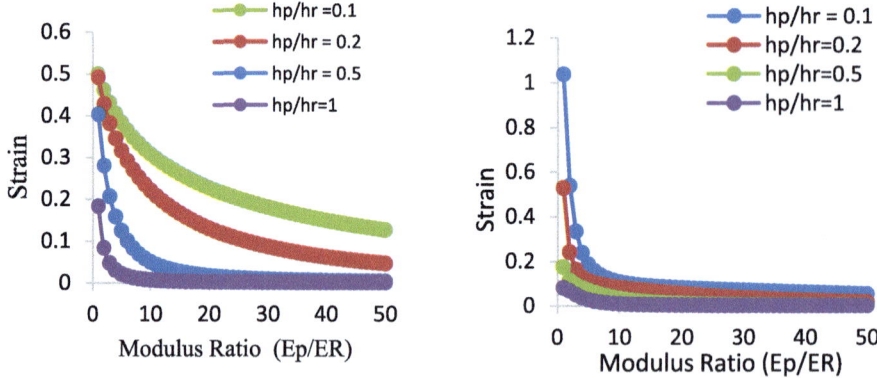

Fig. 2. (a) Average strain of PZT layer. (b) Average strain of retaining wall.

$$n = \frac{E_P}{E_R} = \text{modulus ratio of PZT over retaining wall}$$

$$\mu = \frac{h_P}{h_R} = \text{Thikness ratio of PZT over retaining wall}$$

By solving the Eq. (4) we have:

$$(\in_{PZT})_{avg} = \frac{M}{bE_R h_R^2}\left[\frac{36(1 + \mu)(1 + \mu n)}{3\mu^5 n^3 + 6\mu^4 n^2 + 5\mu^3 n + \mu^2 n^2 + \mu^2 + 4\mu n + 2}\right] \tag{5}$$

Similarly, for the backfill layer, the strain in centroid surface of the RW is

$$\left(\in_{RW}\right)_{avg} = \frac{M}{12bE_R h_R^2}\left[\frac{n^2(6\mu^3 + 4\mu^2) + 3\mu^2 n + 2}{12\mu^5 n^4 + 4n^2(3\mu^4 + 6\mu^3 + 14\mu^2) + 4n(5\mu^3 + 9\mu^5 + 6\mu^4 + \mu) - n\mu^3 + 1}\right]$$

(6)

The plot of the average strain in the PZT with four different thickness ratios is shown in Fig. 2(a). And the plot of the average strain in the backfill with four different thickness ratios is shown in Fig. 2(b).

Figure 2(a) and (b) show that the strain in the beam and the strain in the PZT have opposite sign. Similarly, stress can be obtained from Eq. 2.

Thus, Average stress of the PZT layer is

$$\left(\sigma_{PZT}\right)_{avg} = \frac{36M}{bh_R^2}\left[\frac{n(1 + \mu)(1 + \mu n)}{3\mu^5 n^3 + 6\mu^4 n^2 + 5\mu^3 n + \mu^2 n^2 + \mu^2 + 4\mu n + 2}\right]$$

(7)

Similarly, Average stress of the backfill layer is

$$\left(\sigma_{PZT}\right)_{avg} = -\frac{M}{12bh_R^2}\left[\frac{n^2(6\mu^3 + 4\mu^2) + 3\mu^2 n + 2}{12\mu^5 n^4 + 4n^2(3\mu^4 + 6\mu^3 + 14\mu^2) + 4n(5\mu^3 + 9\mu^5 + 6\mu^4 + \mu) - n\mu^3 + 1}\right]$$

(8)

Figure 3(a) and (b) are the plots of the average stress in the PZT and the average stress in the retaining wall respectively.

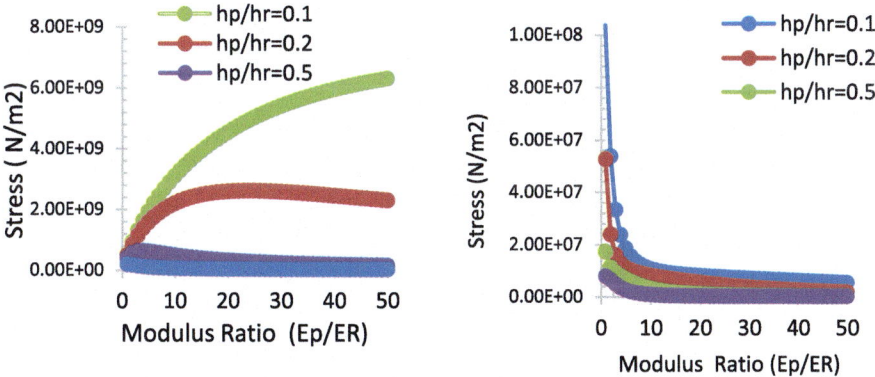

Fig. 3. (a) Average stress in PZT layer. (b) Average stress in retaining wall.

4 Conclusions

The stress strain analysis is performed by varying thickness ratio of the PZT and retaining wall. These graphs indicate that the average strain in PZT layer is decreases when the ratio of modulus n increases. The ratio of modulus of PZT and retaining wall indicates the flexibility of the retaining wall compared to the PZT. Average stress in PZT layer is increases when the ratio of modulus increases. The concept of the PZT indicates that strain in PZT is proportional to strain.

References

1. Dyke, S.J., et al.: Role of control-structure interaction in protective system design. ASCE J. Eng. Mech. **121**(2), 322–338 (1995)
2. Kerboua, M., et al.: Semi active control of civil structures, analytical and numerical studies. Phys. Procedia **55**, 301–306 (2014)
3. Miah, M.S., et al.: Semi-active control for vibration mitigation of structural systems incorporating uncertainties. Smart Mater. Struct. **24**, 05501 (2015)

Ground Improvement by the In-Tube Deep Dynamic Compaction Method

Ping Li[1,2], Jianhui Tang[1,2(✉)], and Chengwen Fan[1,2]

[1] Key Laboratory of Ministry of Education for Geomechanics and Embankment Engineering, Hohai University, Nanjing 210098, China
tangjianhui1994@163.com
[2] College of Civil and Transportation Engineering, Hohai University, Nanjing 210098, China

Abstract. Rammed from the deep layer gradually to the ground surface along the ramming tube can be achieved by the in-tube deep dynamic compaction method, which is a new dynamic compaction construction technique. Based on the field test in Fujian, the reinforcement effect was explored during reinforcement. The test results show that the accumulation of excess pore water pressure during the tamping process can be effectively avoided by the in-tube deep dynamic compaction method, and the excess pore water pressure will dissipate completely after 24 h of dynamic consolidation; The effective depth of foundation treatment can reach more than 13 m depth, the defects of long period of dissipation of excess pore pressure and shallow depth of effective reinforcement in conventional dynamic consolidation method were overcame.

Keywords: In-tube deep dynamic compaction method
Dissipation of excess pore water pressure · Depth of effective reinforcement
Technique test

1 Introduction

Dynamic consolidation method (DCM) is a cost-effective way to treat foundation, which was first proposed by French engineer Menard in 1969 [1]. After years of development, DCM has been widely used in the treatment of gravels, sand, collapsible loess, and other foundation reinforcement [2–4]. However, the following limitations still exist. Firstly, there is still controversy about the reinforcement effect of soft soil foundation (especially muddy soil), whether the excess pore water pressure after tapping can be quickly dissipated is the key factor that determines the success or failure of DCM in such foundation [5]; Secondly, The effective reinforcement depth of dynamic compaction has always been the main problem for deep consolidation of soft soil foundation [6].

Based on the research status of DCM [7, 8], a new technology called the in-tube deep dynamic compaction method was developed in this paper. The conventional dynamic compaction process which tamped down from the ground surface was changed to punning the foundation from the deep layer gradually to the ground surface, so as to achieve the effect of deep reinforcement. In order to explore its reinforcement effect in

© Springer Nature Switzerland AG 2018
W. Wu and H.-S. Yu (Eds.): *Proceedings of China-Europe Conference on Geotechnical Engineering*, SSGG, pp. 983–986, 2018.
https://doi.org/10.1007/978-3-319-97115-5_21

the process of reinforcing soft soil, a field test was carried out, which provided a reference for the design and application of the new method in the consolidation treatment of large-area soft soil foundation.

2   The In-Tube Deep Dynamic Compaction Method

The Process Equipment. As shown in Fig. 1, the process equipment used is the combination of ramming tube and vibrating hammer, the vacuum suction pipe is arranged on both sides of the ramming tube, the upper part of the tube is provided with a feeding inlet and a built-in rammer, the rammer is lifted and detached by opening and closing of the clamp-shaped decoupling device so as to achieve the required ramming energy; In the process of tamping, while ramming side pumping, the excess pore water pressure can be dissipated rapidly. For the flow plastic silt, the dynamic replacement can be carried out by filling coarse particles.

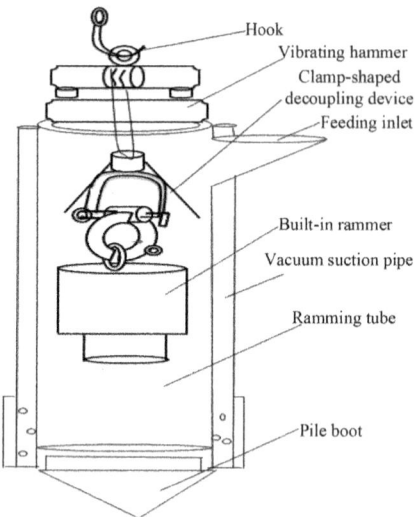

Fig. 1. Structure of the process equipment

The Construction Technology. Firstly, turn on the vibratory hammer and drive the tube into the design reinforcement depth; Secondly, connect the vacuum suction pipe on both sides of the tube and open the vacuum pump to carry out the deep precipitation around the tube; Thirdly, poured into the coarse filler from the feeding inlet and release the rammer to compact the packing to replace the silt layer, this is called the first level ramming; Subsequently, raise the ramming tube about 2 to 3 m, repeat the third step, this is called the second level ramming; By analogy, the single in-tube deep dynamic compaction was completed until the tube was pulled to the ground.

3   Field Test Research

3.1   Project Overview

The test site was located in the coastal area of Fujian Province, the site was originally the coastal beach. The groundwater in the site area was 0.6 m above the surface. According to geological prospecting data, site strata can be divided from top to bottom: ① Plain fill; ② Silty soil; ③ Silty clay; ④ Residual clay; ⑤ Fully weathered granite. The main soil physical and mechanical indexes are shown in Table 1.

Table 1.   Physical and mechanical index of soil strata

Physical and mechanical index	② Layer silty soil	③ Silty clay	④ Layer residual clay
Top buried depth/m	0.60–12.80	6.80–12.80	10.20–21.10
Bulk density/(kN·m^{-3})	16.5	17.7	18.5
Water content/%	45.7	30.3	36.7
Void ratio	1.25	0.89	1.07
Cohesive force/kPa	13.4	22.1	17.2
Internal friction angle/(°)	6.1	15.1	17.7
Compression coefficient/MPa^{-1}	0.85	0.35	0.55

3.2   Test Area Layout

The total area of the test zone was 100 m^2, the filler amount of the area was 5.0 m^3. The test fillers were made of medium coarse sand with 30%–50% of sludge in the field, and the layout of the tamping point was 2.0 m × 2.0 m plum-type. In the test, a group of pore water pressure gauges were embedded to monitor the change of pore water pressure around the tamping point. The buried depth of gauges was 4, 7, 10, 13, 16 and 19 m respectively which 1.0 m from the tamping point center.

4   Analysis of the Change of Pore Water Pressure

The designed depth was 10 m. After step-by-step packing and tamping, the pore water pressure changes at different depths were monitored in 24 h. We can draw the following conclusions from the curve in Fig. 2 which the pore water pressure changes with the number of tamping series and time.

(1)   After the tube entering the design depth of foundation and carrying out the first level ramming, the pore water pressure of 13 m above the depth of the soil has increased obviously, which indicated that the vibration produced by tamping pipe and the effect of compaction on the soil were obvious.

(2)   In the subsequent stage by stage packing ramming process, the excess pore water pressure produced by ramming reduced with the increase of tamping times. This is because the vacuum suction pipes on both sides of the ramming tube was continuously pumping

water around the tamping point, so that the excess pore water pressure produced by tamping can be quickly dissipated.

(3) After the completion of ramming operation, the excess pore water pressure was dissipated after 24 h. The effective depth of foundation treatment can reach more than 13 m depth.

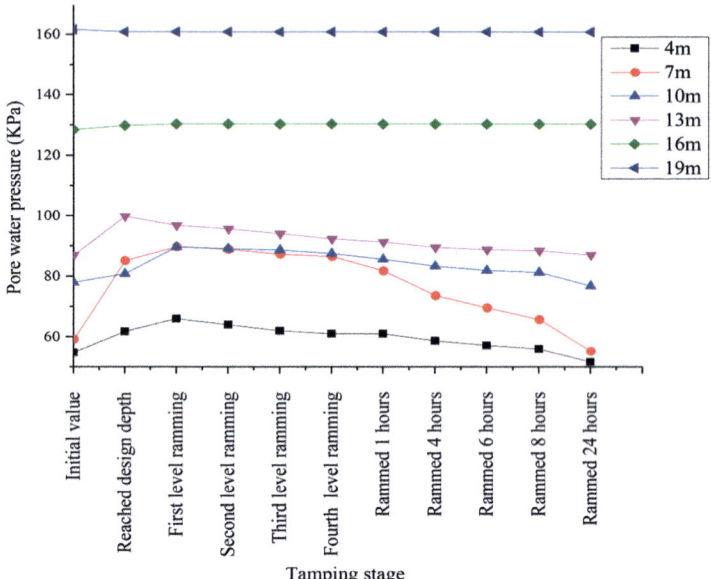

Fig. 2. Variation of pore water pressures under tamping

References

1. Menard, L., Broise, Y.: Theoretical and practical aspect of dynamic consolidation. J. Geotech. Eng. **25**(1), 3–18 (1975)
2. Gao, Z., Du, Y., Huang, X.: Reinforcement mechanism and construction technology of broken stone fills by dynamic consolidation. Chin. J. Rock Mechan. Eng. **32**(2), 377–384 (2013)
3. Zhang, J., Zhang, J., Meng, L.: Soil layer settlement of compacted aeolian sand foundations. Eng. Mech. **31**(Suppl.), 145–148 (2014)
4. Feng, S., Du, F., Shi, Z.: Field study on the reinforcement of collapsible loess using dynamic compaction. Eng. Geol. **185**, 105–115 (2015)
5. Meng, Q., Wang, R., Chen, Z.: Pore water pressure mode of oozy soft clay under impact loading. Chin. J. Rock Soil Mech. **25**(7), 1017–1022 (2004)
6. Jiang, R.: Applicable conditions of dynamic consolidation method in saturated soil foundation. Chin. J. Rock Mech. Eng. **31**(Suppl.), 3197–3202 (2012)
7. Fu, H., Wang, J., Cai, Y.: Experimental study of combined application of electro-osmosis and low-energy dynamic compaction in soft ground reinforcement. Chin. J. Rock Mech. Eng. **34**(3), 612–620 (2015)
8. Sun, T., Liu, S., Wang, Y.: In-situ experimental research on high energy dynamic compaction on soft foundation. Eng. J. Wuhan Univ. **47**(6), 789–793 (2014)

Field Study on Load Transfer Mechanism of XCC Pile Raft Foundation

Ping Li[1(✉)], Chengwen Fan[2], Jianying Lai[2], Xuanming Ding[2], and Ke Cheng[3]

[1] Key Laboratory of Geomechanics and Embankment Engineering of Ministry of Education, Institute of Engineering Safety and Disaster Prevention, College of Civil and Transportation Engineering, Hohai University, Nanjing 210098, China
lipings0110@163.com
[2] College of Civil Engineering, Chongqing University, Chongqing 400045, China
dxmhhu@163.com, newborn1021@foxmail.com, hero2008lai@163.com
[3] School of Mechanical and Power Engineering, Nanjing Tech University, Nanjing 211816, China
ck9392@njtech.edu.cn

Abstract. X-section cast-in-place concrete pile (referred to as XCC pile) is a new type of abnormal section pile developed in geotechnical research institute of Hohai University and patented in China. It has been widely applied in practice for its larger bearing capacity and use of less concrete than the common circular section pile. In this paper the construction techniques are introduced firstly, then comparative field study on load transfer mechanism of X-section and circular section pile rafts are presented. To reveal the bearing characteristics of the XCC pile raft foundations, the load-settlement relationship, side friction and so on are detailed analyzed. The results show that XCC pile raft has superior bearing capability compared with the circular pile raft, especially at the stage of high load pressure. The side friction percentage of XCC pile is larger than that of the circular pile while the tip resistance percentage is smaller.

Keywords: X-section cast-in-place concrete pile · Load transfer mechanism
Field study · Load-settlement relationship · Side friction

1 Introduction

Pile foundation is a tried method to support superstructure and has been widely applied in engineering for a very long history. The cross section of pile is either circular or square commonly. It is known that the perimeter of a circular section pile is less than that of a polygon or square section pile with the same area, which means that the side friction of a circular section pile is smaller than that of a non-circular section pile. Therefore, it is significant to develop abnormal section pile to attain higher bearing capacity and less use of material. By the increase of side surface area for the abnormal section pile, the side resistance of the soil around pile is enhanced, as a result, the bearing capability is increased [1]. X-section cast-in-place concrete pile (referred to as XCC pile) is a new

© Springer Nature Switzerland AG 2018
W. Wu and H.-S. Yu (Eds.): *Proceedings of China-Europe Conference on Geotechnical Engineering*, SSGG, pp. 987–991, 2018.
https://doi.org/10.1007/978-3-319-97115-5_22

type of abnormal section pile developed in geotechnical research institute of Hohai University and patented in China [2, 3], the cross section of which is shown as in Fig. 1.

Fig. 1. The XCC pile

As a new developed abnormal section pile, the bearing characteristics of XCC pile are different from the traditional circular section pile. The bearing capacity of single XCC pile subjected to vertical and lateral loads has been researched by field and large-scale model tests [4] as well as numerical and theoretical analysis [5–8]. XCC pile raft foundation has been widely applied to reinforce soft ground; however, the research about the XCC pile raft is rare. In this paper, field study on load transfer mechanism of XCC pile raft and comparative tests on circular pile raft are presented and the load-settlement relationship, side friction and so on are detailed investigated.

2   Construction Method of XCC Pile

The machine contains a steel casing with X cross section and a valve pile shoe made of steel arc plates. The section size of XCC pile can be determined by three variables: the

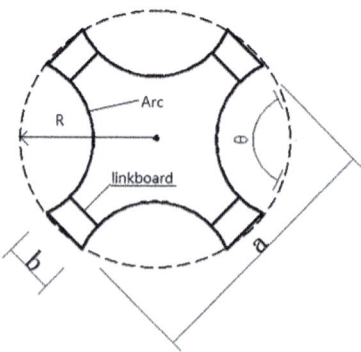

Fig. 2. The size schematic of the steel mould of XCC pile

diameter of circumcircle a, the distance between arc endpoint b and the arc angle θ (as shown in Fig. 2). Differently, the section size of circular section pile can be determined by only one variable R.

XCC pile is more economical than the common circular section pile. For instance, if XCC pile size is assumed as $a = 61.1$ cm, $b = 12$ cm and $\theta = 130°$, the area of XCC pile is 50% smaller than that of circular section pile with the same perimeter($Cx = Cc$, where Cx and Cc are the perimeters of X and circular sections, respectively); similarly, the perimeter of XCC pile is 50% larger than that of circular section pile with the same area ($Sx = Sc$, where Sx and Sc are the areas of X and circular sections, respectively). Therefore, XCC pile offers a better alternative method to circular section pile.

3 Site Condition and Testing Process

The test site locates in Nanjing, where the landform is a floodplain of Yangtze River. The cone penetration test (CPT) and laboratory soil test are carried out here.

4 Test Results and Analyses

4.1 The Load-Settlement Curve

Figure 3 shows the load-settlement relationship. As a result, the settlement of XCC pile raft is smaller than that of the circular pile raft at the same load pressure. The load applied on XCC pile raft is larger than that applied on circular pile raft, the settlement of XCC pile raft is smaller than that of circular pile raft after unloading. XCC pile raft has superior bearing capability compared with the circular pile raft, especially at the stage of high load pressure.

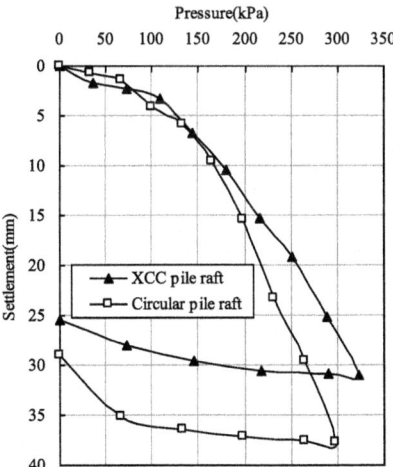

Fig. 3. The load-settlement relationship

4.2 The Distribution of Side and Tip Resistances

Figure 4 presents the percentage of side and tip resistance of two types of piles. There is little difference between XCC pile and circular pile when the load is small. However, the difference is more and more obvious as the increase of load when it is larger than 100 kN. The side area of XCC pile shaft is greater than that of circular pile, therefore, the side resistance of XCC pile is larger than that of circular pile, while the tip resistance of which is smaller. For instance, when the load is 400 kN, the side and tip resistances of XCC pile are 61% and 39%, respectively, while those of circular are 53% and 47%, respectively. As a friction pile, the side friction percentage of XCC pile is larger than that of circular pile while the tip resistance percentage is smaller.

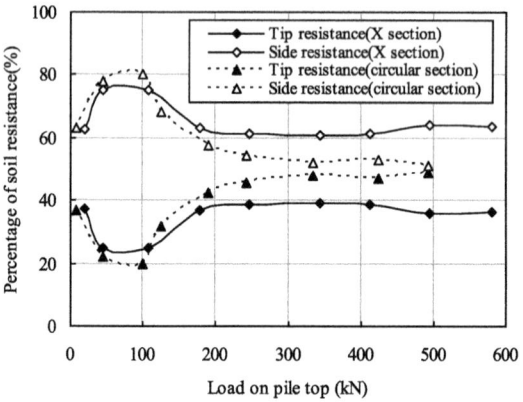

Fig. 4. The percentage of side and tip resistances

References

1. Lin, T.J.: Theoretical discussion on modern special form piles and their mechanical characteristics. Mech. Eng. **20**(6), 7–11 (1998)
2. Liu, H.L.: A sinking steel mould for cast-in-place XCC pile. Chinese patent ZL200720036892.6 (2007)
3. Chen, Y.M., Liu, H.L., Kong, G.Q.: Development and application of X-section cast-in-situ concrete pile. In: Proceedings of the Fourth Japan-China Geotechnical Symposium, pp. 389–392. Okinawa, Japan (2010)
4. Lv, Y.R., Liu, H.L., Ding, X.M., et al.: Field tests on bearing characteristics of X-Section pile composite foundation. J. Perform. Constr. Facil. **26**(2), 180–189 (2011)
5. Liu, H.L., Zhou, H., Kong, G.Q., et al.: XCC pile installation effect in soft soil ground: a simplified analytical model. Comput. Geotech. **62**, 268–282 (2014)
6. Lv, Y.R., Liu, H.L., Ng Charles, W.W., et al.: A modified analytical solution of soil stress distribution for XCC pile foundations. ACTA Geotech. **32**(2), 349–362 (2013)

7. Lv, Y.R., Liu, H.L., Ng, C.W.W., Gunawan, A., et al.: Three-dimensional numerical analysis of the stress transfer mechanism of XCC piled raft foundation. Comput. Geotech. **55**, 365–377 (2014)
8. Lv, Y.R., Ding, X.M., Wang, D.B.: Effects of the tip location on single piles subjected to surcharge and axial loads. Sci. World J. (2013)

Stability Analysis of L-Shape Slurry Trench During Concrete Diaphragm Wall Installation in $c - \varphi$ Soils

Wei Liu[1], Peixin Shi[1(✉)], Qiang Tang[1], and Fei Wang[2]

[1] School of Urban Rail Transportation, Soochow University,
Suzhou 215131, China
pxshi@suda.edu.cn
[2] Suzhou Rail Transit Group Co.Ltd., Suzhou 215131, China

Abstract. In the concrete diaphragm wall (CDW) construction, the L-shaped diaphragm wall is used to form an enclosed retaining structure. Due to the complicated stress change in trenching, the L-shaped trench has the lower stability than the regular rectangular trench. This paper investigates the trench instability during the L-shape slurry trench excavation based on the upper bound analysis. A 3D kinematically admissible mechanism defining the failure for the L-shaped trench is constructed. The safety factor on the trench and corresponding failure pattern are obtained through the upper bound analysis. The influence of trench geometries and the soil properties on the trench stability are discussed by the parameter analysis.

Keywords: Stability · L-shaped trenching · Diaphragm wall
Upper bound analysis

1 Introduction

Installation of the diaphragm wall panel involves excavating a trench filled with bentonite slurry, placing reinforcement cage, and pouring concrete inside the trench. During trench excavation, ground instability may occur due to the ground stress release. The ground instability collapses the trench and generates excessive movement to adjacent structures. The mechanism for the instability of regular rectangular trenching was well investigated [1–4]. Large underground structures are mostly formed with rectangular shape. To avoid panel connection, the panels at the corner of a rectangular excavation are installed by excavating an L-shape slurry trench and placing an L-shape enforcement cage inside the trench. Compared to the regular trenching, the excavation of the L-shape trench involves in significant risk on ground instability due to the complicated stress state at the corner.

   This paper investigates the trench wall collapse during the L-shaped trenching. A 3D kinematically admissible mechanism for this kind failure is constructed. The upper bound analysis on the instability is carried out in the light of the new model. In this analytical approach, the most critical mechanism is found from the numerical optimization. At the meantime, the stability of the trench is obtained.

© Springer Nature Switzerland AG 2018
W. Wu and H.-S. Yu (Eds.): *Proceedings of China-Europe Conference on Geotechnical Engineering*, SSGG, pp. 992–996, 2018.
https://doi.org/10.1007/978-3-319-97115-5_23

2 Mechanism for the Instability of L Shaped Trenching

2.1 The L-Shaped Trenching

Figure 1 shows L-shaped trenching in the diaphragm installation. The L-shaped trench is constructed by intersecting two small rectangular trenches. The corner angle of the L-shaped trench is β. For the simplification, it is assumed that the two components have the uniform length of L. During the excavation, the trench is supported by the bentonite slurry with the unit weight of γ_s. In this paper, the unfavorable situation that both of the slurry level and the groundwater level are identically at the ground surface is considered. The residual support pressure resisting the wall at a depth z is $(\gamma_s - \gamma_w)z$, where γ_w is the unit weight groundwater. The surcharge on the ground surface is σ_s. The soil characterized as a $c - \phi$ material following Mohr-Coulomb criterion and the associated flow, is assumed to be homogenous with the effective weight of γ', cohesion of c' and frictional angle of φ'.

2.2 Upper Bound Analysis

Figure 2 shows the 3D kinematically admissible mechanism in which the failure involved in the movement of the solid prism model. The moving block OABC is a prism. The base of the prism is triangle AOB and on the ground surface. The lateral triangle ABC is the failure surface inclined to the horizontal with the angle of α. H is the height of prism OABC defining the height of failure slope. To satisfy the condition of $\varphi' \leq \theta_d \leq \pi - \varphi'$, the velocity $\vec{V}$ of the prism inclines to the line with the angle of φ'. In the limit analysis, an upper bound on the load can be obtained from the kinematic method [5–7]. By equating the work rate P_e of the external loads applied to the system to the dissipation work rate $P_v(P_e = P_v)$, the upper bound solution is obtained and expressed as follows,

$$F_s = \frac{c'N_c + \gamma_s L N_s}{\gamma' L N_\gamma + \sigma_s}, \tag{1}$$

where,

$$N_c = \frac{\cos\varphi'}{\cos\alpha \, \sin(\alpha - \varphi')}, \tag{2}$$

$$N_s = \frac{1}{3}(1 - \frac{\gamma_w}{\gamma_s}) \tan^2 \alpha \cot(\alpha - \varphi')\cos\frac{\beta}{2} \tag{3}$$

$$N_\gamma = \frac{1}{3}\cos\frac{\beta}{2}\tan\alpha. \tag{4}$$

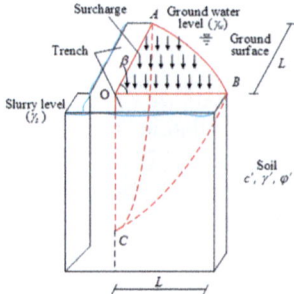

Fig. 1. The L-shaped trenching

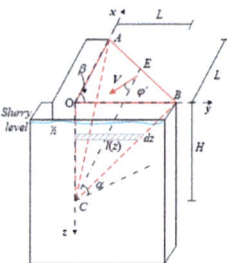

Fig. 2. 3D kinematically admissible mechanism for the instability

3   Results and Discussions

The analysis of stability starts with the particular situation of $\gamma_s = \gamma_w$ and $\sigma_s = 0$. Equation (1) can be simplified as $F_s = c'N_c/\gamma'LN_y$. With the given geometry and soil properties, the safety factor is a function of α. The optimization on the failure mechanism is carried out by searching the optimal α leading to the smallest F_s. The variation of α depending on φ' is shown in Fig. 3. It is clear that α increases with the increase of φ'. The relationship between α and is fitted by the equation of $\alpha = 90°-\varphi'$. The increase of α indicates failure slope becomes steeper.

For a particular trench ($L = 3$ m and $\beta = 90°$), the shape of failure zone is shown in Fig. 4. The increase of φ' results in the shrink of failure zone. According to Fig. 3, as φ' increases from 20° to 40°, α decreases from 70° to 50° and the height H decrease from 5.828 m to 3.674 m in response. The decreases of α and H contribute to the shrink of failure zone.

Figure 5 shows the variation of safety factor depending on φ' for the cohesive soil (i.e. $c' = 5$kPa). Generally, F_s nonlinearly increases with the increase of φ'. When φ' is higher than 30°, the increase rate of F_s starts to go up. The increase of F_s is attributed to the shrink of failure zone in response to the increase of φ'. For the same φ', F_s decrease with the increase of trench length L. The lines of F_s for the different L dispatch as φ' increases.

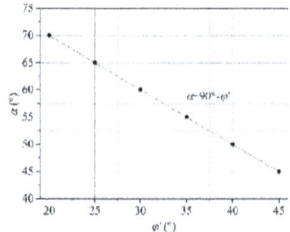

Fig. 3. The optimal α depending on φ'

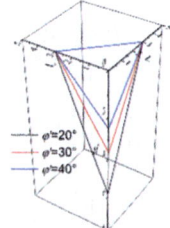

Fig. 4. The shape of failure zone

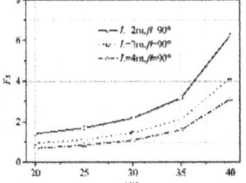

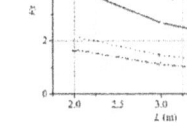

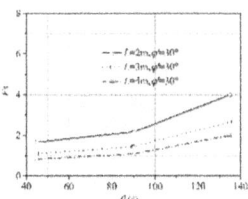

Fig. 5. Variation of F_s depending on φ'

Fig. 6. Influence of trench length L on F_s

Fig. 7. The influence of β on F_s

Figure 6 shows the influence of trench geometry on the stability for a particular c-φ soil (i.e. $c' = 5\text{kPa}$ and $\varphi' = 30°$). As shown in Fig. 6, the trench length makes the negative effect on trench stability. F_s Nonlinearly decreases with the increase of L. The decrease of F_s is fast until L is longer than 3 m. For the same φ', F_s is also influenced by the corner angle β. The increase of F_s with β bigger than 90° is more significant than that with β smaller than 90°. The influence of β on F_s is detailed in Fig. 7. It is found that F_s nonlinearly grows with the increase of β. The increase of F_s with obtuse angle β is faster than that with acute angle β. This means that the risk of instability is much higher when the β is acute angle.

4 Conclusions

The wall collapse during L-shaped trenching is analyzed in this paper. A 3D kinematically admissible mechanism for the instability is proposed. Based on this model, the safer factor on the trench stability is obtained by use of upper bound analysis. Also, the optimal failure pattern and the critical height of failure slope can be determined. The increase of φ' results enhances trench stability and the shrink of failure slope. The critical inclined angle α linearly varies with φ' following the rule of $\alpha = 90°-\varphi'$. The decrease of trench length or increase of corner angle makes a positive effect on the L-shaped trench stability. The charts in the paper may serve as practical tools for L-shaped trench stability assessment.

References

1. Morgenstern, N., Amir-Tahmasseb, I.: The stability of a slurry trench in cohesion-less soils. Géotechnique **15**(4), 387–395 (1965)
2. Fliz, G., Adams, T., Davidson, R.: Stability of long trenches in sand supported by bentonite-water slurry. J. Geotech. Geoenviron. Eng. (ASCE) **130**(9), 915–921 (2004)
3. Fox, P.J.: Analytical solutions for stability of slurry trench. J. Geotech. Geoenviron. Eng. (ASCE) **130**(7), 749–758 (2004)
4. Han, C.Y., Wang, J.H., Xia, X.H., et al.: Limit analysis for local and overall stability of a slurry trench in cohesive soil. Int. J. Geomech. (ASCE) **15**(5), 06014026 (2012)

5. Chen, W.F.: Limit Analysis and Soil Plasticity. Elsevier Scientific Publishing Co., Amsterdam (1975)
6. Michalowski, R.L., Drescher, A.: Three-dimensional stability of slopes and excavations. Géotechnique **59**(10), 839–850 (2009)
7. Leca, E., Dormieux, L.: Upper bound and lower bound solutions for the face stability of shallow circular tunnels in frictional material. Géotechnique **40**(4), 581–606 (1990)

Experimental Studies on Model Single Pile and Pile Groups Subjected to Torque

Sagar Mehra[(✉)] and Ashutosh Trivedi

Delhi Technological University, Delhi 110042, India
sagarmehra36@gmail.com, atrivedi@dce.ac.in

Abstract. Many structures namely tall buildings, offshore platforms, bridge bents and electric transmission towers are subjected to lateral loads of considerable magnitude due to wind and wave actions, ship impacts, or high-speed vehicles. Significant torsional forces can be transferred to the foundation of transmission towers due to the laterally loaded high tension wires. Inadequate consideration of torque in design of the piles against these loads result into disastrous consequences. The design and analysis of pile foundations present a complex problem to the engineers because of several factors that affect the foundation behaviors. Such factors include mode of loading, soil properties, pile geometry, placement and method of construction. The mode and magnitude of loads transferred from the superstructure influences the selection of pile foundation to resist the imposed loads. The present work considers a mechanism of applying torque to a model single pile and model (m × n) pile groups; secondly, to find out experimentally the torque-twist curves of a model single pile and model (m × n) pile groups; and to analyse basis of pile-soil-pile interactions in model (m × n) pile groups subjected to torque. The experiments on a model single pile (1 × 1) and model (m × n) pile groups consisting of (1 × 2), (1 × 3) and (2 × 2) piles were performed to analyse the effect of torque on a model single pile and model (m × n) pile groups. The experiments were performed in loose and dense sand. The results of application of torque on a model single pile were compared with that of model (m × n) pile groups.

Keywords: Single pile · Pile groups · Torque application
Torsional pile groups

1 Introduction

1.1 Torque on a Single Pile

Torsional wind loading on tall buildings has been discussed by many researchers. Boggs et al. (2000) identified sources of torsional loading in terms of building shape, interfering effects of nearby buildings, and dynamic characteristics of structural frame. Tall buildings (Mayer-Kaiser, Miami and Great Plains life, Lubbock, Texas) had suffered permanent damage due to wind action (Vickery 1979).

© Springer Nature Switzerland AG 2018
W. Wu and H.-S. Yu (Eds.): *Proceedings of China-Europe Conference on Geotechnical Engineering*, SSGG, pp. 997–1000, 2018.
https://doi.org/10.1007/978-3-319-97115-5_24

O'Neill (1964) presented the mechanical modal composed of rigid elements connected by torsional springs. These springs can simulate nonlinear twist behavior and the spring constants can be varied to account for variable pile cross section and shear modulus (Fig. 1).

Poulos (1975) presented a solution, for the elastic response of single pile under torque. Randolph (1981) derived solutions for the torsional stiffness of a single pile based on a simple assumption concerning the stress field around a pile undergoing torsion. Chow (1985) presented a discrete element approach in which the pile was modelled as a series of elements and the soil was treated as a series of independent layers, each with a modulus of subgrade reaction.

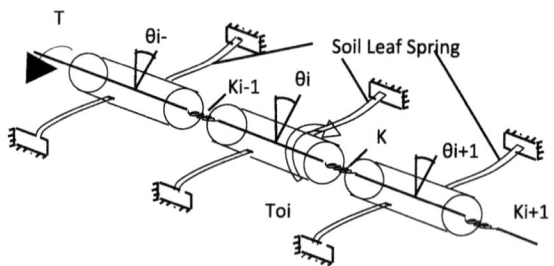

Fig. 1. Discrete element mechanical modal (after O'Neill 1964).

1.2 Torque on Pile Groups

Pile group behavior has been summarized by many researchers such as Poulos (1989), Reese and Van Impe (2001) etc. According to Kong (2006) load sharing among piles in a group is influenced by pile-to-pile interaction, group stiffening effect, load-deformation coupling and soil non-linearity. The response of a pile group subjected to torsion is governed by the interaction between the torsional and lateral behaviour of the individual piles (Fig. 2).

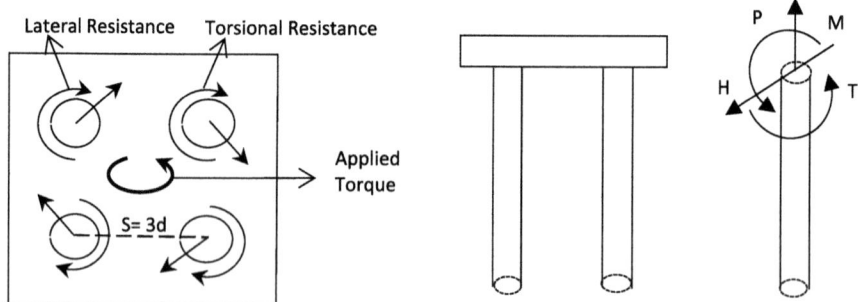

Fig. 2. Movements in (2 × 2) pile group under torque and pile head forces in a torsional loaded pile group.

2   Mechanism of Torque Application

The mechanism of torque application is shown in Fig. 3. It is capable of applying torque to a model single pile and model (m × n) pile groups (after Mehra 2011).

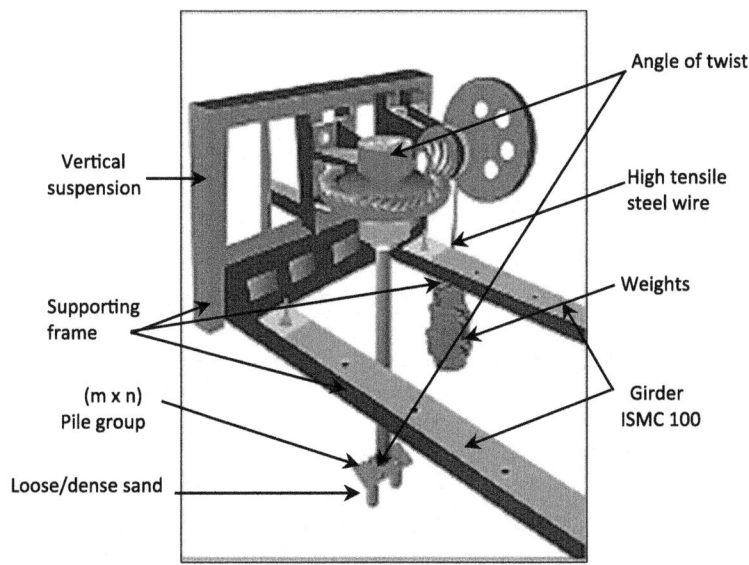

Fig. 3. Mechanism of applying torque on a model single pile and model (m × n) pile groups, where m and n are the number of piles in a group (after Mehra 2011).

3   Result

The results of experiments were compared and plotted as torque applied versus angle of twist as shown in Figs. 4 and 5 on loose and dense sand respectively.

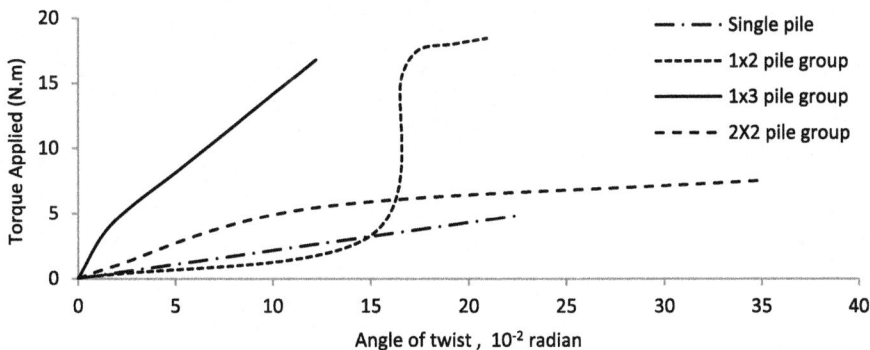

Fig. 4. Plot of applied torque and twist angle in loose sand.

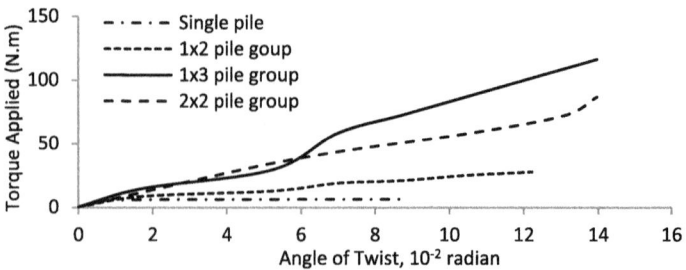

Fig. 5. Plot of applied torque and twist angle in dense sand.

4 Conclusions

The experimental result indicates that model (m × n) pile groups show greater resistance to applied torque than the model single pile. The resistance of a model single pile and model (m × n) pile groups to the application of torque is greater in the dense sand as compared to the loose sand. The tests also reveal that a model single pile and center pile in model (1 × 3) pile group shows similar type of rotational behavior. The behavior of model (1 × 2) pile group and model (2 × 2) pile groups were found to be similar since the outer piles in a group have the peripheral movement. The patterns in the shear zones of pile in the dense and loose sand were found to be different.

References

Chow, Y.K.: Torsional response of piles in non-homogenous soil. J. Geotech. Eng. Div. ASCE **111**(7), 942–947 (1985)

Boggs, D.W., Hosaya, N., Cochran, L.: Sources of torsional wind loading on tall buildings, lessons from the wind tunnel. In: Elgaaly, M. (ed.) Proceedings of the 2000 Structures Congress and Exposition, SEI/ASCE (2000)

Kong, L.G.: Behavior of pile groups subjected to torsion. Ph.D. Thesis, Hong Kong University of Science and Technology, Hong Kong, 339 p (2006)

Mehra, S.: Experimental study on model pile groups subjected to torque. M.E. Thesis, University of Delhi, Delhi, India (2011)

O'Neill, M.W.: Determination of the pile-head torque-twist relationship for a circular pile embedded in clay soil. M.S. thesis, University of Texas, Austin, Texas (1964)

Poulos, H.G.: Torsional response of piles. J. Geotech. Eng. Div. **ASC101**(GT10), 1019–1035 (1975). Proc. Paper 11629

Poulos, H.G.: Pile behaviour theory and application. Geotechnique **39**(3), 365–415 (1989). 29th Rankine Lecture

Randolph, M.F.: Piles subjected to torsion. J. Geotech. Eng. Div. **ASCE107**(GT8), 1095–1111 (1981). Proc. Paper 16424

Reese, L.C., Van Impe, W.F.: Single Pile and Pile Groups Under Lateral Loading. A.A Balkema, Rotterdam (2001)

Vickery, B.J.: Wind effects on building and structures-critical unsolved problems. In: IAHR/IUTAM Practical Experiences with Flow-Induced Vibrations symposium, Karlsruhe, Germany, pp. 823–828 (1979)

Use of Drainage Holes on Reinforced Concrete Pipe Piles to Accelerate Soil Consolidation

Guoxiong Mei[1(✉)], Pengpeng Ni[2], Meijuan Xu[1], and Yanlin Zhao[1]

[1] Key Laboratory of Disaster Prevention and Structural Safety of Ministry of Education,
College of Civil Engineering and Architecture, Guangxi University, Nanning 530004, China
meiguox@163.com
[2] School of Civil and Environmental Engineering,
Nanyang Technological University, Singapore 639798, Singapore

Abstract. This paper investigates the application of permeable piles (reinforced concrete pipe piles with drainage holes) to accelerate soil consolidation during pile driving. To achieve that, finite element models are generated in ABAQUS with infinite element boundary condition. The results of numerical analyses show that the performance of permeable piles can be improved by drilling drainage holes. However, the influence of these openings on the structural performance of permeable piles has not been evaluated before. The structural behavior of permeable piles is investigated in this study using uniaxial compression tests and four point flexure tests. Although drainage holes could reduce the ultimate compressive strength of pile specimens, the measured values were much larger than design specifications. For bending tests, cracks with smaller width were initiated and propagated over an increasingly wider area in permeable piles, and an improved flexural capacity was obtained. These results of the current study show that permeable pile is an attractive alternative to accelerate soil consolidation.

Keywords: Permeable piles · Drainage holes · Soil consolidation

1 Introduction

During pile driving, excess pore water pressures are generated in the vicinity of the pile, and the dissipation of these excess water pressures will take weeks or months especially in sensitive clays. In practice, drainage paths are often introduced near the pile to speed up the soil consolidation process, such as the techniques of prefabricated vertical drains, sand drains, and stone columns. It should be emphasized that the maximum excess pressure occurs at the soil-pile interface. All these measures cannot be installed at locations that are too close to the driven pile.

Some alternative solutions of introducing vertical drainage paths in the soil emerged recently, including pervious concrete ground-improvement piles [1] and permeable pipe piles [2, 3]. Pervious concrete piles have limitations in strength to resist loads in the axial and lateral directions compared to normal concrete piles [1, 4]. Contrarily, permeable pile is made of normal reinforced concrete, with drainage holes drilled around the

© Springer Nature Switzerland AG 2018
W. Wu and H.-S. Yu (Eds.): *Proceedings of China-Europe Conference on Geotechnical Engineering*, SSGG, pp. 1001–1004, 2018.
https://doi.org/10.1007/978-3-319-97115-5_25

circumference the pile as illustrated in Fig. 1. The arrangement of these drainage holes will be opened to enable the dissipation of excess pore pressures, once the driving process is completed. This investigation will demonstrate the effectiveness of permeable piles in accelerating soil consolidation using numerical simulations. Both compression and flexural tests are conducted to check whether opening drainage holes will result in a reduction of strength for permeable piles.

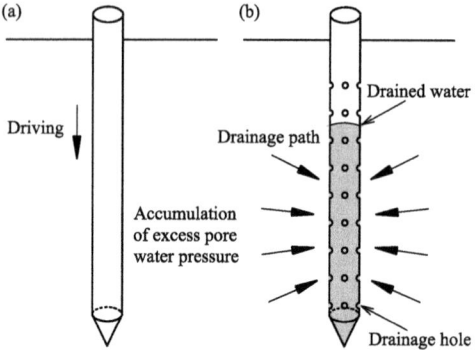

Fig. 1. Schematics of (a) normal pile and (b) permeable pile.

2 Results of Finite Element Analysis

A three-dimensional finite element model is developed in ABAQUS to simulate the behavior of permeable piles, with the incorporation of infinite elements at lateral boundaries. Details of the modelling strategy can be referred to Ni et al. [2]. The average degrees of consolidation (U_{avg}) are calculated for normal and permeable piles embedded in three soils with different hydraulic conductivity (k), Fig. 2. One can see that the soil around permeable piles always has a higher U_{avg} value at a specific time. The consolidation curve moves to the left when drainage holes are opened for permeable piles, which demonstrates that the soil consolidation process is accelerated.

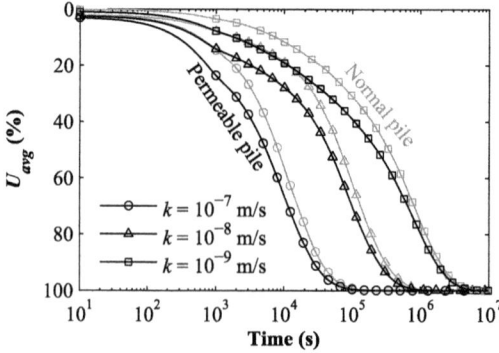

Fig. 2. Comparison of U_{avg} between normal and permeable piles.

3   Results of Uniaxial Compression Tests

Two series of uniaxial compression tests (different concrete strength of C50 and C80) have been performed on permeable pile specimens. Details of the testing program can be found in Ni et al. [5]. Figure 3 depicts the correlation between the strength reduction and the opening ratio of drainage holes. The ultimate compressive strength decreases monotonically with openings, until a plateau of strength reduction of 80% is reached. Calculations are carried out to determine the minimum required compressive strength for normal piles, which is found to be much less than the measured values for permeable piles. Drainage holes only reduce the ultimate compressive strength by a maximum of 20%, and permeable piles still fulfill the design needs.

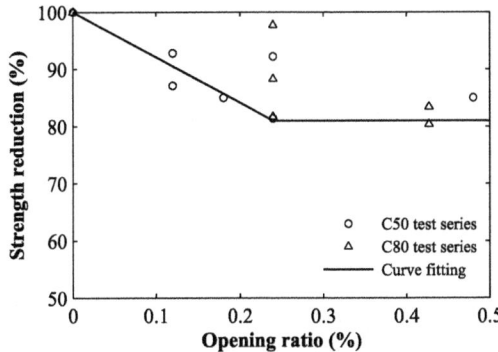

Fig. 3.   Reduction in ultimate compressive strength due to the opening of drainage holes.

4   Results of Four Point Flexural Tests

Four point flexural tests are performed on six prototype-scale permeable piles. Interested readers can find more details regarding the testing program in Ni et al. [5]. The patterns of cracks in the concrete at the end of the tests are plotted in Fig. 4 for six piles. Cracks in the normal pile are more concentrated, whereas cracks in the permeable pile spread more over a wider area. The widths of all cracks are mapped, and it is found that the widest crack always occurs in the normal pile. All cracks first initiate from the weak zone in the pile. For permeable piles, the spread of cracks through multiple weak zones can help to redistribute the flexural moment.

The variations of bending moment with the applied load are drawn in Fig. 5. As expected, the measured bending moments from both normal and permeable piles are comparable at lower loads, which follow a linear pattern of theoretical solution. With the increase of the applied load, the measured values deviate from the analytical solution. At the same load in a later stage, the bending moment of permeable pile is lower than that of normal pile, which explains that less significant cracks are observed for permeable pile. This also demonstrates that drilling drainage holes can help permeable piles to withstand a higher level of flexure.

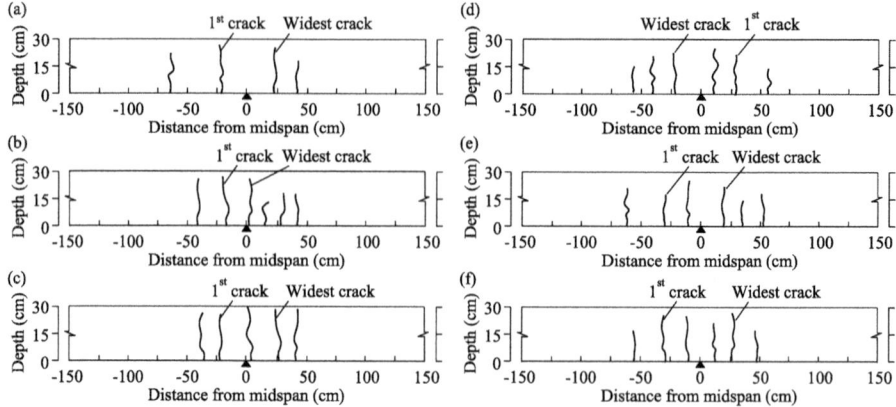

Fig. 4. Distribution of cracks in normal pile: (a), (b) and (c), and permeable pile: (d), (e) and (f).

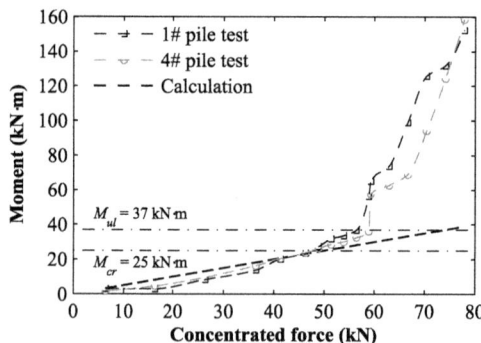

Fig. 5. Measured bending moment for normal (1#) and permeable (4#) piles.

References

1. Suleiman, M.T., Ni, L., Raich, A.: Development of pervious concrete pile ground-improvement alternative and behavior under vertical loading. J. Geotech. Geoenviron. Eng. **140**(7), 04014035 (2014)
2. Ni, P., Mangalathu, S., Mei, G., Zhao, Y.: Permeable piles: an alternative to improve the performance of driven piles. Comput. Geotech. **84**, 78–87 (2017)
3. Ni, P., Mangalathu, S., Mei, G., Zhao, Y.: Laboratory investigation of pore pressure dissipation in clay around permeable piles. Can. Geotech. J. (2017). https://doi.org/10.1139/cgj-2017-0180
4. Suleiman, M.T., Ni, L., Raich, A., Helm, J., Ghazanfari, E.: Measured soil-structure interaction for concrete piles subjected to lateral loading. Can. Geotech. J. **52**(8), 1168–1179 (2015)
5. Ni, P., Mangalathu, S., Mei, G., Zhao, Y.: Compressive and flexural behaviour of reinforced concrete permeable piles. Eng. Struct. **147**, 316–327 (2017)

Ageing Effect on Pull-Out Capacity of Driven Micropiles in Dunkirk Sand

Imane Salama[1(✉)], Christophe Dano[1], Matias Silva[1,2], and Miguel Benz Navarrete[3]

[1] Univ. Grenoble Alpes, CNRS, Grenoble INP, 3SR, 38000 Grenoble, France
{imane.salama,christophe.dano}@3sr-grenoble.fr
[2] Department of Civil Engineering, Universidad Tecnica Federico Santa Maria, Valparaiso, Chile
[3] Sol Solution Company, BP 178, ZA des Portes de Riom Nord, 63204 Riom, France

Abstract. In order to better understand the origin of ageing on driven micropiles in sand, a series of field pull-out tests on small diameter driven piles were carried out in Dunkirk. The micropiles were about 51 mm in diameter, 8 mm wall thickness and 2 m embedment. The micropiles were made of different steel types (mild steel or stainless steel) with different shaft roughness. Sand characteristics, field penetrometer tests and pile tests are summarized. Tension tests were performed at different intervals of time after micropiles installation in attempt to investigate the influence of: (i) physiochemical effects, by testing both stainless and mild steel piles; (ii) the early stages of ageing. A significant increase of capacity was observed in the case of mild steel micropiles, regardless of the shaft roughness, while no ageing effects were observed on stainless steel micropiles. Some micropiles have been pre-loaded in compression at different energy levels. These results suggest the great influence of physiochemical effects (corrosion) for small diameter piles driven in sand. In the case of a prior compressive plastification, the loss of the benefit of ageing is obvious.

Keywords: Ageing · Sand · Micropiles

1 Introduction

The substantial increase of soil properties and bearing capacity of geotechnical structures with time, independently of any consolidation process, refer to the ageing process. An increase of both stiffness and strength of sand specimens with time, under constant effective stress, was showed by Daramola [2] through triaxial tests. Despite several investigations [1, 3, 5, 9, 10], the origin of ageing remains an open question. Three possible mechanisms [5] behind this phenomenon are suggested: (i) higher stationary radial effective stresses acting on pile shaft and induced by the stress redistribution; (ii) dilatative shaft radial stress contributing to growth in capacity [1, 8]; and (iii) physiochemical processes such as corrosion [11].

In the aim of better understanding these mechanisms in sands, a series of field pull-out tests on small diameter open-ended driven piles has been carried out in Dunkirk (France). A summary of the soil parameters [5, 6] is shown in Table 1. Dynamic

© Springer Nature Switzerland AG 2018
W. Wu and H.-S. Yu (Eds.): *Proceedings of China-Europe Conference on Geotechnical Engineering*, SSGG, pp. 1005–1008, 2018.
https://doi.org/10.1007/978-3-319-97115-5_26

penetrometer Panda2® carried out in the closeness of micropiles indicated a denser central zone in the pile test area. The tip resistance q_c increased from 0 to 40 MPa from surface to 1.5 m BGL before reducing between 1.8 and 2 m to approximately 35 MPa [7]. Below 1 m a spatial scatter in q_d is observed and it goes below q_c: from 1 to 2 m q_d is about 10 MPa to 25 MPa in the soft zone and about 20 MPa to 40 MPa in the undisturbed zone.

Table 1. Soil parameters and ground conditions.

	Unit	
Water table BGL	m	4–4.7
Description		Dense to very dense
Origin		Marine hydraulic fill
Dry unit weight (γ_d)	kN/m^3	17.5
Water content	%	5–7
S_r	%	25–40
D_{50}	mm	0.26

2   Testing Programme

The testing campaign commenced in January 2016 and was completed in December 2016. 35 steel (mild and stainless steel) open-ended piles were installed. The piles had an Outside Diameter (OD) of 51.0 mm and 50.6 mm (mild steel and stainless steel piles respectively), a wall thickness of 8 mm and 7.5 mm (mild steel and stainless steel piles respectively), and a length of 2.2 m so that the tip of the pile was embedded 2 m BGL. A compact driving machine, the Sol Solution Grizzly tool was used to drive all the piles. The piles were subjected to static load tension tests at intervals of approximately 1, 14, 28, 90, 175, 272 and 315 days after installation, with a loading device comprising: (i) a reaction frame consisting of two IPE 270 steel beams and two HEB 160 steel beams placed on timber plates; (ii) a manual Enerpac hydraulic jack (HBM C6A) inducing tension in the pile; (iii) three Linear Variable Displacement Transformers (LVDT) transducers, HBM WA200 mm. The loading was applied by increments of about 10% of the estimated capacity, each of them being kept constant for 15 min, until reaching the failure of the pile. Failure, here, is considered as the maximum value of the load during the test and the estimated capacity is calculated following the ICP method [4].

3   Results and Conclusions

Smooth mild steel piles showed a tensile capacity of about 30 kN after installation and a clear increase of capacity gradually to a mean value of 70 kN 315 days after driving (see Fig. 1). Nonetheless, the capacity after installation was larger for rough piles (about 50 kN) than smooth ones (30 kN) and it reached higher values at long term (up to 90 kN). After extraction, all the mild steel piles exhibited a crust of sand attached (glued) to the pile shafts. For stainless steel piles, the tension capacity varied in a small scatter from

25 to 30 kN whatever the deadline can be. It does not show any increase as for the mild steel piles. Despite a larger initial roughness, the shot-blasted rough stainless steel piles have almost the same capacities than smooth piles. No crust of sand was observed attached to the pile shaft after extraction (see Fig. 1).

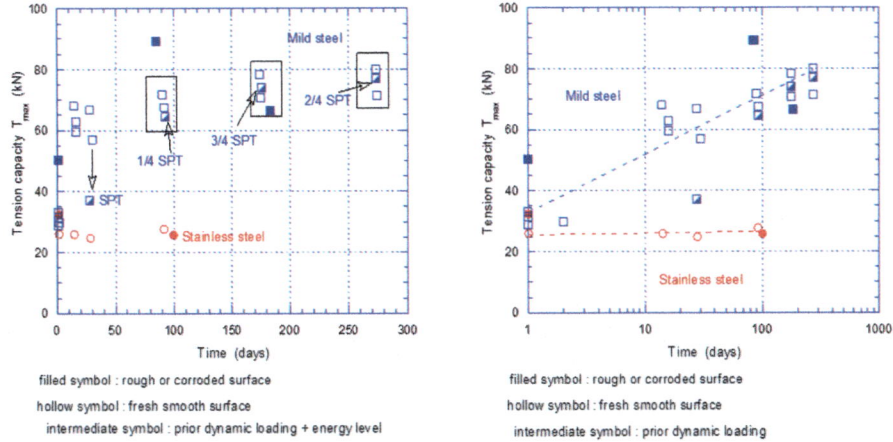

Fig. 1. Evolution of capacity with time.

Obviously, for small piles, physiochemical mechanisms play an important role in the set-up process. Oxidation and cementation that attaches (glues) sand grains to the pile induce a shift of the shearing plane into the sand and increase constrained dilatancy. Effects of chemistry such as corrosion were evident as no ageing was found in stainless piles versus ageing in mild steel piles. Stiffness does not follow a similar evolution.

Some piles were submitted, before tension tests, to a dynamic compressive load and compared to intact piles at the same age of testing, in order to verify the durability of the set-up benefices. Piles re-driven with a fraction of the SPT energy did not present any significant decrease of the tension capacity. In contrast, piles strongly re-driven at the SPT energy level showed an important decrease of their capacity to a value close to the capacity directly after installation (see Fig. 1). Two of these piles were let in place after compression load and re-tested some months later to see if set-up could be re-activated but no evolution of the tension capacity was observed (see Fig. 1).

Acknowledgements. The authors want to acknowledge the partners of the project, Sol Solution company for their interest in the project, their practical help and fruitful discussions. Many thanks to European partners: Prof. Jardine (Imperial College of London), Dr. Carroll and Dr. Carotenuto (Norwegian Geotechnical Institute), Prof. Gavin (TU Delft), as well as Grand Port Maritime de Dunkerque and Dong Energy for the access to the site. Finally, we want to thank 3SR Laboratory (Geomechanics team) for the financial support.

References

1. Chow, F.C.: Investigations into Displacement Pile Behaviour for Offshore Foundations. Ph.D. Thesis, Department of Soil Mechanics, Imperial College, London, UK (1997)
2. Daramola, O.: Effect of consolidation age on stiffness of sand. Geotechnique **30**(2), 213–216 (1980)
3. Gavin, K., Igoe, D., Kirwan, L.: The effect of ageing on piles in sand. Spec. Ed. Ice J. Geotech. Eng. Geotech. Chall. Renew. Energy Proj. **166**(2), 122–130 (2013)
4. Jardine, R. J., Standing, J. R., Chow, F. C.: Field research into the effects of time on the shaft capacity of piles driven in sand. In: Proceedings of ISFOG 2005 on Frontiers in Offshore Geotechnics, pp. 705–710 (2005)
5. Jardine, R., Standing, J., Chow, F.: Some observations of the effects of time on the capacity of piles driven in sand. Geotechnique **56**(4), 227–244 (2006)
6. Jardine, R.J., Standing, J.R.: Field axial cyclic loading experiments on piles driven in sand. Soils Found. **52**(4), 723–736 (2012)
7. Lehane, B. M., Lim, J. K., Carotenuto, P., Nadim, F., Lacasse, S., Jardine, R. J., Van Dijk, B. F. J.: Characteristics of unified databases for driven piles. In: Proceedings of the 8th International Conference on Offshore Site Investigation and Geotechnics, London, UK (2017)
8. Lehane, B., Jardine, R., Bond, A., Frank, R.: Mechanisms of shaft friction in sand from instrumented pile tests. J. Geotech. Eng. **119**(1), 19–35 (1993)
9. Lim, J.K., Lehane, B.M.: Characterisation of the effects of time on the shaft friction of displacement piles in sand. Geotechnique **64**, 476–485 (2014)
10. Schmertmann, J.: The mechanical aging of soils. J. Geotech. Eng. **117**(9), 1288–1330 (1991)
11. White, D., Zhao, Y.: A model-scale investigation into 'set-up' of displacement piles in sand. In: Sixth International Conference on Physical Modelling in Geotechnics, Hong Kong. Taylor & Francis (Eds.) (2006)

The Joint Inspector - A New Method of Quality Control for Diaphragm Walls

Nikolaus Schneider[✉]

GuD Geotechnik und Dynamik Consult GmbH, Darwinstr.13, 10589 Berlin, Germany
schneider@gudconsult.de

Abstract. The construction of diaphragm walls requires a high level of technical experience to cope with one of the most challenging tasks in geotechnical engineering, the interlock of discreet panels to a continuous wall. The joint inspector is a new tool that supports the quality check of diaphragm walls during construction by providing real-time data to initiate corrective actions during construction, if necessary.

Keywords: Diaphragm walls · Joint elements · Stop ends · Quality control Water stops

1 What Are Diaphragm Walls and How Are They Constructed?

Diaphragm walls are built underground to a depth of 100 meters below ground level. They are designed to either transfer load into the ground or to retain horizontal load in the case of a deep excavation pit (Fig. 1).

A mixture of dry bentonite powder, a type of clay with a special microscopic structure, and water provide a slurry that can be poured into an open trench. This allows to open long and deep trenches in which reinforcement cages and concrete can be installed. By applying a special technique of pouring concrete into the trench, tremie piping, the diaphragm wall can be built in a quality that is comparable to reinforced concrete walls in the upper structure.

The most challenging process step in constructing diaphragm walls is to connect the separate trenches into a continuous wall. The trenches are excavated and concreted as independent construction units. For the final concept, however, these independent units must be joined to form a continuous wall. Two types of stop ends have established themselves over time. Joint construction, which remain permanently in the diaphragm wall and elements that are taken out during construction (Figs. 2 and 3).

© Springer Nature Switzerland AG 2018
W. Wu and H.-S. Yu (Eds.): *Proceedings of China-Europe Conference on Geotechnical Engineering*, SSGG, pp. 1009–1012, 2018.
https://doi.org/10.1007/978-3-319-97115-5_27

Fig. 1. Diaphragm wall as retaining wall exposed to water pressure

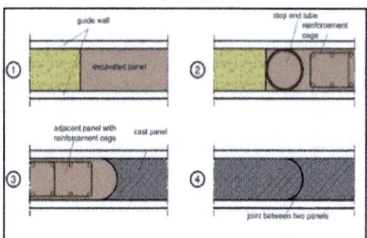

Fig. 2. Stop ends remaining in the wall   **Fig. 3.** Retrievable stop ends

2   What Type of Construction Failure Is Likely to Occur and What Are the Consequences?

The tremie pipe concrete, provides enormous horizontal pressure on the joint construction element, since it is designed to have liquid properties. The possibilities for construction faults arise there, where the side way excavation has an over brake. The stop end is pushed into the over brake and concrete flows around it. The adjacent panel consequently does not achieve the designed connection to previous panel. The connection is disturbed by excessive concrete that was pushed through. These joints, disturbed by excessive concrete, are only visible after the retaining walls have been excavated (Figs. 4 and 5).

Fig. 4. Stop end with unremoved concrete   **Fig. 5.** Excessive concrete at stop end

3   What Is the Joint Inspector and How Does It Work?

The joint inspector is an innovative quality control method to provide a reliable reading on site during construction. During the desanding phase, the joint inspector is attached to the frame of the mechanical grab and lowered down into the trench to measure the distance between the joint inspector and the exposed joint element of the primary panel (Fig. 6).

Fig. 6. Joint inspector attached to a mechanical grab

The arms are released, once the final depth is achieved. They read the distance to the exposed stop end of the primary panel (Figs. 7 and 8). The results are transmitted via blue tooth to the computer of the site engineer, who can take the decision to further remove excessive concrete or not.

Fig. 7. Distance readers   **Fig. 8.** Reset of the joint inspector

The readings provide several graphs, i.e. distance profiles between the joint inspector and the exposed stop end, the approximated inclination of the stop end and areas of excessive concrete for each arm. All values taken are referenced in depth (Figs. 9 and 10).

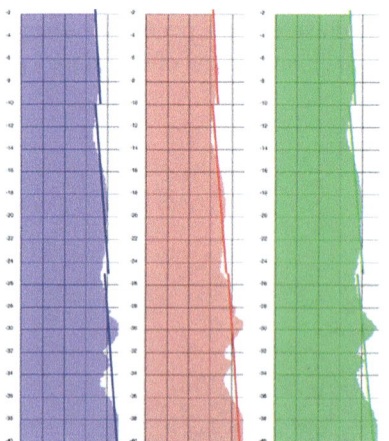

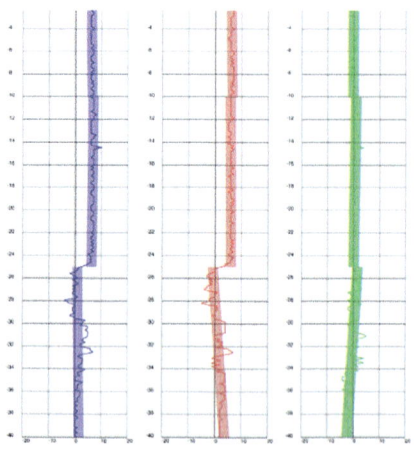

Fig. 9. Distance profiles   **Fig. 10.** Areas of excessive concrete

4   Conclusions

Mostly all construction failures of diaphragm walls occur along the transition between two panels. Presently, quality checks on the joints are possible only, after the wall is built ("cross hole sonic logging") or during excavation. The joint inspector enables the site engineers to take readings during construction and check on uncontrolled concrete on the stop ends. Based on the results, he can decide to remove the concrete and clean the joint to ensure the design quality of these joints in diaphragm walls. Taking the joint inspector into the construction process reduces the risk of failure joints and saves significantly cost of repair work.

Analysis of Cofferdam Stability
for Foundation of a Bridge Pillar

Monika Sulovska[✉] and Jakub Stacho

Department of Geotechnics, Slovak University of Technology, Bratislava, Slovakia
monika.sulovska@stuba.sk

Abstract. Temporary cofferdams for foundations of bridge pillars are ranked among the demanding constructions, and they often require high resources for their construction. The newly planned bridge over the Danube River in the Komarno city between Slovakia and Hungary has been designed as an asymmetric cable-stayed bridge with single pillar. The cofferdam made of double – row sheet pile wall of ground plan dimensions of 20 × 44 m is designed for foundation of the pillar. A narrow space between sheet piles will be filled by an atypical concrete.

The paper focuses on stability analysis of the cofferdam, which construction differs to typically proposed constructions of excavation pits for founding of the bridge pillars. It was necessary to verify stability of the cofferdam in several design situations, such as construction process of the cofferdam, injection of the subsoil from the cofferdam, which is backfilled by the soil and the process of excavating the pit in several phases with gradual installation of struts. The task was solved using an analytical method and numerical method – FEM. Modeling the behavior of the construction of the cofferdam using different methods allowed wide analysis of results and more reliable verification of the stability of the construction as well as the design of individual elements.

Keywords: Cofferdam · Sheet piles · Stability analysis

1 Introduction

The foundation of a pillar of a new bridge in the Danube River requires construction of a cofferdam. The selection of the most appropriate type of the cofferdam was based on consideration of all technological and economic aspects. The stability of the cofferdam was verified in each construction phase. The design also included verification of positive and negative impacts of water level in the riverbed as well as verification of the stability in the case of the flood.

The geological conditions in the area of the cofferdam were presented by [1]. The upper layer of a thickness of 1 m consists of gravels. The following layer of a thickness of 3 m consists of fine-grained soils, silts gravely and silts with medium plasticity. Layers of sands with fine soils and cemented sands are situated below these fine-grained soils.

© Springer Nature Switzerland AG 2018
W. Wu and H.-S. Yu (Eds.): *Proceedings of China-Europe Conference on Geotechnical Engineering*, SSGG, pp. 1013–1016, 2018.
https://doi.org/10.1007/978-3-319-97115-5_28

2 Design of the Cofferdam

The cofferdam has ground plan dimensions of 44×20 m and it's designed of sheet piles (type VL 606) in two rows. The main rectangular shape of the cofferdam is extended by two triangular segments which reduce impact of flowing water. The inner sheet piles have a length of 19 m. An embedded length is 6.5 m below the riverbed. The part remaining will resist to loading from the water and earth pressures. An extended outer row is designed as an overflow protection in the case of the flood. The inner and outer rows are connected by rods. The space between sheet piles of a thickness of 0.6 m will be filled by concrete. The construction will be braced by struts in two levels. The jet grouting technology will be used for creating two improved layers in the subsoil. The first one of a thickness of 2 m will be created in the depth corresponding to the base of sheet piles and the second one at the bottom of the excavation inside the cofferdam. The upper layer has a thickness of 1 m. While the lower layer will prevent water from entering into the excavation pit, the other one will increase global stiffness of the construction.

Typical cross section for all static calculations was created in the middle of the cofferdam. The design of the cofferdam was made in such a way as to best represent the different construction phases:

- The assessment of the concrete filling between inner and outer sheet piles,
- The assessment of the load impact which is caused by the piling rig,
- The assessment of the cofferdam stability during all excavation phases,
- The assessment of the cofferdam during the flood situation.

Two different computational methods were used for analysis of the cofferdam [2]:

- Analytical model based on the dependent pressure method,
- Numerical modeling - FEM.

An analytical solution based on the dependent pressure method was done using the FINE Geo5 geotechnical software. The software allows verification of retaining structures for different excavation phases. The software includes only some basic models which can be used. The cofferdam was analyzed using three different models:

1. Diaphragm wall model - two rows of sheet piles with concrete inside were replaced by a diaphragm wall of a thickness of 0.6 m;
2. Double sheet pile model - stiffness of two rows of sheet piles were taken into account without modeling of concrete between them;
3. Single sheet pile model - simplified computational model with only one row of sheet piles.

The Plaxis software was used for numerical modeling of the construction. The geometry of the cofferdam as well as the soil and material properties were the same for both computational methods. The stability analysis of the cofferdam was verified in following phases: installation of sheet piles, concreting the space between sheet piles, backfill inside the cofferdam, load from the piling rig, jet grouting of the subsoil, pumping water and excavation of the soil from the cofferdam, installation of struts (in corresponding phase of excavation). The analysis showed that the biggest horizontal deformations are obtained in the phase, where the cofferdam with backfill is loaded by the piling rig. The horizontal deformations of inner sheet piles calculated using different computational methods and models are shown in Fig. 1 for this phase. Due to large deformations, the stabilizing stone backfill from the outer side of the cofferdam was required. The horizontal deformations were reduced to about half, see Fig. 2.

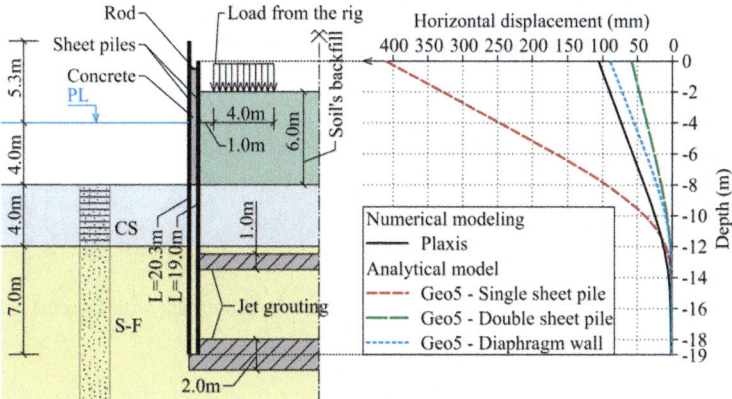

Fig. 1. Horizontal deformations of sheet piles calculated using different computational methods and models - without outer stone backfill

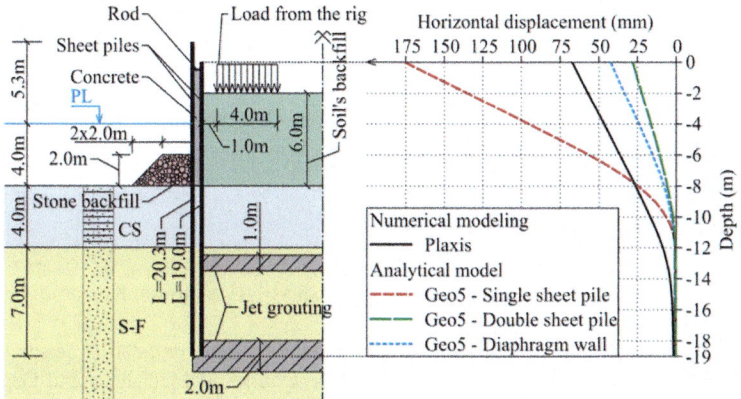

Fig. 2. Horizontal deformations of sheet piles calculated using different computational methods and models - with outer stone backfill

The sheet piles were braced by struts during excavation phases, which caused that horizontal deformations were not increased furthermore. The biggest deformations were obtained in the phase, when the cofferdam with backfill is loaded by the rig and in the phase, which simulates the flood situation. The total deformations from these phases determined by FEM are shown in Fig. 3. These phases were the most critical for stability of the cofferdam and they were crucial to designing individual elements.

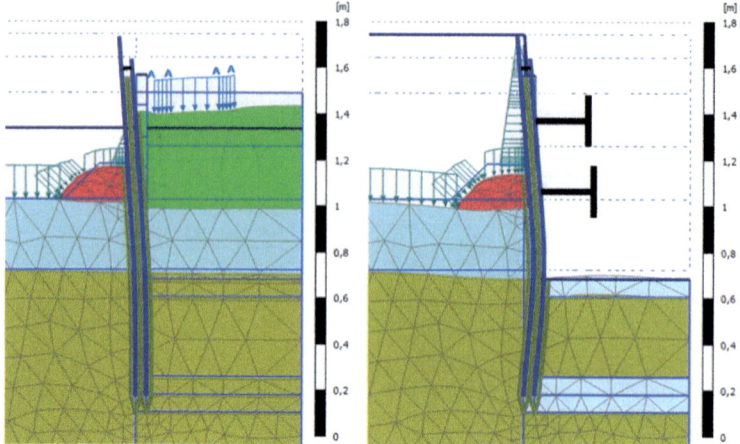

Fig. 3. Deformations of the cofferdam in phase when the cofferdam with backfill is loaded by the rig (left) and in the flood situation (right)

3   Conclusion

The construction of the cofferdam in the middle of the Danube River represented the difficult task. The design was based on extensive analysis using different computational methods such as numerical modeling and analytical calculations. Numerical modeling was considered the most accurate, because allows modeling of all elements of the construction. The most accurate results of analytical solutions were obtained in the case, when the double sheet pile wall is replaced by the diaphragm wall. The solution with single row of sheet piles was too conservative.

References

1. Pozsar, L.: Final report of engineering geological survey. Consortium Komarno – Komarom, 113 p. (2014)
2. Turcek, P., Sulovska, M., Ladicsova, E.: Evaluation of bridge foundation design over river Danube in Bratislava. In: 16th European Conference on Soil Mechanics and Geotechnical Engineering, pp. 799–804. ICE Publishing, Edinburg UK (2015)

Dynamic Response of Embedded Block Foundation Under Vertical Vibration

Kavita Tandon[(✉)], Rohit Ralli, B. Manna, G. V. Ramana, and M. Datta

Indian Institute of Technology Delhi, New Delhi 110016, India
kavita.tandon2008@gmail.com

Abstract. The dynamic response of footing depends on several influencing parameters such as the shape and size of the foundation, effect of embedment, static load and dynamic excitation intensity. Here, efforts have been made to experimentally investigate the dynamic response of embedded block foundation under vertical vibration. Field experiments and analytical investigations using elastic half-space theory are carried out to study the dynamic response of embedded block foundation of aspect ratio $L/B = 1.5$ under vertical vibration for three different embedment depths ($D = 0$, 0.25 and 0.5 m) under static load of 6.6 kN. The block vibration tests are carried out using a Lazan-type mechanical oscillator under four different eccentric moments ($W.e = 0.2, 0.8, 1.4$ and 2.0 Nm). The frequency amplitude response of the block foundation obtained from elastic half-space theory are compared with the response obtained from the experimental investigations and it is observed that the maximum resonant amplitude decreases and resonant frequency increases as the embedment depth increases.

Keywords: Dynamic response · Vertical vibration · Resonant amplitude
Resonant frequency · Embedment effect

1 Introduction

The design of footings exposed to dynamic loads and the prediction of dynamic response of such footings resting in/on soil deposits are of major importance. The theory of surface footings neglects the fact that footings are partly or fully embedded and overestimates the real response of the foundation. The effect of partial embedment of the footings in the soil, on vertical vibrations has been recognized for many years experimentally [1, 2] and theoretically [3]. The design of such footings is of considerable interest so as to place the foundation on a good bearing stratum and to provide a convenient block top level for operating the machine it supports. Gazetas and Stokoe [2] experimentally evaluated the dynamic response of model footings embedded at various depths. Saran et al. [4] conducted field tests on embedded block foundation under vertical and coupled mode of vibrations. In order to determine the response of soil-foundation system subjected to dynamic loads experimental investigations are carried out for a model block foundation in the field and using elastic half-space theory proposed by Velestos and Wei [1] and Novak and Beredugo [3].

© Springer Nature Switzerland AG 2018
W. Wu and H.-S. Yu (Eds.): *Proceedings of China-Europe Conference on Geotechnical Engineering*, SSGG, pp. 1017–1021, 2018.
https://doi.org/10.1007/978-3-319-97115-5_29

2   Experimentation

The vertical block vibration tests were carried out on M30 concrete block of dimension 0.9 m × 0.6 m × 0.5 m casted in-situ resting in/on soil for different eccentric moment ($W.e = 0.221, 0.868, 1.450$ and 1.944 Nm) with a static load (W_s) of 6.6 kN. The average density of the soil in the field was 16.72 kN/m³ with an average N-value of 9 up to a depth of 3.0 m and the shear wave velocity was 216.42 m/sec. The shear wave velocity (V_s) of the soil was obtained from the correlation [5] as given in Eq. 1.

$$V_s = 86.0\,N^{0.42} \text{ m/sec} \tag{1}$$

The vibration tests were done for three embedment depths of $D = 0, 0.25$ and 0.5 m and time-acceleration response were measured using accelerometers connected with data acquisition system for different frequencies. The complete experimental setup is shown in Fig. 1.

Fig. 1. Experimental setup of block foundation under vertical vibration.

3   Elastic Half-Space Theory

The dynamic response of block foundation for surface footings is determined using the approximate analytical solution proposed by Velestos and Wei [1] and Novak and Beredugo [3] for embedded foundations. The theory considers soil as an elastic half-space and stiffness and damping constants of soil-foundation system are considered as frequency dependent.

4   Results and Discussions

The dynamic response of the embedded foundation obtained for block vibration test indicates that footing responses are greatly affected by the depth of embedment into the soil and the excitation intensity. The dynamic field test results were compared with the calculated values based on elastic half-space theory and the results are discussed below.

4.1   Effect of Excitation Force Level

The typical frequency-amplitude curve depicting the effect of excitation force for both field testing and elastic half-space theory is shown in Fig. 2. It can be seen from Fig. 2 that the maximum resonant amplitude of vibration increases notably and the resonant frequency drops slightly with increasing exciting force level for both surface and embedded footings. The experimental resonant frequencies match with good agreement with elastic-half theory. The experimental resonant amplitudes are considerably larger than the theoretical values.

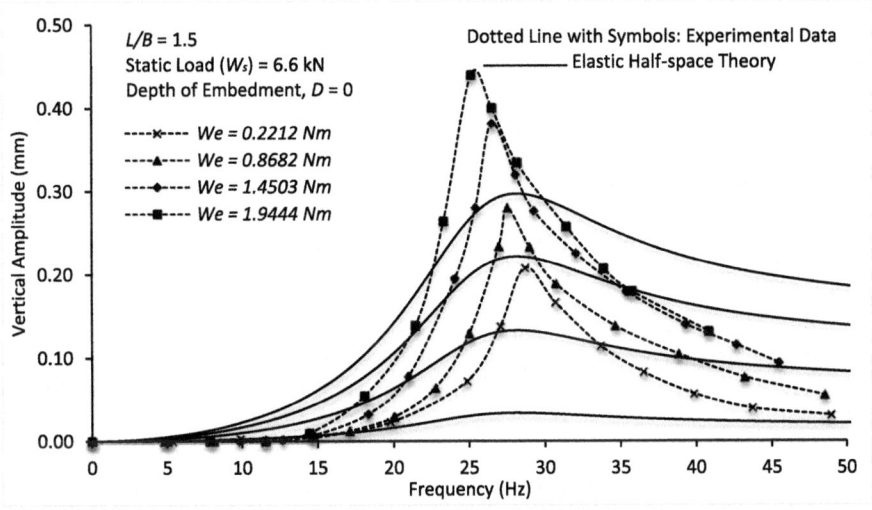

Fig. 2. Response curves of a surface foundation under vertical vibration for different eccentric moments.

4.2   Effect of Embedment Depth

The typical frequency-amplitude curves for block foundation subjected to vertical vibration for both field testing and elastic half-space theory is shown in Fig. 3. It can be seen from Fig. 3 that the maximum resonant amplitude of block foundation under vertical vibration decreases and the frequency increases significantly with the increasing depth of embedment. It is observed from the experimental investigation that the decrease in resonant amplitude is about 20 to 35% for the embedment depth of 0.25 m and 30 to 50% for the embedment depth of 0.5 m, whereas, increase in resonant frequency is found to be 5 to 20% for different depths of embedment as compared to surface block foundation.

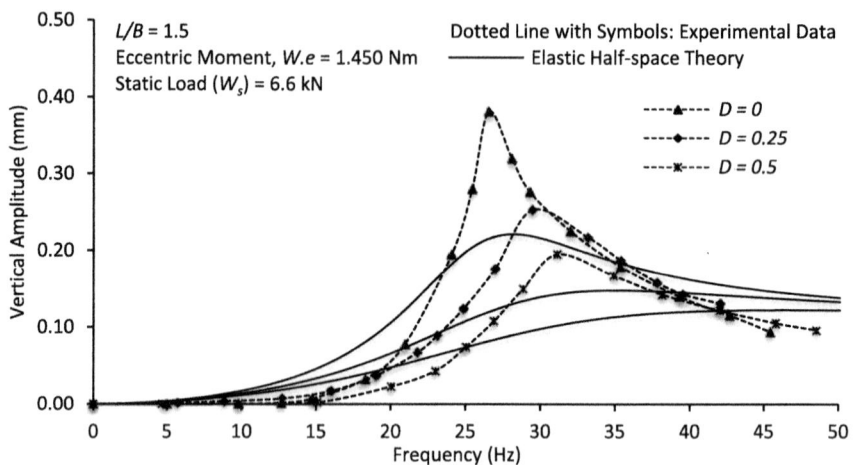

Fig. 3. Response curves under vertical vibration for different embedment depths.

5   Conclusions

Based on the results of block vibration tests carried out on model block foundation in the field and analysis using elastic half-space theory, the following conclusions are drawn:

1. The experimental resonant frequencies match with good agreement with elastic-half theory. The experimental resonant amplitudes are considerably larger than the theoretical values.
2. For a constant excitation level, the resonant amplitude decreases and resonant frequency increases with the increase in embedment depth of foundation.
3. The maximum resonant amplitude of vibration increases notably and the resonant frequency drops slightly with increasing exciting force level for both surface and embedded footings.

References

1. Velestos, A.S., Wei, Y.T.: Lateral and rocking vibration of footings. J. Soil Mech. Found. Eng. ASCE **97**(SM9), 1227–1248 (1971)
2. Gazetas, G., Stokoe, K.H.: Free vibration of embedded foundations: theory versus experiment. J. Geotech. Eng. **117**(9), 1382–1401 (1991)
3. Novak, M., Beredugo, Y.O.: Vertical vibration of embedded footings. J. Soil Mech. Found. Div. ASCE **98**(SM12), 1291–1310 (1972)

4. Saran, S., Ranjan, G., Vijayvargiya, R.C.: Embedment effects on foundations under vertical vibrations. In: First International Conference on Recent Advances in Geotechnical Engineering and Soil Dynamics, Missouri, pp. 749–754 (1981)
5. Hanumantharao, C., Ramana, G.V.: Dynamic soil properties for microzonation of Delhi, India. J. Earth Syst. Sci. **117**(2), 719–730 (2008)

Free Vibration Calculations of an Euler-Bernoulli Beam on an Elastic Foundation Using He's Variational Iteration Method

Alima Tazabekova, Desmond Adair$^{(\boxtimes)}$, Askar Ibrayev, and Jong Kim

Nazarbayev University, Astana 010000, Kazakhstan
dadair@nu.edu.kz

Abstract. Free vibration characteristics for an Euler-Bernoulli beam supported using a simple Winkler linear elastic foundation are calculated. The method of solution is by He's variational iteration method developed for various boundary (end) conditions. The beam's natural frequencies and mode shapes are obtained, with rapid convergence noted during the calculations. The calculation method is tested using a clamped-clamped beam. In this paper a robust and efficient algorithm is also given, based on He's method, which can be easily modified for more complicated elastic foundations.

Keywords: Free vibrations · He's variational method · Winkler foundation

1 Introduction

Beams resting on elastic foundations have a wide application in modern engineering, including railway engineering, but pose technical problems in structural design [1]. The Winkler model of elastic foundation is one of the simplest, where the vertical displacement of the beam is assumed to be proportional to the contact pressure at an arbitrary point [2]. A variety of investigations on free vibration, buckling and stability behavior of Winkler foundation beams have been conducted by researchers [3–6]. He's variational iteration method is a modification of a general Lagrange multiplier method [7] and has been used as a powerful tool for calculating free vibration [8, 9]. In this paper, we proceed to investigate the free vibrations of an Euler-Bernoulli beam resting on an elastic foundation using the relatively new and more efficient method by He [10].

2 Governing Equation and Application of He's Method

The equation of motion for transverse vibrations of a uniform Euler-Bernoulli beam resting on a Winkler elastic foundation, as shown on Fig. 1, can be written as

$$\frac{\partial^2}{\partial x^2}\left(EI(x)\frac{\partial^2 w(x,t)}{\partial x^2}\right) + k_w(x)w(x,t) + \rho A(x)\left(\frac{\partial^2 w(x,t)}{\partial t^2}\right) = 0, \quad 0 < x < l. \quad (1)$$

© Springer Nature Switzerland AG 2018
W. Wu and H.-S. Yu (Eds.): *Proceedings of China-Europe Conference on Geotechnical Engineering*, SSGG, pp. 1022–1026, 2018.
https://doi.org/10.1007/978-3-319-97115-5_30

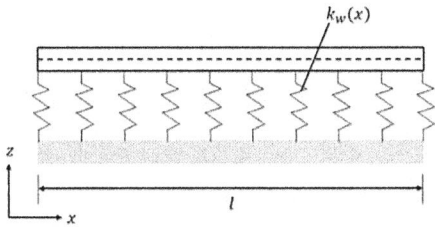

Fig. 1. Euler-Bernoulli beam on Winkler foundation.

Using $w(x,t) = W(x)h(t)$ and without loss of generality Eq. (1) is now made non-dimensional to give

$$\frac{\mathrm{d}^4 W(X)}{\mathrm{d}X^4} - PW(X) = 0, \quad o < x < 1, \tag{2}$$

where P is the eigenvalue of the problem and is equal to

$$P = \frac{(k_w(X) - \rho A\omega^2)l^4}{EI}. \tag{3}$$

The correctional function can now be obtained

$$W_{n+1}(X) = W_n(X) + \int_0^X \lambda \left(\frac{\mathrm{d}^4 W_n(t)}{\mathrm{d}t^4} - PW_n(t)\right) \mathrm{d}t, \tag{4}$$

and the Lagrange multiplier λ can be found as

$$\lambda = \frac{(t - X)^3}{6}. \tag{5}$$

To start the iterations associated with Eq. (4) the $W_0(X)$ term is needed, which is represented as a Maclaurin series of the first four terms, and the solution is found as $W(X) = \lim_{K \to \infty} W(X)_k$. The boundary conditions can also be written in dimensionless form

$$\left[\alpha_{r3} \frac{\mathrm{d}^3 w(X)}{\mathrm{d}x^3} + \alpha_{r2} \frac{\mathrm{d}^2 w(X)}{\mathrm{d}x^2} + \alpha_{r1} \frac{\mathrm{d}w(X)}{\mathrm{d}x} + \alpha_{r0} W(X)\right]\Bigg|_{x=0} = 0, \quad r = 1, 2,$$

$$\left[\beta_{r3} \frac{\mathrm{d}^3 w(X)}{\mathrm{d}x^3} + \beta_{r2} \frac{\mathrm{d}^2 w(X)}{\mathrm{d}x^2} + \beta_{r1} \frac{\mathrm{d}w(X)}{\mathrm{d}x} + \beta_{r0} W(X)\right]\Bigg|_{x=1} = 0, \quad r = 1, 2. \tag{6}$$

For Eq. (6), the second of the boundary conditions can be rewritten as

$$\sum_{j=0}^{3} f_{rj}^{[k]}(P)W^{(j)}(0) = 0, \; r = 1, 2. \tag{7}$$

Here, $f_{rj}^{[k]}$ are polynomials of P with respect to K. On solving the first boundary condition of Eqs. (6) and (7) simultaneously for the non-trivial solutions $W^{(j)}(0)$ $(j = 0, 1, 2, 3)$, the ith eigenvalue $P_i^{[k]}$ corresponding to K can be obtained, and the number of iterations M is decided from

$$\left| P_i^{[n]} - P_i^{[n-1]} \right| \leq \varepsilon. \tag{8}$$

3 Numerical Example: Clamped-Clamped (C-C) Beam

The following are results for a C-C beam. Important to this study is the efficiency of the method, which is demonstrated in Fig. 2. The variational iteration method proved extremely fast with convergence achieved after very few iterations. The linear modulus used in Fig. 2 and Table 1 took the form $k_w = K_{w0}(1 - \alpha X)$.

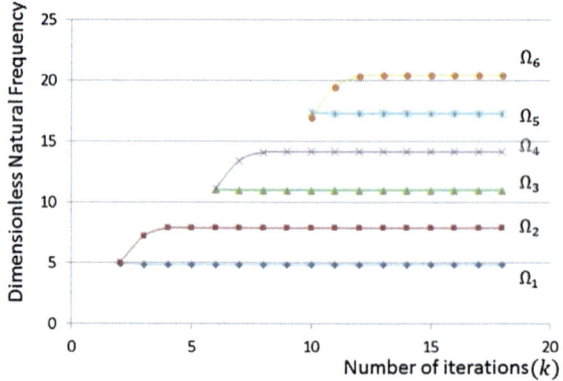

Fig. 2. Convergence for the first six natural frequencies when $k_{w0} = 50, \alpha = 0.2$.

The frequency parameters for the C-C beam with a linear elastic modulus are shown in Table 1. These values compare well with the equivalent found in the literature [9].

The mode shape functions shown in Fig. 3 are obtained using the eigenvalues and a polynomial formed in terms of X. The second diagram in Fig. 3 shows the effect of increasing the slope of the linear function on the natural frequency. Basically the natural frequency falls linearly.

Table 1. Frequency parameters for C-C beam with linear foundation modulus ($\alpha = 0.2$).

k_{w0}	Ω_1	Ω_2	Ω_3	Ω_4	Ω_5	Ω_6
1	4.73217	7.85367	10.99578	14.13718	17.27933	20.39929
10	4.75116	7.85785	10.99730	14.13792	17.28016	20.39638
50	4.83294	7.87633	11.00407	14.14103	17.28250	20.41832
100	4.92965	7.89925	11.01250	14.14508	17.28361	20.42077
200	5.10758	7.94451	11.02930	14.15305	17.28789	20.39431
2000	6.92482	8.65221	11.31955	14.29380	17.36672	20.46750

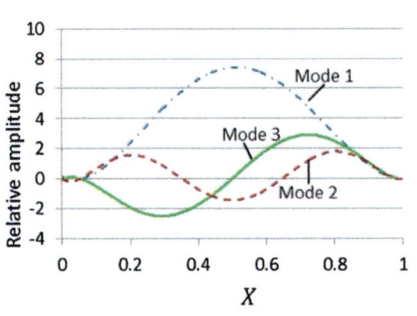
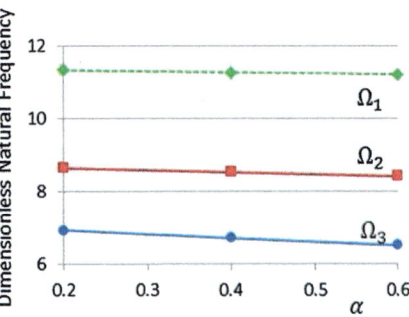

Fig. 3. Mode shapes for the first three natural vibrations and the effect of the linear foundation parameter, α on the natural frequencies.

References

1. Zheng, D.Y., Cheung, Y.K., Au, F.T.K., Cheng, Y.S.: Vibration of multi-span non-uniform beams under moving loads by using modified beam vibration functions. J. Sound Vib. **212** (3), 455–467 (1998)
2. Hetenyi, M.: A general solution for the bending of beams on an elastic foundation of arbitrary continuity. J. Appl. Phy. **21**, 55–58 (1950)
3. Eisenberger, M., Clastornik, J.: Vibrations and buckling of a beam on a variable Winkler elastic foundation. J. Sound Vib. **115**, 233–241 (1987)
4. Ding, Z.: A general solution to vibrations of beams on variable Winkler elastic foundation. Comput. Struct. **47**, 83–90 (1993)
5. Eisenberger, M., Yankelevsky, D.Z., Clastornik, J.: Stability of beams on elastic foundations. Comput. Struct. **24**, 135–140 (1986)
6. Farghaly, S.H., Zeid, K.M.: An exact frequency equation for an axially loaded beam-mass-spring system resting on Winkler elastic foundation. J. Sound Vib. **185**, 357–363 (1995)
7. He, J.-H.: Variational iteration method for autonomous ordinary differential systems. Appl. Math. Comput. **114**(2/3), 115–123 (2000)

8. Tari, H., Ganji, D.D., Babazadeh, H.: The application of He's variational iteration method to nonlinear equations arising in heat transfer. Phy. Lett. A **363**(3), 213–217 (2007)
9. Dehghan, M., Shakeri, F.: Application of He's variational iteration method for solving the Cauchy reaction-diffusion problem. J. Comput. Appl. Math. **214**(2), 435–446 (2008)
10. He, J.-H.: Variational iteration method - A kind of non-linear analytical technique: Some examples. Int. J. Non-Linear Mech. **34**(4), 699–708 (1999)

Vertical Capacity of Large and Deep Barrette Pile for Bangkok Subsoils

Wanchai Teparaksa[1(✉)], Mike Sinkinson[2], and Jirat Teparaksa[3]

[1] Chulalongkorn University, Bangkok 10330, Thailand
wanchai.te@chula.ac.th
[2] Thai Bauer Co., Ltd., Bangkok, Thailand
[3] Strategia Engineering Consultants Co., Ltd., Bangkok 10400, Thailand

Abstract. Recently, barrette piles have been used to replace conventional bored piles in order to increase allowable vertical pile capacity in the limited space projects such as high-rise building projects and elevated train projects. The basic of construction equipment, the slurry used during construction and construction time between barrette pile and bored pile construction are totally different. Therefore, their design parameters for pile capacity estimation should be different. In this research, pile load test was carried out on the fully-instrumented barrette piles. The test result can be separated into friction of each soil layer and end bearing. By considering stress-strain relationships, adhesion factor of clay layer, friction factor of sand layer and end bearing factor of sand layer can be calculated. These factors were found to be able to estimate ultimate pile capacity. Based on these factors, the estimated pile capacity was compared with another barrette pile tests. This estimation agreed well with the test results.

Keywords: Barrette pile · Pile capacity · Adhesion factor · Friction factor

1 Introduction

Recently, the barrette pile has been used to replace the conventional bored pile in Bangkok city since it provides higher capacity in the limited space. The construction process of a barrette pile is the same as that of a diaphragm wall but it is different from wet process bored pile. The bentonite slurry is used as borehole stabilized agent during boring of the barrette which is different from wet process bored pile which polymer based slurry is used (Teparaksa and Teparaksa 2012). The bentonite slurry creates the cake film along the shaft of borehole and sedimentation at pile toe which leads to a decrease in shaft friction and end bearing (Teparaksa 1994, 2000, Teparaksa et al. 1999). This paper presents the behavior of deep barrette pile constructed in Bangkok subsoil based on the recent fully instrumented test piles.

© Springer Nature Switzerland AG 2018
W. Wu and H.-S. Yu (Eds.): *Proceedings of China-Europe Conference on Geotechnical Engineering*, SSGG, pp. 1027–1031, 2018.
https://doi.org/10.1007/978-3-319-97115-5_31

2 Soil Condition

The Bangkok subsoils consists of 13–16 m thick of soft to medium clay and followed by a stiff clay layer to about 21–28 m deep. The first dense silty sand layer is encountered below stiff to hard silty clay. The very stiff silty clay is alternated with the second dense silty sand layer at about 45–55 m deep. Generally, pile foundation of superstructure is penetrated in this second very dense sand layer. The piezometric level or phreatic surface of Bangkok aquifer is drawdown from −23.0 m below ground surface in 1995 and increased to −13.0 m in 2016.

3 Shaft Friction Behavior of Deep Barrette Pile

The deep barrette pile has been used for many years in Bangkok city; however, only 12 tested deep barrette piles with fully instrumentation had been collected. The barrette pile tip mostly penetrated in the second very dense silty sand layer at the depth of 50–60 m. All barrette piles were not based grouted. The unit pile shaft friction (f_s) can be estimated from the following equations;

$$f_s = \alpha\, S_u (\text{for clay layer}) \text{ and } f_s = \beta\, \sigma'_v (\text{for sand layer}) \tag{1}$$

where

f_s = Unit pile shaft friction (kN/m^2)
S_u = Undrained shear strength of clay (kN/m^2)
σ'_v = Effective overburden pressure in drawdown condition (kN/m^2)
α = Adhesion factor for clay
β = Friction factor for sand

α and β value can be determined from load transfer curve derived from instrumented test barrette piles which consists of vibrating wire strain gauges and extensometers installed near to the boundary of each soil layer. The α-value derived from

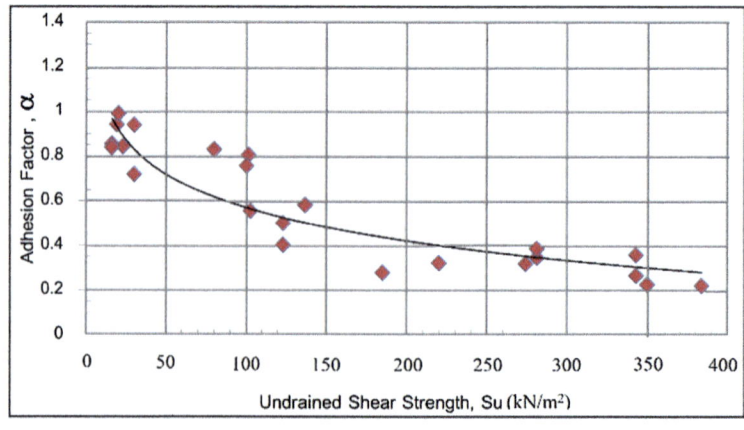

Fig. 1. Adhesion factor and undrained shear strength of barrette pile.

mobilized skin friction of all collected barrette pile test is presented against undrained shear strength of clay layer in Fig. 1. The relationship between β-value and effective angle of internal friction (Φ') of the sand layer derived from instrumented pile load test is presented in Fig. 2.

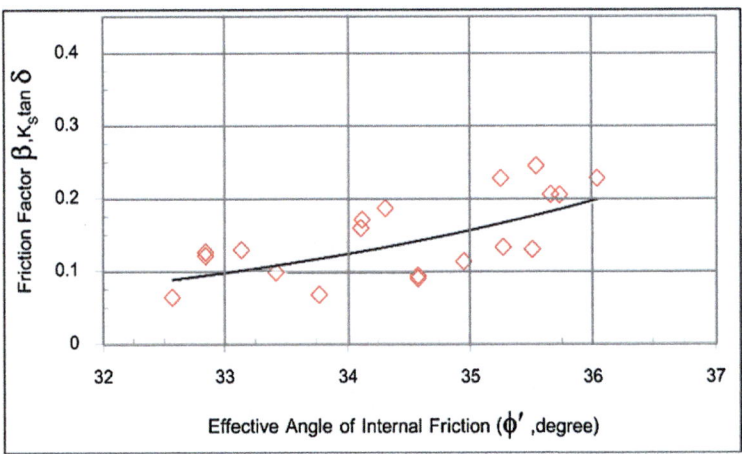

Fig. 2. Friction factor and angle of internal friction of barrette pile

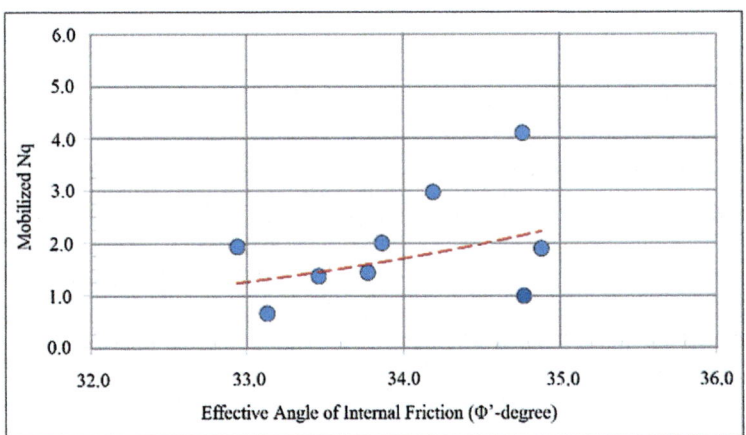

Fig. 3. N_q-parameter of barrette pile

4 End Bearing Behavior of Deep Barrette Pile

The bearing capacity of a barrette pile with tip penetrated in the sand layer generally derived from the same approach as bored pile as Eq. 2;

$$q_b = N_q \, \sigma'_v \tag{2}$$

where

q_b = Unit end bearing (kN/m^2)
N_q = End bearing coefficient
σ'_v = Effective overburden pressure in drawdown condition (kN/m^2)

The mobilized N_q-parameter of tested barrette pile is based on the pile deflection at the point where the tested load closes to the yield point of Butler & Hoy or Mazurkiewicz methods defined by Fellenius (1980) as presented against effective angle of internal friction of sand layer in Fig. 3.

5 Back Analysis of Barrette Pile Capacity Without Instrumentation

To verify the proposed α, β and N_q parameter for determination of ultimate capacity of deep barrette pile, pile capacity of two tested barrette piles without instrumentation, BP-T1 and BP-T2, was estimated and compared with pile load test. As present, BP-T2 is the deepest barrette pile (71.0 m deep) in Thailand. Table 1 presents the result of comparison. It was found that the calculated or estimated ultimate pile capacity agreed well with test results.

Table 1. Comparison of predicted and tested barrette pile capacity.

No	Project	Dimension (m)	Length (m)	Max. test load (tons)	Back analysis of pile capacity (tons)
BP-T1	MRT blue line extension	0.80 × 2.70	55.0	2,750	3,170
BP-T2	Landmark waterfront	1.00 × 2.70	71.0	5,150	5,124

6 Conclusions

The behavior of deep fully instrumented barrette pile was investigated by full scale load tests in Bangkok subsoils. The skin friction factor α and β value for barrette pile is determined from unit skin friction of strain gauges and extensometers installed in the test piles. The end bearing coefficient (N_q) was quite low due to the sedimentation at pile toe as a result of bentonite slurry. The verification of all parameters was carried out by predicting the ultimate capacity of another barrette pile test and comparing with its test results.

References

Fellenius, B.H.: The analysis of results from routine pile load test. Ground Eng. **13**(6), 19–31 (1980)

Teparaksa, W.: Newly developed toe grouted bored pile in soft Bangkok clay, performance and behavior. In: Proceedings of the International Conference on Design and Construction of Deep Foundation, pp. 1337–1351. FHWA, Orlando, USA (1994)

Teparaksa, W., Thasananipan, N., Anwar, M.A.: Base grouted of wet process bored pile in Bangkok subsoils. In: Proceedings of the 11th Asian Regional Conference on Soil Mechanics and Geotechnical Engineering, Seoul, vol. 1, pp. 269–272 (1999)

Teparaksa, W.: Estimating ultimate capacity of deep bored pile in Bangkok subsoils in line with global research trend. In: Proceedings of Annual Conference of the Engineering Institute of Thailand, pp. 41–56 (2000)

Teparaksa, W., Teparaksa, J.: Capacity of deep barrette piles with time effect. In: Proceedings of the 18th Southeast Asian Conference (18SEAGC) and Inaugural AGSSEA Conference (IAGSSEA), Singapore (2012)

Application Examples of Elastic Support Layers for High-Rise Buildings in China

Alexander Tributsch[1(✉)], Silke Appel[1], and Bertram Grass[2]

[1] GuD Geotechnik und Dynamik Consult GmbH, Darwinstr. 13, 10589 Berlin, Germany
`tributsch@gudconsult.de`
[2] Getzner Werkstoffe GmbH, Herrenau 5, 6706 Bürs, Austria

Abstract. Public transport systems excite ground vibrations that are transferred to nearby buildings and may exceed the comfort level or legal limit values for residential use. The paper aims to present application examples in China, where the assessment was based on Chinese and German standards.

Keywords: Rail traffic vibrations · Vibration assessment · Vibration prognosis

1 Introduction

Public transport systems, such as subway and suburban trains or tramways, emit vibrations into the ground below. These vibrations are transferred to the floors/ceilings of nearby buildings via soil, foundations and walls. Vibration prognoses are often conducted by a scheme provided in [1], based on piecewise transfer functions from the excitation source to the soil below, to the soil outside a neighboring building, to the foundation and finally to the floors/ceilings. Excitation spectra and/or transfer functions may be measured or simulated to predict the vibration immissions inside the building. The assessment process proves the admissibility or reveals the necessity of vibration mitigating measures.

One effective way of reducing the vibration immissions inside a building is to decouple it from the ground by an elastic layer. It is necessary to carefully investigate the effect of the elastic layer by taking into account the excitation characteristics, the building's eigenmodes/-frequencies, and the parameters of the elastic layer (e.g. permissible compressive load, …). A wrong design of an elastic support system could impair the system performance. Thus, the design process requires numerical simulations of the dynamic building response and the soil pressure below the foundation.

2 Assessment

2.1 Vibrations

Chinese Standards. According to Chinese Standard GB10071-88, "Measurement method of environmental vibration of urban area" [2], a so called VL_{zmax}-value has to

© Springer Nature Switzerland AG 2018
W. Wu and H.-S. Yu (Eds.): *Proceedings of China-Europe Conference on Geotechnical Engineering*, SSGG, pp. 1032–1035, 2018.
https://doi.org/10.1007/978-3-319-97115-5_32

be determined and compared to the acceptable values, listed in Chinese Standard GB10070-88 [3].

In order to calculate VL_{zmax}, the vibration acceleration level VAL is calculated based on the following equation [2]

$$VAL = 20\,log\frac{a}{a_0} \qquad (1)$$

where a is the RMS vibration acceleration in m/s² and a_0 is the reference acceleration of 10^{-6} m/s². This operation is performed for 1/3 octave bands to yield vibration levels VL_i for each frequency band i. Subsequently, the so-called z-weighting by W_k-values according to ISO 2631-1 [4] is applied. The quadratic or energetic sum of the weighted third-octave spectral values results in the VL_{zmax}-value.

The tolerable vibrations listed in GB 10070-88 range from 65 dB for special residential areas at night to 75 dB for industrial areas at daytime. For average residential areas the limit values of 67 dB and 70 dB are provided for nighttime and daytime, respectively.

The measurement point is defined at the ground surface outside a building. In our understanding, the assessment of vibrations at individual floors of a building is not intended by the standards mentioned above.

German Standards. In Germany, the standard DIN 4150-2 [5] is typically used to assess vibration levels in residential environments. Based on high pass filtered data and an exponential moving average, so-called KB-values are calculated. The assessment focuses on maximum values (KB_{Fmax}) and on time averaged values (KB_{FTr}).

For typical residential areas in cities with residential and commercial use, a maximum KB_{Fmax} of 0.30 is allowed at night. If KB_{Fmax} is between 0.15 and 0.30, an averaged value over 8 h at night is calculated and must be not greater than 0.07.

Two main differences between vibration assessment in China and in Germany can be found. On the one hand, KB-values are directly calculated from vibrations velocities, while VL-values are calculated from accelerations. On the other hand, the definition of the assessed point is different. While Chinese standard GB 10071-88 uses a measurement point outside a building, DIN 4150-2 relates to the immission point at the floor, i.e. directly where vibrations are perceived by persons.

2.2 Secondary Airborne Noise

Chinese standard JGJ/T170-2009 [6] is used to evaluate secondary noise immissions. The assessed value L_{Aeq} should be below 35 dB(A) to 42 dB(A), depending on the regional classification.

Secondary airborne noise, radiated by vibrating ceilings, walls, etc. is often difficult to separate from primary noise in measurements. Thus, the secondary noise levels are commonly calculated by empirical relations based on the measured or predicted floor/ceiling vibrations. The A-weighting of the calculated third octave spectra considers the characteristics of the perceived noise by the human ear.

3 Application Examples

Although several similar projects in Beijing were investigated, in the following, only one exemplary application is presented in more detail. In order to assess vibrations in several high-rise buildings and evaluate the necessity of vibration mitigation measures under consideration of a projected metro line, a dynamic simulation of the situation was necessary.

The new metro tunnel will be constructed at about 15 m below ground level and with a minimum distance of about 20 m to the investigated building assembly. Measured vibration responses at a similar, existing tunnel utilized to calculate the free field response at the construction site, see Fig. 1. One of the high-rise buildings was selected as a typical structure and modelled by finite elements, see Fig. 1, middle/right. The calculated free field excitation was applied at spring-dashpot-elements below the foundation. The following effects are taken into account:

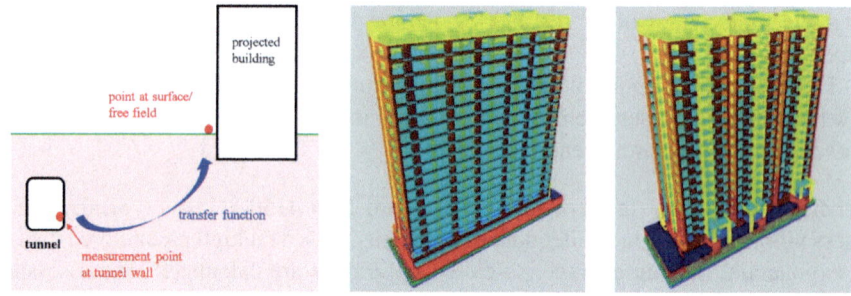

Fig. 1. Sketch of the situation (left), Finite element model, south/north facade (middle/right).

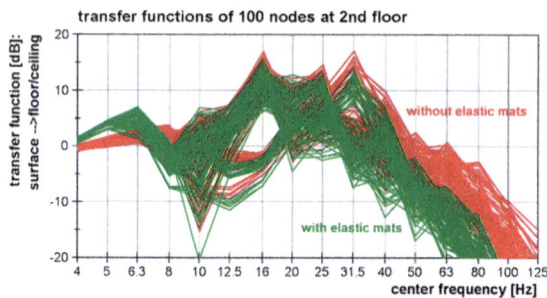

Fig. 2. Transfer functions of 100 exemplary finite element nodes with/without elastic mats.

- 3-dimensional wave propagation in soil
- vibration amplitude decay and delayed arrival time depending on distance to tunnel
- iteratively determined, load-dependent discrete spring-dashpot-elements for soil
- frequency tuned, load-dependent discrete spring-dashpot-elements for elastic layer.

The numerical model of the structure allowed to determine transfer functions from free field to the building floors/ceilings and to simulate the effect of an elastic layer below the foundation. For frequencies above 20 Hz the vibration reduction can be read from the comparison of the red and green curves in Fig. 2. By consecutively applying the transfer functions "tunnel to free field" and "free field to floors" onto the measured excitation spectra, a prognosis of floor vibrations was obtained. The assessment lead to the VL_{zmax}-values illustrated in Fig. 3 for the foundation slab. In the upper stories, vibrations were assessed on basis of German standards with predicted KB_{Fmax}-values below 0.10. Secondary airborne noise was calculated from the predicted floor vibrations. The noise levels were well below the limits because of relatively low excitation at frequencies above 50 Hz and the higher mitigation capacity of the elastic layer in the higher frequency range.

Fig. 3. Predicted VL_{zmax}–values [dB] at the foundation slab.

4   Conclusions

Based on numerical simulations, a concept for decoupling several buildings in Beijing by an elastic layer made of elastomeric materials by company Getzner Werkstoffe GmbH, Austria, was developed to reduce rail traffic immissions in residential buildings. Depending on ground water levels, materials Sylomer® and Sylodyn® were recommended for installation. For an optimized design of the decoupling layer, the numerical simulations considered the excitation and soil characteristics as well as the properties and the support structure of the buildings. The assessment of predicted vibrations in the insulated and uninsulated structural model supported decision making and highlighted the necessity of vibration mitigation measures.

References

1. DB Netz AG: Ril 820.2050 - Erschütterungen und sekundärer Luftschall. (Vibrations and Secondary Airborne Noise) (2017)
2. Chinese Standard GB10171-88: Measurement Method of Environmental Vibration of Urban Area (1988)
3. Chinese Standard GB10070-88: Standard of Environmental Vibration in Urban Area (1988)
4. International Organization for Standardization: ISO 2631-1:1997, Mechanical Vibration and Shock - Evaluation of Human Exposure to Whole-Body Vibration (1997)
5. DIN Deutsches Institut für Normung e.V.: German Standard DIN 4150-2: 1999-06. Available at Beuth Verlag Gmbh, 10772 Berlin, Germany (1999)
6. Chinese Standard JGJ/T170-2009: Standard for Limit and Measuring Method of Building Vibration and Secondary Noise Caused by Urban Rail Transit (2009)

CPT-Based Approach to Study the Load-Displacement Behavior of Driven Piles by the New Method of Stress Characteristics

Fatemeh Valikhah[1], Abolfazl Eslami[1], and Mehdi Veiskarami[2(✉)]

[1] Department of Civil and Environmental Engineering,
Amirkabir University of Technology, Tehran, Iran
[2] Department of Civil and Environmental Engineering, Shiraz University, Shiraz, Iran
mveiskarami@shirazu.ac.ir, mveiskarami@gmail.com

Abstract. In view of principles of the theory of plasticity, the stress field is not independent of the displacement and/or deformation fields. Therefore, a more reliable analysis will be achieved if the bearing capacity and load-displacement behavior of piles are analyzed simultaneously. In this study, a new analytical-numerical method has been proposed to estimate the bearing capacity and axial load-displacement behavior of driven piles in granular soils using CPT records. For this purpose, the method of stress characteristics is used to analyze the stress field below and around the pile and in effect, the failure mechanism. This failure mechanism is then used by implementation of the kinematical approach of the limit analysis to compute the displacement field. This procedure is employed in a step-wise manner to gradually calculate the stress and displacement field as the pile is assumed to penetrate into the ground. One should note that the mobilization of the friction angle is linked to the gradual increase in shear strains in the field. This is done by making use of the CPT results which are both continuous and reliable in comparison to standard laboratory tests often conducted on disturbed samples at discrete intervals. Hence, the step-wise procedure is expected to give rise to a complete load-displacement behavior of driven piles shown in a practical case.

Keywords: Pile · CPT · Load-displacement · Stress characteristics method
Displacement field

1 Introduction

In this work, an analytical-numerical study has been conducted to estimate the bearing capacity and axial load-displacement behavior of driven piles in granular soils using CPT records, which often assumed to be a "model pile" (Eslami and Fellenius 1997). The ingredients of the computational procedure and assumptions behind it are presented in the next sections.

© Springer Nature Switzerland AG 2018
W. Wu and H.-S. Yu (Eds.): *Proceedings of China-Europe Conference on Geotechnical Engineering*, SSGG, pp. 1036–1040, 2018.
https://doi.org/10.1007/978-3-319-97115-5_33

2 Proposed Approach

The procedure comprises three different ingredients: First, the stress state at every point around the pile were computed using the slip lines equations. Then, an admissible velocity field has been found corresponding to the failure mechanism already obtained by the stress characteristics. Construction of the velocity field, or equivalently, the displacement increments is done by using the velocity hodograph corresponding to a multi-rigid blocks mechanism, enclosed by slip lines (Veiskarami et al. 2014). Figure 1 shows the failure pattern obtained by the method of stress characteristics and the velocity field acting on the slip lines.

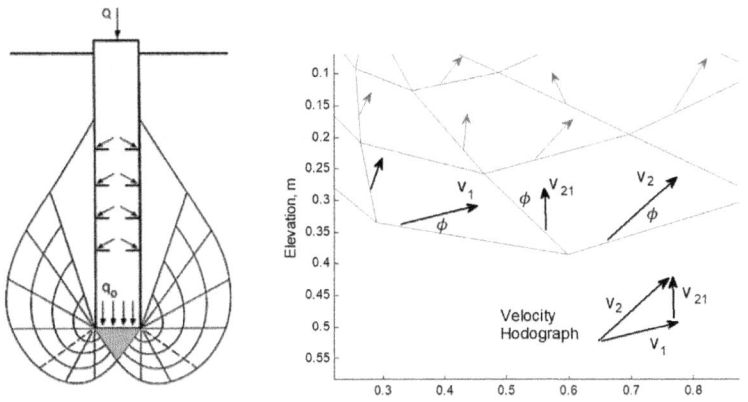

Fig. 1. Failure mechanism around the pile and the velocity vectors acting on the slip lines.

Having known the velocity field, the maximum (major principal) shear strain at each point can be computed. The soil shear strength is assumed to be a function of the maximum shear strain and the residual shear resistance of the soil. This relationship can be found by direct use of CPT data based on which, the residual shear strength of the soil can be found. The collected database used for calibration the new equation in presented in Table 1.

Table 1. Case study records (Schneider 2007).

No.	Pile shape	Material	L (m)	B (mm)	No.	Pile shape	Material	L (m)	B (mm)
1	Square	Concrete	11	253	11	Round	Concrete	7.5	280
2	Square	Concrete	15	253	12	Round	Concrete	11.5	280
3	Square	Concrete	19.5	610	13	Round	Concrete	15.5	280
4	Round	Concrete	8	280	14	Square	Concrete	12.8	235
5	Round	Concrete	16	280	15	Square	Concrete	15.2	406
6	Round	Steel	19	660	16	Round	Steel	9.2	273
7	Round	Steel	6.8	356	17	Round	Steel	8.9	457
8	Round	Steel	6.1	457	18	Round	Steel	12	457
9	Round	Steel	6.9	356	19	Round	Steel	15	457
10	Round	Steel	11	813	20	Round	Steel	14.2	305

A hyperbolic relationship between $\sin \phi_{mob.}$ and the maximum shear strain can be assumed (Lade and Duncan 1975; Clark 1998). Therefore, the following equation has been chosen as a basis for the functional dependency of the mobilized friction angle:

$$\sin \phi_{mob.} = \frac{\gamma}{a + b\gamma} \tag{1}$$

In this equation, a and b can be considered as representatives of mechanical parameters, i.e. the modulus of elasticity, E and the critical state (or the residual) friction angle, $\phi_{c.s.}$. For instance, a is some measure of E and $b = 1/\sin \phi_{c.s.}$.

3   Verification

In order to verify the obtained load-displacement trend via proposed approach, two actual cases are studied (Fellenius et al. 2017). The Pile B2 was constructed with a continuous flight auger and Pile C2 by full displacement equipment. Both piles are 445 mm width and 9.5 m length. The load-displacement curve of these two piles are predicted by the proposed approach and presented in Fig. 2. As it is shown, the results obtained by the proposed approach are in acceptable agreement with the measured load-displacement curve for the piles.

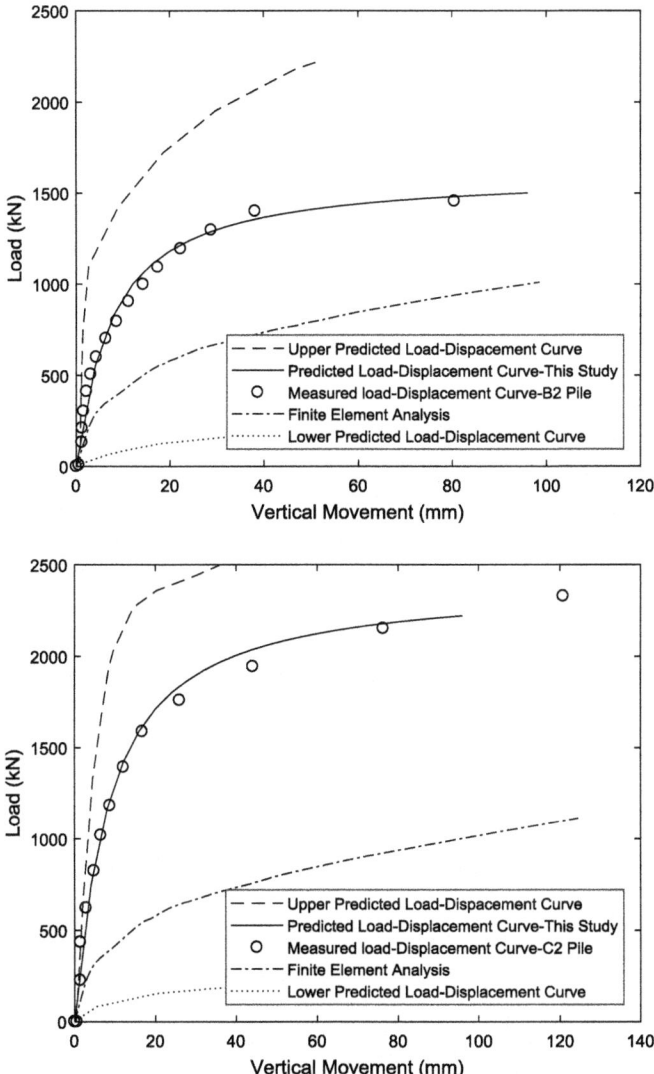

Fig. 2. Predicted load-displacement responses of Bolivian piles using proposed approach and comparison with the other results.

4   Conclusions

In this paper, a method with a theoretical basis based on CPT records has been proposed to predict the load-displacement behavior and the bearing capacity of driven piles in sand. The results have been verified with the pile load test data including 20 driven piles in sand. The load-displacement behavior of two selected piles is predicted by use of the proposed method. The results demonstrate better agreement with the measured load-displacement curves.

References

Clark, J.I.: The settlement and bearing capacity of very large foundations on strong soils. 1996 R.M. Hardy Keynote address. Can. Geotech. J. **35**, 131–145 (1998)

Eslami, A., Fellenius, B.H.: Pile capacity by direct CPT and CPTu methods applied to 102 case histories. Can. Geotech. J. **34**(6), 886–904 (1997)

Fellenius, B.H., Massarsch, K.R., Herrera, M.T.: Resutls of the static loading tests carried out at the Bolivian Experimental Site for Testing Piles (B.E.S.T). In: 3rd Bolivian International Conference on Deep Foundations, 27–29 April 2017, Santa Cruz de la Sierra, Bolivia, vol. 3 (2017)

Lade, P.V., Duncan, J.M.: Elastoplastic stress–strain theory for cohesionless soil. J. Geotech. Eng. ASCE **101**(10), 1037–1053 (1975)

Schneider, J.A.: Analysis of Piezocone Data for Displacement Pile Design, Ph.D. thesis University of Western Australia (2007)

Veiskarami, M., Kumar, J., Valikhah, F.: Effect of the flow rule on the bearing capacity of strip foundations on sand by the upper-bound limit analysis and slip lines. Int. J. Geomech. ASCE **14**(3), 04014008 (2014)

On Trench Construction of Diaphragm Wall in Medium-Coarse Sand: Slurry Composition and Construction Optimization

Jingyu Wang[1(✉)], Jianguo Liu[1], Longlong Fu[1], Weitao Ye[1], and Guangwei Xu[2]

[1] Key Laboratory of Road and Traffic Engineering, Ministry of Education, Tongji University, Shanghai, China
1632402@tongji.edu.cn
[2] CCCC Tunnel Engineering Co., Ltd., Beijing 100102, China

Abstract. On basis of the slurry preparation of diaphragm wall construction in Jinshan Metro Station in Fuzhou, firstly, laboratory tests are conducted to obtain 3D characteristic figures which exhibit the effects of silt soil and bentonite content on slurry properties including density, viscosity and water loss. Furthermore, influence of slurry density, viscosity on filter cake forming process under different slurry pressure is analyzed through laboratory tests. The test results indicate that: the 3D characteristic figures can give a variety of slurry proportion combinations of specific density, viscosity and water loss. The final unit water filtration first drops then starts to increase as slurry density increases. As for viscosity and slurry pressure, the final unit water filtration decreases as viscosity increases and it increases as slurry pressure increases. In addition, the influence of viscosity on final unit water filtration is less than that of density. Finally, optimization of trench construction is proposed based on filter cake formation time obtaining from the test results and construction conditions.

Keywords: Medium-coarse sand layer · Diaphragm wall
Slurry constituents proportion · Filter cake formation · Trench construction

1 Introduction

In trench construction, instability of groove wall often occurs in the surface soil [1] or in the shallow soil with depth of about 5–15 m. Meanwhile, a large thickness of the sand layer will also affect the stability of groove wall [2]. Slurry preparation, which is very important to the deformation control of groove wall, directly affects the trench construction [3] and is one of the keys to the construction of diaphragm wall.

The slurry properties are closely related to the proportion of slurry material, and have a significant impact on the filter cake formation. In this study, taking the construction of diaphragm wall of Jinshan Metro Station in Fuzhou as background, the relationship between slurry proportion and slurry property, and is studied by laboratory tests and the influence of interaction between slurry density and viscosity on filter cake formation is comprehensively analyzed.

© Springer Nature Switzerland AG 2018
W. Wu and H.-S. Yu (Eds.): *Proceedings of China-Europe Conference on Geotechnical Engineering*, SSGG, pp. 1041–1045, 2018.
https://doi.org/10.1007/978-3-319-97115-5_34

2 Project and Geology

The enclosure structures in Jinshan Metro Station are diaphragm walls of 800 mm in thickness and 38 m in depth. The geotechnical parameters are shown in Table 1.

Table 1. Geotechnical parameters in Jinshan Metro Station

Soil layer	Thickness/m	Natural unit weight (kN/m^3)	Water content/%	Void ratio
Fill	4.5	18.5	–	–
Medium-coarse sand	14	19	26.7	0.667
Silt soil	23	17	45.2	1.266
Gravel	–	21	–	–

3 Laboratory Tests of Slurry Proportion

Different slurry materials and proportion have a great impact on slurry properties. Based on controlling variable method, 29 groups of laboratory tests are designed using bentonite, silt soil, CMC to study the effects of various materials on slurry properties by measuring density, viscosity and water loss of each group. Regression equations and three-dimensional characteristics figures of the effects of slurry materials on slurry properties are obtained from the tests results. The results indicate that:

(1) Slurry density which is mainly controlled by silt soil increases with the increase of bentonite and silt soil.
(2) Slurry viscosity increases as the amount of bentonite and CMC increase.
(3) Slurry water loss decreases with the increase of the amount of CMC and the concentration of particles.

Regression equations and 3D figures are shown in (1)–(3) and Figs. 1, 2 and 3.

$$F_1(x, y) = 0.9978 + 4.375 \times 10^{-4}x + 1.485 \times 10^{-4}y + 3 \times 10^{-7}x^2 \tag{1}$$

$$F_2(x, y) = 1.54 - 1.629 \times 10^{-2}x - 8.505 \times 10^{-4}y + 5.554 \times 10^{-4}x^2 \tag{2}$$

$$F_3(x, y) = 8.591 - 6.611 \times 10^{-2}x - 2.005 \times 10^{-3}y + 1.94 \times 10^{-4}x^2 \tag{3}$$

Where $F_1(x, y)$, $F_2(x, y)$ and $F_3(x, y)$ represent density, viscosity and water loss of slurry. x and y represent quality of silt soil and bentonite.

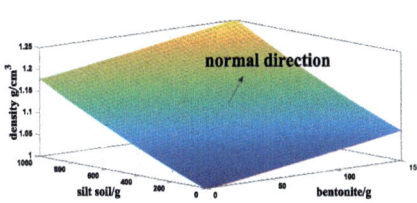

Fig. 1. 3D figure of density

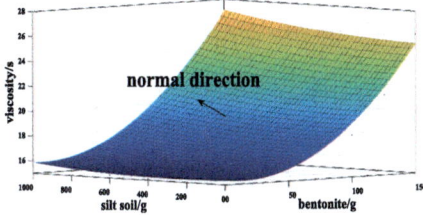

Fig. 2. 3D figure of viscosity

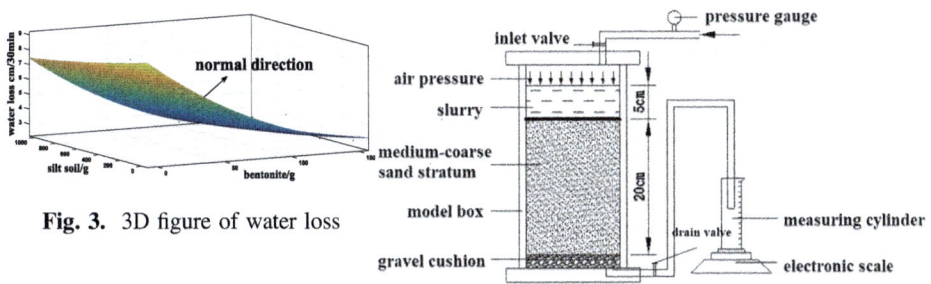

Fig. 3. 3D figure of water loss

Fig. 4. Filter cake formation test device

4 Laboratory Tests of Filter Cake Formation

Slurry properties become decisive for the formation of slurry protection that has a great influence on the stability of groove wall.

4.1 Brief Introduction of Tests

A filter cake formation test device is made using plexiglass which is a cylinder with an inner diameter of 14 cm and a height of 30 cm (as shown in Fig. 4).

The lateral earth pressure is:

$$p = k_0(\gamma_1 h_1 + \gamma_2 h_2) + \gamma_w h_w \tag{4}$$

where γ_1, γ_2, γ_w and h_1, h_2, h_w represent unit weight and thickness.

It can be calculated according to Eq. (4) that the lateral earth pressure of medium-coarse sand layer is from 0.052 Mpa to 0.285 Mpa which means slurry pressure can be divided into three levels of 0.1, 0.2, 0.3 Mpa.

A total of 16 groups of tests are designed depending on the actual situation (see Table 2) and each density level corresponds to four viscosity levels.

Table 2. Classification of slurry density and viscosity.

Soil layer	Slurry density (g/cm³)	Slurry viscosity(s)	Slurry materials
Medium-coarse sand layer	1.05	18	Na-bentonite
	1.12	22	Silt soil
	1.18	26	Carboxymethyl cellulose (CMC)
	1.25	30	

4.2   Analysis of Test Results

Comparing Fig. 5(a)–(d), it can be found that for the same viscosity, slurry with the density of 1.05 g/cm³ has the maximum amount of unit water filtration which is the amount of water permeating through mud protection per unit area; slurry with the density of 1.12 g/cm³ and 1.18 g/cm³ have a smaller unit water filtration. The increase mainly occurs before filter cake formation. The results also indicate that:

(1) For the same density, the larger the viscosity, the smaller the final unit water filtration is; the larger the slurry pressure, the more obvious this effect is.

(2) For the same kind of slurry, the final unit water filtration increases with the increase of slurry pressure.

(3) For the same viscosity and slurry pressure, as density keeps increasing, the final unit water filtration first drops then starts to increase.

(4) The influence of viscosity on final unit water filtration is less than that of density.

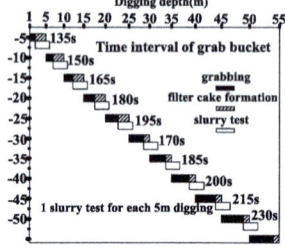

Fig. 6. Construction bar chart

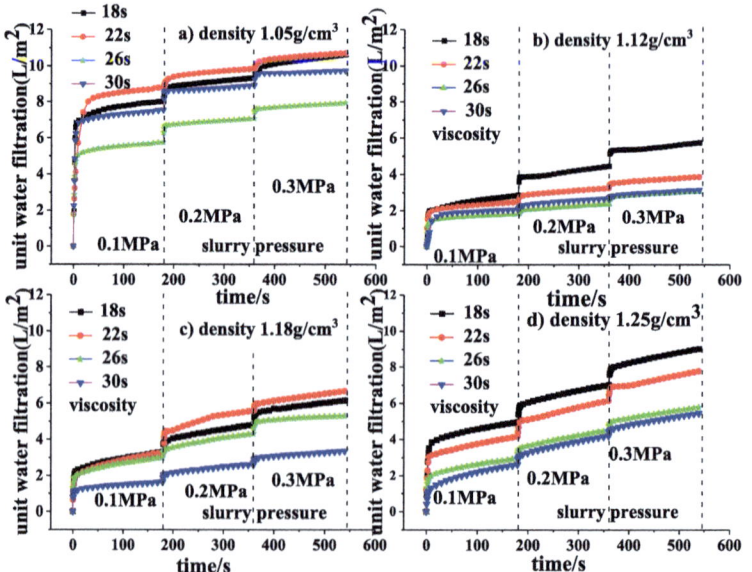

Fig. 5. The results of unit water filtration

5   Optimization of Trench Construction of Diaphragm Wall

In this case study, the filter cake formation time is considered as the time when the amount of unit water filtration tends to be stable or the slope of the curve doesn't exceed 0.1 ml/s. A construction bar chart (see Fig. 6) designed by considering the filter cake formation time and grabbing time using circulating slurry is obtained, in which all soil layers are considered under the worst conditions (medium-coarse sand layer) and actual situation of construction machine and depth of groove wall are also taken into account.

References

1. Morgenstern, N., Amir-Tahmasseb, I.: The stability of a slurry trench in cohesionless soils. Geotechnique **15**(4), 387–395 (1965)
2. He, C.: Construction technology of diaphragm wall in Shiziyang Tunnel. Tunn. Constr. **s1**, 429–432 (2010)
3. Tsai, J.S.: Stability of weak sublayers in a slurry supported trench. Can. Geotech. J. **34**(2), 189–196 (1997)

Tunneling Induced Building Settlement Based on Variational Principle

Xinjiang Wei[1,2(✉)], Xiao Wang[1], Xingfu Yu[2], and Lianying Zhou[2]

[1] College of Civil Engineering and Architecture,
Zhejiang University, Hangzhou 310058, China
weixj@zucc.edu.cn
[2] Zhejiang University City College, Hangzhou 310015, China

Abstract. Construction of tunnels often causes uneven settlement of the above buildings, resulting in engineering accident such as the skew or even collapse of buildings. Assuming that the vertical displacement distribution mode of buildings is built, then the variational control equation is established. A calculation method of building settlement caused by construction of double line parallel tunnels based on variation method is put forward. Through two engineering examples, it is concluded that the building settlement curve calculated by this method is in agreement with the measured value, which verifies this method.

Keywords: Building settlement · Tunnels · Variational principle

1   Introduction

A number of approaches have been used to predict building settlement caused by construction of shield tunnels. These solutions, such as theoretical analytical solution [1], measured analysis method [2] and finite element method [3]. In this paper, taking the energy variation method of vertical displacement of underground pipelines as reference [2], a method to predict the building settlement caused by construction of tunnels is proposed and verified by engineering examples.

2   Calculation Method of Building Settlement

2.1   Calculation Model

The location of the building is assumed to above the center line of the double line tunnels, where R is the radius of the tunnel, h is the depth of the tunnel, and i_b is the width coefficient of the settlement trough of the building (see Fig. 1). To ensure the accuracy, the width of the settlement trough is considered to be $20i_b$.

© Springer Nature Switzerland AG 2018
W. Wu and H.-S. Yu (Eds.): *Proceedings of China-Europe Conference on Geotechnical Engineering*, SSGG, pp. 1046–1050, 2018.
https://doi.org/10.1007/978-3-319-97115-5_35

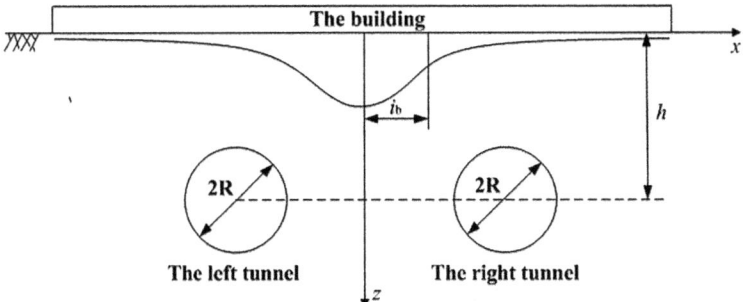

Fig. 1. Diagram of building settlement curve

2.2 Ground Surface Displacement Caused by Crossing of Double Line Parallel Tunnels

Assuming that the left tunnel starts to excavate firstly, the solution in article [4] is used to calculate the ground surface settlement caused by construction of tunnels:

$$
\begin{aligned}
u_z = u_1 + u_2 &= R^2 \cdot \left(\frac{h}{(x+l/2)^2 + h^2} \right) \cdot \frac{4Rg_1 - g_1^2}{4R^2} \cdot B_1 \cdot \exp\left(\frac{(x+l/2)^2 \ln \lambda_1}{(h+R)^2} \right) \\
&+ R^2 \cdot \left(\frac{h}{(x-l/2+b)^2 + h^2} \right) \cdot \frac{4Rg_2 - g_2^2}{4R^2} \cdot B_2 \cdot \exp\left(\frac{(x-l/2+b)^2 \ln \lambda_2}{(h+R)^2} \right)
\end{aligned}
\tag{1}
$$

In which l is the distance between two tunnels, b is the distance between the maximum settlement value of the right tunnel and the vertical axis of the right tunnel, η_1, η_2 are the soil loss rate of the left and the right tunnel respectively, the value of η is determined in journal [5], g_1, g_2 are the equivalent soil loss parameters for the left and the right tunnel respectively, B_1, λ_1 are the calculation parameter of the left tunnel, B_2, λ_1 are the parameters of the right. Details of how to value these parameters can be found in [4].

2.3 The Matrix Representation of the Vertical Displacement of the Building

The shape of the vertical displacement curve of the building can be assumed to conform to the normal distribution curve. According to the Fourier series expansion of the normal distribution curve, two independent finite matrices are used to represent v_b:

$$
v_b = \{X_n\}\{a\}
\tag{2}
$$

$$
\{X_n\} = \left\{ 1, \cos\frac{\pi x}{L}, \cos\frac{2\pi x}{L}, \cos\frac{3\pi x}{L}, \cdots, \cos\frac{n\pi x}{L} \right\}
\tag{3}
$$

$$
\{a\} = \{a_0, a_1, a_2, a_3, \cdots, a_n\}^T
\tag{4}
$$

Where v_b is the vertical displacement of building caused by tunnels excavation, L is the half width of building settlement trough, {a} is the nonlinear parameter of vertical displacement of the building.

Assuming that i_b is equal to the width coefficient of the settlement trough on the surface of the soil, then i_b can be obtained according to [6]:

$$i_b = 0.5h \tag{5}$$

2.4   Energy Equation

The building is considered as a horizontal elastic foundation beam, its length is 2L, width is 1 m, and the height is equal to d, meaning the depth of the building foundation. The total potential energy is determined with:

$$\prod = U + W = \int_{-L}^{L} \frac{1}{2} E_b I_b \left(\frac{d^2 v_b}{dx^2} \right)^2 dx + \int_{0}^{l} \frac{1}{2} k (u_z - v_b)^2 dx \tag{6}$$

In which E_b is the elastic modulus of the building, I_b is the moment of inertia to the axis y in the cross section of the building, U is the bending strain energy of the building, $u_z - v_b$ is the relative displacement of the soil and the building, k is the back force modulus of foundation bed, W is the work done by the soil displacement to the building. According to the principle of minimum potential energy, the extreme values of the total potential energy can be calculated with:

$$\frac{\partial \prod}{\partial a_i} = 0 \; i = 1, 2, \ldots, n + 1 \tag{7}$$

in which a_i is the element of {a}. {a} can be obtained by formula (7), and v_b can be calculated according to formula (2). The above calculation method can be solved with Mathematica, and the calculation accuracy can be satisfied by preserving the ten-order matrix in the results.

3   Engineering Examples

Two engineering examples is listed to be analyzed. Chong Si building of Shanghai Xuhui high school in thesis [7] and Electrical Education building of Wuhan University of Technology in article [8] are both located above the construction area of the double line shield tunnels, and the left tunnel started to excavate in both examples (see Table 1). In Table 1, the value of k is referenced by article [7, 8], and the value of E_b is referenced by article [9, 10].

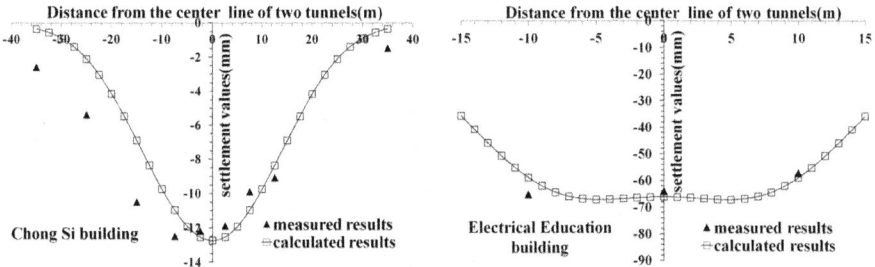

Fig. 2. Comparison diagram of building settlement value and measured settlement value.

The Chong Si building is about 66 m long, and the Electrical Education building is 30 m long, the measured value of their settlement is both included in Fig. 2, which shows that the settlement values of buildings obtained by this method are basically consistent with the measured values. The result of Chong Si building also illustrates that the settlement trough of the building is narrower than that of the measured value, but still shows the characteristic of Gauss districation. The settlement trough is in a transition from a "v" shape settlement trough to a "w" shape settlement trough as seen on the result of Electrical Education building.

Table 1. Corresponding parameters of each engineering example

Parameter	h/m	l/m	η_1/%	η_2/%	b/m	k/kN/m^3	E$_b$/kN/m^2	I$_b$/m^4	R/m	d/m
Chong Si building	22.70	16.50	0.74	0.74	4	15000 [7]	50000 [9]	0.28	3.1	1.5
Electrical Education building	13.77	16.34	1.1	1.1	1	17000 [8]	40000 [10]	1.15	5.5	2.4

4 Conclusions

Assuming the vertical displacement distribution mode, the method to predict the building settlement caused by construction of double line shield tunnels is proposed in this research. The method is verified through two engineering examples. This study shows that the shape of the building settlement trough is similar to the shape of Gauss distribution when the distance of the tunnels is close enough, and in a transition from a "v" shape to a "w" shape when the distance of tunnels is further, changing to "w" shape at last.

References

1. Loganathan, N., Poulos, H.G.: Analytical prediction for tunneling-induced ground movement in clays. J. Geotech. Geoenviron. Eng. **124**(9), 846–856 (1998)
2. Liu, X.Q., Liang, F.Y., Zhang, H., et al.: Energy variational solution for settlement of buried pipeline induced by tunneling. Rock Soil Mech. **35**(supp. 2), 217–222, 231 (2014). (in Chinese)
3. Ding, L., Wu, X., Zhang, L., et al.: How to protect historical buildings against tunnel-induced damage: a case study in China. J. Cult. Herit. **16**, 904–911 (2015)
4. Wei, G., Pang, S.Y., Zhang, S.M.: Prediction of ground deformation induced by double parallel of ground deformation induced by double parallel shield tunneling. Disaster Adv. **6** (13), 91–98 (2013)
5. Wei, G.: Selection and distribution of ground loss ratio induced by shield tunnel construction. Chin. J. Geotech. Eng. **32**(9), 1354–1361 (2010). (in Chinese)
6. Mair, R.J., Taylor, R.N., Bracegirdle, A.: Subsurface settlement profiles above tunnels in clays. Geotechnique **43**(2), 315–320 (1993)
7. Ding, Z.: Prediction of deformation and study on the influence of shield tunnel on adjacent buildings. Ph.D. dissertation, Zhejiang University (2004). (in Chinese)
8. Wu, Y.G., Huang, B., Zhang, T., et al.: Amendment and assessment on laboratory results of subgrade coefficient. Port Waterw. Eng. **7**, 27–30 (2010). (in Chinese)
9. Ouyang, W.B., Ding, W.Q., Xie, D.W.: Calculation method for settlement due to shield tunneling considering structure stiffness. Chin. J. Undergr. Space Eng. **9**(1), 155–160 (2013). (in Chinese)
10. Qu, T.J., Xu, R.H., Shi, Y.X.: Experimental study on influence of ratio of reinforcement to modulus of elasticity of reinforced concrete component. Concrete **9**, 113–115, 119 (2014). (in Chinese)

Measured and Predicted Response of a Post-grouted Pile in Cohesionless Soil

Xiong Xiao[(✉)], Shanyong Wang, Scott Sloan, and Daichao Sheng

ARC Centre of Excellence for Geotechnical Science and Engineering,
Faculty of Engineering and Built Environment, University of Newcastle,
Australia, Newcastle, Australia
Xiong.xiao@uon.edu.au

Abstract. Although compaction grouting beneath the pile tips has been proven to improve the vertically loaded capacity of piles, its design is still largely based on empirical experience and lack of rational design guide. An analytical model that relates the tip resistance to the pressure to expand a spherical cavity for prediction of pile tip bearing capacity is presented. The proposed approach prediction matches quite well with tests results. However, more tests should be done to confirm the correctness of this method. In this paper, a new laboratory setup for investigating the effect of compaction grouting on pile capacity was designed and assembled. This apparatus allows a model pile to be driven into or buried in the sand sample and then a low mobility grout is delivered through a grouting tube inside the model pile into a membrane that is used to prevent grout fracture the sand sample. Then soil stress and pore pressure change are monitored by soil pressure and pore pressure transducer buried in the sample. Pile load test is conducted after the grout have been cured and the pile penetration resistance is measured by the load cell.

Keywords: Post-grouted pile · Cavity expansion · Compaction grouting
Pile load test · Bearing capacity

1 Introduction

Pile foundations are largely used to support the load of the superstructure like bridges, skyscrapers. The fast city expansion has caused increasing amount of infrastructure facility to be constructed on soft and unstable soils. Due to the high mobilized skin and tip resistance, driven pile is one of the widely used foundation choices. However, the tough requirement of minimizing the construction noise and vibration forced us to abandon this technique and find an alternative ground treatment method. The drilled shaft become popular because of its less intrusive nature. Nevertheless, some researches show the jetting/installing process will cause soil disturbance and resulting a load capacity decreases. Many researchers have investigated using compaction grouting technique to densify the disturbed soil and mobilize tip resistance at smaller displacement. In 2006, Mullins et al. briefly illustrated the postgrouting process and discussed parameters affecting the performance based on some previous field test results [6]. Then Pooranampillai et al. conducted some grouted shaft tests in a circular

© Springer Nature Switzerland AG 2018
W. Wu and H.-S. Yu (Eds.): *Proceedings of China-Europe Conference on Geotechnical Engineering*, SSGG, pp. 1051–1054, 2018.
https://doi.org/10.1007/978-3-319-97115-5_36

steel chamber with 2.4 m in diameter and 2.7 m in high [7]. Their test results show the tip capacity increase more significant in 50% relative density sand than 70% relative density sample. In 2013, Thiyyakkandi et al. presented both experimental and FEM modelling of grouting process and axial loading response of various dimension jetted and grouted piles [9]. Fang et al. emphasized the prestressing effect of the post grouting in smaller the vertical displacement to mobilise the tip resistance and demonstrated this influence of preloading on a field case [5]. However, all of these studies were limited to analysis the influence factors or present the performance of post-grouted pile tests without giving a reasonable design methodology. In this study, cavity expansion theory is introduced to study the densification behaviour of compaction grouting process first and then tip resistance of post-grouted shaft is calculated based on the its relationship with the limit pressure of cavity expansion. In order to verify the proposed method, some small-scale physical modelling test will be conducted. The laboratory setup of the apparatus is presented herein.

2   Analysis Based on Cavity Expansion Theory

According to the previous researches as mentioned in Sect. 1, grout injection pressure is one of the most important parameters in the post-grouted pile design. One photograph of a post-grouted pile after excavation demonstrated the grout bulb at the pile tip is more or less like a spherical shape and grout zones formed along the pile similar to a cylindrical shape [9]. Then it is reasonable to use cavity expansion theory to estimate the injection pressure. Carter et al. and Yu & Houlsby are two-milestone researches presented analytical solution of cavity expansion in elastic-perfect-plastic medium [2, 12]. Then Cao et al. and Chen & Abousleiman given the elastoplastic solution of cylindrical cavity expansion in modified cam-clay materials [1, 3, 4]. However, the method described in these studies are limit in analysing the dilation behaviour of sand and overconsolidated soil in large cavity deformation scenario.

Lately, Li et al. proposed a unified solution to drained expansion of a spherical cavity in clay and sand [10]. Yao et al.'s unified hardening parameter based critical state model (UHP model) was used in its study. This model can capture both of the dilatancy and peak strength of sand [11]. With the self-similarity assumption, the three differential equations are developed and conditions at elastic-plastic boundary are used to solve these as an initial value problem. These equations are developed based on 1. Equilibrium between radial and tangential stress around the cavity; 2. Constitutive model linking the stresses and strains using the elastic and plastic potential function; 3. The consistency between the stresses according to the yield function.

The estimate grouting pressure and radial stress distribution determined by Li et al.'s analytical solution is then compared with the measured radial stresses around the spherical grout bulb during tip grouting of 0.203-m-square pile in Thiyyakkandi et al.'s tests. The cavity radius after grouting is 0.254 m. The soil parameters used in analytical study are $\varphi_c = 31^0$, $\varphi_p = 36^0$, $v_0 = 1.83$, $\lambda = 0.0374$, $\kappa = 0.008$, $\mu = 0.3$, $G_0 = 5049$, $p_0 = 40\,\text{kPa}$. In Fig. 1a, it can be seen that the measured stresses match reasonably well with the analytical stress distribution. It should be noted that the radial stress diminished to 20% of the cavity stress at r/a = 3 and 10% of the cavity stress at r/a = 5. As shown in Fig. 1b, the injection pressure will increase dramatically at the beginning of injection, during which a large pressure increase will result in a small volume increment. 90% of the limit pressure will be reach when the cavity expands to about 2 times of initial radius. In order to prevent shear failure happens between grout and pile, a ratio of 2.5 is suggested as the size of grout zone.

In 1994, Randolph et al. presented a correlation between the spherical cavity limit pressure p_L and ultimate end bearing pressure q_b, which is used to estimate the tip resistance of the pile. Then the ultimate tip load is obtained by multiplying tip capacity with the grout bulb tip area [8].

$$q_b = [1 + \tan \phi_c \tan(45^0 + \phi_c/2)]p_L \qquad (1)$$

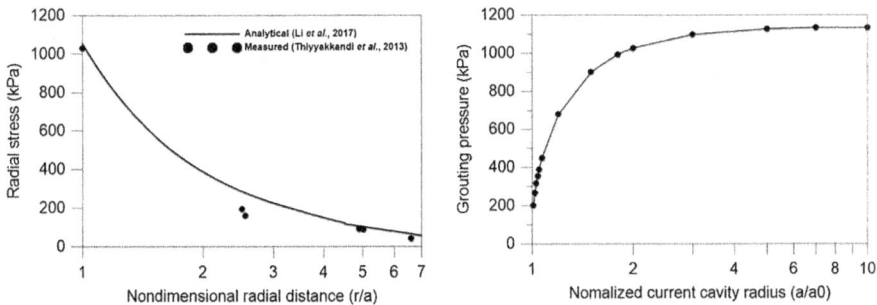

Fig. 1. (a) Stress distribution with radial distance and (b) cavity expansion curve

3 Laboratory Setup

Three pressure/volume controllers are connected to a standard large 1000 kPa triaxial cell (sample size: 0.3 m diameter and 0.6 m high) to control the cell pressure, the back pressure and the injection pressure. The model pile is placed at the top of the sample with its tip through the sample top cap and embedded into the centre of specimen. Then the pile driven process is simulated by upwards bottom piston of the load frame. GDSLab is used to record the triaxial cell displacement, the pressure/volume controllers, the volume gauge and the pore pressure transducer reading.

The compaction grouting system is prepared by rolling over standard 12.7 mm outer diameter. medical latex tubing over the model pile. During injection, the pressure or volume is measured and controlled using the pressure/volume controller.

In this test the sand sample is prepared by raining the sand sample into the sample former that sits on top of the triaxial cell base. The dry pluviation techniques is used to make the sand sample. Sand is raining into the cylindrical mould from a disc filter placed 600 mm over the top of the sample. The density of the prepared sample is almost the same lying between 1500 and 1550 kg/m^3. Some miniature pressure strain gauges and pore pressure transducers are buried in the specific place of the sample to measure the stress distribution.

4 Conclusions

A design method based on cavity expansion theory for grouted pile is presented in this paper. The comparison between its estimates and the rare experiment measurements are reasonable. However, more tests result still needed to confirm the correctness of the proposed method. So, labor test will be performing in the setup described in this paper to further study the behavior of grouted pile.

References

1. Cao, L.F., The, C.I., Chang, M.F.: Undrained cavity expansion in modified Cam clay. Geotechnique **51**, 323–334 (2001)
2. Carter, J.P., Booker, J.R., Yeung, S.K.: Cavity expansion in cohesive frictional soils. Géotechnique **36**, 349–358 (1986)
3. Chen, S.L., Abousleiman, Y.N.: Exact undrained elasto-plastic solution for cylindrical cavity expansion in modified Cam-clay soil. Geotechnique **62**(5), 447–456 (2012)
4. Chen, S.L., Abousleiman, Y.N.: Exact drained solution for cylindrical cavity expansion in modified Cam clay soil. Géotechnique **63**(6), 510–517 (2013)
5. Fang, K., Zhang, Z., Zhang, Q., Liu, X.: Prestressing effect evaluation for a grouted shaft: a case study. Proc. ICE Geotech. Eng. 1–9 (2013)
6. Mullins, G., Winters, D., Dapp, S.: Predicting end bearing capacity of post-grouted drilled shaft in cohesionless soils. J. Geotech. Geoenviron. Eng. **132**(4), 478–487 (2006)
7. Pooranampillai, S., Elfass, S., Vanderpool, W., Norris, G.: Large scale laboratory testing of low mobility compaction grouts for drilled shaft tips. Geotech. Test. J. **33**(5), 13 (2010)
8. Randolph, M.H., Dolwin, J., Beck, R.: Design of driven piles in sand. Geotechnique **44**(3), 427–448 (1994)
9. Thiyyakkandi, S., MaVay, M., Bloomquist, D., Lai, P.: Measurement and predicted response of a new jetted and grouted precast pile with membranes in cohesionless soils. J. Geotech. Geoenviron. Eng. **139**(8), 1334–1345 (2006)
10. Li, L., Li, J., Sun, D., Gong, W.: Unified solution to drained expansion of a spherical cavity in clay and sand. Int. J. Geomech. **17**(8), 04017028 (2017)
11. Yao, Y.P., Sun, D.A., Matsuoka, H.: A unified constitutive model for both clay and sand with hardening parameter independent on stress path. Comput. Geotech. **35**(2), 210–222 (2008)
12. Yu, H.S., Houlsby, G.T.: Finite cavity expansion in dilatant soils: loading analysis. Geotechnique **41**(2), 173–183 (1991)

Parameter Identification of the High-Fill Foundations Using Response Surface Method

Ming Xu$^{(\boxtimes)}$, Dehai Jin, and Erxiang Song

Department of Civil Engineering, Tsinghua University, Beijing 100084, China
mingxu@mail.tsinghua.edu.cn

Abstract. The response surface method (RSM) is introduced to identify the model parameters of a synthetic cross-section of high-fill foundations. Four different initial exploration regions are employed to show the robustness of the RSM. The results show that, the method can effectively search the satisfactory parameters. Thus, it can be applied for parameter identification in realistic engineering problem once enough field monitor data is obtained.

Keywords: Parameter identification · High-fill foundation
Deformation prediction · Response surface method

1 Introduction

Deformation prediction is necessary for the high-fill foundations. In general, the model parameters used in numerical analysis are determined according to the lab tests, which might underestimate the true deformation. This is not safe for the operation of such structures. Thus, identifying the satisfactory model parameters based on the in situ monitoring data is necessary. In this paper, response surface method [1] is employed to solve the parameter identification (PI) problem. A synthetic cross-section is built in FLAC 2D, and the numerical results of specified parameters (i.e., exact solution) are assumed as the filed data. Based on this, RSM searches for the parameters. Four initial exploration regions are used to investigate the robustness of the RSM.

2 Response Surface Method and Material Model

2.1 Response Surface Method

Response surface method (RSM) is proposed by Box and Wilson, which is initially used in the field of chemistry and chemical industry [2]. Usually, this approach searches for the satisfactory parameters by the following three steps.

Step 1: First-order analysis. In an initial exploration region, the first-order polynomial is formulated as follows

$$y^{(1)} = \beta_0^{(1)} + \beta_1^{(1)} x_1 + \beta_2^{(1)} x_2 + \ldots + \beta_k^{(1)} x_k \tag{1}$$

© Springer Nature Switzerland AG 2018
W. Wu and H.-S. Yu (Eds.): *Proceedings of China-Europe Conference on Geotechnical Engineering*, SSGG, pp. 1055–1058, 2018.
https://doi.org/10.1007/978-3-319-97115-5_37

where $\beta_i^{(1)}$ ($i = 0, \ldots, k$) are the coefficients, which can be obtained by the least-square regression. Here, a fitting tolerance (R_{ft}) is defined. If R^2 of Eq. (1) is smaller than R_{ft}, the searching procedure can be continued. Otherwise, the exploration region is contracted to half of previous region, and a new first-order polynomial analysis is required.

Step 2: Steepest descent searching. The gradient of $y^{(1)}$ is chosen as the searching direction. Note that, this method just provides the searching direction and the actual step size is determined by the designer The procedure is continued along the chosen direction until no further decrease in response is observed. Step 1 and Step 2 can be executed until the designer arrive in the vicinity of optimal region. At this moment, a higher-order polynomial is required.

Step 3: Second-order analysis. The second-order polynomial is formulated as follows

$$y^{(2)} = \beta_0^{(2)} + \sum_1^k \beta_i^{(2)} x_i + \sum_1^k \beta_{ii}^{(2)} x_{ii}^2 + \sum_i \sum_j \beta_{ij}^{(2)} x_i x_j \tag{2}$$

where $\beta_0^{(2)}$, $\beta_i^{(2)}$, $\beta_{ii}^{(2)}$ and $\beta_{ij}^{(2)}$ are the coefficients, which can be determined by the least square regression. Equation (2) is rewritten in matrix notation,

$$y^{(2)} = \beta_0^{(2)} + x' b + x' B x \tag{3}$$

The optimal solution of Eq. (3) is formulated as: $x_0 = -B^{-1} b / 2$. The second-order analysis can be executed until satisfactory parameters are obtained.

2.2   Material Model

Rockfills have been extensively used for the construction of the high-fill foundations. In this paper, the modified Burgers model proposed by Jin and Xu is introduced to simulate the creep behaviors of rockfills. In brief, there are eleven parameters in the modified Burgers model. K, n, R_f, φ_0 and $\Delta\varphi$ are for describing the instant deformation. K_1, n_1, K_2, n_2, K_3 and n_3 are for describing the creep deformation.

3   Back-Analysis of Model Parameters

3.1   FDM Model

A synthetic cross section of the high-fill foundation (Fig. 1) is built in FLAC 2D, which has a prototype from the Yuanshangzi Valley of the Jiuzhai-Huanglong Airport, the mountainous area of Sichuan Province in China. The bedrock is elastic with a Poisson's ratio of 0.3 and Young's modulus of 30 GPa. The filling material is described by the modified Burgers model. The construction process is simulated in six steps. The maximum height of the filling body is 90 m, and the slope ratio is 1:2.

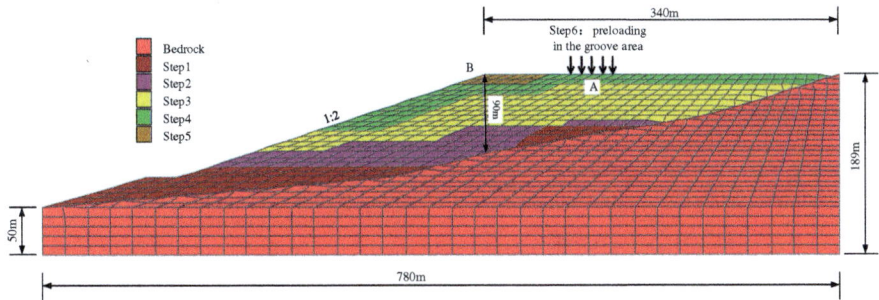

Fig. 1. Numerical mesh of synthetic cross-section of high-fill foundations

3.2 Objective Function

In this paper, the objective function is formulated as,

$$\mathbf{F}(\mathbf{p}) = \sum_{r=1}^{m} \sum_{s=1}^{n} \sqrt{\left(\frac{u_{rs}^{h}(\mathbf{p}, t_s) - u_r(t_s)}{u_r(t_s)}\right)^2} \tag{4}$$

where $\mathbf{p} = [p_1, p_2, ..., p_k]^T$ is a $(k \times 1)$ vector of the unknown parameters; $u_r(t_s)$ is the monitoring deformation of location r at time s; and $u_{rs}^{h}(\mathbf{p}, t_s)$ is the corresponding numerical deformation. m and l are the numbers of the monitoring locations and time points for the back analysis, respectively. In addition, a convergence tolerance (F_{ct}) is defined to judge the results of PI result is satisfactory.

3.3 Back-Analysis Results

To investigate the efficiency of RSM, the field data are not used here since the exact solution is unknown. Instead, specified parameters are pregiven, and the corresponding calculation results are assumed as the in situ monitoring data. Two parameters (K_1 and K_2) are identified to graphically represent the convergence process of the PI. The exact solution is assumed to be $K_1 = 1.550$ and $K_3 = 7.800$. The other nine model parameters are listed in Table 1. The convergence tolerance (F_{ct}) and the fitting tolerance (R_{ft}) are set to be 2% and 90%, respectively.

Table 1. The other nine parameters of the modified Burgers model

K	n	R_f	φ_0	$\Delta\varphi$	n_1	K_2	n_2	n_3
800	0.50	0.90	44.0	8.20	0.75	5.50	0.75	0.75

Four different sets of initial exploration region (Case 1–4) are selected to investigate the robustness of the RSM. The results of PI for the four cases are displayed in Fig. 2. The satisfactory parameters for the four cases are estimated to be $(K_1, K_3) = (1.548, 7.840)$, $(1.553, 7.817)$, $(1.551, 7.817)$, and $(1.552, 7.781)$. Based on these results, the estimation errors are $-1.3‰ -1.9‰$ in K_1, and $-2.4‰ -5.1‰$ in K_3, consistent with the exact solution. The above results reveal that the RSM can effectively search the satisfactory parameters of the high-fill foundations with high precision.

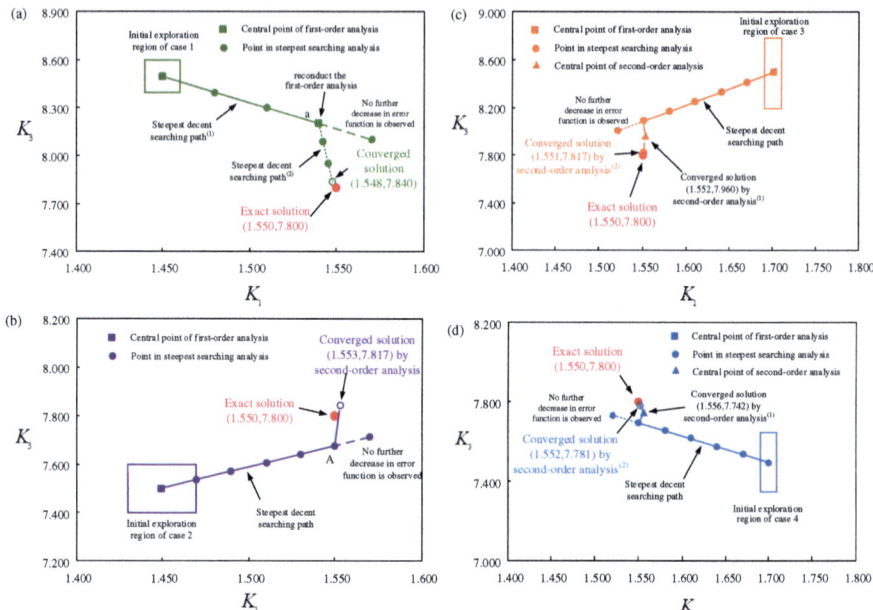

Fig. 2. Searching procedure of PI for: (a) Case 1, (b) Case 2, (c) Case 3, and (d) Case 4.

4 Conclusion

In this paper, a synthetic cross-section of high-fill foundations is built in FLAC 2D to investigate the efficiency and robustness of RSM. The satisfactory parameters are obtained by three steps and the results reveal that RSM can work quite accurately and efficiently. Therefore, the approach can be applied to predict the deformation of a realistic engineering problem once enough field monitor data are obtained.

References

1. Box, G.E.P., Wilson, K.B.: On the experimental attainment of optimum conditions. J. Roy. Stat. Soc.: Ser. B (Methodol.) **13**(1), 1–45 (1951)
2. Hill, W.J., Hunter, W.G.: A review of response surface methodology: a literature survey. Technometrics **8**(4), 571–590 (1966)
3. Jin, D., Xu, M.: Parameter correction method of burgers model for coarse-grained soil considering confining pressure. Eng. Mech. **12**, 017 (2016). (in Chinese)

Distribution Patterns of Horizontal Foundation Modulus Under Pre-excavation Dewatering

Xiu-Li Xue, Miao-Kun Li, and Chao-Feng Zeng[(✉)]

Hunan Provincial Key Laboratory of Geotechnical Engineering for Stability
Control and Health Monitoring, School of Civil Engineering,
Hunan University of Science and Technology, Xiangtan 411201, Hunan, China
cfzeng@hnust.edu.cn

Abstract. Pre-excavation dewatering (PED) can cause a significant wall deflection. How to predict this deformation is an urgent problem to be solved. The elastic foundation beam method is widely adopted to predict the wall deflection. A key issue that need to be addressed in the use of this method is to obtain the distribution of the horizontal foundation modulus. However, the distribution pattern of horizontal foundation modulus under PED is not clear. In this study, a numerical model is established and verified, and a series of numerical simulations are carried out to investigate the distribution of horizontal foundation modulus along depth under conditions of different excavation width and depth. The numerical results are compared with those computed by the 'm method', which is a traditional excavation design method recommended in the Chinese Code. The results indicate that under the common conditions of PED, the horizontal foundation modulus decreases greatly compared with the values determined by m method, and the larger the excavation width and dewatering depth, the greater the decrease extent. When using the elastic foundation beam method to predict the PED-induced wall deflection, the deflection will be underestimated if the m method is used to determine the horizontal foundation modulus.

Keywords: Deep excavation · Pre-excavation dewatering
Horizontal foundation modulus · Finite element analysis · M method

1 Introduction

As to deep excavation in soft ground with high groundwater level, the whole construction processes of excavation mainly consist of 7 stages [1], which are the construction of retaining wall, pre-excavation dewatering (PED), staged excavation, staged dewatering, construction of slabs or struts, removal of struts and backfill and the groundwater level recovery. The PED is an important construction stage before excavation to check the completion quality of dewatering wells (i.e., the discharge flow rate and drawdown in dewatering well need to be tested and should meet the design requirement before excavation) [1, 2]. Considering that the retaining wall already existed before PED, the retaining wall is expected to be affected by PED and appear deformation response.

© Springer Nature Switzerland AG 2018
W. Wu and H.-S. Yu (Eds.): *Proceedings of China-Europe Conference on Geotechnical Engineering*, SSGG, pp. 1059–1063, 2018.
https://doi.org/10.1007/978-3-319-97115-5_38

Based on field measurements, Zeng et al. [3] found that the deformation of retaining structure and surrounding environment induced by PED can reach centimeter level. Zeng et al. [4, 5] found that the responses of retaining wall to PED are totally differ under different parameters. Under the demand of strict deformation control in current deep excavation, the PED should be included in the excavation design. The designer should be able to predict the deformation caused by the whole process (including the PED stage) of deep excavation. At present, the elastic foundation beam method is widely adopted to predict the excavation-induced wall deflection, and the "m method" is recommended in the Chinese Code [6] to calculate the horizontal foundation modulus (k_h). However, the code does not deal with how to analyze the PED-induced wall deflection. If the elastic foundation beam method is still used, how to calculate k_h under PED is still unclear, which brings uncertainty in excavation design.

In this study, a two-dimensional (2D) finite element (FE) model considering soil-fluid coupling is developed to simulate a PED process reported in literature [3]. The model is verified by the test and numerical results reported in literature [3]. Then, a series of numerical simulations are carried out to investigate the distribution of horizontal foundation modulus along depth during PED under conditions of different excavation width and depth.

2   Numerical Analysis

2.1   Model Setup and Verification

In this section, there is a total of 9 FE models with different excavation width (denoted as b and $b = 20$, 60 and 120 m are considered) and dewatering depth (denoted as H_d and $H_d = 11$, 16 and 21 m are considered) are established. Figure 1 shows the mesh of the FE model with excavation width of 20 m. Considering the symmetry of this question, only half of the excavation is modelled. In the vertical direction, the model is set to 50 m in depth. According to the actual field conditions of the excavation in literature [3], the strata in the model are divided into 9 layers.

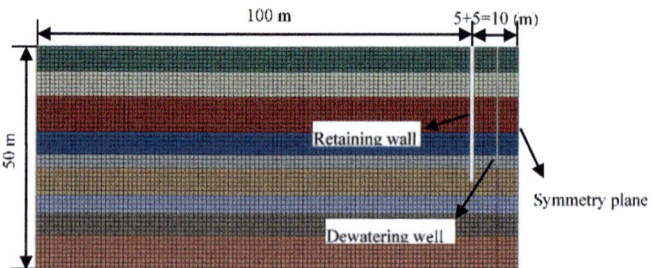

Fig. 1. Finite element model for pre-excavation dewatering of pit with 20 m in width.

The soil behavior is assumed to obey the constitutive theory of modified Cam-Clay (MCC) during PED. The main parameters of each soil layer can be seen in literature [3]. The linear elastic model is adopted for the retaining wall and dewatering well. The Young's moduli of the retaining wall and dewatering well are 30 and 210 GPa, respectively. The initial water level is assumed to be located at the ground surface. The lateral soil boundaries are supported by rollers and recharged by a constant head at the ground surface. The bottom of the soil is pinned and impermeable. The function of dewatering well is simulated by applying a seepage boundary on a zone where the screen of the dewatering well is located. In this paper, the drainage-only flow (DOF) seepage boundary will be set on the screen zone [2]. The FE modelling are conducted according to the PED process reported in literature [3], i.e., let all the dewatering wells operate simultaneously and the dewatering time is 21 days. The calculated results of the above 9 cases with different b and H_d are compared with the results in literature, as shown in Fig. 2. It can be seen that the calculated maximum wall deflections are close to the results reported by Zeng et al. [3] under the same conditions, which means the use of the 2D numerical models established in this study to analyze the problem of PED-induced deformation is reasonable. Besides, the results from Fig. 2 indicate that with the dewatering depth (H_d) and excavation width (b) increased, the maximum wall deflection induced by PED is more seriously.

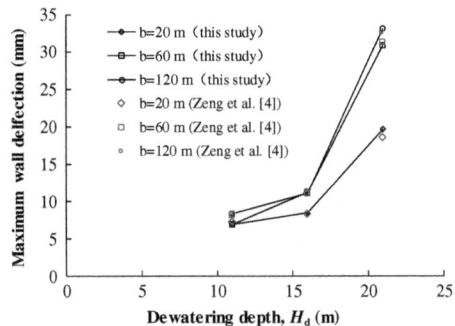

Fig. 2.   Calculated wall deflections and their comparisons with results in literature [3]

2.2   Distribution of Horizontal Foundation Modulus (K_h)

In Chinese Code, when analyzing the excavation-induced wall deflection, k_h is assumed to increase with depth linearly, i.e., $k_h = mz$, where z = soil depth and m = scale factor (m values depend on the soil property and can be obtained via table lookup). However, during PED, the distribution of k_h is totally different. Figure 3 presents the distribution of k_h/m along depth at the end of dewatering under different dewatering depth (H_d) and excavation width (b). Apparently, k_h does not increase linearly with depth, and is smaller than the values determined by m method. This is because the seepage force around the dewatering well during PED can cause the soils inside the excavation to develop inward horizontal displacement (i.e., towards the direction away from the retaining wall), and thus the resistance from inner soil to retaining wall will somewhat decrease [3]. When using the elastic foundation beam method to predict the PED-induced wall deflection, the deformation will be underestimated if the m method is used to determine the k_h. Besides, as to the case with larger H_d and b, k_h is basically much smaller in the range of H_d. This is why greater wall deflection occurs in the case with larger H_d and b (see Fig. 2).

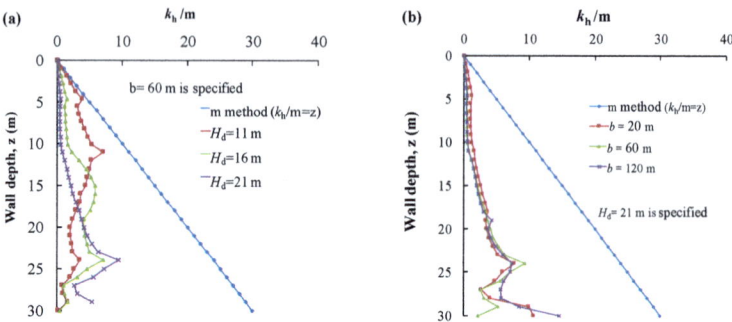

Fig. 3. Distribution of k_h/m along depth under different: (a) depth (H_d) and (b) excavation width

3   Concluding Remarks

(1) Under the common conditions of PED, the horizontal foundation modulus (k_h) decreases greatly compared with the values determined by m method. When using the elastic foundation beam method to predict the PED-induced deformation of the retaining structure, the deformation will be underestimated if the m method is used to determine the horizontal foundation modulus. (2) The excavation width (b) and dewatering depth (H_d) influence the distribution of k_h along depth. As to the case with larger H_d and b, k_h is basically much smaller in the range of H_d.

Acknowledgements. The research is funded by the National Natural Science Foundation of China under Grant No. 51708206 and 11602083, the Natural Science Foundation of Hunan Province under Grant No. 2016JJ6044, the China Postdoctoral Science Foundation under grant No. 2018M633298, and the Special Funding Project for Postdoctoral Research in Guangxi under grant No. BH2018054. These financial supports are gratefully acknowledged.

References

1. Zeng, C.F., Xue, X.L., Zheng, G., Xue, T.Y., Mei, G.X.: Responses of retaining wall and surrounding ground to pre-excavation dewatering in an alternated multi-aquifer-aquitard system. J. Hydrol. **559**, 609–626 (2018)
2. Zheng, G., Zeng, C.F., Diao, Y., Xue, X.L.: Test and numerical research on wall deflections induced by pre-excavation dewatering. Comput. Geotech. **62**, 244–256 (2014)
3. Zeng, C.F., Zheng, G., Xue, X.L.: Wall deflection induced by pre-excavation dewatering in large-scale excavations. Chin. J. Geotech. Eng. **39**(6), 1012–1021 (2017)

4. Zeng, C.F., Xue, X.L., Zheng, G.: A parametric study of lateral displacement of support wall induced by foundation pre-dewatering in soft ground. Rock Soil Mech. **38**(11), 3295–3303 + 3318 (2017)
5. Zeng, C.F., Xue, X.L., Zheng, G.: Effect of soil permeability on wall deflection during pre-excavation dewatering in soft ground. Rock Soil Mech. **38**(10), 3039–3047 (2017)
6. China Academy of Building Research. Technical Specification for Retaining and Protection of Building Foundation Excavations, JGJ 120-2012. China Architecture and Building Press, Beijing (2012)

New Method for Determining Foundation Bearing Capacity Based on Plate Loading Test

Guanghua Yang[1,2(✉)], Yan Jiang[1,2], Chuanbao Xu[1,2],
Zhiyun Li[1,2], Fuqiang Chen[1,2], and Kai Jia[1,2]

[1] Guangdong Research Institute of Water Resources and Hydropower, Guangzhou, China
1084242143@qq.com
[2] The Geotechnical Engineering Technology Center of Guangdong Province, Guangdong
Research Institute of Water Resources and Hydropower, Guangzhou 510610, Guangdong, China

Abstract. A new method for determining the bearing capacity of the foundation based on the plate loading test is proposed in this paper. Assuming that the p (pressure)-s (settlement) curve of the foundation is a hyperbolic equation, the method back calculates the strength and deformation parameters of the soil from the plate loading test and then derives the p-s relationship of specific foundation by using a 'Tangent Modulus Layer-wise Summation Method'. Eventually, based on the predicted p-s curve of the specific foundation, the bearing capacity of the foundation can be determined by the strength safety and foundation settlement, which we call it the 'double control principle'.

Keywords: Tangent modulus layer-wise summation method
Foundation bearing capacity · Plate loading test

1 Introduction

Generally, the foundation bearing capacity is derived by reducing the ultimate foundation bearing capacity to as much as 2–3 times (factor of safety k); and the ultimate foundation bearing capacity is obtained according to the Limit Equilibrium Method based on the soil strength parameter c and φ, accompanied with associated boundary conditions such as foundation size and buried depth. The allowable bearing capacity of the foundation can be determined if the associated settlement meets the requirements of the upper structure. Generally, the strength of foundation is determined by the Limit Equilibrium Method or the elasto-plastic solution.; And the settlement is obtained by elastic solution [1] or the Layer-wise Summation Method [2]; however, the commonly used but rough calculation theory, in which the strength of soil is not directly linked to deformation, is not able to really meet the need of nowadays engineering, especially the accuracy of settlement prediction. One of the most essential reason of inaccuracy is that the parameters for calculation is obtained by laboratory tests, which is inherently deviated from reality. Hence, inaccurate parameters make accurate settlement prediction difficult.

© Springer Nature Switzerland AG 2018
W. Wu and H.-S. Yu (Eds.): *Proceedings of China-Europe Conference on Geotechnical Engineering*, SSGG, pp. 1064–1067, 2018.
https://doi.org/10.1007/978-3-319-97115-5_39

It is generally believed that the most reliable method for determining the foundation bearing capacity is in-site plate loading test. However, the problem of reasonably determining the bearing capacity of the foundation by the plate loading test is still unsolved due to the fact that the size and depth of the test are different from real foundation. As the soil is a non-linear material, it is difficult to define a linear part in the p (pressure)-s (settlement) curve of the test. Therefore, the code for foundation design in China applies a 'Settlement Ratio Method' to determine the bearing capacity of the foundation. According to this method [2], for a test using a plate of 0.25–0.5 m^2, the value of bearing capacity is determined as the pressure when s(settlement)/b (width or diameter of foundation) = 0.01–0.015, at the meantime, the value is not allowed to be greater than half of maximum load of the plate loading test. The method above provides the characteristic value of bearing capacity, and for any specific foundations, the allowable bearing capacity can be obtained by empirical modification based on the width and depth of the foundation. However, the bearing capacity determined by this method does not guarantee that the settlement of the actual foundation satisfies the requirements of the upper structure, furthermore, the settlement ratio ($s/b = 0.01$–0.015) has a range rather than a specific value, which indicates the value varies with different individuals. Therefore, even though the 'Settlement Ratio Method' is relatively useful method, it is still far from perfection.

2   New Method in Determining the Allowable Bearing Capacity Based on the Plate Loading Test

Based on the plate loading test, a new method to determine the bearing capacity of the foundation is proposed in this paper. The new method back calculates the strength and deformation parameters of the soil through the curve of plate loading test; Then, derive pressure-settlement relationship of specific foundation based on the back-calculated soil parameters; and eventually the safety of factor of foundation strength can be determined according to the derived pressure-settlement relationship; the allowable foundation bearing capacity can be derived by considering the allowable settlement of upper structure. The method is illustrated by the case as follows:

A plate loading test with a circular plate of diameter D equals 0.5 m, the foundation soil is silty clay, the corresponding relationship curve of pressure p and settlement s is shown in Fig. 1.

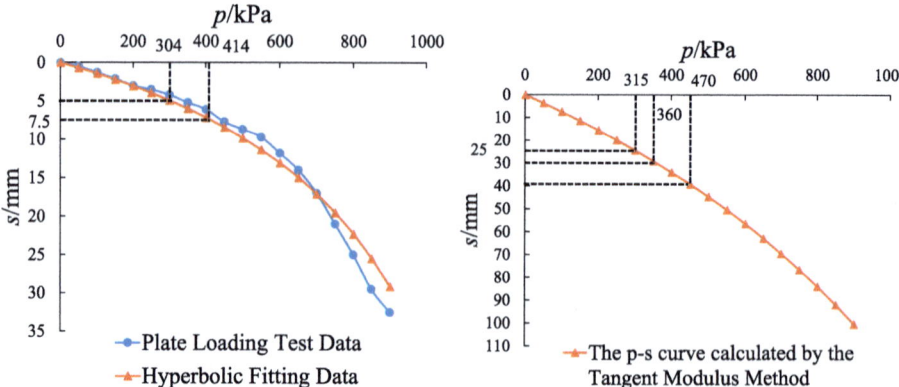

Fig. 1. The *p-s* curve of a plate loading test

Fig. 2. The *p-s* curve calculated by the Tangent Modulus Method (0 burial depth and width *B* equals 2 m)

Assume that the *p-s* curve is a hyperbolic equation [3]:

$$p = s/(a + bs) \tag{1}$$

And the value of a and b can be obtained: $a = 0.01316$, $b = 0.00066$, according to the assumption of hyperbolic Equation, there are: $a = 1/k_0$, $b = 1/p_u$, k_0 is the initial inclination of the test curve. When $s \to 0$, settlement can be calculated according to the Boussinessq Solution, then we have:

$$a = B(1 - \mu^2)\omega/E_0 \tag{2}$$

The ultimate bearing capacity of the foundation can be derived when b is solved. The value of φ obtained by laboratory test is relatively stable compared to the value of c, so if the value of φ is determined by laboratory tests, the value of c can be back calculated according to the bearing capacity of the foundation. In this case:

$$E_{t0} = B(1 - \mu^2)\omega/a = 30.4 \, \text{MPa}, \quad p_u = 1/b = 1505.0 \, \text{kPa}$$

Assuming $\phi = 20°$, then $c = 70.2$ kPa. Considering safety aspect, the values of c, φ is back calculated by the maximum load of the test of p equals 900 kPa, although p_u reaches as much as 1505.0 kPa. Taken p equals 900 kPa as the ultimate value of plate load test, and when $\phi = 20°$, the cohesion c is 59 kPa. It is conservative when applied to real foundation.

For the plate loading test curve, based on hyperbolic fitting equation, different characteristic values corresponding to different settlement ratios can be obtained according to the Chinese code: When $s/b = 0.01$, $p_{0.01} = 304$ kPa, safety factor: $k = 900/304 = 2.96$. When $s/b = 0.015$, $p_{0.015} = 414$ kPa, safety factor: $k = 900/414 = 2.17$.

In this way, different values of bearing capacity taken by settlement ratios judged by different people, and the settlement of the real foundation is also unclear when using different values, so the method is not perfect.

Our method is to calculate the p-s curve of specific foundation by using a 'Tangent Modulus Layer-wise Summation Method' [3, 4] when the values of E_{t0}, c, and φ are obtained, then the bearing capacity of specific foundation is determined according to the p-s curve calculated and the strength as well as deformation (so called 'double control principle').

For example, for a foundation with 0 burial depth and width B equals 2 m, the p-s curve calculated by the Tangent Modulus Method is shown in Fig. 2.

Considering the safety aspect, when using $\varphi = 20°$, $c = 59$ kPa, the calculated ultimate bearing capacity is $p_u = 982$ kPa.

According to the code of foundation design in China, for a foundation with no burial depth and width $B = 2$ m, no depth and width modification is needed. It can be seen from the (Fig. 2) that when the foundation allowable settlement is $s = 25$ mm, the corresponding bearing capacity is $p = 315$ kPa, and the safety factor is: $k = 982/315 = 3.1$; when the foundation allowable settlement is $s = 30$ mm, then the base stress is 360 kPa, the safety factor is $k = 982/360 = 2.7$; when the foundation allowable settlement is $s = 40$ mm, the foundation bearing capacity $p = 470$ kPa, and the foundation safety factor $k = 982/470. = 2.1$. In this way, we can directly determine the reasonable bearing capacity of the foundation based on the predicted p-s curve of the specific foundation and the 'double control principle' of strength safety and foundation settlement. This is obviously a more scientific and reasonable method.

3 Conclusion

The new method for determining the bearing capacity of the foundation is to obtain the soil parameters by the in-situ plate loading test, then use the Tangent Modulus Method to calculate the p-s curve of real foundation, and eventually based on the p-s curve, the bearing capacity of real foundation is determined according to the 'double control principle' of strength safety and foundation settlement. This is a new and more scientific method to determine the bearing capacity of foundation.

References

1. Das, B.M.: Shallow Foundations: Bearing Capacity and Settlement. CRC Press, Boca Raton (2009)
2. National Standards Compilation Group of the People's Republic of China. GB50007–2011 Foundational Design Code for Building Foundations. China Architecture and Building Press, Beijing, China (2011)
3. Yang, G.: Undisturbed soil tangent modulus method for calculation of nonlinear settlement of foundation. Chin. J. Geotech. Eng. **11**, 1927–1931 (2006)
4. Yang, G.-H., Luo, Y.-D., Zhang, Y.-C., Wang, E.-Q.: Application of the tangent modulus method in nonlinear settlement analysis of sand foundation. In: Proceedings of the 18th International Conference on Soil Mechanics and Geotechnical Engineering, Paris, pp. 3483–3486 (2013)

Numerical Investigation of T-Shaped Soil-Cement Column Supported Embankment Over Soft Ground

Yaolin Yi[1]([⊠]), Pengpeng Ni[1], and Songyu Liu[2]

[1] School of Civil and Environmental Engineering, Nanyang Technological University, Singapore 639798, Singapore
yiyaolin@ntu.edu.sg
[2] Institute of Geotechnical Engineering, Southeast University, Nanjing 210096, China

Abstract. Deep mixing is a technique to stabilize soft soils *in situ* by mixing with cementitious binders to form columns. An innovative soil-cement column is proposed, with an enlarged column cap at shallow depth, and as such the column shape is analogous to a letter "T". This paper presents a numerical investigation on the performance of T-shaped soil-cement column supported embankment over soft ground. The sensitivity of differential settlement between soil and column, as well as soil and column stresses in the middle of the embankment to the diameter and the height of the enlarged column cap is systemically studied.

Keywords: T-shaped column · Embankment · Numerical modelling

1 Introduction

The surcharge load induced by highway embankments can result in excessive settlement in soft ground, and the deep cement mixing method is often used to stabilize soft soils. Liu et al. [1] originally conceived the idea of increasing the diameter of column cap at shallow depth for deep mixing soil-cement columns as illustrated in Fig. 1, and the field performance of a highway embankment section supported with T-shaped columns was encouraging compared to another section above conventional columns with a fixed diameter [1, 2]. Yi et al. [3] conducted laboratory modelling of T-shaped and conventional column-treated soft ground under embankment loading, and found that the differential settlement of T-shaped column-treated ground was significantly lower. As a further work, this study employs three-dimensional numerical model to investigate the impacts of the diameter and the height of the enlarged column cap on the performance of T-shaped soil-cement column supported embankment.

2 Numerical Modelling

An embankment over soft soils is studied using a commercially available computer program, FLAC-3D. Due to symmetry, a half embankment model is established, and the dimensions of the model are given in Fig. 1. A fine mesh is defined near the T-shaped

© Springer Nature Switzerland AG 2018
W. Wu and H.-S. Yu (Eds.): *Proceedings of China-Europe Conference*
on Geotechnical Engineering, SSGG, pp. 1068–1071, 2018.
https://doi.org/10.1007/978-3-319-97115-5_40

soil-cement columns and the meshes become coarser at a far distance from the embankment. The bottom boundary is fixed in all degrees of freedom, and the lateral boundaries are restrained in the normal direction. The groundwater table is assumed at the ground surface, where a fully drained boundary is adopted. The embankment fill, soft clay, bearing stratum and soil-cement columns are all characterized as Mohr Coulomb materials with parameters under the drained conditions (see Fig. 1). This is to simulate the long-term performance of the embankment. The soil-cement column has a fixed D_2 of 0.5 m and L of 12 m. All columns are arranged in a squared pattern with a spacing of S = 1.8 m. The beneficial effect of T-shaped column with three D_1 values (0.75, 1.0 and 1.25 m) and three L_1 values (2, 4 and 6 m) on the performance of embankment compared to the conventional fixed diameter column is investigated.

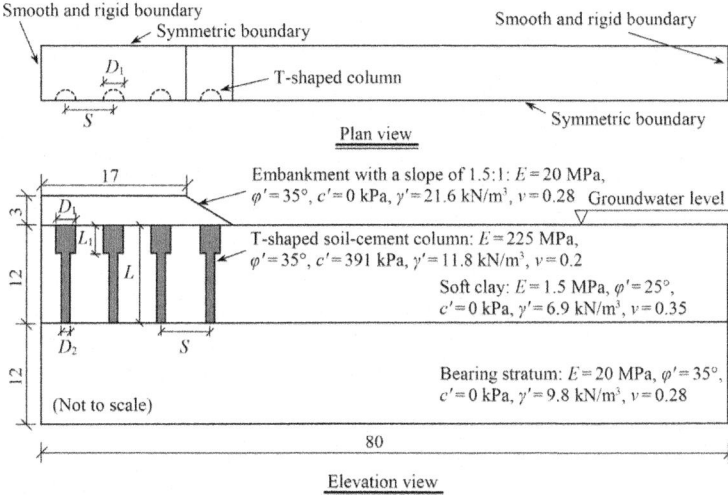

Fig. 1. Schematics of a T-shaped soil-cement column supported embankment (unit: m).

3 Numerical Results

In the middle of the embankment, the maximum settlement of soil occurs at the ground surface, and it decreases with depth as presented in Fig. 2a. For T-shaped columns, the enlarged column cap can help to reduce the maximum soil settlement significantly by approximately 50% compared to the conventional column. When the enlarged column cap has a larger diameter and a higher height, the soil settlement is smaller. All studied scenarios have the same area replacement ratio below 8 m, where the profile of soil settlement with depth becomes similar, indicating that the enlarged column cap does not influence the magnitude of soil settlement at greater depths. The variations of column settlement with depth are plotted in Fig. 2b. The magnitude of column settlement calculated for both T-shaped and conventional columns is similar. In general, a low column settlement is obtained when the column cap has a higher L_1 and a lower D_1.

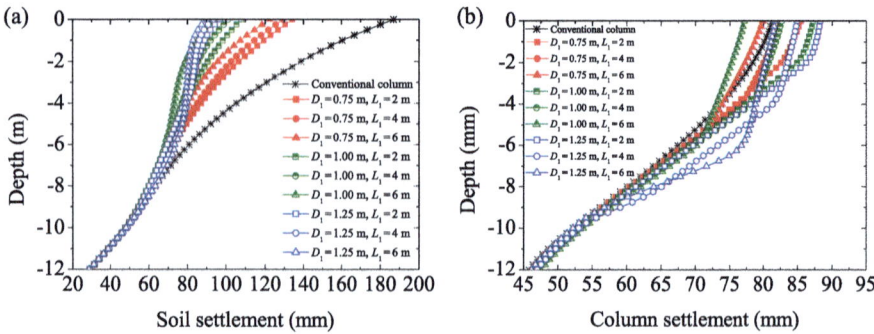

Fig. 2. Variations of settlement with depth: (a) soil, and (b) column.

The variations of differential settlement between soil and column with depth are calculated in Fig. 3a. The soil between columns settles more than the settlement of column itself at shallow depths for both T-shaped and conventional columns. It should be noted that the differential settlement is decreased significantly by introducing an enlarged column cap, which enables load redistribution and minimizes the detrimental effect of negative skin friction mobilized at the soil-column interface. With the increase of depth, the differential settlement becomes positive. A higher positive settlement demonstrates that the column tip penetrates the bearing stratum. As presented in Fig. 3b, the maximum differential settlement is inversely proportional to the D_1 value. The influence of L_1 on the mobilized differential settlement is relatively negligible. In design, a higher value of D_1 should be selected, but the value of L_1 can be minimized.

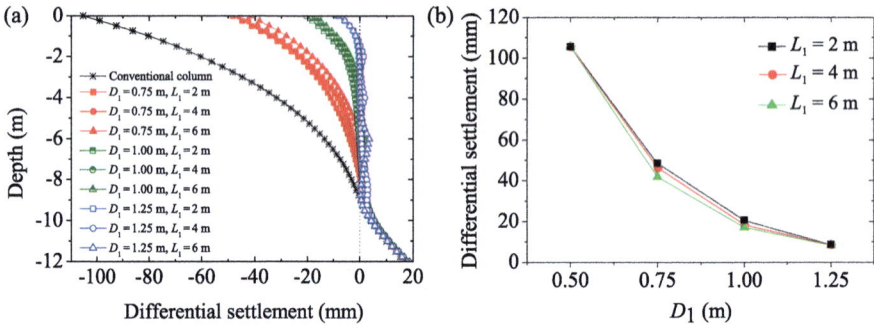

Fig. 3. Differential settlement: (a) profile with depth, and (b) influence of column dimensions.

The profiles of column stress with depth in the middle of the embankment are plotted in Fig. 4a. The equivalent unconfined compressive strength of the column is 1000 kPa. For the conventional column, the column stress increases almost linearly with depth, and the peak value is much less than the strength, indicating an overdesign of the column in design. For T-shaped columns, a lower stress is mobilized in the column cap, and a sudden increase of stress can be observed below the column cap. The column cap should be designed to have a larger cap diameter, but a smaller cap height to fully utilize the strength along most of the column height. The profiles of soil stress between columns with depth in the middle of the embankment are plotted in Fig. 4b. At the ground surface, the soil stress is significantly lower than the column stress due to the arching effect occurred in the embankment. The column cap of T-shaped columns acts as a load transfer platform, which further reduces the soil stress compared to the conventional column-treated ground. This effect is mainly controlled by the value of D_1 while nearly not influenced by the value of L_1 (Fig. 4b).

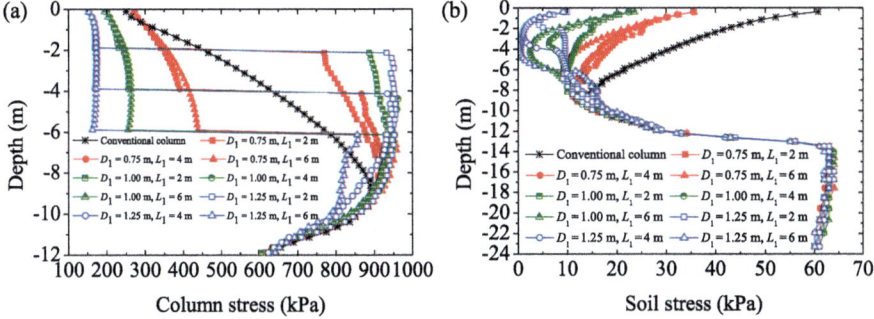

Fig. 4. Variations of (a) column stress, and (b) soil stress with depth.

4   Conclusions

The numerical study shows that the use of T-shaped column can significantly reduce the differential settlement between soil and column under embankment loading compared to the conventional column. The differential settlement is inversely proportional to the cap diameter, while the cap height has a negligible effect. Hence, a larger cap diameter and a smaller cap height are recommended for practical design.

References

1. Liu, S.Y., Du, Y.J., Yi, Y.L., Puppala, A.J.: Field investigations on performance of T-shaped deep mixed soil cement column–supported embankments over soft ground. J. Geotech. Geoenviron. Eng. **138**(6), 718–727 (2012)
2. Yi, Y., Liu, S., Puppala, A.J., Xi, P.: Vertical bearing capacity behaviour of single T-shaped soil–cement column in soft ground: laboratory modelling, field test, and calculation. Acta Geotech. **12**(5), 1077–1088 (2017)
3. Yi, Y., Liu, S., Puppala, A.J.: Laboratory modelling of T-shaped soil–cement column for soft ground treatment under embankment. Géotechnique **66**(1), 85–89 (2016)

Experiment Study on Wave Induced Excessive Pore Pressure Around Near-Sea Foundation Pit

Hong-wei Ying[1,2(✉)] [iD], Ding-ye Xu[1,2] [iD], and Cheng-wei Zhu[1,2] [iD]

[1] Research Center of Coastal and Urban Geotechnical Engineering,
Zhejiang University, Hangzhou, China
ice898@zju.edu.cn
[2] MOE Key Laboratory of Soft Soils and Geoenvironmental Engineering,
Zhejiang University, Hangzhou, China

Abstract. Model test of coastal excavations is based on the excavation of Gongbei tunnel located in the Ling-ding Ocean, which is a part of the famous Hongkong-Zhuhai-Macau Bridge project. A series of model tests are carried out in the ZJU multifunctional wave-current flume to study on the effects of wave factors on the pore pressure response around clayey silt excavation. It is shown that the attenuation of the amplitude of the oscillatory and accumulated excess pore pressure are related to the wave height, the wave period and the water depth. Oscillatory excess pore and accumulated excess pore pressure along with maximum seepage path are affected by various wave factors.

Keywords: Near-sea excavation · Wave factors
Oscillatory excess pore pressure · Accumulated excess pore pressure

1 Introduction

In the near-sea excavation, the groundwater is affected by waves, tides and other dynamic water environment. Present experimental researches on the pore pressure response mainly focus on seabed and offshore structures under wave conditions [1–4], but there are few reports on effects of wave action on near-sea excavations. This paper based on former experiment [5] aims to investigating the law of pore pressure response around the offshore foundation pit under different wave factors. The emphasis is put on the amplitudes of the oscillatory and maximum cumulated pressure around foundation pit in this study.

2 Model

Model is determined by theory of similar scale. The test model is 2.3 m long, 0.6 m wide and 1.3 m high, the detailed size is shown in Fig. 1. Pore pressure transducers were arranged on different sections of the model to record pore pressure changes during the

© Springer Nature Switzerland AG 2018
W. Wu and H.-S. Yu (Eds.): *Proceedings of China-Europe Conference
on Geotechnical Engineering*, SSGG, pp. 1072–1076, 2018.
https://doi.org/10.1007/978-3-319-97115-5_41

test process, as shown in Fig. 2. The analysis of P2 and P12 pore pressure are not intro-duced due to the breakage of these two devices during the test.

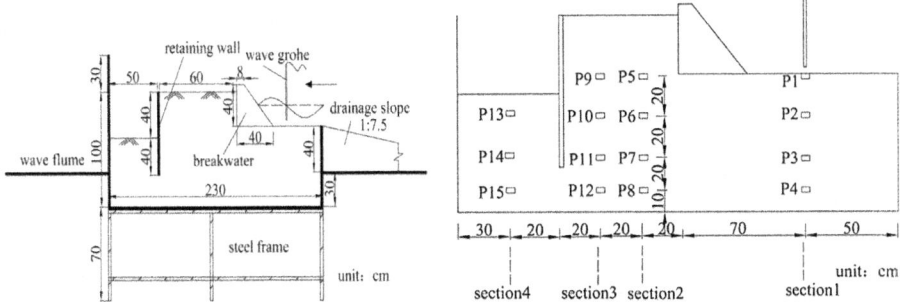

Fig. 1. Test model (left)

Fig. 2. Pore pressure transducers distribution (right)

3   Results Analysis

3.1   Excess Pore Pressure Distribution Under Wave Heights Variations

The regularity of distribution of oscillatory pressure in the same vertical section at different wave heights is consistent (Fig. 3), and the amplitude of the oscillating

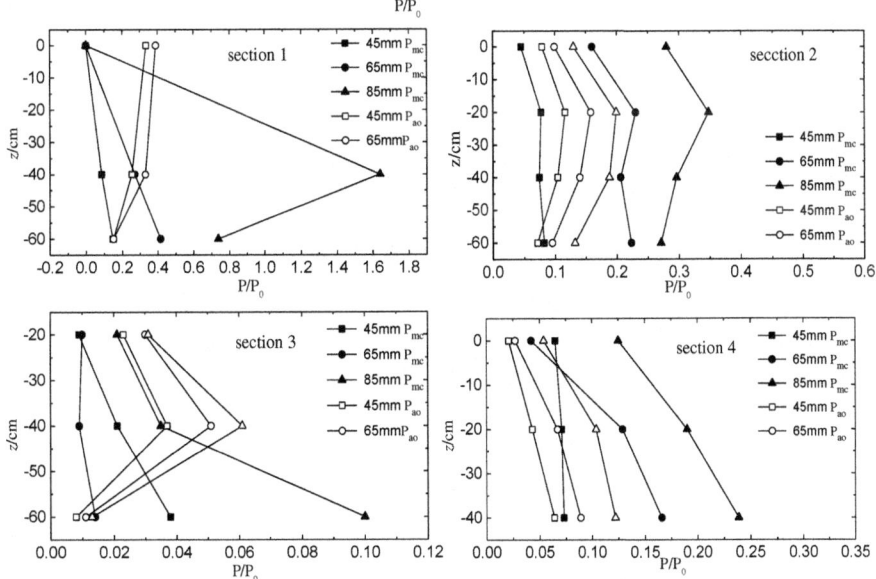

Fig. 3. Excess pore pressure distribution in different sections under wave height variations (filled symbols: Pmc, open symbols: Pao)

excessive pore pressure (Pao) and the maximum cumulative pore pressure (Pmc) of the same measuring point augment with the increase of wave height. It can be seen that Pao and Pmc along with the seepage path increase with the raise of the wave height. It can be seen that the amplitude of the oscillating excessive pore pressure and the maximum cumulative excess pore pressure along with the seepage path increase with the raise of the wave height.

3.2   Excess Pore Pressure Distribution Under Wave Period Variations

When the wave height and water depth are constant, the longer the wave period, the smaller Oscillating pore pressure amplitude and the maximum cumulative excess pore pressure (Fig. 4). P1 has been damaged during this test, so no longer analyzed here. Although the accumulated pore pressure inside the foundation pit is less affected by the periodic variation of the waves, the accumulation trend of the excess pore pressure is identical at different wave periods. Oscillating pore pressure amplitude under shorter wave period attenuations along with the flow path more quickly, and the cumulative pore pressure along with maximum seepage path attenuation changes slowly.

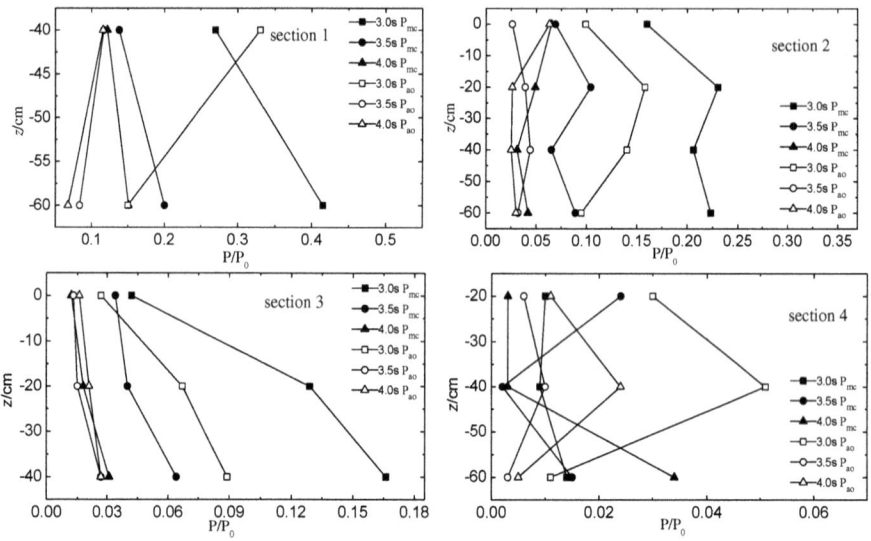

Fig. 4. Excess pore pressure distribution in different sections under wave period variations (filled symbols: Pmc, open symbols: Pao)

3.3   Excess Pore Pressure Distribution Under Water Depth Variations

As shown in Fig. 5, the amplitude of the oscillatory pressure and the maximum value of the cumulative excess pore pressure of each measuring point on the active side of the foundation pit under different water depths show consistent trend, which indicates both decrease significantly with the increase of water depth, while pore pressure on the passive side was less affected. The smaller the test water depth is, the faster the amplitude of the oscillating excess pore pressure and the maximum cumulative excess pore pressure decrease along the seepage path.

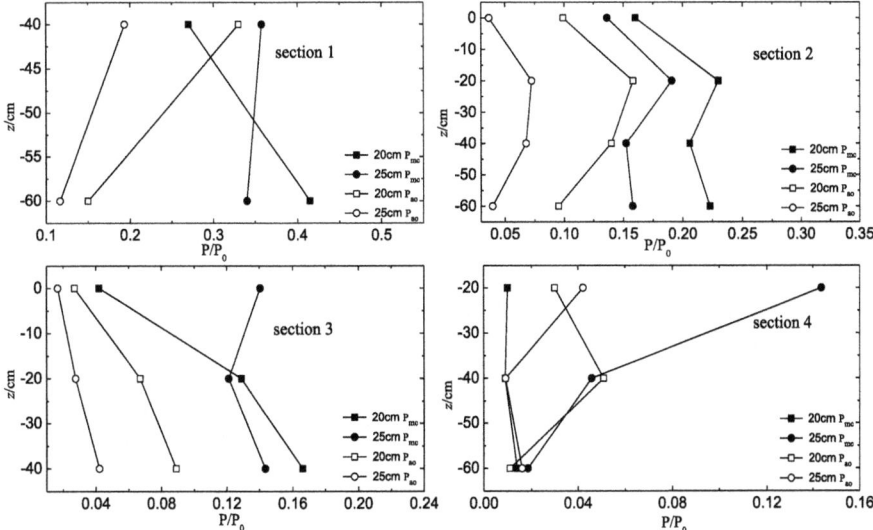

Fig. 5. Excess pore pressure distribution in different sections under water depth variations (filled symbols: Pmc, open symbols: Pao)

4   Conclusion

When the wave period and the water depth is constant, the amplitudes of the oscillatory excess pore pressure and the maximum value of cumulated pore pressure in the active zone of the foundation pit augment with the increase of the wave height. The maximum cumulated pressures decrease with the increase of the wave period in the active zone of the pit, while is less affected in the passive zone. The increase of water depths leads to the decrease of the amplitude of the oscillatory excess pore pressure and the maximum cumulated pore pressure in the active zone of the excavation.

References

1. Bennett, R.H., Hulbert, M.H., Meyer, M.M.: Fundamental response of pore-water pressure to micro fabric and permeability characteristics: Eckernförde Bay. Geo-Mar. Lett. **16**(5), 182–188 (1996)
2. Zen, K., Yamazaki, H.: Mechanism of Wave-Induced Liquefaction and Densification in Seabed. J. Jpn. Soc. Soil Mech. Found. Eng. **30**(4), 90–104 (2008)
3. Sudhan, C.M., Sundar, V., Rao, S.N.: Wave induced forces around buried pipelines. Ocean Eng. **29**(5), 533–544 (2002)

4. Sui, T., Zhang, J., Zheng, J., Zhang, C.: Modeling of wave-induced seabed response and liquefaction potential around pile foundation. In: ASME 2013 International Conference on Ocean, Offshore and Arctic Engineering, vol. 1, pp. 161–170 (2013)
5. Ying, H., Sun, W., Zhu, C.: Experiment research on response of excess pore pressure to wave around near-sea excavation. Rock Soil Mech. **37**(s2), 187–194 (2016). (in Chinese)

Centrifugal Model Tests on Working Behavior of Composite Foundation Reinforced by Rigid Piles with Caps Under Embankment

Jian-Lin Yu[1,2(✉)], Jia-Nan Zhong[1,2], Jun-Yuan Li[1,2],
Ri-Qing Xu[1,2], and Xiao-Nan Gong[1,2]

[1] Coastal and Urban Geotechnical Engineering Research Center,
Zhejiang University, Hangzhou 310058, China
yujianlin72@126.com
[2] Zhejiang Urban Underground Space Development Engineering Research Center,
Zhejiang University, Hangzhou 310058, China

Abstract. Centrifuge model tests are carried out to study the composite foundation under embankment which is reinforced by rigid piles with caps on the layered ground that the upper layer was soft clay and the lower layer was stiff sand. Variable acceleration loading method, constant acceleration loading method and gasbag loading method are applied to simulate the construction process, operation process and failure process of embankment. Working behavior, which including displacement of embankment and ground, internal force and deformation of piles in different position, soil stress and pore water pressure between piles, and failure modes of composite foundation are analyzed. Results show that the ground deformation is presented as settlement, horizontal displacement and toe heave. The pile-soil stress ratio in the middle of embankment (25.0) is greater than that in slope shoulder (17.4). Failure modes of piles in different position under embankment are different. To the piles under the middle of embankment, the most likely failure mode is compression failure. The piles under slope shoulder may be subjected to press-bending failure or overturning failure, and the piles under slope toe may occur tension-bending failure or overturning failure.

Keywords: Embankment · Composite foundation reinforced by rigid piles
Centrifugal model tests · Working behavior · Failure mode

1 Introduction

Because of the higher demanding for bearing capacity, settlement and stability of composite foundation under embankment of expressways and high-speed railways, rigid piles are increasingly applied to the treatment of soft soil foundation. However, in engineering practice, the overall instability failure of embankment may be appeared even if reinforced by rigid piles composite foundation [1]. Relevant specifications [2] recommend that the limit equilibrium method to analyze the overall stability of embankment

© Springer Nature Switzerland AG 2018
W. Wu and H.-S. Yu (Eds.): *Proceedings of China-Europe Conference on Geotechnical Engineering*, SSGG, pp. 1077–1080, 2018.
https://doi.org/10.1007/978-3-319-97115-5_42

which assuming shear failure of all piles. But the research results of Broms (1999) and Kitzume (2005) show that this method seriously overestimates the stability of embankment, and there are other failure modes in piles [3–5]. Therefore, how to reasonably judge the failure mode of composite foundation improved by rigid piles under embankment and accurately analyze the overall stability of embankment is still a problem to be solved.

Centrifuge model test with the high-speed rotation of centrifuge can create a stress field with the same stress level as prototype, which can reproduce the traits of prototype and effectively simulate force characteristics and deformation modes of prototype under the original gravity field by reducing the model size [6]. This paper designed and completed a centrifugal model test of composite foundation improved by rigid piles under embankment. The variable acceleration loading method, constant acceleration loading method and gasbag loading method are applied to simulate the construction process, operate process and failure process of embankment, for analyzing the behavior and failure mode of composite foundation under embankment.

2   Centrifuge Model Test

In the experiment, model piles are made of aluminum alloy hollow tube. The principle is that the model piles and prototype piles need to satisfy similarity of axial stiffness and bending stiffness. On the improved area, 6 columns with length of 200 mm (20 m in prototype) are arranged from the center of embankment to the toe, and the space between columns is 50 mm. The square caps of pile are made of aluminum alloy plates with border lengths of 20 mm and thick of 3.7 mm.

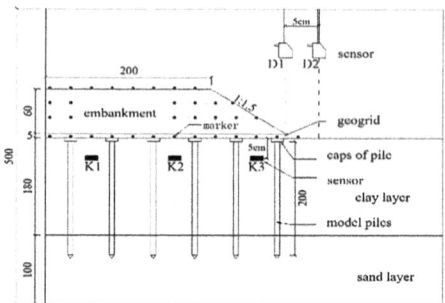

Fig. 1. A section of the experimental model.

Figure 1 shows a section of the experimental model. In the experiment, an embankment with height of 60 mm was built on a ground with two layers including a soft clay layer depositing on a stiff sand layer. The thickness of clay layer is 180 mm and the sand layer is 100 mm. The clay used in the test was Kaolin clay. Kaolin power was mixed with tap water in a vacuum mixer to produce uniform slurry with water content of 120%. The clay slurry was poured into the model box, then pre-consolidated for 48 h by vertical pressure of 9.8 kN/m^2. After completing the pre-consolidation, the model clay ground was subjected to high centrifugal acceleration of 100 g to consolidate by enhanced self-weight for 6 h and overloading consolidate for 3 h. The sand used in test was Fujian standard sand, the relative density was 60%. And the embankment was made by coarse sand (d_{50} = 0.5 mm). The geogrid was made by plastic gauze, tensile strength was 1.738 kN/m.

3   Test Results and Discussion

3.1   Pile Failure Mode

The failure mode and failure time of piles in different positions under embankment are different. The piles under the center of embankment are mainly subjected to vertical loads and may be damaged by pressure. The piles under the slope shoulders mainly bear vertical load and horizontal load of soil and may be subjected to press-bending failure or overturning failure. The piles under the toe of slope not only subjected to horizontal load, but also bear the upward pulling force of the soil uplift. The piles under slope toe may occur tension-bending failure or overturning failure. Due to the low strength of the geogrid, there are spiny failure on the top of piles under embankment. Therefore, increasing the tensile strength of reinforced cushion can control the spiny failure of the piles and allow the piles to share more of the pressure, reduce the settlement and improve the stability of embankment.

3.2   Ground Deformation

The PIV method was used to investigate the movement of ground. Figure 2 shows the displacement of ground and embankment. The sliding deformation of ground was obvious when the embankment instability. There were settlement and horizontal displacement under embankment, and vertical uplift outside the toe. The farther away from the center of the embankment, the greater horizontal displacement of ground. Therefore, the deformation of ground mainly includes settlement, horizontal displacement and toe heave.

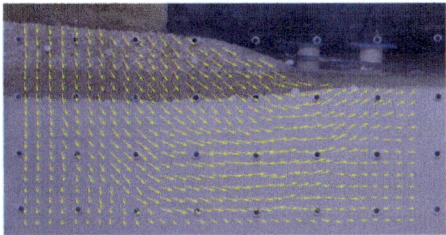

Fig. 2. The displacement trend of ground and embankment.

3.3 Stress Ratio of Pile-Soil

During the process of layered filling of embankment and loading at the top of embankment, the maximum pile-soil stress ratio in the center of embankment is 25, and that in slope shoulder is 17.4. The result shows that the piles near the center of embankment can effectively share more load from embankment.

3.4 Excess Pore Water Pressure

With the increase of centrifugal acceleration and load from the top of embankment, the excess pore water pressure in ground rise rapidly. Keeping the steady load at 100 g after 180 min (2 years in prototype), the excess pore water pressure gradually dissipated, and the dissipation rate gradually slow down.

4   Conclusions

Through the centrifugal model tests on working behavior and failure pattern of composite foundation reinforced by rigid piles with caps under embankment, the major conclusions drawn from this study are summarized below:

(1) Under the load of embankment, the ground deformation is presented as settlement, horizontal displacement and toe heave.
(2) The pile-soil stress ratio in the center of embankment is greater than that in slope shoulder, indicating that the piles under the center of embankment can share more load from embankment effectively.
(3) There are significant differences in the failure mode of piles in different positions under embankment. Therefore, it is inappropriate to use the limit equilibrium method which assuming shear failure of all piles to analyze the stability of composite foundation reinforced by rigid piles under embankment.

References

1. Liu, J.F.: Stability analysis of flow-slide of embankment on rigid-piles composite ground. Geotech. Inves. Surv. J. **6**, 17–22 (2013)
2. JTG D30-2004. Specification for design of highway subgrades. Industry Standard of China
3. Broms, B.: Can lime/cement columns be used in Singapore and Southeast Asia, p. 214. Nanyang Technological University and NTU-PWD Geotechnical research Centre (1999)
4. Kitazume, M., Okano, K., Miyajima, S.: Centrifuge model tests on failure envelope of column type deep mixing method improved ground. Soils Found. **40**(4), 43–55 (2000)
5. Kitazume, M., Maruyama, K.: Collapse failure of group column type deep mixing improved ground under embankment. In: Proceedings of the International Conference on Deep Mixing, pp. 245–254. ASCE (2005)
6. Editorial Committee of The Principles and Engineering Applications of Rock and Soil Centrifugal Simulation Technology.: Principles and Engineering Applications of Rock and Soil Centrifugal Simulation Technology. Yangtze River Press, Wuhan (2011)

On Performance of Anchored Retaining Piles and Stability of Excavations

Xiaoyi Yuan, Longzhu Chen[✉], and Chunyu Song

Shanghai Jiao Tong University, Shanghai 200240, China
lzchen@sjtu.edu.cn

Abstract. Accurate calculation of lateral displacements and internal forces of retaining structures and analysis of stability against sliding has great significance. This paper presents an analytic method to calculate displacements of anchored retaining piles and analyzes excavation stability with the modified mechanical model specified in the Chinese code of excavation. This research work is financially supported by the National Natural Science Foundation of China with Grant No. 51379122.

Keywords: Analytic calculation · Anchored retaining pile
Stability of excavation

1 Calculation of Internal Forces and Displacements of Retaining Piles and Stability Analysis

Assume that there are M layers of soil and N composite bolts above the pit bottom, and the retaining structure meets the following assumptions:

(1) The calculation is a plane strain problem, and the calculative width of the retaining pile is b_1;
(2) Retaining piles are linear elastic vertical beams with constant sections, and the top end of the pile bears no shear force or moment;
(3) Earth pressure acting on the upper part of the pile (related to the pit bottom) is governed by Rankine's earth pressure theory, and that on the lower part is governed by the elastic foundation beam assumption;
(4) The effects of the pile's vertical displacements are ignored.

The calculation model of retaining pile is shown in Fig. 1.

The equilibrium differential equation of the pile above the pit bottom can be expressed using Singular Function Method (see Ref. [1]) in Eq. (1):

$$EI\frac{d^4y}{dz^4} = q(z) \tag{1}$$

© Springer Nature Switzerland AG 2018
W. Wu and H.-S. Yu (Eds.): *Proceedings of China-Europe Conference on Geotechnical Engineering*, SSGG, pp. 1081–1084, 2018.
https://doi.org/10.1007/978-3-319-97115-5_43

Where, EI is the flexural rigidity of the retaining pile, $q(z) = \sum\limits_{i=1}^{M} q_i(z)$ $+ \sum\limits_{j=1}^{N} p_j(z)$, $p_j(z) = k_j y_j \langle z - h_j \rangle^{-1}$, k_j is equivalent horizontal stiffness of the composite bolts, $q_i(z) =$ $b_1 \Big[q_{i1} \langle z - z_{i1} \rangle^0 - q_{i2} \langle z - z_{i2} \rangle^0$ $+ \frac{q_{i2} - q_{i1}}{z_{i2} - z_{i1}} \big(\langle z - z_{i1} \rangle^1 - \langle z - z_{i2} \rangle^1 \big) \Big].$ The equilibrium differential equation below the pit bottom is Eq. (2):

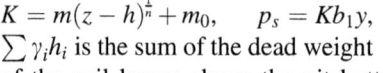

$$EI \frac{d^4 y}{dz^4} = -Kb_1 y + p_h b_1 \qquad (2)$$

where,

$p_h = (\sum \gamma_i h_i + q_0)\overline{K}_a - 2\bar{c}\sqrt{\overline{K}_a},$
$\overline{K}_a = \tan^2(45° - \overline{\varphi}/2),$
$K = m(z - h)^{\frac{1}{n}} + m_0, \quad p_s = Kb_1 y,$
$\sum \gamma_i h_i$ is the sum of the dead weight

Fig. 1. Calculation model of anchored retaining pile

of the soil layers above the pit bottom, q_0 is ground overload, $\bar{c}$ and $\overline{\varphi}$ are average cohesion and internal friction angle weighted by the thickness of soil layers above the pit bottom, respectively.

Solutions of Eqs. (1) and (2) can be expressed in a single formula in Eq. (3) using Singular Function Method:

$$y = Y_1(z)\langle z - h \rangle^0 + Y_2(z)\Big(1 - \langle z - h \rangle^0\Big) \qquad (3)$$

Where, $Y_1(z) = y_b A[\alpha(z-h)] + \frac{\varphi_b}{\alpha} B[\alpha(z-h)] + \frac{M_b}{\alpha^2 EI} C[\alpha(z-h)] + \frac{Q_b}{\alpha^3 EI} D[\alpha(z-h)]$

$+ \frac{p_h b_1}{\alpha^4 EI} E[\alpha(z-h)]$ $Y_2(z) = \frac{1}{EI}\Big[I_1(z) + \sum\limits_{j=1}^{n} I_{2,j}(z) y_j \Big]$, $A(\alpha x) = 1 + \sum\limits_{s=1}^{\infty} \frac{(-1)^s n^{4s}}{\{s\}_{R=0}^{!!}} (\alpha x)^{\frac{4n+1}{n}s}$,

$B(\alpha x) = \alpha x + \sum\limits_{s=1}^{\infty} \frac{(-1)^s n^{4s}}{\{s\}_{R=n}^{!!}} (\alpha x)^{\frac{4n+1}{n}s+1}$, $C(\alpha x) = \frac{(\alpha x)^2}{2} + \sum\limits_{s=1}^{\infty} \frac{(-1)^s n^{4s}}{2\{s\}_{R=2n}^{!!}} (\alpha x)^{\frac{4n+1}{n}s+2}$, $D(\alpha x) =$

$\frac{(\alpha x)^3}{6} + \sum\limits_{s=1}^{\infty} \frac{(-1)^s n^{4s}}{6\{s\}_{R=3n}^{!!}} (\alpha x)^{\frac{4n+1}{n}s+3}$, $E(\alpha x) = \frac{(\alpha x)^4}{24} + \sum\limits_{s=1}^{\infty} \frac{(-1)^s n^{4s}}{24\{s\}_{R=4n}^{!!}} (\alpha x)^{\frac{4n+1}{n}s+4}$.

Let $A'(\alpha x) \sim E'(\alpha x)$ be derivatives of $A(\alpha x) \sim E(\alpha x)$ of αx, and $A(\alpha l_d) = A$, $A'(\alpha l_d) = A'$ and so forth, $W = \frac{1}{6}\sum\limits_{i=1}^{M} b_1(z_{i2} - z_{i1})[q_{i1}(3h - z_{i2} - 2z_{i1}) + q_{i2}(3h - 2z_{i2}$ $-z_{i1})]$, $T_0 = AB' - A'B$, $T_1 = BC' - B'C$, $T_2 = BD' - B'D$, $T_3 = BE' - B'E$, $T_4 =$

$$A'C - AC', \; T_5 = A'D - AD', \; T_6 = A'E - AE', \; P = \tfrac{1}{2}\sum_{i=1}^{M} b_1(q_{i1} + q_{i2})(z_{i2} - z_{i1}), \; f(z) =$$

$$\sum_{i=1}^{M} b_1\left[\tfrac{q_{i1}}{24}\langle z - z_{i1}\rangle^4 - \tfrac{q_{i2}}{24}\langle z - z_{i2}\rangle^4 + \tfrac{q_{i2}-q_{i1}}{120(z_{i2}-z_{i1})}\left(\langle z - z_{i1}\rangle^5 - \langle z - z_{i2}\rangle^5\right)\right], \quad T_7 = f'(h),$$

$T_8 = f(h)$, and the lateral displacement and rotation of the bottom of the retaining pile
be δ and θ, $m_0 = 0$, n is a positive integer, then:

$$I_1(z) = f(z) - T_8 + z\left(\tfrac{WT_4}{\alpha T_0} + \tfrac{PT_5}{\alpha^2 T_0} + \tfrac{p_h b_1 T_6}{\alpha^3 T_0}\right) + \tfrac{W(T_1 - \alpha T_4 h)}{\alpha^2 T_0} + \left(T_7 + EI\tfrac{\alpha \delta A' - \theta A}{T_0}\right)(h - z)$$

$$+ \tfrac{P(T_2 - \alpha T_5 h)}{\alpha^3 T_0} + \tfrac{p_h b_1(T_3 - \alpha T_6 h)}{\alpha^4 T_0} + EI\tfrac{\alpha \delta B' - \theta B}{\alpha T_0},$$

$$I_{2,j}(z) = k_j\left[\tfrac{(T_1 - \alpha T_4 h)(h - h_j)}{\alpha^2 T_0} + \tfrac{T_2 - \alpha T_5 h}{\alpha^3 T_0} + \tfrac{(h - h_j)^2(2h + h_j)}{6}\right] + \tfrac{1}{6}k_j\langle z - h_j\rangle^3$$

$$+ zk_j\left[\tfrac{\alpha T_4(h - h_j) + T_5}{\alpha^2 T_0} - \tfrac{(h - h_j)^2}{2}\right], \quad \varphi_b = \tfrac{1}{\alpha^3 EI T_0}\left[\alpha^2 M_b T_4 + \alpha Q_b T_5 + p_h b_1 T_6\right] - \tfrac{\alpha \delta A' - \theta A}{T_0},$$

$$y_b = \tfrac{1}{\alpha^4 EI T_0}\left[\alpha^2 M_b T_1 + \alpha Q_b T_2 + p_h b_1 T_3\right] + \tfrac{\alpha \delta B' - \theta B}{\alpha T_0}.$$

Here y_j are lateral displacements of retaining pile at the positions of composite
bolts, and their values can be obtained through solving Eq. (4):

$$[U_{ij}]\,\vec{y} + EI[E]\,\vec{y} = \vec{H} \tag{4}$$

where, $U_{ij} = -I_{2,j}(h_i)$, $\vec{y} = [y_1, y_2, \ldots y_N]^T$, $[E]$ is unit matrix, $\vec{H} = [I_1(h_1),$
$I_1(h_2), \ldots I_1(h_N)]^T$.

Easy to prove that if shear force and moment acted on the bottom of pile were 0, the
formula would have similar formation except that $T_0 \sim T_6$ should be replaced by their
2nd order derivatives of αl_d, and terms about δ and θ should be deleted.

When $m_0 \neq 0$ and n is positive integer, the solution has the same formation as
above, but the expressions of $A(\alpha x) \sim E(\alpha x)$ cannot be given in analytic form.

A modified formula is proposed in Eq. (5) based on the analytical solution established above and Swedish Slice Method to analyze slip stability of the excavation:

$$K_{s,i} = F_a/F_p \tag{5}$$

where, $K_{s,i}$ is safety factor of the ith slip arc, $F_p = b_1 r_i \sum (q_j d_j + \Delta G) \sin \theta_j$, $F_a =$
$b_1 r_i \sum \{c_j l_j + [(q_j d_j + \Delta G)\cos \theta_j - u_j l_j]\tan \varphi_j\} + b_1 r_i \sum R'_k[\cos(\theta_k + \alpha_k) + \psi_v]/s_{x,k}$
$-M_p + Q_p D_p$, $l_j = b_j/\cos \theta_j$, $\psi_v = 0.5 \sin(\theta_k + \alpha_k)\tan \varphi$, c_j and φ_j are cohesion and
friction angle of the jth soil slice at slip arc, d_j is the width of the soil slice, θ_j is the angle
between the normal direction of the slip arc at the midpoint and vertical direction, q_j is
attached load on soil slice, ΔG is dead weight of the soil slice, u_j is water pressure, R'_k is the
uplift capacity of the bolts, α_k is the inclination of kth bolt, θ_k is the angle between the
normal direction of the slip arc at the intersection point of the bolt and vertical direction,
$s_{x,k}$ is the lateral distance between adjacent bolts, r_i is radius of the slip arc, M_p and Q_p are
moment and shear force of the retaining pile at intersection point of slip arc and retaining
pile respectively, and D_p is the vertical distance between the circle center of slip arc and
the intersection point of slip arc and retaining pile.

2 Validation of the Formula

The formula was applied in an excavation engineering in Wenzhou, Zhejiang, China. The excavation was in thick soft clay ground, and composition of the soils were mainly mud and mud silt. The excavation had a depth of 6.2 m, while the pile was 26 m in length. The retaining pile had a section diameter of 0.8 m, and center distance of adjacent piles was 1.1 m. The piles were connected by top beams which had a cross section dimension of 1.1 m × 0.8 m. Two composite bolts were placed along the retaining pile with a vertical distance of 1.4 m and 3.9 m from the top of the pile respectively, and lateral distance between adjacent bolts was 2.0 m.

Assume that the ground overload was 20 kPa, and the groundwater level was 1 m lower than the ground surface nearby the pit and 1 m lower than the pit bottom inside the pit. Moment and shear force acting on the bottom of pile was set to 0 while lateral displacement and rotation of the pile bottom was set unconstrained. The results are shown in Fig. 2. The calculated results showed good agreement with the measured data.

The slope stability of Wenzhou excavation engineering was analyzed using the method established above, and the results are shown in Fig. 3. It indicates that the slip arc of the minimum safety factor (0.774) may not pass the bottom of retaining pile as checked with the Chinese code.

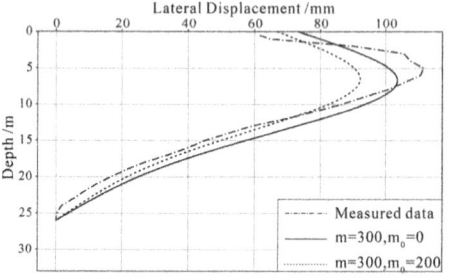

Fig. 2. Calculated relative lateral displacement of Wenzhou excavation.

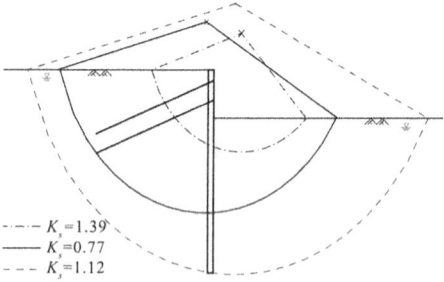

Fig. 3. Slope stability analysis of Wenzhou excavation engineering.

Reference

1. Beer, F.P., Jounston, E.R., Dewolf, J.T., Mazurek, D.F.: Mechanics of Materials, 6th edn. McGraw-Hill Companies, Inc., New York (2012)

Responses of Retaining Wall to Pre-excavation Dewatering Under Staggered Layout of Dewatering Wells

Chao-Feng Zeng[(✉)], Zhi-Cheng Yuan, Xiu-Li Xue, and Wei Hu

Hunan Provincial Key Laboratory of Geotechnical Engineering for Stability Control and Health Monitoring, School of Civil Engineering, Hunan University of Science and Technology, Xiangtan 411201, Hunan, China
cfzeng@hnust.edu.cn

Abstract. Pre-excavation dewatering (PED) is an important stage in deep excavation in soft ground with high groundwater level. The PED-induced deformations of retaining wall and surrounding environment can reach centimeter level. However, the method to control this deformation has not been yet proposed by the traditional pit design theory and relevant research. In this paper, a countermeasure, which is called as staggered layout of dewatering well, is proposed. In this method, it is recommended that deeper wells and smaller spacings of neighboring wells are employed near the retaining wall. The numerical results show that this method can effectively limit the PED-induced wall deflection.

Keywords: Deep excavation · Environmental impact
Pre-excavation dewatering · Staggered layout of dewatering well
Finite element analysis

1 Introduction

Pre-excavation dewatering (PED) is an important construction stage before excavation to check the completion quality of dewatering wells. The discharge flow rate and drawdown in dewatering wells need to be tested and should meet the design requirement before excavation [1, 2]. Based on field measurements and numerical simulations, Zeng et al. [3, 4] found that the PED can induce obvious wall deflection, which can reach centimeter level. Zheng et al. [2] pointed out that the seepage force induced by PED will cause the horizontal soil displacement and eventually lead to the wall development. However, in practice, it is usually considered that no wall deflection exists before excavation. The method to control PED-induced deformation has not been systematically studied. In this paper, a two-dimensional (2D) finite element (FE) model using the Biot consolidation theory is developed in a FE software (ABAQUS) to simulate a PED process reported in literature [3]. The model is verified by the test and numerical wall

© Springer Nature Switzerland AG 2018
W. Wu and H.-S. Yu (Eds.): *Proceedings of China-Europe Conference on Geotechnical Engineering*, SSGG, pp. 1085–1089, 2018.
https://doi.org/10.1007/978-3-319-97115-5_44

deflections reported in literature [3]. Then, a countermeasure, which is called as staggered layout of dewatering well, is proposed. Through series of numerical simulations, the effect of this method in controlling the PED-induced wall deflection is investigated.

2   Numerical Model and Verification

Figure 1 shows the mesh of the 2D FE model with excavation width (b) of 120 m. For simplification, only half of the excavation is modeled. In the vertical direction, the model is set to 50 m in depth. The depth of the retaining wall and dewatering wells are 33 m and 27 m, respectively, which complies the description in literature [3].

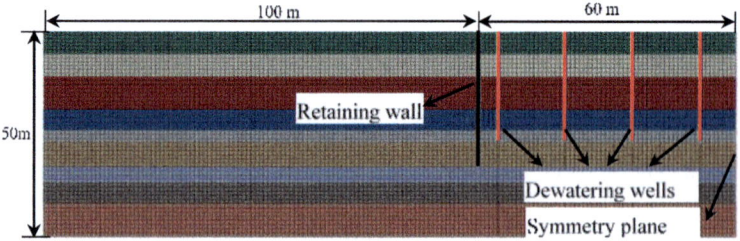

Fig. 1. Mesh of the FE model.

The soil behavior is assumed to obey the constitutive theory of modified Cam-Clay during PED. The linear elastic model is adopted for the wall and wells. The main parameters of each soil layer, wall and wells and boundary conditions can be seen in literature [3]. The function of dewatering well is simulated by applying a seepage boundary on a zone where the screen of the dewatering well is located [2]. The FE modeling is conducted according to the PED process reported in literature [3], i.e., let all the dewatering wells operate simultaneously and the dewatering time is 21 days. Figure 2 shows the comparison between the calculated results of PED with different dewatering depth (H_d) with the results under the same condition in literature [3]. It can be seen that

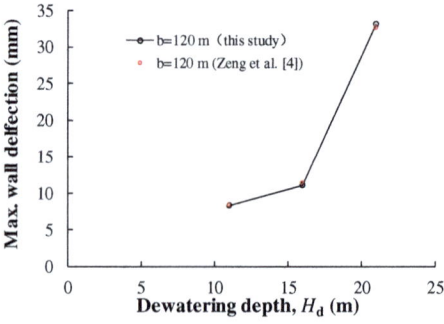

Fig. 2. Calculated wall deflections and their comparisons with results in literature [3].

the calculated maximum wall deflections are close to the results reported by Zeng et al. [3]. Besides, the results from Fig. 2 indicate that with H_d increased, the maximum wall deflection is much greater. When an unacceptable deformation occurs, countermeasures should be adopted.

3   Effect of Staggered Layout of Dewatering Wells

Figure 3 shows the comparison between the conventional layout and the staggered layout of dewatering wells. In the proposed method, it is proposed that deeper wells and smaller spacings of neighboring wells are employed near the retaining structure. In the following, 4 cases of PED with both the conventional and staggered layout of dewatering wells are simulated through the FE model established above. The dewatering time is 21 days. Table 1 presents the basic information about the 4 cases.

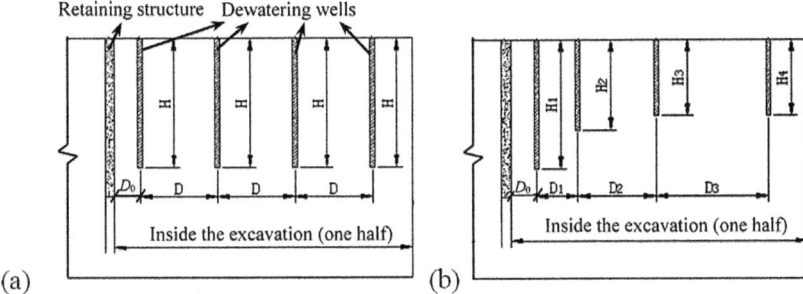

(a)   (b)

Fig. 3. Layout of dewatering wells: (a) conventional layout; and (b) staggered layout.

Table 1. Basic information about calculation cases.

Case No.	Depth, H_i (m)	Spacing, D_i (m)	Note
1	$H_1 = H_2 = H_3 = H_4 = 21$	$D_1 = D_2 = D_3 = 15.7$ m	Conventional layout
2	$H_1 = 23; H_2 = 20; H_3 = H_4 = 19$	$D_1 = D_2 = D_3 = 15.7$ m	Staggered Layout 1
3	$H_1 = H_2 = H_3 = H_4 = H_5 = 21$	$D_0 = 1; D_1 = 5; D_2 = 9; D_3 = 14; D_4 = 20$	Staggered Layout 2
4	$H_1 = 23; H_2 = 20; H_3 = H_4 = H_5 = 19$	$D_0 = 1; D_1 = 5; D_2 = 9; D_3 = 14; D_4 = 20$	Staggered Layout 3

Figure 4 shows the distribution of pore-water pressure (p_w) at excavated side of the wall after PED under Cases 1–4. p_w distribution in the four cases is basically the same, and the drawdown can all reach approximately 20 m, which indicates that the dewatering intensity in all the cases are identical.

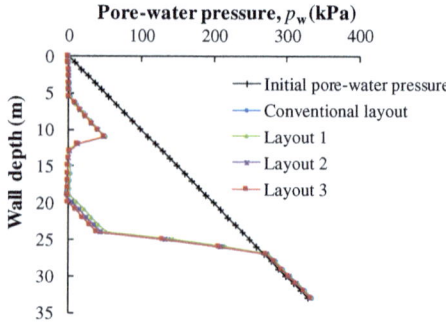

Fig. 4. p_w distribution at excavated side of the wall after PED under different cases.

Figure 5 shows the PED-induced wall deflection (δ_w) under different cases. Apparently, δ_w in the cases with staggered layout of dewatering wells is smaller than that in the conventional layout of wells. By using the Layout 3, the maximum δ_w can be reduced from 32 mm to 18 mm, and the reduction extent reaches 44%. It should be noted that the spacing and depth of dewatering wells can be further adjusted according to the proposed method to achieve a more effective reduction of PED-induced wall deflection. Besides, the adjustment of spacing between wells is more effective compared to the adjustment of well depth.

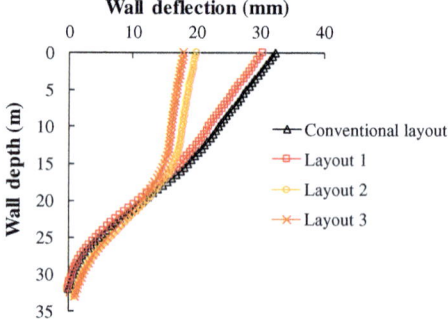

Fig. 5. PED-induced wall deflection under different cases

4   Concluding Remarks

(1) A staggered layout of dewatering wells is proposed as a countermeasure to the PED-induced wall deflection. In this method, the deeper wells and smaller spacings of neighboring wells are employed near the retaining structure. The numerical simulation shows that this method can effectively limit the PED-induced wall deflection.

(2) The spacing and depth of dewatering wells can be further adjusted according to the proposed method to achieve a more effective reduction of PED-induced wall de-flection.

Besides, the adjustment of spacing between wells is more effective compared to the adjustment of well depth.

Acknowledgements. The research is funded by the National Natural Science Foundation of China under Grant No. 51708206 and 11602083, and the Natural Science Foundation of Hunan Province under Grant No. 2016JJ6044. These financial supports are gratefully acknowledged.

References

1. Zeng, C.F., Xue, X.L., Zheng, G., Xue, T.Y., Mei, G.X.: Responses of retaining wall and surrounding ground to pre-excavation dewatering in an alternated multi-aquifer-aquitard system. J. Hydrol. **559**, 609–626 (2018)
2. Zheng, G., Zeng, C.F., Diao, Y., Xue, X.L.: Test and numerical research on wall deflections induced by pre-excavation dewatering. Comput. Geotech. **62**, 44–56 (2014)
3. Zeng, C.F., Zheng, G., Xue, X.L.: Wall deflection induced by pre-excavation dewatering in large-scale excavations. Chin. J. Geotech. Eng. **39**(6), 1012–1021 (2017)
4. Zeng, C.F., Xue, X.L., Zheng, G.: Effect of soil permeability on wall deflection during pre-excavation dewatering in soft ground. Rock and Soil Mech. **38**(10), 3039–3047 (2017)

Experimental Study of Temperature Effect on Binding Properties of Resin Anchors

Zha Wenhua[1,2,3(✉)] and Liu Zaobao[3]

[1] Key Laboratory of Safety and High-efficiency Coal Mining,
Anhui University of Science and Technology, Huainan 232001, China
whzha@126.com
[2] School of Mining and Safety Engineering,
Anhui University of Science and Technology, Huainan 232001, China
[3] Laboratory of Mechanics of Lille, University of Lille, 59650 Villeneuve d'Ascq, France

Abstract. This study analyzes the temperature effect on the anchoring behavior of the resin anchoring agent and bolt. Fully resin anchoring specimens consisting of surrounding rock-resin anchoring agent-bolt were designed. Through push-out experiment of specimens at 25 °C, 30 °C, 40 °C, 45 °C and 55 °C gradient temperature of, variation laws of anchor force and anchoring interfacial shear stress response rules along the anchor length direction were obtained. Results showed that, with temperature increase, maximum anchoring force and the average anchoring interface shear stress became smaller, Moreover, the failure mode of the anchoring interface gradually changed from brittle failure to plastic failure similar to rock. Under same temperature, the shear stress shows obvious non-uniform distribution, and the shear stress gradually transferred away from the load side. With the distance from load side increased, the shear stress became smaller. The interfacial shear stress tends to power function distribution with the increase of temperature.

Keywords: Resin anchoring interface · Temperature · Anchoring force
Shear stress distribution

1 Introduction

Rallying bolt support is widely used in mine roadway support. The load is transferred by the two interface shear stress between the anchor body and the resin anchoring agent and between the anchoring agent and the surrounding rock. The two interface shear stress distribution directly affects Anchor anchoring effect. The force transmission mechanism is very complicated and has many influencing factors. The mechanical transmission of rock bolts involves the properties of rocks, anchoring agents and reinforcing steel materials as well as the interaction of materials in the process of stress. The force transmission mechanism is very complicated [1–4] and has many influencing factors [5–7]. With the occurrence of high-temperature thermal damage in mines, Anchor anchoring performance will be affected.

© Springer Nature Switzerland AG 2018
W. Wu and H.-S. Yu (Eds.): *Proceedings of China-Europe Conference on Geotechnical Engineering*, SSGG, pp. 1090–1093, 2018.
https://doi.org/10.1007/978-3-319-97115-5_45

At present, it is mainly designed as a steel pipe anchoring system, which uses the indoor pull-out test or push-out test to analyze the distribution characteristics of shear stress at the anchoring interface [8–10]. In this paper, The distribution of the sliding displacement and shear stress at the inter-face between anchor and anchoring agent is analyzed under the influence of temperature, which will provide the basis for determining the relevant parameters of theoretical analysis and numerical simulation.

2   Experimental Method and Material

With RMT-150B test system, push-out test of anchor system is carried out. Using the GD-65/150 high and low temperature equipment to control the temperature of the anchor specimen.

The precast mixed surrounding rock were made from crushed mudstone at a water-cement ratio of 1:4 and a cement-to-mudstone ratio of 1:1. The mudstone is taken from the 1000 m deep 11501 mining face in Zhuji West Mine. Bolt is a Φ22 L-longitudinal non-longitudinal rebar with a length of 300 mm, of which 250 mm length was used for full-length anchoring. The bolt grooved on both sides, and paste 6 strain gauges on each side symmetrically. The anchoring material is a mixture of epoxy resin and curing agent, which ratio is 2:1.

3   Results and Discussion

3.1   Relationship Between Load and Displacement Slip Under Temperature

Distribution curves of the push-out load and displacement under temperature of 25 °C, 30 °C, 40 °C, 45 °C and 55 °C are shown in Fig. 1. The curve showed two stages of ascent and descent. Before reaching the peak intensity, the curves showed similar changes as the temperature increased. However, after the peak strength, the anchoring strength decreased rapidly before 45 °C and behaved action similar to rock brittle failure. After 45 °C, The reduction of anchoring strength decreases gradually and behaved action similar to rock plastic failure.

Take the average bond stress as the interfacial bond strength. Curve of peak push-out load, the interfacial bond strength, and the displacement corresponding to the peak intensity for various temperature as shown in Fig. 2. The anchoring strength gradually decreases with the increase of temperature. After 45 °C, the rate of load change increased from a liner 0.625 kN/°C to 0.92kN/°C. The above results show that under test conditions, when the temperature is higher than 45 °C, the ability of the anchor strength to resist load decreases, the anchor is easy to failure. Therefore, in anchoring project under high temperature environment, should pay attention to the selection of the anchor material and avoid the temperature effect of anchoring performance.

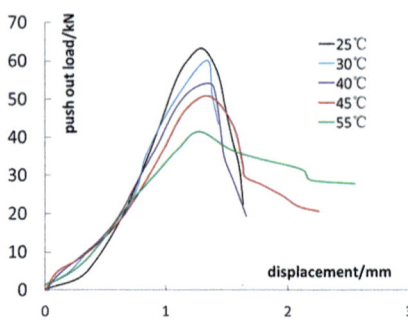
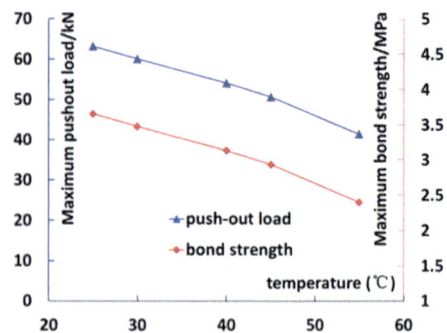

Fig. 1. Load–displacement curves for various temperature.

Fig. 2. Push-out load and maximum bond strength varies with temperature.

3.2 Influence of Temperature on Interfacial Shear Stress Transfer

The relationship between strain on the right and time under 25 °C temperatures as shown in Fig. 3, which shows that along the length of the anchorage, the shear stress response has a certain time difference and transfers from the load side to the interior along the anchorage length. After the strain value of R1 near the load side reaches the maximum, the other test points reach the peak strain until the anchor fails.

Variation of the peak strain along the anchorage length at 25 °C, 30 °C, 40 °C as Shown in Fig. 4, which shows that strain at both sides is not consistent at 25 °C. Under different temperature, stress concentration is formed on the load side, and the strain value is the largest. Then along the anchorage length, the shear stress distribution around the bolt is not uniform. At 25 °C, the logarithmic function fitted with the largest R-squared value of 0.9125 and 0.7542 respectively. At 30 °C and 40 °C, the power function fitted with the largest R-squared value is of 0.9333 and 0.9232 respectively. The interfacial shear stress is mainly logarithmic and power function distribution, and tends to power function distribution with the increase of temperature.

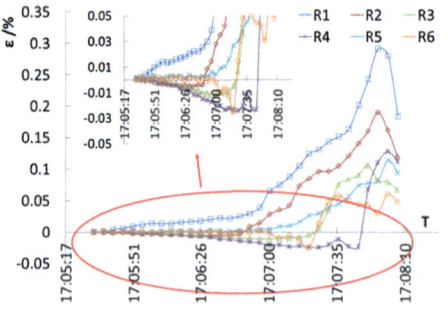
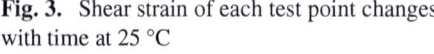
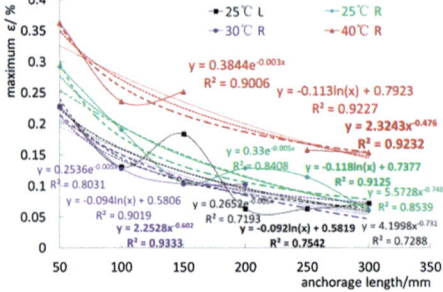

Fig. 3. Shear strain of each test point changes with time at 25 °C

Fig. 4. Variation of interfacial maximum shear strain at different temperatures

4   Conclusions

As the temperature increases, the bond strength between the anchor rod and the anchoring agent decreases. At high temperature, the influence of temperature on the anchorage performance should be emphasized. With the temperature increase, the distribution of shear stress changes from logarithmic to power function distribution.

The anchoring performance of resin anchors is affected by many factors, such as surrounding pressure, deformation of surrounding rock, mining disturbance and anchoring material properties, which are the research contents to be carried out in the future.

Acknowledgements. This work was partially supported by the Natural Science Foundation of China (NSFC) through the grant number 51474005, 51504006, and by the Key projects of Domestic and foreign visits to study for Outstanding young talents in universities through the grant number gxfxZD2016074.

References

1. Phillips, S.: Factors Affecting the design of anchorages in rock. Cementation Research Ltd., London (1970)
2. Freeman, T.J.: The behaviour of fully-bonded rock bolts in the Kielder experimental tunnel. Tunn. Tunn. **6**(10), 37–40 (1978)
3. Zou, J., Li, L., Yang, X., Wang, Z., Zhao, L.: Study on load transfer mechanism for span-type anchor based on the damage theory. J. China Railw. Soc. **12**(29), 84–88 (2007)
4. Deb, D., Das, K.C.: Modelling of fully grouted rock bolt based on enriched finite element method. Int. J. Rock Mech. Min. Sci. **2**(48), 283–293 (2011)
5. K111c, A., Yasar, E., Atis, C.D.: Effect of bar shape on the pullout capability of fully-grouted rock bolts. Tunn. Undergr. Space Technol. **1**(18), 1–6 (2003)
6. Hu, B.: Study on Mechanical Properties of Resin Capsules for Full-length Pre-stressed Anchor Bolts. China Coal Research Institute CCRI, Beijing (2011)
7. Fan, J.Q., Dong, H.X., Gao, Y.H., Lou, M.L.: Experimental study of shear stress distribution in internal anchoring section of a full-length grouting anchor. J. Exp. Mech. **29**(2), 250–256 (2014)
8. Li, F.H., Quan, X.H., Jia, Y., Wang, B., Zhang, G.B., Chen, S.Y.: The experimental study of the temperature effect on the interfacial properties of fully grouted rock bolt. Appl. Sci. **7**(4), 327–337 (2017)
9. Mark, C., Compton, C.S., Dolinar, D.R., Oyler, D.C.: Field performance testing of fully grouted roof bolts. SME Annu. Meet. and Exhib. **2**, 24–26 (2003)
10. Karanam, U., Rao, M., Dasyapu, S.K.: Experimental and numerical investigations of stresses in a fully grouted rock bolts. Geotech. Geol. Eng. **3**(23), 297–308 (2005)

An Innovative Bolt Fastener for Steel Tube Bracing in Deep Excavations

Mingju Zhang, Meng Yang[✉], Pengfei Li, and Dechun Lu

The Key Laboratory of Urban Security and Disaster Engineering,
Beijing University of Technology, Beijing 100124, China
yangmeng88@emails.bjut.edu.cn

Abstract. In order to improve the active node structure in steel tube bracing system, an innovative Bolt Fasten Wedge (BFW) active node is proposed in this study. Firstly, the BFW active node was researched on its working principle, function and structure by theoretical derivation and numerical simulation. Then, the elastic and ultimate bearing capacity of the BFW active node were obtained by laboratory loading tests. Finally, the bearing capacity and stiffness of the BFW active node were compared with those of many different types of commonly used steel tube bracings and the applicability of the BFW active node was evaluated. The results show that the BFW active node has good bearing capacity and strong stiffness. In conclusion, the BFW active node is of simple structure, reasonable stress, large bearing capacity and strong stiffness which meet the requirements of deformation and stability for braced excavations.

Keywords: Braced excavation · Steel tube bracing · Bolt Fasten Wedge
Active node · Laboratory loading tests

1 Introduction

In the subway construction, the steel tube inner supporting system is currently the preferred form because of its light weight, convenient construction, recycling and reusing for retaining structure with inner supports in excavation engineering. In the steel tube inner supporting system, the active node is one of the key components to control the stability and deformation of excavations [1, 2]. The commonly used active node is implemented by inserting steel wedges into a gap formed by welded steel plates (see Fig. 1).

This kind of node is easy to be operated in-situ but has many defects which does not satisfy the design principle of "nodes should be stronger than components". A number of scholars pointed out the problem by case study and theoretical analysis as Zhang [3], Yang [4], Li [5], Chowdhury [6], etc. Based on that, a series of new structural forms based on different principles have been proposed, just like beam string type [7], screw bar type, hydraulic pressure type, hoop type [8], etc. These inventions have not been tested, calculated, or applied in practice. A new type of active node, which is simple in principle and easy to operate, needs to be developed, calculated and tested.

© Springer Nature Switzerland AG 2018
W. Wu and H.-S. Yu (Eds.): *Proceedings of China-Europe Conference
on Geotechnical Engineering*, SSGG, pp. 1094–1097, 2018.
https://doi.org/10.1007/978-3-319-97115-5_46

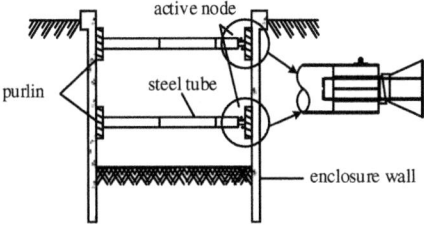

Fig. 1. The commonly used active node.

2   Proposed Bolt Fasten Wedge (BFW) Active Node

A Bolt Fasten Wedge (BFW) active node consists of 1 wedge seat with flange, 2 splints, 6 high strength bolts, 2 polish round rods and end plates, the full size model is shown in Fig. 2(a). In any state, we can fasten the bolts so as to adjust the length, the different adjustment statuses are shown in Fig. 2(b–d).

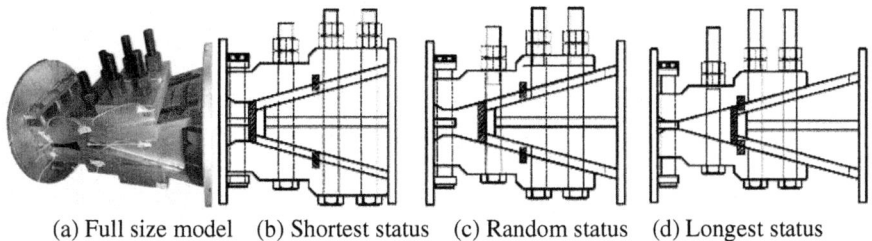

(a) Full size model   (b) Shortest status   (c) Random status   (d) Longest status

Fig. 2. The BFW active node model and its different adjustment status.

The structural design principle of the BFW active node is decomposing the force under static equilibrium, then the multiple rows of high strength bolts clamping the wedge to bear the axial force.

3   Mechanical Analysis

3.1   Calculations for Bearing Capacity and Stiffness

The design bearing capacity and stiffness of the BFW active node were calculated respectively.

Firstly, the design parameters were determined, like the angle of the contact surface of the splint, the number of bolts, the size of the node, etc.

Then, the force relationship was calculated by a series of formulas, like the key formulas: $N \leq N_{bt}2n \tan(15° + \delta)/[1 - \mu \tan(15° + \delta)]$, $K_d = F/\Delta L = EA_d/h$.

Finally, with the calculation, the design bearing capacity is 4490 kN and the stiffness is 2442 kN/mm.

3.2   Numerical Simulation

The finite element model was established by using ABAQUS and simulated under the axial and eccentric loads respectively. The load-displacement curves are shown in Fig. 3. Eccentricity load is divided into 3 cases: along the bolts' axis (X-X), vertical the bolts' axis (Y-Y), between the two directions (45°). Under the eccentric load, the deformation of near load is different from that of far load, so 2 curves are used to show that.

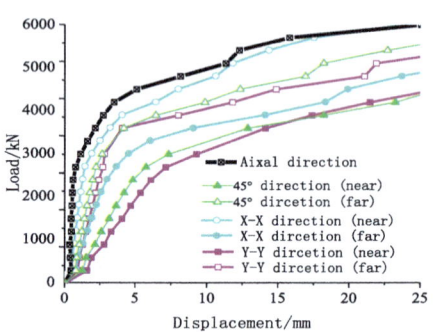

Fig. 3.   The load-displacement curves through FEA.

The comparison of simulation results shows that under the eccentric compression, the bearing capacity does not weaken obviously but the stiffness is significantly reduced. The order of eccentric stiffness is X-X direction, biaxial eccentric direction, Y-Y direction.

3.3   Elastic Performance and Ultimate Bearing Capacity

3 full size BFW active node test specimens were processed to carry out tests under different operating conditions. The elastic test was followed by the destructive test.

The adjustable length of the elastic test is set to 3 statuses (50/100/150 mm). Figure 4(a) shows the results of absolute displacement (solid) and relative displacement (hollow) under the axial direction load, the stiffness is 2002 kN/mm.

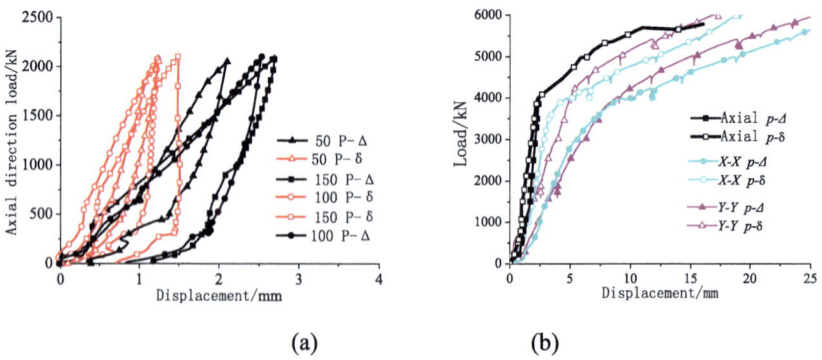

(a)   (b)

Fig. 4.   Results of elastic loading tests and destructive compression tests.

The adjustable length of the destructive test is set to the most unfavorable condition (150 mm). The specimens were continuously loaded slowly and uniformly until they were destroyed. Figure 4 (b) shows the results of absolute displacement (solid) and relative displacement (hollow) under the axial and eccentric direction loads, the ultimate bearing capacity is 5865kN. Applicability Analysis.

The ultimate bearing capacity and stiffness of steel tube bracings of different types and spans were calculated. The ultimate bearing capacity and the stiffness have been obtained through the mechanical property test. By comparison, it is found that the BFW active node is superior to all Φ609 and Φ630 steel bracings in both bearing capacity and stiffness. Part of the large span Φ800 steel bracings need a stronger BFW active node to fit.

4   Conclusion

The BFW active node was developed based on the inadequacy of the present. The bearing capacity and stiffness are excellent, the local stress concentration appears, and the bolts are the weak position, through calculation, analysis, numerical simulation.

Through elastic test and destructive test, the adjustable length is larger, the property is weaker. Under the axial compression, the property is better.

Both the bearing capacity and stiffness of the BFW active node are larger than many steel tube bracings which meets the design principle of "nodes should be stronger than components".

References

1. Jovaševića, S., Correiaab, J.A.F.O., Pavlović, M.: Global fatigue life modelling of steel half-pipes bolted connections. Procedia Eng. **160**(8), 278–284 (2016)
2. Hao, J., Sun, X., Xue, Q.: Research and applications of prefabricated steel structure building systems. Eng. Mech. **34**(1), 1–13 (2017)
3. Zhang, K., Li, J.: Accident analysis for "08.11.15" foundation pit collapse of Xianghu station of Hangzhou metro. Chinese. J. Geotech. Eng. **32**(S1), 338–342 (2010)
4. Yang, X.: Several issues in design, construction and monitoring of foundation pits. Chin. J. Rock Mech. Eng. **31**(11), 2327–2333 (2012)
5. Li, H., Wang, G.: Causes and suggestion on deep foundation excavation accident in some metro station. Construction Technology **39**(03), 56–82+62 (2010)
6. Chowdhury, S.S., Deb, K., Sengupta, A.: Estimation of design parameters for braced excavation: numerical study. Int. J. Geomech. **13**(3), 234–247 (2012)
7. Park, J.-S., Joo, Y.-S., Kim, N.-K.: New earth retention system with prestressed wales in an urban excavation. J. Geotech. Geoenviron. Eng. **135**(11), 1596–1604 (2009)
8. Zhang M., Li J., Yuan Y.: Study on mechanical properties of the wedge-disconnectable coupling structure in internal support system of excavation engineering. In: Proceedings of the IACGE International Symposium on Geotechnical and Earthquake (IACGE 2016), pp. 223–230. IACGE, Beijing, China (2016)

Seepage Stability Calculation of Excavation Base due to Groundwater Level Fluctuation

Lisha Zhang[1,2,3] (ID), Hongwei Ying[2,3(✉)] (ID), Di Wang[2,3] (ID),
Kanghe Xie[2,3] (ID), and Cheng-wei Zhu[2,3] (ID)

[1] Department of Civil Engineering, Zhejiang University City College, Hangzhou, China
[2] Research Center of Coastal and Urban Geotechnical Engineering,
Zhejiang University, Hangzhou, China
ice898@zju.edu.cn
[3] MOE Key Laboratory of Soft Soils and Geoenvironmental Engineering,
Zhejiang University, Hangzhou, China

Abstract. Based on leaky theory, the analytical solution of exit gradient in the excavation base near the retaining structure was derived due to groundwater fluctuation, and a simplified method was proposed to calculate seepage stability of the excavation base considering unsteady seepage. The validity of analytical solution was verified and the influential factors of exit gradient in the excavation base were analyzed based on idealized case studies. The results showed that the influence factor of exit gradient was described by the dimensionless factor, which was positively correlated with the coefficient of permeability and the constrained modulus of the soil, but was negatively correlated with the angular frequency of groundwater level variation and the squared of total thickness of the fine-grained soil in the analytical model. The results also imply that the exit gradient variation was asynchronous with the groundwater level variation when the permeability is low, such situation should be avoided that the exit gradient did not reduce timely and efficiently through dewatering process to prevent seepage damage of the excavation project.

Keywords: Seepage stability of the excavation base
Groundwater level fluctuation · Unsteady seepage · Exit gradient

1 Introduction

According to Chinese Codes [1, 2] and Handbooks [3, 4], seepage stability of the excavation base was always analyzed under steady seepage condition. However, the excavations are under unsteady seepage, when the groundwater level is varied due to raining, pumping and dewatering, or water level variation of the lakes and rivers nearby. The deformation and stability of the excavation unsteady seepage were conducted by numerical methods [5, 6], rarely the dealing with seepage stability of the excavation base under unsteady seepage was proposed.

This paper introduced a simplified method for seepage stability calculation of the excavation base, considering the influence of unsteady seepage.

© Springer Nature Switzerland AG 2018
W. Wu and H.-S. Yu (Eds.): *Proceedings of China-Europe Conference on Geotechnical Engineering*, SSGG, pp. 1098–1101, 2018.
https://doi.org/10.1007/978-3-319-97115-5_47

2   Analytical Solution

The passive zone is assumed to be a hypothetical passive zone under the active zone [7] to describe the seepage flow along the retaining structure, as shown in Fig. 1, for the continuous see page at the bottom of the retaining structures. Then the unsteady seepage induced by the variation of groundwater lever along the retaining structure can be simplified into one dimensional problem based under assumptions that: (1) the foundation pit stretches along the coast with large size; (2) the construction duration of the final excavation phase is relatively longer than other typical engineering projects; (3) the penetration of the retaining structure is over the excavation depth.

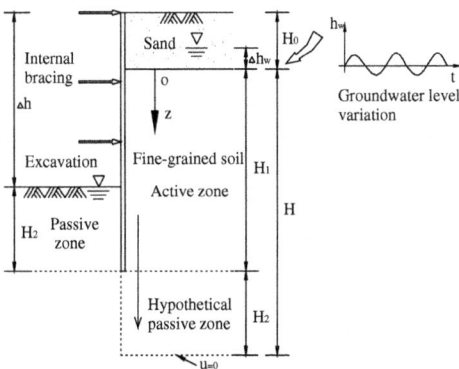

Fig. 1. Simplified calculation model for seepage stability of the excavation base due to groundwater level variation.

For the typical coastal ground, a highly permeable stratum (i.e., sandy layer) overlays an fine-grained soil layer. Only a slight difference exists between the natural unit weight and the saturated unit weight of sand, the total stress in fine-grained soils generally varies slightly as the groundwater fluctuates in the sandy layer, and can be regarded as constant. Therefore, variations in the groundwater table in the top sandy layer can be considered equivalent to the variations in the excess pore water pressure at the upper boundary of the fine-grained soil layers, which results in a change in the pore pressure. Moreover, the evolution of excess pore pressure with time in the fine-grained layer generally does not reflect changes that occur at the boundary.

Accordingly, the governing equation and other solution conditions used to obtain the excess pore water pressures in fine-grained soil model can be characterized as follows:

$$c_{vs}\partial^2 u / \partial z^2 = \partial u / \partial t \tag{1}$$

$$u\big|_{z=0} = \gamma_w \cdot f(t) \tag{2}$$

$$u\big|_{z=H} = 0 \tag{3}$$

$$u|_{t=0} = 0 \tag{4}$$

where u is the excess pore pressure; z and t are the spatial coordinate and time, respectively; c_{vs} is the coefficients of the swelling or consolidation, and is generally accepted to be constant during this process [8, 9]; γ_w is the unit weight of water; H is the total thickness of the fine-grained soil model.

The analytical solution of $u(z, t)$ in Eq. (1) can be derived by Fourier technique and Duhamel theorem, and the answers are in Conte and Troncone [9]'s and Ying et al. [10]'s study. Based on the definition of hydraulic gradient [11], analytical solution of exit gradient i_e in the excavation base near the retaining structure due to groundwater fluctuation was derived:

$$i_e = \triangle\, i_e + i_{e0} = -1/\gamma_w \cdot \partial u(z,t)/\partial z|_{z=H} + \left(\triangle\, h_w + h_1 - h_2\right)/H \tag{5}$$

where $\triangle\, i_e$ is the unsteady seepage component induced by groundwater variation, and i_{e0} is the steady seepage component; $\triangle\, h_w$ is the distance between the steady groundwater level and the top the fine-grained soil model; h_1 is the thickness of the active zone, and h_2 is the thickness of the passive zone.

Consequently, the seepage stability can be analyzed by the following equation:

$$Ki_e \leq i_{cr} \tag{6}$$

where K is the safety factor; i_{cr} is the critical exit gradient.

3   Discussion

3.1   Verification for the Analytical Solution of Exist Gradient

The verification result in Fig. 2 shows that if the coefficient of permeability of the excavation soil was large enough, the analytical solution results of exit gradient considering unsteady seepage were in agreement with the traditional results under the steady seepage assumption, which verified the validity of the analytical solution.

3.2   The Influence Factors of Exist Gradient

The dimensionless influence factor of exit gradient θ is defined by $\theta = c_{vs}/\omega H^2$, which was positively correlated with the coefficient of permeability and the constrained modulus of the soil, but was negatively correlated with the angular frequency of groundwater level variation and the squared of total thickness of the fine-grained soil in analytical model. Figure 3 shows that the curves of exist gradient i_e changes with dimensionless time t/T, and such change is influenced by the dimensionless factor θ significantly. An increase in the dimensionless parameter θ leads to a decrease in amplitude attenuation and phase shift of the exit gradient. The results also implies that the exit gradient variation was asynchronous with the groundwater level variation when the permeability is low, such situation should be avoided that the exit gradient did not reduce timely and efficiently through dewatering process to prevent seepage damage of the excavation project.

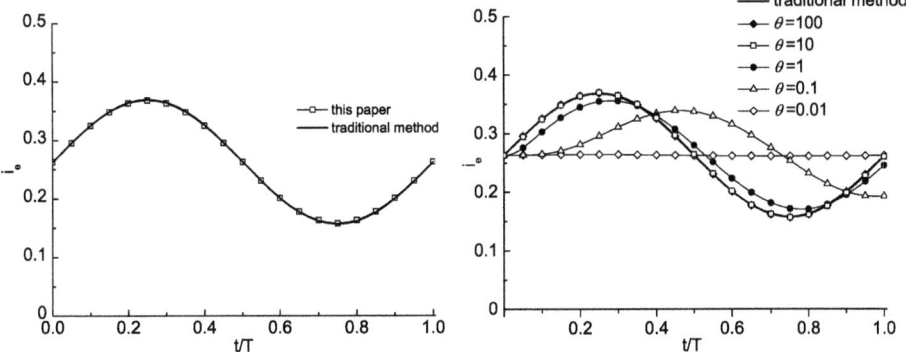

Fig. 2. Verification of analytical solution for exit gradient.

Fig. 3. Variation curves of exit gradient against time with different θ.

References

1. The Professional Standards Compilation Group of People's Republic of China: GB50007-2011 Code for design of building foundation. China Architecture and Building Press, Beijing (2012)
2. The Professional Standards Compilation Group of Shanghai: DG/TJ08-61-2010 Technical code for excavation engineering. [s. n.], Shanghai (2010)
3. Gong, X.N., Gao, Y.C.: Deep Excavation Engineering Design and Construction Handbook. China Architecture and Building Press, Beijing (1998)
4. Liu, G.B., Wang, W.D.: Excavation Engineering Handbook. China Architecture and Building Press, Beijing (2010)
5. Liu, J.: The deformation characteristics analysis of deep foundation pit while dewatering and excavation considered simultaneously. Shanghai Jiao Tong University, Shanghai (2010)
6. Zhang, Q.W.: Stable analysis of foundation pit retaining structure under unsteady seepage. Zhengzhou University, Zhengzhou (2007)
7. Li, Y.Q., Ying, H.W., Xie, K.H.: On the dissipation of negative excess pore water pressure induced by excavation in soft soil. J. Zhejiang Univ. (Eng. Sci.) **6A**(3), 188–193 (2005)
8. Leroueil, S.: Natural slopes and cuts: movement and failure mechanisms. 39th Rankine Lecture. Géotechnique **51**(3), 197–243 (2001)
9. Conte, E., Troncone, A.: Soil layer response to pore pressure variations at the boundary. Géotechnique **58**(1), 37–44 (2008)
10. Ying, H.W., Zhang, L.S., Xie, K.H., et al.: Pore and earth pressure response to groundwater fluctuation out of foundation pit. J. Zhejiang Univ. (Eng. Sci.) **48**(3), 492–497 (2014). (in Chinese)
11. Terzaghi, K., Peck, R.B., Mesri, G.: Soil mechanics in engineering practice. Wiley, Canada (1996)

Design Charts and Simplified Approach for the Bearing Capacity of Strip Footings on Sand Overlying Clay

Haizuo Zhou[1,2,3], Gang Zheng[1,2,3(✉)], and Jiapeng Zhao[1,2]

[1] School of Civil Engineering, Tianjin University, Tianjin 300072, China
zhenggang1967@163.com
[2] Key Laboratory of Coast Civil Structure Safety, Ministry of Education,
Tianjin University, Tianjin 300072, China
[3] State Key Laboratory of Hydraulic Engineering Simulation and Safety,
Tianjin University, Tianjin 300072, China

Abstract. Native soils are often deposited in layers. The evaluation of ultimate bearing capacity of shallow foundations on sand overlying clay has been relevant topic. This problem is complex as it is highly dependent on geometric and geotechnical properties of a given soil. Moreover, the conventional methods provide oversimplified approaches, yielding unreliable results. In this study, the bearing capacity and failure mechanism of footings placed on sand overlying clay are evaluated using discontinuity layout optimization (DLO). By introducing a reduction coefficient, a set of design charts that can be directly applied to the classical bearing capacity formulation is presented. A regression-based approach, relevant to a wide range of geometric and strength parameters, is then developed to accurately capture the bearing capacity. The model uncertainty for the proposed method is investigated with the artificial data and results from literatures. The model bias is characterized as a lognormal random variable with a mean of 1.126 and a coefficient of variation (COV) of 0.186.

Keywords: Sand · Clay · Bearing capacity · Simplified approach

1 Introduction

Establishing the ultimate bearing capacity of shallow foundations has long been an important component of geotechnical engineering. The load-spread method [1] and the punching-shear method [2] are the two most common analytical design models for shallow foundations on sand overlying clay. The load-spread method assumes that the footing load develops at a certain dispersion angle α_p from sand layer to underlying clay stratum, and the shear resistance of the sand is neglected. The value of α_p is a determinant factor, but it does not have a precise physical interpretation. The punching-shear method provides a useful insight into the behavior of sand. It is assumed that the failure slip comprising a vertically sided block beneath the footing that punched into the clay, and the punching-shear coefficient (K_s) is determined by a chart or an interpolation method.

© Springer Nature Switzerland AG 2018
W. Wu and H.-S. Yu (Eds.): *Proceedings of China-Europe Conference on Geotechnical Engineering*, SSGG, pp. 1102–1105, 2018.
https://doi.org/10.1007/978-3-319-97115-5_48

For the load-spread method, an underestimation of bearing capacity is observed for large sand thickness as the shearing resistance of sand is neglected. The punching-shear method gives a conservative solution for a footing without embedment, whereas an overestimation is found for large sand thickness [3]. Therefore, both methods provide oversimplified approaches. Tang et al. [4] proposed a regression equation to modify the load-spread method and the punching shear method in calculating the bearing capacity associated with onshore foundations. A database of centrifuge tests for the peak resistance of flat-footing and spudcan footing was used for verification.

The design chart [5] and the empirical method [6] can provide efficient tools for practitioners to determine the bearing capacity of foundations when Terzaghi's theory cannot adopted. The objective of this investigation is to present direct solutions for bearing capacity of sand overlying clay, particularly focuses on the strip footing. The accuracy of the developed approach is verified with artificial data and results in literatures.

2   Method and Results

A rigorous analysis in the framework of upper-bound limit state plasticity, known as discontinuity layout optimization (DLO), capable of accurately estimating the ultimate load of strip footings, is adopted in this study. As a highly efficient tool for directly obtaining the ultimate load and critical collapse mechanism of geotechnical stability issues, it works without assuming prescribed failure geometry, enabling pre-assuming a slip surface and determination of both a critical mechanism of collapse and limit load. A reduction coefficient R is introduced to quantify the influence of voids on the ultimate bearing capacity. R is expressed as

$$R = \frac{q_{u,v}}{q_u} = \frac{q_{u,v}}{c \cdot N_c + 0.5 \cdot \gamma \cdot B \cdot N_\gamma} \tag{1}$$

where $q_{u,v}$ is the bearing capacity obtained from DLO analysis, and q_u is the bearing capacity obtained from the classical bearing capacity solution for a single layer of homogeneous soil. Figure 1 illustrates the design charts of the bearing capacity ratio R for sand overlying clay. The R value increases with H/B as the sand contributes to larger bearing capacity. For cohesionless soil with $\varphi < 42°$, R finally increases up to 1.0, indicating the influence of the clay layer on the limit load disappears and the sand layer fails in a classical Prandtl-type failure mechanism. The R value cannot approach 1.0 when friction angles $\varphi > 48°$ because a larger frictional strength enlarges the slip surface. The minimum depth at which the failure surface develops completely within the sand layer and the clay layer does not contribute to the bearing capacity is denoted as a critical depth $(H/B)_{cri}$. This value increases with the increasing φ value, whereas it decreases with the soil cohesion $c/\gamma B$. Note that an increase in frictional strength leads to a large decrease in R value. Relative phenomena will be further elaborated with the failure mechanism.

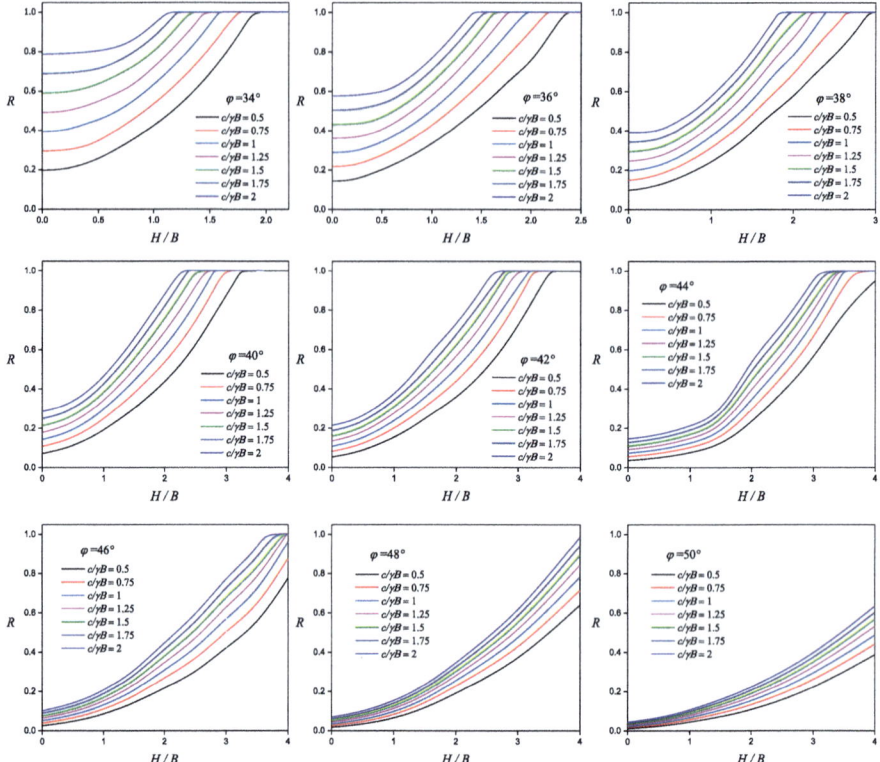

Fig. 1. Design charts of the bearing capacity ratio R for strip footings on sand overlying clay.

From the above design charts, the R shows a nonlinearly increasing relationship with the sand thickness, which is described by a power function; the intercept R_0 representing the bearing capacity ratio between sand and clay layers. A simplified model that relates R to the three variables (H/B; φ; and $c/\gamma B$) through regression analyses. Thus, a simple equation is developed:

$$R = \left(\alpha_1 X + \alpha_2 Y + \alpha_3\right) Z^{\left(\alpha_4 X^{(-1)} + \alpha_5\right)} + R_0 \tag{2}$$

where $X = \tan\varphi$; $Y = c/\gamma B$; $Z = H/B$; $R_0 = \dfrac{cN_c}{0.5\gamma BN_\gamma}$; and the value of each coefficient for Eq. (2) determined using MATLAB through the least-square regression as $\alpha_1 = -0.1357$; $\alpha_2 = 0.0111$; $\alpha_3 = 0.1692$; $\alpha_4 = 1.2210$; $\alpha_5 = 0.6953$. The proposed approach is first validated with the results in the design charts. The coefficient of determination R^2 is 0.986, and all the estimated values are within $\pm20\%$ of the 1:1 line. Further validation of the developed simplified model is carried out using a total of 81 data in previous literatures. The mean trend line is close to the 45° line, and the R^2 is found to be 0.922, as shown in Fig. 2. The p value from the Kolmogorov-Smirnov goodness-of-fit test (KS test) is 0.907, which is much higher than 0.05, demonstrating the lognormal distribution for BF appears to be reasonable. The mean and coefficient of variation (COV) of *BF* are 1.126 and 0.186, respectively. In addition, the standard deviation (SD) is 0.21, as shown in Fig. 3. The comparison demonstrates the accuracy of the proposed model.

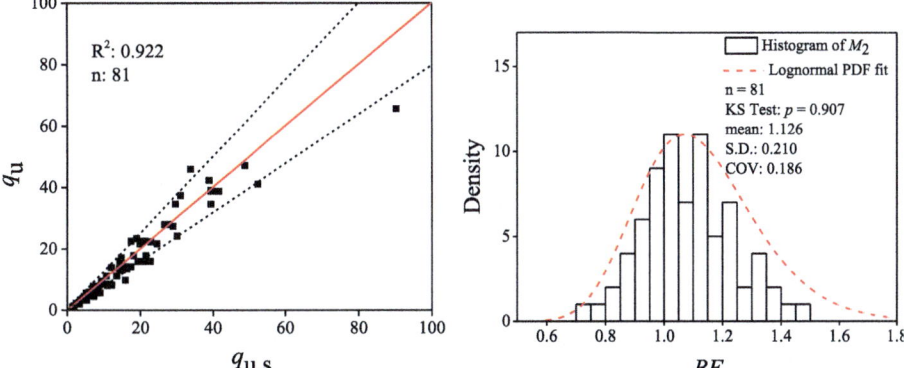

Fig. 2. Performance of simplified approach   **Fig. 3.** Histogram of model bias factor (BF)

3   Conclusion

In this paper, a set of reduction coefficients R for strip footings placed on dense sand overlying clay is presented as design charts. The critical failure mechanism is further discussed to elaborate the relative phenomena. A simplified semi-empirical model, relevant to a wide range of geometric and soil strength parameters, is developed to predict the ultimate load. Moreover, the model uncertainty of the proposed approach is investigated using the artificial data and the results of previous literatures. Verification demonstrates the developed model could be applicable for the evaluation of bearing capacity.

References

1. Yamaguchi, H.: Practical formula of bearing value for two layered ground. In: Proceedings of the 2nd Asian Reginal Conference on Soil Mechanics and Foundation Engineering, Japanese Society of Soil Mechanics and Foundation Engineering, Tokyo, vol. 1, pp. 176–180 (1963)
2. Hanna, A.M., Meyerhof, G.G.: Design charts for ultimate bearing capacity of foundations on sand overlying soft clay. Can. Geotech. J. **17**(2), 300–303 (1980)
3. Okamura, M., Takemura, J., Kimura, T.: Centrifuge model tests on bearing capacity and deformation of sand layer overlying clay. Soils Found. **37**(1), 73–88 (1997)
4. Tang, C., Phoon, K.K., Zhang, L., Li, D.Q.: Model uncertainty for predicting the bearing capacity of sand overlying clay. Int. J. Geomech. **17**(7), 04017015 (2017)
5. Zhou, H., Zheng, G., Yin, X., Xu, X., Zhang, T., Yang, X.: Bearing capacity of strip footings on c–φ soils with square voids. Acta Geotech. **13**, 747–755 (2018)
6. Ahmadi, M.M., Mofarraj Kouchaki, B.: New and simple equations for ultimate bearing capacity of strip footings on two-layered clays: numerical study. Int. J. Geomech. **16**(4), 06015014 (2016)

Field Tests on Behavior of Pre-bored Grouted Planted Pile in Soft Soil Area with Existing Pile Foundation

Jiajin Zhou[1] ⓘ, Xiaonan Gong[1(✉)] ⓘ, and Rihong Zhang[2,3] ⓘ

[1] Zhejiang University, Hangzhou 310058, China
13906508026@163.com
[2] Ningbo University, Ningbo 315000, China
[3] ZCONE High-tech Pile Industry Holdings Co., Ltd., Ningbo 315000, China

Abstract. This paper presents a practical renewal project in soft soil area, and the site foundation should be improved owing to the increasing load from super-structure. The installation of "new" pile foundation was a great challenge due to the existing square piles in the foundation. A group of field tests were conducted to investigate the compressive bearing capacity of the PGP pile and bored pile, and the applicability of these two piles in soft soil area with existing pile foundation was also studied. The results show that the ultimate unit skin friction of the PGP pile is about 1.22 times the ultimate unit skin friction of the bored pile. The PGP pile was chosen as the foundation of this renewal project, and the existing square piles around were scarcely disturbed during the PGP pile installation process.

Keywords: Pre-bored grouted planted pile · Bored pile · Existing pile foundation
Field test

1 Introduction

The pre-bored grouted planted (PGP) pile is a new type of composite foundation recently developed in China mainly utilized in deep soft soil areas. Zhou et al. (2013, 2016, 2017) have investigate the load transfer mechanism of PGP pile in soft soil areas based on field and model tests. The specific installation process of the PGP pile is shown in Fig. 1. This figure shows that the installation process of the PGP pile can be mainly divided into the following four steps: (1) Drilling; (2) Enlarged base construction; (3) grouting along the pile hole; (4) Precast pile planting.

This paper presents a practical renewal project in soft soil area with plenty of existing square piles. A group of field tests of the PGP piles and bored piles were conducted to compare the applicability, the compressive bearing capacity and economical aspect of the two piles, moreover, the successful application of the PGP pile in this renewal project was also introduced.

© Springer Nature Switzerland AG 2018
W. Wu and H.-S. Yu (Eds.): *Proceedings of China-Europe Conference on Geotechnical Engineering*, SSGG, pp. 1106–1110, 2018.
https://doi.org/10.1007/978-3-319-97115-5_49

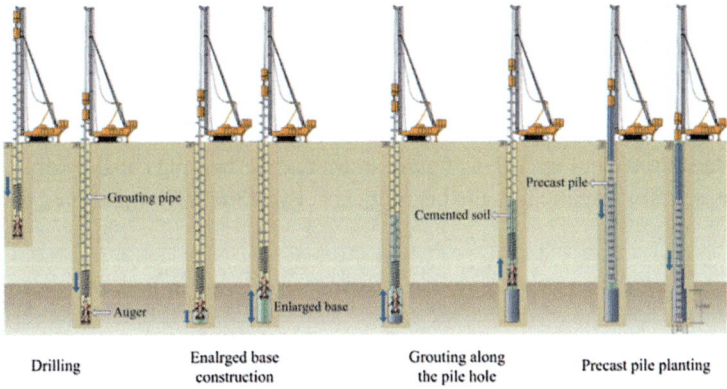

Fig. 1. PGP pile installation process.

2   Field Tests of PGP Pile and Bored Pile

The test piles were installed in the Fourth Stage Project of Wenzhou Power Plant. The Fourth Stage Project (mainly consists of 2×660 MW supercritical coal-fired units) was constructed just on the site of the First Phase (2×135 MW coal-fired units) for promoting the capacity of the Power Plant. The foundation of the power plant, as a result, should also be strengthened as the applied load from the superstructure increasing. The existing precast concrete square piles in the site brought significant difficulty for the installation of "new" piles.

Six test piles were installed in the test site, including four PGP piles and two bored piles. The four PGP piles could be divided into two categories: the 900 mm PGP pile (TP3 and TP4) and the 700 mm PGP pile (TP1 and TP2). Two 1000 mm diameter bored piles (TP5 and TP6) were also installed in the test site, and the six test piles were all 61 m long.

Fig. 2. PGP pile installation process.

The PGP pile installation process is shown in Fig. 2. This figure shows that the precast pile was planted into the pile hole with its own weight, and no additional load was needed in the precast pile planting process.

Static load tests were carried out according to the local Technical Code for Testing of Building Foundation Piles (CABR 2014). The load-displacement responses of the test piles are shown in Fig. 3. This figure shows that the bearing capacity of the 900 mm PGP pile is almost identical to the bearing capacity of the 1000 mm bored pile.

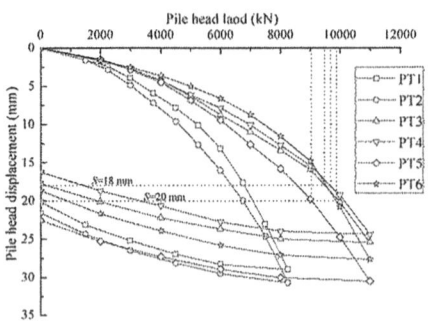

Fig. 3. Load-displacement responses of test piles.

The shaft and base resistances cannot be separated completely in the present study as lack of sensors along the pile shaft, nevertheless, the shaft resistance of the test pile can be approximately estimated. The shaft capacity is mobilized at much smaller displacements (typically 0.5 to 2.0% of the pile diameter) than the tip capacity (5 to 10% of the pile base diameter) (Fleming et al. 2009). The pile head displacement of the test pile needed to fully mobilize the shaft capacity is assumed to be 2.0%D, considering the large elastic shortening of the pile shaft. The calculated average skin friction of the PGP pile is about 1.22 times the skin friction of the bored pile, when pile head displacement reaches 2%D.

3   Comparison on Two Types of Piles

The compressive bearing capacity of the 900 mm PGP pile, according to the field test results, is similar to the bearing capacity of the 1000 mm bored pile. The materials of the two pile foundations needed for the Fourth Stage Project of Wenzhou Power Plant foundation are arranged as shown in Fig. 4. This figure shows that the materials needed for the PGP pile scheme are all less than that of the bored pile scheme. Moreover, the slurry emission in PGP pile installation process is largely reduced compared with that in bored pile installation process.

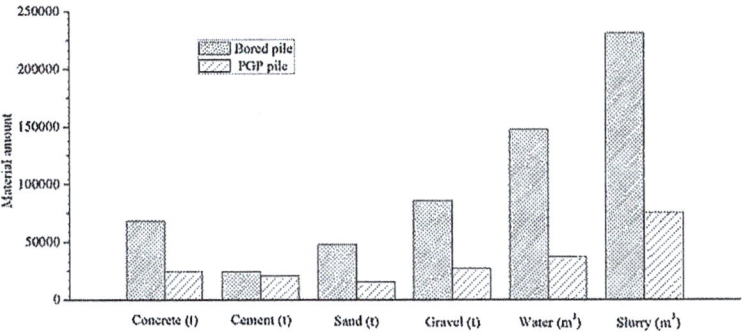

Fig. 4. Materials needed for two pile foundation schemes.

The PGP pile was selected to be the pile foundation of the Fourth Stage Project of Wenzhou Power Plant. The construction site after PGP pile installation is presented in Fig. 5. This figure shows that the PGP piles were located among the existing square piles. The lateral displacements of the square piles were measured during the PGP pile installation, and the measured lateral displacements are pretty limited, and finally, the square piles are all straight after the PGP pile installation. The behavior of the existing square pile was scarcely influenced by the installation of the PGP pile, based on the field test results before and after the installation of the PGP pile.

(a)Layout of PGP piles and existing square piles; (b)Sketch of PGP pile and existing square pile

Fig. 5. Scene of construction site.

References

CABR (China Academy of Building Research). Technical Code for Testing of Building Foundation Piles. JGJ106–2014. China Architecture and Building Press, Beijing (2014) (in Chinese)

Fleming, K., Weltman, A., Randolph, M., Elson, K.: Piling Engineering, 3rd edn. Taylor & Francis, London (2009)

Zhou, J.J., Wang, K.H., Gong, X.N., Zhang, R.H.: Bearing capacity and load transfer mechanism of a static drill rooted nodular pile in soft soil areas. J. Zhejiang Univ. Sci. A (Appl. Phys. Eng.) **14**(10), 705–719 (2013)

Zhou, J.J., Gong, X.N., Wang, K.H., Zhang, R.H., Yan, T.L.: A model test on the behavior of a static drill rooted nodular pile under compression. Marine Georesources Geotechnol. **34**(3), 293–301 (2016). 150429065916002

Zhou, J.J., Gong, X.N., Wang, K.H., Zhang, R.H., Yan, J.J.: Testing and modeling the behavior of pre-bored grouting planted piles under compression and tension. Acta Geotech. **12**, 1061–1075 (2017)

Effect of Vertical Shaft Resistance on the Lateral Behavior of Large-Diameter Pile Foundation

Ming-xing Zhu[1]([⊠]) [iD], Hong-qian Lu[1], Guo-liang Dai[2],
and Zhi-hui Wan[2]

[1] China Energy Engineering Group Jiangsu Power Design Institute,
Nanjing 211102, China
phd_mxingzhu19856l@vip.163.com
[2] Southeast University, Nanjing 210096, China

Abstract. Current research gradually recognizes that resisting moment M_s, induced by vertical shaft resistance developed on the passive side of pile shaft, has non-ignorable influence on the lateral bearing characterize of large-diameter pile embedded in stiff soil layers. This work firstly presents numerical solution for resisting moment which is suitable for any type of side friction models. Accordingly, a series of analytical expressions for resisting moment versus slope are established with hardening and softening τ-s curve models (i.e., shaft resistance). Furthermore, the comparison of case study indicates that the influence of shaft resisting moment cannot be ignored for large-diameter pile embedded in stiff soil material. Finally, parametric study is performed and results reveal that for hardening τ-s curve, resisting moment M_s will increase with increase of pile diameter, equivalent limit friction $\tau_{u,eq}$ and as decreasing critical displacement s_{eu}; for softening τ-s model, M_s will increase with increasing pile diameter, ratio of residual-critical displacement to peak displacement and ratio of residual shaft friction to maximum shaft friction.

Keywords: Pile foundation · Shaft resisting moment · Hardening τ-s model
Softening τ-s model

1 Introduction

A growing number of in-situ tests gradually shows that the lateral bearing capacity of testing results is larger than that of calculated results by means of conventional p-y method, and this difference will become more significant as increasing vertical shaft resistance. [4] defined this moment effect caused by vertical shaft resistance as resisting moment. [5] has established the fitting solutions of resisting moment for socketed pile, which is only suitable for soft rock. Based on the numerical simulations, [3] have proposed empirical formula of resisting moment for pile embedded in clay soil and pile responses by theoretical approach considering resisting moment induced by vertical shaft resistance match well with results of numerical simulations. However, proposed resisting moment model can only apply to clay soil.

© Springer Nature Switzerland AG 2018
W. Wu and H.-S. Yu (Eds.): *Proceedings of China-Europe Conference on Geotechnical Engineering*, SSGG, pp. 1111–1114, 2018.
https://doi.org/10.1007/978-3-319-97115-5_50

Based on the proposed numerical solution for resisting moment, this work aims to establish analytical models of resisting moment when vertical shaft resistance calculated by hardening and softening τ-s models, respectively. Furthermore, case study is performed to reveal the contribution of vertical skin friction to laterally loaded pile.

2 Solutions for Shaft Resisting Moment Per Unit Length

2.1 Numerical Solutions

Ashour et al. [1] assumed that vertical skin friction would develop on the front-side (i.e., the passive side) of the pile shaft. Thus, by dividing semi-circumference arc of an arbitrary section of shaft into $2n$ identical parts [5], the shaft resisting moment caused by vertical shaft resistance can be described as

$$M_s = \frac{\pi r^2}{n} \sum_{i=1}^{n} \left[\tau_i(\theta) \sin\left(\frac{(2i-1)\pi}{4n} \right) \right] \tag{1}$$

in which M_s/(kN·m/m) is the shaft resisting moment per unit length caused by vertical shaft resistance; r is pile radius; θ is the shaft section slope; $\tau_i(\theta)$ is the average friction of ith arc part (after dividing), which can be calculated by a given τ-s model and the relative displacement s_i ($=x_i \times \theta$, $x_i = r \times \sin[(2i-1)\,\pi/(4n)]$).

2.2 Analytical Solutions

Based on the hardening τ-s model (seeing Fig. 1a), the analytical model of shaft resisting moment M_s could be calculated by the following equation:

$$\frac{M_s}{M_{su,eq}} = \begin{cases} \frac{\theta}{\theta_{ref}} & \left(0 \le \frac{\theta}{\theta_{ref}} \le 0.785 \right) \\ A \times \left(\frac{\theta}{\theta_{ref}} \right)^{-2.142} + \left[\frac{k_2}{k_1} \times \frac{\theta}{\theta_{ref}} + \left(1 - \frac{k_2}{k_1} \right) \right] & \left(\frac{\theta}{\theta_{ref}} > 0.785 \right) \end{cases} \tag{2}$$

where $M_{su,eq} = \tau_{u,eq} \times d^2/2$; $\theta_{ref} = 8s_{eu}/(\pi d)$; $A = 0.128(k_2/k_1-1)$; k_1 and k_2 are shear stiffness coefficients of initial stage and hardening stage, respectively; s_{eu} and $\tau_{u,eq}$ ($= k_1 \times s_{eu}$) are critical displacement and equivalent limit friction.

The results from Eq. 2 are identical to that of Eq. 1 (Fig. 1b), which verifies the correctness of proposed analytical solutions. Moreover, M_s will increase with increasing d, $\tau_{u,eq}$ and as decreasing s_{eu} (Fig. 1b). Similarly, for the softening τ-s model (Fig. 1c), the analytical model of shaft resisting moment M_s could be calculated by

$$\frac{M_s}{M_{su,eq}} = \begin{cases} \frac{\theta}{\theta_{ref}} & \left(0 \le \frac{\theta}{\theta_{ref}} \le 0.785\lambda \right) \\ \left(B_1 \times \left(\frac{\theta}{\theta_{ref}} \right)^2 + B_2 \times \frac{\theta}{\theta_{ref}} + B_3 \right) / \left(\frac{\theta}{\theta_{ref}} + B_4 \right) & \left(0.785\lambda < \frac{\theta}{\theta_{ref}} \le 0.785\beta\eta \right) \\ C_1 \times \left(\frac{\theta}{\theta_{ref}} \right)^{C_2} + C_3 & \left(0.785\beta\eta < \frac{\theta}{\theta_{ref}} \right) \end{cases} \tag{3}$$

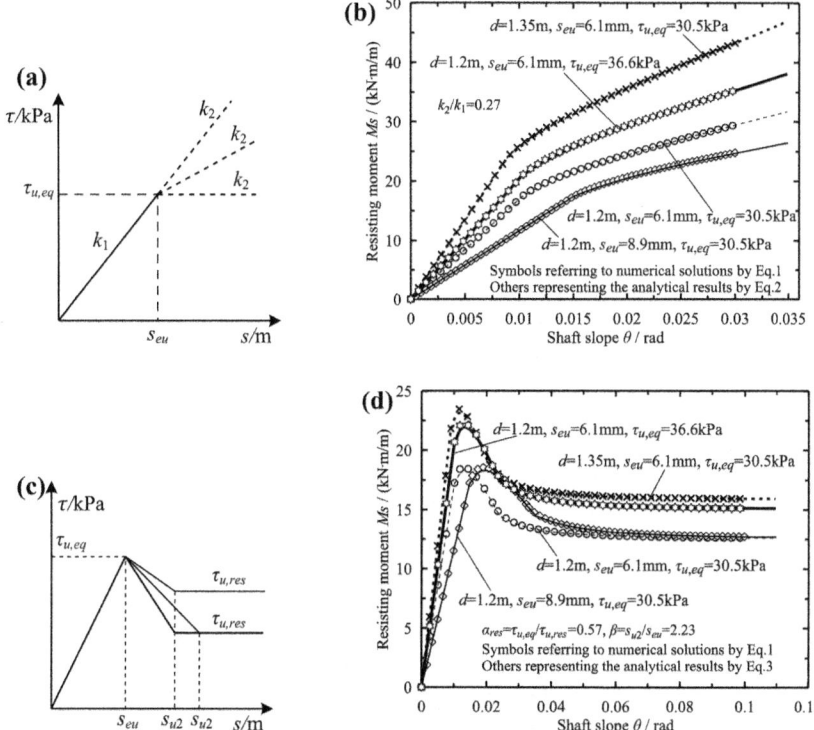

Fig. 1. **a** Hardening model of skin friction, **b** $M_s - \theta$ curves for hardening model, **c** softening model of skin friction, **d** M_s-θ curves for softening model.

in which $\lambda = (0.0726\beta + 0.726) + (0.0116\beta^2 - 0.1118\beta + 0.3439)\, \alpha_{res}$; α_{res} is the ratio of residual strength $(\tau_{u,res})$ to peak friction $(\tau_{u,eq})$; β is the ratio of residual-critical displacement (s_{u2}) to peak displacement (s_{eu}); $\eta = (0.0412\alpha_{res}^2 + 0.0412\alpha_{res} -0.022)\beta + 1.1$; $C_1 = (-0.076\beta^{2.9} - 0.41)\alpha_{res} + (-0.08\beta^{2.88} + 0.29)$, $C_2 = (-0.04\beta -0.25)\alpha_{res} - (0.09\beta + 2.16)$, $C_3 = 0.993\alpha_{res} -0.006$ and parameters $B_1 \sim B_4$ are calculated by

$$\begin{cases} B_1 = ((-0.00523\beta + 1.052)/(\beta - 0.95))\alpha_{res} - (0.0115\beta + 1.01)/(\beta - 0.95) \\ B_2 = ((-0.0229\beta - 1.63)/(\beta - 0.95))\alpha_{res} + (1.044\beta + 0.5407)/(\beta - 0.95) \\ B_3 = (0.604/(\beta - 0.95))\alpha_{res} - (0.554\beta + 0.09)/(\beta - 0.95) \\ B_4 = (-0.54\beta^2 + 0.11\beta - 0.34)/(\beta^2 + 0.61\beta - 0.613) \end{cases} \quad (4)$$

The results from Eq. 3 are identical to that of Eq. 1 (Fig. 1d), which verifies the correctness of proposed analytical solutions. Moreover, M_s will increase with increasing d, β and α_{res} (seeing Fig. 2). Furthermore, [2] had performed an in-situ test for laterally loaded pile embedded into stiff clay soil. Theoretical analysis is calculated by transfer matrix approach [6] and corresponding comparison is shown in Fig. 3, which implies that predictions considering resisting moment match better with field data.

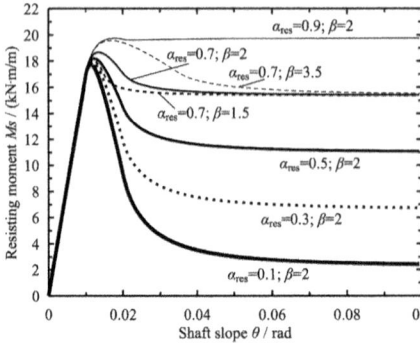

Fig. 2 Impact analysis of α_{res} and β

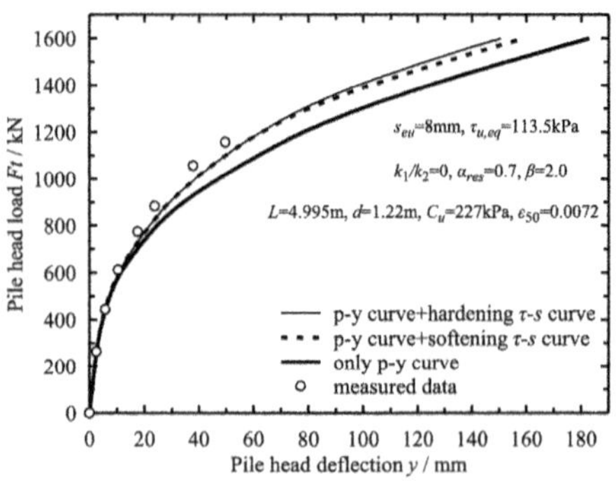

Fig. 3 Comparison between predictions and test

3 Conclusions

To investigate the contribution of vertical shaft resistance to lateral bearing capacity of pile foundation, this work deduces analytical expressions of shaft resisting moment for hardening and softening τ-s models, respectively. Case study implies that resisting moment has a significant influence on laterally loaded pile embedded in stiff soil.

Parametric studies show that for hardening τ-s curve, resisting moment M_s will increase with increase of d, $\tau_{u,eq}$ and as decreasing s_{eu}; for softening τ-s model, M_s will increase with increasing d, $\beta(= s_{u2}/s_{eu})$ and $\alpha_{res}(= \tau_{u,res}/\tau_{u,eq})$.

Acknowledgements. The authors would like to acknowledge financial support from China Postdoctoral Science Foundation (2017M611955), Jiangsu Province Postdoctoral Science Foundation (1701028B) and Science & Technology Project of JSPDI (32-JK-2016-003).

References

1. Ashour, M., Helal, A.: Contribution of vertical skin friction to the lateral resistance of large-diameter shafts. J. Bridge Eng. **19**(2), 289–302 (2014)
2. Bhushan, K., Fong, P.T., Haley, S.C.: Lateral load tests on drilled piers in stiff clays. J. Geotech. Eng. Div. **105**(8), 969–985 (1979)
3. Karapiperis, K., Gerolymos, N.: Combined loading of caisson foundations in cohesive soil: finite element versus Winkler modeling. Comput. Geotech. **56**, 100–120 (2014)
4. Lam, I.P., Martin, G.R.: Seismic Design of Highway Bridge Foundations. Springfield, Virginia (1986)
5. Mcvay, M.C., Niraula, L.: Development of P-Y Curves for Large Diameter Piles/Drilled Shafts in Limestone for FBPIER. University of Florida, Florida (2004)
6. Zhu, M.X., Zhang, Y.B., Gong, W.M.: Generalized solutions for axially and laterally loaded piles in multilayered soil deposits with transfer matrix method. Int. J. Geomech. **17**(4), 04016104 (2017)

Part VIII: Underground Construction

Tunnelling in Urban Environments: Protecting Sensitive Buildings

Eduardo E. Alonso[✉]

Department of Civil and Environmental Engineering,
Universitat Politècnica de Catalunya, Barcelona, Spain
eduardo.alonso@upc.edu

Abstract. The paper stresses the difficulty of estimating the potential damage induced by tunneling on sensitive buildings. This situation justifies the adoption of protective measures. Among a wide range of solutions two comforting techniques are described: Compensation grouting and protection walls. In the first case attention is given to the response of pile foundations to the tunnel volume loss. A semi-analytic solution, which includes a few fundamental solutions, is outlined. It is an alternative to expensive numerical analysis which may prove to be accurate enough in practice. The tunnel-protection wall interaction was also analyzed by a similar simplified solution. Main dimensionless parameters controlling the wall performance are identified and their relevance is discussed.

Keywords: Analytical solutions · Compensation grouting · Damage
Protection walls · Tunneling · Volume loss

1 Introduction

Singular and sensitive buildings are often encountered by tunnel construction in city environments. A typical case is found when a tunnel boring machine, excavating at cover ratios in the range 0.5 to 3, goes underneath or close to historic, ancient or artistic monuments.

In standard practice, the risk of damaging a structure by a mechanized excavation follows a few well established steps:

(a) Tunnelling-induced volume loss is estimated on the basis of past experience in a given soil profile and an estimated performance of the tunnelling machine.

(b) The volume loss generates a displacement field which can be approximated by simple rules which go back to the pioneering work of Peck (1969). Soil settlement and horizontal strains provide the basic information to estimate the risk of inducing a given damage to the affected structure (Fig. 1).

(c) Under a few simplifying assumptions (the affected building is characterized by a thick beam which follows the "green field" motion induced by the tunnel; damage is associated to a critical maximum tensile strain) the damage plots provided by

© Springer Nature Switzerland AG 2018
W. Wu and H.-S. Yu (Eds.): *Proceedings of China-Europe Conference on Geotechnical Engineering*, SSGG, pp. 1117–1127, 2018.
https://doi.org/10.1007/978-3-319-97115-5_51

Boscardin and Cording (1989) or Burland (1995) provide an estimate of the expected building damage.

The summarized steps can be substituted by more accurate procedures. If the structure-tunnel interaction is introduced in step (c) (Potts and Addenbrooke 1997; Franzius et al. 2005), the estimated damage level is typically reduced.

Steps (a) and (b) may be substituted by a numerical analysis which attempts to reproduce the tunnelling operations and introduces the actual soil profile and its geotechnical characterization. Unfortunately, transforming tunnelling operations into a 3D numerical model is a difficult task and the experience indicates that design predictions are often far from observations.

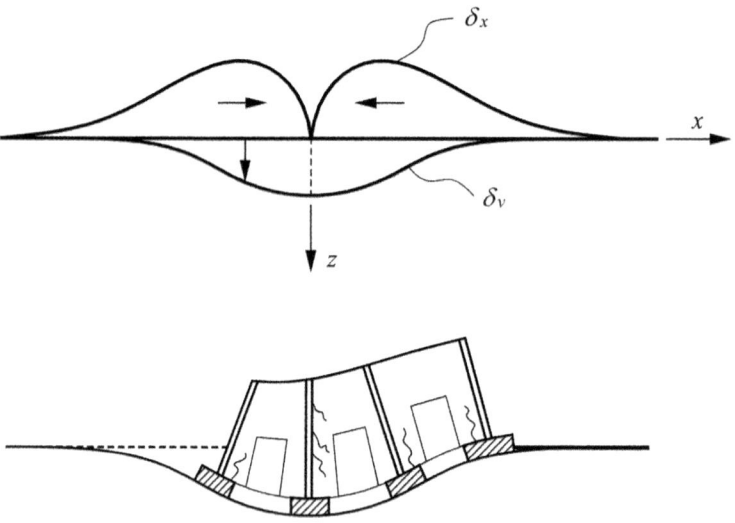

Fig. 1. Classical pattern of surface settlements and horizontal displacements due to tunnel construction

Nevertheless, numerical analysis is a step ahead in standard practice when complex structures are involved. One example is given in Fig. 2. The plot shows the calculated settlement profile and horizontal displacements in a particular cross section of Terminal 2 of Barcelona Airport. The figure shows the irregular, non-symmetric shape of the settlement trough and horizontal displacements expected from the excavation by an Earth Pressure Balance Shield (EPBS) machine, of a railway tunnel connecting the two terminals of the airport. The presence of a massive and rigid foundation caisson and the pile foundations of the terminal structure modify the smooth classical Gauss distribution of settlements and associated horizontal displacements.

Numerical procedures have a chance to be accurate if they are validated against field measurements, typically, green field data. This is obviously not the case of the design stage. Therefore, singular buildings require a particular attention, which often results in designing protection measures.

Two techniques, among other possibilities will be described: Compensation grouting and protective walls. In both cases, simplified semi-analytical calculation procedures will be presented and discussed. They are useful to highlight the critical aspects controlling design and they also provide a convenient and even accurate calculation procedure.

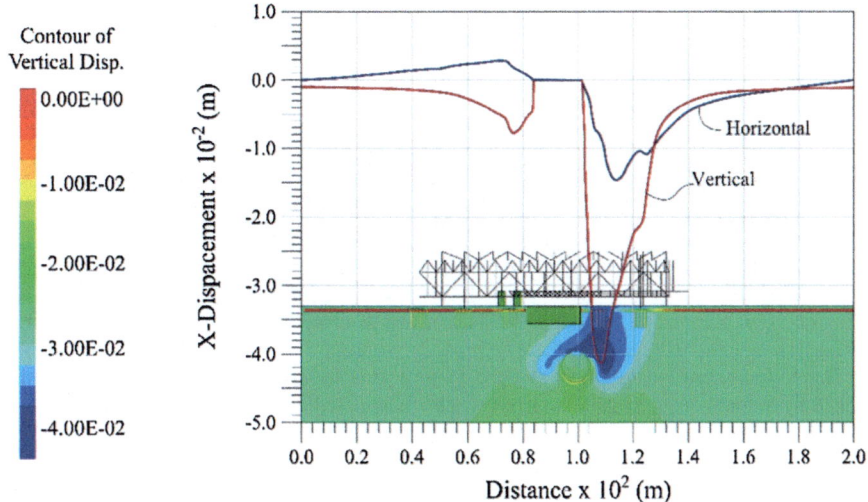

Fig. 2. FE calculation of settlements and horizontal displacements for the EPBS excavation of a railway tunnel under Barcelona Airport Terminal 2

2   Compensation Grouting

Compensation grouting techniques began to be used in many cities around the world in 1990–1995. The following papers describe case records in Lisbon (Simic and Gittoes 1996), Antwerp (Schweiger et al. 2004), London (Harris et al. 1994); and Rome (Kummerer and Sciotti 2013).

Figure 3 shows the layout of the compensation grouting of a sensitive building in Bilbao in a complicated soil profile, because of the underlying karstic marls and limestones. This was a case of building settlement induced by uncontrolled seepage towards the tunnel and the resulting subsidence. Also shown in the figure is the position of an array of injection boreholes (TAM: "Tubes à manchette") under building foundations intended to compensate the expected settlement.

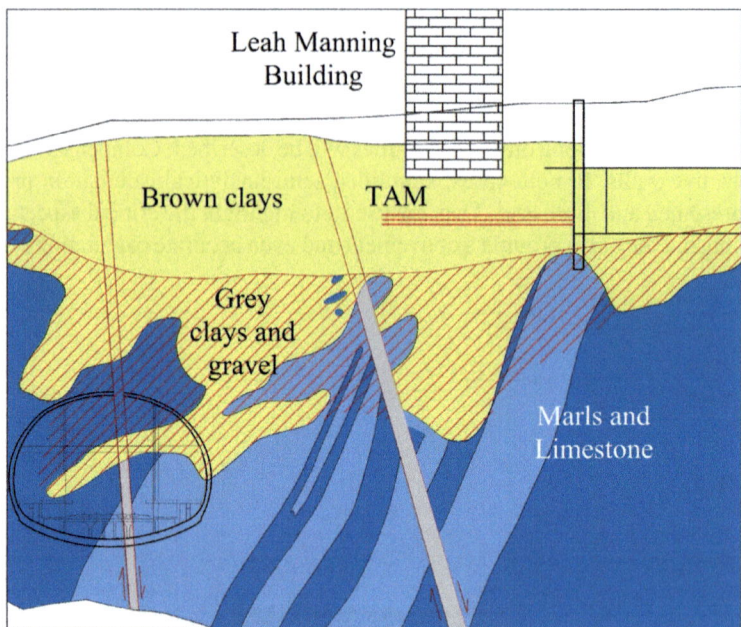

Fig. 3. A compensation grouting scheme to protect a building against expected settlement induced by tunnel boring excavation in a complex karstic soil profile. Bilbao, Spain

Figure 4(a) describes some basic concepts of compensation grouting. In impervious clayey materials the injected grout induces a soil fracture, which is controlled initially by the state of the stress. Eventually, the modification of the stress field, as a result of the injection, leads to a horizontal propagation of cracks and a vertical heave. In pervious sandy soils a "compaction grouting" is more likely. Figure 4(b) summarizes the phenomenon of generation of an impervious "filtered grout" and the associated soil expansion as grouting pressure increases. In applications the end result in both cases is a volume increase in the vicinity of the grouting lines which will deform the soil and the existing foundations. The "efficiency" of this procedure, defined as the ratio of surface heave (relevant for shallow foundations) and injected volume, is very often a small number in the vicinity of 0.1.

A case of interest, which is often found in cities, are the pile foundations. The "interaction" between the deforming soil and the pile groups is more complex if compared with the case of shallow foundations. This case could be analysed by a simplified procedure.

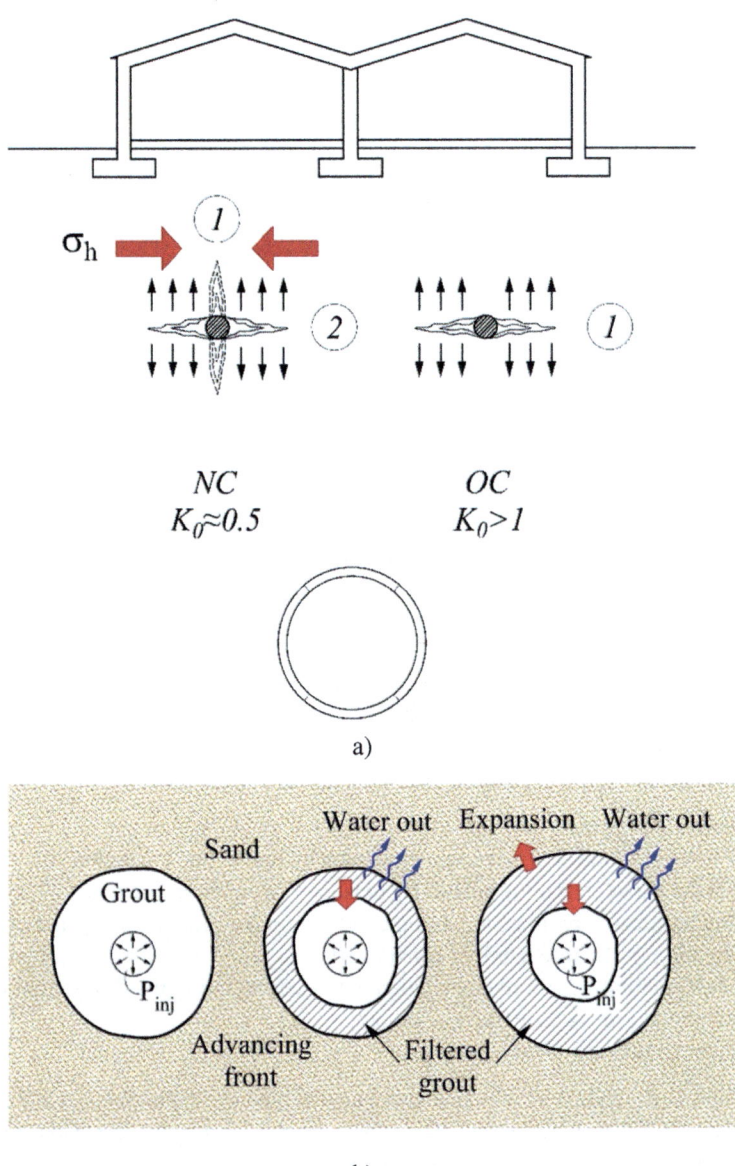

a)

b)

Fig. 4. (a) Fracturing in impervious soil induced by injection in Normally Consolidated and Over-consolidated soil; (b) Compaction grouting in pervious soils. Development of a filtered grout and volume expansion

2.1   Pile Foundations Affected by Volume Changes at Depth

This case is illustrated in Fig. 5. A pile group supporting building loads suffers the effect of a volume decrease or increase at an arbitrary location in the soil.

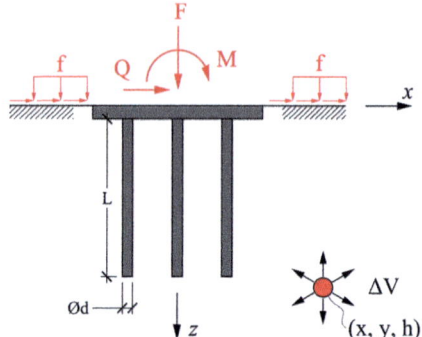

Fig. 5. The problem of the response of a pile group in the presence of a soil volume loss or volume expansion

A model was built using three fundamental solutions: Boussinesq (1885) and Mindlin (1936) for the point load on the surface and inside an elastic half-space and Sagaseta (1987) for a volume change at a point in a half-space. The solution is sought by enforcing the compatibility of soil and piles displacements at a number of discretization points. The following additional assumptions were made:

- Superposition applies;
- Shear forces on pile shafts are approximated by the Mindlin solution;
- Interaction between piles is solved by means of Mindlin solution;
- Horizontal stresses on piles are assumed to act on the planar rectangular equivalent cross section of the pile element considered;
- Piles deformations and forces are calculated by integrating the differential equation governing the pile as a structural element;
- The pile cap was assumed rigid.

The outlined procedure was first used to estimate the behaviour of a 3×3 deep foundation of Ø1.5 m bored piles, 20 m long, capped by a rigid slab. This foundation supported a 50 m high railway viaduct pillar. The soil expanded at depth because of anhydrite- related swelling mechanism. The case is described in Alonso et al. (2015). Table 1 shows a comparison of measured and calculated inclination of pile cap. The model was calibrated against a "green field" heave at surface. The agreement is quite good having in mind the complexity of the real case. This procedure was not yet applied to pile foundations affected by volume loss but it is an interesting simplified procedure to tackle the complicated case of tunnel-foundation interaction.

Table 1. Pont de Candí bridge. Comparison of calculated and measured pile cap rotations (in sexagesimal degrees of arc)

Direction of inclination	Calculated (Candí)	Measured (21 Nov/2007 to 25 Apr/2008)
x	0.0155	0.0176
y	0.0119	0.0191

3 Protective Walls: Crossing Barcelona by the High Speed Railway Line to France

Deep walls isolating a sensitive building from the tunnel excavation are often a convenient protection. Peck (1969) in his well-known state of the art on "Deep excavations and tunnelling on soft ground" describes this technique with reference to a real case. This procedure was selected to reduce soil displacements in the vicinity of the Unesco World Heritage "Sagrada Familia" church. The bored tunnel, 5 km long, crosses the "Art Nouveau (Modernism)" quarter of Barcelona where a large number of modernist homes dating from the early decades of 20th Century are a distinctive feature of the city. Streets are relatively narrow (~20 m) and the tunnel (11.5 m of excavation diameter with cover thickness of around 20 m) is capable of inducing noticeable surface displacements in a band 70 to 80 m wide, centred on the tunnel axis.

The possibility of inducing some damage on the Sagrada Familia church raised a major concern in the city. Figure 6 shows the protective wall designed. A 230 m long wall of Ø1.5 m, 40 m long excavated piles, separated at 2 m interval, was built. Piles were capped by a rigid beam and a counterfort buried concrete block was added to reduce

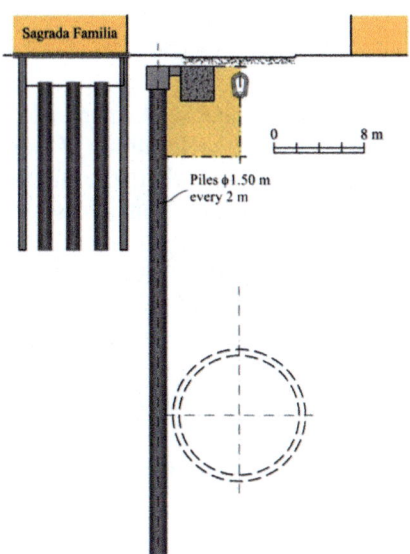

Fig. 6. Pile wall for the protection of the Gloria façade of the Sagrada Familia church.

horizontal displacement of pile heads. Comprehensive FE analysis helped to design and to estimate the performance of this solution. A simplified analysis of the tunnel-soil-wall interaction was also performed to isolate the main variables controlling the efficiency of the adopted solution.

The theoretical problem is illustrated in Fig. 7. The deformation field induced by tunnel excavation results in a set of forces acting on the pile's shaft and tip. Those forces are resisted by the piles. They may be determined using fundamental solutions for line loading inside an elastic half space and the displacement field associated with tunnel excavation. Additional hypothesis of the analysis performance is: compatibility of soil and pile deformations at points of the pile shaft, superposition principle applies, the elastic solution derived by Melan (1932) for a line load inside de soil allow the calculation of soil displacements; the Longanathan and Poulos (1998) solution for tunnel induced displacement is used and equilibrium and displacements compatibility is formulated in the vertical direction. Ledesma and Alonso (2017) describe in detail the procedure.

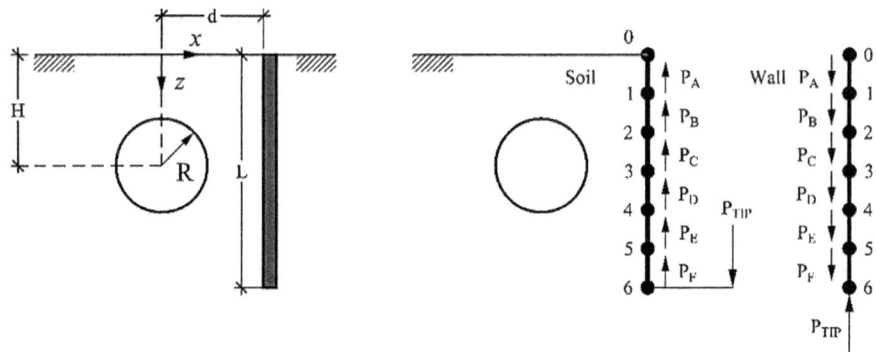

Fig. 7. Geometry defining the tunnel-protection wall interaction and solution scheme

The analysis indicates that the tunnel-wall interaction is controlled by a few dimensionless variables (see Fig. 7 for the definition of geometrical variables: $\Pi_1 = H/R$, $\Pi_2 = d/R$, $\Pi_3 = L/R$, $\Pi_4 = E_s/(E_w \cdot A_w)$ where the stiffness ratio Π_4 is calculated in terms of soil and wall stiffness (E_s, E_w) and the cross section area of the wall per unit length (A_w).

Bilotta and Russo (2011) defined the wall efficiency by the relationship $\eta = (s_{\text{green field}} - s_{\text{bored wall}})/s_{\text{green field}}$, where the surface settlements s refer to the point in the soil immediately behind the wall in the protected area. η varies between 0 (no effect of wall) and 1 (maximum protection; no settlement behind the wall). Figure 8 shows the effect of the pile length to tunnel radius (Π_3) on the wall efficiency for two cases. A reference case defined by $\Pi_1 = 3$; $\Pi_2 = 2$; $\Pi_4 = 0.0025$ and a volume loss of 1%, and the wall actually built for the protection of the Sagrada Familia assuming a volume loss of 1%. Figure 8 indicates an efficiency of 0.6 for the Sagrada Familia wall.

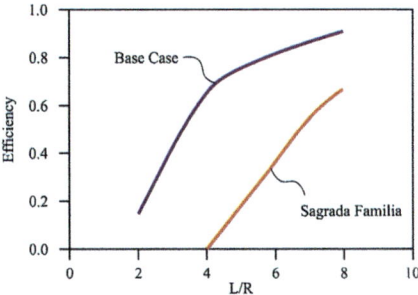

Fig. 8. Calculated efficiencies of a base case and the Sagrada Familia protecting wall.

A small strain elastoplastic finite element analysis (Plaxis) of the wall for an estimated volume loss of 0.5% resulted in the settlement trough shown in Fig. 9. Maximum settlements vary between 5 and 6.5 mm (the difference is explained by the application or not of the church loading). Figure 9 shows also measured settlements. They are negligible under the street and essentially zero in the protected area. This irregular settlement trough corresponds to a very low estimated volume loss: 0,04%.

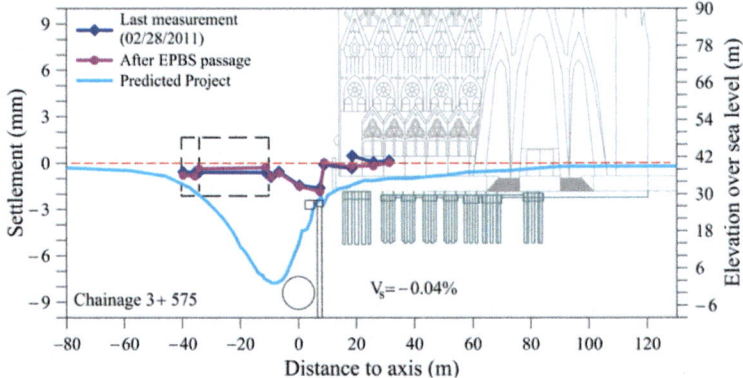

Fig. 9. FE calculations of surface settlement trough ("Predicted Project") and actual measurements. They imply a volume loss of 0.04%

4   Conclusions and Lessons Learnt

Simplified solutions used in practice to estimate the expected damage induced by tunnelling are only a first approximation when dealing with the structural complexity of singular buildings.

The estimation of building performance, even if a joint tunnel excavation-structural analysis is carried out, has a limited accuracy unless a previous knowledge of tunnelling deformations under similar soil conditions and excavation procedures is available. Given this uncertainty, "external" protection of singular buildings is often adopted. The paper describes two methods: compensation grouting and protection walls. Both are amenable to numerical analysis. However, semi-analytical methods are useful for a better understanding of deformation mechanisms and a first estimation of field performance. Modern shield tunnelling and, in particular, Earth Pressure Balance Shied machines are capable, under a rigorous control and experience, to reduce volume loss to very small amounts. Medium to high soil stiffness help to achieve small volume loss targets, often not larger than 0.1%. Accurate controlled excavation is therefore the first desirable situation for the protection of singular buildings.

Acknowledgements. The contribution of Engineers S. Sánchez and E. V. Silva from Geoconsult SA, L. Prieto from Rodio-Kronsa SA and M. Sondon is acknowledged with thanks.

References

Alonso, E.E., Sauter, S., Ramon, A.: Pile groups under deep expansion: a case history. Can. Geotech. J. **52**, 1111–1121 (2015)

Bilotta, E., Russo, G.: Use of a line of piles to prevent damages induced by tunnel excavation. J. Geotech. Geoenvironmental Eng. **137**(3), 254–262 (2011)

Boscardin, M.D., Cording, E.J.: Building response to excavation-induced settlement. J. Geotech. Eng. ASCE **115**(1), 1–21 (1989)

Boussinesq, J.: Applications des potentiels à l'etude de l'equilibre et du mouvement des solides élastiques. Gauthier-Villard, Paris (1885)

Burland, J.B.: Assessment of risk of damage to buildings due to tunnelling and excavation: invited special lecture. In: Proceedings of the 1st International Conference on Earthquake Geotechnical Engineering, Tokyo (1995)

Franzius, J.N., Potts, D.M., Burland, J.B.: The response of surface structures to tunnel construction. In: Proceedings of the Institution of Civil Engineers. Geotechnical Engineering, vol. 158(GE1), pp. 1–15, June 2005

Harris, D.I., Mair, R.J., Love, J.P., Taylor, R.N., Henderson, T.O.: Observations of ground and structure movements for compensation grouting during tunnel construction at Waterloo station. Géotechnique **44**(4), 691–713 (1994)

Kummerer, C., Sciotti, A.: Compensation Grouting with shallow and deep foundations - case study from the Metro B1 in Rome. In: Proceedings of the 18th International Conference on Soil Mechanics and Geotechnical Engineering, Paris (2013)

Ledesma, A., Alonso, E.E.: Protecting sensitive constructions from tunnelling: the case of World Heritage buildings in Barcelona. Géotechnique **67**(10), 914–925 (2017)

Loganathan, N., Poulos, H.G.: Analytical prediction for tunneling-induced ground movements in clays. J. Geotech. Geoenvironmental Eng. ASCE **124**(9), 846–856 (1998)

Melan, E.Der: Spannungszustand der durch eine Einzelkraft im Innern beanspruchten Halbscheibe. Z. Angew Math Mech. **12**, 343–346 (1932)

Mindlin, R.D.: Force at a point in the interior of a semi-infinite solid. J. Appl. Phys. **7**, 195–202 (1936)

Peck, R.B.: Deep excavation and tunnelling in soft ground: state-of-the-art report. In: Proceedings of the 7th International Conference on Soil Mechanics and Foundation Engineering, Mexico City, 1969, 225–290 (1969)

Potts, D.M., Addenbrooke, T.I.: A structure's influence on tunnelling-induced ground movements. In: Proceedings of the Institution of Civil Engineers. Geotechnical Engineering, vol. 125, pp. 109–125 (1997)

Sagaseta, C.: Analysis of undrained soil deformation due to ground loss. Géotechnique **37**(3), 301–320 (1987)

Schweiger, H.F., Kummerer, C., Otterbein, R., Falk, E.: Numerical modelling of settlement compensation by means of fracture grouting. Soils Found. **44**(1), 71–86 (2004)

Simic, D., Gittoes, G.: Ground behaviour and potential damage to buildings caused by the construction of a large diameter tunnel for the Lisbon metro. In: Mair, R., Taylor, N. (eds.) Geotechnical Aspects of Underground Construction in Soft Ground, Balkema, Rotterdam, pp. 745–750 (1996)

Validation of an In-Situ Stress Level Rating Method

Fei Chen[1] and Jianhui Deng[2(✉)]

[1] School of Architecture and Civil Engineering, Chengdu University,
Chengdu 610106, China
[2] State Key Laboratory of Hydraulics and Mountain River Engineering,
Sichuan University, Chengdu, China
jhdeng@scu.edu.cn

Abstract. In-situ stress is an important factor that controls the behavior of rock mass. Since 1990s, a few methods have been suggested worldwide to rate the in-situ stress level so as to facilitate underground rock engineering design. Among them, the method by Hoek et al. (1995) and Martin et al. (1999) is widely used. The method is based on the ratio of the maximum principal in-situ stress to the uniaxial compression strength of rocks. After clarifying some concepts, 25 case histories of large-scale underground excavations constructed in hard rocks in China are studied to validate the method. The results show that the rating method is effective in general to delineate high stress conditions and the rating can be improved if the dry uniaxial compression strength (UCS) of rocks is used.

Keywords: Hard rock · In-situ stress rating · Validation · UCS

1 Introduction

In-situ stress is an important factor that controls the behavior of rock masses. In general, the failures of hard rocks are structurally controlled under low stress and stress-induced under high stress (Sun 1993; Hoek et al. 1995). Hence, the determination of the in-situ stress level is important for hard rock engineering.

In the past few decades, a number of methods have been suggested for in-situ stress level rating, among which the method by Hoek et al. (1995) and Martin et al. (1999) is widely used in the design of underground rock engineering structures (Table 1). The method is based on the ratio of the maximum principal in-situ stress to the uniaxial compression strength (UCS) of rocks, i.e.

$$RSS = \sigma_1 / UCS \tag{1}$$

where RSS is rock stress state; σ_1 is the maximum principal in-situ stress.

During the past 20 years, many hydropower projects have been completed in high in-situ stress grounds in China. These projects provide a wealth of data for rock mechanics research. After clarifying some concepts related to stress rating, data collected from 25 well-documented case histories are used to validate the method.

© Springer Nature Switzerland AG 2018
W. Wu and H.-S. Yu (Eds.): *Proceedings of China-Europe Conference
on Geotechnical Engineering*, SSGG, pp. 1128–1131, 2018.
https://doi.org/10.1007/978-3-319-97115-5_52

Table 1. In-situ stress level rating suggested by Hoek et al. (1995) and Martin et al. (1999).

Stress state	Low in-situ stress	Intermediate in-situ stress	High in-situ stress
RSS	<0.15	0.15–0.4	>0.4

2 Basic Considerations

Firstly, a clear definition is needed for high stress. High stress is often used to describe a stress state under which a rock may experience stress-induced failure. For hard rocks, the physical and failure characteristics are listed in Table 2.

Table 2. Qualitative description of high stress in hard rocks.

	Physical characteristics	Failure characteristics
1	Structurally compact rock mass	Core disking
2	Extremely low permeability	Rockburst
3	Low unit-absorption rate	Onion skin, spalling, slabbing or bulking in sidewalls
4	Equivalent modulus of rock block and rock masses	Buckling of seams
5	Fresh rock	Borehole or tunnel breakout
6	/	Zonal disintegration
7	/	Shear rupture

The essence of rock failure under high stress is the failure of rock blocks, either dynamically as rockburst or pseudo-statically as sidewall spalling or bulking. This suggests that high in-situ stress can be defined qualitatively as a stress state under which there exist failures of rock blocks. The definition of high stress emphasizes the existence of high-stress-induced failures, but ignores their frequency and intensity. The three stress states listed in Table 1 can be qualitatively described in Table 3.

Table 3. Qualitative description of the in-situ stress rating.

Stress state	Low in-situ stress	Intermediate in-situ stress	High in-situ stress
Description	No high-stress-induced rock failure	High-stress-induced rock failure exists locally or mildly during site investigation and may worsen during construction	High-stress-induced rock failure is noticeable, either occurring frequently in a large extent or with increased intensity during investigation

Failure of soft or heavily fractured rocks under high stress loading, such as squeezing, is not considered here. Furthermore, gas outburst is also not considered in Table 2.

Secondly the selection of the two parameters in Eq. (1) needs to be consistent. (1) the measured in-situ stress should be selected at the location where a cavern is located. (2) the selection of the UCS and the in-situ stress should be consistent, i.e., they should be from the same rock. (3) the average UCS is suggested to be used for stress rating due to the variability of the test data. (4) the moisture content of rock specimens needs to be considered. Both saturated UCS (UCSsat) and dry UCS (UCSdry) are used in this paper to reveal the influence of moisture content on stress rating.

Thirdly excavation disturbances need to be considered, especially at the construction stage. Due to its complexity the discussion is restrained to in-situ stress.

3 Case Histories and Validation

Case histories are selected according to the availability and integrity of related data with regard to detailed description of rock mass behaviors during site investigation, in-situ stress measurement, comprehensive laboratory test data etc.

Altogether 25 cases are selected. The qualitative ratings shown in Table 4 were based on the rock failure characteristics (Table 2) and the high-stress-failures observed during site investigation, while the quantitative ratings are evaluated according to the RSS value. If both the quantitative and the qualitative ratings agree with each other, then the ratings are considered as consistent. The consistence rates, defined as the ratio of the number of consistent cases to all cases, are 92% and 84% for using UCSdry and UCSsat to calculate RSS respectively. In general, the method can give a good prediction of the in-situ stress levels. However, the prediction is better if UCSdry is used.

Table 4. In-situ stress level ratings for 25 underground powerhouse caverns and tunnels.

SN	Project	Main rocks	σ_1 (MPa)	UCS$_{dry}$ (MPa)	UCS$_{sat}$ (MPa)	Qualitative rating	RSS		Quantitative rating	
							dry	sat.	dry	sat.
1	BHT_L	Basalt	18.90	144.69	116.00	I	0.13	0.16	L	I
2	BHT_R	Basalt	24.70	144.69	116.00	I	0.17	0.21	I	I
3	XLD_L	Basalt	19.10	270.00	231.00	L	0.07	0.08	L	L
4	XLD_R	Basalt	20.49	270.00	231.00	L	0.08	0.09	L	L
5	XJB	Sandstone	12.60	153.86	118.44	L	0.08	0.11	L	L
6	JP I	Marble	35.70	87.40	66.00	H	0.41	0.54	H	H
7	JP II-H	Marble	16.79	95.14	85.20	I	0.18	0.20	I	I
8	JP II-T	Marble	63.09	107.51	88.00	H	0.59	0.72	H	H
9	GD	Basalt	35.17	188.60	151.10	I	0.19	0.23	I	I
10	ET	Basalt	38.40	263.90	189.70	I	0.15	0.20	I	I
11	SJK	Granite	28.96	139.00	97.30	I	0.21	0.30	I	I
12	HZY	Limestone	36.43	155.41	125.05	I	0.23	0.29	I	I
13	CHB	Granite	31.96	185.12	137.72	I	0.17	0.23	I	I
14	HJP	Diorite	23.23	175.90	146.00	L	0.13	0.16	L	I
15	DGS	Granite	19.28	103.00	81.60	I	0.19	0.24	I	I
16	PBG	Granite	27.30	147.20	100.10	I	0.19	0.27	I	I

(*continued*)

Table 4. (*continued*)

SN	Project	Main rocks	σ_1 (MPa)	UCS_{dry} (MPa)	UCS_{sat} (MPa)	Qualitative rating	RSS		Quantitative rating	
							dry	sat.	dry	sat.
17	LXW	Granite	22.87	157.00	110.00	I	0.15	0.21	I	I
18	XLD1	sandstone	3.46	107.40	60.00	L	0.03	0.06	L	L
19	XW	gneiss	26.7	173.61	134.34	I	0.15	0.20	I	I
20	NZD	Granite	8.27	143.69	120.30	L	0.06	0.07	L	L
21	SBY	Limestone	5.62	82.50	59.50	L	0.07	0.09	L	L
22	JB	granite	14.21	148.30	102.40	L	0.10	0.14	L	L
23	LT	Sandstone	12.93	183.00	155.00	L	0.07	0.08	L	L
24	LBG	Dolomite	17.00	102.20	82.80	L	0.17	0.21	I	I
25	XJ	Tuff	14.09	121.00	83.00	L	0.12	0.17	L	I

_L = The left powerhouse; _R = The right powerhouse; _H = The powerhouse; _T = The headrace tunnel; L = low stress;
I = intermediate stress; H = high stress; BHT = Baihetan; CHB = Changheba; DGS = Dagangshan; GD = Guandi; ET = Ertan;
HJP = Huangjinping; JB = Jiangbian; JP I = Jinping I; JP II = Jinping II; LBG = Lubuge; LT = Longtan; LXW = Laxiwa;
NZD = Nuozhadu; PBG = Pubugou; SBY = Shuibuya; SJK = Shuangjiangkou; XJ = Xianju; XJB = Xiangjiaba;
XLD1 = Xiaolangdi; XLD = Xiluodu; XW = Xiaowan

4   Conclusions

After clarifying the concept of high stress, 25 case histories are studied to validate the stress level rating scheme by Hoek et al. (1995) and Martin et al. (1999), and the following conclusions can be drawn.

(1) The rating scheme can provide a very good indication of the in-situ stress level.
(2) The stress level rating can be improved if dry state UCS is used.

Additional case histories including soft rocks can be studied to further validate the rating method.

Acknowledgement. The support of the National Science Foundation of China under grant 41772322 is greatly appreciated.

References

Hoek, E., Kaiser, P.K., Bawden, W.F.: Support of underground excavations in hard rock. Balkema, Rotterdam (1995)

Martin, C.D., Kaiser, P.K., McCreath, D.R.: Hoek–Brown parameters for predicting the depth of brittle failure around tunnels. Can. Geotech. J. **36**, 136–151 (1999)

Sun, G.Z.: Engineering Geology and Geological Engineering. Seismological Press, Beijing (1993). (in Chinese)

Dynamic Response of Underground Structure Under Bidirectional Shaking in Layered Liquefiable Ground

Renren Chen[ID], Rui Wang[ID], and JianMin Zhang[(✉)]

Department of Hydraulic Engineering, National Engineering Laboratory for Green and Safe Construction Technology in Urban Rail Transit, Tsinghua University, Beijing 100084, China
zhangjm@mail.tsinghua.edu.cn

Abstract. Seismic liquefaction can cause severe damage to underground structures in saturated sandy soil, and vertical ground motion is considered as an important factor for such damage. This paper investigates the bidirectional seismic response of underground structures in layered liquefiable grounds using a high fidelity finite element calculation method. A unified plasticity model for large post-liquefaction shear deformation of sand is used for the liquefiable soil to fully capture the behavior of saturated sand. A technique of combining beam elements with quadrilateral elements is developed to simulate reinforced concrete structures. Shallow buried underground structures in level and sloping layered liquefiable ground are analyzed under horizontal (H-input) and bidirectional (HV-input) motions, respectively. Comparison of the results show that the effect of vertical seismic excitation is significant for vertical axial force, but has little effect for the shear force, bending moment, and shear de-formation of the structure.

Keywords: Underground structure · Soil liquefaction · Seismic response
Layered liquefiable ground · Bidirectional shaking

1 Introduction

With the rapid development of infrastructure in China, more and more under-ground structures are under construction in seismically active with liquefiable soil. Past earthquake cases have shown significant damages to underground structures and facilities in liquefiable soil, such as in the Kobe earthquake (1995) [1], and in the Tohoku earthquake (2011) [2]. The seismic performance of underground structures in liquefiable ground has come under scrutiny.

Vertical ground motion has been suggested to have a significant influence on the seismic response of underground structures [3]. However, its effect on under-ground structures in layered liquefiable ground has not been studied in detail [4–6].

This paper compares the effect of horizontal and bidirectional excitation for underground structures in level and sloping layered liquefiable ground. The response of ground and structure are studied.

© Springer Nature Switzerland AG 2018
W. Wu and H.-S. Yu (Eds.): *Proceedings of China-Europe Conference on Geotechnical Engineering*, SSGG, pp. 1132–1135, 2018.
https://doi.org/10.1007/978-3-319-97115-5_53

2 Numerical Analyses Details

The analysis is a plane strain analysis. The ground is assumed to be 35 m thick saturated layered soil, with a 6-m liquefiable layer passing through the under-ground structure. The hypothetical underground structure is a typical one-story and two-span subway station, with the top slab 6 m below ground surface. 0.5 m × 0.5 m square cross section central pillars with 3 m spacing are used.

The bottom of the model is constrained to follow the input motion and is un-drained. Tied boundaries are used at the two lateral boundaries to simulate the free-field motion, which are also undrained. The water level is assumed to be at the ground surface, with free drainage boundary.

Before the dynamic analysis, the application of gravity and excavation is conducted to obtain the initial stress field [7]. Dynamic analysis is then carried out, followed by another 100 s of calculation after the earthquake motion to allow for excess pore pressure dissipation. The scaled acceleration time history (with peak value 2.2 m/s^2) from the El Centro motion is used as the H-input, and the vertical component is 0.17 g and 0.62 g in bidirectional conditions.

Liquefiable soil is modeled using a three-dimensional plasticity model, which is capable in analyzing large post-liquefaction deformation of sand [8, 9]. Both level and sloping ground is analyzed. Non-liquefiable soil is modelled using elastic perfect-plastic Mohr-Coulomb model for simplification. Soil-structure interface is modelled using the Clough-Duncan interface model. The underground structure is modelled using combined elements that consist of both quadrilateral elements and nonlinear fiber beam elements, which could be effective to reflect the dynamic response of reinforced concrete.

The shear stress-strain relationship, stress path, excess pore pressure ratio in the far field soil are compared. The displacement, deformation and internal forces in the structure are analyzed. Besides, different slopes (from 0° to 4°) and vertical ex-citation component (0.17 g and 0.62 g) are investigated.

3 Dynamic Response Discussion

The basic response of the far-field soil at point M and N (Fig. 1) is presented in Fig. 2. The excess pore pressure ratio (EPPR) at −8 m reaches 1.0 towards the end of the input motion, suggesting that the soil achieves liquefaction. Although the soil below −12 m does not fully liquefy, significant increase in excess pore pressure is observed. The EPPR accumulated with intensive vibration in HV-input condition, compared with the H-input condition. However, the backbone of the EPPR time history curves are the same. Zero effective stress is reached at −8 m depth, while the vertical effective stress decreases but does not reach zero at −12 m. The shear stress-strain relationship and the vertical effective stress is only minorly affected by the vertical excitation.

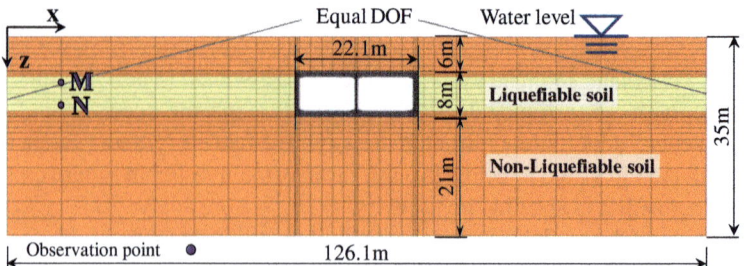

Fig. 1. Finite element analysis model setup.

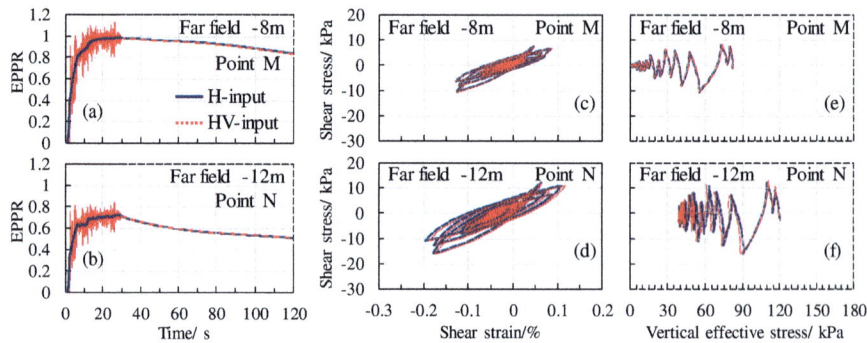

Fig. 2. Basic response of far field soil in the liquefiable interlayer.

Figure 3 presents the time histories of axial force, shear force, and moment of the structure in the central pillar. The axial forces in the central pillar change intensively in HV-input condition. But the bidirectional excitation has almost no effect on the time history of shear forces and moments in the central pillar. The shear force and moment response in walls has similar trends to the pillar.

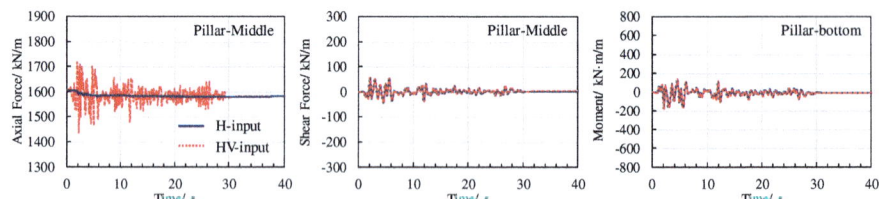

Fig. 3. Time histories of axial force, shear force and moment in the central pillar.

Table 1 shows the maximum structure drift ratio in different conditions. It shows that the drift ratios increase with the increased slope angle but is barely affected by different levels of vertical input motion.

Table 1. Maximum structure drift ratio during the ground motion.

	Angle/°	0	0.5	1	1.5	2	2.5	3	3.5	4
Drift ratio/%	H-input	0.093	0.100	0.116	0.133	0.178	0.243	0.317	0.404	0.504
	HV ($a_v =$ 0.17 g)	0.092	0.100	0.116	0.134	0.181	0.247	0.323	0.408	0.508
	HV ($a_v =$ 0.62 g)	0.092	0.096	0.112	0.136	0.185	0.253	0.330	0.417	0.518

4   Concluding Remarks

Shallow buried underground structures in level and sloping layered liquefiable ground are analyzed under horizontal (H-input) and bidirectional (HV-input) motions, respectively. The comparison analysis shows that the vertical excitation can increase the axial force in central pillar, with little influence on the shear force, moments, and shear deformation of the underground structure.

Acknowledgements. The authors would like to acknowledge the National Natural Science Foundation of China (No. 51708332 and No. 51678346) for funding the work in this paper.

References

1. Samata, S., Ohuchi, H., Matsuda, T.: A study of the damage of subway structures during the 1995 Hanshin-Awaji earthquake. Cem. Concr. Compos. 19(3), 223–239 (1997)
2. Tokimatsu, K., Tamura, S., Suzuki, H., et al.: Building damage associated with geotechnical problems in the 2011 Tohoku Pacific Earthquake. Soils Found. 52(5), 956–974 (2012)
3. Du, X., Ma, C., Lu, D., et al.: Collapse simulation and failure mechanism analysis of the Daikai subway station under seismic loads. China Civ. Eng. J. 50(1), 53–62 (2017). (in Chinese)
4. Yang, D., Naesgaard, E., Byrne, P.M., et al.: Numerical model verification and calibration of George Massey Tunnel using centrifuge models. Can. Geotech. J. 41(5), 921–942 (2004)
5. Madabhushi, S.S.C., Madabhushi, S.P.G.: Finite element analysis of floatation of rectangular tunnels following earthquake induced liquefaction. Indian Geotech. J. 45(3), 233–242 (2015)
6. Bao, X., Xia, Z., Ye, G., et al.: Numerical analysis on the seismic behavior of a large metro subway tunnel in liquefiable ground. Tunn. Undergr. Space Technol. 66, 91–106 (2017)
7. Wang, R., Fu, P., Zhang, J.M.: Finite element model for piles in liquefiable ground. Comput. Geotech. 72, 1–14 (2016)
8. Wang, R., Zhang, J.M., Wang, G.: A unified plasticity model for large post-liquefaction shear deformation of sand. Comput. Geotech. 59, 54–66 (2014)
9. Wang, R.: Single Piles in Liquefiable Ground: Seismic Response and Numerical Analysis Methods. Springer, Heidelberg (2016)

Non-destructive Technology for Underground Utility Mapping: A Case Study

Kulin Dave[1(⊠)] and Silky Agrawal[2]

[1] Institute of Technology, Nirma University, Ahmedabad, India
kulkulin.kd.kd@gmail.com
[2] GeoCarte Radar Technology Pvt. Ltd., IIT Gandhinagar, Gandhinagar, India

Abstract. This article presents a case study of underground utility mapping survey carried out for a proposed Bridge site in Ahmedabad, Gujarat. The main focus was on mapping possible underground metallic as well as non- metallic utilities, viz, water line, sewer line, gas pipeline, cables etc. using Ground Penetrating Radar (GPR). With the growing needs and ever pinching demand for progressive infrastructure, the pre-existing infrastructure needs to be preserved. GPR is one such tool that can be used for non-destructive survey for utility mapping, concrete inspection, road inspection, archaeology, environmental assessment and many more. An area of around 15000 m^2 is scanned over the bitumen road surface of stretch 700 m length and 22 m width. A GSSI GPR system SIR-3000 was used equipped with 400 MHz ground coupled antenna to cover the depth of penetration up to 3 m. The data was collected at spacing of 6 m across the road where as 2.5 m along the road to map possible utilities in both the directions. In comparison to traditional methods of excavation and trenching which involve a huge loss in terms of time, money and man power GPR can prove to be cost efficient, less invasive and more reliable. GPR survey is expected to be a more efficient technique for underground survey especially in infrastructure projects which involve huge quantum of money in excavation and filling and where cost cutting is a must.

Keywords: Utility mapping · Ground Penetrating Radar

1 Introduction

Ground Penetrating Radar (GPR) is one of the useful non-destructive geophysical methods for subsurface investigation. It is based on Electromagnetic technique and provides high resolution subsurface profiles which are used to detect subsurface features. The depth of penetration for a GPR ranges from a few centimeters to larger depth of around 15–20 m. The success of the GPR survey to a great extent depends on soil condition like, clay content, sediment mineralogy, moisture content, depth of target, electromagnetic interference [1] etc. A GPR system is made up of three main components i.e. the control unit, the antenna and the power supply. GPR equipment can be run with a variety of power supplies ranging from small rechargeable batteries to vehicle batteries and normal 110/220-V. The control unit contains the electronics which triggers the pulse of radar energy that the antenna sends into the ground. It also has a

© Springer Nature Switzerland AG 2018
W. Wu and H.-S. Yu (Eds.): *Proceedings of China-Europe Conference on Geotechnical Engineering*, SSGG, pp. 1136–1139, 2018.
https://doi.org/10.1007/978-3-319-97115-5_54

built-in computer and hard disk to store data for examination after fieldwork. The antenna receives the electrical pulse produced by the control unit, amplifies it and transmits it into the ground or other medium at a particular frequency. Antenna frequency is one major factor in depth of penetration [2]. The higher the frequency of the antenna, the shallower into the ground it will penetrate. High frequency of waves gives better and clearer resolution for given depth.

2   Scope and Layout of Study Area

The GPR survey has been carried out on an area of about 14700 m^2 at Ajit mill circle along Lal Bahadur Shastri road, Ahmedabad, Gujarat, India of dimension 700 m × 21 m. The scope of work includes the mapping of all possible underground utilities up to the depth of 2.5–3 m using non-destructive GPR technology. The purpose of the project is to prevent the damage to the existing utilities during the construction phase of a proposed flyover bridge. The area has been surveyed using 400 MHz centre frequency antenna mounted with SIR 3000 GPR system to cover the depth with fair resolution. The results after analysis and interpretation of the survey executed have been discussed in subsequent sections. The layout of the surveyed area is as shown in Fig. 1. The data is collected in grid of 6 m spacing across the road and 2.5 m spacing along the road. The data was acquired in four patches of around 350 m stretch and 10.5 m wide with proper barrication. For ease of field work and in order to have minimal disturbances to the traffic flow as well as to the data acquisition, survey was executed in night time from 10:00 PM to 6:00 AM.

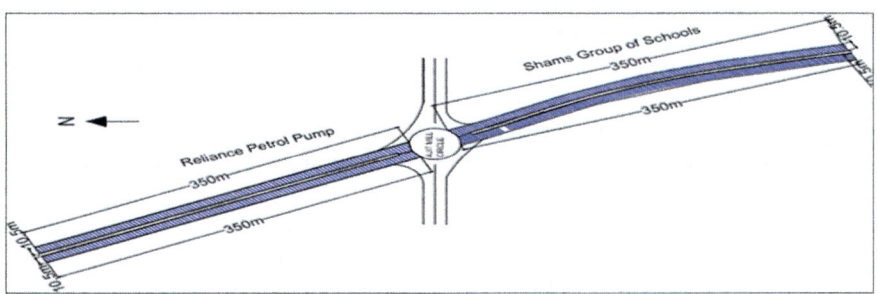

Fig. 1. Layout of the area surveyed

3   Methodology

3.1   Data Acquisition

With 400 MHz antenna, data was collected in distance mode with time window of 60 ns, scans per unit 100 and transmission rate of 100 kHz, with antenna in mono static mode. The stacking of 30 was applied during data acquisition for better data resolution and quality. The survey wheel was calibrated every time on the field to ensure least possible error in location of any underground utility. Similar specifications were applied for all the patches throughout the survey. In all around 265 profiles were collected.

3.2   Data Processing

The acquired data from the field was in raw form and could not be interpreted directly. Therefore, the raw data was processed adequately to make it more interpretable using commercially available software RADAN 7 [3]. For 400 MHz, primarily, time zero correction and background removal were applied to the whole acquired data. Besides, band pass filters of 50 MHz–500 MHz for 400 MHz antenna have been applied, which were decided based on the dominant frequency content obtained from spectrum in RADAN. Range gain was applied along the depth appropriately to amplify the weak deeper reflections using the commercial software RADAN (GSSI). As the data quality was good, we could see clear hyperbolas in subsurface profiles. To calculate velocity of the EM waves in the ground surface, the dielectric constant is estimated using kir-choff's migration in RADAN by fitting Ghost hyperbola. The dielectric of the medium is obtained to be around 7. The velocity corresponding to (7) dielectric constant is 1.13×10^8 m/sec. The estimated velocity is fed to obtain the depth of the observed features. The whole data collected using respective antennas has been processed with respective specifications in similar fashion for all the patches.

4   Interpretation of Features

After the processing of the acquired data, each and every profile is analyzed to observe the feature reflections from the underground objects. The observed features of specific reflection pattern and depth are marked on its locations in the plan view. The repeating points of similar reflection pattern at particular depth were joined by a line to obtain the alignment of the utility. The interpreted utilities in a given patch if found similar to some in earlier surveyed areas are joined throughout their length to get proper alignment. The interpreted features (utilities) are marked on the plan view in AutoCAD, as shown in the Fig. 2. The whole of the surveyed area is divided into four patches that are patch 1 (road beside reliance petrol-pump) patch 2 (road beside omega factory) patch 3 (road beside shams group of schools) and patch 4 (road beside Alfalah masjid) respectively. In patch 1 total 4 utilities were found which were named as U1, U2, U3 and U4 at depths of 0.8 m, 0.9 m, 0.9 m and 1.3 m respectively. The average distances of utilities from the divider were 8.5 m, 6.5 m, 5.3 m and 4.3 m respectively. The repeating reflections which were traced throughout the patch for utility U1 are as shown in Fig. 2. Similar approach was followed in identifying and marking reflections for other utilities as well. In total 4 utilities in patch 2 named U5, U6, U7 and U8 at depths 1.2 m, 0.9 m, 1 m, 1 m and at distances 5.1 m, 6.2 m, 6.9 m, 8.5 m respectively from the divider were found. In patch 3 (that is the road beside the school) utilities U1, U3 and U4 from patch 1 are found to be continued but utility U2 is terminated at Ajit mill circle itself may be due to change in alignment. Moreover the distances from divider of utilities U1, U3 and U4 are 8 m, 4.1 m and 6.2 m respectively in patch 3 which signifies that through its chain age utility U4 has crossed utility U3. Similarly, in patch 4 the pipelines U5, U6, U8 and U9 from patch 2 have continued but utility U7 has discontinued at the circle itself. The distances of utilities U5, U6 and U8 are 5.4 m, 6.5 m and 8.5 m respectively from the divider and the new found utility U9 has a distance of 7.5 m and depth 0.9 m. After marking utilities one more feature i.e. manholes can also be identified through this technique. They have typical continuous reflections of finite width with alternate high and low reflections throughout the depth of the cross section. They are marked with blue dots as shown in Fig. 2.

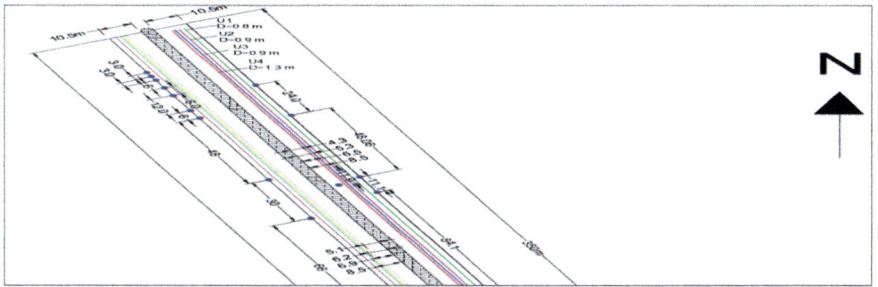

Fig. 2. Section of the road stretch with utility markings in AutoCAD

5 Conclusion

A non-destructive survey for a proposed bridge site has been conducted. The survey depth and data quality are mainly governed by the type of work, selection of antenna frequency, soil conductivity and moisture content of the underlying layer, input parameters while data acquisition and post processing of raw data. A sound knowledge of data interpretation is also must because sometimes there is clutter present in the data such as buried wire scraps, boulders and small metal objects which though give hyperbolic reflections but should not be confused with utilities. This is a very cost-efficient and a reliable non-destructive method which if put into practice can surely avoid breaking and damaging of pipelines during excavation. This in turn reduces the repair and rehabilitation cost reducing the overall cost of the infrastructure project.

References

1. Natural Resources Conservation Service Soils under United States Department of Agriculture (USDA). https://www.nrcs.usda.gov/wps/portal/nrcs/detail/soils. Accessed 1 Feb 2019
2. Smith, D.G.: Smith: Ground penetrating radar: antenna frequencies and maximum probable depths of penetration in Quaternary sediments. J. Appl. Geophys. **33**, 93–100 (1995)
3. Geophysical Survey Systems, Inc. (GSSI). https://www.geophysical.com/software. Accessed 1 Feb 2019

Numerical Investigation on the Influence of the Excavation Rate on the Mechanical Response of Deep Tunnel Fronts in Cohesive Soils

Luca Flessati[(✉)] and Claudio di Prisco

Politecnico di Milano, Piazza Leonardo da Vinci 32, Milano, Italy
{luca.flessati,claudio.diprisco}@polimi.it

Abstract. In conventional tunnelling, under difficult ground conditions, the fronts are commonly stabilized by employing expensive and time-consuming soil reinforcement techniques. The design of these soil reinforcement techniques is commonly based on the analysis of unreinforced front mechanical response.

In a previous paper, the authors numerically analysed the mechanical response of deep tunnel fronts under undrained conditions and suggested the employment of a non-dimensional front characteristic curve. In this paper, the same approach is employed to study the influence of the excavation rate on the front response. The soil behaviour is modelled by employing the Modified Cam Clay model.

1 Introduction

When tunnels are excavated under difficult ground conditions, both the cavity and the front are commonly supported by reinforcing the tunnel cavity or the advance core. The design of the front reinforcements is largely based on prevision of the unreinforced front mechanical response. In the past, the mechanical response of the front was studied theoretically [1–3], experimentally [4–9] and numerically [8–12].

In this paper, the authors numerically analyse the mechanical response of deep tunnel fronts excavated in cohesive soils and, in particular, the influence of the excavation rate on the system response. The numerical results will be analysed by employing the front characteristic curve, relating the total stress applied on the front (σ_f) and the average front displacement (u_f).

2 Numerical Model

To study the hydro-mechanical coupled response of the front, the authors performed a series of 3D FEM numerical analyses by means of the commercial

W. Wu and H.-S. Yu (Eds.): *Proceedings of China-Europe Conference on Geotechnical Engineering*, SSGG, pp. 1140–1143, 2018.
https://doi.org/10.1007/978-3-319-97115-5_55

code Midas GTS NX (http://en.midasuser.com/). The geometry of the numerical model is represented in Fig. 1. The tunnel cross section is circular and the diameter is hereafter named D. The cover diameter ratio (H/D) is 5. The tunnel is assumed to be excavated in a homogeneous cohesive soil stratum. The soil behaviour is modelled by employing the Modified Cam Clay model. The constitutive parameters are λ (virgin loading line inclination), κ (unloading-reloading line inclination), M (critical state line slope), e_0 (initial void ratio) and ν (Poisson's ratio). The soil is assumed to be hydraulically homogeneous and isotropic (the permeability is hereafter named k). The rigid tunnel lining is assumed to fill the gap between the lining and the front usually realized during the construction.

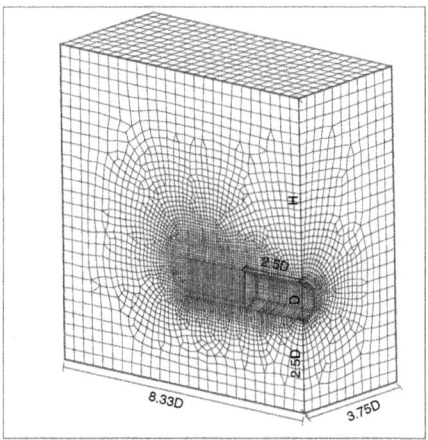

Fig. 1. Numerical model

The numerical analyses were performed according to the two following phases:

1. the initial effective state of stress is obtained by keeping constant the geometry and increasing gravity under drained conditions. The initial pore water pressure distribution is assumed to be hydrostatic and the water table coincident with the ground surface;
2. the initial total geostatic horizontal pressure applied to the front (σ_{f0}) is progressively nullified during an unloading time period named t_u.

3   Numerical Results

In the following, the front characteristic curves are plotted in the non-dimensional $Q_f - q_f$ plane ([10,11]), where the non dimensional variables are defined as:

$$Q_f = \left(1 - \frac{\sigma_f}{\sigma_{f0}}\right)\frac{\sigma_{f0}}{S_u^*} \qquad q_f = \frac{u_f}{u_{fr,elu}}\frac{\sigma_{f0}}{S_u^*}, \tag{1}$$

where S_u^* is the equivalent undrained strength for the system [10], whereas $u_{fr,elu}$ is the elastic undrained front displacement corresponding to $\sigma_f = 0$ ([11]). Both S_u^* and $u_{fr,elu}$ are estimated by following the approach proposed in [10]. According to [11] the use of Q_f and q_f is particularly convenient since, under undrained conditions, in the $Q_f - q_f$ plane the front characteristic curve does not depend on the soil properties and the system characteristic length.

To study the influence of the excavation rate, the authors performed a large series of numerical analyses by changing the unloading rate. The numerical results are reported in Fig. 2a (for $D = 12m$, $\nu = 0.3$, $\kappa = 0.05$, $\lambda = 0.25$, $M = 1$ and $e_0 = 0.5$). As was expected, by increasing the unloading time, the front displacement increases. Moreover, for small t_u values the system response is characterized by a stable "hardening" response ([11]), whereas for large t_u the system response is unstable.

By following the approach proposed in [11], a non-dimensional excavation rate is introduced:

$$\Upsilon = \frac{\gamma_w D^2}{kK t_u}, \tag{2}$$

being γ_w is the water unit volume weight whereas K is the elastic bulk modulus evaluated at the tunnel depth axis at the beginning of the unloading process. In Fig. 2b the results obtained for a fixed Υ value and different D, k and soil mechanical properties are reported. The numerical curves are practically superimposed. This may be justified by observing that, during front unloading the material points in the advance core get very soon the critical state at which plastic hydro-mechanical coupling is absent. For this reason the authors decided to employ the elastic bulk modulus in the non-dimensional excavation rate definition even in the elastic-plastic case.

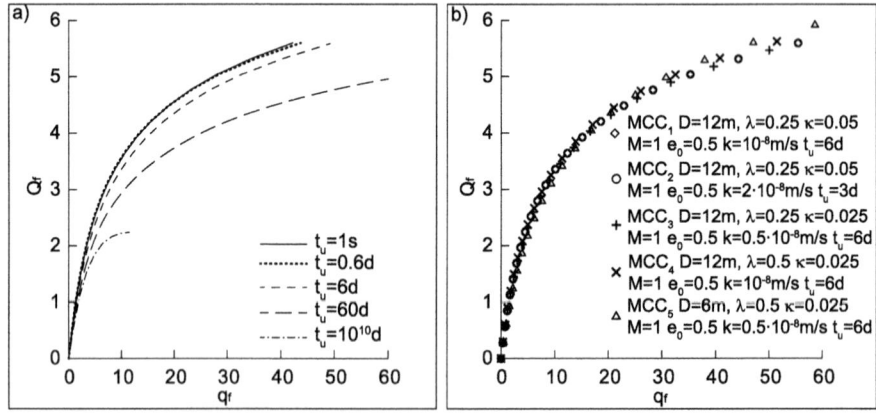

Fig. 2. (a) Influence of the unloading rate and (b) Characteristic curves for a fixed Υ value

4 Concluding Remarks

In this paper the authors show the results of a series of 3D FEM numerical analyses aimed at reproducing the response of deep tunnel fronts in cohesive soils. The results are obtained by assuming a strain hardening elastic-plastic constitutive model for the soil. The authors analyse the influence of the unloading rate: by decreasing the unloading rate, the front displacement significantly increases. Finally, the authors show that for a fixed value of a suitable non-dimensional excavation rate the front characteristic curve is unique independently on the soil mechanical/hydraulic properties and on the system characteristic length.

Acknowledgements. This research was funded by Rocksoil S.p.A. and Maccaferri S.p.A. within the framework of an experimental/numerical program aimed at defining innovative design solutions for the front reinforcements by means of fibreglass tubes.

References

1. Davis, E.H., Gunn, M.J., Mair, R.J., Seneviratine, H.N.: The stability of shallow tunnels and underground openings in cohesive material. Géotechnique **30**, 397–416 (1980)
2. Horn, N.: Horizontaler erddruck auf senkrechter absulussflächen von tunnelröhren. Landeskonferenz der Ungarishen Tiefbauindustrie, pp. 7–16 (1961)
3. Leca, E., Dormieux, L.: Upper and lower bound solutions for the face stability of shallow circular tunnels in frictional material. Gotechnique **40**, 581–606 (1990)
4. Broms, B.B., Bennermark, H.: Stability of clay at vertical openings. J. Soil Mech. Found. Div. **93**, 71–94 (1967)
5. di Prisco, C., Flessati, L., Frigerio, G., Castellanza, R., Caruso, M., Galli, A., Lunardi, P.: Experimental investigation of the time-dependent response of unreinforced and reinforced tunnel faces in cohesive soils. Acta Geotech. **13**(3), 651–670 (2018)
6. Juneja, A., Hegde, A., Lee, F.H., Yeo, C.H.: Centrifuge modelling of tunnel face reinforcement using forepoling. Tunn. Undergr. Space Technol. **25**, 377–381 (2010)
7. Mair, R.J.: Centrifugal Modelling of Tunnel Construction in Soft Clay. Cambridge University, Cambridge (1979)
8. Sterpi, D., Cividini, A.: A physical and numerical investigation on the stability of shallow tunnels in strain softening media. Rock Mech. Rock Eng. **37**, 277–298 (2004)
9. Yoo, C., Shin, H.K.: Deformation behaviour of tunnel face reinforced with longitudinal pipes: laboratory and numerical investigation. Tunn. Undergr. Space Technol. **18**, 303–319 (2003)
10. Di Prisco, C., Flessati, L.: A numerical tool for estimationg deep tunnel front displacements: the role of the excavation rate in cohesive soils. Acta Geotechnica (submitted for publication)
11. di Prisco, C., Flessati, L., Frigerio, G., Lunardi, P.: A numerical exercise for the definition under undrained conditions of the deep tunnel front characteristic curve. Acta Geotech. **13**(3), 635–649 (2018)
12. Vermeer, P.A., Ruse, N., Marcher, T.: Tunnel heading stability in drained ground. Felsbau **20**, 8–18 (2002)

Choosing Optimal Parameters of Mine Air Conditioning Systems

A. F. Galkin[1]([⊠]) and I. V. Kurta[2]

[1] Saint Petersburg Mining University, Saint Petersburg, Russia
afgalkin@mail.ru
[2] Ukhta State Technical University, Ukhta, Russia
ivkurta@mail.ru

Abstract. Regulation of thermal regime of mines, mine workings and underground constructions of the North is conducted both with the purpose of securing stability of the mine workings constructed in permafrost rocks as well as creating comfortable working conditions for the mine workers. The regulation process itself is quite costly and requires significant economic expenditure. In order to decrease these costs, it is suggested to use different classes of air conditioning systems based on special heat-retaining workings (HRW). In this work, we have considered the task of choosing the optimal parameters of borehole system used in the mine air conditioning system. The aim of the work has been to determine the optimal number of boreholes and their length depending on the differing air expenditure in the system. Various economic indicators influencing the efficiency of the system, such as the cost of the electric and thermal energy, and the cost of constructing the boreholes, have been considered as well. It is established that usage of boreholes significantly increases the energy and economic efficiency of mine air conditioning systems. With conditions characteristic of the mines of the North, the usage of boreholes allows to increase energy efficiency of the mine systems three times.

Keywords: Thermal regime · Air conditioning · Boreholes · The North
Energy efficiency

1 Introduction

One of the methods of increasing the energy and economic efficiency of mine air conditioning systems is the intensification of heat-exchange processes through bringing additional volume of rocks into the process [1–4]. The work [5] has presented a new idea: to use boreholes together with the HRW as an additional link in the system chain. Bore systems of heat exchange are used in geothermal technologies, in mining for thermal drainage and for weakening the rocks. Permafrost boring is a fairly well studied process to which a large volume of research works was devoted. As a result, thermal regulation systems using boreholes, the basic version of which is the system described in [5], are prospective and technologically possible.

In this work, we have considered the task of choosing the optimal parameters of borehole system used in the mine air conditioning system. The aim of the work has

© Springer Nature Switzerland AG 2018
W. Wu and H.-S. Yu (Eds.): *Proceedings of China-Europe Conference on Geotechnical Engineering*, SSGG, pp. 1144–1148, 2018.
https://doi.org/10.1007/978-3-319-97115-5_56

been to determine the optimal number of boreholes and their length depending on the differing air expenditure in the system. Various economic indicators influencing the efficiency of the system, such as the cost of the electric and thermal energy, and the cost of constructing the boreholes, have been considered as well.

2 Optimal Parameters of the Boreholes Algorithms

Constructing the objective function of total expenditures for air heating:

$$Z = Z_1 - Z_2, \text{rub.}/\text{year} \tag{1}$$

where Z_1 – costs of heating air in a traditional way (heating with an industrial air heater); Z_2 – costs of mine air conditioning using the boreholes; Z – the economic effect of using the boreholes method of mine air conditioning.

The aim is to maximize the given objective function, that is, find the length of a borehole and their number to receive the maximum economic effect. The costs of air conditioning with the use of boreholes consist of the following parts:

$$Z_2 = z_3 + z_4 + z_5, \text{ rub.}/\text{year}, \tag{2}$$

where Z_3 – costs of constructing the boreholes; Z_4 – costs of overcoming the aerodynamic resistance of the boreholes; Z_5 – costs of additional air heating to the necessary temperature (+2 °C).

The formulas for calculation of individual cost components have the following form (rub./year):

$$Z_1 = \frac{G c_p \gamma \cdot \tau \cdot Z_T (T_\Pi - T_B)}{1163 \cdot 10^3}; \quad Z_3 = (E + E_1) Z_\Pi \cdot l \cdot n; \quad Z_4 = \frac{a' G^3 Z_3 \cdot \tau \cdot l \cdot n}{51 \cdot \pi^2 \cdot R^5};$$
$$Z_5 = \frac{G c_p \gamma \cdot \tau \cdot Z_T (T_\Pi - T_1)}{1163 \cdot 10^3}; \tag{3}$$

Air temperature at the borehole exit is calculated according to the formula:

$$T_1 = T_2 + (T_B - T_e) \cdot \exp(-Al), °C \quad \text{where} \quad A = \frac{2K_\tau \cdot \pi \cdot R \cdot n}{G c_p \cdot \gamma}, \ 1/M. \tag{4}$$

The coefficient of non-stationary heat exchange κ_τ is calculated according to formulas included in [4], where the heat transfer coefficient can be found according to the formula:

$$\propto = 2,7\varepsilon \left(\frac{\gamma G}{n \pi R^2}\right)^{0,8} \cdot \left(\frac{1}{R}\right)^{0,2}, B_T / (M^2 \cdot \kappa). \tag{5}$$

Here and further the following designations are used: T_B – cold air temperature, °C; T_1 – air temperature at the borehole exit, °C; l – the length of the borehole, m; R – borehole radius, m; n – number of boreholes; τ – aeration time, h; γ – air density, kg/m^3; Z_3 – price of the electric energy, rub/kWh; α' – coefficient of aerodynamic resistance of the borehole, kgs $\cdot$ s^2/m^4; η – energy conversion efficiency; c_p – total heat capacity of moist air, J/kg $\cdot$ K; Z_T – price of a unit of thermal energy considering the energy conversion efficiency of a ventilation device, rub/Gcal; E – normative coefficient, 1/year; ε – coefficient of surface roughness; Z_Π – expenses of construction of one meter of a borehole, rub./m; E_1 – a part of total expenses for the operation of the boreholes, fractions of a unit; h_{nped} – maximum depression in the system, kgs/m^2.

This task will be converted into a task of minimization with constraints.

Minimization of $Z' = -Z(l, n)$ with constraints:

$$a_1 \leq l \leq b, \quad a_2 \leq n \leq b_2, \quad h(l, n) \leq h_{nped},$$

where a_i, b_i – are constants, i = 1, 2.

The resulting optimization task with simple constraints on l, n variables and a smooth non-linear constraint function will be transformed into a task of non-constrained optimization task using the penalty method:

$$Z'' = Z' + \left[\max\{0, h(l, n) - h_{nped}\}\right]^2 \quad \text{with conditions} \quad a_1 \leq l \leq b_1, \quad a_2 \leq n \leq b_2.$$

In order to solve the optimization task, we will apply the Hooke-Jeeves method of multi-dimensional minimum search with the use of one-dimensional Brent minimization [6]. Multivariate calculations were conducted with the price of construction of a borehole Z_Π = 10 rub./m and 20 rub./m; the duration of aeration of the borehole τ = 720 h, 1440 h, 2160 h; air usage in the system G = 10, 20, 30, 40 m^3/s.

An analysis of the results is conducted. It is established that the optimal parameters l_{opt}, n_{opt} are independent of the air heating temperature. The expenses of air heating using the boreholes are calculated, with the desired air temperature T_Π = + 2 °C. The total expenses with optimal parameters constitute 40.7 thousand rub., of them capital expenses are 17.5 thousand rub (43%); the expenses of overcoming the aerodynamic resistance – 574 rub (1.4%), expenses of heating air to the required temperature (+2 °C) – 22.6 thousand rub (55.6%). The air temperature at the borehole exit reaches –9.4 °C, that is, the system increases the outside air temperature by 30.6 °C.

The energy efficiency of the proposed system in comparison with heat-exchanging workings will be assessed. A system of heat-exchanging workings increases the temperature of outside air during the coldest period of the year by, on average, 10 °C. In case of air supply through the boreholes this value will be equal to:

$$t_k'' = T_B + (t_k' - T_B) \cdot \frac{G_0'}{G''}, \, °C, \tag{6}$$

where t_k' – air temperature at the mine working exit during the coldest month with the usage of all the supplied air G_0 via mine workings, °C; T_B – average temperature of outside air in the mine workings in case of availability of boreholes, m^3/s. With the total air expenditure of 45 m³/s and supply of 30 m³/s through the boreholes the temperature at the mine workings' exit will be equal to $t_k'' = -10, °C$, that is, the temperature difference in this case constitutes roughly 30 °C.

Temperature of the mixture at the exit from the mine workings and boreholes will be:

$$t_{CM} = \frac{t_k'' \cdot G'' + t_k \cdot (G_0' - G'')}{G_0'} = \frac{-10 \cdot 15 - 9,4 \cdot (45 - 15)}{45} = -9,6, °C. \qquad (7)$$

This way, it is possible to increase energy efficiency of the system three times using the boreholes. For this example, the optimal length of boreholes is 283 m and their count 37.

The construction of boreholes of such length presents a difficulty, therefore, it is more effective it construct a set of shorter boreholes from parallel mine workings. The total number of boreholes needs to be chosen in such a way that their total depression is lower than the depression of the mine workings. A minimum number of boreholes that needs to be constructed parallel to the mine working is determined by the formula:

$$n \geq \frac{G_0' - G''}{G''} \sqrt{\frac{\alpha_c \cdot U_c S_B^3 l_c}{\alpha_B U_B \cdot S_c^3 \cdot l_B}} \qquad (8)$$

Substituting numerical values for the considered case, we will receive n = 116. Because $n > n_{opt}$, it is necessary to choose an optimal number, and the level of depression needs to be determined taking the characteristics of the existing ventilator into account. In this case, the necessary number of boreholes will be determined based on the formula:

$$n \geq (G' - G'') \frac{\alpha_c U_c l_c}{S_c^3 \cdot h_d}, \qquad (9)$$

If even in this case $n > n_{opt}$, such a number is chosen which it is technically possible to achieve, and with a fixed n parameter, the optimization task is solved again, finding the optimal parameters of air expenditure and borehole length [6].

3   Conclusions

The main conclusions are the following:

1. An algorithm and a program for solving the task of choosing the optimal parameters of a system of heat regime regulation using mine workings and boreholes.
2. It is established that usage of boreholes significantly increases the energy and economic efficiency of mine air conditioning systems. With conditions character-istic of the mines of the North, the usage of boreholes allows to increase energy efficiency of the mine systems three times.

References

1. Galkin, A.F.: Integrated use of mine openings in criolithic zone. Metall. Min. Ind. **2**, 312–315 (2015)
2. Ying, M.W., Chu, Y.: Geothermal preheating of mine intake air. Trans. Inst. Min. Met. **A94**, 189–194 (1985)
3. Galkin, A.F.: Thermal control mining system design. Metall. Min. Ind. **4**, 396–399 (2015)
4. Dyad'kin, Yu.D.: Osnovy gornoi teplofiziki dlya shakht i rudnikov Severa (Basics of Mining Themophysics for Underground Mines in the North, in Russian). Nedra, Moscow (1968)
5. Kim V.P.: Ispolzovanie nizkopotencialnyh istochnikov tepla na shahtah Severa (Use of low-potential heat sources in the mines of the North, in Russian), Yakutsk (1991)
6. Bazaar, M., Shetty, K.: Nelineynoe programmirovanie. Teoriya I algoritmy (Nonlinear programming. Theory and algorithms, in Russian). Mir, Moscow (1983)

Monitoring System for Previous Convergence Measurements in Tunnel Construction

Stephan Großwig[1(✉)] and Maria-Barbara Schaller[2]

[1] GESO GmbH & Co. Projekt KG, Loebstedter Strasse 50,
07749 Jena, Germany
stephan.grosswig@geso.eu
[2] GGB mbH, Leipziger Straße 14, 04571 Roetha, Germany

Abstract. During excavation and operation of geotechnical projects such as tunnels, their deformation affects other infrastructure facilities. Therefore, continuous monitoring is required to minimize potential hazards. This task can be realized with the methods of DSTS - technology. Fiber optic sensor technology has progressed at a rapid pace over the last decade. Strain measurement with a distributed Brillouin scattering based sensor system provides excellent opportunity for geotechnical monitoring. Using this technology, it will be possible to create the wide metrological basis for an evidence-based risk assessment in geotechnology. This makes it possible, among other things, to optimize the construction process in such a way that a balance is created between improving safety and reducing costs.

Keywords: Monitoring system · Tunnel construction

1 Distributed Brillouin Scattering Based Sensing Technology

One class of Brillouin-based sensors is based on the Brillouin loss technique, whereby two counter-propagating laser beams, a pulse and a CW, exchange energy through an induced acoustic field. When the beat frequency of the laser beams equals the acoustic (Brillouin) frequency, v_B, the pulsed beam experiences maximum amplification from the CW beam. By measuring the depleted CW beam and scanning the beat frequency of the two lasers, a Brillouin loss spectrum centered about the Brillouin frequency is obtained.

The sensing capability of Brillouin scattering arises from the dependence of the Brillouin frequency, v_B, on the local acoustic velocity and refractive index in glass, which has a linear strain dependence through

$$v_B(T_0, \varepsilon) = C_\varepsilon(\varepsilon - \varepsilon_0) + v_{B0}(T_0, \varepsilon_0) \tag{1}$$

where C_ε are the strain coefficient, and ε_0 and T_0 are the strain and temperature corresponding to a reference Brillouin frequency v_{B0}. By varying the spatial resolution, it can provide the scale of material strain measurement and structural strain monitoring.

Spatial information along the length of the fiber can be obtained through optical time domain analysis (OTDA) by measuring propagation times for light pulses

© Springer Nature Switzerland AG 2018
W. Wu and H.-S. Yu (Eds.): *Proceedings of China-Europe Conference on Geotechnical Engineering*, SSGG, pp. 1149–1151, 2018.
https://doi.org/10.1007/978-3-319-97115-5_57

traveling in the fiber. This allows continuous distributions of the strain to be monitored. The spatial resolution (gauge length) can be varied according to the application required, even after the fibers have been installed in the structure, by simply altering the length of the light pulse used. These systems offer unmatched flexibility of measurement locations and the ability to monitor a virtually unlimited number of locations simultaneously.

2   Monitoring System for Previous Convergence Measurements in Tunnel Construction

Over the past few years the number of tunneling projects in Europe has increased considerably. The construction of tunnels is still one of the most dangerous construction projects. Although the security measures have improved significantly in recent years, but during tunneling there are still numerous incalculable risks. Cause of accidents are mostly loose sand, gravel, and mud masses within solid rock layers that penetrate during excavation in the tunnel and very often lead to fatal accidents. The geology of the mountain range represents often a major problem for the tunnel builders, because certain geotechnical factors are often impossible to estimate by the geologist or civil engineers (Fig. 1).

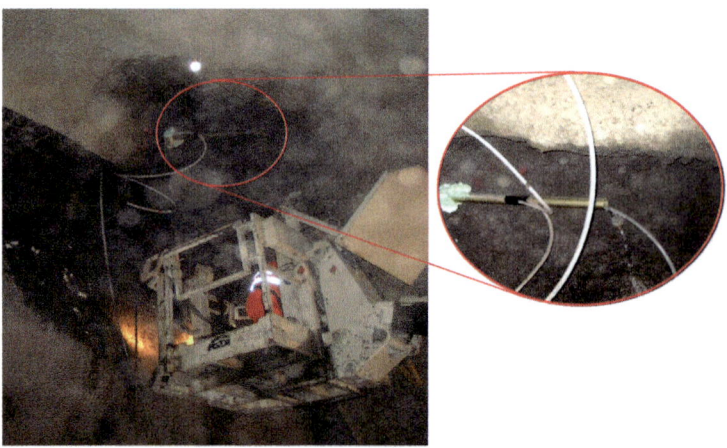

Fig. 1. Situation during the instrumentation process

The objective of an ongoing R&D project is to develop a method for previous convergence measurements in tunnel new construction or renovations tunnel. The system is intended to allow the measurement of convergences (movements of the rock) qualitatively and quantitatively already during mined excavation, to assess and determine appropriate measures to secure the mountain and for correcting the forward speed.

Thus, an amount to increase working safety will be made. To achieve this goal it is planned to install a measuring system during mined excavation along the tunnel axis using the DSTS System.

The Brillouin – technology makes it possible to determine differences in temperature and expansions in fiber optic cables. The innovation of this R&D project is to fix a fiber optic cable rigidly to a support medium (tube) und the instrumentation in a borehole for connection to the surrounding rock. Possibly occurring temperature fluctuations must be compensated for in the measurements.

A large series of experiments also in situ have been performed statistically supported. The results will be presented. A Reliable monitoring system for in situ sensing and in situ assessment of the condition of a structure is offered at the market (Fig. 2).

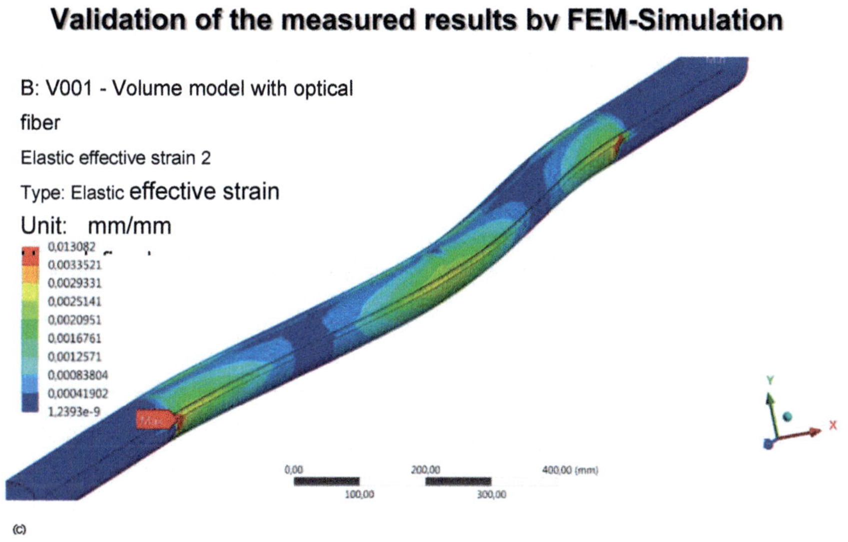

Validation of the measured results bv FEM-Simulation

Fig. 2. FEM simulation of the optical fibres behavior

3 Summary and Outlook

A Reliable monitoring system for in situ sensing and in situ assessment of the condition of a structure is offered at the market. Using the Software GKSpro®, the new system enables the detection of complicated geometrical subjects, their visualization and therefore the control and efficient interpretation.

The consistent linking of measurement technology and complex in-situ measurement data processing permits the preliminary approval of proportionality and necessity of expensive and complex reconstructions. Therefore, an up-to-date decision by the responsible authorities is given.

Numerical Investigation of Impact of Non-circular Tunneling in Sensitive Soft Clay Layers

Tadashi Hashimoto[1], Yujian Liu[1(✉)], Teruo Nakai[1],
Hossain Md. Shahin[2], Yaohong Zhu[3], and Zibo Dong[3]

[1] Geo-Research Institute, Osaka 5400008, Japan
liu@geor.co.jp
[2] Islamic University of Technology, Gazipur, Bangladesh
[3] Ningbo Rail Transit, Ningbo, China

Abstract. Large section non-circular shield tunnel has been used for a new metro line where the underground space was not enough for the traditional parallel twin circular shield tunnel in the central area of Ningbo city in China. Ground of the construction site contains layers of very soft sensitive clay having a typical mechanical behavior of stress softening and others due to the destruction of the soil structure. In consequence, long-term consolidation settlement and inclination of the adjacent buildings could occur. In this paper, 2D soil-water coupling elastoplastic FE-analyses have been carried out to predict the impact of tunneling on the pile foundation of adjacent buildings, using FEMtij-2D program. As the constitutive model of soils, the Subloading t_{ij} model has been used which properly consider typical consolidation and shear behaviors of various kind of soils under general three-dimensional stress conditions.

Keywords: Non-circular shield tunnel · FEM
Soil-water coupling elastoplastic

1 Introduction

Ningbo city is one of the large harbor cities in the east of China with over 7.6 million populations and an extensive Metro network is necessary to meet the demand of the current traffic volume. The ground contains particularly soft clay layers with high sensitive property which causes some typical mechanical behaviors of stress softening and others due to destruction [1]. In addition to this troublesome geotechnical condition, Metro line No. 4 is planned to underpass the central area of the city, where dense old buildings with pile foundations restrict the space for the metro tunnels. The common used parallel twin circular tunnels method takes extra width because it needs a certain gap between the parallel tunnels. On the other hand, a double-track large diameter circular tunnel takes too much cross-sectional area that could induce any impact to the sensitive clay. Therefore, a double-track tunnel with 10.6×6.04 m^2 non-circular section shield tunnel was chosen to this project.

© Springer Nature Switzerland AG 2018
W. Wu and H.-S. Yu (Eds.): *Proceedings of China-Europe Conference
on Geotechnical Engineering*, SSGG, pp. 1152–1155, 2018.
https://doi.org/10.1007/978-3-319-97115-5_58

A field test was conducted in a branch into the maintenance area. As the result, the non-circular shield tunnel is successfully constructed in the soft clay ground. The settlement of ground surface in construction period can be controlled in an acceptable range. However, relatively large displacement in long terms could occur as the monitor data of the ground settlement show no convergence even some months after the boring machine passed through [2].

In this study, a prediction analysis using 2D soil-water coupling elastoplastic Finite Element Method (FEM) based on a constitutive model - Subloading t_{ij} model, was conducted to evaluate the impact on the pile foundation of adjacent buildings from this kind of non-circular shield tunneling in the soft clay ground for a long period.

2 Analysis and Results

The ground condition, the location of the metro tunnel and pile foundations of the adjacent building of the simulation model are shown in Fig. 1.

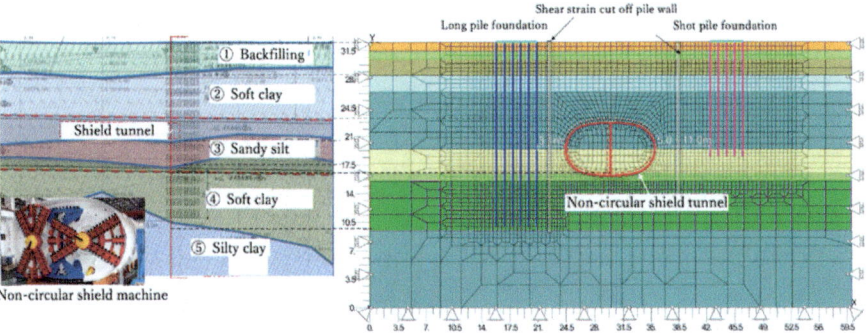

Fig. 1. Ground condition and analysis model

In order to determine the mechanical parameter of the structured soft clay, conventional consolidation tests and consolidated undrained triaxial compression tests on undisturbed samples were conducted. The undisturbed samples were obtained using thin-walled tube sampler, cushion packaged samples were transported to the lab carefully to avoid vibration acted on the samples as much as possible. Parameters which are necessary to the Subloading t_{ij} model can be partially determined directly by the tests mentioned above. Those include compression index λ and swelling index κ; initial void ratio, e_0, void ratio at the NCL for p = 98 kPa, N, and the principal stress ratio at critical state in triaxial compression - $R_{cs} = (\sigma_1/\sigma_3)_{CS(comp)}$ [3].

There are also β, which is responsible for the shape of the yield surface, $a^{AF} = a^{IC}$ which are the density parameters, and ω group including $b^{AF} = b^{IC}$, which mainly reflects the bonding property of structured clay. These parameters can be determined by simulating the conventional consolidation test and consolidated undrain triaxial compression test simultaneously, which means this constitutive model can take the consolidation and shear deformation into consider, simultaneously. Figure 2 shows the results of parametric simulations and all the parameters are listed in Table 1.

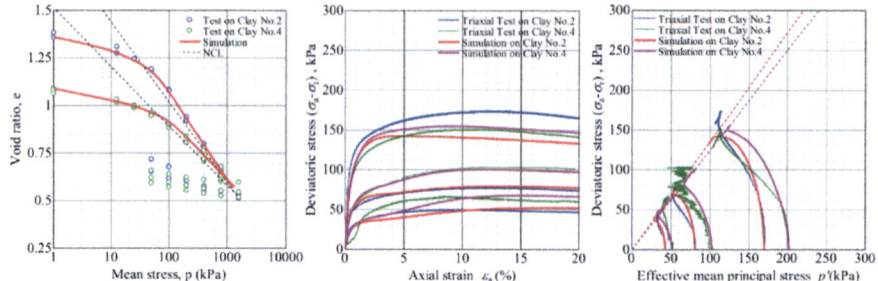

Fig. 2. Simulations on consolidation tests and triaxial compression tests

Table 1. Parameters of soil

Name	No.	γ_t (kN/m³)	e –	λ –	κ –	N –	Rcs –	β –	a –	b –	ω –	k_s (cm/sec)×10⁻⁶	k_y (cm/sec)×10⁻⁶
backfill soil	①₁ₐ	16.8	1.36	0.18	0.018	1.02	3.50	1.9	80	3.50	0.35	5.200	4.633
clay	①₂	16.8	1.36	0.18	0.018	1.02	3.50	1.9	80	3.50	0.35	5.200	4.633
muddy clay	①₃ᵦ	18.8	1.36	0.18	0.018	1.02	3.50	1.9	80	3.50	0.35	0.229	0.214
clay	②₁	18.5	1.36	0.18	0.018	1.02	3.50	1.9	80	3.50	0.35	3.997	3.557
muddy clay	②₂ᵦ	17.0	1.36	0.18	0.018	1.02	3.50	1.9	80	3.50	0.35	0.299	0.221
sandy slit	③₁ₐ	18.8	0.90	0.12	0.001	1.10	3.15	2	50	0.00	–	100.700	68.233
muddy clay	④₁ᵦ	17.7	1.09	0.13	0.022	0.90	3.15	1.7	230	3.00	0.40	0.405	0.351
clay	④₂ₐ	18.1	1.09	0.13	0.022	0.90	3.15	1.7	230	3.00	0.40	3.406	3.041
slit like clay	⑤₁ᵦ	18.8	1.12	0.18	0.012	1.00	3.20	1.7	400	3.00	2.80	5.901	4.718

In this study, the impact due to the shield tunneling is simulated by an impact ratio corresponds to the magnitude of the stress-release as percentage of the initial stress. When the impact ratio is controlled under 10%, ground surface settlement is 1.3 cm, however, it will be 3.23 cm after 400 days. The surface settlement is 4.83 cm for impact ratio of 20% and it will be 9.15 cm after 400 days, as seen in Fig. 3.

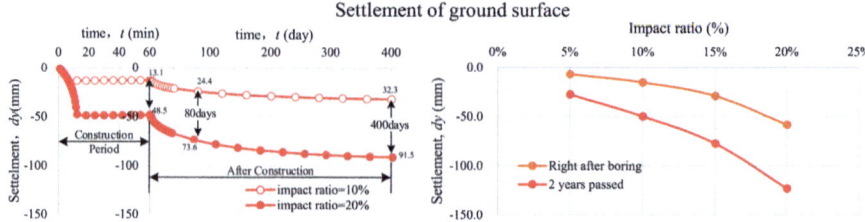

Fig. 3. Settlement of ground surface

The effect to the short piles is more significant than that to the long piles even though the distance of the short piles is further than that of the long piles. The relative depth between the pile-tips and the tunnel plays major role which can be verified by the distribution of maximum shear strain in Fig. 4.

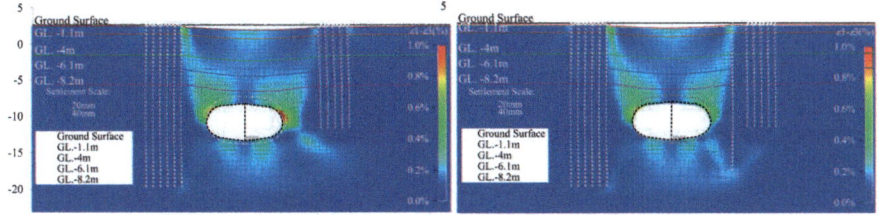

Fig. 4. Shear strain distribution shows pile sheet reduced the impact on the pile-foundations

3   Conclusion

In such soft clay, the settlement of ground surface during the construction period is about half of that caused by long-term consolidation.

The settlement and inclination of the building where the pile tips are deeper than the invert of the tunnel are smaller than those with the shorter pile, even though their positions are closer to the tunnel.

Analyses with shut-off pile-wall which is installed between the tunnel and the existing building are also carried out, to ensure the effect of the countermeasure. The simulations show that the wall which is deeper than the spring line of the tunnel is effective to reduce the settlement and inclination of the building. It can be concluded that the countermeasure - shut-off pile-wall which is installed between the tunnel and the existing building is an effective method to reduce the de-formation of the adjacent pile foundation.

References

1. Chen, C., Ye, G., Zhu, Y., Hashimoto, T.: Depositional history, laboratory tests and constitutive model of Ningbo soft clay. Chin. J. Geotech. Eng. (2018, Submitted)
2. Zhu, Y.H., Zhu, Y.F., Huang, D., et al.: Development and application of key technologies to quasi rectangular shield tunneling. Tunn. Constr. **37**(9), 1055–1062. (in Chinese)
3. Nakai, T., Shahin, H.M., Kikumoto, M., et al.: A simple and unified three-dimensional model to describe various characteristics of soils. Soils Found. **51**(6), 1149–1168 (2011)

Numerical Analysis of Prefabricated Column-Base Connections in Tunnels

Haixi Jiang[1], Longjin Li[1], Jie Cao[2], and Haitao Yu[1(✉)]

[1] Department of Geotechnical Engineering, Tongji University, Shanghai, China
yuhaitao@tongji.edu.cn
[2] China Jikan Research Institute of Engineering Investigations and Design, Co., Ltd., Xi'an, China

Abstract. Large shield tunnels with double decks inside have been developed rapidly for highway tunnels in China, and prefabricated structural members are applied frequently for their internal structures. Based on the engineering practice of Zhuguanglu tunnel in Shanghai, the mechanical performance of two different column-base connections is studied through three-dimensional numerical analyses. Results showed that: the specimen with grouted splice sleeve connection had the same load-bearing capacity as the specimen with cast-in-site connection. Plastic hinge appeared at the bottom of the columns in both cases, while the upper region of the grouted splice sleeve connection shows significant stress and severe plastic damage indicating a second plastic hinge.

Keywords: Prefabricated structural members
Grouted splice sleeve connection · Cast-in-site connection

1 Introduction

Large shield tunnels with double decks inside have been developed rapidly for highway tunnels in China. Due to the limitation of space, it is hard to organize the construction procedure for the cast-in-site method. Therefore, it is more efficient to apply prefabricated members. A large shield tunnel with double decks, named Zhuguanglu Tunnel is in construction in Shanghai. The outer diameter of the tunnel is 14.00 m, and the thickness of segments is 0.6 m. Most of the internal components of the tunnel are prefabricated. The connection between components has a great influence on the load-bearing capacity of prefabricated structures. The connection should be convenient for construction; meanwhile, it must be robust enough to maintain its integrity under seismic loading [1]. This paper studies the performance of the column-base connection in Zhuguanglu Tunnel. Because of the successful precedents, grouted-splice-sleeve-connected longitudinal reinforcements are used. Studies [2, 3] have shown that the bearing capacity and seismic performance of the columns with grout sleeves connecting longitudinal reinforcements are close to the cast-in-site columns. Three-dimensional finite element models were built for the specimen with grouted splice sleeves and cast-in-site specimen.

© Springer Nature Switzerland AG 2018
W. Wu and H.-S. Yu (Eds.): *Proceedings of China-Europe Conference on Geotechnical Engineering*, SSGG, pp. 1156–1159, 2018.
https://doi.org/10.1007/978-3-319-97115-5_59

2   Three-Dimensional Finite Element Models

Two specimens are designed and modeled. The cross-section of the column is a square with each width of 500 mm. Longitudinal steel of the columns in two specimens consist of 10 bars with diameter of 28 mm distributed along the circumferential direction of the cross-section. Transverse stirrups of the columns have the diameter of 14 mm and are arranged at a spacing of 100 mm within the range of 700 mm from the bottom of the columns at a spacing of 150 mm.

The column-base connection is the only difference between the two specimens. For the cast-in-site specimen, the column and the base are cast together; by contrast, the other specimen is cast separately. For the specimen with grouted splice sleeves, 10 sleeves are embedded at the end of the precast column. Correspondingly, 10 steel bars are extended from the base.

As shown in Fig. 1, the concrete members are represented by eight-node brick elements. The steel reinforcement is modeled by two node truss elements. The steel bars are embedded inside the concrete. The concrete damage plasticity model and the bilinear elasto-plastic model are used to simulate the elastic-plastic behavior of the concrete and reinforcement, respectively.

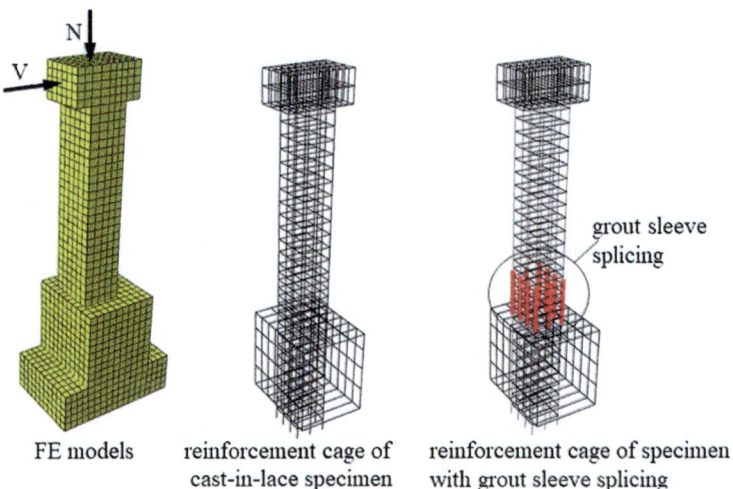

FE models   reinforcement cage of   reinforcement cage of specimen
             cast-in-lace specimen   with grout sleeve splicing

Fig. 1.   Finite element models of the specimens.

The total length of the sleeves adopted in Zhuguanglu Tunnel is 560 mm, and the inner and outer diameters are 51 mm and 70 mm, respectively. In order to study the mechanism of the grouted splice sleeves, a uniaxial tensile test is carried out. Results of the test shows that the ultimate failure mode is the rupture of the rebar outside the sleeve. During the entire test process, deformation within the sleeve section is very small and approximately linear. Therefore, linear elastic truss elements are used to simulate the grouted splice sleeves. The axial tensile stiffness of the sleeves is k = 1553.2 kN/mm.

The equivalent elastic modulus, E, of the sleeves is 481.71 GPa, which could be obtained by k = EA/l (A is the area of cross-section of the sleeves, and l is the length of the sleeves). The downside of the base is fixed in the vertical direction. The cast-in-site column is perfectly tied to the base. For the grouted splice sleeve connection, hard contact and penalty friction contact are used to simulate the contact between the column and the base in normal and tangential directions, respectively. Axial force (N = 1500 kN) is applied at the top of the columns and remains constant; then, monotonic increasing lateral loading is applied on the side face.

3   Results and Analyses

Numerical results show that tensile damage of the cast-in-site specimen initiates from the bottom of the column and develops upward gradually. The tensile damage of concrete in the sleeve is smaller than that of the cast-in-site specimen because the sleeve restricts the tensile cracking.

The stress of rebars near the bottom of the column is largest in both cases, and the longitudinal reinforcement has yielded on one side. However, for the specimen with grouted splice sleeve, the stress of rebars is great at the upper part of the sleeve area, too, while the stress within the sleeve is very small.

Figure 2 shows the curves of lateral force-displacement on the top of the columns. It could be seen that the F-u curves of the two specimens are very similar. The yield loading is about 280 kN. Both curves could be divided into three parts. Firstly, when the lateral force F is less than 120 kN, the two curves are almost identical and shows linear increment with the lateral force. Secondly, when F is greater than 120 kN and smaller than the yield loading, the stiffness of the specimens decreases. Lateral displacement of the specimen with grouted splice sleeve is smaller in response to the same increment of lateral force. After the yielding phase, the two curves show similar trends.

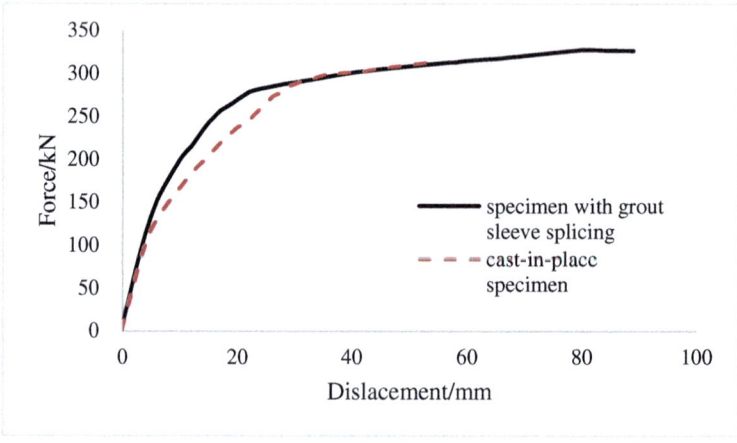

Fig. 2. Force-displacement curve.

4   Conclusions

Following conclusions could be obtained through numerical analyses:

The bearing capacity of the specimen with grouted splice sleeve connection is similar to that of the cast-in-site specimen. When the lateral loading reaches the yield loading (280 kN), longitudinal reinforcement of both specimens yield at the bottom of the columns. The yield loading is slightly larger than the calculated value (251 kN) according to the *Code for design of concrete structures (GB50010)*.

Plastic hinge appears at the bottom of the cast-in-site column. For the specimen with grouted splice sleeve connection, apart from the plastic hinge at the bottom of the column, significant stress and severe damage occur at the upper region of the grouting sleeves, which may be a second plastic hinge.

Acknowledgements. The research has been supported by the National Key Research and Development Plan of China (2017YFC1500703), the National Natural Science Foundation of China (51678438 & 51478343 & 51778487), the Shanghai Rising-Star Program (17QC1400500), and the Shanghai Committee of Science and Technology (16DZ1200302 & 16DZ1201904 & 18DZ1205 103). The authors acknowledge the support from the Fundamental Research Funds for the Central Universities of China.

References

1. Marsh, M.L., Stanton, J.F., Wernli, M.: NCHRP Report 698: Application of Accelerated Bridge Construction Connections in Moderate-to-High Seismic Regions (2011)
2. Wei, H., Xiao, W., Wang, Z.: Experimental study on seismic performance of precast bridge pier with grouted splice sleeve. J. Tongji Univ. (Nat. Sci.) **07**, 1010–1016 (2016)
3. Su, Q., Xie, Z., Lu, S.: Experimental research on grout sleeve splicing of rebars technology in prefabricated pier. Prestress Technol. **05**, 11–16 (2013)

Experimental Study on the Properties of Geocell-Reinforced Embankments

Lihua Li, Feilong Cui, Zhi Hu$^{(\boxtimes)}$, and Henglin Xiao

Hubei University of Technology, Wuhan 430068, Hubei, China
huzhi@hbut.edu.cn

Abstract. A series of model experimental tests were conducted to compare the performance of reinforced embankments with geocells. Several design factors of reinforced embankments were studied in tests, these design factors include embedment depth of geocell, number of geocell layers. Experimental results suggested that, compared with unreinforced embankments, reinforced embankments effectively improved bearing capacity, reduced vertical and lateral displacements, and more uniformly distributed additional stresses. These characteristics indicate that the reinforced embankments can effectively enhance the stability of embankments. This study also revealed an optimal embedment depth for the geocell layers. The bearing capacity of geocell-reinforced embankments increased with the number of reinforcement layers, meanwhile the vertical and lateral displacements decreased.

Keywords: Geocell · Reinforced embankment · Bearing capacity
Stability · Reinforcement effect

1 Introduction

Various planar reinforcement, such as geogrid, geotextiles, welded wire mesh and steel strips, are often used to improve soil strength and reduce its compressibility [1–7]. In recent years, three-dimensional (3D) (i.e., cellular) reinforcement materials such as geocell and waste tires tend to offer superior performance than two-dimensional (2D) (i.e., planar) reinforcement materials, particularly when soft soil is encountered. Krishnaswamy et al. [8] and Zhang et al. [9] reported that the bearing capacity of embankments significantly increases due to the inclusion of geocell. Huang et al. [10] compared the reinforcement effects of geocell with those of single-layer and double-layer geogrid, and it is concluded that geocell reinforcement achieved the best results in terms of the deformation mitigation. Tafreshi and Dawson [11] conducted model tests of a geocell-reinforced embankment slope under cyclic loading, and they found that geocell reinforcement can effectively reduce the cumulative deformation. Latha et al. [12] found that the geocell-supported embankment can effectively improve the stability of embankment.

This study outlines an experimental program used to compare the behaviors of embankments constructed with different types of reinforcement schemes.

© Springer Nature Switzerland AG 2018
W. Wu and H.-S. Yu (Eds.): *Proceedings of China-Europe Conference on Geotechnical Engineering*, SSGG, pp. 1160–1163, 2018.
https://doi.org/10.1007/978-3-319-97115-5_60

2 Method and Materials

Model scale experiments were conducted inside a chamber, which was framed by timber and braced steel plates as shown in Fig. 1. The embankment was built using clean river sand, and the geocell was made from polymer with high tensile strength connected by rivets. One side of the model box was built with thick plexiglass for observation of embankment deformation. 5 mm-diameter steel balls, indicated by A1 to D4, were placed in the embankment backfill, adjacent to the plexiglass, in order to measure the displacement of soil during the experiment. A vertical load was applied to the embankment via a loading plate. The loading was terminated when significant settlement of loading plate was observed. To test the additional stress in embankment, six earth pressure cells (Full Scale = 1.5 MPa) were placed at locations B1, B3, C1, C3, D1, and D3. By applying the symmetry of testing model, the earth pressure and displacements diagrams were depicted directly.

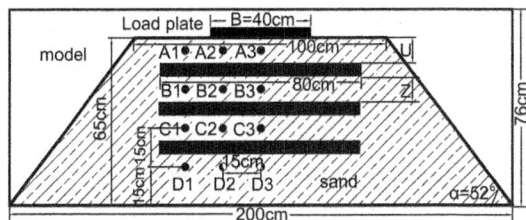

Fig. 1. Schematics of reinforced embankment systems.

3 Results and Discussion

Test results show that the settlement of embankment slope increases with the load increase, and the settlement rate also increases gradually, especially on the later stage of loading. When the settlement of embankment slope rapidly increases and many cracks appear on its surface, the corresponding applied load can be defined as the bearing capacity of embankment. From Fig. 2a, the bearing capacity for unreinforced and reinforced embankments can be the load at settlement levels of s/B = 0.03 and s/B = 0.05, respectively (s is the settlement of loading plate and B is the width of the loading plate).

Compared to unreinforced embankments, the bearing capacity of geocell-reinforced embankments increases by 143%, 157%, and 114% with the embedment depth of 0.125B, 0.25B, and 0.375B, respectively. We may attribute this to the geocell's confinement effects, which restrain the soil's lateral deformation. Additionally, the geocell's net effects and the vertical friction between the soil and the geocell's sidewalls also enhance the bearing capacity and reduce the settlement.

Figures 2b–d shows the vertical stress at locations B1 and B3, vertical displacements at locations B1, B2, and B3 and lateral displacements at locations A1, B1, C1, and D1 under a 1.5 MPa load. Parabolic distributions of vertical stress and

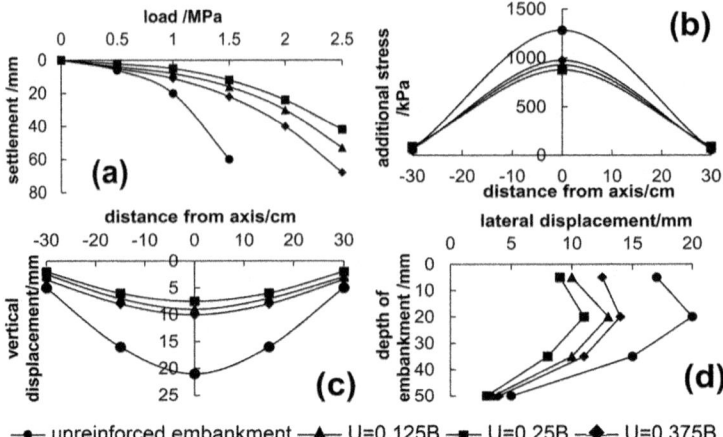

Fig. 2. Effect of embedment depth on (a) load-settlement curves (b) vertical stresses (c) vertical displacement and (d) lateral displacement.

displacement in the embankment are observed. In the reinforced embankment, the distribution of stress and displacement is more uniform than in the unreinforced embankment. With increasing geocell embedment depth, the peak values of stress and displacement firstly decrease and then increase. When the geocell embedment depth is 0.25B, the peak stress and displacement both reach the minimum value. These results suggest that 0.25B is an optimal embedment depth for geocell layers.

To examine the effects of the number of reinforcement layers on the embankment, load-settlement curves and induced responses are presented in Fig. 3, where vertical

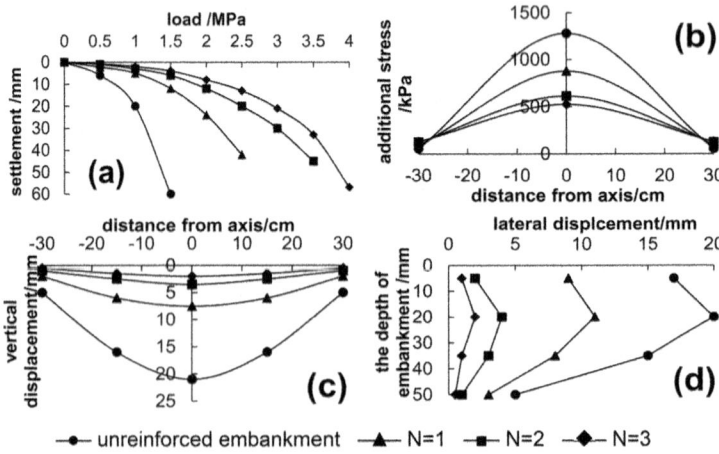

Fig. 3. Effect of number of reinforcement layers on (a) load-settlement curves (b) vertical stresses (c) vertical displacement and (d) lateral displacement.

spacing between reinforcement layers Z is set as 0.25B. From the load-settlement curves, the bearing capacity of reinforced embankment increases with the number of reinforcement layers and is larger than that of unreinforced embankment. The observation locations for induced responses are the same with the study of embedment depth. It can be obtained that the peak values of additional stress, vertical displacement, and lateral displacement decrease with an increasing number of reinforcement layers.

Although geocell's reinforcement effectiveness increases with an increasing number of reinforcement layers, additional reinforcement layers do not make major contributions in reducing the embankment's overall settlement, since the embankment's deformation mainly concentrates in the top layer, and the restrain effects are not fully mobilized. As the performance improvement of reinforced embankment slope between 2 and 3 reinforcement layers is not obvious, the optimal reinforcement layer is chosen as 2.

Acknowledgments. The authors would like to extend their gratitude to Hubei Provincial Department of Education project (No. D20151402), the National Natural Science Foundation of China (No. 51678224, 521778217), National Program on Key Research Project of China (No. 2016YFC0502208) and the Open Research Fund of State Key Laboratory of Geomechanics and Geotechnical Engineering, Institute of Rock and Soil Mechanics, Chinese Academy of Sciences (No. Z014011).

References

1. Dash, S.K., Rajagopal, K., Krishnaswamy, N.R.: Behaviour of geocell-reinforced sand beds under strip loading. Can. Geotech. J. **44**(7), 905–916 (2007)
2. Alamshahi, S., Hatal, N.: Bearing capacity of strip footings on sand slopes reinforced with geogrid and grid-anchor. Geotext. Geomembr. **27**(3), 217–226 (2009)
3. Azzam, W.R., Nasr, A.M.: Bearing capacity of shell strip footing on reinforced sand. J. Adv. Res. **6**(5), 727–737 (2015)
4. Hegde, A., Sitharam, T.G.: 3-Dimensional numerical modelling of geocell reinforced sand beds. Geotext. Geomembr. **43**(2), 171–181 (2015)
5. Xiao, H.W., Liu, Y.: A prediction model for the tensile strength of cement-admixed clay with randomly orientated fibres. Eur. J. Environ. Civ. Eng. (2016)
6. Dash, S.K., Rajagopal, K., Krishnaswamy, N.R.: Performance of different geosynthetic reinforcement materials in sand foundations. Geosynth. Int. **11**(1), 35–42 (2004)
7. Cicek, E., Guler, E., Yetimoglu, T.: Effect of reinforcement length for different geosynthetic reinforcements on strip footing on sand soil. Soils Found. **55**(4), 661–677 (2015)
8. Krishnaswamy, N.R., Rajagopal, K., Madhavi Latha, G.: Model studies on geocell supported embankments constructed over a soft clay foundation. Geotech. Test. J. **23**(2), 45–54 (2000)
9. Zhang, L., Zhao, M., Shi, C., Zhao, H.: Bearing capacity of geocell reinforcement in embankment engineering. Geotext. Geomembr. **28**(5), 475–482 (2010)
10. Huang, G.J., Zhang, Q.L., Yu, X.J., Luo, Q., Cai, Y.: Comparison of settlement control of reinforcement layer. Chin. J. Geotech. Eng. **23**(5), 598–601 (2001). (in Chinese)
11. Tafreshi, S.N.M., Dawson, A.R.: A comparison of static and cyclic responses of foundations on geocell reinforced sand. Geotext. Geomembr. **32**, 55–68 (2012)
12. Latha, G.M., Rajagopal, K., Krishnaswamy, N.R.: Experimental and theoretical investigations on geocell-supported embankments. Int. J. Geomech. **6**(1), 30–35 (2006)

Study on the Joint Bending Stiffness of Large-Diameter Shield Tunnel: A Case Study

Yu Liang and Linchong Huang[(✉)]

Sun Yat-sen University, Guangzhou 510275, China
hlinch@mail.sysu.edu.cn

Abstract. Large-diameter shield tunnel is wildly adopted to accommodate the increasing traffic demands in China. As the diameter of the tunnel segmental lining increases, the external loading condition also becomes more complicated, resulting in highly non-uniform and nonlinear internal forces. Depend on the loading condition, the segments joints exhibit different mechanical behaviors. To obtain an accurate evaluation of the internal forces in the segmental lining, it is of great importance to precisely characterize the joint stiffness in prior. In this paper, the internal forces of the tunnel segments, as functions of the joint stiffness, are derived based on force method. Various types of external loading, including the grout buoyancy force, are considered in the derivation. Based on the evaluated loading condition, the mechanical behaviors of the joints are categorized into different cases, in which the effective bending stiffness of the joints is derived accordingly. The coupled nonlinear equations of the joint stiffness and internal forces are solved via fixed-point iteration method. The approach proposed in this work is applied on a large-diameter road tunnel project in China. The calculation results show a good agreement with field measured data and numerical results. It is found that the joint bending stiffness differs from each other greatly at different locations, which is influenced significantly by outer loads. The joint bending stiffness of tunnel can be quickly obtained with this method mentioned in this work.

Keywords: Shield tunnel · Segmental lining · Joint bending stiffness
Internal force · Fixed-point iteration

1 Introduction

Large-diameter shield tunnel is wildly adopted to accommodate the increasing traffic demands in China. As the tunnel gets larger, the lining segments are going to be subject to more complicated external loading. The various external loadings would result in the highly non-uniform and nonlinear internal forces distributed in the segmental lining. In addition, the existence of the segment joints further complicates the internal force distribution. The effective joint bending stiffness, which characterizes the bending resistance of the joints, is an important factor in the design of tunnel segmental lining. It is of great importance to precisely characterize the bending stiffness of the segment joints.

The approaches to characterize the effective bending stiffness of the joints can be divided into four groups: analytical solution [1–3], numerical simulation [4–6],

© Springer Nature Switzerland AG 2018
W. Wu and H.-S. Yu (Eds.): *Proceedings of China-Europe Conference on Geotechnical Engineering*, SSGG, pp. 1164–1167, 2018.
https://doi.org/10.1007/978-3-319-97115-5_61

experimental loading test [7–9], and empirical knowledge. Though it is the most favored and precise approach to obtain the joint stiffness, the experimental loading test approach is expensive and project-dependent. The numerical simulation, such as the FEM modeling, could be computational expensive, due to the mesh issue to accommodate the small scale of joints and large scale of segments. The theoretical analysis turns to be the most convenient approach to characterize the joint stiffness. But the theoretical models mentioned above are usually case-wise, and involve complex counting process.

The aim of this work is to develop a reliable and computational cheap approach to characterize the effective bending stiffness of the segment joints under various loading conditions. In this work, general and computational cheap contact models are developed. The proposed approach is validated based on an example of a realistic large-diameter tunnel in China, in which the location-wise joint stiffness is successfully obtained.

2 Solutions Based on Fixed-Point Iteration Method

Based on the fixed-point iteration method, the nonlinear equations about joint stiffness (k1, k2, ⋯, k5) could be obtained. The detailed implementation follows the flowchart shown in Fig. 1.

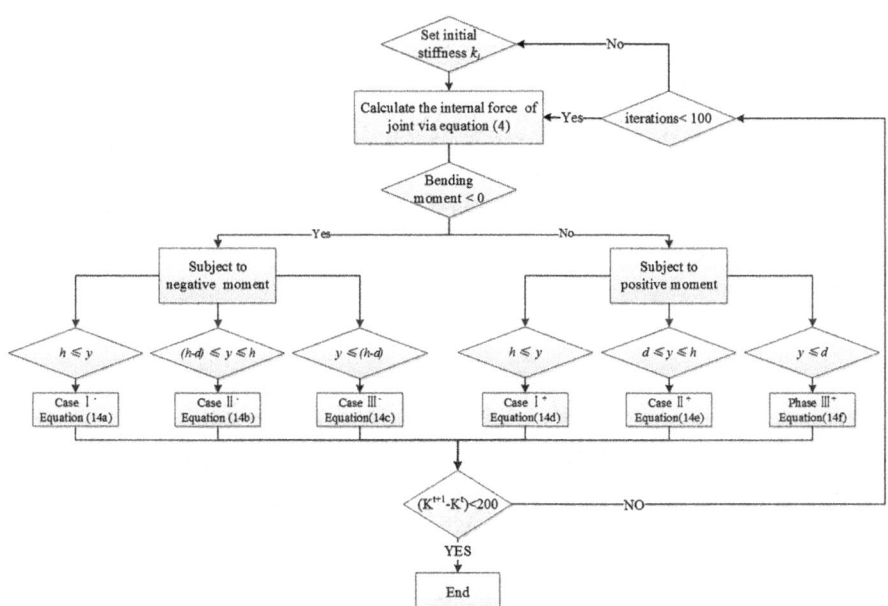

Fig. 1. The flowchart of the fixed-point method in solving the coupled equations of joint stiffness and internal forces.

3 Engineering Validation

The proposed method is applied to a realistic large-diameter underground shield tunnel in China, to evaluate the joint stiffness and internal force in the segmental lining. In this example, the grout buoyancy force is included. The initial guess of the joint stiffness is estimated based on the experimental result in Guo et al. [8]. Convergence criterion reached after 13th iteration. The results for each iteration are presented in Table 1. Though with the same initial guess for all the joints, the effective joint stiffness can converge to corresponding solution rapidly.

Table 1. The results of the bending stiffness of the joints

	K1	K2	K3	K4	K5
M (kN·m)	379.45	−68.24	−258.94	778.02	−18.98
N (kN)	989.54	1305.50	1647.30	1252.20	790.07
k (kN·m/rad)	131700	5697	12663	83469	2872
Iteration No.	13				

4 Comparison and Validation

To verify the iteration results, the field measurement of the radial pressure of the selected segments had been carried out, based on a large-diameter tunnel in China. Furthermore, numerical beam-spring models were adopted (Case 1 and Case 2), to analyze the influence of diversity of joint bending stiffness. Case 1 assumed that the joint bending stiffness varied with different loads (based on the results in Table 1), and Case 2 assumed that the joint bending stiffness was unified and equaled to 105 kN·m·rad. The bending moment along segments are shown in Fig. 2, accordingly with measured data, numerical and iteration method. We can conclude that the value of joint

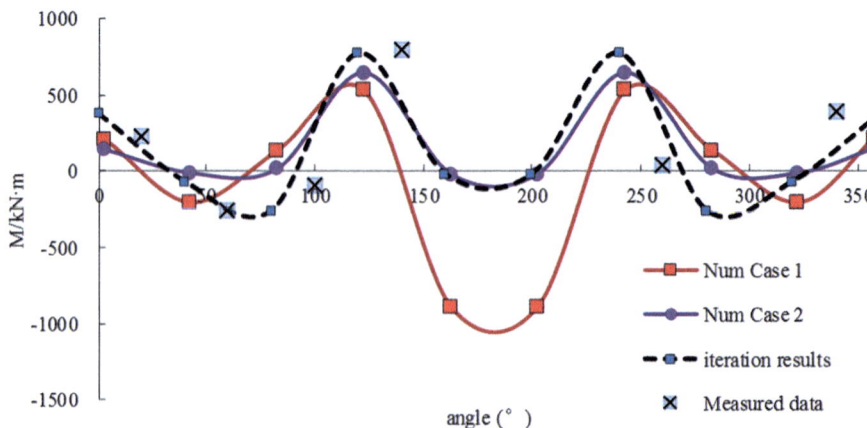

Fig. 2. The distribution curve of bending moment with different method

bending stiffness has great influence on the distribution of internal force. The location-difference of joint bending stiffness will adjust the segments deformation and reduce the fluctuation of bending moment. The calculation results with iteration method are closed to the field measured data, which has proved the reliability of iteration method. In addition, the solving speed of iteration method is dramatically faster than that of numerical method, hence applicable to different calculation conditions.

5 Summary

The various internal forces at the joints would result in different types of joint behaviors. As such the joints exhibit different effective joint stiffness, which in turn will impact the internal force distribution. Thus, the determination of the joint stiffness should be coupled with the internal forces at the joints. In this case, the joint bending stiffness is observed to have great influence on the bending moment in the lining. The bending stiffness of the joints subject to positive bending moment is much greater than that of the joints subject to negative bending moment. The proposed framework, based on force method and fixed-point iteration method, is general and computationally cheap, as well as incorporates with various external loading condition. The calculation results with iteration method are closed to the field measured data, which has proved the reliability of it. This framework can be readily applied to other tunnel project with similar lining structure.

References

1. Wood, A.M.: Discussion: the circular tunnel in elastic ground. Géotechnique 25(1), 115–127 (1975)
2. Teodor, I.: Prefabricated lining, conceptional analysis and comparative studies for optimal solution. Int. J. Rock Mech. Min. 38(3), 136–137 (1995)
3. Jiang, H.S., Hou, X.Y.: Theoretical study of rotating stiffness of joint in shield tunnel segments. Rock Mech. Rock Eng. 23(9), 1574–1577 (2004)
4. Cavalaro, S.H.P., Blom, C.B.M., Walraven, J.C., Agrado, A.: Structural analysis of contact deficiencies in segmented lining. Tunn. Undergr. Sp. Tech. 26(6), 734–749 (2011)
5. Li, Z.L., Song, K., Wang, F., Wright, P., Tsuno, K.: Behavior of cast-iron tunnel segmental joint from the 3d FE analyses and development of a new bolt-spring model. Tunn. Undergr. Sp. Tech. 41(1), 176–192 (2014)
6. Do, N.A., Dias, D., Oreste, P., Djeran, M.I.: Three-dimensional numerical simulation for mechanized tunnelling in soft ground: the influence of the joint pattern. Acta Geotech. 9(4), 673–694 (2014)
7. Koyama, Y.: Present status and technology of shield tunneling method in Japan. Tunn. Undergr. Sp. Tech. 18(2), 145–159 (2003)
8. Guo, R., He, C., Feng, K., Xiao, M.Q.: Bending stiffness of segment joint and its effect on segment internal force for underwater shield tunnel with large cross-section. China. Rai. Sci. 34(5), 46–53 (2013)
9. Li, Z.L., Soga, K., Wright, P.: Behavior of cast-iron bolted tunnels and their modelling. Tunn. Undergr. Sp. Tech. 50, 250–269 (2015)

Coal Pillar Design by Numerical Simulation to Protect an Inclined Tunnel

Baoguo Liu[(⊠)], Yi Qi, Xiaomeng Shi, and Yan Wang

Key Laboratory of Urban Underground Engineering of Ministry of Education,
Beijing Jiaotong University, Beijing 100044, China
bgliu@bjtu.edu.cn

Abstract. A finite element model simulation was conducted to evaluate the influence of coal mining on the performance of two parallel 7.3 m diameter inclined shield tunnels located approximately 15 m above a coal seam in China. The tunnels will be excavated by a tunnel boring machine and supported by a segmented concrete liner; this is the first application of this method in a coal mine in China. The 2D finite element model was constructed to study the influence of the coal mining on the stresses and displacements in the tunnel liner. The modelling was also performed to determine the allowable distance that mining could encroach on the tunnel location and thus the optimal size of the coal pillar to leave below the tunnels.

Keywords: Inclined TBM tunnel · Finite element simulation · Coal mining

1 Introduction

With the increasing depletion of shallow coal resources, many mines in China are now extracting coal from depths greater than 600 to 800 m [1]. At these depths in a coal mine, high in-situ stresses, high earth temperatures, and high water pressures can occur. These factors increase the tunnel support and maintenance requirements and increase the risk for rockbursts, water inrush, and large rock deformations [2, 3]. Mining shafts or tunnels used to access deep coal seams have a high cost and a long construction period. Typically, vertical shafts have been used for this purpose. However, the use of an inclined tunnel excavated with a tunnel-boring machine (TBM) is being considered for a coal mine in China. It is believed that a TBM will significantly shorten the construction period because simultaneous installation of support (concrete liner segments) can occur while the tunnel advances.

2 Project Overview

The New Street coal mine is operated by Shenhua Company in the province of Inner Mongolia, in China. This area is in a tectonically stable region that has not experienced any destructive earthquakes. The mine has a total area of 734.5 km^2, with a total coal resource 13287 Mt. To exploit new coal resources within the mining area, there are plans to excavate two inclined 7.3 m diameter tunnels from the ground surface to a

W. Wu and H.-S. Yu (Eds.): *Proceedings of China-Europe Conference on Geotechnical Engineering*, SSGG, pp. 1168–1171, 2018.
https://doi.org/10.1007/978-3-319-97115-5_62

depth of 692 m. These tunnels will be approximately 6.3 km long and inclined at 6°. The tunnels will be driven parallel to each other and spaced 60 m apart. One tunnel will be used to transport people underground while the other will be used to transport coal out of the mine. These tunnels will be excavated with a tunnel boring machine using the earth pressure balance shield method and semi-earth pressure balance shield method. This will be the first time a TBM has been used to excavate an inclined tunnel in a Chinese coal mine. Therefore, considerable research has been conducted to study the feasibility and risks associated with this project [4].

The inclined tunnels will start at elevation 1324 m and end at elevation 632 m above sea level. The first coal seam is 1.5 to 3.5 m thick and is about 670 m beneath the ground surface. Where the tunnels intersect the coal, the seams have a dip of up to 3°. Laboratory testing was used to measure the physical properties of the different rock types located near the coal seam and the values are listed in Table 1. The vertical stress at the upper coal seam is estimated to be 15 MPa based on the depth of the coal below the ground surface.

While two parallel tunnels are planned for the coal mine, the physical models were simplified by constructing only one of the tunnels and taking advantage of symmetry of the geometry. The distance between the edge of the numerical model and the tunnel corresponds to one half the distance between the tunnels in the field. In this paper, only the results from the model simulating a 15 m separation between the tunnel and the coal seam are presented as illustrated in Fig. 1.

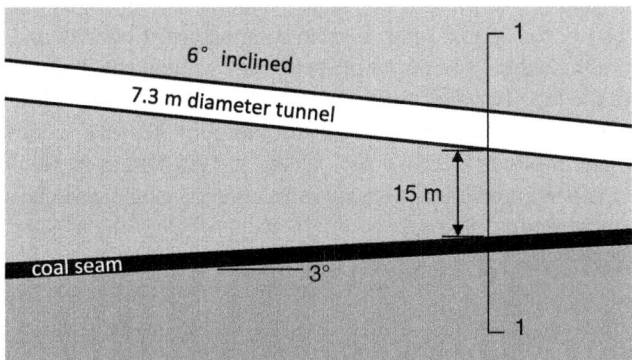

Fig. 1. Relative position of coal seam and inclined tunnel.

3   Finite Element Simulation by RS2

Numerical models are often used to examine the influence of nearby excavations on a tunnel. For example, Gong et al. [5] used a finite element model to study the influence of mining areas under existing road tunnels. The results showed that the mining was found to influence the tunnel structures when the mining approached within a horizontal distance of 30 m to the tunnel for a coal seam that was 20 m below the tunnels. In another example, Wang et al. [6] used Flac3D to study the forces and displacements at an existing tunnel that was affected by a new tunnel that crossed underneath. During a preliminary study of this project, a UDEC [7] model was built with roughly estimated rock properties to study the stress distribution around a future coal pillar [8]. This work found that there was a saddle-shaped high stress zone around the pillar and as the distance increased between the coal seam and the inclined tunnel getting, a wider coal pillar was required to keep tunnel vertical displacement in a safe range.

A finite element program was used to analyze the physical model results to provide further insight into the tunnel behaviour and the predicted stresses within the tunnel liner. The program RS2 [9] was used. A 2D finite element model was constructed to represent the physical model. The model closely replicated the dimensions of the physical model. It is consisted of 81448 nodes and 39874 triangular 6-noded elements. Mohr-coulomb model and Plane strain simulation were selected. Five types of rock with different material properties were modelled as well as the tunnel liner. Just like the physical model, the numerical model consisted of 16 layers of different rock materials. The bedding planes between each rock type were simulated with joint elements with open ends and with a normal stiffness 100 GPa/m and shear stiffness 10 GPa/m. The mechanical properties of the different materials used in the finite element model are listed in Tables 3 and 5. The modelling was done to replicate the physical model, not the coal mine, which is why the values in these tables were used. As mentioned before, a fine layer of black mica powder was used in physical model to reduce the cohesive strength between layers representing different rock types. In the finite element model, the contacts between the rock layers were represented by joint elements. The rock and the joints were allowed to yield if the Mohr-Coulomb strength was exceeded.

Figure 2 shows the model after 8 excavation stages. Roller boundary conditions were applied to the sides and bottom of the model while a constant stress of 15 MPa was applied to the top. To allow for more realistic behaviour associated with the yield of the rock near the excavated coal seam, a Voronoi joint network is chosen to simulate the coal layer and the four layers of rock above it. [10] The average joint length was 0.02 m with open-end conditions. An irregular polygon shape was chosen to simulate the materials representing the rock.

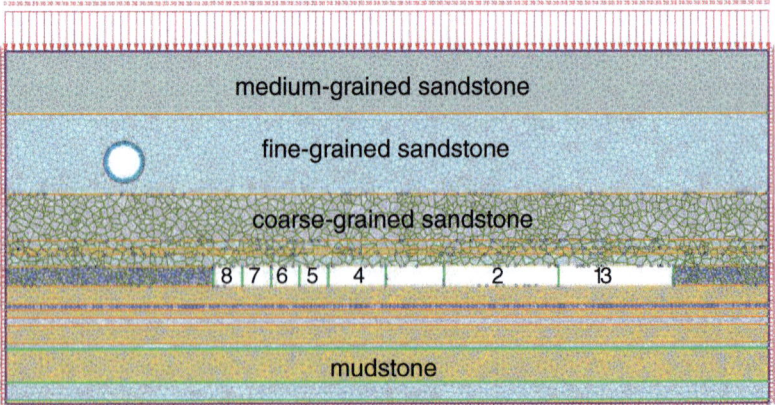

Fig. 2. RS2 finite element model

In this numerical model, a Young's modulus reduction method is used to simulate the coal excavation procedure. For each coal block that was excavated (8 blocks), the Young's modulus of the coal was reduced to 0 in three steps, to simulate the coal excavation. The modelling was run in 11 stages: application of the overburden stress, tunnel excavation, liner installation, and eight coal block excavations.

References

1. He, M.C., Xie, H.P., Peng, S.P., Jiang, Y.D.: Study on rock mechanics in deep mining engineering. Yanshilixue Yu Gongcheng Xuebao/Chin. J. Rock Mech. Eng. **24**, 2803–2813 (2005)
2. Kulatilake, P.H., Wu, Q., Yu, Z., Jiang, F.: Investigation of stability of a tunnel in a deep coal mine in China. Int. J. Min. Sci. Technol. **23**, 579–589 (2013)
3. He, M.C.: Present situation and prospect of rock mechanics in deep mining engineering. In: Proceedings of the 8th Chinese Conference on Rock Mechanics and Engineering, pp. 88–94 (2004)
4. Hou, G.-Y., Liang, R., Gong, Y.-F., Liu, L., Tian, L.: Risk analysis and trend prediction of long inclined-shaft construction in coalmine by TBM. Yantu Lixue/Rock Soil Mech. **35**, 325–331 (2014)
5. Gong, J.W., Xia, C.C., Lei, X.W., Zhou, D.H., Gao, S.J.: 3D numerical analysis of influence of construction of underlying mining area on existing road tunnel structures. Yantu Gongcheng Xuebao/Chin. J. Geotech. Eng. **35**, 120–125 (2013)
6. Wang, D.Y., Yuan, J.X., Zheng, Y., Zhu, Y.Q.: Application of WSS in mined tunnel crossing under the existing shield tunnel. Adv. Mater. Res. **838**, 1341–1345 (2013)
7. Itasca Consulting Group, Inc., UDEC Universal Distinct Element Code, Version 4.0. User's Manual, Minneapolis (2014)
8. Qi, Y., Liu, B., Ma, Q.: Determination of the width of protective coal pillar under inclined shaft with segment structure affected by mining action. Beijing Jiaotong Daxue Xuebao/J. Beijing Jiaotong Univ. **38**(4), 160–165 (2014)
9. Rocscience Inc. RS2 (2014). https://www.rocsciencecom/rocscience/products/rs2
10. Collon, P., Steckiewicz-Laurent, W., Pellerin, J., Laurent, G., Caumon, G., Reichart, G.: 3D geomodelling combining implicit surfaces and Voronoi-based remeshing: a case study in the Lorraine Coal Basin (France). Comput. Geosci. **77**, 29–43 (2015)

On Stability of Shallow Tunnel by Model Test and Numerical Simulation

Nader Moussaei[1,3(✉)], Mostafa Sharifzedeh[2,3], Kourosh Sahriar[2,3], and Mohammad Hossein Khosravi[1,3]

[1] Tehran University, Tehran, Iran
n.moussaei@ut.ac.ir
[2] Curtin University, Perth, Western Australia, Australia
[3] Amirkabir University of Technology, Tehran, Iran

Abstract. The geometry and mechanical properties of the discontinuity as well as opening dimensions are the govern parameter in the instability of the shallow underground tunnels. Presented research is relate to the analyses such structure using small scale experimental and numerical models. In the tested models beddings dip varies from 0 to 90°. Layer thickness is considered 16 mm. The tunnel width and height are considered 10, 12, and 14 times as big as the bedding thickness. Also, the joint spacing is one to two times as big as the bedding thickness. Based on the 72 model results the extension of the collapsed zone measured and its variation analyzed. Among other considered parameters, the bedding dip has notable effect on collapsed zone for shallow underground spaces.

Keywords: Shallow underground structure · Physical model · Collapsed zone

1 Introduction

Generally, stratified structures have continuously extended, and they have a planar geometry. Moreover, these structures do not sustain tensile stress in the direction perpendicular to bedding plane, and shear strength of the bedding surface is lower than intact rock [1].

Generally, all previous parametric studies using physical modelling [2, 3] have two defects. First is that they are not applied the influence of various effective factors in these models. The second is that they are considered one aspect of instability mechanism like maximum unsupported span or roof failure mechanism. To overcome these short coming, in this research, a physical model is proposed to determine the effect of bedding dip, joint spacing and tunnel dimensions on tunnel instability in stratified Structure with crosscut joint sets (masonry structure). For this purpose, a testing apparatus were developed, and tests were performed under the full-face tunnel excavation condition.

© Springer Nature Switzerland AG 2018
W. Wu and H.-S. Yu (Eds.): *Proceedings of China-Europe Conference on Geotechnical Engineering*, SSGG, pp. 1172–1176, 2018.
https://doi.org/10.1007/978-3-319-97115-5_63

2   Test Setup

The strain less steel was used as framework with the dimensions of 1000×1100 mm^2 (Fig. 2). The stratified structure models are made from 1500 blocks on average. The model material must be representative of the real material [1]. Medium-Density Fiberboard (MDF) is considered as a model material.

2.1   Experiment Design

The discontinuity dip and spacing, and Tunnel height and span are considered as main variables and experiments designed to capture reasonable range of these variables. The inclination of the bedding was considered from horizontal to vertical consisting; 0, 30, 60 and 90° with thickness of 16 mm. The discontinuities spacing was determined based on layer thickness from 1 and 2 times of bedding thickness. The continuity factor (the ratio of tunnel diameter to block diameter) is between 6 and 15 for discontinuous area; therefore, tunnel dimensions are considered as 160, 192, and 224 mm. Among 72 model tests were need to study the failure mechanisms, only 16 model simulated as small scale model using Taguchi's experimental design method, and the remain states modeled using numerical modelling.

2.2   Numerical Model

To verify the accuracy of physical modeling results, numerical modeling was also performed using Universal Distinct Element Code (UDEC) [4]. Numerical model geometry and material properties is simulated similar to the physical model (Fig. 1 and Table 1). Constitutive model for block is considered as an elastic-isotropic model; because the stress level is low and all blocks in physical model behave like rigid material. Mohr-Coulomb constitutive model assumed for discontinuities surface to allow them to shear failure. It should be noted that, numerical model was simulated under plain strain condition.

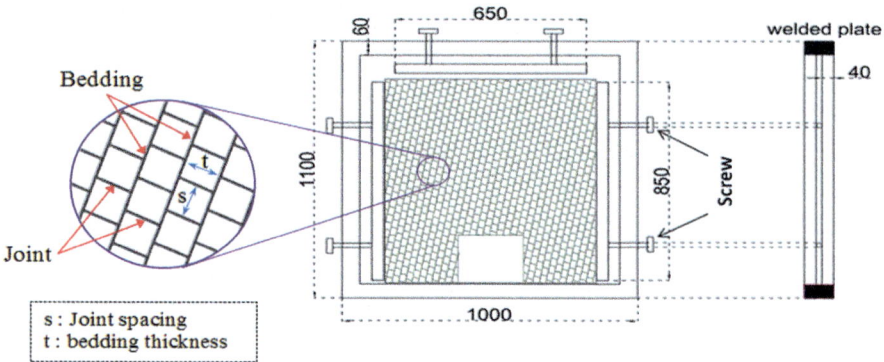

Fig. 1. Test framework. (a): Front view (b): Side view (dimension in mm).

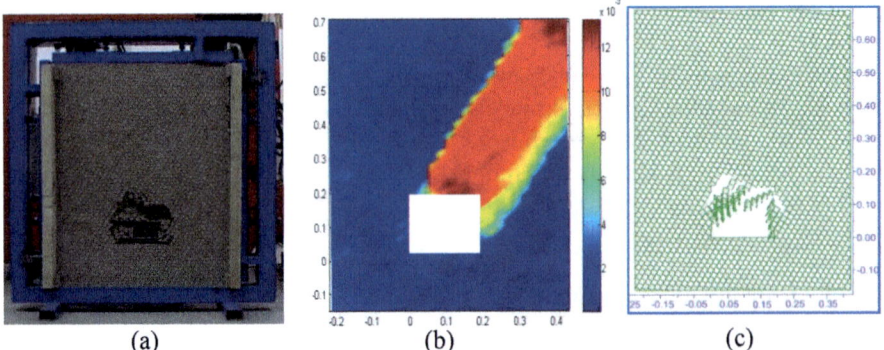

(a)   (b)   (c)

Fig. 2. Model test; (a) physical model, (b) image processing (displacement contour), (c) numerical analysis.

2.3   Model Test Results

Figure 1 shows the model failure, image velocimetry analysis, as well as numerical model for four different bedding dips. Based on 72 model tests the area of the collapsed zone have been measured. Based on the Fig. 2 the effect of the bedding dip and tunnel dimensions variation on collapsed zone is notable. The increasing bedding dip from zero to 30° makes no sharp variation in collapsed zone. On the other hand, the huge area of collapsed zone was created when bedding dip increased to the 60°; but, collapsed zone is dwindled in vertical bedding. This reduction also has affect the variation of the collapsed zone area versus tunnel height. This event is called multicollinearity in

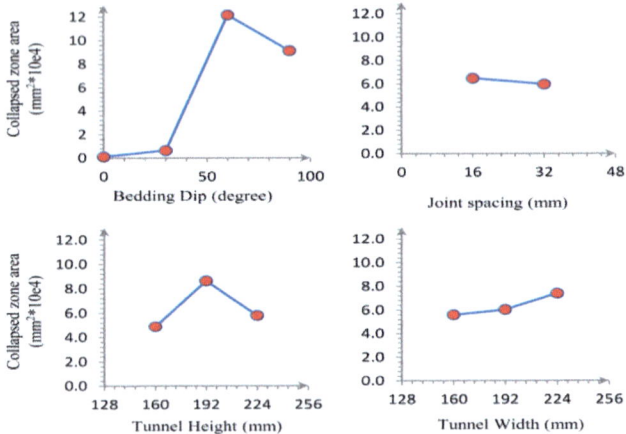

Fig. 3. Model test; (a) physical model, (b) image processing (displacement contour), (c) numerical analysis.

the regression model and it is due to highly correlated data. Three main reason control the variation of the collapsed zone versus bedding dip. The extent of potential area to collapse as showed in Fig. 3. The decreasing compressive strength with increasing the angle between discontinuity plane and axial load as uttered by jaeger and cook (1979) is the second reason (Fig. 3). Third parameter is the decreasing shear strength of the bedding surface with increasing the bedding dip due to the decreasing normal component of the block weight (Figs. 4 and 5).

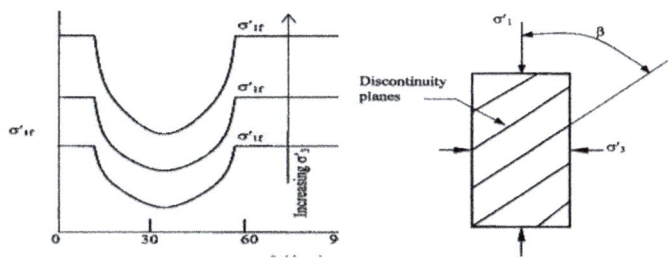

Fig. 4. Model test; (a) physical model, (b) image processing (displacement contour), (c) numerical analysis [5].

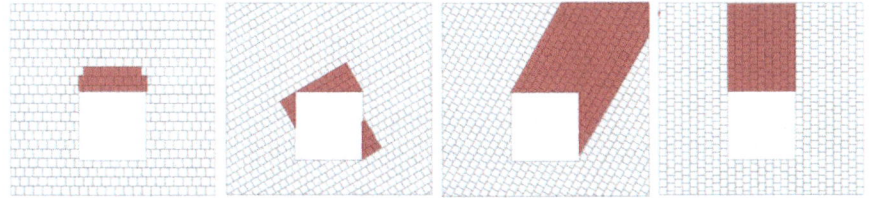

Fig. 5. Model test; (a) physical model, (b) image processing (displacement contour), (c) numerical analysis.

3 Conclusions

Presented research contain 72 shallow tunnel models done through physical small scale model as well as numerical model. The extension of the collapsed zone measured for different model with various tunnel dimensions, rock mass block size as well as bedding dip. Among other mentioned parameter bedding dip has notable effect on instability of the shallow underground structure bored in masonry structure.

References

1. Brady, B.H.G., Brown, E.T.: Rock Mechanics for Underground Mining, 2nd edn. Kluwer Academic Publisher, London (1993)
2. Moyo, T., Stacey, T.R.: Mechanisms of rockbolt support in jointed rock masses. In: Potvin, Y. (ed.) Proceedings of the Sixth International Seminar on Deep and High Stress Mining, pp. 91–103. Australian Centre for Geomechanics, Perth (2012)
3. Fuenkajorn, K., Phueakphum, D.: Physical model simulation of shallow openings in jointed rock mass under static and cyclic loadings. Eng. Geol. **113**(1–4), 81–89 (2010)
4. Itasca: UDEC—Universal Distinct Element Code, Minneapolis. Itasca Consulting Group (2004)
5. Zhang, L.: Engineering Properties of Rock, 4. Elsevier Science, London (2006)

Semmering Base Tunnel, Austria

Michael Proprenter[(✉)] and Enrico Soranzo

iC consulenten ZT GmbH, Schoenbrunnerstr. 297, 1120 Vienna, Austria
M.Proprenter@ic-group.org

Abstract. Semmering Base Tunnel, a major infrastructure project in Europe, consists of two 27.3 km single-track tubes, numerous cross passages and a complex underground emergency and ventilation station with two 420 m deep ventilation shafts. Intermediate access points include a 1.2 km long access tunnel with two 220 m deep sub-surface shafts for connection to the main tunnels, temporary 100 m deep shafts from the surface and large caverns at the shaft bases. The dimensions of the larger caverns are in the range of 20 m by 18 m. The maximum overburden is approximately 950 m in difficult geological conditions with large fault zones and extensive water inflow, which requires complex heading concepts, extensive support measures and partially special ground improvement works such as grouting. Sophisticated numerical 2D- and 3D-calculations were carried out to verify the adequacy of the designed solutions.

Construction using the NATM and TBM method is ongoing from 2014 until 2026.

Keywords: NATM and TBM · Shafts and caverns · Tunneling

1 Introduction

Semmering Base Tunnel, with a length of more than 27 km, represents the major structure of the Austrian section of the Baltic-Adriatic Railway Corridor, which connects the Baltic Sea (Gdansk in Poland) with the Adriatic Sea (Bologna in Italy). The tunnel is located 80 km south of Vienna, Austria.

The existing railway line crossing the mountain range has a steep gradient and small radii along tunnels and viaducts and does not correspond to the requirements for freight and passenger transport. After finalization of the new tunnel system, the existing railway line will remain in place as a fallback in case of maintenance works in the new tunnels and for local traffic purposes. The line is also a designated world heritage site due to its engineering excellence and therefore an often visited tourist attraction.

The tunnel system with a total length of 27.3 km includes two parallel single track tubes connected by numerous emergency cross passages. A centrally located emergency station, including two ventilation shafts, forms the core of the final tunnel system (see Fig. 1).

The total investment costs for the project are around 3.3 bil. EUR, with 1.5 bil. EUR related to the civil construction works. Pre-construction works started in 2011 and the start of operation is envisaged for 2026.

© Springer Nature Switzerland AG 2018
W. Wu and H.-S. Yu (Eds.): *Proceedings of China-Europe Conference on Geotechnical Engineering*, SSGG, pp. 1177–1180, 2018.
https://doi.org/10.1007/978-3-319-97115-5_64

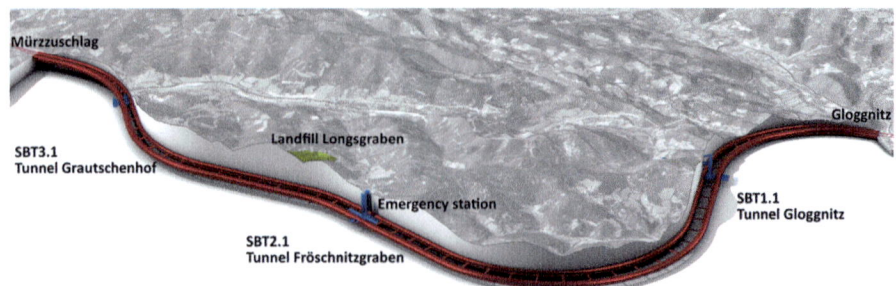

Fig. 1. Tunnel system and construction lots (source: ÖBB).

2   Underground Conditions

From the geological point of view, the project is located in an area of intense tectonic imbrication. The main geological units are the so-called Greywacke zone, the Semmering unit and the Wechsel unit.

Frequently changing conditions include rocks of metasediments such as phyllite, schist, quartzite, locally sulphate rocks and metasandstone of various tectonic units as well as carbonate rocks and highly fractured metamorphic crystalline schist and gneiss. The units are separated by numerous distinct fault zones. Especially the fault zones consisting of cataclastic fault material with high overburden are a challenge to the excavation works. In the carbonate rock formations, a water inflow of up to 300 l/s is expected.

3   Tunnel and Cavern System

The tunnel system with a total length of 27.3 km consists of two parallel single track tubes, which are connected by 56 emergency cross passages. The maximum spacing of the cross passages is 500 m. The cross passages act as emergency connections in case of an accident and accommodate technical rooms for electrical and communication purposes as well as maintenance installations. The excavation area of the main tunnels is about 75 m^2 and 35 m^2 for the cross passages.

In addition, due to national and international safety regulations, an underground emergency station with various caverns, connection tunnels and two, more than 420 m deep, ventilation shafts is located at the center of the tunnel alignment. The caverns in the emergency station provide space for ventilation and rescue purposes as well as for the accommodation of electrical and mechanical installations. The length of the emergency station is close to 1 km. In case of an accident trains shall stop at the emergency station and passengers shall be moved to a safe area and board a rescue train after arrival in the non-affected tunnel tube.

The spacing of the two tunnel tubes is around 40 m, which is enlarged at the underground emergency station as well as in massive fault zones only. At the portals the spacing is reduced to minimize land acquisition costs. The alignment of the tunnels is based on the foreseen maximum train velocity of 230 km/h with a maximum inclination of 8.4‰.

The cross sections for the tunnels, cross passages and caverns are designed with a drained, double shell lining including a waterproofing system and sidewall drainages (see Fig. 2). The cast in-situ concrete for the inner lining of the main tunnels is unreinforced except in areas of fault zones or intersections with cross passages.

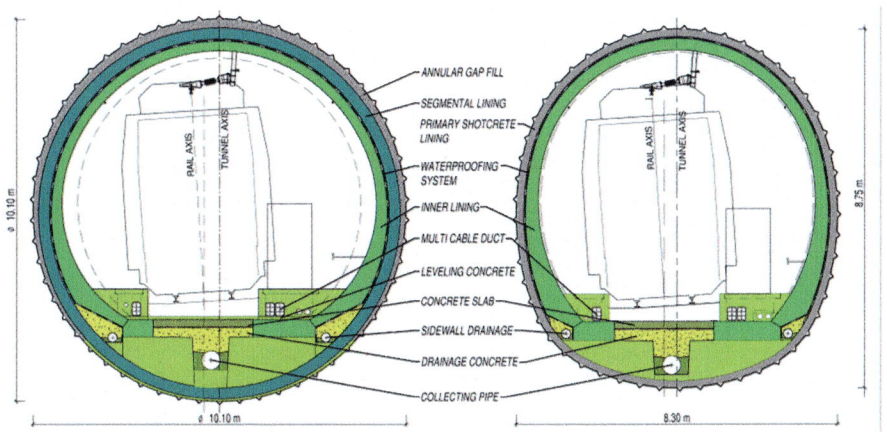

Fig. 2. Main tunnel tubes (TBM and NATM)

The track bed system includes a slab system for the majority of the tunnel. Underneath residential areas, a mass-spring system is foreseen as to minimize vibration and noise emissions by the rail traffic.

4 Construction Procedure

Due to the given timeline and logistical reasons, the Austrian Federal Railways decided to tender the underground works in three separate construction lots designated SBT 1.1, SBT 2.1 and SBT 3.1 (see Fig. 1). For the cut and cover and open cut tunnel section with a total length of 650 m, as well as the modernization of the adjacent railway station in Mürzzuschlag at the southern portal, an additional construction lot is foreseen.

In addition to the two permanent ventilation shafts at the emergency station, four temporary shafts with depths ranging between 100 and 220 m are required to provide access to the individual underground sections.

Due to the given access conditions, two shafts must be built underground, starting at the end of a 1.2 km long temporary access tunnel. At the top of the subsurface shafts and at the bases of all shafts, temporary caverns and connection tunnels provide space for site installations, areas for machinery assembly and material handling underground (see Fig. 3).

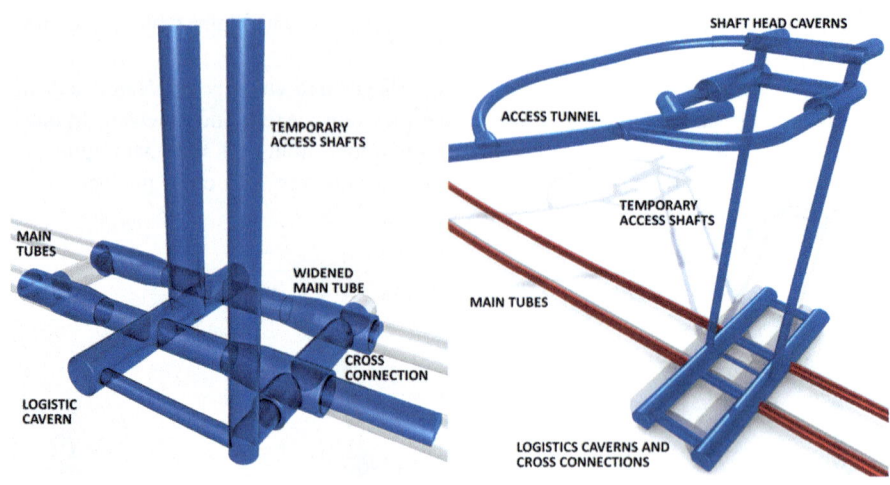

Fig. 3. Temporary shafts and caverns.

The maximum dimensions of the caverns are in the range of 20 m by 18 m. After finalization of the construction works, the access tunnel and the temporary caverns will be backfilled.

For the construction of the tunnels both construction methods, NATM and TBM, are applied. The NATM method will be used for a total length of 19 km in the northern and southern stretches of the tunnels. Based on the expected geological conditions, the TBM method will be applied only in the center part of the tunnels over a length of 8 km. The TBM assembly takes place in the underground caverns located in the emergency station. During the main construction phase, a total of 12 NATM and two TBM headings are excavated parallel at the same time, which requires a sophisticated and detailed logistics concept, at the surface as well as underground.

The support measures for the NATM method includes reinforced shotcrete, lattice girders and rock bolts as well as forepoling pipes for overhead protection. The segmental lining for the TBM tunnels consist of a 6+0 system with a minimum thickness of 30 cm.

References

1. Wagner, O.K., Proprenter, M.: Semmering Base Tunnel – Large Caverns in Challenging Conditions. WTC Bergen, Norway (2017)
2. Wagner, O.K., Proprenter, M.: Semmering Base Tunnel - 17 miles of SEM and TBM tunneling under challenging conditions in Austria, RETC San Diego, USA (2017)
3. Proprenter, M.: Semmering base tunnel – logistic challenges. In: 11th International Tunnelling and Underground Structures Conference, ITA Slovenia, Ljubljana, Slovenia (2017)

Track Differential Settlement in a Culvert-Embankment Transition Zone Due to Adjacent Shield Tunneling: A Case Study

Yao Shan, Yi Lu[✉], Li Su, and Longlong Fu

Key Laboratory of Road and Traffic Engineering, Ministry of Education,
Tongji University, Shanghai 201804, China
1632489@tongji.edu.cn

Abstract. Shield tunneling beneath an existing railroad leads to the track differential settlement, which will directly affect the operation safety of the railroad and the ride comfort of the passengers, especially in culvert-embankment transition zones. In this paper a case study is taken to investigate the track differential settlement distribution discipline due to an adjacent shield tunnel construction based on a tunnel construction project in Jinan, China. A double - line shield tunnel is planned to be constructed beneath an existing railroad between Beijing and Shanghai. A three-dimension finite element method combined with field monitoring data is employed to investigate the influence of the shield tunneling on the track differential settlement in a culvert-embankment transition zone. The results show that the construction of shield tunnel underpassing railway will cause uneven settlement of the railway subgrade and bridge transition culvert section, and by monitoring the feedback in real time, controlling the construction parameters and taking protective measures, the deformation of railway bridge can meet the control requirements.

Keywords: Shield tunnel
The railway subgrade and bridge culvert transition section · Deformation
Filed monitoring · Control measures

1 Introduction

By the rapid development of urban rail transit in china, shield tunneling under pass the railway in complex geological condition occurs frequently. when the shield tunnel is constructed beneath a culvert-embankment transition zones, there is very strict control requirement of settlement especially the differential settlement in a culvert-embankment transition zone. To ensure the safety of railway, we must reduce the deformation of foundation and the track coursed by the construction process of shield tunnel. The previous researches mainly used numerical simulation to study the deformation of railway subgrade or frame Bridges under shield tunneling, and rarely involved the difference settlement between railway subgrade and bridge culvert transition section [1–4]. In this paper a three-dimension finite element method combined with field monitoring data is employed to investigate the influence of the shield tunneling on the track differential settlement in a culvert-embankment transition zone.

© Springer Nature Switzerland AG 2018
W. Wu and H.-S. Yu (Eds.): *Proceedings of China-Europe Conference on Geotechnical Engineering*, SSGG, pp. 1181–1185, 2018.
https://doi.org/10.1007/978-3-319-97115-5_65

2 Engineering Situation

The left line of Jinan rail transit line R1 Yu Fu he station ∼ between Wang Fu Zhuang station tunnel is planned to constructed beneath the Beijing-shanghai railway frame Bridges section, and the right line of shield tunnel under railway subgrade section. The track structure is composed of ballast-track, and the speed of the train is 120 km/h. The net distance between the left line shield tunnel and the railway frame bridge culvert structure is 7.0–7.7 m, and the structure arch of the right line is 16.8 m deep. The frame bridge was built in 2010. It is a reinforced concrete frame bridge with a span of 10.5 m + 15.0 m + 15.0 m + 10.5 m. The middle two holes are motor vehicles, both sides are non-motorized lanes, with a net structure of 8.1 m, and the clearance height is 3.0 m, and the urban pipeline space is reserved under the road (Fig. 1).

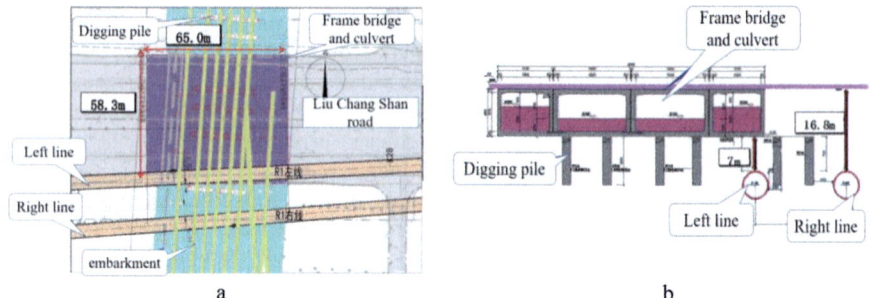

a   b

Fig. 1. Shield tunnel underpass the transition section of frame culvert and subgrade

3 Numerical Calculation

Solid elements are adopted to simulate the formation and railway subgrade. The constitutive model adopts the moori-coulomb model, the shell element simulation is adopted for the pipe, and the solid element model is adopted for the railway frame bridge, and the linear elastic model is adopted for the mechanical model. Calculation model around the boundary using roller constraints, under the surface with fixed constraint, on the surface using free constraints, considering the railway roadbed, frame Bridges with slip, may have created between formation and orbital in Bridges, establish contact surface between soil and the railway track.

3.1 Analysis of Numerical Simulation Results

The wall which is closest to the left line shield center, which is about 1.02 m. The maximum lateral displacement of the wall is about 0.625 mm. With the farther and farther away from the shield center location, the horizontal displacement of the wall decreases rapidly. The settlement of one of track structure is shown in Fig. 2 below and the maximum settlement is 3.654 mm.

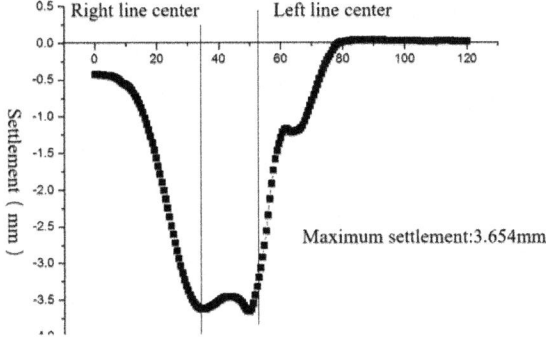

Fig. 2. The settlement of track structure

4   Field Monitoring

According to the relevant regulations, the monitoring distance is 2 times the depth of the shield segment outward. Considering the integrity of the frame culvert and the overall monitoring of the frame culvert, the mileage of this monitoring project is K475+885 to K475+998 from Beijing to Shanghai. The corresponding specific mileage of total 19 monitoring points are shown in Table 1 below:

Table 1. Three - dimensional monitoring point layout table

Mileage	Point name	Point name	Remark
K475+885	L1	L8	Side wall
K475+898	L2	L9	Mid-wall
K475+915	L3	L10	Mid-wall
K475+930	L4	L11	Mid-wall
K475+943	L5	L12	Side wall
K475+970	L6	L13	Subgrade
K475+997	L7	L14	Subgrade
K475+910	L15		Catenary mast
K475+963	L16	L18	Catenary mast and subgrade
K475+984	L17	L19	Catenary mast and subgrade

In the monitoring area, the distance between the adjacent lines above the axis of the shield tunnel is 5 m, and each line is designed at six places with a point interval of 20 m.

4.1   Monitoring Data Analysis

The construction sequence is to push the right line first and then push the left line. The relative distance in the diagram is calculated from the small mileage side of the frame bridge to the large mileage direction.

When the shield tunneling is completed, the settlement gradually becomes stable, and the culvert-embarkment transition section has a great differential settlement. The settlement of the embarkment is about two times of that of the culvert, and the settlement of the track structure presents a large settlement trough, with the maximum settlement of 0.9 mm (Figs. 3 and 4).

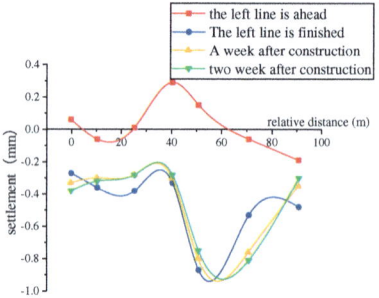

Fig. 3. Settlement of culvert and subgrade

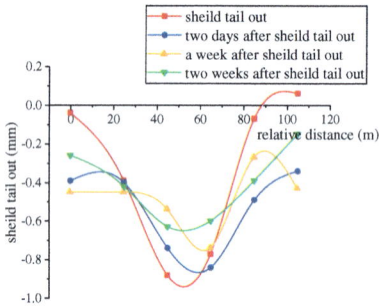

Fig. 4. Deformation of track structure

5   Conclusions

The results of numerical simulation and field monitoring show that the deformation caused by shield tunnel under-passing through the culvert-embarkment transition section is controllable through reasonable construction control measures. The results of numerical simulation are larger than the field monitoring, and the law of settlement deformation is consistent, which can provide reference for similar projects in the future. The settlement of the embarkment is about two times of that of the culvert and the differential settlement of culvert-embarkment transition section cannot be ignored.

References

1. Wang, Q., Xie, X.Y.: Study of settlement through over quadruple-tube parallel shield tunnel crossing railway track. Chin. J. Rock Mech. Eng. **36**(S2), 4235–4243 (2017)
2. Yan, Ch.: Analysis on safety in construction of shield tunnel undergoing through frame bridge on existing railway. Railw. Stand. Des. (7), 84–87(2011)
3. Zhang, H.Zh.: Settlement analysis and control research in metro shield tunnel tunneling underneath existing railway. Beijing Jiaotong University (2015)
4. Lu, Y.J.: Analysis and measures for risk of metro shield tunneling crossing underneath tracks of railway station. Chin. J. Undergr. Space Eng. **9**(6), 1412–1418 (2013)

Effects of Fault Zones on Failure Mechanism of Tunnel, an Experimental Study

Ba Thao Vu[1,2], Hehua Zhu[2(✉)], Qianwei Xu[2,3], and Xiaoying Zhuang[2,4]

[1] Department of Geotechnical Engineering, Hydraulic Construction Institute,
Vietnam Academy for Water Resources, Hanoi, Vietnam
[2] State Key Laboratory of Disaster Reduction in Civil Engineering,
College of Civil Engineering, Tongji University, Shanghai, China
zhuhehua@tongji.edu.cn
[3] Urban Rail Transit and Railway Engineering Department,
Tongji University, Shanghai, China
[4] Institute of Structural Mechanics, Leibniz University Hanover,
Hanover, Germany

Abstract. This paper presents a large-scale experiment for studying the effects of fault zones on the failure process and failure pattern of tunnel. Firstly, appropriate synthetic materials representing fault zones and rock mass were developed. The mix of fine sand and grease was adopted to model the fault zones. Secondly, an experimental model was then performed to investigate the effects of fault zones on the progressive failure of tunnels. It was found that the presence of fault zones caused the asymmetry of convergent deformations and failure patterns. The fault zone itself failed initially and became a slip surface, which then led to the failure of surrounding rock masses and eventually caused the collapse of the tunnel. Failure mechanisms of tunnels with a fault zone obtained by model tests are helpful for the implementation of optimum support designing.

Keywords: Synthetic material · Fault zone · Progressive failure
Tunnel

1 Introduction

Physical model tests have been one of the most effective means to investigate the progressive failure of rock mass surrounding the tunnel due to the model being tested till collapse state. This study aimed to investigate only the most unfavorable parameters of the fault zones on the instability of the tunnel. The effects of various parameters of fault zones which are dip, locations, in situ stress states, shear strength of the fault zones, on the instability of the tunnel were summarized based on the results of some typical previous studies [1–3]. They found that the most unfavorable parameters of the fault zones are: the fault zone passes through the roof with the dip of 50°; thickness of the fault is 0.9 m; the stress ratio is 0.25.

© Springer Nature Switzerland AG 2018
W. Wu and H.-S. Yu (Eds.): *Proceedings of China-Europe Conference on Geotechnical Engineering*, SSGG, pp. 1186–1191, 2018.
https://doi.org/10.1007/978-3-319-97115-5_66

2   Experimental Procedures

2.1   Similar Materials

The rock mass of Grade IV, based on the design codes of road tunnel in China, has been chosen as a prototype for this study. Proportion of raw material has been selected: Barite powder : Gypsum : Fine sand : Laundry detergent : Water = 12:4:1.8:0.33:1. Three kinds of materials, namely, talcum powder, fine sand, and a mix of fine sand-grease were used to develop the similar material for fault zone through a series of large-scale interface shear tests. Development process and characteristics of the model material is presented in another article [4].

2.2   The Test Rig and Hydraulic Loading System

The tests were conducted using a large-scale two-dimensional test rig. The test rig consists of a stiff modular frame and a hydraulic loading system. The load system is able to generate non-uniform loads and automatically control. Dimensions of the model test are 2.0 m × 2.0 m × 0.4 m. Tunnel profile is located at the center of the model. A transparent glass plate with dimension of 0.8 m × 0.8 m is installed at the center to observe the deformation of the tunnel (Fig. 1). The referred prototype for the tests was two-lane cross sections tunnel; span and height of tunnel were 11.4 m and 5.1 m, respectively. By scaling down with the ratio of 1:30, the tunnel span and height were 0.38 m and 0.17 m, respectively. The initial overburden depth of tunnel, H was 30 m. Particle Image Velocimetry (PIV) technique was adopted for measuring the deformation of rock mass surrounding the tunnel [5].

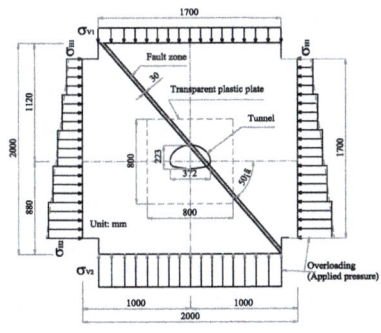

Fig. 1. Sketch and photo of model test of tunnel with fault zone (unit: mm).

2.3   Test Scheme

To investigate the whole failure process of tunnel surrounding rock mass, the model test process was divided into two stages. In the first stage, the actual excavation process and the load applied on the model equivalent to the initial overburden depth of 30 m is tested. The second stage is conducted to evaluate the behavior of surrounding rock mass under the higher in-situ stresses by increasing the overloads step by step.

The test scheme were: (1) Model construction: The model material was placed and then compacted homogeneously inside the test frame. (2) Apply loads equivalent to the overburden depth. (3) Excavation: The excavation sequence was: left part – right part – middle upper part – middle lower part. (4) Overloading process: The overloading was increased gradually and the tunnel finally collapsed. The stress ratio was increased simultaneously with unchanged ratio, as it guaranteed that the lateral pressure coefficient remained unchanged during overloading process.

3 Results and Analyses

The results revealed that the fault zone has a significant effect on the progressive failure of the tunnel. Compared to other tests where the tunnel was derived through homogeneous medium [6], the failure process of surrounding rocks differed completely due to the presence of the fault zone. The failure process can be classified into five main stages: (1) Fault zone failure and slip surface formation, (2) Initiation of crack near the fault zone, (3) Failure wedges initiation, (4) Recurring failure process of the wedges and, (5) Collapse, as seen in Fig. 2.

Stage 1: Fault Zone Failure and Slip Surface Formation. The presence of fault zone led magnitudes of tunnel strain to increase significantly. After complete excavation, the tunnel strain at the crown and floor were up to 2.3% and 0.4%, respectively, whereas negligible in the case without fault, as seen in [6]. The effects of the fault zone on the instability of the tunnel were displayed more evidently with additionally applied overloading. Fortunately, the fault zone failing and then becoming a slip surface can be identified through the shear band that was attained by PIV technique. This shear band confirmed that the material inside the fault zone had failed and acted as a slip surface.

Stage 2: Initiation of Crack Near Fault Zone. When the overloading increased to 90 m, the tunnel strain at the crown rose to 3.1%. The crack first initiated underneath the fault zone in the crown area. Then, numerous micro-cracks occurred above the fault zone at the right corner of the tunnel. With an increase in applied stress, the crack at the crown propagated downwards in a direction perpendicular to the fault zone and eventually met the tunnel periphery. A potential failure zone was sequentially formed.

Stage 3: Initiation of Failure Wedges. The surrounding rocks in the crown region were divided into three major deformation zones: the first one was located upon the fault and the other appeared below the fault. The third zone, named as potential failure zone, formed within a region covered by the initial crack, fault and tunnel boundary; it tended to move into the opening. While applied pressure was increasing, the potential failure zone was gradually falling into the opening. It is named "failure wedge" in this study. As soon as the failure wedge formed, rock masses close to the fault in the crown region became unstable and then resulted in the failure wedge to expand upwards.

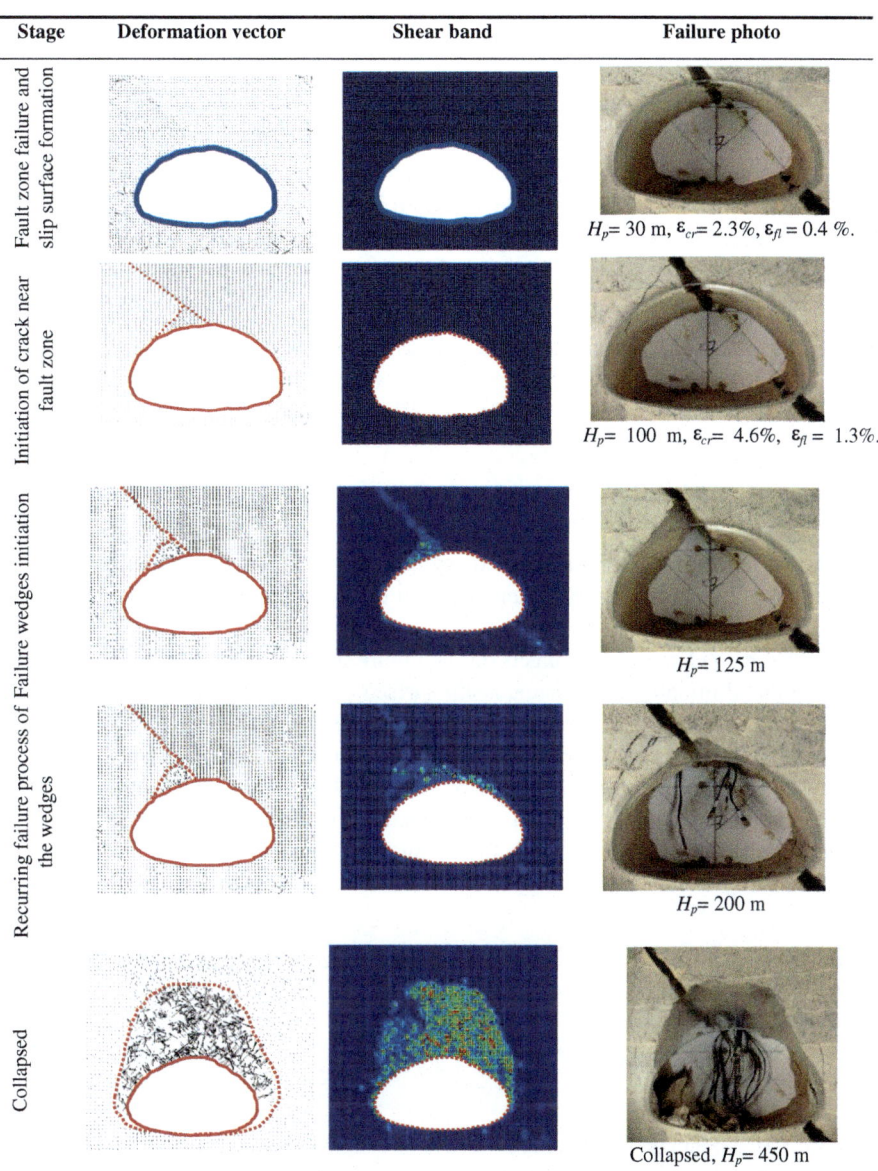

Stage	Deformation vector	Shear band	Failure photo

Fig. 2. Failure process and failure patterns of tunnel.

Stage 4: Recurring Failure Process of the Failure Wedge. A tendency of initiation and propagation of the recurring failure wedges at the crown can be summarized as follows:

- Crack firstly initiated underneath the fault zone in the crown region and then propagated downwards in a perpendicular direction to the fault zone.

- A failure wedge occurred when a closed region was formed by initial crack, fault zone and tunnel periphery.
- The failure wedge expanded upwards and symmetric geometry was established; it remained stable unless further overloading was applied.

From this pattern, it can be concluded that the initial failure wedge underneath the fault zone played a significant role in controlling the failure process of the whole surrounding rocks. To improve the stability of tunnel in case a fault zone passes through the crown, a reasonable rock bolt support system is required and installed at initial failure wedge.

Stage 5: Collapsed. The tunnel collapsed when the overloading increased to 450 m. The entire failure process of tunnel can be seen visibly from the pattern of displacement vector and shear zone that attained by the PIV technique.

4   Conclusions

- The presence of a fault zone caused asymmetric convergent deformation and failure pattern of tunnel where cracks and the failure zone occurred around the intersections of the fault and tunnel boundary.
- Fault zone had significant effects on the failure process of the tunnel. The fault zone itself failed initially and became a slip surface, which then led to the failure of rock masses close to it and eventually caused the collapse of tunnel.
- Failure process and failure pattern of tunnels with a fault zone attained through the present experiment are helpful for the implementation of optimum support designing. It can be suggested that a local support system needs to be installed around the intersections of the fault and tunnel boundary to prevent the failure of the fault itself and the rock masses close to it.

Acknowledgements. This study was financially supported by Natural Science Foundation of China (No. 41130751).

References

1. Hao, Y.H., Azzam, R.: The plastic zones and displacements around underground openings in rock masses containing a fault. Tunn. Undergr. Space Technol. **20**(1), 149–161 (2005)
2. Huang, F., Zhu, H.H., Xu, Q.W., Cai, Y.C., Zhuang, X.Y.: The effect of weak interlayer on the failure pattern of rock mass around tunnel - scaled model tests and numerical analysis. Tunn. Undergr. Space Technol. **35**, 207–218 (2013)
3. Zhang, Z.Q., Chen, F.F., Li, N., Swoboda, G., Liu, N.F.: Influence of fault on the surrounding rock stability of a tunnel: location and thickness. Tunn. Undergr. Space Technol. **61**, 1–11 (2017)
4. Vu, B.T., Zhu, H.H., Xu, Q.W., Dinh, T.H.M.: Development and application of model material for tunnel model test in soft rock mass. In: Proceedings of the 6th International Symposium on In-situ Rock Stress, Sendai, pp. 876–885 (2013)

5. White, D., Take, W.: Particle Image Velocimetry (PIV) software for use in geotechnical testing. Cambridge University Engineering Department, Technical report CUED/D-SOILS/TR322 (2002)
6. Vu, B.T., Zhu, H.H., Xu, Q.W., Zhuang, X.Y.: Physical model tests on progressive failure of unsupported tunnel in soft rock. In: Proceedings of the 26th KKHTCNN Symposium on Civil Engineering, Singapore (2013)

Field Study on the Behavior Super-Long and Large-Diameter Grouted Drilled Shafts

Zhihui Wan[1], Guoliang Dai[1(⊠)], Weiming Gong[1], and Mingxing Zhu[2]

[1] Key of Laboratory for RC and PRC Structure of Education Ministry, School of Civil Engineering, Southeast University, Nanjing 210096, China
daigl@seu.edu.cn
[2] China Energy Engineering Group Jiangsu Power Design Institute, Nanjing 211102, China

Abstract. This paper presents bi-directional O-cell loading tests on two super-long and large-diameter drilled shafts with diameter of 2 m and lengths of more than 110 m. The field test results show that the ultimate bearing capacity of the shaft after post-grouting at the shaft tip and side is 1.95 times that of the shaft without post-grouting, and the total side resistance and the base resistance of the shaft after grouting at the shaft tip and side are 183% and 280–314% larger than that of the shaft without post-grouting, respectively. Consequently, the combined tip-and-side grouting has a significantly effect to improve the bearing characteristics of super-long and large-diameter drilled shaft. As the load reaches the maximum test load, the proportion of the load carried by the shaft base is not more than 20% before and after combined grouting. Furthermore, the relationship between side resistance and shaft-soil relative displacement for super-long and large-diameter drilled shaft is described by the hyperbolic model, and this model is also used to capture the base resistance-base displacement relationship. In addition, the base displacement required to fully mobilize the base resistance for the shaft after combined grouting is about 1.0% of shaft diameter.

Keywords: Super-long and large-diameter drilled shaft · Post-grouting Osterberg cell method · Base resistance · Side resistance

1 Introduction

With the continuous development of high-rise buildings and large bridges, the diameter and length of drilled shafts are increasing constantly, whereas super-long and large-diameter drilled shaft with a length of more than 50 m and a diameter of more than 2.0 m are also becoming more common. However, the problems of super-long and large-diameter drilled shafts are mainly manifested in the difficulty of construction, and quality is difficult to be guaranteed. For the above-mentioned issues, post-grouting technique has been gradually applied to the construction of super-long drilled shafts, especially in important projects [1, 2]. At present, post-grouting technique has become an effective method to enhance the bearing capacity of shaft and ensure the quality of

© Springer Nature Switzerland AG 2018
W. Wu and H.-S. Yu (Eds.): *Proceedings of China-Europe Conference on Geotechnical Engineering*, SSGG, pp. 1192–1196, 2018.
https://doi.org/10.1007/978-3-319-97115-5_67

shaft foundation [3, 4]. However, owing to the complexity of the mechanism for post-grouting technique, there is a need to analyze the response of grouted drilled shaft with super-long and large-diameter.

Based on two in-situ loading tests of Shishou Yangtze River Highway Bridge engineering, the behaviors and performance of combined tip-and-side grouting drilled shaft with super-long and large-diameter are studied by comparing the field test results of combined grouting shaft and shaft without grouting. The behaviors of side and base resistance under axial load are further investigated on the basis of the obtained data of strain gauges.

2   Project Description and Soil Conditions

Shishou Yangtze River Highway Bridge crosses the Yangtze River, which is located in Shishou, Central China. The proposed bridge will adopt long-span (which has a length of 820 m) of cable stayed bridge. The length of the main bridge is 1.45 km. According to the geological drilling at the bridge site, the results show that no bedrock is found within 180 m of the exploration. The soil layer at the site is composed of silty clay, fine sand and gravel. The design lengths of the main pylon shafts are up to 120 m. To investigate the behavior of super-long and large-diameter drilled shafts, it is necessary to test shafts in the proposed bridge site. The tested drilled shafts TS5 and TS6 were 2.0 m in diameter, had lengths of 110 m and 115 m, respectively.

3   Test Results

After the test shafts were formed, the shafts TS5 and TS6 were conducted to combined tip-and-side grouting. The detailed combined post-grouting sequences were discussed by [2]. The bi-directional O-cell test was applied in the test shafts, and double-level O-cell was mounted in the shafts for this test. The detailed about the double-level O-cell tests were described by [1]. In the following, based on the results of field test, we will study the behaviors and performance of super-long and large-diameter drilled shafts before and after post-grouting.

3.1   Load-Displacement Responses

The test results of the bi-directional O-cell before and after post-grouting were equivalently converted into the load-displacement curves of conventional static load test. Detailed equivalent conversion of the bi-directional O-cell test was described by [1]. Figure 1 presents the equivalent load-displacement responses of the shafts before and after combined grouting. It can be observed that the equivalent load-displacement curves of the shafts TS5 and TS6 before and after combined grouting have obvious inflections, and the equivalent displacement of the shafts increases gradually with increasing the equivalent load. [2] proposed that the load prior to failure can be regarded as the ultimate bearing capacity of a single shaft. Therefore, the ultimate bearing capacity of the shafts TS5 and TS6 before and after combined grouting are

52.22, 50.90, 101.44, and 99.12 MN, respectively. Consequently, the bearing capacity of combined grouting shaft is about 1.95 times than that of the shaft without grouting.

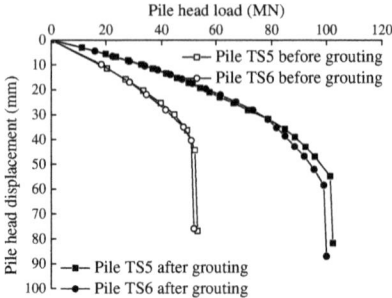

Fig. 1. Equivalent load versus displacement curves of the test shafts.

3.2   Mobilized Side and Base Resistance

Figure 2 shows the side resistance-relative displacement relationship and the base resistance-base displacement relationship for the shafts TS5 and TS6. It can be found that the side resistance-relative displacement relationship and the base resistance-base displacement relationship for the shafts are described by the hyperbolic model, as well as have a higher accuracy. It can also be observed from Fig. 2b that the base displacements required to fully develop the base resistance for the shafts before and after combined grouting are 11.39, 6.73, 20.01, and 22.24 mm, corresponding to 0.6%, 0.3%, 1.0%, and 1.1% of the shaft diameter, respectively. It is noted that the base displacement needed to mobilize base resistance for the shaft before grouting is smaller than that after grouting.

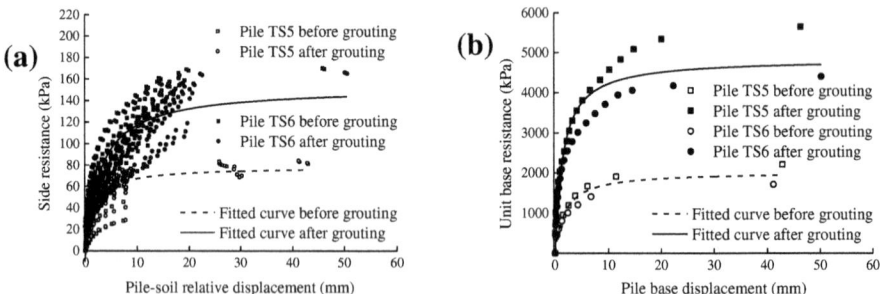

Fig. 2. (a) Measured and fitted relationship between side resistance and shaft-soil relative displacement, (b) Measured and fitted relationship between base resistance and base displacement.

The ultimate bearing capacity, total shaft resistance, base resistance of each shaft, and their corresponding increased range are summarized in Table 1. The results in

Table 1 show that the ultimate bearing capacity of the shafts after combined grouting in extra-thick fine sand layer is increased by about 95%, the total side resistance is increased by about 83%, and the base resistance is increased by 179.96–213.70%. At the ultimate load of the shafts TS5 and TS6, the ratios of the base resistance to the ultimate bearing capacity for the shafts before combined grouting are 8.67% and 13.97%, respectively, and increases to 11.47% and 16.53% after combined grouting, respectively. The test results show that the function of super-long and large-diameter drilled shaft after combined grouting still shows a friction shaft.

Table 1. Test values of ultimate bearing capacity, total side resistance, base resistance and their increase proportion.

Pile no.	Before combined grouting		After combined grouting		$\frac{Q_{su}'-Q_{su}}{Q_{su}}$ (%)	$\frac{Q_{bu}'-Q_{bu}}{Q_{bu}}$ (%)	$\frac{Q_{u}'-Q_{u}}{Q_{u}}$ (%)
	Total side resistance Q_{su} (kN)	Mobilized base load Q_{bu} (kN)	Total side resistance Q_{su}' (kN)	Mobilized base load Q_{bu}' (kN)			
TS5	46235	5989	84677	16768	83.14	179.96	94.25
TS6	48486	4414	85273	13846	83.44	213.70	94.74

4   Conclusions

The paper presents the results of the bidirectional O-cell loading tests on two super-long and large-diameter drilled shafts before and after combined grouting. The test results demonstrate that the bearing capacity of combined grouting shaft is obviously larger than that of the shaft without grouting under the same conditions. At the ultimate load level, the total side resistance and base resistance for the shafts after combined grouting are increased by about 83% and 179.96–213.70%, respectively. Meanwhile, the proportion of the load carried by the shaft base after combined grouting is not more than 20%. The combined grouting drilled shaft with super-long and large-diameter is functioning as a friction shaft. Moreover, the relationship between side resistance and shaft-soil relative displacement for the shafts before and after combined grouting is described by the hyperbolic model, and the relationship between base resistance and base displacement for the shafts before and after combined grouting is also addressed in this model. Additionally, the base displacement needed to fully mobilize the base resistance for the shafts after combined grouting is about 1.0% of shaft diameter.

Acknowledgements. The authors would like to acknowledge financial support from the National Natural Science Foundation of China (51478109; 51678145) and China Postdoctoral Science Foundation (2017M611955).

References

1. Dai, G., Gong, W., Zhao, X., et al.: Static testing of pile-base post-grouting piles of the Suramadu bridge. Geotech. Test. J. **34**(1), 34–49 (2010)
2. Wan, Z., Dai, G., Gong, W.: Full-scale load testing of two large diameter drilled shafts in coral-reef limestone formations. Bull. Eng. Geol. Environ. **10**, 1–17 (2017)
3. Bruce, D.A.: Enhancing the performance of large diameter piles by grouting. Ground Eng. **19**(4), 9–15 (1986)
4. Mullins, G., Dapp, S., Frederick, E., et al.: Post grouting drilled shaft tips: Phase I. Research Report University of South Florida, Tampa, USA, pp. 127–156 (2001)

Numerical Modeling of the Effects of Ground Fissure Dislocation on Metro Tunnel

Ming Wu[⊠], Jianbing Peng, Yahong Deng, and Xin Liu

School of Geological Engineering and Geomatics,
Chang'an University, Xi'an 710054, China
d05wuming@zju.edu.cn

Abstract. The formation of ground fissures is caused by tectonization activities such as earthquake or by human engineering activities such as excessive pumping of underground water. The latter has become a major cause of ground fissures formation in Xi'an, China. These ground fissures are still active and has become an obstacle for metro construction. In this paper, Xi'an metro line 3 was taken as a practical example to analyze the dislocating effects on shallow buried tunnels. The method proposed in Chinese tunnel design code was employed to evaluate the safety of tunnel liner and was implemented by using the programming language embedded within FLAC^{3D}. The results show that tunnel passing crossing ground fissures cannot serve well without multiple deformation joints and that the failure scopes of the tunnel liner are approximately 23 m in the case. In addition, an engineering countermeasure was proposed to mitigate the dislocating effects on tunnel.

Keywords: Ground fissure · Shallow tunnel · Numerical analysis

1 Introduction

Since 1950, there have been approximately fourteen ground fissures discovered in Xi'an, China, which have caused considerable damage to buildings, infrastructure and underground facilities. In response to this, a code was established to offer guidelines for practitioners and to make it compulsory that buildings cannot be built over ground fissures. The Metro, however, being designed as a transport web to cover the entire city, inevitably intersects with ground fissures. Thus, it is very important to study the dislocating effects of ground fissures on tunnels. Peng [4] focused on obstacles that arose from the perpendicular crossing of ground fissures by the Metro Tunnel and proposed several countermeasures to mitigate the dislocating effects of ground fissures on the tunnel based on a series of large-scale model tests. Unfortunately, some new phenomena have emerged during the construction of a new planned Metro line in Xi'an where the tunnel crosses the ground fissures at a small angle of less than 20°. In this case, the effects on the tunnel exerted by the dislocation of ground fissures are much larger than in cases where the tunnel intersects with ground fissures perpendicularly. In this paper, numerical methods are employed to investigate the dislocating effects that occur where the tunnel crosses ground fissures at a small angle.

© Springer Nature Switzerland AG 2018
W. Wu and H.-S. Yu (Eds.): *Proceedings of China-Europe Conference on Geotechnical Engineering*, SSGG, pp. 1197–1200, 2018.
https://doi.org/10.1007/978-3-319-97115-5_68

2   Numerical Analysis

2.1   Model

The numerical program FLAC3D [1] was employed to simulate the dislocating effects on the tunnel. Figure 1 shows the numerical model and mesh configurations for the simulation of the interaction between soil and tunnel. The total length, width and height of the numerical model were 350 m, 300 m and 100 m, respectively. The axis of the tunnel runs along the y-direction (from 0 m to 350 m), within which the lengths of the tunnel in the hanging wall and the footwall are 250 m and 100 m, respectively. The tunnel crown is 17 m below ground level, running along the z-direction. The section of tunnel is horseshoe-shaped with a width of 10 m and a height of 9.9 m. The ground fissures are inclined from the horizontal plane with a dip angle of 70°, intersecting with the tunnel axes with a 20° skew angle.

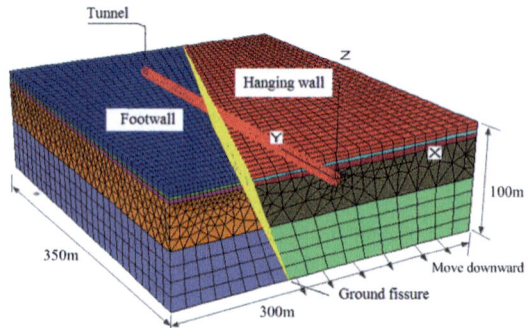

Fig. 1. Numerical model.

2.2   Simulation of Ground Fissures

To simulate the dislocating effects of ground fissures on the tunnel, an interface element was used in FLAC3D. The stress-strength relationship in the interface is in accordance with the Mohr-Coulomb strength criterion. Our research group has published a number of strength parameters on soils within the zone of ground fissures [3]. Herein, the obtained strength parameters were used.

2.3   Parameters and Conditions

The constitutive model for the soil is ideal-elastoplastic, with a Mohr-Coulomb yield condition and a non-associative flow rule. The tunnel is linearly elastic, which is simulated with the structural element "shell" in FLAC3D. Based on monitoring data of ground fissure dislocation over a number of decades, we [5] predict that the maximum magnitude (vertical movement) of a ground fissure is 0.5 m over a period of 100 years. To numerically simulate the situation described above, a vertical displacement of 0.1 m was added to the bottom of the model at each step of the computation.

2.4   Safety Magnitude of Liner

After all computation steps are accomplished and the internal forces of the liner are obtained, it is important to appraise the safety of the liner. According to the Chinese Code [2], liners are considered as eccentric compression members to determine the strength force, and the magnitude of safety is defined as the ratio of computed force to strength force. A magnitude of safety within an element that is less than 1 means that the liner has collapsed. These procedures were implemented using a programming language embedded within FLAC3D.

3   Results

(1) Moment: To shed light on the results, tow typical sections along the tunnel axes were selected to draw contours of internal forces. Regarding the contour of moment, the moment is drawn on the side in which the liner resists tensile stress. Figure 2 demonstrates that the maximum moment of approximately 1065 kN•m appears in the section of y = 205 m in the lower right corner of the tunnel.

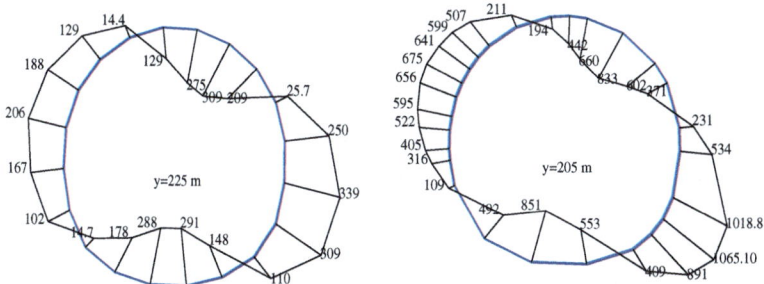

Fig. 2. Moment on typical section with hanging wall moving 0.5 m (kN·m)

(2) Axial force: It can be seen that the liner is under a state of eccentric compression. All internal forces are summarized in Table 1, where the magnitude of the moment increases as the dislocation of the ground fissure increases. It can also be seen from Table 1 that, as the hanging wall moves 0.1 m vertically, the liner installed in the footwall is subjected to tensile forces, indicating that the liner is at the brink of failure.

(3) Magnitude of safety: It can be seen from Table 1 that the minimum magnitude of safety in each section is approximately 7, which is considerably larger than the value of 1.7 suggested in the code. The minimum magnitude of safety decreases as the vertical movement of the hanging wall increases. Finally, it can be concluded that, as the vertical movement of the hanging wall increases, the failure of eccentric compression tends to occur in the liner installed in the hanging wall and adjacent to the ground fissure, which appears in the lower right corner of the tunnel.

Table 1. Computation results

Vertical movement (m)	Maximum moment (kN·m)		Tunnel in hanging wall	
	Section y = 205 m	Section y = 195 m	Minimum axial force (kN)	Minimum magnitude of safety
0	115	93	29	7.2
0.1	287	239	−82	6.5
0.2	476	420	−126	3.23
0.3	666	598	−286	1.88
0.4	864	771	−294	1.38
0.5	1065	970	−386	1.09

4   Conclusions

(1) As the vertical movement of the hanging wall increases, the tensile forces occur firstly in the liner installed at the footwall. When the vertical movement of the hanging wall increases to 0.5 m, the axial force in the liner installed at the hanging wall approaches the maximum tensile forces, indicating that failure has occurred in the liner.

(2) The surface of a tunnel intersecting with the ground fissures will failure when the dislocation of a ground fissure occurs, so it is necessary to take measures to mitigate the failure.

References

1. Peng, J., Huang, Q., Hu, Z., Wang, M., Li, T.: A proposed solution to the ground fissure encountered in urban metro construction in Xi ' an, China. Tunn. Undergr. Space Technol. Incorporating Trenchless Technol. Res. **61**, 12–25 (2017)
2. Itasca Consulting Group. FLAC Fast Lagrangian Analysis of Continua, Version 4.0. User's Manual, Available (2009)
3. JTG D70-2004 Code for design of road tunnel. China Communications Press, Beijing (2004)
4. Wu, M., Peng, J., He, k., Meng, s.: Numerical analysis of dislocation of ground fissure on the metro tunnel paralleling to the ground fissure's strike. J. Eng. Geol. **23** (2015)
5. Wu, M., Peng, J.: Numerical analysis of dislocation of ground fissure on themetro tunnel paralleling to the ground fissure's strike. J. Eng. Geol. **23** (2015). http://doi.org/10.13544/j.cnki.jeg.2015.05.029. (in Chinese)

Drift Ratio Limit for the Seismic Design of Underground Structures

Mian Xiao⬤, Renren Chen⬤, Rui Wang⬤, and Jianmin Zhang$^{(\boxtimes)}$

Tsinghua University, Beijing 100084, China
18686628006@163.com

Abstract. The interlayer drift ratio of underground structures is an important seismic design index in China's Seismic Design Code. However, existing elastic and elasto-plastic limit of drift ratio are adopted from ground structures, and those of underground structures have yet to be comprehensively studied. Based on the design practice of underground structures in China, a set of finite element pushover analysis is conducted on underground structures in this paper to investigate the seismic deformation of six common subway stations. A technique of combining beam elements with quadrilateral elements is utilized to capture the realistic elasto-plastic behavior of structural components. Computation results show that 1/550 and 1/1000 are appropriate elastic drift ratio limiting values for underground structures two stories or lower and underground structures three stories or higher, respectively. The elasto-plastic drift ratio limit of 1/250 that is adopted from China's Code for Seismic Design of Building (CCSDB) is a conservative limiting value. The influence of buried depth and soil stiffness is also analyzed.

Keywords: Underground structure · Drift ratio · Elastic limit
Elasto-plastic limit

1 Introduction

In the seismic design of structures, interlayer drift ratio is one of the most important indexes in quantifying the seismic performance [1]. However, comprehensive studies have yet to be conducted on the drift ratio for underground structures. In 2010, China's Code for Seismic Design of Buildings (CCSDB) adopts the same elasto-plastic drift ratio limit of 1/250 for underground structures. There is few experimental or numerical validation of these limiting values in previous research for underground structures. As a result, the drift ratio, which is important for evaluating the seismic deformation for underground structures, remains unclear for practical design.

In this paper, a set of pushover analysis is performed on six underground subway stations to obtain their elastic and elasto-plastic drift ratios [2]. The elastic drift ratio is the drift ratio at which the first plastic hinge occurs in the structure, the elasto-plastic drift ratio is the drift ratio at which the structure or part of the structure becomes a mechanism.

© Springer Nature Switzerland AG 2018
W. Wu and H.-S. Yu (Eds.): *Proceedings of China-Europe Conference on Geotechnical Engineering*, SSGG, pp. 1201–1205, 2018.
https://doi.org/10.1007/978-3-319-97115-5_69

2 Numerical Analysis Detail

Plane strain pushover analyses are performed on six station structures through the OpenSees finite element platform. Station 1, 2 and 3 with reinforced concrete column are constructed using the cut-and-cover method, with the burial depths of 2 m and 4 m. Station 4, 5 and 6 with concrete-filled steel tube column and arched crown are constructed using the undercutting method, with the burial depths of 5 m, 10 m and 15 m. Each structure has three spans apart from station 1, which has just two spans. Station 1, 2 and 4 has two stories, while station 3 and 5 are three-storied. Station 6 is a triple-arched structure with only one story. Besides, the dynamic Young's modulus of surrounding soil is set as 10, 150 and 300 MPa in the analyses. These three different dynamic Young's moduli represent the stiffness of soft clay, loose sand and dense sand in a strain range between 0.3% and 1.5% [3].

As shown in Fig. 1, the surrounding soil is considered elastic isotropic at a depth of 50 m, discretized with the quadrilateral elements, while structural components are modeled by combining quadrilateral elements and beam elements with fiber section in order to simulate the realistic elasto-plastic behavior of reinforced concrete structure [5]. The compressive strength of concrete is 16.7 MPa and the yield strength of steel is 335 MPa. The columns are modeled according to their geometric size, and the thickness of soil elements and structure elements is set up to be the same as the spacing of columns.

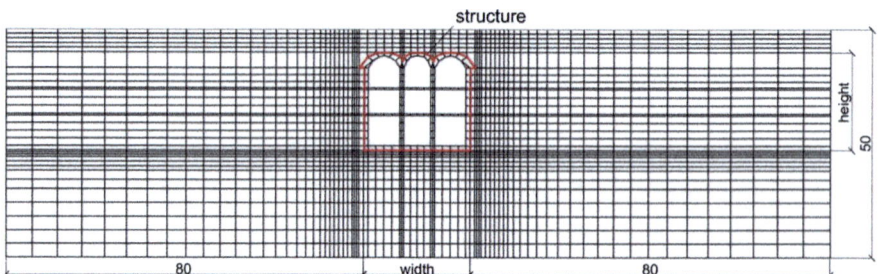

Fig. 1. Mesh configuration of soil and structure for one station.

Analysis starts with a geostatic step to calculate the structure's initial internal force before pushover loading. After the initial step, an inverse-triangular horizontal displacement toward right is applied to both sides of soil boundary to perform pushover analysis [4]. The internal force and nodal displacement of the structures are recorded to determine the occurrence of plastic hinge and the failure of the structures.

3   Computation Results

Typically, plastic hinge first occurs at the left bottom of the lowest story's outer wall, and when plastic hinge occurs at the top of the same wall, the structure becomes a mechanism. At these two moments, each layer's horizontal drift is recorded to calculate the maximum elastic and elasto-plastic drift ratio. Station 1 to 5 tend to end up with overall failure, that is to say, with both ends of the lowest story walls and columns failed. Station 6 ends up with local failure as the left side arch becomes a mechanism.

In Fig. 2, it shows that the higher an underground structure is, the lower its elastic and elasto-plastic drift ratio are. 1/460 and 1/612 are the minimum elastic drift ratio in underground structures two stories or lower and underground structures three stories or higher, respectively. 1/151 is the minimum elasto-plastic drift ratio in all these structures. Therefore, 1/550 and 1/1000 are chosen to be the limiting values for underground structures two stories or lower and underground structures three stories or higher, respectively (in Fig. 2). Based on CCSDB, this paper conservatively takes 1/250 as the drift ratio's elasto-plastic limit.

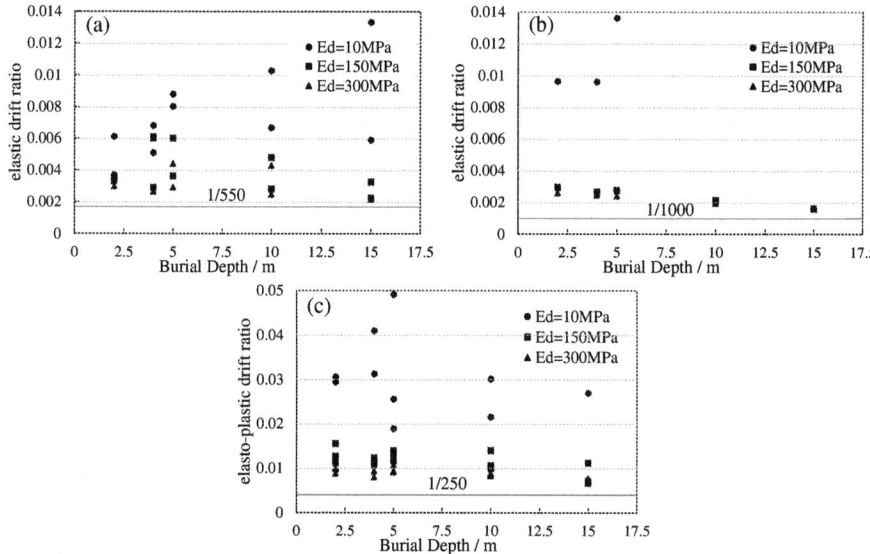

Fig. 2. Drift Ratio of different conditions. (a) elastic drift ratio for underground structures two stories or lower, (b) elastic drift ratio for underground structures three stories or higher, (c) elasto-plastic drift ratio underground structures

Figure 3(a) shows the effects of burial depth on the elastic and elasto-plastic drift ratio of station 6. In most cases, the elastic and elasto-plastic drift ratio decrease with the increased burial depth. However, this trend reverses in extremely soft soil. On the one hand, as the burial depth increases, an underground structure would be designed with stronger strength of components, which would lead to a higher drift ratio. On the other hand, the stiffness of the structure would usually increase as it is designed stronger, which would lead to a lower drift ratio. When the surrounding soil is extremely soft, the deformation of the underground structure is mainly controlled by its strength, but not the interaction between the surround soil and the structure, which is mainly controlled by their relative stiffness. Therefore, in extremely soft ground, the deeper an underground structure is buried, the higher its elastic and elasto-plastic drift ratio could be.

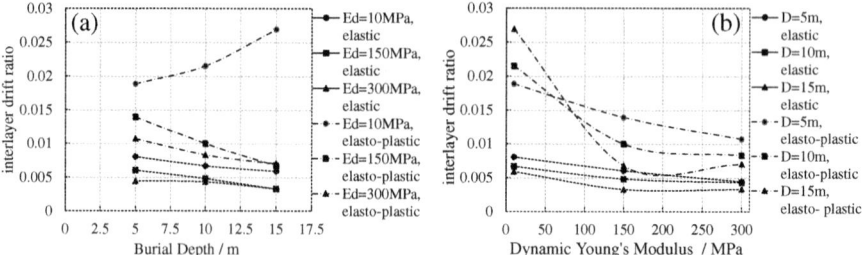

Fig. 3. Effects of burial depth and dynamic Young's modulus of surrounding soil on drift ratio. (a) burial depth, (b) dynamic Young's modulus.

Figure 3(b) shows the effects of dynamic Young's modulus of surrounding soil on the drift ratio of station 6. It indicates that the larger dynamic Young's modulus usually results in lower elastic and elasto-plastic drift ratio. The reason is that the interaction force produced by unit deformation of hard surrounding soil could be larger than that by the soft one. As a result, underground structures generally fail at a lower interlayer drift ratio when the dynamic Young's modulus of surrounding soil is larger.

4   Conclusions

- For the stations in this study, the two ends of the lowest story walls are the weakest part under horizontal seismic load.
- 1/550 and 1/1000 are appropriate elastic drift ratio limiting values for underground structures with two stories or lower and underground structures with three stories or higher, respectively. The elasto-plastic drift ratio limit of 1/250 that is adopted from China's Code for Seismic Design of Building is a conservative limiting value.

References

1. Dong, Z.F., Wang, J.J., Yao, Y.C., Su, J.S.: Research on the seismic performance index system of urban mass transit underground structures. Earthq. Eng. Eng. Dyn. **34**(Suppl.), 699–705 (2014). (in Chinese)
2. Li, B.: Theoretical Analysis of Seismic Response of Underground Subway Structures and Its Application. Tsinghua University Doctoral Dissertation (2005). (in Chinese)

3. He, R.C.: Dynamic triaxial test on modulus and damping. Chin. J. Geotech. Eng. **19**(2), 39–48 (2009). (in Chinese)
4. Hashash, Y., Hook, J.J., Schmidt, B.: Seismic design and analysis of underground structures. Tunn. Undergr. Space Technol. **16**(4), 247–293 (2001)
5. Chen, R.R., Yao, Y., Wang, R., Zhang, J.M.: Three-dimensional finite element analysis of underground structures dynamic response in liquefiable soil. In: Advances in Soil Dynamics and Foundation Engineering: Geoshanghai International Congress, pp. 572–578 (2014)

Numerical Analysis on Interaction Between Closely Spaced Shield Tunnels

Congcong Xiong[1]([✉]), Binglong Wang[1], Zhi Liu[2], Weitao Ye[1],
Wei Zhao[1], and Jianbing Long[3]

[1] Key Laboratory of Road and Traffic Engineering of the Ministry of Education,
Tongji University, Shanghai 201804, China
`1632498@tongji.edu.cn`
[2] Shenzhen Metro Group Co., Ltd., Shenzhen 518173, China
[3] Ningbo Rail Transit Group Co., Ltd., Ningbo 315010, China

Abstract. This paper presents a three-dimensional numerical model simulating a small-spacing overlapped shield tunnel driving course. Using this model, the influences of shield tunnel construction parameters on ground settlement, and existing shield tunnel axis floating are studied. The driving process of the shield will bring inevitable impacts upon the surrounding stratum and the existing tunnel. Therefore, the determination of construction parameters of shield tunnel, including support pressure in the face, decrease of soil modulus caused by shield machine, grouting pressure and properties of lining grouting slurry is a big problem. The Hardening soil (HS) model as it is implemented in the finite element code PLAXIS 3D, is used to simulate construction parameters and tunneling construction process. It is shown based on the analysis of construction parameters that when construction parameters were controlled reasonably, the ground settlement and deformation of existing tunnel lining induced by shield tunnel construction can be decreased effectively. The first tunnel construction produced disturbance on surrounding soil making the following tunnel longitudinal and transverse settlement value increasing than the first hole. Influence of the following tunnel construction to the first of the hole was a process (arrival, passes through this process), which is illustrated in the transverse settlement curve, the first cave settlement increased first and then decreased.

Keywords: Overlapped shield tunnel · Numerical simulation
Sensitive parameters identification · Ground movement · Axis floating

1 Introduction

The driving process of the shield will bring inevitable impacts upon the surrounding stratum and the existing tunnel. The settlement along the tunnel axis is generally non-uniform. The excessive differential settlement of the metro tunnel not only leads to structural damages (segment cracks and even breakage of the rail in Fig. 1), but also impairs the serviceability of the metro line [1, 2].

The driving process of the shield will bring inevitable impacts upon the surrounding stratum and the existing tunnel. Therefore, the determination of construction parameters of shield tunnel, including support pressure in the face, decrease of soil

© Springer Nature Switzerland AG 2018
W. Wu and H.-S. Yu (Eds.): *Proceedings of China-Europe Conference on Geotechnical Engineering*, SSGG, pp. 1206–1210, 2018.
https://doi.org/10.1007/978-3-319-97115-5_70

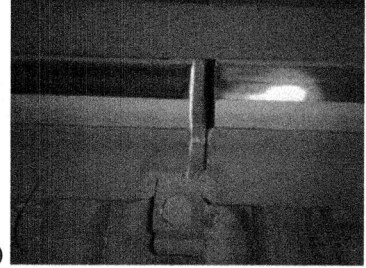

(a)   (b)

Fig. 1. Typical structural damages in operational metro tunnel induced by the construction disturbance: (a) segment damage; (b) breakage of the rail.

modulus caused by shield machine, grouting pressure and properties of lining grouting slurry is a big problem [6]. It is shown based on the analysis of construction parameters that when construction parameters were controlled reasonably, the ground settlement and deformation of existing tunnel lining induced by shield tunnel construction can be decreased effectively.

The fast development of urban underground traffic has brought more construction projects of small-spacing tunnels. How is mutual influence between neighborhood shield tunnel construction? However, the study of the interaction of small-spacing overlapped shield tunnel is rarely seen. Therefore, based on the background of new shield tunnel of Shenzhen Metro Line 5, this paper presents a three-dimensional numerical model simulating a small-spacing overlapped shield tunnel driving course.

2 Project Description

The Shenzhen City is located in southern China. The landform of Shenzhen is the Marine alluvial plain where thick muddy silty clay layer and muddy clay layer are widely distributed. The groundwater level is located approximately 1.0 m below the ground surface. Compared with the soft soils in Shanghai, the soft muddy clay soils in Shenzhen possess even worse engineering proper-ties, e.g., high porosity, high water content, high compressibility, high rheology, and medium sensitivity. The soils are very vulnerable to the impact of construction disturbances.

3 Analysis of the Sensitivity to Construction Parameters

Detailed description of the relevant physical subsystems needed to be modelled within the simulation of mechanized tunneling by means of slurry shield tunnel boring machine (slurry shield TBM) is given in [3].

Figure 2 shows the layout of the FEM mesh of a three-dimensional numerical model simulating a small-spacing overlapped shield tunnel driving course. The FEM model has a length of 310 m, a width of 150 m and a depth of 45.5 m. The left and right boundaries are fixed in the horizontal direction, and the bottom boundary is fixed in both horizontal and vertical directions. The soil parameters in the FEM model follow the actual profile as shown in Fig. 3. Considering that the bolt connections in the tunnel linings can reduce the overall rigidity, the stiffness of the plate elements is reduced by multiply a reduction factor of 0.75 [4, 5].

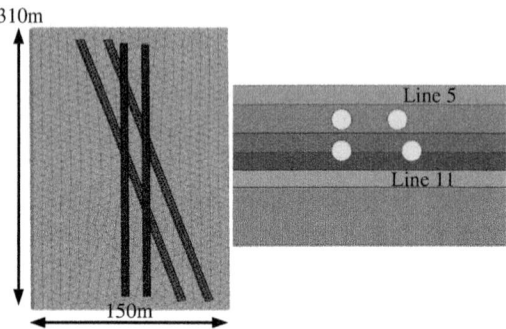

Fig. 2. Dimensions of the main FE-model (left) and the cross section of 5L-11L (right).

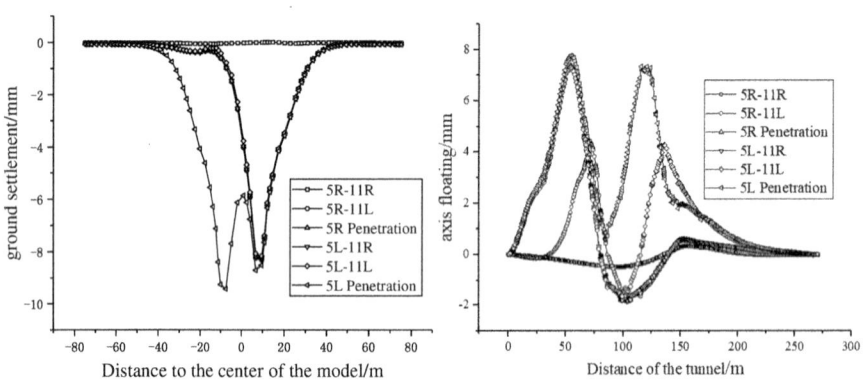

Fig. 3. Ground settlement (left) and Axis floating(right).

The parameters of the linear elastic model assigned to the plate elements are given in Table 1. The slurry shield TBM is 9 m long, simulated via plate elements. Plate elements are also used for the concrete tunnel lining. The parameters of the linear elastic model assigned to the plate elements are given in Table 1. The action of the grout pressure on the surrounding subsoil is simulated via nonuniformly distributed load applied to the soil elements by deactivating the plate elements at the place grouting acts.

Table 1. The parameters of the linear elastic model.

Parameter	Lining	TBM	UNIT
d	0.3	0.35	[m]
E	30000	210000	[MPa]
γ	24	38	[KN/m^3]
v	0.10	0.30	[-]

4 Results

When the right line of line 5 is completed, the maximum surface deformation takes place near the right line. The elastic modulus of slurry is 2.0 MPa, 4.0 MPa and 6.0 MPa respectively, and the deformation is 8.31 mm, 6.62 mm and 5.80 mm respectively. When the left line is completed, the surface deformation of the right side basically stays the same. The main deformation is the surface deformation on the left line. The elastic modulus of the slurry is 2.0 MPa, 4.0 MPa and 6.0 MPa respectively, and the deformation is 9.43 mm, 7.41 mm and 6.46 mm respectively. Compared with the influence of 2.0 MPa, 4.0 MPa and 6.0 MPa on the surface deformation after the condensation of the slurry, it is known that the change of the elastic modulus of the slurry set in the model has great influence on the ground settlement.

5 Conclusions

It is shown based on the analysis of construction parameters that when construction parameters were controlled reasonably, the ground settlement and deformation of existing tunnel lining induced by shield tunnel construction can be decreased effectively. The first tunnel construction produced disturbance on surrounding soil making the following tunnel longitudinal and trans-verse settlement value increasing than the first hole. Influence of the following tunnel construction to the first of the hole was a process (arrival, passes through this process), which is illustrated in the transverse settlement curve, the first cave settlement increased first and then decreased.

References

1. Wang, X.B., Wang, R.L., Liu, J.H.: Disposal method of unequal settlement of metro tunnel in operation in Shanghai soft ground. J. Shanghai Jiaotong Univ. **46**(01), 26–31 (2012)
2. Zhou, S.H., Di, H.G., Xiao, J.H., Wang, P.X.: Differential settlement and induced structural damage in a cut-and-cover subway tunnel in a soft deposit. J. Perform. Constr. Facil **30**(5), 04016028 (2016)
3. Schanz, T., Vermeer, P.A., Bonnier, P.G.: The hardening soil model: formulation and verification. Beyond 2000 in Computational Geotechnics - 10 Years of PLAXIS (1999)

4. Zhu, W.: Japanese Standard for Shield Tunneling (translated version). China Architecture and Building Industry Press (2001)
5. Fang, Y., He, C.: Numerical analysis of parallel shield tunnel construction affecting on the existent one. Rock Soil Mech. **28**(7), 1402–1406 (2007)
6. Huang, H., Zhang, L., Yang, X.Q.: Experimental study of mechanical properties of cemented-soil. J. Taiyuan Univ. Technol. **31**(6), 705–709 (2000)

Hydraulic Response on Shield Tunnel Face Subjected to Water Level Fluctuation

Hongwei Ying[1,2(✉)] ⓘ, Hua Wei[1,2] ⓘ, Jinhong Zhang[3],
and Chengwei Zhu[1,2] ⓘ

[1] Research Center of Coastal and Urban Geotechnical Engineering,
Zhejiang University, Hangzhou 310058, China
ice898@zju.edu.cn
[2] MOE Key Laboratory of Soft Soils and Geoenvironmental Engineering,
Zhejiang University, Hangzhou 310058, China
[3] Zhejiang Province Institute of Architectural Design and Research,
Hangzhou 310008, China

Abstract. Subaqueous shield tunnels in offshore or coast conditions are constructed in a complex environment. The tunnel face stability is highly affected by tides and waves. The pore water pressure response near the tunnel excavation face is very important during the construction of the tunnel. Based on the Qingchun Road Tunnel in Hangzhou, a three-dimensional numerical model was established and the pore water response on tunnel surface to tide impact is studied. It is shown that the water level fluctuation is easier to propagate in silty soil with the increase of the soil permeability or the decrease of the soil compressibility. The seepage force on the tunnel face varies with time and the maximum value is smaller than that calculated in static water condition with the highest tidal level. The characteristics of the total head, pressure head and hydraulic gradient at different locations near the tunnel face are investigated. The phenomenon of phase lag is more obvious as the depths increase. It is also found that the phase difference on the same horizontal plane is almost equal.

Keywords: Water level fluctuation · Tunnel face · Seepage field
3D FEM

1 Introduction

Subaqueous shield tunnels in offshore or coast conditions are constructed in a complex environment, seepage flow is unfavorable for the stability of the tunnel face. The pore water pressure response near the tunnel face is very important during the construction of the tunnel. The tunnel face stability is highly affected by tides and waves, bringing new scientific problems under the condition of water level fluctuation. Previous works concerning the tunnel face stability tend to focus only on static situations under a constant water level. In studies of shield tunnel face stability concerning seepage, numerical methods for the seepage field are widely used. Seepage forces from numerical results can be applied to different theoretical analysis approaches like limit equilibrium method or limit analysis method [1–4, 6].

© Springer Nature Switzerland AG 2018
W. Wu and H.-S. Yu (Eds.): *Proceedings of China-Europe Conference
on Geotechnical Engineering*, SSGG, pp. 1211–1215, 2018.
https://doi.org/10.1007/978-3-319-97115-5_71

2   Numerical Model

A three-dimensional numerical model is established and the pore water response on tunnel face to tide impact is studied. In order to analyze the hydraulic response on shield tunnel face subjected to water level fluctuation, the characteristics of the Qingchun Road Tunnel in Hangzhou are considered in this study (see Fig. 1(a)). Due to the vertical symmetry plane, the computational domain consists of half of the system. Figure 1(b) shows the size and the coordinate axis of the model, and the finite element mesh adopted for the calculations. The excavated length is 50 m. A no-flow boundary condition is prescribed to the tunnel wall (impervious lining), as well as to the surrounding and lower boundary of the model. A constant pressure head $h_p = 0$ is prescribed to the tunnel face. For the sake of simplicity, a harmonically fluctuating water level is prescribed to the upper boundary of the model to simulate the tide. The average water level H_w is 15 m. According to the historic hydrologic records [5], the tide with a cycle, T, of 0.5 day, and an amplitude, A, of 5 m remains unchanged in this study. Moreover, it is assumed that the drainage of the tunnel does not cause any change in the fluctuation of the water table.

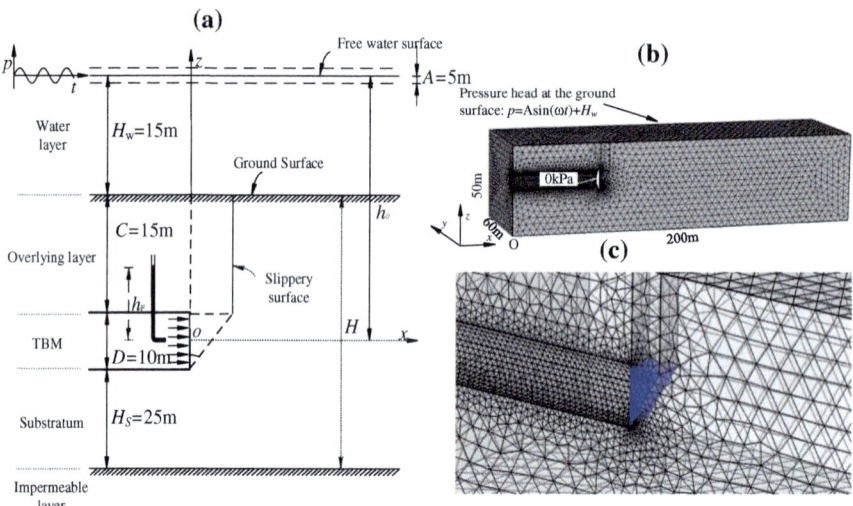

Fig. 1. (a) Analysis model of pore pressure response during excavation of shield tunnel with water level fluctuation. (b) Finite element mesh. (c) The 'wedge-prism' in the numerical model

3   Effect of Soil Properties

It is found from the numerical analysis that the hydraulic response around the tunnel is related to the soil's hydraulic conductivity (K) and volumetric specific storage (S_s). The fluctuation of the horizontal seepage force in the wedge, F_x (Fig. 1(c)), is observed during the study. Different K and S_s are used in several combinations in the computation. Parts of the results are plotted in Fig. 2(a) and (b) respectively. The fluctuation is easier to propagate in soil with the increase of the soil permeability or the decrease of the volumetric specific storage. The seepage force on the tunnel face varies with time and the maximum value is smaller than that calculated in static water condition with the highest tidal level.

(a)   **(b)**

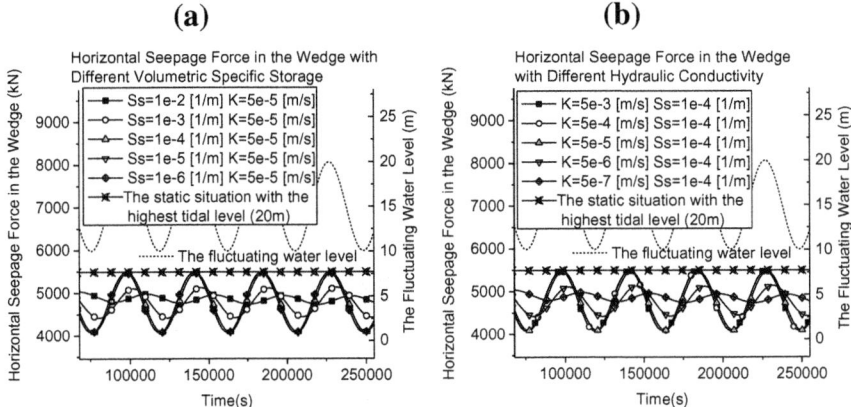

Fig. 2. (a) Seepage force in the wedge with different volumetric specific storage. (b) Seepage force in the wedge with different hydraulic conductivity

4   Case Study

The soil layer which the Qingchun Road Tunnel crosses, consisting of silty soil, clay and other kinds of soil [5], is simplified to a homogeneous sandy silty soil layer. The parameters in the case study are set as follows: $K = 5 \times 10^{-6}$ m/s, $S_s = 1 \times 10^{-4}$ 1/m, and the porosity, $n = 0.42$. Two series of points are observed in order for the study of the fluctuation of hydraulic characteristics. One series is on the central axis of the tunnel (line A), and the other one is on the vertical line on the symmetry plane 2.5 m away from the tunnel face (line B) (Fig. 3). The characteristics of the total head, pressure head and hydraulic gradient at different locations are investigated. Two typical figures are mapped in the text to show the total water head with time at different points (Fig. 4(a) and (b)). Phase lag and amplitude decay of seepage are found in the case study. The phase lag at different points on the same horizontal plane is almost identical (line A), while the phase lag is more obvious as the depth increases (line B) and this characteristic is not affected by the drainage condition of the tunnel face, indicating that the phase lag is caused by the obstacles during the process of the downward propagation of fluctuation. Several other conclusions can be drawn in the last section.

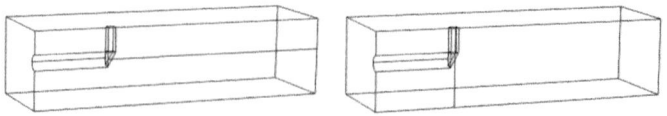

Fig. 3. Two typical lines in the model for the case study

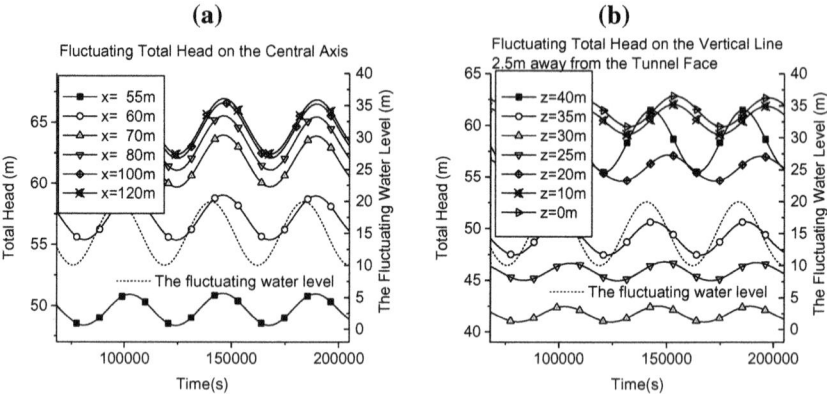

Fig. 4. (a) Fluctuating total head on the central axis. (b) Fluctuating total head on the vertical line 2.5 m away from the tunnel face

5   Conclusion

1. The fluctuation is easier to propagate in soil with the increase of the soil permeability or the decrease of the volumetric specific storage. The seepage force on the tunnel face varies with time and the maximum value is smaller than that calculated in static water condition with the highest tidal level.
2. The phenomenon of phase lag and amplitude decay exists in the fluctuation of all hydraulic elements. The phase lag at different points on the same horizontal plane is almost identical (line A), while is more obvious as the depth increases (line B). The amplitude of the fluctuation is relatively small near the tunnel face or in deeper area.
3. The seepage force near the tunnel face fluctuates in a complex way. The value of it needs to be reasonably decided in the theoretical calculation of the limiting support force in the shield tunnel construction subjected to water level fluctuation.

References

1. Anagnostou, G., Kovári, K.: The face stability of slurry-shield-driven tunnels. Tunn. Undergr. Space Technol. **9**(2), 165–174 (1994)
2. Anagnostou, G., Kovári, K.: Face stability conditions with earth-pressure-balanced shields. Tunn. Undergr. Space Technol. **11**(2), 165–173 (1996)

3. Leca, E., Dormieux, L.: Upper and lower bound solutions for the face stability of shallow circular tunnels in frictional material. Géotechnique **40**(4), 581–606 (1990)
4. Lee, I.M., Nam, S.W., Ahn, J.H.: Effect of seepage forces on tunnel face stability. Can. Geotech. J. **40**(40), 342–350 (2003)
5. Cungang, L.: Research on Shield Tunnelling-induced Ground Surface Heave and Subsidence and Behavior of Underwater Shield-driven Tunnels Subject to Tidal Bores. Zhejiang University, Hangzhou (2014). (in Chinese)
6. Perazzelli, P., Leone, T., Anagnostou, G.: Tunnel face stability under seepage flow conditions. Tunn. Undergr. Space Technol. Inc. Trenchless Technol. Res. **43**, 459–469 (2014)

Undrained Stability of Tunnel with a Longitudinal Gradient

Fei Zhang[(⊠)] [ID], Guangyu Dai, Hu Wang, Yufeng Gao, and G. H. Lei

Key Laboratory of Ministry of Education for Geomechanics
and Embankment Engineering, Hohai University, Nanjing, China
feizhang@hhu.edu.cn

Abstract. This paper involves the effects of longitudinal gradient into the stability analyses of tunnel face in undrained clay. A kinematical approach of limit analysis based on continuous velocity field is employed to estimate the upper-bound solution of supporting pressure on tunnel face. The obtained results demonstrate significant influences of the longitudinal gradient on the stability of tunnel face below an inclined ground surface. When tunnels are excavated below a horizontal ground surface, the influences are minor and then can be ignored in practice.

Keywords: Tunnel · Face stability · Limit analysis

1 Introduction

Shield-driven tunnel generally has a longitudinal gradient, which is less than 3% for vehicle driving safety. Such a small longitudinal gradient has negligible influences on the stability of tunnel face and its effects are always excluded in most stability analyses of tunnel face. Recently, many shield-driven tunnels in construction have great longitudinal gradient, which can exceed 5% and even reach 10% especially for some traffic tunnels and cable tunnels crossing a river or sea. Based on the kinematical approach of limit analysis, Zhao et al. [5] adopted the failure mechanism with multiple elliptical cone to investigate the influences of the tunnel inclination angles on the supporting pressure of the tunnel face. Their calculated results demonstrate the significant effects for both of the active and passive failure modes but are limited to the tunnels in cohesive frictional soils. The purpose of this study is to explore the effects of the longitudinal gradient on the stability of tunnel face in purely cohesive soils. For undrained failures in clay, a continuous velocity field proposed by Mollon et al. [3] is employed here to obtain the upper bound solutions of support pressure on tunnel face.

2 Continuous Velocity Field and Formulations

The generated continuous velocity field for 3D undrained failures of tunnel face by Mollon et al. [3] has the shape of a torus, as shown in Fig. 1. A maximum velocity flow line is located at a distance L_1 from the centre of the tunnel face. In a curvilinear coordinate system (r, θ, β), the axial component of the velocity (v_β), perpendicular to

© Springer Nature Switzerland AG 2018
W. Wu and H.-S. Yu (Eds.): *Proceedings of China-Europe Conference
on Geotechnical Engineering*, SSGG, pp. 1216–1220, 2018.
https://doi.org/10.1007/978-3-319-97115-5_72

the circular cross sections, Π_β, decreases from the maximum velocity flow line to the circular boundary in a parabolic way. In an arbitrary plane Π_β, the radial distance from the point of the maximum velocity to the boundary can be expressed as

$$R_{\max}(\theta, \beta) = R_{\max}(\theta, \beta_{TF}) \cdot \frac{R(\beta)}{R_i} \tag{1}$$

$$R_{\max}(\theta, \beta_{TF}) = \frac{2L_1 \cos\theta + \sqrt{D^2 - 2L_1^2 + 2L_1^2 \cos 2\theta}}{2} \tag{2}$$

$$R(\beta) = R_i + \frac{\beta - \beta_{TF}}{\beta_G - \beta_{TF}}(R_f - R_i) \tag{3}$$

where D = tunnel diameter; C = tunnel cover depth; β_{TF} = inclination angle of tunnel face; β_G = inclination angle of ground surface; $R_i = D/2 + L_1$; $R_f = C\sin\beta_G/\sin(\beta_G - \beta_{TF}) + D/2 + L_1$. Details of the continuous velocity field can be found elsewhere (e.g. [2–4]). When the tunnel face has a longitudinal gradient, as shown in Fig. 1, the inclination angles of the tunnel face and the ground surface are considered to obtain the expression describing the velocity field. Following the same symbols and the procedure by Klar and Klein [2] and Zhang et al. [4], the axial and radial components of the velocity can be obtained. For the orthoradial component of the velocity, $v_\theta = 0$. Using the kinematically admissible velocity field can obtain an upper bound on the limit pressure of tunnel face σ_t, by equating the rate of work of external forces to the rate of internal energy dissipation. In the continuous velocity field, the sum of the rates of work done by the soil weight, the supporting pressure σ_t on the tunnel face (plane for $\beta = \beta_{TF}$) and a possible uniform surcharge (σ_s) acting on the ground face (plane for $\beta = \beta_G$) can be expressed and then equal to the rate of the internal energy dissipation

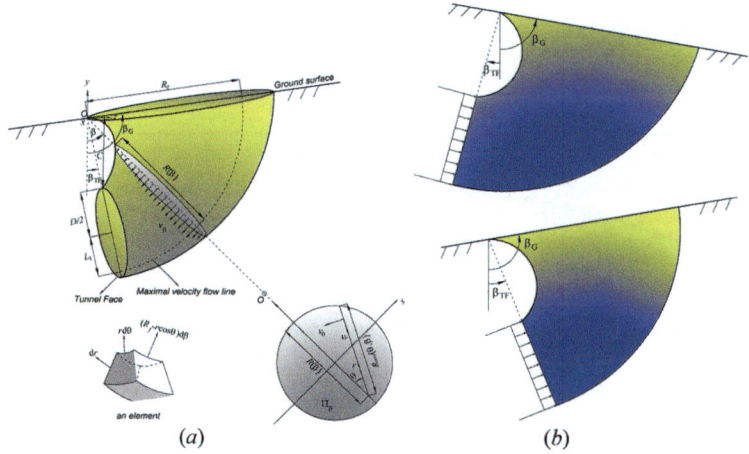

(a)   (b)

Fig. 1. Continuous velocity field for the undrained failure of the tunnel face with a longitudinal gradient (following Mollon et al. [3])

within the volume V. Based on the balance equation, the supporting pressure σ_t for collapse of tunnel face can be explicitly obtained as

$$\sigma_t = \gamma D N_\gamma + \sigma_s N_s - s_u N_c \tag{4}$$

where the dimensionless coefficients N_γ, N_s, and N_c can be obtained from the balance equation; s_u is the undrained shear strength. The coefficient N_s is equal to 1.0. Following a dimensionless form of the stability number introduced by Davis et al. [1], the stability number $N = [\sigma_s - \sigma_t + \gamma(C + D/2)]/s_u$ is also adopted here to show the upper-bound solutions for both of collapse and blow-out. Giving the values of the $\gamma D/s_u$, C/D, angles β_{TF} and β_G, the least upper-bound solutions of N needs to be found by an optimization procedure. Only one independent variable L_1/D is considered in the minimization procedure.

3   Results and Discussions

For shield-driven tunnels below the horizontal ground surface, the obtained stability number is plotted versus the ratio of C/D for several values of the inclination angle $\beta_{TF} = \pm 10°$, $\pm 5°$ and $0°$, as shown in Fig. 2. The negative and positive values of β_{TF} represent downward and upward movements of the tunnel boring machine, respectively. Both of collapse and blow-out failures of the tunnel face are considered. The value of $\gamma D/s_u$ has insignificant effects on the results because of $N_\gamma \approx C/D + 0.5$. The upward movement yields more critical value of stability number, which implies that neglecting the effects of the longitudinal gradient may underestimate the supporting pressure for tunneling in upward. The underestimation increases with the reducing C/D and reaches 10% for $\beta_{TF} = \pm 10°$ and $C/D = 0.5$. Figure 3 illustrates the stability numbers for tunnels excavated in the inclined ground surface. The results demonstrate the significant effects of $\gamma D/s_u$ on the stability number. As the value of $\gamma D/s_u$ increases, the influences of the longitudinal gradient become more profound. For collapse of

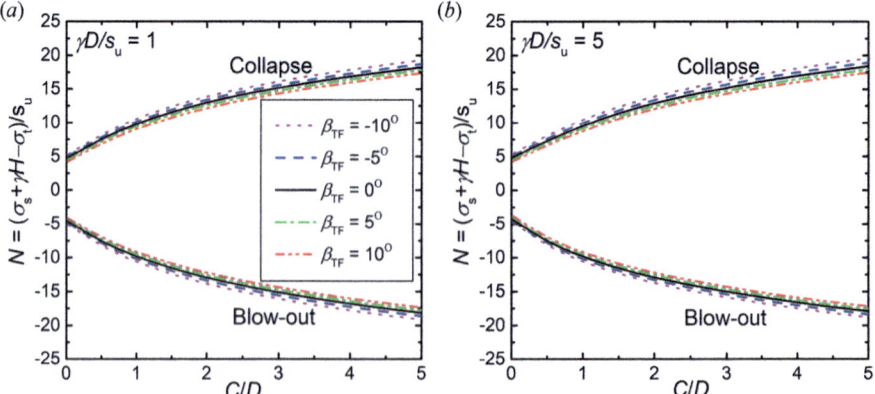

Fig. 2. Stability numbers for tunnel face in horizontal ground surface.

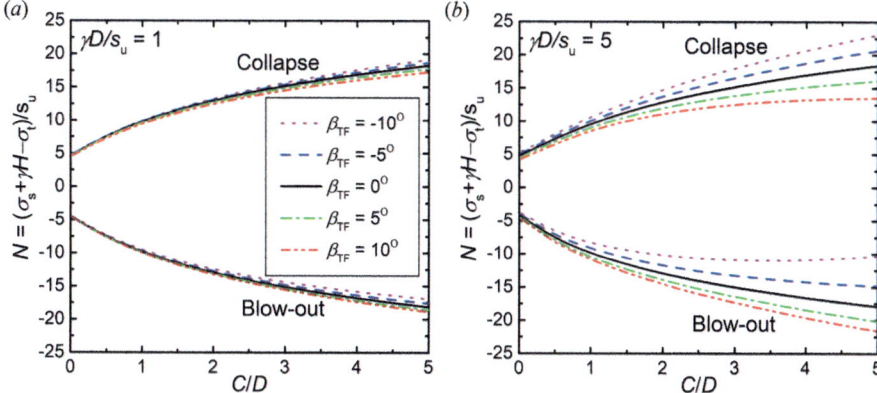

Fig. 3. Stability numbers for tunnel face in inclined ground surface.

tunnel face, the upward movement of tunneling requires greater supporting pressure. Conversely, the downward movement needs smaller pressure against the blow-out of the tunnel face. It implies that, neglecting the effects of the longitudinal gradient may lead to blow-out failures of the tunnel face in downward. The overestimation increases with the increasing C/D and exceeds 20% for $\gamma D/s_u = 5$, $\beta_{TF} = -10°$ and $C/D = 5$.

4 Conclusions

Based on the kinematic approach of limit analysis, the effect of the longitudinal gradient is involved in stability analyses of tunnel face in undrained clay. For tunneling below the horizontal ground surface, the longitudinal gradient has negligible influence on the upper-bound solution of the supporting pressure on tunnel face. When the tunnel is excavated in an inclined ground surface, the effect of the longitudinal gradient cannot be ignored. Its induced difference on the supporting pressure becomes more significant with the increasing value of $\gamma D/s_u$ and tunnel depth C. For tunneling in downward movement, neglecting the effects of the longitudinal gradient could overestimate the supporting pressure for blow-out failure. On the contrary, it results in underestimation of the supporting pressure on tunnel face against collapse when tunneling is in upward. The longitudinal gradient should be paid more attentions on face stability of tunneling in the inclined ground surface, especially for large-diameter shield-driven tunnels.

References

1. Davis, E.H., Gunn, M.J., Mair, R.J., Seneviratine, H.N.: The stability of shallow tunnels and underground openings in cohesive material. Géotechnique **30**(4), 397–416 (1980)
2. Klar, A., Klein, B.: Energy-based volume loss prediction for tunnel face advancement in clays. Géotechnique **64**(10), 776–786 (2014)

3. Mollon, G., Dias, D., Soubra, A.: Continuous velocity fields for collapse and blowout of a pressurized tunnel face in purely cohesive soil. Int. J. Numer. Analyt. Methods Geomech. **37**(13), 2061–2083 (2013)
4. Zhang, F., Gao, Y.F., Wu, Y.X., Zhang, N.: Upper-bound solutions for face stability of circular tunnels in undrained clays. Geotechnique **68**(1), 76–85 (2018)
5. Zhao, L., Li, D., Li, L., Yang, F., Cheng, X., Luo, W.: Three-dimensional stability analysis of a longitudinally inclined shallow tunnel face. Comput. Geotech. **87**, 32–48 (2017)

Shaking Table Test of Tunnel-Shaft Junction in Shield Tunnel

Jinghua Zhang, Haitao Yu$^{(\boxtimes)}$, and Yong Yuan

Department of Geotechnical Engineering, Tongji University, Shanghai, China
yuhaitao@tongji.edu.cn

Abstract. In order to have a better understanding of the seismic behavior of the tunnel-shaft junction in shield tunnel, a shaking table test precisely targeted on this part of the tunnel is conducted. A flexible wall barrel is used as the model container. The structure model is made of concrete. The model soil is a combination of saw dust and dry sand. Both the longitudinal and the radial joints are simulated in the test. This paper presents the design of the test and the maximum openings of the circumferential joints.

Keywords: Shield tunnel · Shaking table test · Tunnel-shaft junction

1 Introduction

The seismic performance of underground structures has drawn much attention, especially after the collapse of Daikai station during the great Hanshin earthquake in 1995 [1]. It is of great importance to protect underground structures from seismic hazard. Although critical joints, such as connecting passages and working shafts, are particularly vulnerable to earthquakes [2], previous studies on the seismic response of shield tunnel are mainly focused on the analysis of regular tunnels. Most researchers have applied numerical methods to solve this problem [3]. This paper will present a shaking table test of the tunnel-shaft junction in shield tunnel.

2 Shaking Table Test

2.1 Flexible Wall Barrel

As shown in Fig. 1, a flexible wall barrel is used as the model container. The operating space of the barrel is a cylinder with the diameter of 3.0 m and the height of 1.8 m. The side wall of the container is made of 5 mm thick rubber. There are steel wires wrapping around the rubber to provide horizontal confinement. The diameter of the steel wire is 4 mm. There is one ring every 60 mm on the vertical direction. This design would allow the soil to give shear deformation at different depths. The top of the container is a rectangular steel frame. Because there are roller balls at the top of the supporting pillars, the frame could go in any directions in the horizontal plane.

© Springer Nature Switzerland AG 2018
W. Wu and H.-S. Yu (Eds.): *Proceedings of China-Europe Conference on Geotechnical Engineering*, SSGG, pp. 1221–1225, 2018.
https://doi.org/10.1007/978-3-319-97115-5_73

Fig. 1. Flexible wall barrel.

2.2   Model Soil

Table 1 shows the similitude ratios of the test. The similitude ratio of dimension is 1:20. Concerning the model soil, the density and the maximum shear modulus both should be decreased compared with the original soil. The model soil is a combination of dry sand and saw dust [4, 5]. Since the original ground is horizontally layered, two kinds of model soil are used to simulate the stratified ground. As shown in Fig. 2, Model Soil 1 has the mass ratio of 1/2 (sand/saw dust), and Model Soil 2 is 1/15.

Table 1. Similitude ratios of the shaking table test.

Parameters	Dimension	Mass	Time	Density	Acceleration	Shear modulus
Similitude ratios	0.05	4.3E−5	0.2	0.34	1.25	0.021

2.3   Tunnel Model

The tunnel ring model designed for this test is shown in Fig. 3. According to the similitude ratio in Table 1, the outer and inner diameters of the ring model are 58 cm and 53 cm, respectively. The width of the model is 10 cm. There are three steel rings embedded inside the model for two purposes. Firstly, the whole model is held up together by the steel rings, so they are working like the circumferential bolts at the radial joints. Secondly, by adjusting the stiffness of the steel rings, the transverse stiffness of the ring model could be controlled. There are eight aluminum sheets in the ring model. They can separate one segment from another, leaving 8 radial joints in the model. Figure 4 is the planar view of the model test. The tunnel model is made of 17 ring models.

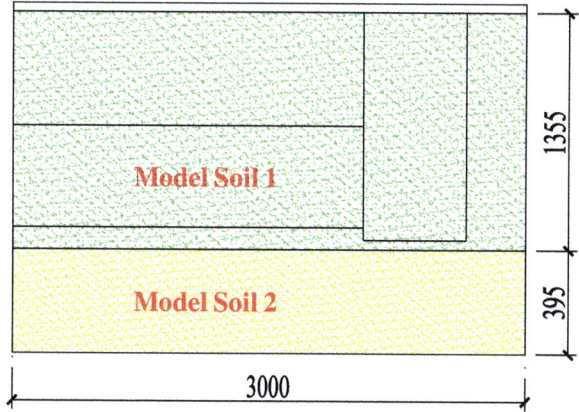

Fig. 2. Cross section of the model soil. (Unit: mm)

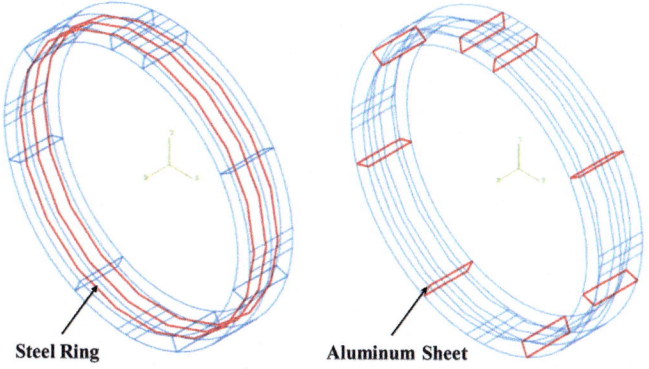

Fig. 3. Tunnel ring model.

2.4   Maximum Joint Openings

Figure 5 shows the maximum circumferential joint openings of the measured joints. The horizontal axis is the distance from the measured joints to the shaft model. The vertical axis is the maximum opening at each measured joint. There are 4 curves from 4 cases. It is obvious that the joint openings increase rapidly coming near the shaft model. And in all cases, the largest opening appears at joint J1, which is the contacting joint between the tunnel and the shaft. The maximum joint openings tend to be stable after joint J6. The maximum joint openings at J1 could be 40 times that of J6.

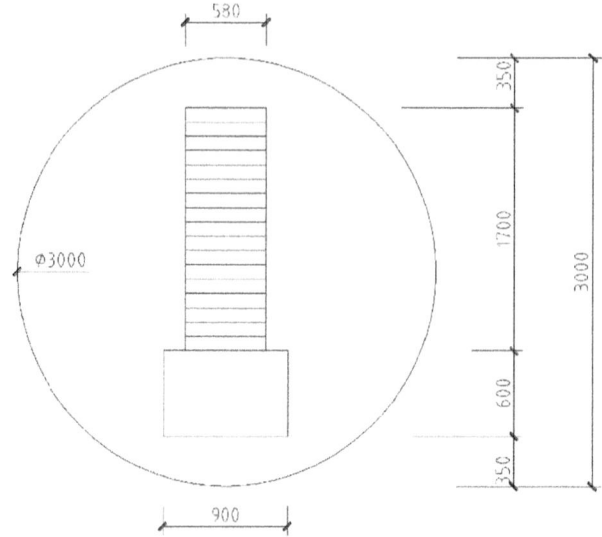

Fig. 4. Planar view of the model test. (Unit: mm)

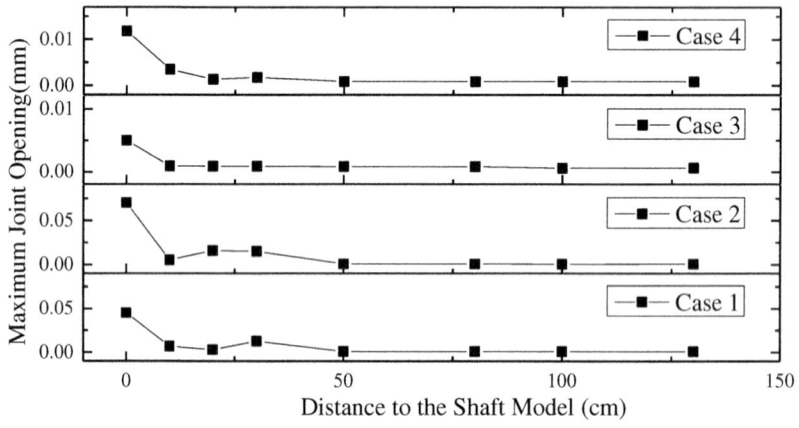

Fig. 5. Maximum joint openings of the measured joints.

Acknowledgements. The research has been supported by the National Key Research and Development Plan of China (2017YFC1500703), the National Natural Science Foundation of China (51678438 & 51478343 & 51778487), the Shanghai Rising-Star Program (17QC1400 500), and the Shanghai Committee of Science and Technology (16DZ1200302 & 16DZ1201904 & 18DZ1205103).

References

1. Uenishi, K., Sakurai, S.: Characteristic of the vertical seismic waves associated with the 1995 Hyogo-ken Nanbu (Kobe), Japan earthquake estimated from the failure of the Daikai Underground Station. Earthq. Eng. Struct. Dynam. **29**(6), 813–821 (2000)
2. Koizumi, A.: Seismic damages and case study for shield tunnel (Translated by W. Zhang, D. Yuan), 1st edn. China Architecture and Building Press, Beijing (2009). (in Chinese)
3. Zhao, W., He, X., Chen, W.: Analysis of seismic damage of segments and joints at the junction of shield tunnel and shaft. Chin. J. Rock Mech. Eng. **31**(2), 3847–3854 (2012). (in Chinese)
4. Yan, X., Yu, H., Yuan, Y.: Multi-point shaking table test of the free field under non-uniform earthquake excitation. Soils Found. **55**(5), 985–1000 (2015)
5. Bao, Z., Yuan, Y., Yu, H.: Multi-scale physical model of shield tunnels applied in shaking table test. Soil Dyn. Earthq. Eng. **100**, 465–479 (2017)

Nonlinear Analysis on Buried Pipelines Effected by Tunnelling

Chenrong Zhang[1,2(✉)] and Haili Li[1,2]

[1] Department of Geotechnical Engineering,
Tongji University, Shanghai 200092, China
zcrong33@tongji.edu.cn
[2] Key Laboratory of Geotechnical and Underground Engineering of Ministry
of Education, Tongji University, Shanghai, China

Abstract. Tunnel excavation may have impact on adjacent pipelines. Ignoring soil nonlinearity, the analysis of responses of pipelines under tunnel excavation will exhibit conservative results. A Winkler subgrade reaction model is developed, in which soil nonlinearity is considered based on the soil stiffness degradation model and the soil shear strain along the pipeline. The soil shear strain for the tunnel-soil-pipeline interaction is evaluated from two aspects. One is the tunnel excavation induced soil strain from the free soil movements. The other is the pipeline-soil interaction induced soil strain, based on a multi-layer-disc elastic model for a laterally loaded pile. The rationality of the Winkler based method in considering soil nonlinearity for the problem of tunnel effects on adjacent pipeline is proved against the published elastic continuum solution.

Keywords: Tunnel excavation · Pipelines · Soil nonlinearity
Winkler subgrade reaction model

1 Introduction

Tunnel excavation induced free soil movements lead to extra stress and deformation on existing pipelines, as illustrated in Fig. 1. Since the buried pipeline is subjected to excavation induced soil movement, the elastic analysis ignoring soil nonlinearity will overestimate the responses of pipelines, such as larger maximum bending moments [1, 2]. In view of this, Vorster et al. [1] presented an equivalent linear elastic continuum approach to take account of soil nonlinearity by evaluating soil stiffness from an average deviatoric strain in the free soil movement field. Marshall et al. [2] introduced an "out of plane" shear argument into Vorster [1]'s method, and verified it by analyzing results from centrifuge model tests. The modification of Vorster's formulation in estimating the tunnel excavation induced soil average strain is given by Klar et al. [3], to rationally consider shape parameters of free soil settlement.

In this paper, a Winkler based subgrade reaction model is used for the analysis of a buried pipeline under tunnel excavation, in which the calculation of soil average deviatoric strains with three components along the pipeline is introduced from free soil movements, as well as a multi-layer-disc elastic model to consider soil nonlinearity for

W. Wu and H.-S. Yu (Eds.): *Proceedings of China-Europe Conference
on Geotechnical Engineering*, SSGG, pp. 1226–1229, 2018.
https://doi.org/10.1007/978-3-319-97115-5_74

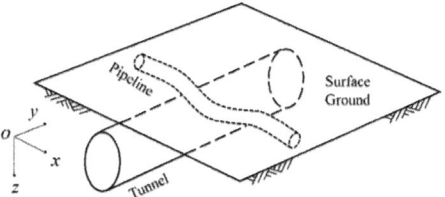

Fig. 1. Schematic graph of the problem

pipeline-soil interaction. The verification of the proposed approach is given against an elastic continuum solution.

2 Analytic Approach

The governing equation for the deflection of a pipeline $w(x)$ induced by tunnelling is given as follows:

$$E_p I_p \frac{d^4 w(x)}{dx^4} + kw(x) = kS_v(x) \tag{1}$$

where $E_p I_p$ is the pipeline bending stiffness, k is passive Winkler subgrade modulus, the expression of which is given by Yu et al. [4], $S_v(x)$ is the vertical free soil movement at pipeline level, which is based on the modified Gaussian curve by Vorster et al. [1] as

$$S_v = \frac{n}{(n-1) + \exp[\alpha(\frac{x}{i})^2]} S_{max} \tag{2}$$

where n is a shape function of parameter α, i is the distance from tunnel centerline to the inflection point of the curve.

According to the stiffness degradation curve, soil nonlinearity is represented as a reduced modulus based on the soil strain from tunnel excavation and pipeline-soil interaction. Therefore, the reduced soil modulus changes the value of Winkler subgrade modulus in Eq. (1), resulting a smaller deflection and maximum bending moment of pipeline.

To simplify the calculation of global shear strain due to tunnelling, Vorster et al. [1] and Klar et al. [3] only considered the shear strain $|\varepsilon_{xz}|$ by free soil movements and an average shear strain $|\varepsilon_{xz}|$ is suggested over an interval of $2.5i$ to give a constant reduced stiffness of soil. Obviously, since all six components of deviatoric strain contribute positively to the shear strain γ, any omitting of them gives a higher soil stiffness and hence more conservative results. Besides, the stiffness changes along the pipeline. In this paper, the 2D global shear strains in the plane of xoz (Fig. 1) are considered, which means $\varepsilon_y = 0$, $\varepsilon_{xy} = 0$, $\varepsilon_{zy} = 0$. The engineering shear strain γ, equaling to the diameter of the Mohr circle of strain [2], is then given as:

$$\gamma = \sqrt{(\varepsilon_x - \varepsilon_z)^2 + 4\varepsilon_{zx}^2} \tag{3}$$

in which, the calculation of ε_{zx} is the same as that in Vorster et al. [1], ε_z and ε_x are derived as

$$\varepsilon_x = \frac{\partial S_u}{\partial x} = -\frac{2nx^2\alpha \exp\left[\alpha\left(\frac{x}{i}\right)^2\right]}{Z_R i^2 \left\{n-1+\exp\left[\alpha\left(\frac{x}{i}\right)^2\right]\right\}^2} S_{max} + \frac{nS_{max}}{Z_R\left\{n-1+\exp\left[\alpha\left(\frac{x}{i}\right)^2\right]\right\}} \tag{4}$$

$$\varepsilon_z = \frac{\partial S_v}{\partial z} = \frac{\partial i}{\partial z}\frac{2n\alpha\exp[\alpha(\frac{x}{i})^2]}{\left\{n-1+\exp[\alpha(\frac{x}{i})^2]\right\}^2}\frac{x^2}{i^3}S_{max} + \frac{n}{\left\{n-1+\exp[\alpha(\frac{x}{i})^2]\right\}}\frac{\partial S_{max}}{\partial z} \tag{5}$$

For the interaction between pipeline and soil, the mechanism of mobilized shear strain around the pipeline is similar to a 2D horizontal plane analysis of a laterally loaded pile, modelling as a rigid disc moving in the nonlinear soil continuum [5]. Using the mobilized strength design method (MSD) and two-layer-deformational disc model, Klar et al. [5] related the shear strain around the pipeline γ_s to the displacement of the inner rigid disc δ_r as

$$\gamma_s = \beta\frac{\delta_r}{r_0} \tag{6}$$

in which the shearing factor β is 1.3, r_0 is pile radius. The result was applied by Marshall et al. [2] to analyze the centrifuge model tests, giving a good performance when compared with Vorster's method [1]. Since soil strain field based on a single layer elastic ring around the rigid disc could not describe the nonuniform distribution of shear strain around the laterally loaded pile, Yu et al. [6] further extended the single-layer disc to multi-layer discs, in which the shearing factor is modified as 0.8, the value used in this paper to calculate the soil shear strain due to pipeline-soil interaction.

Based on the above analysis, the equivalent shear strain value γ_{eq} for the Winkler analysis of the pipeline, is obtained as a combination of shear strain by free soil movement γ and that by pipeline-soil interaction γ_s, given as [3]

$$\gamma_{eq} = \sqrt{\gamma^2 + \gamma_s^2} \tag{7}$$

An interactive procedure is needed in the nonlinear calculation of buried pipeline under tunnel excavation, for the reduced soil stiffness in fact depends on the deflection of the pipeline.

3   Comparison with the Elastic Continuum Solution

The comparisons of normalized maximum bending moments between the results by the present Winkler analysis and those by elastic continuum analysis using Vorster's method and Klar's method are given in Fig. 2, along with the Winkler analysis ignoring soil nonlinearity. The case parameters in this calculation is corresponding to the centrifuge tests in Marshall et al. [2] for test 2. It shows that the present method gives a better prediction of the bending moment with a rational consideration of soil nonlinearity.

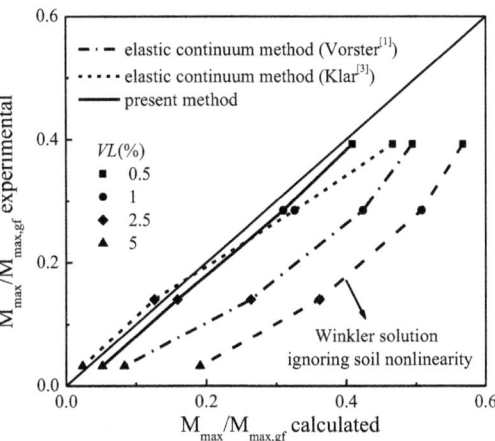

Fig. 2. Comparison between maximum bending moments of different volume loss

4   Conclusion

In this paper, a Winkler analysis is presented to investigate soil nonlinearity on response of buried pipeline under tunnel excavation. The procedure to calculate soil shear strain by tunnelling and pile-soil interaction is introduced respectively, to obtain the reduced soil stiffness for the Winkler subgrade modulus. A comparison with the elastic continuum solutions proved the rationality of the analysis in this paper.

References

1. Vorster, T.E.B., Klar, A., Soga, K., Mair, R.J.: Estimating the effects of tunneling on existing pipelines. J Geotech. Geoenviron. Eng. **131**(11), 1399–1410 (2005)
2. Marshall, A.M., Klar, A., Mair, R.J.: Tunneling beneath buried pipes: view of soil strain and its effect on pipeline behavior. J. Geotech. Geoenviron. Eng. **136**(12), 1664–1672 (2010)
3. Klar, A., Elkayam, I., Marshall, A.M.: Design oriented linear-equivalent approach for evaluating the effect of tunneling on pipelines. J. Geotech. Geoenviron. Eng. **142**(1), 04015062 (2016)
4. Yu, J., Zhang, C.R., Huang, M.S.: Soil-pipe interaction due to tunnelling: assessment of winkler modulus for underground pipelines. Comput. Geotech. **50**(5), 17–28 (2013)
5. Klar, A.: Upper bound for cylinder movement using elastic fields and its possible application to pile deformation analysis. Int. J. Geomech. ASCE **8**(2), 162–167 (2008)
6. Yu, J., Huang, M.S., Li, S., Leung, C.F.: Load-displacement and upper-bound solutions of a loaded laterally pile in clay based on a total-displacement-loading EMSD method. Comput. Geotech. **83**, 64–76 (2017)

Adequate Numerical Simulation of Tail Void Grouting for Tunneling in Saturated Soil

Chenyang Zhao$^{(\boxtimes)}$, Arash A. Lavasan, and Tom Schanz

Chair of Foundation Engineering, Soil and Rock Mechanics,
Ruhr-University Bochum, Universitaetstr. 150, 44801 Bochum, Germany
{Chenyang.Zhao,Arash.AlimardaniLavasan,Tom.Schanz}@rub.de

Abstract. In this research, a 3D numerical simulations of tunneling in saturated soil are conducted. The tunnel shield and lining segments with different diameters are explicitly modeled during the progressive excavation. Special attention is paid to accurate simulation of the grout hydration in the tail void where a constitutive model that accounts for the time dependent stiffness is developed to describe the hardening behavior of grout mortar in the coupled hydro-mechanical analysis. Additionally, the effect of grouting pressure distribution along the lining segments is investigated as well. The results reveal that neglecting the evolution of stiffness may lead to underestimation of surface settlements. The results also indicated the insignificant impact of the shape of the annular gap on the soil deformations and lining forces.

1 Introduction

With the development of computational techniques, numerical simulation methods have become a powerful tool for reliable prediction of tunneling induces soil deformations and tunnel structure forces (Shah et al. 2017). Franzius and Potts (2005) summarized a number of 3D finite element studies. They pointed out the effects of employed constitutive model, geometry, mesh dimension, excavation method on the model responses. Zhao et al. (2015) conducted a case study of Western Scheldt tunnel using the finite element method, and the numerical results well match the real measurements. Lavasan et al. (2017) numerically investigated the influence of important factors that affect the system behavior for tunneling in saturated soil, special attention was paid to the infiltration of fine particles due to tail void grouting. However, the volume loss due to conical shape of shield and grout hardening was neglected. Therefore, this research aims to take into account the volume loss and grout hardening in the numerical simulation of tunneling in saturated soil.

© Springer Nature Switzerland AG 2018
W. Wu and H.-S. Yu (Eds.): *Proceedings of China-Europe Conference on Geotechnical Engineering*, SSGG, pp. 1230–1233, 2018.
https://doi.org/10.1007/978-3-319-97115-5_75

2   Numerical Simulation of Mechanized Tunneling

In this study, the tunneling process is modeled using finite element software *Plaxis*. The 3D model geometry is shown in Fig. 1. The TBM head has a diameter of $D = 8.5$ m with an overburden depth of $1D$ at the tunnel crown. In order to model the overcut at TBM head, 0.1% volume loss is applied at the front of TBM shield using contraction factor method. This contraction linearly increases to 0.5% at the tail of the shield to model the conicity shape of TBM shield. The outer diameter of the lining segments is assumed to be 8.2 m. Therefore, there is a gap between the lining and the surrounding soil (see Fig. 1) where grout material is injected. By varying the ratio of d_1 and d_2, influence of shape of annular gap between lining and surrounding soil on the ground movements can be investigated. The ground water level is assumed at the ground surface.

The isotropic hardening soil (HS) model is employed to describe the soil behavior during the tunneling process. The TBM shield and lining segments are modeled by elastic shell elements. The corresponding material properties are same as those in Lavasan et al. (2017).

The lining element has a length of 1.5 m which is identical to the advancement distance of the TBM in each excavation step. A uniformly distributed grouting pressure of 200 kPa is assumed to be applied as the mechanical pressure to the tail void with the length of one ring (single-grouting). According to the experimental study of Dias and Bezuijen (2015), the grouting pressure and gradient are the maximum directly after the shield and they gradually decrease along the newly installed lining segments. Due to this reason, the grouting pressure is defined in an alternative way that the grouting pressure gradually decrease to the steady water pressure within the length of 4 lining segments (multi-grouting).

In order to model the stiffness evolution of grout, a constitutive model with time dependent stiffness (TDE) is developed based on CEB-FIP code (1990):

$$E(t) = E_{28} \left[\exp \left(s_{\text{stiffness}} \left(1 - \sqrt{\frac{t_{28}}{t}} \right) \right) \right]^{0.5}, t \leq t_{28}$$

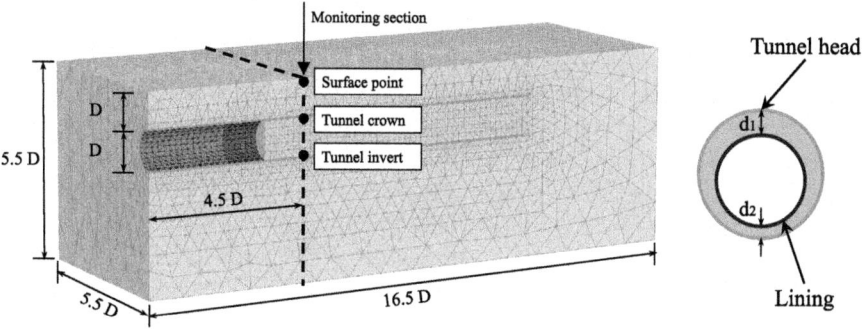

Fig. 1. Geometry and mesh discretization of the 3D model

where $s_{stiffness} = 0.2$ controls how fast the stiffness increases with time. The stiffness of the grout is assumed to be constant after 28 days as $E_{28} = 5250$ kPa.

Table 1 shows the numerical scenarios employed in this study. In addition to the TDE model, linear elastic (LE) model is also applied to model the grout behavior with constant stiffness of E_{28}. The soil permeability of the soil is assumed to be 10^{-9} m/s, therefore, the effect of infiltration can be neglected (Lavasan et al. 2017). In case of low excavation speed, drain analysis is utilized to model the tunneling process. When the excavation speed is high, consolidation analysis is employed to capture the hydro-mechanical coupling behavior.

Table 1. Description of the employed numerical scenarios

Details	Scenario 1	Scenario 2	Scenario 3	Scenario 4	Scenario 5	Scenario 6
Model of grout	TDE	LE	TDE	TDE	TDE	TDE
Grouting area	Multi	Multi	Single	Multi	Multi	Multi
Volume loss	$d_1 = 2d_2$	$d_1 = 2d_2$	$d_1 = 2d_2$	$d_1 = d_2$	$d_1 = 2d_2$	$d_1 = 2d_2$
Analysis type	Drained	Drained	Drained	Drained	Consolidation	Consolidation
Permeability of lining	–	–	–	–	impermeable	permeable

3   Results and Discussions

The tunneling induced ground movements at the monitoring section are shown in Figs. 2(a) and (b). As seen, when TDE model is applied instead of LE in drained analysis, larger settlement is observed at ground surface due to gradually increasing stiffness. By modeling the grouting pressure along the lining segments, the surface settlement is decreased. While the shape of the annular gap seems not influencing on the surface settlements according to scenarios 1 and 4. Due to the fact that dissipation of excess pore pressure takes time, less surface settlements are generated by the end tunnel excavation in the consolidation analyses. Furthermore, in case that lining segments are permeable, the excess pore pressure in the near field around the tunnel can be rapidly dissipated, which results in higher settlements at tunnel crown.

By analogy, Figs. 2(c) and (d) presents the lining forces by the end of excavation for different scenarios. When the grout is modeled as constant stiffness E_{28}, maximum axial forces and bending moments are obtained due to the higher stresses around the tunnel under less soil deformations. On the contrary, in the consolidation analysis, the generated excess pore pressure decreases the effective stress. Subsequently, the axial forces and bending moments are reduced.

Acknowledgements. This research has been supported by the German Research Foundation (DFG) through the Collaborative Research Center (SFB 837).

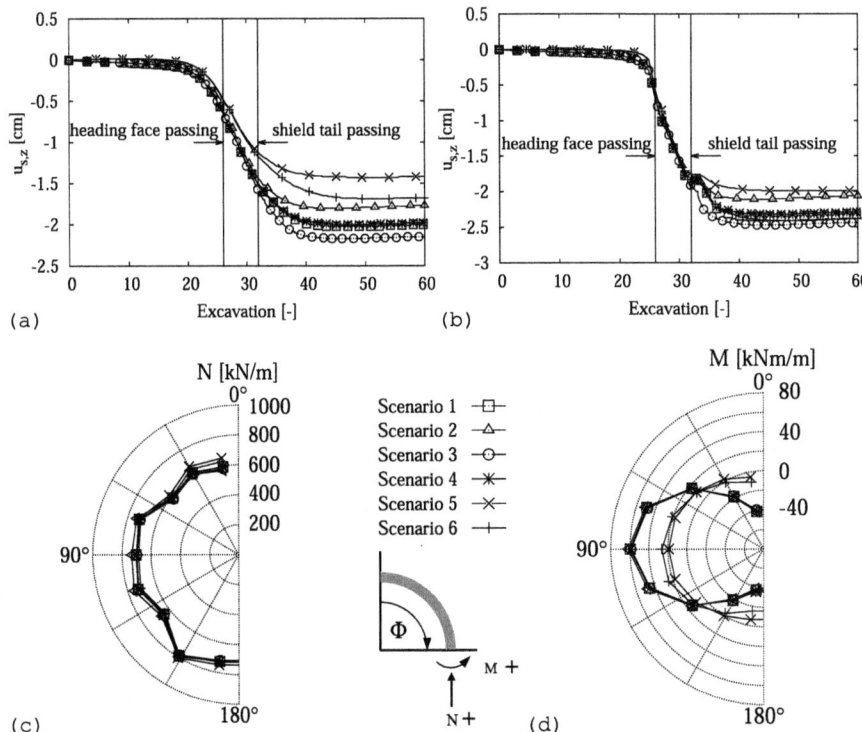

Fig. 2. Comparison of the model responses for (a) surface displacement, (b) vertical displacement at tunnel crown, (c) lining axial forces, (d) lining bending moments

References

CEB-FIP code: Design code-comite Euro-international du Beton. Thomas Telford (1990)

Dias, T., Bezuijen, A.: TBM pressure models-Observations: Theory and Practice. Geotechnical Synergy in Buenos Aires (2015)

Franzius, J., Potts, D.: Influence of mesh geometry on three-dimensional finite-element analysis of tunnel excavation. Int. J. Geomech. **5**(3), 256–266 (2005)

Lavasan, A., Zhao, C., Barciaga, T., Schaufler, A., Steeb, H., Schanz, T.: Numerical investigation of tunneling in saturated soil: the role of construction and operation periods. Acta Geotech. **13**(3), 671–691 (2018)

Shah, R., Zhao, C., Lavasan, A.A., Peila, D., Schanz, T., Lucarelli, A.: Influencing factors affecting the numerical simulation of the mechanized tunnel excavation using FEM and FDM techniques. In: EURO:TUN 2017, Innsbruck, Austria (2017)

Zhao, C., Lavasan, A.A., Barciaga, T., Zarev, V., Datcheva, M., Schanz, T.: Model validation and calibration via back analysis for mechanized tunnel simulations: the Western Scheldt tunnel case. Comput. Geotech. **69**, 601–614 (2015)

Torque Fluctuation and Penetration Analysis of Shield Tunnel in Rock–Soil Mixed Ground

Yu Zhao[✉], Quanmei Gong, Runlai Zhang, and Jie Xia

Key Laboratory of Road and Traffic Engineering of the Ministry of Education,
Tongji University, Shanghai 201804, China
yuzhao@tongji.edu.cn

Abstract. Tunnelling in rock–soil interface mixed ground has often been faced with cutterhead torque fluctuation, a reasonable penetration rates to reduce the torque fluctuation and improve tunnelling efficiency is of great significance. In this paper, the cutting torques model was established. Torque changes with the rotation angle of cutterhead has been studied when the cutterhead in rock–soil interface with different hard rock area ratios (Fc). The results show that the average torque of cutterhead increases with the increase of Fc. The torque fluctuation cannot be eliminated, and torque fluctuation reaches its maximum when the Fc is about 10%–20%, which most likely to cause blockage and damage for the cutterhead. Such compound stratum sections should be avoided as far as possible when designing of tunnel lines. In the case of the same average torque, the penetration does not decrease linearly with the increase of Fc because of torque fluctuation. The penetration decreases slightly when the Fc reaches 10%–20%, and decreases obviously when the Fc exceeded 30%. The reason is that the penetration needs to be reduced to decrease the torque as well as torque fluctuation; when Fc reaches 60%, the penetration reaches the lowest; when Fc exceeds 60%, the penetration increases slightly to improve the tunnelling efficiency because torque fluctuation is reduced.

Keywords: Shield tunnel · Rock–soil interface · Torque fluctuation
Penetration

1 Introduction

Rock–Soil Interface (RSI) mixed ground refers to the tunnel face which is composed of more than two layers which greatly differ from each other in mechanical [1]. And this ground is one of the most difficult conditions for shield tunneling method [2]. As the cutters move over the RSI they are taking up highly variable loads and the result of such loading is torque fluctuation, which makes the tunneling parameter difficult to choose. Many researchers concentrate on shield machine performance prediction based on field data [3]. However, few literatures focus on the interaction between the cutter and the RSI, as well as lack of penetration analysis. Against this background. This research concerning on the torque fluctuation in RSI mixed ground and relationship between penetration and the hard rock area ratios considering the torque fluctuation based on theoretical analysis and validated by field date, which is rarely reported.

© Springer Nature Switzerland AG 2018
W. Wu and H.-S. Yu (Eds.): *Proceedings of China-Europe Conference on Geotechnical Engineering*, SSGG, pp. 1234–1238, 2018.
https://doi.org/10.1007/978-3-319-97115-5_76

2 Total Cutter Torque

When the geological condition is determined, the frictional resistance torque of the cutterhead can be regarded as a constant. Considering the fluctuation of torque in the rotary cutting process, we only calculate the torque of the cutting tool. For disc cutter force, the individual cutter force can be calculated by the CSM model. Where F_r is the tangential load per disc cutter (see Fig. 1):

$$F_R = \left[\sigma_c h^2 + \frac{4\tau\phi h^2 (S - 2h\tan\theta/2)}{D(\phi - \sin\phi\cos\phi)} \right] \tan\theta/2 \qquad (1)$$

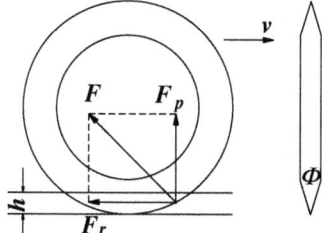

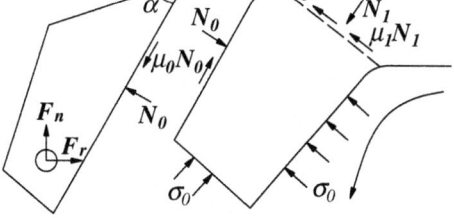

Fig. 1. Force of disc cutter   **Fig. 2.** Force of soft ground tool

For soft ground tools force, a new model has been conducted considering the influence of the earth pressure (see Fig. 2):

$$F_t = BN_0 \qquad (2)$$

$$N_0 = \frac{A(CF_1\cos\theta - \sigma_0\cos\alpha F_2 + \sigma_0\sin\alpha F_3) - M(CF_1\sin\theta + \sigma_0\sin\alpha F_2 + \sigma_0\cos\alpha F_3)}{AB - MN} \qquad (3)$$

where the A, B, M, N can be expressed as: $A = \mu_1\sin\theta - \cos\theta$, $B = \sin\alpha + \mu_0\cos\alpha$, $M = \sin\theta + \mu_1\cos\theta$, and $N = \cos\alpha - \mu_0\sin\alpha$ with F_t being the tangential load per soft ground tool. The EPB shields are equipped with single and double disk cutters and soft ground tools. To simplify, it is also generally assumed that each kind of tools take the same tangential load in the same geology and therefore the total cutting torque will be:

$$T_1 = \sum_{1}^{i} F_{Ri} r_i + \sum_{1}^{j} F_{Tj} r_j \qquad (4)$$

where r_i is the distance between the disc cutter i and the center of cutterhead, r_j is the distance between the soft ground tools j and the center of cutterhead.

3 Characteristics of Torque Fluctuation

At present, the torque is empirically determined or build model through experiment. However, both two methods can only obtain a fixed value and cannot reflect the torque fluctuation when tunnelling in RSI. Combining with geological conditions and taking into account of cutterhead structure and the interaction between cutterhead and RSI, a composition calculation method and evaluation index are presented.

3.1 Torque Fluctuation with Cutterhead Rotating

The hard rock area ratios (F_c) is defined as the area of the hard rock (S_{hard}) divided by the total area of the excavation face (S_{total}) (Figs. 3 and 4).

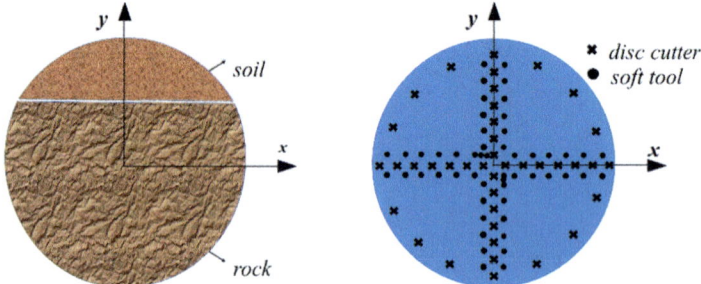

Fig. 3. Geological conditions   **Fig. 4.** Tool distribution

The upper part is in residual soil and the lower part is in weathered granite with strength of 50 MPa. The simulation results are shown as Fig. 5 which selected a typical Herrenknecht shield as the object.

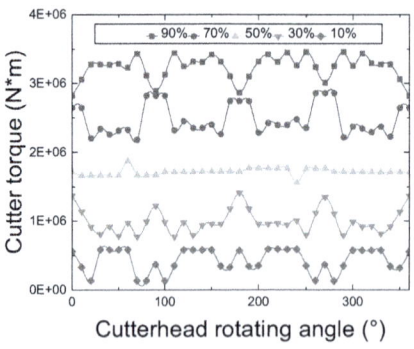

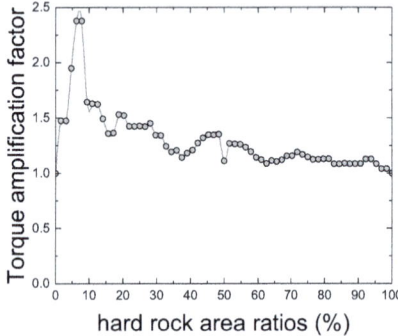

Fig. 5. The torque fluctuation with cutter-head rotating in different F_c.

Fig. 6. The relationship between torque amplification factor and F_c.

This interface brings great torque fluctuation. And fluctuating loads will not only cause vibration of cutterhead, but also cause blockage accident during shield tunneling. Torque amplification factor (η_d) is introduced as $\eta_d = T_{max}/T_{avg}$. Torque amplification factor is shown as Fig. 6. The fluctuation value changed from 1 to 2.4 and reach the peak at 8%. Tóth has conducted that the most significant drop of penetration rate can be observed where the F_c ranges in 0–30% [1]. This is because the torque fluctuation intensely at this time and the penetration needs to be reduced to decrease the vibration.

3.2   The Measured Torque Fluctuation

The RSI mixed ground has been encountered in Guangzhou Metro Line 21. The η_d fluctuates between 1.1–5.2 and mainly distributed between 1.2–1.8, with an average of 1.5, which is in good agreement with the previous analysis (see Fig. 7).

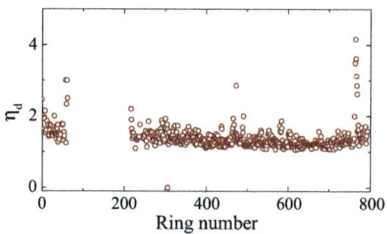

Fig. 7. The torque amplification factor based on field data

4   Penetration Analysis Based on Field Data

The penetration and torque has been analyzed as shown in Figs. 8 and 9 that based on field data of tunnelling in RSI mixed ground in Guangzhou.

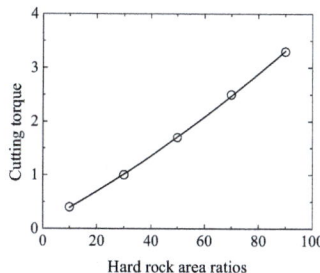
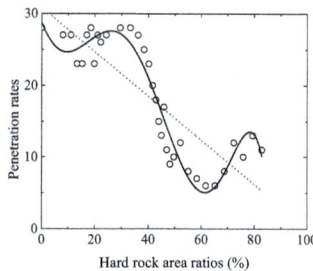

Fig. 8. Relationship between torque and F_c.

Fig. 9. Relationship between penetration and F_c.

When penetration as a constant, the average torque of cutterhead increases with the increase of Fc as shown in Fig. 8. the penetration needs to be reduced to decrease the torque, which is generally considered that penetration decreases linearly with increasing of Fc. Actually, the penetration is fluctuating based on field data analyses as shown in Fig. 9, the main reason is the torque fluctuation causes the shield driver to adjust the thrust to change the penetration and reduce the fluctuation of the torque.

References

1. Tóth, Á., Gong, Q., Zhao, J.: Case studies of TBM tunneling performance in rock–soil interface mixed ground. Tunn. Undergr. Space Technol. **38**(9), 140–150 (2013)
2. Ma, H., Yin, L., Gong, Q., Wang, J.: TBM tunneling in mixed face ground Problems and solutions. Int. J. Min. Sci. Technol. **25**(4), 641–647 (2015)
3. Vergara, I.M., Saroglou, C.: Prediction of TBM performance in mixed-face ground conditions. Tunn. Undergr. Space Technol. **69**, 116–124 (2017)

Semi-analytical Solutions on Seepage Field of Twin Tunnels

Chengwei Zhu[1,2,3]($\boxtimes$) (iD), Hongwei Ying[1,2,3] (iD), Xiaonan Gong[1,2,3] (iD),
Huawei Shen[1,2,3] (iD), and Xiao Wang[1,2,3] (iD)

[1] Research Center of Coastal and Urban Geotechnical Engineering,
Zhejiang University, Hangzhou 310058, China
zhuchengwei@zju.edu.cn
[2] Key Laboratory of Soft Soils and Geoenvironmental Engineering
of Ministry of Education, Zhejiang University, Hangzhou 310058, China
[3] Engineering Research Center of Urban Underground Development,
Hangzhou 310058, China

Abstract. Based on the governing equation of steady-state seepage, the semi-analytical solutions of seepage field of twin tunnels considering the effect of lining is derived rigorously using the technique of conformal mapping. These solutions can provide the values of total hydraulic head at an arbitrary point in the seepage field and the tunnel water ingress. Discussion is given about the effects of the tunnel distance on the distribution of the total hydraulic head outside the twin tunnels and the tunnel water ingress. It is found that simplifying the twin tunnels as a single tunnel in analysis could overestimate the hydraulic head around tunnels and the tunnel water ingress.

Keywords: Twin tunnels · Semi-analytical solution · Conformal mapping
Tunnel distance

1 Introduction

Some research has been done in terms of seepage field of single tunnel. The common analytical methods in the previous literature include mainly the following two ways: (1) Method of images [1–3], (2) Mapping method [4–7]. Noted that the majority of the existent research focused on single tunnel, but most practical projects adopted twin tunnels rather than single tunnel because of the limitation of geology condition or technical means. Compared with single tunnel, the existence of twin tunnels would influence the seepage field around either tunnel. This paper illustrates the derivation of the semi-analytical solutions for hydraulic head distribution and water ingress of twin tunnels, which could take into account the properties of the lining layer such as the permeability and thickness of the lining layer. Moreover, discussion is given about the effect of tunnel distance on the distribution of the total hydraulic head outside the twin tunnels and water ingress into tunnels.

© Springer Nature Switzerland AG 2018
W. Wu and H.-S. Yu (Eds.): *Proceedings of China-Europe Conference
on Geotechnical Engineering*, SSGG, pp. 1239–1243, 2018.
https://doi.org/10.1007/978-3-319-97115-5_77

2 Semi-analytical Solutions

As shown in Fig. 1, the whole domain is divided into two parts, i.e., the aquifer (I) and the lining layer (II). The upper tunnel has the internal and external radii of r_1 and R_1, respectively, with a burial depth of h_1. The nether tunnel has the internal and external radii of r_2 and R_2, respectively, with a burial depth of h_2. Above the mud line is the water layer with a thickness of H. The mud line is chosen as the elevation reference datum. This paper adopts the following assumptions:

(1) The aquifer and lining layer are isotropous porous medium;
(2) The water flow obeys the Darcy's law;
(3) The water pressure on the internal surface of the tunnels is constant, i.e., u_j

The governing equation for such a problem is shown below [8]:

$$\partial^2 \phi / \partial x^2 + \partial^2 \phi / \partial y^2 = 0 \tag{1}$$

where $\phi = y + p/\gamma_w$, y is the elevation head, p is the water pressure and γ_w is the unit weight of water.

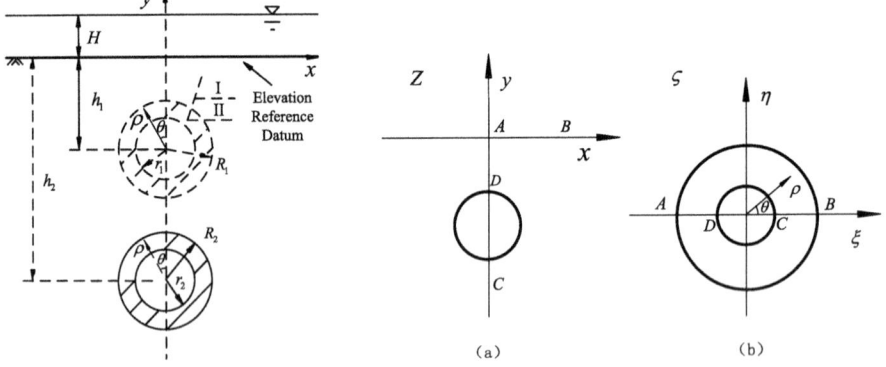

Fig. 1. Schematic diagram.   **Fig. 2.** Conformal mapping.

Figure 1 and the basic assumptions could provide the following two boundary conditions:

$$\phi|_{y=0} = H \tag{2}$$

$$\phi|_{\sqrt{x^2 + (y+h_j)^2} = r_j} = y + h_{uj} \tag{3}$$

where $h_{uj}, j = 1, 2$, is the pressure head on the internal surfaces of the twin tunnels.

Equation (4) could map the semi-infinite aquifer into a ring domain in Fig. 2(b) conformally,

$$\zeta = \xi + \eta i = (z + ia_m)/(z - ia_m) \tag{4}$$

where z is an arbitrary point in plane - Z, $a_m = \sqrt{h_m^2 - R_m^2}$ and i is the imaginary unit; ζ is the corresponding point in the mapped domain of the point z; ξ and η are the Cartesian coordinates in plane - ς.

According to the property of conformal mapping, Eq. (1) has the following form in plane $-\varsigma$

$$\partial^2 \phi / \partial \xi^2 + \partial^2 \phi / \partial \eta^2 = 0 \tag{5}$$

The expression of the hydraulic head in the aquifer is shown as Eq. (6)

$$\phi_1 = H + \sum_{m=1}^{2} \left(B_m \ln \rho_m + \sum_{n=1}^{\infty} C_{mn} \left(\rho_m^n - \rho_m^{-n} \right) \cos n\theta_m \right) \tag{6}$$

where A_m, B_m and C_{mn} are unknowns to be determined; ρ_m and θ_m are the polar coordinates in plane $-\varsigma$.

Similarly, the general solution to Eq. (1) in the lining layer in plane $- Z$ is shown below

$$\phi_{\mathrm{II}j} = D_j + E_j \ln \rho_j + \sum_{n=1}^{\infty} \left(F_{jn} \rho_j^n + G_{jn} \rho_j^{-n} \right) \cos n\theta_j \tag{7}$$

Where $\phi_{\mathrm{II}j}$ represents the hydraulic head in the lining layer; D_j, E_j, F_{jn} and G_{jn} are coefficients related with B_m and C_{mn}; ρ_j and θ_j are the polar coordinates in plane $- Z$, whose directions are shown in Fig. 1.

The values of B_m and C_{mn} could be inferred by expanding the seepage continuity condition on the interface between domain I and domain II, namely, Eq. (8) as Fourier series and comparing each term's coefficient [9].

$$k_s / k_{\mathrm{l}j} \cdot \partial \phi_1 / \partial \rho = \partial \phi_{\mathrm{II}j} / \partial \rho \tag{8}$$

Moreover, the expression of water ingress could be derived from Eq. (7), as shown below

$$Q_j = \int_0^{2\pi} k_{\mathrm{l}j} \cdot \partial \phi_{\mathrm{II}j} / \partial \rho \big|_{\rho = R_j} R_j d\theta = 2\pi k_{\mathrm{l}j} E_j \tag{9}$$

3   Discussion

Figure 3 illustrates that the hydraulic head on the interface increases as s becomes larger. Compared with single tunnel, twin tunnel system could reduce the water pressure imposed on the lining layer to a great degree. In addition, the effect induced by the nether tunnel is much more obvious at the invert contrast to the crown. The hydraulic head outside of the nether lining layer is influenced greatly no matter it is at the crown or the invert. Especially, when the twin tunnels get closer, the maximum hydraulic head outside the nether lining layer moves from the crown to the invert.

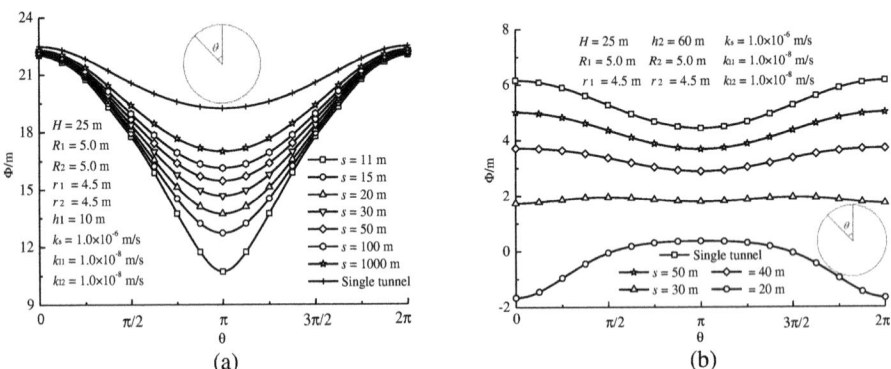

Fig. 3. Hydraulic distribution on exterior circumference of linings (a) Upper tunnel; (b) Nether tunnel.

From Figure 4, an increasing water ingress is observed for both tunnels when the tunnel distance becomes larger, which means ignoring the effect of twin tunnels could overestimate the water ingress into tunnels. For example, when $s = 11$ m, $Q_1 = 1.381$ m$^3 \cdot$day^{-1} per meter while the water ingress is 1.585 m$^3 \cdot$day^{-1} per meter

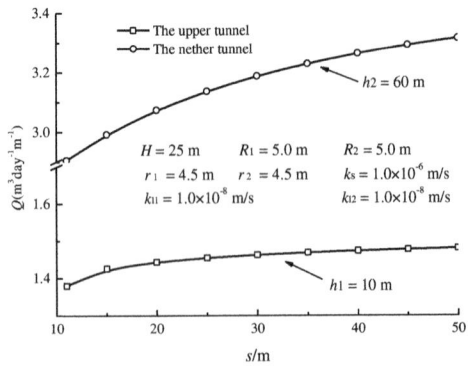

Fig. 4. Effect of tunnel distance on water ingress.

for single tunnel, and the ratio between these two values is 87.1%; $Q_2 = 2.905$ $m^3 \cdot day^{-1}$ per meter while the water ingress is 3.365 $m^3 \cdot day^{-1}$ per meter for single tunnel, and the ratio between these two values is 86.3%.

References

1. Harr, M.E.: Groundwater and Seepage. McGraw-Hill, New York (1962)
2. Ying, H.W., Zhu, C.W., Gong, X.N.: Analytical solution on seepage field of underwater tunnel considering grouting circle. J. Zhejiang Univ. (Eng. Sci.) **50**(6), 1018–1023 (2016). (in Chinese)
3. Joo, E.J., Shin, J.H.: Relationship between water pressure and inflow rate in underwater tunnels and buried pipes. Géotechnique **64**(3), 226–231 (2014)
4. Tani, M.: Circular tunnel in a semi-infinite aquifer. Tunn. Undergr. Space Technol. **18**(1), 49–55 (2003)
5. Kolymbas, D., Wagner, P.: Groundwater ingress to tunnels–the exact analytical solution. Tunn. Undergr. Space Technol. **22**(1), 23–27 (2007)
6. Zhu, C.W., Ying, H.W., Gong, X.N.: Analytical solutions for seepage fields of underwater tunnels with arbitrary burial depth. Chin. J. Geotechn. Eng. **39**(11), 1984–1991 (2017). (in Chinese)
7. Park, K.H., Owatsiriwong, A., Lee, J.G.: Analytical solution for steady-state groundwater inflow into a drained circular tunnel in a semi-infinite aquifer: a revisit. Tunn. Undergr. Space Technol. **23**(2), 206–209 (2008)
8. Arjnoi, P., Jeong, J.H., Kim, C.Y., et al.: Effect of drainage conditions on porewater pressure distributions and lining stresses in drained tunnels. Tunn. Undergr. Space Technol. **24**(4), 376–389 (2009)
9. Ying, H.W., Zhu, C.W., Gong, X.N.: Tide-induced hydraulic response in a semi-infinite seabed with a subaqueous draSined tunnel. Acta Géotechnica, 1–9 (2017)

Analytical Solutions for Seepage Field of Underwater Tunnel

Chengwei Zhu[1,2,3(✉)] ⓘ, Hongwei Ying[1,2,3] ⓘ, Xiaonan Gong[1,2,3] ⓘ,
Huawei Shen[1,2,3] ⓘ, and Xiao Wang[1,2,3] ⓘ

[1] Research Center of Coastal and Urban Geotechnical Engineering,
Zhejiang University, Hangzhou 310058, China
ice898@zju.edu.cn
[2] Key Laboratory of Soft Soils and Geoenvironmental Engineering
of Ministry of Education, Zhejiang University, Hangzhou 310058, China
[3] Engineering Research Center of Urban Underground Development,
Hangzhou 310058, China

Abstract. Based on the patterns of the hydraulic head outside the lining layer obtained from numerical simulation, a new boundary condition for the tunnel is proposed that the hydraulic head appears sinusoidal distribution on the lining's external circumference. The analytical solutions about the hydraulic head and water ingress are derived rigorously, which could consider the effect of the lining property such as the lining thickness and permeability. The new solution is compared with another three classic analytical solutions in the existent literature. It is found that the traditional assumption that the lining layer is an equipotential surface could overestimate the water ingress especially when the tunnel is buried shallowly.

Keywords: Analytical solution · Sinusoidal distribution · Burial depth

1 Introduction

All tunnels could be regarded as drained because of its cracks in the lining layer, including the assembly gaps of the lining segments and crevices originating from tunnel deformation [1–3]. The majority of the existent literature based on the assumption that the lining layer is hydraulic equipotential [4–6]. This process is acceptable when the tunnel is buried deeply, but some errors could occur if the tunnel has a shallow burial depth, which is common in urban tunnel construction.

This paper proposed a new boundary condition based on the numerical results from the software COMSOL that the hydraulic head on the external circumference of the lining layer is distributed sinusoidal. The analytical solutions about the hydraulic head and water ingress into the tunnel are derived rigorously. A discussion is given about the effect of burial depth on the water ingress calculated from the proposed solution and another three different methods in the previous literature.

© Springer Nature Switzerland AG 2018
W. Wu and H.-S. Yu (Eds.): *Proceedings of China-Europe Conference
on Geotechnical Engineering*, SSGG, pp. 1244–1248, 2018.
https://doi.org/10.1007/978-3-319-97115-5_78

2 Analytical Solutions

As illustrated in Fig. 1, the whole research area is divided into two parts, i.e., the lining layer (I) and the aquifer (II). The lining layer has the internal and external radii of r and R, respectively, with burial depth of d. Here the burial depth is defined as the distance between the tunnel center and mud line. Above the aquifer is the water layer with a thickness of H and the mud line is chosen as the elevation reference datum. The whole paper is based on the following assumptions.

(1) The aquifer and the lining are isotropic porous medium;

(2) The water flow within the seepage field obeys the Darcy's law;

(3) Based on the numerical result obtained from the software COMSOL, as shown in Fig. 2, it is assumed that the hydraulic head distribution pattern is sinusoidal.

(4) The water pressure on the internal surface of the lining layer is constant, namely, u.

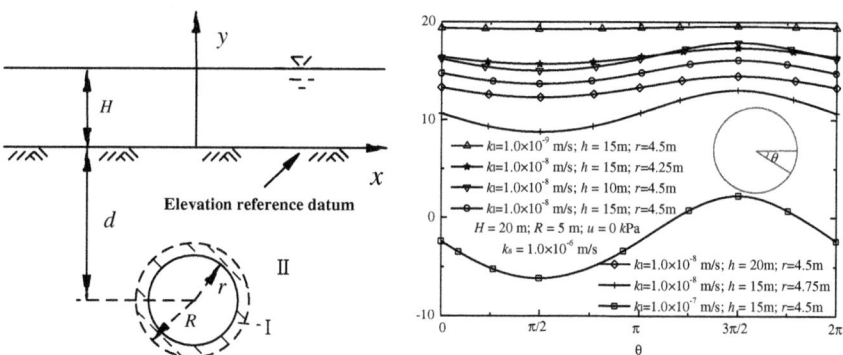

Fig. 1. Schematic diagram

Fig. 2. Distribution pattern of ϕ outside the lining

The governing equation for such a problem is shown as Eq. (1)

$$\nabla^2 \phi = 0 \tag{1}$$

Based on Fig. 1 and assumptions, the following three boundary conditions could be obtained,

$$\phi|_{y=0} = H \tag{2}$$

$$\phi|_{x^2 + (y+d)^2 = R^2} = Ay + B \tag{3}$$

$$\phi|_{x^2 + (y+d)^2 = r^2} = y + h_u \tag{4}$$

Taking into considering Eqs. (3 and 4), the general solution to Eq. (1) in the lining layer, in polar coordinates is shown below:

$$\phi_1 = C_1 \ln(\rho/r) - d + h_u + \left(C_2\rho + (1 - C_2) \cdot r^2/\rho\right) \sin\theta \tag{5}$$

As the conformal mapping function, Eq. (6) could transform the semi-infinite aquifer in plane - Z as the ring domain in plane $-\varsigma$ and facilitate the derivation to a great extent.

$$\varsigma = \xi + \eta i = (Z + ia)/(Z - ia) \tag{6}$$

where i is the imaginary unit, ξ and η are the Cartesian coordinates in plane $-\varsigma$, and $a = \sqrt{d^2 - R^2}$.

Based on the property of the conformal mapping, Eq. (1) in plane $- \zeta$ has a new form:

$$\partial^2\phi/\partial\xi^2 + \partial^2\phi/\partial\eta^2 = 0 \tag{7}$$

The general solution considering Eq. (2) is given as follows

$$\phi_{II} = H + C_3 \ln\rho + \sum_{n=1}^{\infty} C_4(\rho^n - \rho^{-n}) \cos n\theta \tag{8}$$

The four unknowns, C_i, could be derived from the seepage continuity condition on the interface between the aquifer and the lining layer, as shown below [7].

$$C_1 = \frac{H - h_u + d + (a - d)C}{\ln(R/r) - k_l/k_s \cdot \ln\alpha} \tag{9}$$

$$C_3 = \frac{H - h_u + d + (a - d)C}{\ln(R/r) \cdot k_s/k_l - \ln\alpha} \tag{10}$$

$$C_4 = -2aC\alpha^{2n}/\left(\alpha^{2n} - 1\right) \tag{11}$$

$$C = C_2 + (1 - C_2) \cdot r^2/R^2 \tag{12}$$

$$C_2 = \frac{\Gamma_2 + (k_1 - \Gamma_1)r^2/R^2}{\Gamma_1(R^2 - r^2)/R^2 + k_1(R^2 + r^2)/R^2} \tag{13}$$

$$\Gamma_1 = -\frac{k_s(a - d)}{\ln(R/r) \cdot (k_s/k_l) - \ln\alpha} \cdot \left(\frac{1}{R} + \frac{\rho'}{\alpha}\right) + 2k_s a \sum_{n=1}^{\infty} \frac{n\alpha^{n-1}(\alpha^{2n} + 1)}{\alpha^{2n} - 1}\rho' \tag{14}$$

$$\Gamma_2 = k_s \frac{H - h_u + d}{\ln(R/r) \cdot (k_s/k_l) - \ln\alpha} \left(\frac{1}{R} + \frac{\rho'}{\alpha}\right) \tag{15}$$

$$\rho' = -2a \cdot (d + R + a)^{-2} \tag{16}$$

The water ingress into the tunnel could be obtained from Eq. (8), as shown below (Fig. 3)

$$Q = 2\pi k_l \frac{H - h_u + d + (a - d)C}{\ln(R/r) - k_l/k_s \cdot \ln \alpha} \tag{17}$$

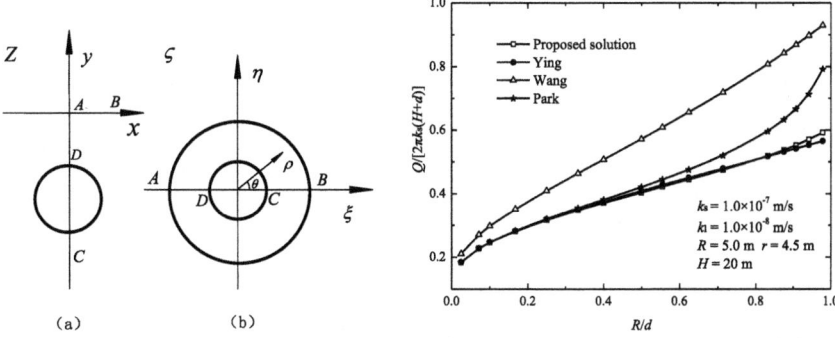

Fig. 3. Original and mapping domains   Fig. 4. Relationship between Q and d

3 Discussion

Many scholars have studied the effect on water ingress induced by burial depth based upon different assumptions. The analytical solution proposed in this paper is compared with three classic solutions in the existent literature, as shown in Fig. 4 where the burial depth varies from R to $40R$.

Figure 4 illustrates that the relative water ingress $Q/[2\pi k_s(H + d)]$ calculated from the four methods increases with declining burial depth, while Wang [8] has a large difference from the others. The result from the proposed solution is close to Ying [9], no matter the tunnel is buried shallowly or deeply. When the tunnel has a great burial depth (e.g., $R/d < 0.4$), Park [6] is similar with the result from the proposed solution; however, when $R/d > 0.4$, the gap between these two methods becomes wider. Thus, it could be concluded that, in terms of water ingress, Wang [8] based on the theory of shaft overestimates water inflow; Park [6] with the assumption that the lining layer is equipotential surface has a good precision when the tunnel is buried deeply ($R/d < 0.4$), but the estimation diverges with the result from the proposed solution gradually with decreasing burial depth; In general, Ying [9] based on the method of images has the best agreement with the new solution.

References

1. Lee, I.M., Nam, S.W.: The study of seepage forces acting on the tunnel lining and tunnel face in shallow tunnels. Tunn. Undergr. Space Technol. **16**(1), 31–40 (2001)
2. Arjnoi, P., Jeong, J.H., Kim, C.Y., et al.: Effect of drainage conditions on porewater pres-sure distributions and lining stresses in drained tunnels. Tunn. Undergr. Space Technol. **24**(4), 376–389 (2009)
3. Cao, Y., Jiang, J., Xie, K.H., et al.: Analytical solutions for nonlinear consolidation of soft soil around a shield tunnel with idealized sealing linings. Comput. Geotech. **61**, 144–152 (2014)
4. Tani, M.: Circular tunnel in a semi-infinite aquifer. Tunn. Undergr. Space Technol. **18**(1), 49–55 (2003)
5. Kolymbas, D., Wagner, P.: Groundwater ingress to tunnels–the exact analytical solution. Tunn. Undergr. Space Technol. **22**(1), 23–27 (2007)
6. Park, K.H., Owatsiriwong, A., Lee, J.G.: Analytical solution for steady-state groundwater inflow into a drained circular tunnel in a semi-infinite aquifer: a revisit. Tunn. Undergr. Space Technol. **23**(2), 206–209 (2008)
7. Verruijt, A.: Complex variable solutions of elastic tunneling problems. Delft University of Technology, Interfacultaire werkgroep Gebruik van de Ondergrondse Ruimte (GOR), Centrum Ondergronds Bouwen (COB) (1996)
8. Wang, J.Y.: Once more on hydraulic pressure upon lining. Mod. Tunn. Technol. **40**(3), 5–9 (2003). (in Chinese)
9. Ying, H.W., Zhu, C.W., Gong, X.N.: Analytical solution on seepage field of underwater tunnel considering grouting circle. J. Zhejiang Univ. (Eng. Sci.) **50**(6), 1018–1023 (2016). (in Chinese)

Part IX: Environmental Geotechnics

Using a Complementary Evapotranspiration Relationship to Estimate Surface Suction for Soil-Atmosphere Interaction Analysis

Hossein Assadollahi[1,2(✉)] and Hossein Nowamooz[1]

[1] ICUBE, UMR7357, CNRS, INSA, Department of Civil Engineering
and Energies, 67000 Strasbourg, France
hossein.assadollahi@insa-strasbourg.fr
[2] DETERMINANT R&D SARL - Department of Engineering and Consulting,
75008 Paris, France

Abstract. The estimation of soil surface condition from climatic parameters is of great importance in soil-atmosphere analysis. This paper deals with the estimation of the surface suction of soils using meteorological data. The rate of evapotranspiration (AE/PE) is related to the soil surface suction by the Wilson equation and it is generally used when the potential evapotranspiration (PE) and the soil surface suction are known while hydrological models allow the calculation of the evapotranspiration rate (AE/PE) by introducing complementary relationships. In this paper, a complementary hydrological model was combined with the Wilson equation in order to estimate the soil surface suction in time. To test the validity of the results, the approach was applied to the Toulouse region in the south of France in 3 different years (1999, 2003, 2004). The 2003 year was distinguished as a high intensity drought period in France that has caused many damages by triggering shrinkage and swelling in clays. By comparing the estimated total surface suction for these three years, the 2003 year showed higher values compared to the two other years as expected.

Keywords: Soil-atmosphere analysis · Surface suction · Evapotranspiration

1 Introduction and Methodology

Extreme drought and humidification cycles affect the soil surface physical parameters resulting in considerable damages mostly on lightweight Civil Engineering constructions like residential or industrial buildings, roads, embankments, etc. Clayey soils are mainly sensitive to these climatic cycles and can easily shrink and swell over time which results in structural damages of constructions. In order to evaluate these climatic changes, soil-atmosphere analysis is mainly used. Many authors have studied the soil's coupled behavior while subjected to climatic changes over time using soil-atmosphere analysis [1–3]. In these studies, the soil suction is normally deduced by using a coupled hydro-thermal simulation which goes through applying water balance and energy balance concepts as natural boundary conditions at the surface of the considered soil medium. This approach could be simplified for estimating the soil surface suction from meteorological variables. Generally, the soil surface suction is related to the relative

© Springer Nature Switzerland AG 2018
W. Wu and H.-S. Yu (Eds.): *Proceedings of China-Europe Conference*
on Geotechnical Engineering, SSGG, pp. 1251–1255, 2018.
https://doi.org/10.1007/978-3-319-97115-5_79

evapotranspiration by the equation proposed by Wilson et al. [4, 5] which was derived by laboratory experiments on different soil samples. On the other hand, hydrological models are capable of estimating the relative evapotranspiration term which is defined by the ratio of actual and potential evapotranspiration (AE/PE). Thus, the combination of these two approaches would give the soil surface suction variation over time. This approach was applied to a region in the south of France known for its risk of shrink-swell characteristic and extreme climate. Results are described in the last section. Many authors have studied the evapotranspiration process from both saturated and unsaturated surfaces by proposing physical and empirical methods [6–8]. The potential evapo-transpiration (PE) is defined as the amount of evapotranspiration that would occur if water availability is sufficient while the actual evapotranspiration (AE) is the amount of possible evapotranspiration from a surface for the given meteorological data and the soil water availability. Bouchet [9] demonstrated that as a surface dried from initially moist conditions, the potential evaporation increased while the actual evaporation was decreasing. He then defined an equilibrium parameter WE (wet surface evapotranspi-ration) that represented the amount of evapotranspiration when the actual and potential evapotranspiration were equal. The complementary relationship (CR) between AE and PE can take the following form $AE + PE = 2WE$. Some authors have evaluated and proposed different methods usually based on this concept [10–14]. In this study the complementary relationship known as the advection aridity model (AA), and the wet surface evapotranspiration proposed by Brutsaert and Stricker [11] were used:

$$PE = \frac{\Delta}{\Delta + \gamma} \frac{R_n}{\lambda} + \frac{\gamma}{\Delta + \gamma} 0.0026(1 + 0.54\,u_2)(e_0 - e_a);\tag{1}$$

$$WE = \alpha \frac{\Delta}{\Delta + \gamma} \frac{R_n}{\lambda}\tag{2}$$

where R_n is the net solar radiation calculated by the FAO 56 model. The coefficient γ (the psychrometric constant relating the partial pressure of water in air to the air temperature) is equal to 66 (Pa/°C). Δ is the slope of saturation vapour pressure curve at the air temperature (kPa/°C). λ is the latent heat, u_2 is the wind speed at 2 m elevation (m/s) and e_a and e_0 are the vapour pressure of the air and the saturation vapour pressure at the air temperature, respectively and α is assumed to be constant equal to 1.26. Following the Bouchet's complementary relationship the actual evapotranspiration could be expressed as a function of WE and PE. On the other hand, Wilson et al. [5] proposed the following equation to relate the effect of soil suction to the relative evapotranspiration expressed as A = AE/PE. The equation was based on the Kelvin's thermodynamic law and derived by measuring the actual evapotranspiration (AE) from a bare soil surface and the potential evapotranspiration (PE) from an adjacent water surface at the same time:

$$\frac{AE}{PE} = \left[\frac{\exp\left(\frac{\psi Mg}{RT}\right) - RH}{1 - RH}\right]\tag{3}$$

In this equation, ψ is the total surface suction, RH is the relative humidity (%), M is the molar masse of water (M = 18.016 g/mol), R is the gas constant (R = 8.3143 J/mol/°K), T is the temperature (°K) and g is 9.81 m/s². This equation also showed that the soil suction measured at the surface of the soil is independent of the type of soil and has a unique relationship with the rate of evapotranspiration. Therefore, climatic data such as air temperature, relative humidity, global solar radiation and wind speed are the only parameters in order to estimate the surface suction using Eqs. 1 to 3.

2   Results and Discussion

The proposed approach was applied to the Toulouse site's meteorological data in three different years (1999, 2003 and 2004). The 2003 drought period was chosen in this study because France (and especially south of France) was severely hit especially during this year. Based on the findings of the proposed approach, the relationship between the complementary relationship (AE/PE) and the total soil surface suction were moreover investigated during these years. The minimum relative humidity was almost equal to 20% and the maximum air temperature was nearly 40 °C in August (32 °C mean temperature) for 2003. Figure 1(a) shows the results of the calculated actual and potential evapotranspiration using Eqs. 1 and 2. It can be observed that there is a significant difference between AE and PE and it becomes higher as we approach dry periods (July–October).

It is this difference between the actual and potential evapotranspiration that triggers suction variations during the year. The more difference between AE and PE, the more suction is generated according to (3). Figure 1(b) shows the results of the calculated suction values for these chosen years. It is clearly observed that the 2003 year shows higher suction values mainly in august due to the fact that the difference between the AE and PE is higher which makes the relative evapotranspiration (AE/PE) a small value corresponding to a high suction value. All calculated suction values for the 2003 year versus the corresponding relative evapotranspiration (AE/PE) showed that the data respected the Wilson equation limits corresponding to a minimum and maximum relative humidity of 0.19 and 0.76 respectively. It should be noted that the surface suction is lower for both 1999 and 2004 year in August due to the fact that the climatic parameters such as relative humidity and temperature are higher and lower respectively for these two years which corresponds to a higher value of relative evapotranspiration. It can be concluded that the soil surface suction could be derived without the use complex simulations.

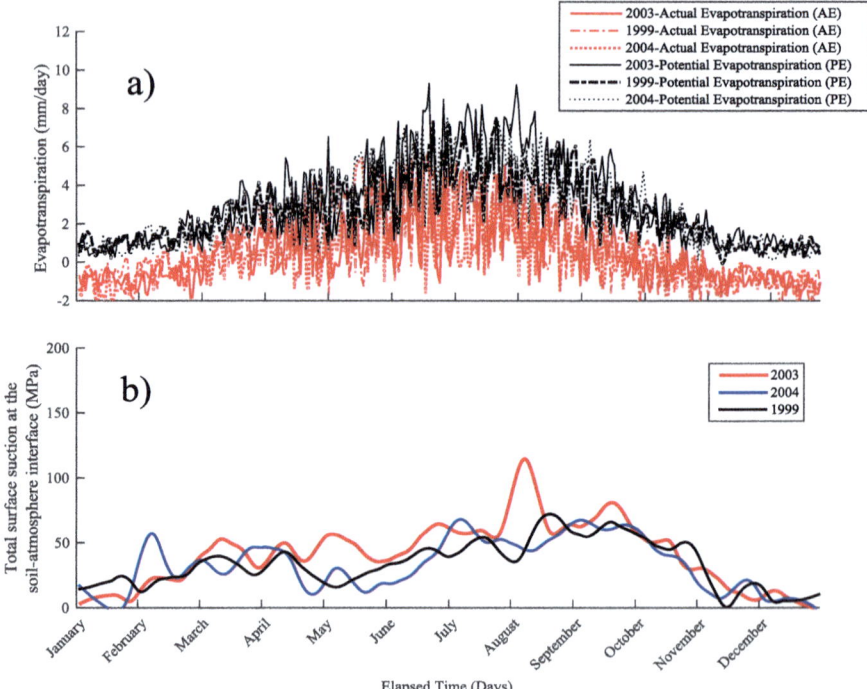

Fig. 1. (a) The calculated actual and potential evapotranspiration from the unsaturated soil surface followed by (b) The total soil surface suction at the soil-atmosphere interface for Toulouse-Blagnac site.

References

1. Ta, A.: Etude de l'interaction sol-atmosphère en chambre environnementale, Ph.D. thesis, Ecole des Ponts Paris-Tech, France (2009)
2. Hemmati, S., Gatmiri, B., Cui, Y.J., Vincent, M.: Thermo-hydro-mechanical modelling of soil settlements induced by soil-vegetation-atmosphere interactions. Eng. Geol. **139–140**, 1–16 (2012)
3. Adem, H.H., Vanapalli, S.K.: Soil-environment interactions modeling for expansive soils. Environ. Geotech. **3**(3), 178–187 (2015)
4. Wilson, G.W., Fredlund, D.G., Barbour, S.L.: Coupled soil-atmosphere modelling for soil evaporation. Can. Geotech. J. **31**, 151–161 (1994)
5. Wilson, G.W., Fredlund, D.G., Barbour, S.L.: The effect of soil suction on evaporative fluxes from soil surfaces. Can. Geotech. J. **34**, 145–155 (1997)
6. Penman, H.L.: Natural evaporation from open water, bare soil and grass. Proced. R. Soc. Lond. Ser. A **193**, 120–145 (1958)
7. Monteith, J.L.: Evaporation and the environment. The State and Movement of Water in Living Organisms. In: Proceedings of the 19th Symposium, Society for Experimental Biology, Swansea. Cambridge University Press, Cambridge, pp. 205–234 (1965)
8. Priestley, C.H.B., Taylor, R.J.: On the assessment of surface heat flux and evaporation using large-scale parameters. Mon. Weather Rev. **100**(2), 81–92 (1972)

9. Bouchet, R.J.: Evapotranspiration réelle evapotranspiration potentielle, signification climatique, pp. 134–142. International Association of Scientific Hydrology, Berkeley (1963)
10. Wang, T.J., Zlotnik, V.A.: A complementary relationship between actual and potential evapotranspiration and soil effects. J. Hydrol. **456**, 146–150 (2012)
11. Xu, C.Y., Singh, V.P.: Evaluation of three complementary relationship evapotranspiration models by water balance approach to estimate actual regional evapotranspiration in different climatic regions. J. Hydrol. **308**, 105–121 (2005)
12. Brutsaert, W., Stricker, H.: An advection-aridity approach to estimate actual regional evaporation. Water Resour. Res. **15**, 443–450 (1979)
13. Morton, F.I.: Operational estimates of areal evapotranspiration and their significance to the science and practice of hydrology. J. Hydrol. **66**, 1–76 (1983)
14. Granger, R.J., Gray, D.M.: Evaporation from natural non-saturated surfaces. J. Hydrol. **111**, 21–29 (1989)

Numerical Modeling of Leakage, Transport and Remediation of Mixed DNAPL and LNAPL in the Unsaturated Clayey Soil Underlain by Saturated Sandy Soil

Yuzhang Bi, Yanjun Du$^{(\boxtimes)}$, Xingyuan You, Kaixuan Yuan,
and Jin Ni

Institute of Geotechnical Engineering, School of Transportation,
Southeast University, Nanjing 210096, China
`duyanjun@seu.edu.cn`

Abstract. The mixed pollution of LNAPL and DNAPL is a common problem in environmental geotechnical engineering. Leakage, migration and extraction of pollutants will affect the distribution of pollutants in the unsaturation and the saturation zone. Furthermore, they will also affect the disposal efficiency of pollutants. The paper mainly focused on the migration of LNAPL under the influence of DNAPL. With the software of T2VOC, the distribution of pollutants under the whole process of "leakage-migration-extraction" was studied, especially about LNAPL. The studies showed that LNAPL will penetrate into the saturation zone through the unsaturation zone with the action of DNAPL, while it is only distributed in the unsaturation zone under the pure LNAPL condition. Meanwhile, with the increase of DNAPL quality ratio, the distribution area of LNAPL in the saturated zone increased accordingly. The study also indicated that extraction process will lead to the increase of LNAPL and DNAPL in some area. The results of the study have practical significance in revealing the migration of pollutants under the mixture conditions of LNAPL and DNAPL in engineering area.

Keywords: Whole process · DNAPL and LNAPAL mixture
Organic pollution · Multi-phase flow · T2VOC

1 Introduction

The contaminated site is a region with hazardous substances due to accumulation, storage, treatment and disposal, which can do harm to human's health and the environment or have potential risk. In general, in the condition that no other pollutants, LNAPL phase pollutant always floats above groundwater. However, under most conditions of practical engineering, LNAPL phase pollutant and DNAPL phase pollutant can coexist at the same time, which can also lead to a series of abnormal migration of LNAPL phase pollutant under the influence of DNAPL phase pollutant. In summary, the current research has the following two problems: (1) the study on the migration of pollutants only focused on the separate stage of leakage, migration and

© Springer Nature Switzerland AG 2018
W. Wu and H.-S. Yu (Eds.): *Proceedings of China-Europe Conference on Geotechnical Engineering*, SSGG, pp. 1256–1259, 2018.
https://doi.org/10.1007/978-3-319-97115-5_80

extraction. However, there was no report on the distribution of pollutants under the whole process of "leakage-migration-extraction"; (2) the selected pollutants were either LNAPL phase or DNAPL phase, but there were few reports about the pollutants under the mixed condition.

2   Case Study

The target site for this study is contaminated by a chemical plant in Changzhou (see Fig. 1(d)). The plant is above the unsaturated zone, and below the unsaturated zone is the saturation zone. The upper layer is clay (3–6 m), and the lower is silty sand (9–29 m) which is confined aquifer. This study follows the predecessor's values and the specific calculation parameters [1, 2]. The numerical cases are set in Table 1.

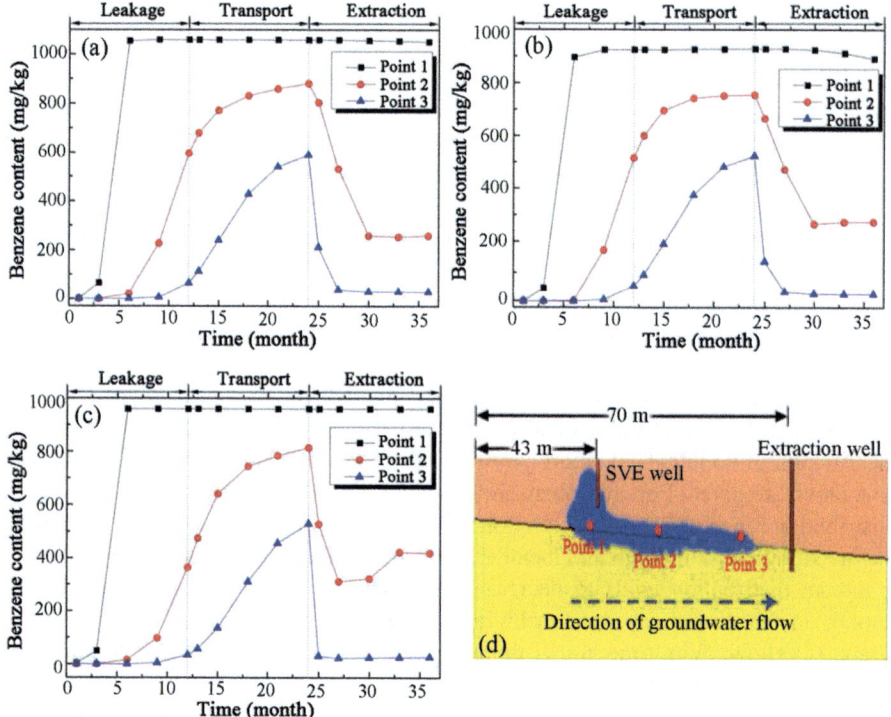

Fig. 1. Variation diagram of benzene along with time in (a) case I, (b) case II, (c) case III, (d) observation point arrangement

Table 1. Simulation examples setting

Case	LNAPL phase	DNAPL phase	Simulation of leakage	Simulation of migration	Simulation of extraction
I	Benzene	None	√	√	√
II	Benzene	Chlorobenzene	√	√	√
III	Benzene	Carbon tetrachloride	√	√	√

3   Result Analysis

In order to understand the pollutant distribution in the process of leakage - free migration - extraction, three observation points were set up in this study to monitor the concentration of benzene pollutants. Figure 1 shows the concentration change of benzene pollutants under three different cases. It can be seen from the figure that the trend of concentration is nearly the same as that during leakage and free migration processes: the concentration of observation point 1 increases with time going, and eventually maintains a stable value consistently; benzene concentrations at observation point 2 and 3 increases over time. The main difference exists in the extraction stage: under condition I, the concentration of observation point 1 has almost no change, but the concentrations of observation point 2 and 3 decreases firstly and then remain stable status. Under case II, the concentration of observation point 1 has a slight decrease, and the observation point 2 has a decrease firstly and then a slight increase, while the observation point 3 decreases firstly and then remains unchanged. The change law of case III is similar to that of case II, but the increase of observation point 2 is greater. It can be seen that anomalous phenomena emerge in some observation sites; to be specific, extraction will lead to an increase in the concentration of pollutants in some part of the site.

In order to explain this phenomenon, this paper deliberately analyzed the process of DNAPL distribution in the extraction stage. As shown in Fig. 2(a), the migration and distribution of pollutants of chlorobenzene and carbon tetrachloride at various stages of extraction are given. For chlorobenzene pollutants: With the progress of extraction, the distribution range of high concentration area (red area) increases gradually and reaches the maximum at 6th month and then decreases gradually. And the total area of the final pollutant distribution tends to decrease significantly compared with that of the first month. In the case of carbon tetrachloride, the area of the highest concentration in the graph is yellow. With time going, the area increases gradually till reaching the maximum at the 6th month and then decreasing gradually. It can be seen that when the DNAPL phase is extracted, the pollutants will flow in a certain direction under the outside pressure, which eventually leads to the concentration increasing in a certain area at a certain moment. The LNAPL will also produce increased levels in some of the monitoring points in Fig. 2(b) under the effects of the DNAPL.

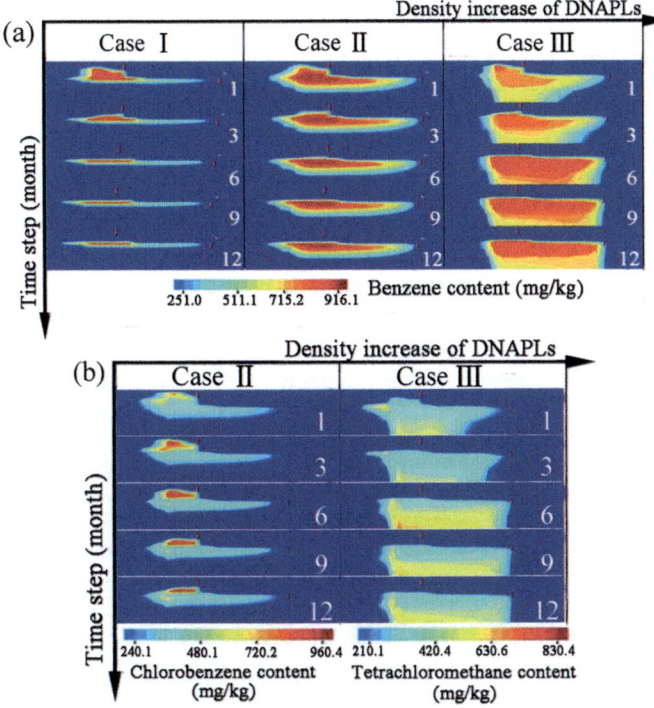

Fig. 2. Left: Migration distribution of LNAPL in three cases while extracting; Right: Migration distribution of DNAPL in three cases while extracting

4 Conclusions

When LNAPL and DNAPL are mixed, LNAPL enters the saturation zone with the DNAPL carrying. As the mass fraction of DNAPL increasing, more LNAPL penetrate into the saturated zone, and the polluted area in the saturation zone gets larger. In the extraction stage, LNAPL pollutants in some areas will have an increase in concentration. This is because DNAPL phase concentrated in a certain area under the pressure of extraction, which brought the LNAPL to the same area.

Acknowledgements. This research was supported by the Fundamental Research Funds for the Central Universities (Postgraduate Research & Practice Innovation Program of Jiangsu Province) (Grant KYCX17_0130).

References

1. Shi, X.Q., Jiang, B.L., Wu, J.C., et al.: Numerical analysis of the effect of leakage rate on dense non-aqueous phase liquid transport in heterogonous porous media. Adv. Water Sci. **23** (3), 376–382 (2012)
2. McCray, J.E., Falta, R.W.: Numerical simulation of air sparging for remediation of NAPL contamination. Groundwater **35**(1), 99–110 (1997)

Long Root Grasses in Pyroclastic Soils: Vegetation Growth and Effects on Induced Soil Suction

Vittoria Capobianco[(⊠)], L. Cascini, and V. Foresta

University of Salerno, Fisciano, Italy
vcapobianco@unisa.it

Abstract. Pyroclastic soil covers in Campania region (South Italy), usually of 2-3 m depth, are systematically affected by rainfall induced shallow landslides in wet season, causing catastrophic consequences. This paper introduces an experimental study aimed to investigate the use of vegetation as a sustainable practice for stabilizing these soils. The first step of the study is focused on both *(i)* the growth of perennial *graminae* grass species with long roots in a 1D column and *(ii)* its effect along depth on induced soil suction during evapotranspiration in wet season. Results show that the effect of roots on increasing soil suction is observable in shallowest layers and it decreases with depth. Moreover, the presence of roots can change the initial soil suction conditions in soil during wet season, when rainfall induced flow-like landslides systematically occur.

Keywords: Long roots · Pyroclastic soils · Suction

1 Introduction

Pyroclastic soils are widely diffused all over the world and those produced by the Somma-Vesuvius volcano are often in unsaturated conditions [1], covering the shallowest layers of slopes in Campania region (South Italy). During the wet season, when soil suction is low, they are systematically involved in rainfall induced shallow flow-like landslides that can reach great distances causing loss of life and economic damages to structures or infrastructures. Structural passive control works, such as dissipative basins and/or brindles, have been usually adopted as risk mitigation measures for these phenomena [2], even if they are expensive and require frequent maintenance.

The use of indigenous vegetation can represent an alternative sustainable bio-engineering practice for stabilizing shallow pyroclastic covers through hydro-mechanical reinforcement. However, how indigenous grasses with long roots (up to 2 m), typically used in bio-engineering practices for contrasting surface erosion, would grow and affect the induced soil suction in pyroclastic soils during evapotranspiration need to be studied. In this study the growing effect of perennial *graminae* species on suction changes of pyroclastic soil during evapotranspiration in wet season is quantified, compared to the soil suction in no-vegetated soil.

© Springer Nature Switzerland AG 2018
W. Wu and H.-S. Yu (Eds.): *Proceedings of China-Europe Conference on Geotechnical Engineering*, SSGG, pp. 1260–1263, 2018.
https://doi.org/10.1007/978-3-319-97115-5_81

2   Materials and Methods

Perennial *graminae* grass species with fine and fasciculate long roots (up to 2 m) has been seeded inside a 1D transparent plexiglass column (200 cm high and inner diameter of 190 mm) filled with pyroclastic soil belonging to class 'B' [3] trough the moist tamping method with a dry density of 12.03 kN/m³ (corresponding to 53.5% of porosity) and gravimetric water content of 10%. Minitensiometers measuring soil suction were installed along the column respectively at 30 cm, 60 cm, 120 cm and 180 cm depth (Fig. 1). In addition, another 1D instrumented column, only filled with bare soil, was built up as control.

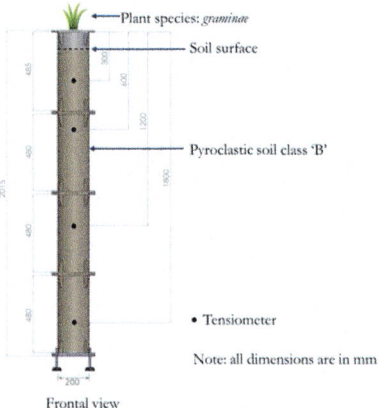

Fig. 1. Schematization of experimental set-up and instrumentation of vegetated column: frontal view. All dimensions are in mm.

The test started in January and the vegetated column was irrigated with the same amount of water every two days for the whole growth period. During the first vegetative year, the average root depth and height of foliage have been monitored trough a graduated scale every month: the average between the four longest roots observed at the four sides of the column was the monthly root depth; the average between the length of five different leaves chosen randomly among all the leaves was the monthly height of foliage. Before the drying test, both vegetated (V) and no-vegetated (NV) soil columns were saturated by applying water from the upper part, even if soil full saturation was never reached because of the lack of an additional water pressure involved in saturation process. Then the columns were exposed for 15 days to the atmospheric conditions for the evapotranspiration process, under the rainout shelter to protect them from natural rainfall. The drying tests in wet season were conducted in April of two consecutive years.

3 Results and Discussions

In Fig. 2(a) the monthly root depth and height of foliage are reported for the first vegetative year. It can be observed that starting from the third month, with the beginning of the spring, the root increased its growth by reaching almost 2 m depth after 8 months, whereas the height of foliage reached 60 cm of length. A strong linear correlation was observed between root depth and height of foliage (Fig. 2b), and, according with agronomical considerations, the ratio between the hypogeum (roots) part and the part above the ground level (foliage) was higher than 3.

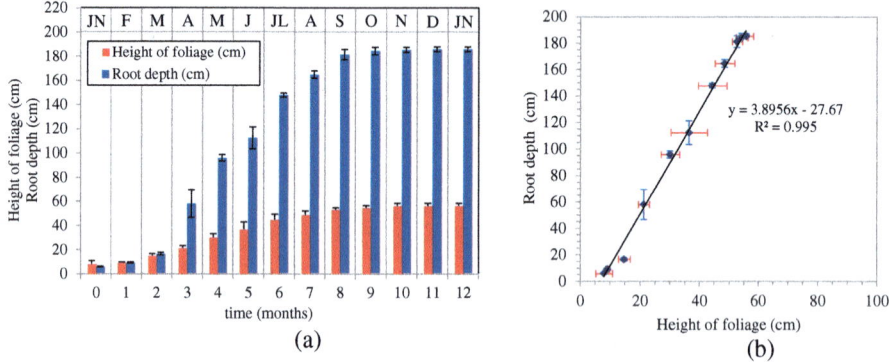

Fig. 2. (a) Monthly measured and (b) correlation law of height foliage and root depth in the first vegetative year.

Figure 3 shows the daily soil suction (s) measured along depth during drying tests conducted in April on no-vegetated soil column (NV_D_A in Fig. 3) and on vegetated column respectively in April of the first vegetative year (V_D_A1 in Fig. 3) and of the second year (V_D_A2 in Fig. 3). After 15 days of drying the induced soil suction in V_D_A1 increased by 13 kPa compared to NV_D_A only at 30 cm (row A), whereas in V_D_A2 it increased respectively by 18 kPa, 7 kPa, 3 kPa at 30 cm (row A), 60 cm (row B) and 1.8 cm (row D) of depth compared to NV_D_A. The daily suction increment (Δs), as the average of the amount of soil suction increased day by day (kPa/d), was calculated for each depth and reported for each test. This latter represents the velocity of suction increasing due to evapotranspiration and it can be observed that it decreases with depth for both NV and V columns. The effect of the presence of roots on Δs is observed in shallowest layer (row A), where in both V_D_A1 (3.1 kPa/d) and V_D_A2 (3.5 kPa/d) it was almost the double of that recorded in NV column (1.7 kPa/d). The daily suction increment at row B was the same between NV_D_A and NV_D_A1, this might be due to the length of roots, which did not yet reach the depth of row B, as consequence the suction increment for vegetated soil is only observable within the rooted zone, in contrast with other studies [4]. Conversely, in totally vegetated soil column (Fig. 3c) Δs was higher than that measured in both NV and V column in the first year, at all depths. Nevertheless, Δs recorded at lowest depth (row

D) in V_D_A2 (0.4 kPa/d) was slightly higher than that recorded in NV_D_A (0.1 kPa/d). This means that the presence of roots significantly changes the induced soil suction during evapotranspiration at shallowest layers.

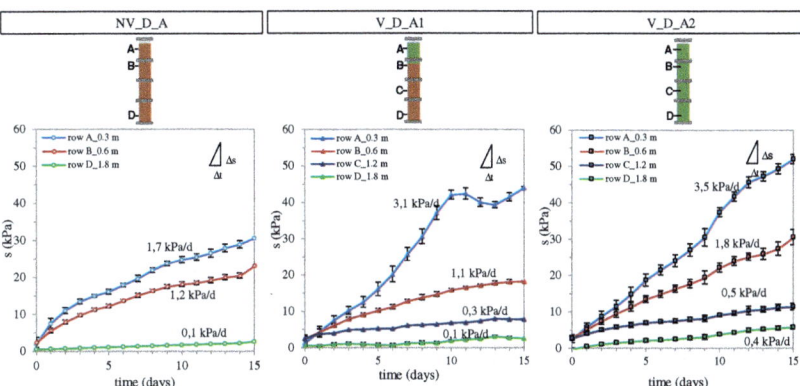

Fig. 3. Daily soil suction during 15 days drying on (NV_D_A) no-vegetated column, (V_D_A1) vegetated column in the first year, (V_D_A2) vegetated column in the second year.

4 Conclusions

This study investigated the effects of long roots grasses growth on induced soil suction of pyroclastic soil during the wet season. During one vegetative year roots can reach about 2 m depths after 8 months, whereas the height of foliage can reach 60 cm of length. The effect of roots on induced soil suction is observable within the root zone and it decreases with depth. The daily suction increment at 30 cm depth in vegetated soil is the double of no-vegetated soil, this means that the velocity of suction increasing due to evapotranspiration is doubled compared to no-vegetated soil. As consequence the presence of roots in shallow pyroclastic covers can change the soil suction condition in wet season, in which rainfall induced flow-like landslides systematically occur.

Acknowledgement. The authors would like to acknowledge Prati Armati S.R.L. for providing the perennial *graminae* species used in this experimental work.

References

1. Damiano, E., Olivares, L., Picarelli, L.: Steep-slope monitoring in unsaturated pyroclastic soils. Eng. Geol. **137**, 1–12 (2012)
2. VanDine, D.F.: Debris flow control structures for forest engineering. Res. Br., BC Min. For., Victoria, BC, Work. Pap, 8 (1996)
3. Bilotta, E., Cascini, L., Foresta, V., Sorbino, G.: Geotechnical characterisation of pyroclastic soils involved in huge flowslides. Geotech. Geol. Eng. **23**(4), 365–402 (2005)
4. Ng, C.W.W., Woon, K.X., Leung, A.K., Chu, L.M.: Experimental investigation of induced suction distribution in a grass-covered soil. Ecol. Eng. **52**, 219–223 (2013)

A Simple Model for Estimating Shear Strength of Root-Soil Composite

Jiulong Ding[(⊠)], Faning Dang, and Songhe Wang

Institute of Geotechnical Engineering,
Xi'an University of Technology, Xi'an, China
jiulong_ding@126.com

Abstract. The shear strength of soils is an important index to evaluate the stability of slopes reinforced with roots. This paper based on the common direct shear test results gives a quantitative analysis of the effect of roots on the stability of the slope. A simple model for the shear strength of root-soil composite was proposed based on the strength data for both the root and soil material. The influence of the root system was investigated with a new soil strength enhancement factor of β to represent the strengthening of shear strength. The influence of root diameter and root density on the shear strength of root-soil composite was discussed. Results show that within a certain range of root reinforced soils, the shear strength of the root-soil composite gradually increased at higher root densities. After reaching the peak value, the root density in the root soil had no significant effect on the shear strength. The small and densely distributed roots exhibit larger effect on the shear strength of root-soil system than that of the large and sparse roots. The results may provide guidance for stability estimating and engineering design of slopes.

Keywords: Shear strength · Root-soil composite · Slope protection

1 Introduction

The slope stability of soils is an important problem in geotechnical engineering. Plants slope protection technology has been widely used worldwide, and can improve soil shear strength. It is of great realistic significance to carry out the study of plant roots reinforcement effect on slopes (e.g. Wu et al. (1979), and Waldron et al. (1977). Research shows that plant roots have the function of reinforcing and anchoring and improving the shear strength of soil so as to improve the stability of slope. Researches in this area is a hot research topic. (e.g. Zhang et al. 2010; Abdullah et al. 2017).

Since the slope stability is highly dependent on the shear strength of soil, Scholars have conducted researches in terms of experimental, theoretical and established models. For the experiment. Ghestem et al. (2014), Zhang et al. (2010) and others carried out the direct shear tests, triaxial tests and in situ tests for common herbs and woody plants, such as Cynodon dactylon, Ophiopogogon japonicas, Ricinus ommunist. etc. Studies have shown that plant roots can significantly improve the shear strength of the soil by increasing the cohesion of the root-soil composite.

© Springer Nature Switzerland AG 2018
W. Wu and H.-S. Yu (Eds.): *Proceedings of China-Europe Conference on Geotechnical Engineering*, SSGG, pp. 1264–1268, 2018.
https://doi.org/10.1007/978-3-319-97115-5_82

Based on the More-Coulomb Criterion, the theoretical model such as Wu - Waldron - Model (WWM) has been put forward by Wu et al. (1979) and Wu Waldron (1998), and Root - Bundle - Model (RBM) put forward by Hidalgo et al. (2001). These theoretical models have laid a theoretical foundation for the study of composite soil. In this paper, based on the previous studies, the research on the shear strength of root-soil composite was conducted from the perspective of soil mechanics. The research results have certain theoretical and practical significance for guiding vegetation revetment.

2   A Soil Composite Shear Strength Physical Model

In geotechnical engineering, the shear strength of either plain soil sample or soil–root composite, can be expressed by Mohr-Coulomb law (Waldron 1977):

$$\tau = c + \sigma_n \tan \varphi \tag{1}$$

Where, ϕ is the internal friction angle and c is the cohesion. Studies have shown effects on root-soil composite shear strength is increased by increasing the shear strength of cohesion, rather than the friction angle (Wu et al. 1979). Therefore, cohesion is equal to a summary of root-soil composite cohesion. Wu et al. (1979) suggested that the root shear stress changes to the root tension stress through the friction between root - soil. Equation (2) written:

$$\tau = c_s + \sigma_n \tan \varphi + c_r \tag{2}$$

where C_s is soil cohesion, and C_r is the root cohesion. Among the C_r:

$$c_r = T_R(RAR)(\sin \omega + \cos \omega \tan \varphi) \tag{3}$$

Where: T_R root tensile strength; ω is the angle between the root and the normal line when the soil is damaged, RAR is the root cross-sectional area ratio which is mainly used to quantify the influence of root on the shear strength of soils. The expression:

$$RAR = \frac{A_R}{A} \tag{4}$$

where, A is the total cross-sectional area of the root system on the shear surface; A_R is the soil occupied by the root system area of cut surface; further Eq. (2) written:

$$\tau = c_S + k \cdot T_R \cdot RAR + \sigma_n \tan \varphi \tag{5}$$

Where, k is the calculation coefficient for ω φ is the internal friction angle.

3   Improved Physical Model

Schwarz et al. (2011) compared root complex in situ soil testing, the experimental results measured in situ with the result of the calculation WWM, Wu overestimate root Waldron Model 60%–100%. Thus, the WWM models overestimate the enhancement effect of root on the shear strength of soils due to the increase in cohesion caused by the root tensile stress. When the root-soil composite shear failure, if the root tensile strength is not enough the root will be broken off, the part of the roots that enhances the shear strength of the soil is the tensile strength of the root system, In the other case, the root system will be pulled out without being broken. Root complex occurred under shear damage and the root system has not been pulled off. Therefore, substituting root tensile strength for Eq. (5) will overestimate its enhancement effect, to explore the roots for improving the soil affect root complex strength. The shear stress of the root-soil composite is divided into two parts: the soil-borne part and the root-borne part. Therefore, the Eq. (2) Written:

$$\tau = \tau_s + \tau_r = c + \sigma_n \tan \varphi + c_R \tag{6}$$

Similar to the *RAR*, under extreme equilibrium conditions, the root-soil stress ratio of the average shear stress of the soil that defines the root shear stress is R_S, that is:

$$R_S = \frac{\tau_r}{\tau_S} \tag{7}$$

Where: τ_r for the root shear strength; τ_s shear strength of soil The density of root will affect the shear strength of root-soil composite. Considering the density of root in root-soil composite, substituting formula (4) into Eq. (6) can be rewritten as:

$$\tau = RAR \cdot \tau_r + \beta(1 - RAR)\tau_s \tag{8}$$

Where reinforcement coefficient of roots for root-soil composites. For the formula (8), if the root system with almost no soil exists, the *RAR* is close to 1, The value of root-soil composite, τ is close to the root shear strength. The root-soil composite shear stress is tensile strength of roots. The law is consistent with WWM Model. Another extreme case, the RAR is close to zero, because almost no effect of root-soil compaction t, the shear strength is 1, therefore, the shear strength close to the soil itself shear strength. Between the above two cases, the shear strength of the root-soil composite is related to the values of RAR and β. Substitute (8) into (9):

$$\tau = RAR \cdot R_S \tau_S + \beta(1 - RAR)\tau_s \tag{9}$$

It is difficult to accurately measure the shear stress of root system in the root-soil composite. Through the previous analysis if we instead shear strength of the root of the tensile strength, it will overestimate the shear strength of the root-soil composite. According to the formula (9), the shear strength and the shear strength of plain soil can be calculated by the method of comparative test. then the R_S can be calculated.

According to Gray and Lott (1983) and others on tensile testing of root soil, β is an empirical value related to plant root diameter.

4   Validation of Improved Models

The effect of root content on the shear strength of the soil-root composite was investigated by Ji (2013). It was found that with the increase of (or *RAR*) root density, the influence on the shear strength of it is increased for C_R and less for internal friction angle. The formula (9) reflects the law (Table 1).

Table 1. Effects of roots on root-soil combinations.

RAR	C	C_R	φ	$\Delta\varphi$
Soil	18.05	0	23.34	0
0.03	24.37	6.32	25.14	1.8
0.04	31.59	7.22	25.39	0.25
0.05	39.71	8.12	26.81	1.42

5   Conclusion

Through the formula quantitative analysis of the roots to enhance the shear strength of the slope, Improved the WWM model. The shear stress of root-soil composite is divided into soil shear stress and root shear stress. Reveals the influence of root content on the shear strength of soil by affecting the *RAR* value and β to represent the strengthening of shear strength. Improved model compared with the WWM model, the effect of the enhanced part of cohesion on the enhancement of its enhanced estimate can be estimated by reducing the root tensile strength. The research results have certain theoretical and practical significance for vegetation revetment.

References

Wu, T.H., Iii, M.K., Swanston, D.N.: Strength of tree roots and landslides on Prince of Wales Island, Alaska. Can. Geotech. J. **16**(1), 19–33 (1979)

Waldron, L.J.: The shear resistance of root-permeated homogeneous and stratified soil. J. Soil Sci. Soc. Am. **41**(5), 843–849 (1977)

Zhang, C.B., Chen, L.H.: Triaxial compression test of soil-root composites to evaluate influence of roots on soil shear strength. Ecol. Eng. **36**(1), 19–26 (2010)

Abdullah, M.F., Kasmin, H.: The Soil-Root Strength Performance of Alternant Hera Ficoidea and Zoysia Japonica as Green Roof Vegetation. vol. 103, p. 04018 (2017)

Wu, T.H., Watson, A.: In situ shear tests of soil blocks with roots. Can. Geotech. J. **35**(4), 579–590 (1998)

Hidalgo, R.C., Kun, F., Herrmann, H.J.: Bursts in a fiber bundle model with continuous damage. Phys. Rev. E Stat. Nonlinear Soft Matter Phys. **64**(2), 066122 (2001)

Schwarz, M.: Pullout tests of root analogs and natural root bundles in soil: experiments and modeling. J. Geophys. Res. Earth Surf. **116**(F2), 1–14 (2011)

Gray, D.H., Lott, J.: Radial versus parallel tie arrays in earth backfills. J. Geotech. Eng. **109**(7), 982–986 (1983)

Ghestem, M., Veylon, G., Bernard, A., Vanel, Q., et. al: Influence of plant root system morphology and architectural traits on soil shear resistance. Plant Soil, **377**(1–2), 43–61 (2014)

Ji, X.: Study on ecological slope stability based on vegetation root distribution pattern. Doctoral dissertation, Nanjing Forestry University, pp. 45–47 (2013)

Field Pilot Scale Ex-Situ S/S of Electroplating Industrial Contaminated Soil Using Two Novel Binders

Yasong Feng[1], Yanjun Du[1(✉)], Weiyi Xia[1], and Krishna R. Reddy[2]

[1] Institute of Geotechnical Engineering,
Southeast University, Nanjing 210096, China
fengyasongys@126.com, duyanjun@seu.edu.in
[2] Department of Civil and Materials Engineering, University of Illinois
at Chicago, 842 West Taylor Street, Chicago, IL 60607, USA

Abstract. A field test was conducted to evaluate the use of GM and KMP to stabilize zinc (Zn) and chloride (Cl) contaminated soil at an abandoned industrial electroplating plant site in this study. The stabilized field soil was cured for 1, 3, 7, and 28 days and tested for dry density, dynamic cone penetration, soil pH and leachability. The results showed that the soil pH and leachable concentrations of Zn and Cl were well below their corresponding remediation goals. Furthermore, the KMP stabilized soil possesses superior performance in terms of higher dry density and strength in the early curing stage (7 days) and also lower dynamic cone penetrometer index and leachable Zn. While GM exhibited superior immobilization of Cl in the contaminated soil.

Keywords: Contaminated site · Heavy metals · Chloride · Stabilization
Solidification · Leachability · Strength

1 Introduction

Solidification/stabilization technique is widely used for treating heavy metal-contaminated soils. Previous studies focused on the high-alkaline cementitious materials such as Portland cement (PC) and lime-fly ash blend [1]. However, PC and lime production is associated with consumption of intensive energy and natural resources. In addition, the presence of certain heavy metals and chloride in the soils could significantly retard hydration of cement-based binders and consequently affect strength development and heavy metals immobilization in stabilized soils [2, 3]. Recently, low-cost industrial by-products and phosphate-containing materials have been successfully used in solidification/stabilization of heavy metal contaminated soils [4, 5]. However, no studies exist that evaluate these binders in treating soils with relatively high concentrations of mixed Zn and Cl.

In the present study, a field test was conducted to investigate the performance of GM and KMP stabilized industrial site soil contaminated with Zn and Cl. The physico-chemical and strength properties and contaminants leaching characteristics were evaluated by determining dry density (ρ_d), dynamic cone penetrometer index (DCP), and leachability of Zn and Cl.

© Springer Nature Switzerland AG 2018
W. Wu and H.-S. Yu (Eds.): *Proceedings of China-Europe Conference on Geotechnical Engineering*, SSGG, pp. 1269–1273, 2018.
https://doi.org/10.1007/978-3-319-97115-5_83

2 Field Pilot Tests

The field tests were performed at an abandoned industrial electroplating plant site in Jiangsu Province, China. Field pilot test was conducted before undertaking full-scale remediation. The average soil pH is 4.23. The leachable Zn and Cl are 8555 and 2050 mg/L, respectively. According to the results from site investigation and environmental risk analysis, the remediation goals were set as: (1) the soil pH should be higher than 5, and (2) the leached Zn and Cl concentrations of the stabilized soil after 28 days of curing should be lower than 100 mg/L and 500 mg/L, respectively.

GM is a mixture of GGBS and reactive MgO in a dry mass ratio of 9:1, while KMP is a mixture of APR, KH2PO4, and reactive MgO in a dry mass ratio of 1:1:2. Preliminary tests suggested that the water content of 23% and binder dosage of 6% yield relatively low contaminant leachability and also high strength of stabilized soils with GM and KMP. An excavator with a standard bucket was used to excavate the contaminated soil and mix the contaminated soil and binder. After thorough mixing, the admixed soils were compacted at the excavated sites and covered with PE sheets.

Field tests and sampling were performed at 1, 3, 7 and 28 d curing. Soil bulk density was determined as per JTG E60/T0923. The DCP test was conducted as per ASTM D6951/D6951M. The leachable Zn was measured as per USEPA Method 6010C. The leachable Cl was determined based on the China GB/T 50123-1999. The soil pH measurement was conducted as per ASTM D4972.

3 Results and Discussion

3.1 Dry Density Tests

Figure 1 shows the dry density variation of the stabilized soil with curing time. Compared to the original soil, an approximate increase of 10% to 12% is observed after 28 d curing. The dry density of the KMP stabilized soil is slightly higher than that of GM stabilized soil regardless of the curing time.

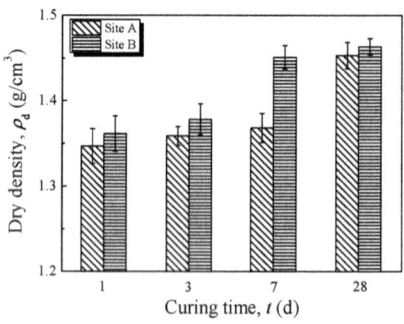

Fig. 1. Variation in the soil dry density with curing time (Number of replicate = 3, COV < 1.4%)

3.2   DCP Tests

Figure 2 presents the variations in the DCP cumulative blow counts, R_s and average R_s with penetration depths. The average R_s of the GM and KMP stabilized soils exhibit a slight increase in the period of 0 d to 3 d. The average R_s of the GM and KMP stabilized soils show different increase patterns in the period of 3 d to 28 d. For example, the remarkable average R_s increases of GM and KMP stabilized soils occur at 7 d to 28 d and 3 d to 7 d, respectively. Finally, the average R_s values of the GM and KMP stabilized soil are very close at 28 d.

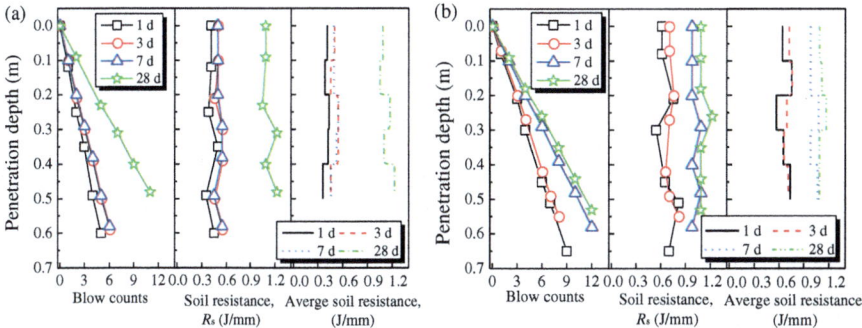

Fig. 2. Variations in blow counts, soil resistance (R_s) with penetration depth and curing time: (a) GM stabilized soil and (b) KMP stabilized soil (Number of replicate = 1)

3.3   Soil PH Tests

Figure 3 shows the pH variation of the field site soil with curing time. It is found that the pH of the GM or KMP stabilized soil significantly increases as compared to the original soil. The pH of the GM stabilized soil is slightly greater than that of the KMP stabilized soil regardless of the curing time. Overall, both the GM and KMP stabilized field soils meet the soil pH requirement of its value to be higher than 5.

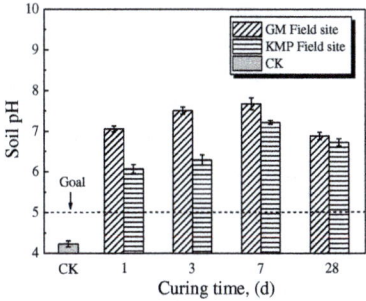

Fig. 3. Variation in the soil pH with curing time (Number of replicate = 3, COV < 1.9%)

3.4   Leachable Zn and Cl

Figure 4 shows the leachable Zn and Cl concentrations of the stabilized soils from the field test. It is found that the leachable Zn and Cl concentrations of KMP stabilized soil meet the remediation goal after 1 d, and the leached Zn and Cl concentrations of GM stabilized soil meet the remediation target after 1 d and 3 d curing, respectively.

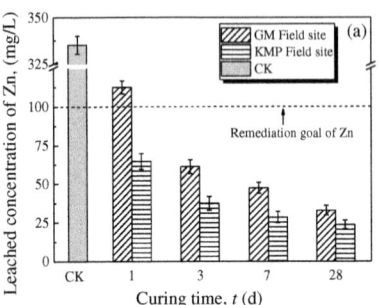 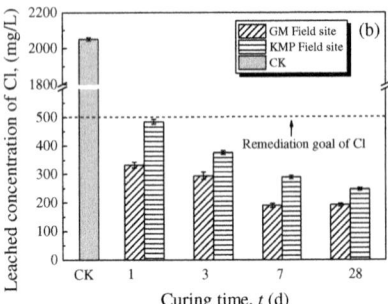

Fig. 4. Variations in the leached Zn and Cl concentration with curing time: (a) Zn and (b) Cl (Number of replicate = 3, COV < 3.3%)

4   Conclusions

Some conclusions can be drawn as follows:

(1) GM and KMP addition can increases the soil pH to 6.75 to 6.96 after 28-day curing as compared to the original value of 4.23.
(2) With GM and KMP stabilization, the dry density has an increase of 11% to 12%. The stabilized soils possess a higher soil resistance relative to the original contaminated soil. The early strength improvement of the KMP stabilized soil is more significant compared to GM stabilized soil.
(3) The leached Zn and Cl concentrations in the soils are significantly reduced with stabilization by using KMP and GM. After 28-day curing, the TCLP leached concentrations in GM and KMP stabilized soils are well below their remediation goals. GM shows superior immobilization of Cl, while KMP is better on immobilizing Zn within the testing conditions presented in this study.

References

1. Sharma, H.D., Reddy, K.R.: Geoenvironmental Engineering: Site Remediation, Waste Containment, and Emerging Waste Management Technologies. Wiley, Hoboken (2004)
2. Du, Y.J., Jiang, N.J., Liu, S.Y., et al.: Engineering properties and microstructural characteristics of cement-stabilized zinc-contaminated kaolin. Can. Geotech. J. **51**(3), 289–302 (2014)

3. Saussaye, L., Boutouil, M., Baraud, F., et al.: Influence of chloride and sulfate ions on the geotechnical properties of soils treated with hydraulic binders. Road Mater. Pavement Des. **14** (3), 551–569 (2013)
4. Jin, F., Al-Tabbaa, A.: Evaluation of novel reactive MgO activated slag binder for the immobilization of lead and zinc. Chemosphere **117**, 285–294 (2014)
5. Du, Y.J., Wei, M.L., Reddy, K.R., et al.: New phosphate-based binder for stabilization of soils contaminated with heavy metals: leaching, strength and microstructure characterization. J. Environ. Manag. **146**, 179–188 (2014)

Three-Dimensional Numerical Analysis of the Air Phase Flow During Air Sparging in Sands

Z. B. Liu$^{(\boxtimes)}$, S. Y. Liu, Z. L. Chen, Y. Wang, L. L. Lu, and G. Y. Du

School of Transportation, Southeast University, Nanjing 210096, Jiangsu, China
seulzb@seu.edu.cn

Abstract. The simulator TOUGH2 was adopted to study the air phase flow during in situ air sparging in sands. It was found that the vertical profile of upward air flow is in the shape of water drop before reaching the groundwater table. As the air phase flows upwards, it also moves horizontally, especially when the air reaches the vadose zone. Finally, the zone of influence in air sparging connects the saturated zone with vadose zone and forms a U-shaped area. The air phase saturation degree is symmetrically distributed around sparging well. The air phase saturation degree decreases with the increase of the distance to the sparging well in the horizontal direction. From the sparging point up to the ground, the air phase saturation degree increases first and then decreases.

Keywords: Numerical simulation · Air phase flow · Air sparging

1 Introduction

As a remediation method of contaminated sites, the in situ air sparging technique has been widely noticed due to its simple, efficient and economic. The compressed air is injected into saturated zone below the contaminated area. With the help of buoyancy, the air will bring the volatile contaminants upward to the unsaturated zone, where the contaminants are removed by soil vapor extraction system. In addition, the oxygen in the air will also do good to the aerobic degradation of organic contaminants [1]. The distribution of air flow plays an important role in remediation efficiency of the air sparging system. As the whole process is deep in the saturated zone, it cannot be directly observed. Many laboratory tests were conducted by researchers in the past [2]. But there exist some problems such as the scale of the model chamber, measurement techniques, and cost. Numerical simulation has unique advantages such as the adjustable parameters, independent of practical conditions, and visualization of results. For example, Gao et al. [3] studied the air and water phase flow from the angle of microstructure with dynamic two phase flow model. However, the simulation of the whole process from initial air sparging until final steady state of air flow has seldom been studied. TOUGH2 is a multi-dimensional numerical model for simulating the coupled transport of water, vapor, air, and heat in porous and fractured media. It is adopted here to simulate the three-dimensional air flow patterns during air sparging, which does not consider the existence of contaminants.

W. Wu and H.-S. Yu (Eds.): *Proceedings of China-Europe Conference on Geotechnical Engineering*, SSGG, pp. 1274–1277, 2018.
https://doi.org/10.1007/978-3-319-97115-5_84

2   Governing Equations

In the two phase flow, the following equation of relative permeability-saturation degree was adopted in this research [4].

$$k_{rw} = \left(\frac{S_w - S_{wr}}{1 - S_{wr}} \right)^{n_1} \tag{1}$$

$$k_{rg} = \left(\frac{S_g}{1 - S_{wr}} \right)^{n_1} \tag{2}$$

where k_{rw} and k_{rg} are the relative permeabilities of water phase and air phase respectively, S_w and S_g are the saturation degrees of water and air, S_{wr} is the residual saturation degree, the power index $n_1 = 3$. In the simulation, the residual saturation degree of the sand is designated as 0.15. In addition, the Capillary pressure-Saturation relationship is important to characterize the multiphase flow during air sparging. Here the van Genuchten's equation is adopted here.

$$P_{cgw} = \frac{\rho_w g}{\alpha_{gw}} \left[\left(\frac{S_w - S_m}{1 - S_m} \right)^{-1/m} - 1 \right]^{1/n_2} \tag{3}$$

where P_{cgw} is the capillary pressure between air and water ($ML^{-1}T^{-2}$), α_{gw} and S_m the material parameters of corresponding porous media, ρ_w is the density of liquid (ML^{-3}), g is the gravitational acceleration (LT^{-2}), $m = 1 - 1/n_2$ is an empirical parameter.

3   Numerical Simulation

3.1   Computational Model and Parameters

A three-dimensional cylindrical model was established to learn the whole process of air phase flow from its initial injection into sand layer until the final steady state during air sparging. The dimension of the model is 5 m in diameter and 10 m in height ($-5 \leqslant x \leqslant 5, -5 \leqslant y \leqslant 5, -8 \leqslant z \leqslant 2$). The water table is 2 m below the ground surface. The soil is meshed into many polygons. The air phase saturation degree is 0 at the bottom is 0, and 0.75 at the top layer of the unsaturated area. The top surface of the model is connected to the atmosphere. The bottom and lateral surfaces are impermeable boundaries. The main parameters used in the numerical simulation are listed in Table 1. The screening point of the sparging well is at x = 0, y = 0, z = −5.8–6.0 m, and the air flow rate is 1.0 g/s.

Table 1. Main parameters for the numerical simulation by TOUGH2.

Parameters	Value
Porosity	0.35
Intrinsic permeability, $k_{xx} = k_{yy} = k_{zz}$, /m^2	5.0×10^{-11}
Soil density, kg/m^3	2650
Temperature, /°C	20
Residual saturation degree of liquid phase, S_{wr}	0.15
n_1 (in K-S equation)	3
n_2 (in P-S equation)	2
S_m	0
α_{gw}	5

3.2   Results and Analysis

It can be seen from the air flow contour (see Fig. 1) that the air phase mainly runs upwards from the screening point. It also moves horizontally in some degree. Before it reaches the water table, the whole area of air phase is in the shape of a water drop. About 20 min later, it reaches the unsaturated zone. Since the water content of the unsaturated area is quite low, its air permeability is relatively high. Thus the air phase laterally moves to even larger area. The test results from the model tests of Culligan et al. [5] also proved that the moving area of air phase was gradually developing into a shape of inverted cone from the very beginning. After 24 h, the horizontal movement of air flow reaches its maximum area. Finally, the steady state of the whole air flow region forms a U shaped area. The main influencing zone of the air flow is from –7 m until 1 m.

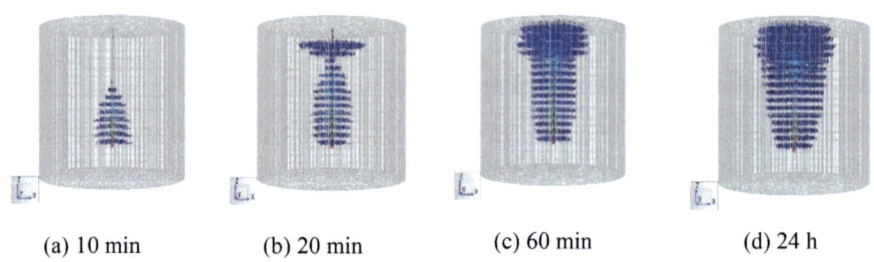

(a) 10 min   (b) 20 min   (c) 60 min   (d) 24 h

Fig. 1. The air flow contour during sparging.

For the calculation results of the model, a typical cross sections y = 0 is selected to draw the contour map of saturation degree (see Fig. 2). It is indicated that the air phase saturation degree displays a centrosymmetric distribution with the center of sparging well. The saturation degree decreases gradually from the central axis towards outside area, which has been verified by filed measurement [6]. The variation of the air phase saturation degree from 0–5 m right above the sparging point is shown in Fig. 3. It can be seen that the value of air phase saturation degree first goes up from 0–0.56 m, and then decreases.

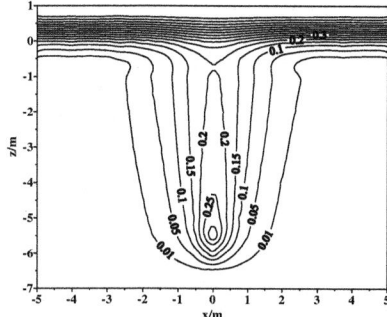

Fig. 2. Contour map of the air phase saturation degree (cross section of y = 0)

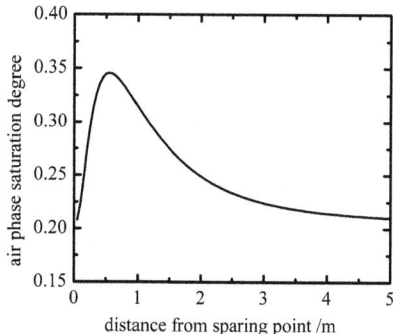

Fig. 3. Variation of the air phase saturation degree in the vertical direction

4 Conclusion

The main conclusions are summarized as follows. The geometric shape of the whole air phase flow seems like a water drop before the air flow reaches the water table. The whole steady air phase flow distributes in a U-shaped area. The saturation degree of air phase gradually decreases from the center axis of the sparging well towards surrounding area. The authors are grateful for the financial support from the National Natural Science Foundation of China (41672280, 41330641).

References

1. Nyer, E.K., Palmer, P.L., Carman, E.P., et al.: In Situ Treatment Technology, 2nd edn. CRC Press, Boca Raton (2001)
2. Hu, L., Meegoda, J., Du, J., et al.: Centrifugal study of zone of influence during air-sparging. J. Environ. Monit. **13**(9), 2443–2449 (2011)
3. Gao, S., Meegoda, J.N., Hu, L.: A dynamic two-phase flow model for air sparging. Int. J. Numer. Anal. Methods Geomech. **37**(12), 1801–1821 (2013)
4. Fatt, I., Klikoff, W.A.: Effect of fractional wettability on multiphase flow through porous media. J. Pet. Technol. **11**(10), 71–76 (1959)
5. Culligan, P.J., Zhu, Y., Germaine, J.T.: Numerical simulation of in situ air-sparging. In: Proceedings of Geo-Frontiers 2005, Austin. Waste Containment and Remediation (2005)
6. Schima, S., Labrecque, D.J., Lundegard, P.D.: Monitoring air sparging using resistivity tomography. Groundw. Monit. Remediat. **16**(2), 131–138 (1996)

Long-Term Performance of a Three-Layer Capillary Barrier Cover System in Humid Climates

Jian Liu[1]([⊠]), Yuedong Wu[1], Rui Chen[2], and C. W. W. Ng[3]

[1] Key Laboratory of Ministry of Education for Geomechanics and Embankment Engineering, Hohai University, Nanjing 210098, China
geoliujian@163.com
[2] Shenzhen Key Laboratory of Urban and Civil Engineering for Disaster Prevention and Mitigation, Shenzhen School, Harbin Institute of Technology, Shenzhen 518055, China
[3] Department of Civil and Environmental Engineering, The Hong Kong University of Science and Technology, HKSAR, Kowloon, Hong Kong

Abstract. An innovative three-layer (silt/gravelly sand/clay) capillary barrier cover system has recently been proposed for minimizing rainfall infiltration in humid climates. Its performance has been preliminarily verified by both physical and numerical modelling subjected to a short extreme rainfall. However, the long-term performance of this three layer cover system is not yet clear in humid climates, such as Hong Kong and Singapore, where an annual rainfall of over 2000 mm is expected. In this study, the long-term feasibility and effectiveness of this three-layer cover system are investigated. A series of numerical simulations were carried out based on a calibrated numerical model. It was found that the long-term feasibility and effectiveness of this three-layer cover system are investigated. A series of numerical simulations were carried out based on a calibrated numerical model even k_s in the clay layer increases to 5×10^{-9} m/s due to the occurrence of cracks, the long-term effectiveness of the three-layer cover system on preventing water into the waste is still satisfactory.

Keywords: Water infiltration · Capillary barrier · Landfill
Long term performance

1 Introduction

Covers commonly placing over landfills have been identified as a crucial component in preventing access to the buried solid waste after their closure. Because of long service life, ease of construction and relatively low cost, the capillary barrier consisting of a fine-grained soil and an underlying coarse-grained soil has been received ever more attention. However, the performance of such capillary barrier in humid climates is unsatisfactory. Recently, a three-layer capillary barrier cover system has been proposed to minimize water percolation into landfills under heavy rainfall [1]. The three-layer cover system consists of a fine-grained soil layer (e.g., silt layer) overlying a coarse-grained soil layer (e.g., gravelly sand layer), which in turn overlies another fine-grained

© Springer Nature Switzerland AG 2018
W. Wu and H.-S. Yu (Eds.): *Proceedings of China-Europe Conference on Geotechnical Engineering*, SSGG, pp. 1278–1281, 2018.
https://doi.org/10.1007/978-3-319-97115-5_85

soil layer such as a compacted clay layer. The results obtained from numerical simulations, as shown in Fig. 1, soil column tests and physical flume model tests show that the percolation under a heavy rainfall was negligible and thus the three-layer cover system performed well subjected to short-term rainfalls. However, the long-term performance of this three layer cover system is not yet clear in humid climates, such as Hong Kong and Singapore, where an annual rainfall of over 2000 mm is expected.

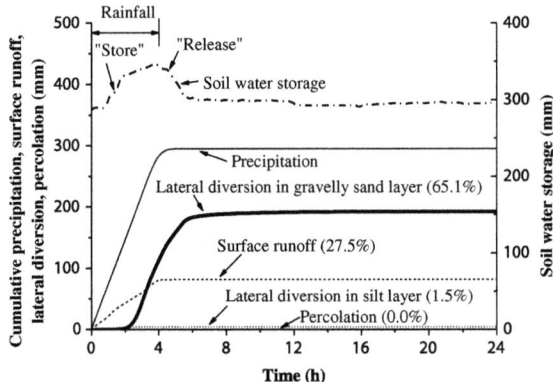

Fig. 1. Water balance analysis in the three-layer cover system

In this study, the long-term feasibility and effectiveness of this three-layer cover system are investigated. A series of numerical simulations were carried out based on a calibrated numerical model. Daily rainfall and evaporation rate lasting one year measured in Hong Kong were chosen to simulate the long-term boundary conditions. The development of pore water pressure, volumetric water content and percolation is investigated. The influence of the coefficient of saturated permeability of bottom clay layer, which may be largely affected by cracks, on water infiltration is discussed.

The cover system used in the numerical simulation consists of a 0.6 m-thick silt layer overlying a 0.2 m-thick sand layer and a 0.8 m–thick clay layer, as shown in Fig. 2. The modelled cover system is 360 m long, whereas the inclination of the slope is 1 V: 6H, which is equivalent to a slope angle of 10°. It is intended to simulate one half of the proposed landfill cover. Hence, the boundary along the lateral upstream side is thus taken as the axis of symmetry and assumed to be impermeable. This implies that the amount of flux that passes through this boundary is equal to zero. On the upper exposed cover surface, an infiltration/runoff boundary is modelled by a potential seepage face. This implies that while the surface is relatively dry (i.e., the pore water pressure is negative), rainfall intensity is applied as the water infiltration rate. Should the water pressure eventually reach zero during a rainfall event, a constant flow boundary is automatically switched to a constant pore water pressure boundary condition. In other words, no ponding is permitted in any analysis. Along the lateral downstream side, the boundary is specified as a drain/soil interface at which water accumulates until the pore water pressure increases to zero, forming a drip face from which the extra water seeps out. For the bottom boundary, pore water pressures are applied according to the water retention curve of the underlying waste. Prior to any transient seepage analysis, the initial pore water pressure at each node of the finite elements along the upper boundary is obtained from a steady state seepage analysis. In this steady state simulation, a very small flux of 0.01 mm/d is applied to the top of boundary. After obtaining the initial steady-state solutions, the rainfall/evaporation conditions are then simulated.

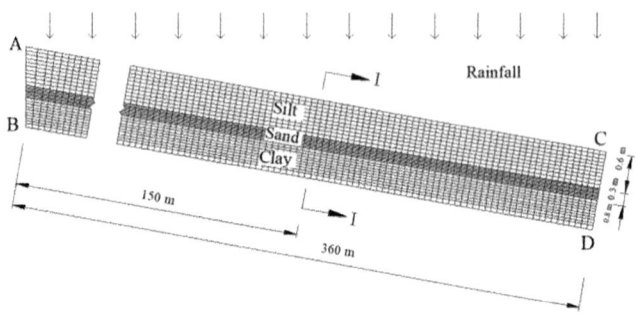

Fig. 2. Finite element mesh used in the transient seepage analysis

Figure 3 shows the development of computed percolation into the waste with four coefficients of saturated permeability (k_s) of the bottom clay layer (i.e., 1×10^{-9}, 2×10^{-9}, 5×10^{-9} and 1×10^{-8} m/s). Correspondingly, the daily rainfall rate is also shown in the figure. As expected, the accumulated percolation increases in wet season (e.g., from May to Oct.) and maintains constant in dry season (e.g., from Jan. to May and Oct. to Dec.). It should be noted that the percolation is almost zero under a heavy rainfall in the dry season (e.g., the daily rainfall is about 100 mm on 6 Feb.). It may be because the initial condition is relatively dry during the dry season and thus the water has not infiltrated through the cover. In other words, there may be no direct relationship between heavy rainfall and percolation. In the wet season, the rainfall is not heavy but occurs almost every day, whereas the percolation increases relatively quickly. It indicates that the small and long rainfall may lead to a big risk on the increase in percolation. As well known, if the cracks occur in the clay layer, k_s of the clay would increase. As the k_s of the clay increases from 1×10^{-9} to 1×10^{-8} m/s, the percolation through the cover increases from 4.5 to 53 mm. When k_s is smaller than 5×10^{-9} m/s, the percolation is smaller than 30 mm/year which is the allowable percolation proposed by Benson et al. [2]. It indicates that even k_s in the clay layer increases to 5×10^{-9} m/s due to the occurrence of cracks, the long-term effectiveness of the three-layer cover system on preventing water into the waste is still satisfactory.

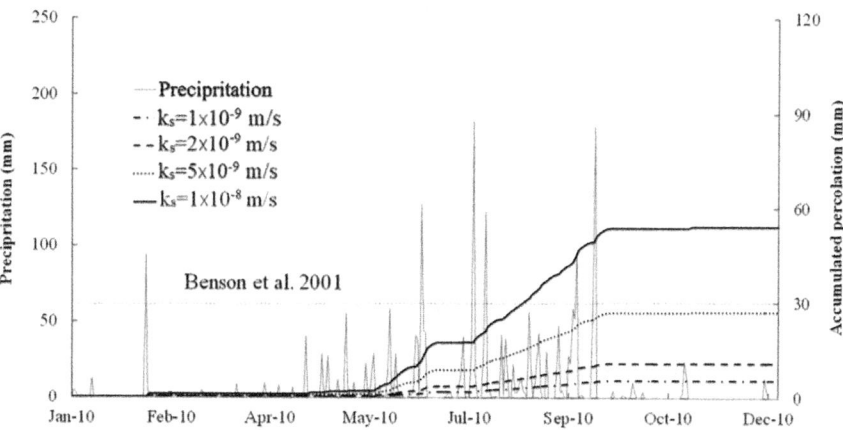

Fig. 3. Development of accumulated percolation with various coefficient of saturated permeability in clay layer

2 Conclusions

In this study, the long-term feasibility and effectiveness of this three-layer cover system are investigated. A series of numerical simulations were carried out based on a calibrated numerical model. Even k_s in the clay layer increases to 5×10^{-9} m/s due to the occurrence of cracks, the long-term effectiveness of the three-layer cover system on preventing water into the waste is still satisfactory.

Acknowledgements. This work was supported by the National Natural Science Foundation of China (Grants No. 51608152), National Higher-education Institution General Research and Development Project (No. 2015B25914) and the Natural Science Foundation of Guangdong Province China (No. 2016A030310368).

References

1. Ng, C.W.W., Liu, J., Chen, R., Xu, J.: Physical and numerical modeling of an inclined three-layer (silt/gravelly sand/clay) capillary barrier cover system under extreme rainfall. Waste Manag. **38**, 210–221 (2015)
2. Benson, C., Abichou, T., Albrigh, W., Gee, G., Roseler, A.: Field evaluation of alternative earthen final covers. Int. J. Phytorem. **3**(1), 105–127 (2001)

NaCl Activation of Steel Slag upon Component Adjustment

Li Liu, Qianwen Liu, Yongfeng Deng[(⊠)], and Yu Zhao

School of Transportation, Southeast University, Nanjing 210096, China
noden@seu.edu.cn

Abstract. The re-utilization of steel slag, a kind of by-products, is relatively rough in China for lack of activity. The adjustment concept of the cement clinker was introduced by adding metakaolin and lime to imitate the suitable cement components. Hereafter the formed composites were activated by NaCl to enhance the strength. The results showed that the strength of the composite slag-based materials arrived at 4.5 MPa after adjustment, furthermore, the maximum strength after NaCl activation is 8.2 MPa, showing great improvement. XRD and SEM results revealed that after adjustment and activation, calcium sili-coaluminate hydrate (C_3ASH), calcium silicate hydrate (C-S-H) and calcium ferrite hydrate were generated for the composite. Among them, the new hydrate of Friedel's salt (Fs) was found after NaCl activation, which is the main reason for the obvious increment of the cementation and strength.

Keywords: Steel slag · Metakaolin · Component adjustment
Chemical activation · Macro strength · Hydration product and microstructure

1 Introduction

According to World Steel Association, the production of crude steel in China accounts for about 50% that of the world per year. Hence the discharge of steel slag, the by-product, is up to 10% of the steel [1], resulting that almost 300 million tons of waste steel have been accumulated for the low recycle ratio (only 10%–21% [2, 3]).

Steel slag with potential cementitious capacity consists of 38%–48% CaO, 10.5%–15.6% SiO_2, 0.9%–6% Al_2O_3 etc., and the main minerals are C_3S, C_2S, C_4AF, RO (CaO, FeO, MnO, MgO solid phase) etc. [4]. Comparing with ordinary Portland cement, the steel slag is lack of active oxides Al_2O_3 and CaO. Thereafter, the high temperature/pressure reconfiguration was applied to improve the activity of steel slag [5], where the regulating components were added [6] to reconstruct the mineral phase, improving its activity and stability. However, the process is of high energy consumption and engineering price. Traditionally, there are another two activating methods, physical and chemical excitation. By comparison, the chemical excitation is the more effective and feasible for ground improvement since the alkalinity of liquid phase accelerates the hydration and hardening process of the slag at atmospheric temperature.

This study aims to investigate the component adjustment by the metakaolin and lime and the activation by NaCl on the slag-based composite for soil modification.

© Springer Nature Switzerland AG 2018
W. Wu and H.-S. Yu (Eds.): *Proceedings of China-Europe Conference on Geotechnical Engineering*, SSGG, pp. 1282–1286, 2018.
https://doi.org/10.1007/978-3-319-97115-5_86

A series of laboratory tests were carried out by tracking the hydrate and pore size distribution to explicit the transformation mechanism of strength and microstructure after adjustment and activation.

2 Materials and Methods

2.1 Materials

The steel slag was pretreated by the wet magnetic separation method from Jiaxing, Zhejiang province. Hereafter the wet steel slag was air dried and then passed through No. 10 sieve. Its chemical component was listed in Table 1. The primary oxides of steel slag are CaO and SiO_2, up to 85%, and the content of Al_2O_3 is limited, less than 5%. The slag content between 0.075 mm and 0.25 mm accounts for 62%, resulting the low activity.

Table 1. Chemical composition of steel slag.

Composition	CaO	Al_2O_3	SiO_2	Fe_2O_3
Proportion (%)	49.83–55.46	0–4.82	28.47–34.28	10.19–14.17

The essential component of Metakaolin (i.e. MK) is anhydrous aluminium silicate by kaolin's dehydrating at the temperature of 600 to 900 °C. High purity MK from BASF was used in this investigation, whose basic properties are shown in Table 2.

Table 2. Basic properties of Metakaolin.

325# (about44 μm) weight of screen residue (w%)	pH	Specific gravity (g/cm³)	SiO_2 (%)	Al_2O_3 (%)
0.1	6	2.5	50–55	25–35

2.2 Methods

Component Adjustment. Cement moduli, including silica modulus (SM), aluminum modulus (IM) and lime saturation factor (KH) was introduced into this investigation. For most cement, SM, IM and KH are between 1.7 and 2.7, between 0.9 and 1.7, between 0.67 and 1.0 respectively. After series of attempts, lime and metakaolin were selected as the supplement for steel slag. The moduli of slag-based composite were designed as: SM = 2.0, IM = 1.6, KH = 0.94. Therefore, the mass fraction of the steel slag, lime and metakaolin of the composite was 30.2%, 60.0% and 9.8% respectively. The proportion of the oxides is listed in Table 3.

Table 3. Chemical composition of slag-based composite.

Component	CaO	SiO$_2$	Al$_2$O$_3$	Fe$_2$O$_3$	Else
Proportion (%)	61.67	20	5.32	3.85	9.16

Experimental Scheme. The reagent NaCl was prepared as solution with concentration of 3%, 5% and 8%. Slag-based composite with and without NaCl activation were prepared as cylinder pieces with a diameter in 50 mm and length in 100 mm, and then cured in the chamber with a temperature of 20 ± 2 °C and a humidity of 95%. After 7 and 28 days' curing, unconfined compression tests were conducted. Hereafter the specimens were freeze-dried for XRD, MIP and SEM tests to clarify the micro-mechanism.

3   Results

3.1   Unconfined Compression Strength (UCS)

The UCS of compacted slag is 0.4 MPa, while the UCS of slag-based composite after 7 and 28 days' curing are 2.7 MPa and 4.5 MPa respectively. After the NaCl activation, the UCS shows significant improvement. When the concentration of NaCl solution is 5%, the UCS after 28 days' curing can arrive at the maximum value, equal to 8.2 MPa. Therefore 5% NaCl solution is the best activation. Hereafter, its microstructure is further studied.

3.2   Microstructure Analysis

The results of UCS illustrate that the strength of the slag was significantly updated after adjustment and activation. To clarify the mechanism, the hydration products and morphology were tracked by XRD and SEM. After 28 days' curing, the main productions of the slag-based composite are C$_3$ASH$_6$, C-S-H and iron-based hydrates. Additionally, for the composite activated by NaCl, Friesdel's salt was generated.

The microstructure of the activated composite after 7 days and 28 days' curing are presented in Fig. 1. The amorphous C-S-H, acicular Aft and irregular hexagonal plates of Fs are revealed, which suppress the pore size and enhances strength. The production of C-S-H further filled the pores in the hydrated composite with the increasing of the curing age. The pore size distributions of hydrated slag-based composite activated by 5% NaCl and without activation are presented in Fig. 2. It is observed that the pore size with NaCl activation is smaller than that without activation, suggesting that activation can effectively promote hydration and enhance strength. Furthermore, the pore diameter decreases with curing periods fir the more productions.

(a) 7 day   (b) 28 day

Fig. 1. SEM of slag-based composite activated by 5% NaCl

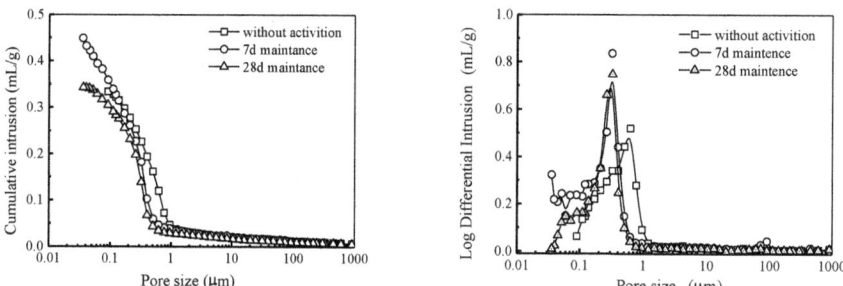

Fig. 2. Pore size distribution of slag-based composite with and without NaCl activation

4   Conclusions

The main conclusions are: (1) after component adjustment, the strength of the composite (2.7 MPa after 7d's, and 4.5 MPa after 28d's curing), is more than that of steel slag compacted at the maximum dry density; (2) the strength of adjustment composite was re-enhanced remarkably by the activation with 5% NaCl solution for the hydration production and Friesdel's salt.

Acknowledgements. This study is supported by the National Natural Science Foundation of China (Grant No. 41572280) and Postgraduate Research & Practice Innovation Program of Jiangsu Province (Grant No. KYCX17-0131).

References

1. Hou, G.H., Li, W., Guo, W.: Microstructure and mineral phase of converter slag. J. Chin. Ceram. Soc. **36**(4), 436–443 (2008)
2. Wang, Q., Bao, L., Yan, P.: Research progress on converter steel slag applied for concrete. Concrete **02**, 53–56 (2009)
3. Huang, Y., Xu, G., Cheng, H., et al.: Analysis on chemical composition, cicro-morphology and phase of typical steel slag. Bull. Chin. Ceram. Soc. **08**, 1902–1907 (2014)

4. Dong, Q., Li, C., Peng, B., et al.: Study of the practicability of asphalt concrete confected by steel slag. J. Wuhan Univ. Technol. **23**(6), 9–13 (2001)
5. Zhao, H., Yu, Q., Wei, J., et al.: Effect of composition and temperature on structure and early hydration activity of modified steel slag. J. Build. Mater. **03**, 399–405 (2012)

Bioengineering for Slope Stabilisation Using Plants: Hydrological and Mechanical Effects

C. W. W. Ng[✉], A. K. Leung, and J. J. Ni

Department of Civil and Environmental Engineering,
Hong Kong University of Science and Technology, Kowloon, Hong Kong
cecwwng@ust.hk

Abstract. The negative impact of climate change calls for more sustainable and environmentally friendly techniques to be developed to improve the engineering performance of our civil infrastructures such as slopes in urban built environments. Soil bioengineering using plants and microorganisms is considered a low-cost and aesthetically pleasant solution for shallow slope stabilisation. Although extensive research has been conducted on the mechanical effects of root reinforcement in past decades, the hydrological effects of plants on shear strength and water permeability of vegetated soil slopes are not clear. This extended abstract presents an all-round, cross-disciplinary research programme consisting of indoor and field experiments, centrifuge testing and theoretical analysis to examine the plant hydrological effects on slope stability. It was revealed that some plant species native to southern China and Europe could preserve a credible amount of suction after heavy rainfall, which is positively correlated with the leaf area index (LAI), the root area index (RAI) and the ratio of root to shoot biomass. The total amount of water infiltration were found to be considerable higher in bare than those in soil covered by grass and tree. Plant-fungus interaction caused a significant increase in root tensile strength, hence potentially the mechanical reinforcement to soil. By developing novel artificial root systems in the centrifuge and deriving new theoretical closed-form stability equations, it was discovered that heart-shaped roots produced stronger stabilisation effects than either tap- or plate-shaped roots. This root architecture preserved higher suction (hence higher soil shear strength) and provided greater mechanical reinforcement effects due to multiple branching. These findings advance the fundamental understanding of plant-soil interaction and its influence on slope bioengineering applications.

Keywords: Soil bioengineering · Matric suction
Root biomechanical properties · Slope hydrology · Slope stability

1 Introduction

In order to combat the negative impact of climate change and leave a better world for future generations, people these days are looking for more environmentally friendly or "green" solutions for slope stabilisation. Public concern has motivated governments and engineers to gradually adopt vegetation as a kind of construction material. Vegetation can potentially offer a more environmentally friendly, cost-effective and

© Springer Nature Switzerland AG 2018
W. Wu and H.-S. Yu (Eds.): *Proceedings of China-Europe Conference on Geotechnical Engineering*, SSGG, pp. 1287–1303, 2018.
https://doi.org/10.1007/978-3-319-97115-5_87

aesthetically pleasant solution for shallow slope stabilisation. However, the current practice considers vegetation mainly for aesthetic purpose and the engineering functions of plant roots have not yet been integrated into the analysis and design of slope stability scientifically.

A well-recognised effect of plants on slope stabilisation is mechanical reinforcement by roots (Wu et al. 1979; Stokes et al. 2009). In past decades, the mechanical effects of root reinforcement have been extensively quantified both experimentally and analytically (Wu et al. 1979; Pollen and Simon 2005; Jotisankasa and Taworn 2016) and have occasionally been considered in slope stability analysis (Ekanayake and Phillips 2002; Mao et al. 2014). In contrast, the hydrological effects of plant transpiration (i.e., increase in soil shear strength but decrease in water permeability) on soil matric suction/moisture content have received relatively little attention, as pointed out by Ng (2017).

In this study, an integrated and complementary research approach was adopted to investigate the influences of hydrological effects on slope hydrology and stability. First, a series of indoor experiments were conducted to quantify the magnitude and distribution of transpiration-induced soil matric suction and then to seek any correlation with plant traits. Effects of plant roots on soil hydraulic properties such as soil water retention curve and water infiltration were measured. How microorganisms (fungi) affect root tensile strength and the mechanical reinforcement of soil was also explored. New theoretical closed-form solutions were derived to estimate the stability of an infinite unsaturated slope reinforced by roots of different architectures under the influence of plant transpiration. In order to study the combined mechanical and hydrological effects of different root architectures on slope hydrology, stability and failure mechanisms in the geotechnical centrifuge, novel artificial root systems were developed. Combining the results obtained from numerical transient seepage and slope stability analyses, the contribution of vegetation to slope stability was quantified in terms of the factor of safety. Moreover, the influences of root architectures on slope failure mechanism were identified. This extended abstract was extracted and translated from the Huangwenxi lecture delivered by the first author (Ng 2017).

2   Hydrological Effects of Plant-Induced Matric Suction

2.1   Effects of Evapotranspiration

To quantify the effects of evapotranspiration (ET; the sum of soil evaporation and plant transpiration) on suction in unsaturated silty sand (see Ng et al. (2014a) for basic soil index properties), a comprehensive test programme was implemented in a plant room at the Hong Kong University of Science and Technology (HKUST). The room has an air conditioning system, which allows for flexible control of the air temperature from 10 °C to 30 °C $\pm$ 1 °C and the air relative humidity (RH) from 40% to 70% $\pm$ 5%. Different parameters that affect ET-induced suction have been studied in this room, including plant species (Ng et al. 2014a; Garg et al. 2015a; Ni et al. 2017), plant spacing (Ng et al. 2016a), plant height (Ng et al. 2018a), soil density (Ng et al. 2014a), soil nutrient (Ng et al. 2018b) and carbon dioxide concentration (Ng et al. 2018b). Some of these results are summarised and reported in Ng et al. (in press).

Figure 1 shows a set of typical test results comparing suction in soil with and without a tree species *Schefflera heptaphylla* native to the southern China including Hong Kong (Ng et al. 2016a). Compacted soil was planted with multiple trees with a spacing of 60 mm. Three replicates were tested. Before rainfall, the vegetated soil samples (test T) induced much higher matric suction than the bare sample (test B), both within and below the root zone. After two hours of rainfall with a constant intensity of 73 mm/h (equivalent to a return period of 10 years in Hong Kong; Lam and Leung 1995), suction was slightly reduced below 150 mm of the bare soil but greatly reduced above that depth. Although suction at the shallow depths of the vegetated samples had mostly vanished after the intense rainfall, suction at depths below 250 mm remained largely unchanged, thanks to tree root-water uptake before the rainfall.

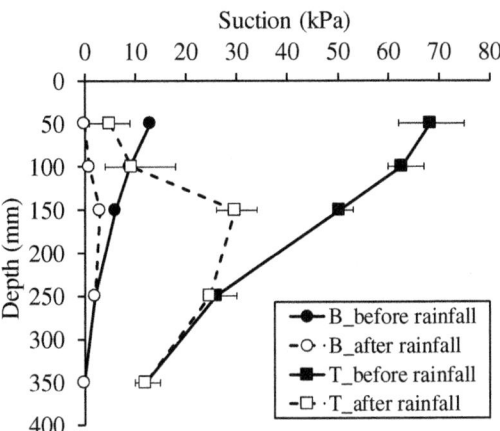

Fig. 1. Distributions of suction with depth before and after intense rainfall. Suction is presented in mean ± standard error of mean (n = 3)

2.2   Correlation Between Suction and Plant Traits

In Fig. 2, suction induced by *S. heptaphylla* in the root zone is correlated with LAI, which is a plant trait that is defined as the total projected area of leaves per unit soil surface area. The measurement methods of LAI are detailed in Ng et al. (2016b). To highlight the effects of tree transpiration, suction is expressed as the mean suction (s) difference between vegetated and bare soil, Δs. It can be seen from the figure that Δs has a strong correlation with LAI ($R^2 = 0.91$). A tree having a higher LAI would intercept a greater amount of radiant energy, meaning that more stomata are available to absorb energy for transpiration (Kelliher et al. 1995).

Another plant trait that is relevant to plant root-water uptake is RAI, which is defined as the ratio of total root surface area for a given depth range to the circular cross-sectional area of soil in the horizontal plane (Francour and Semroud 1992). RAI is also a dimensionless number that indicates the water uptake ability of roots. The

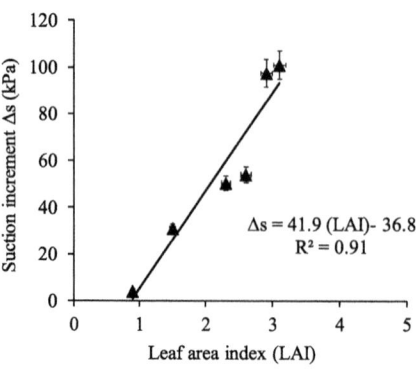 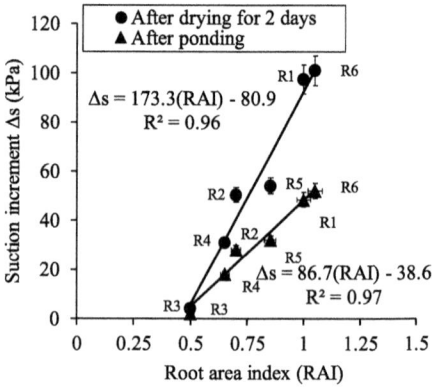

Fig. 2. Relationship between suction and LAI

Fig. 3. Relationship between suction and RAI

measurement methods of RAI are detailed in Garg et al. (2015a). Correlations between Δs and mean peak RAI are shown in Fig. 3. As expected, for a given RAI, Δs measured after ponding is always lower than that after ET because of infiltration. Δs recorded either after ET or ponding is strongly correlated with the RAI ($R^2 = 0.96$).

Boldrin et al. (2017a, b) recently shows that plant root-water uptake was not correlated with the above-ground or below-ground trait alone, but a combination of them. They investigated the effectiveness of 10 woody shrub species native to Europe at the ability of inducing matric suction. These include *Buxus sempervirens* (Bs), *Corylus avellana* (Ca), *Crataegus monogyna* (Cm), *Cytisus scoparius* (Cs), *Euonymus europaeus* (Ee), *Ilex aquifolium* (Ia), *Ligustrum vulgare* (Lv), *Prunus spinosa* (Ps), *Salix viminalis* (Sv) and *Ulex europaeus* (Ue). They defined a new trait, namely the

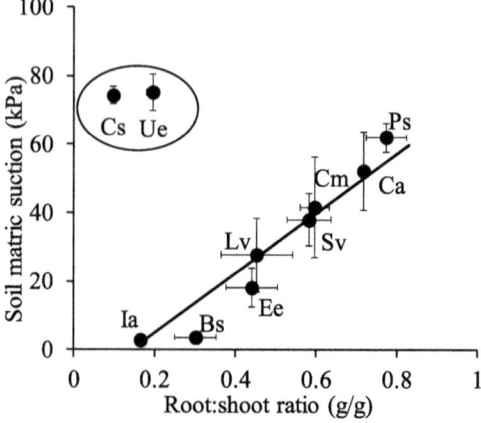

Fig. 4. Relationship of matric suction with root-to-shoot ratio in planted pots. Mean values of species are reported $\pm$ standard error of mean (n = 3). (Boldrin et al. 2017a, b)

root-to-shoot ratio, which is the ratio of root biomass to shoot biomass. Among the traits examined, this ratio has the strongest correlation with induced suction (Fig. 4) for all species except Cs and Ue, which have photosynthetic organs, twigs and thorns that are distinctively different from those of the other eight species.

2.3   Root-Induced Changes in Soil Water Retention Curve

Apart from plant transpiration, the presence of plant roots in soil can alter the soil pore size and its distribution. Ng et al. (2016c) proposed a theoretical model to capture the effects of plant roots on SWRCs. The model assumes that part of the air void in soil is occupied by plant roots. According to the mass-volume relationship and the phase diagram of an unsaturated rooted soil, its void ratio may be expressed as:

$$e = \frac{e_0 - R_v(1+e_0)}{1 + R_v(1+e_0)} \tag{1}$$

where e_0 is the void ratio of bare soil, R_v is the root volume ratio, which is a dimensionless trait defined as the total volume of roots per unit volume of soil. $R_v = 0$ means that there are no plant roots in the soil (i.e., bare soil). R_v is less than $e_0/(1 + e_0)$, as total root volume cannot be larger than the total soil pore size. Depending on the plant type, R_v is a function of depth within the root zone (Ni et al. 2018c).

The void ratio-dependent SWRC equation proposed by Gallipoli et al. (2003) is as follows:

$$S_r = \left[1 + \left(\frac{se^{m_4}}{m_3}\right)^{m_2}\right]^{-m_1} \tag{2}$$

where S_r is the degree of saturation of soil; s is the matric suction; and m_1, m_2, m_3, and m_4 are the model parameters. m_1 and m_2 control the shape of an SWRC (van Genuchten 1980), while m_3 and m_4 are related to the air-entry value (AEV) of the parent soil.

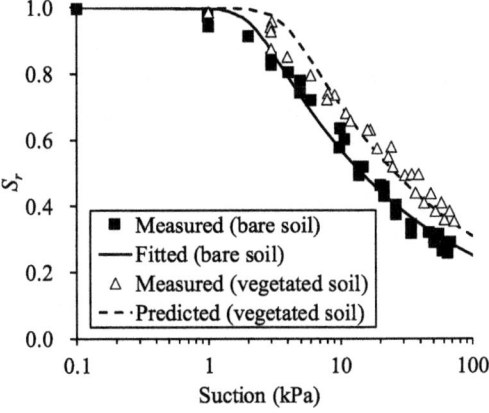

Fig. 5. Measured and predicted SWRCs of bare and vegetated soil

Hence, once the SWRC of the parent soil and R_v are known, the SWRC of rooted soil can be predicted from Eqs. (1) and (2). Figure 5 compares the predicted SWRCs of rooted silty sand with the measurements made by Leung et al. (2015a). The parameters, m_1, m_2, m_3 and m_4 were obtained by best fitting the SWRC of the bare soil. Thus, the good match for the case of bare soil is expected. For a known root parameter R_v of 0.043, the SWRC of rooted soil was predicted and a reasonably good agreement was obtained (Ng et al. 2016c).

2.4   Effects of Vegetation on Water Infiltration

To investigate the vegetation effects on slope hydrology, field experiments were conducted at the HKUST Eco-Park (Fig. 6), including in-situ double-ring infiltration tests (Leung et al. 2015b; Ng et al. 2016c) and field instrumentation and monitoring of a vegetated embankment (Garg et al. 2015b). This extended abstract covers only the results of the infiltration tests, while the plant-induced change in the hydrology of the embankment is interpreted in Garg et al. (2015b). All infiltration tests were conducted on the top flat surface of the embankment. Three treatments were considered, namely bare soil, grass-covered soil and tree-covered soil.

Figure 7 compares the cumulative water infiltration over time between bare, grass-covered and tree-covered soil during ponding. The volume of water that infiltrated the bare soil increased with time but at a decreasing rate. After ponding for 1 h, the volume of water infiltrated increased linearly with time, indicating that a steady state was reached. Similar trends were observed for grass- and tree-covered soil but the rate of water infiltration in the two vegetated covers was considerably smaller than that in the bare soil. It is evident that vegetation reduced the volume of water infiltrated by up to 50% compared to the bare case at the steady state.

Fig. 6. The HKUST Eco-Park

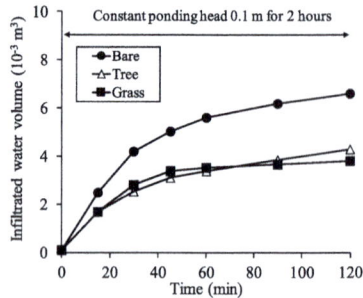

Fig. 7. Plant effects on water infiltration

3 Effects of Fungi on Root Tensile Strength

In addition to the study of the hydrological effects of plants discussed above, mechanical reinforcement by plant roots is also able to enhance slope stability. The contribution of mechanical root reinforcement primarily depends on root tensile strength. Although root biomechanical properties have been investigated over the last several decades (Wu et al. 1979; Pollen and Simon 2005; Boldrin et al. 2017a, b), few studies have focused on the effects of microorganisms on any biological change of root tissues or their biomechanical properties. Arbuscular mycorrhizal fungi (AMF) are one of the most important microorganisms that have been associated with plants since 400 million years ago (Smith and Read 2008). A test programme was designed and implemented to investigate the effects of various fungi (*Rhizophagus intraradices*(Ri), *Funneliformis mosseae* (Fm) *and Glomus aggregatum* (Ga)) on root growth and biomechanical property of Vetiver grass (*Chrysopogon zizanioides*). Sterilized inoculum (autoclave, 121 °C for 2 h) is also added to another five pots as a control (non-mycorrhizal, NM). The proportions of cellulose and hemicellulose contents in the steles were measured and they are summarised in Table 1.

Table 1. Comparison of the percentages of lignin, cellulose + hemicellulose and lipid + hydrosoluble contents in the steles of roots based on different mycorrhizal inoculations.

Treatment	Lignin (%)	Cellulose + hemicellulose (%)	Lipid + hydrosoluble (%)
NM	8.78 ± 3a	67.51 ± 3.65a	23.71 ± 0.97b
Ga	8.12 ± 1.69a	72.47 ± 1.77b	19.41 ± 1.7a
Ri	8.11 ± 1.95a	72.06 ± 2.96b	19.83 ± 2.52a
Fm	6.92 ± 1.19a	73.15 ± 1.9b	19.93 ± 3.06a

Figure 8 shows that AMF treatment increased the root biomass significantly. This led to an increase in root tensile strength (Fig. 9). For the stele diameter ranges of 0.1–0.2 and 0.2–0.3 mm, all AMF treatments significantly increased the root tensile strength. In the range of 0.3–0.4 mm, only the Ri treatment increased the root tensile strength. For the ranges of 0.4–0.5 and 0.5–1.0 mm, both the Ri and Fm treatments enhanced the root tensile strength, but not the Ga treatment. These indicate that AMF treatment can in general enhance the tensile strength of steles, especially the fine ones (0.1–0.3 mm). This is due to the fact that AMF can easily colonize fine roots, but not so many coarse roots (Wu et al. 2015), and exert local effects in the colonized cortical cells (Fiorilli et al. 2009). The AMF structures were only found in fine roots in this study.

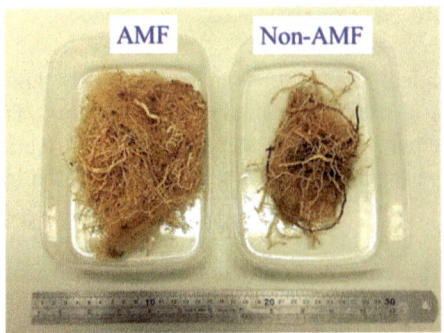

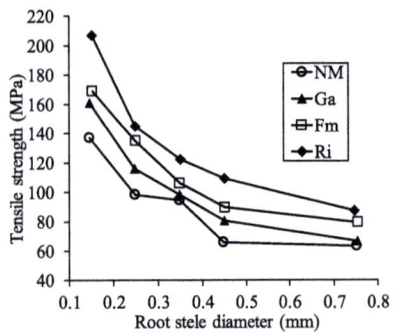

Fig. 8. Comparison of roots with and without AMF treatment

Fig. 9. Effects of AMF treatment on root tensile strength based on the stele diameter

Tensile strength is positively correlated with the cellulose and hemicellulose contents in roots (Genet et al. 2005). Higher percentages of cellulose and hemicellulose contents were found in the treatments with AMF inoculation (Table 1). The proportion of lipid and hydrosoluble contents was reduced in these treatments, and it was compensated only by cellulose and hemicellulose, but not lignin. The responses of plants upon AMF colonization differed across plant species. Nearly half of the 64 plant species inoculated with *GA* showed biomass increments, while the other half showed decrements (Klironomos 2003). It is unclear at this stage whether AMF symbiosis would enhance the proportions of cellulose and hemicellulose contents in all plant species.

4   Theoretical Analysis of the Stability of Vegetated Soil Slopes

In order to estimate the hydrological effects of plant roots on the stability of an infinite slope at a pre-defined slip depth Z, closed-form solutions were derived (Liu et al. 2016). The slope considered in these solutions has an inclination β, thickness H and water table H. The soil is considered perfectly plastic and the soil shear strength can be characterised by effective cohesion c' and effective friction angle ϕ'. Considering that shear stress induced by the self-weight of the vegetated soil at Z must be balanced by mobilising soil shear strength and force equilibrium, the FOS can be expressed as:

$$FOS = \frac{c' - u_w(z)\tan\phi^b}{\left[\gamma_d(H-z) + \gamma_w \int_z^H \theta_w(z)dz\right]\sin\beta\cos\beta} + \frac{\tan\phi'}{\tan\beta} \tag{3}$$

where γ_s is the dry unit weight of the vegetated soil; z is an arbitrary depth referenced to the slope surface; and u_w is the pore-water pressure (PWP); ϕ^b is the angle indicating the rate of increase in shear strength relative to suction (Fredlund and Rahardjo 1993; Ng and Menzies 2007) and θ_w is the volumetric water content. For a vegetated slope, as shown in Sect. 2 above, the soil PWP is significantly affected by plant transpiration (Fig. 1) and the magnitude is a function of root architecture (Leung et al. 2015a; Ng et al. 2016a). Liu et al. (2016) derived closed-formed steady- and transient-state solutions of u_w to capture the effects of plant transpiration for four idealised root architectures, namely exponential, parabolic, uniform and triangular architectures (Fig. 10). Once the plant-induced changes in u_w are known, the change in the FOS can be calculated according to Eq. (3).

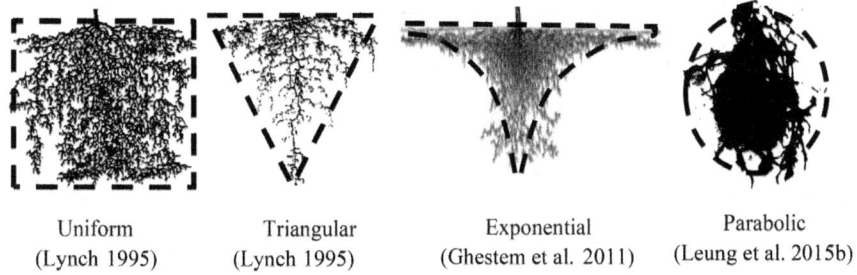

Uniform	Triangular	Exponential	Parabolic
(Lynch 1995)	(Lynch 1995)	(Ghestem et al. 2011)	(Leung et al. 2015b)

Fig. 10. Different idealised root architectures (Liu et al. 2016)

Figure 11 shows the ratio of the FOS of vegetated silty sand slope (with $c' = 10$ kPa and $\phi' = 38°$) to the FOS of bare slope varying with depth during rainfall (with a constant intensity of 181 mm/d for 24 h; equivalent to a 10-year return period in Hong Kong). Any FOS ratio larger than 1.0 indicates plant roots' contribution to slope stability. Before rainfall, water uptake by the exponential roots induced a higher FOS ratio than that by the parabolic roots at shallow depths. After raining for 1 h, the exponential and parabolic cases had similar FOS ratios. At the end of the rainfall event at t = 24 h,

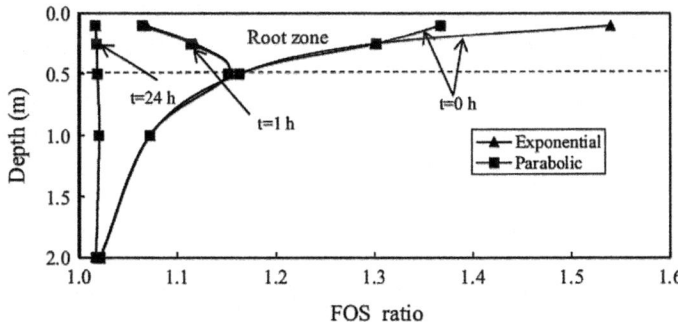

Fig. 11. Ratio of FOS of vegetated to bare slope under a rainfall intensity of 181 mm/day for 24 h (equivalent to 10-year return period)

the hydrological effects of transpiration vanished. This implies that vegetation has a much more profound positive effect to enhance the stability of slope under short and intense rainfall.

5   Plant Effects on Slope Hydrology, Stability and Failure Mechanisms

Although some centrifuge model tests have been carried out to investigate the mechanical effects of plant roots (Sonnerberg et al. 2010), the hydrological effects of plant have rarely been studied. In order to study the combined mechanical and hydrological effects of different root architectures on slope hydrology, stability and failure mechanisms in the geotechnical centrifuge, novel artificial root systems were developed (Ng et al. 2014b, 2016d; in press; Leung et al. 2017). All dimensions reported in this section are converted in prototype scale, unless stated otherwise.

5.1   Artificial Roots for Modelling Root-Water Uptake and Reinforcement

Figure 12 shows the artificial root models of three different architectures created by Ng et al. (2014b). They were developed based on the idealisation and simplification of real roots retrieved from three species—*S. heptaphylla*, *Rhodomyrtus tomentosa* and *Melastoma sanguineum*—that are commonly used for slope rehabilitation and eco-logical restoration in tropical and subtropical regions (Hau and Corlett 2003). These artificial root models were made of a porous material called cellulose acetate (CA), which has a tensile strength and an elastic modulus fairly close to those typically identified in real roots (Stokes and Mattheck 1996). Based on the design of Ng et al. (2014b), these root models connect with a vacuum system, which includes a vacuum chamber that is partially filled with de-aired water. Through a vacuum source con-nected to the chamber, different vacuum pressures can be applied to the CA and hence, different pressures up to -100 kPa can be induced in the water reservoir. Since the CA is in contact with soil, any applied vacuum, and hence reduction in total head inside the root model, would result in water flow from soil to the chamber through the filter. Any decrease in soil moisture would hence induce suction in soil. Leung et al. (2017) showed that the distribution of root area ratio (RAR) with root depth was reasonably captured by the three root models, taking into account the natural variability of plants in the field. The pull-out resistance of these three root models in saturated and unsaturated soil was investigated by Kamchoom et al. (2014). More details of the test results and interpretation can be found in Ng et al. (in press).

5.2   Plant Effects on Slope Hydrology

Figure 13 shows a typical model set-up. Three centrifuge tests were conducted to compare the contributions of three different root architectures—tap, heart and plate—to the hydrology and stability of $45°$ slope models made of completely decomposed granite. Each slope model supported by 15 artificial roots was subjected to a period of

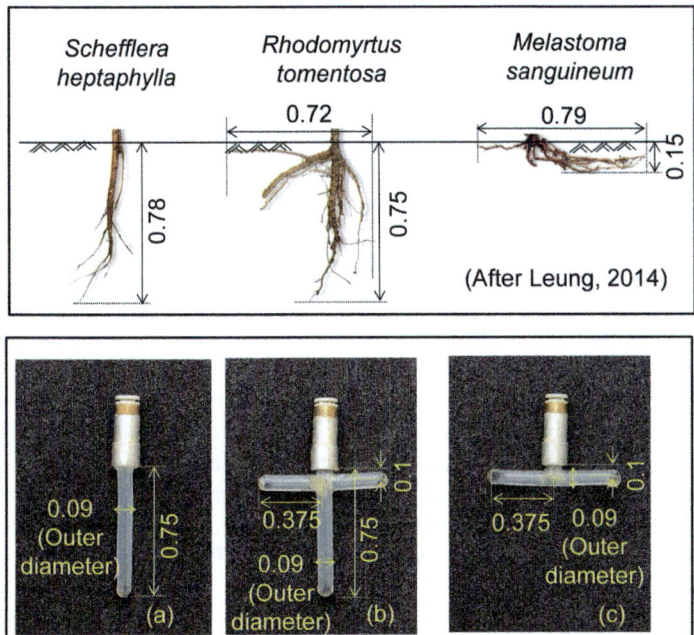

Fig. 12. Idealization and simplification of root systems with representative architectures (unit: m; converted to prototype scale)

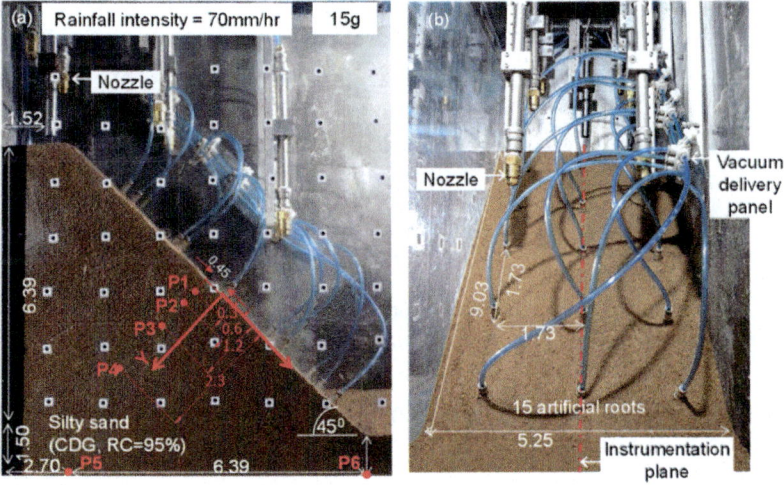

Fig. 13. (a) Side and (b) elevation views of the centrifuge model package and instrumentation (unit: m; converted to prototype scale)

simulated transpiration using the root system, followed by an intense rainfall event with a constant intensity of 70 mm/h for 36 h (i.e., equivalent to a 1000-year return period in Hong Kong). Six pore-pressure transducers (PPTs) were installed to monitor the PWP in each slope. All tests were carried out at 15 g in the geotechnical centrifuge facility at the HKUST. The soil properties, model setup, instrumentation and test procedures are detailed in Leung et al. (2017) and Ng et al. (in press).

Figure 14 compares the measured PWP profiles of the three slopes during rainfall. For the slope supported by tap-shaped roots, after raining for eight hours (i.e., 1000-year return period rainfall in Hong Kong), 2 kPa of suction (i.e., negative PWP) was preserved at the shallower depths, while a positive PWP of about 4 kPa was built up at a depth of 1.2 m. This is consistent with the computed results from the transient seepage analysis performed by Leung et al. (2017) using SEEP/W (Geo-Slope Int. 2009a). Below the root depth where PWP was less affected by transpiration, the PWP profile was distributed almost hydrostatically. In the case of the heart-shaped roots, slightly higher suctions of 2–3 kPa were retained within the root depth. Their additional branches extended the influence zone of transpiration to deeper regions. In the case of the plate-shaped roots which had a much shallower root depth, very little suction was preserved in the root zone, while positive PWP was recorded at all depths and the profile was distributed hydrostatically. The plate-shaped roots did not have a taproot component, and their induced suction thus had less effect at greater depths.

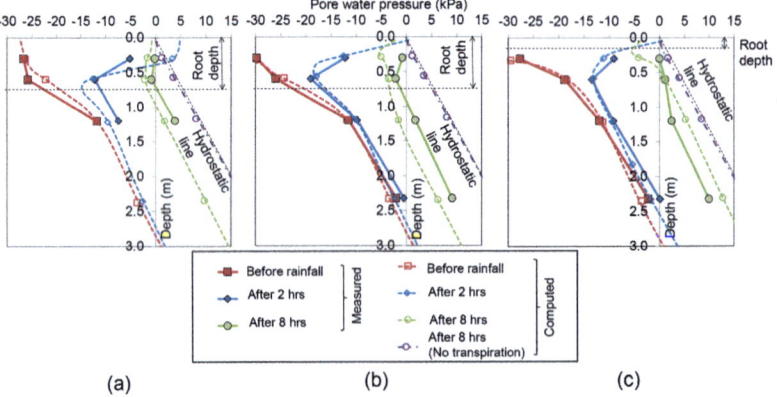

Fig. 14. Distribution of PWP along depth before rainfall and after two and eight hours of rainfall in slope models supported by (a) tap-, (b) heart- and (c) plate-shaped roots

5.3   Effects of Root Architecture and Transpiration on Slope Stability and Failure Mechanism

During and after the three centrifuge tests, no failure was observed. Based on the back-analysed PWP responses, slope stability analysis was conducted using SLOPE/W (Geo-Slope Int. 2009b) for determining the FOS in each case (Fig. 15). Before transpiration, the FOS was similar for all three slopes and exceeded 1.0 (i.e., the slopes were stable). When suction was created by simulating transpiration, the FOS of each slope increased, but not very significantly (less than 4%). This is because transpiration affected mainly the PWP in the top 1.2 m of each slope.

After raining for eight hours, the FOS of all three slopes dropped significantly, following a reduction in PWP upon infiltration. Despite of the reduction in the FOS, all slopes remained stable, as observed in the model tests. The FOS of the slope supported by the heart-shaped roots was 16% and 28% higher than those of slopes supported by the tap- and plate-shaped roots, respectively. The greater stability provided by the heart-shaped roots came from the substantial suction preserved after rainfall and their higher mechanical pull-out resistances than the other two architectures (Kamchoom et al. 2014; Ng et al. in press). If the transpiration effects before the rainfall event were ignored, the values of FOS after rainfall in all three cases dropped below 1.0. Regardless of the root architecture, neglecting the effects of transpiration on slope stability resulted in a significant underestimation of FOS by up to 50%.

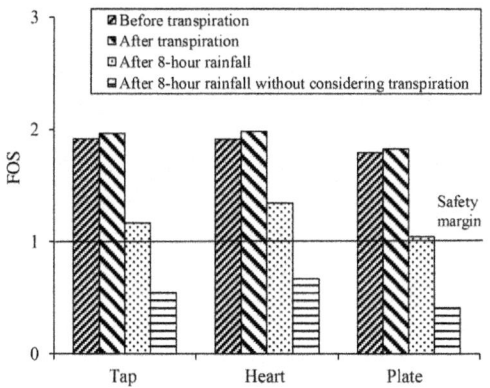

Fig. 15. FOS for slopes supported by different root architectures

In order to study the role of root architecture on the slope failure mechanism, two centrifuge tests were conducted for steeper slopes (i.e., 60°) vegetated with tap- and heart-shaped root models. The two slopes were subjected to an extreme rainfall event with a constant intensity of 70 mm/h for eight hours continuously (equivalent to a rainfall return period of 1000 years in Hong Kong) until failure. Comparing the post-failure geometries between the two slopes in Fig. 16 suggests that shallower slip was formed and a smaller volume of soil failed when the slope was reinforced with heart-shaped roots as compared with the tap-rooted slope. The runout distance from the heart-shaped slope toe was about 9% (4.31 m versus 4.71 m) shorter than that of tap-shaped one. These observations suggest that heart-shaped roots are more effective for stabilising slopes and reducing runout distance than tap- and plate-shaped roots.

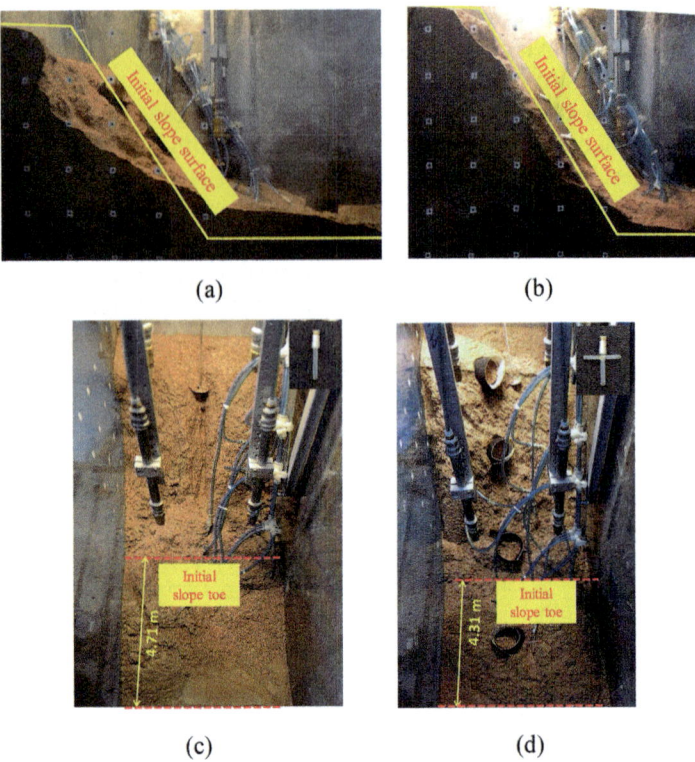

Fig. 16. (a) Side view and (c) plan view of slope reinforced by tap-shaped roots; (b) side view and (d) plan view of slope reinforced by heart-shaped roots

6 Conclusions

Slope bioengineering using plants and microorganisms was studied through a multi-disciplinary and complementary research programme. Various tree, shrub and grass species widespread and native to southern China (*Schefflera heptaphylla* and *Chrysopogon zizanioides*) and Europe with temperate climates (e.g., *Ulex europaeus*, *Salix viminalis*, and *Corylus avellana*) were studied. Some of these species were able to preserve credible soil matric suction during and after intense rainfall. The magnitude of suction preserved was positively correlated with plant traits such as the leaf area index (LAI), the root area index (RAI) and the ratio of root to shoot biomass. Plants not only induce suction via transpiration but also change soil water retention behaviour and infiltration characteristics. The total amount of water infiltration were found to be considerably higher in bare than those in vegetated soil covered by grass and tree. Moreover, the biological interaction of plant roots with symbiotic fungi in soil enhanced root tensile strength significantly.

In order to study the combined hydrological and mechanical effects of plant roots on slope stability, novel artificial root systems with different idealised yet representative root architectures were developed for testing vegetated slopes in the HKUST geotechnical centrifuge facility. New theoretical closed-form solutions were also derived to capture their effects. Heart-shaped roots provided stronger stabilisation effects than either tap- or plate-shaped roots by preserving more matric suction (hence higher soil shear strength and lower water permeability) and contributing greater mechanical reinforcement due to multiple branching. Evidently, ignoring the effects of plant transpiration would underestimate FOS by up to 50%. This is not favourable for developing more sustainable and environmentally friendly cities worldwide.

Acknowledgements. The authors acknowledge research grant 51778166 awarded by the National Natural Science Foundation of China, research grant 2012CB719805 from the National Basic Research Program (973 Program) administered by the Ministry of Science and Technology of the People's Republic of China, and research grant HKUST6/CRF/12R awarded by the Research Grants Council of the Government of the Hong Kong SAR. The second author would also like to acknowledge support from the EU Marie Curie Career Integration Grant for the project 'BioEPIC slope'.

References

Boldrin, D., Leung, A.K., Bengough, A.G.: Correlating hydromechanical properties of vegetated soil with plant functional traits. Plant Soil **416**(1–2), 437–451 (2017a)

Boldrin, D., Leung, A.K., Bengough, A.G.: Root biomechanical properties during establishment of woody perennials. Ecol. Eng. **109**, 192–206 (2017b)

Ekanayake, J.C., Phillips, C.J.: Slope stability thresholds for vegetated hillslopes: a composite model. Can. Geotech. J. **39**, 849–862 (2002)

Fiorilli, V., Catoni, M., Miozzi, L., Novero, M., Accotto, G.P., Lanfranco, L.: Global and cell-type gene expression profiles in tomato plants colonized by an arbuscular mycorrhizal fungus. New Phytol. **184**, 975–987 (2009)

Francour, P., Semroud, R.: Calculation for the root area index in Posidonia oceanica in the Western Mediterranean. Aquatic Bot. **42**(3), 281–286 (1992)

Fredlund, D.G., Rahardjo, H.: Soil Mechanics for Unsaturated Soils. Wiley Inter Science, New York (1993)

Gallipoli, D., Wheeler, S.J., Karstunen, M.: Modelling the variation of degree of saturation in a deformable unsaturated soil. Géotechnique **53**(1), 105–112 (2003)

Garg, A., Leung, A.K., Ng, C.W.W.: Comparisons of soil suction induced by evapotranspiration and transpiration of S. heptaphylla. Can. Geotech. J. **52**(12), 2149–2155 (2015a)

Garg, A., Coo, J.L., Ng, C.W.W.: Field study on influence of root characteristics on suction distributions in slopes vegetated with Cynodon dactylon and Schefflera heptaphylla. Earth Surf. Process. Landf. **40**(12), 1631–1643 (2015b)

Genet, M., Stokes, A., Salin, F., Mickovski, S.B., Fourcaud, T., Dumail, J.F., van Beek, R.: The influence of cellulose content on tensile strength in tree roots. Plant Soil **278**, 1–9 (2005)

Geo-Slope International Ltd.: Seepage Modeling with SEEP/W, An Engineering Methodology, 4th edn. (2009a)

Geo-Slope International Ltd.: Stress-Deformation Modeling with SIGMA/W, An Engineering Methodology, 4th edn. (2009b)

Ghestem, M., Sidle, R.C., Stokes, A.: The influence of plant root systems on subsurface flow: implications for slope stability. Bioscience **61**(11), 869–879 (2011)

Hau, B.C., Corlett, R.T.: Factors affecting the early survival and growth of native tree seedlings planted on a degraded hillside grassland in Hong Kong, China. Restor. Ecol. **4**(11), 483–488 (2003)

Jotisankasa, A., Taworn, D.: Direct shear testing of clayey sand reinforced with live stake. Geotech. Test. J. **39**(4), 608–623 (2016)

Kamchoom, V., Leung, A.K., Ng, C.W.W.: Effects of root geometry and transpiration on pull-out resistance. Géotechnique Lett. **4**(4), 330–336 (2014)

Kelliher, F.M., Leuning, R., Raupach, M.R., Schulze, E.D.: Maximum conductances for evaporation from global vegetation types. Agric. Forest Meteorol. **73**, 1–16 (1995)

Klironomos, J.N.: Variation in plant response to native and exotic arbuscular mycorrhizal fungi. Ecology **84**, 2292–2301 (2003)

Lam, C.C., Leung, Y.K.: Extreme rainfall statistics and design rainstorm profiles at selected locations in Hong Kong. Royal Observatory, Hong Kong (1995)

Leung, T.Y.: The use of native woody plants in slope upgrading in Hong Kong, Ph.D. thesis, The University of Hong Kong (2014)

Leung, A.K., Garg, A., Ng, C.W.W.: Effects of plant roots on soil-water retention and induced suction in vegetated soil. Eng. Geol. **193**, 183–197 (2015a)

Leung, A.K., Garg, A., Coo, J.L., Ng, C.W.W., Hau, B.C.H.: Effects of the roots of *Cynodon dactylon* and *Shefflera heptaphylla* on water infiltration rate and soil hydraulic conductivity. Hydrol. Process. **29**(15), 3342–3354 (2015b)

Leung, A.K., Kamchoom, V., Ng, C.W.W.: Influences of root-induced soil suction and root geometry on slope stability: a centrifuge study. Can. Geotech. J. **54**(3), 291–303 (2017)

Liu, H.W., Feng, S., Ng, C.W.W.: Analytical analysis of hydraulic effect of vegetation on shallow slope stability with different root architectures. Comput. Geotech. **80**, 115–120 (2016)

Lynch, J.: Root architecture and plant productivity. Plant Physiol. **109**(1), 7–13 (1995)

Mao, Z., Bourrier, F., Stokes, A., Fourcaud, T.: Three-dimensional modelling of slope stability in heterogeneous montane forest ecosystems. Ecol. Eng. **273**, 11–22 (2014)

Ng, C.W.W.: Atmosphere-plant-soil interaction: theories and mechanisms, Huangwenxi lecture. Chin. J. Geotech. Eng. **39**(1), 1–47 (2017)

Ng, C.W.W., Menzies, B.: Advanced Unsaturated Soil Mechanics and Engineering. Taylor & Francis, London (2007)

Ng, C.W.W., Ni, J.J., Leung, A.K., Zhou, C., Wang, Z.J.: Effects of planting density on tree growth and induced soil suction. Géotechnique **66**(9), 711–724 (2016a)

Ng, C.W.W., Garg, A., Leung, A.K., Hau, B.C.H.: Relationships between leaf and root area indices and soil suction induced during drying-wetting cycles. Ecol. Eng. **91**, 113–118 (2016b)

Ng, C.W.W., Ni, J.J., Leung, A.K., Wang, Z.J.: A new and simple water retention model for root-permeated soils. Géotechnique Lett. **6**(1), 106–111 (2016c)

Ng, C.W.W., Kamchoom, V., Leung, A.K.: Centrifuge modelling of the effects of root geometry on the transpiration-induced suction and stability of vegetated slopes. Landslides **13**(5), 1–14 (2016d)

Ng, C.W.W., Leung, A.K., Woon, K.X.: Effects of soil density on grass-induced suction distributions in compacted soil subjected to rainfall. Can. Geotech. J. **51**(3), 311–321 (2014a)

Ng, C.W.W., Leung, A.K., Kamchoom, V., Garg, A.: A novel root system for simulating transpiration-induced soil suction in centrifuge. Geotech. Test. J. **37**(5), 1–15 (2014b)

Ng, C.W.W, Wang, Z.J., Leung, A.K., Ni, J.J.: Effects of leaf and root characteristics on the ability of plant root water uptake. Géotechnique (2018a, Accepted)

Ng, C.W.W., Tasnim, R., Capobianco, V., Coo, J.L.: Influences of soil nutrients on plant characteristics and soil hydrological responses. Géotechnique Lett. **8**(1), 1–6 (2018b)

Ng, C.W.W., Woon, K.X., Leung, A.K., Chu, L.M.: Experimental investigation of induced suction distributions in a grass-covered soil. Ecol. Eng. **52**, 219–223 (2013)

Ng, C.W.W., Leung, A.K., Ni, J.J.: Plant-Soil Slope Interaction. Taylor & Francis, USA (in press). ISBN 978-1-13-819755-8

Ni, J.J., Leung, A.K., Ng, C.W.W.: Investigation of plant growth and transpiration induced suction under mixed grass-tree conditions. Can. Geotech. J. **54**(4), 561–573 (2017)

Ni, J.J., Leung, A.K., Ng, C.W.W., Shao, W.: Modelling hydro-mechanical reinforcement of vegetation to slope stability. Comput. Geotech. **95**, 99–109 (2018c)

Pollen, N., Simon, A.: Estimating the mechanical effects of riparian vegetation on stream bank stability using a fiber bundle model. Water Resour. Res. **41**, W07025 (2005)

Smith, S.E., Read, D.J.: Mycorrhizal Symbiosis. Academic Press, London (2008)

Sonnenberg, R., Bransby, M.F., Hallett, P.D., Bengough, A.G., Mickovski, S.B., Davies, M.C. R.: Centrifuge modelling of soil slopes reinforced with vegetation. Can. Geotech. J. **47**, 1415–1430 (2010)

Stokes, A., Mattheck, C.: Variation of wood strength in tree roots. J. Exp. Bot. **47**(5), 693–699 (1996)

Stokes, A., Atger, C., Bengough, A.G., Fourcaud, T., Sidle, R.C.: Desirable Plant root traits for protecting natural and engineered slopes against landslides. Plant Soil **324**(1), 1–30 (2009)

van Genuchten, M.T.: A closed-form equation for predicting the hydraulic conductivity of unsaturated soils. Soil Sci. Soc. Am. J. **44**(5), 892–898 (1980)

Wu, Q.S., Liu, C.Y., Zhang, D.J., Zou, Y.N., He, X.H., Wu, Q.H.: Mycorrhiza alters the profile of root hairs in trifoliate orange. Mycorrhiza **26**(3), 237–247 (2015)

Wu, T.H., Mickinnell III, W.P., Swanston, D.N.: Strength of tree-roots and landslides on Prince of Wales Island. Can. Geotech. J. **16**, 19–33 (1979)

Effect of Liner Consolidation on Contaminant Transport Through a Landfill Bottom Liner System

Hefu Pu$^{(\boxtimes)}$, Jinwei Qiu, Junjie Zheng, and Rongjun Zhang

School of Civil Engineering and Mechanics,
Huazhong University of Science and Technology, Wuhan, China
puh@hust.edu.cn

Abstract. In municipal solid waste landfills, a triple-layer composite liner consisting of a geomembrane liner (GML), a geosynthetic clay liner (GCL) and a compacted clay liner (CCL) is commonly used at the landfill bottom to isolate the leachates from surrounding environment. This paper presents a numerical investigation of the effect of liner consolidation on the transport of a volatile organic compound (VOC), benzene, through the GML/GCL/CCL composite liner system. The numerical simulations were performed using the model CST3. The performed numerical simulations considered the impact of consolidation on contaminant transport for a GML/GCL/CCL liner system. The simulation results indicate that, depending on conditions, consolidation of the GCL and CCL can have significant impact on the transport results of benzene, both during the consolidation process and long after the completion of consolidation. The traditional approach for the assessment of liner performance neglects consolidation of the liners and fails to consider the consolidation-induced transient advection and concurrent changes in material properties and, therefore, can lead to significantly different results.

Keywords: Consolidation · Contaminant transport · Compacted clay liner
Geosynthetic clay liner · Numerical modeling

1 Introduction

For modern sanitary landfills, installation of liner systems at the landfill bottom is required to control the landfill leachates so that the harmful contaminants will not be readily released into the surrounding environment. The bottom liner systems typically consist of a composite liner containing a geomembrane liner (GML) underlain with either a geosynthetic clay liner (GCL) or a compacted clay liner (CCL) or with both. Previous studies by the authors (Pu et al. 2016a and 2016b on the ASCE Journal of Geotechnical and Geoenvironmental Engineering) have investigated the contaminant transport through the GML/CCL and the GML/GCL composite liner systems. The present study is focused on the GML/GCL/CCL composite liner system. Conventionally, performance of these liner systems regarding contaminant containment was evaluated using advective-diffusive models which assume that the liners are rigid during landfill service life such that consolidation of liners and its impact on the

© Springer Nature Switzerland AG 2018
W. Wu and H.-S. Yu (Eds.): *Proceedings of China-Europe Conference
on Geotechnical Engineering*, SSGG, pp. 1304–1307, 2018.
https://doi.org/10.1007/978-3-319-97115-5_88

performance of the liner systems is neglected. In reality, however, vertical stress will be applied on these bottom liners due to placement of the waste materials. Since the municipal solid wastes can sometimes be higher than a hundred meters, such vertical stress can be very high - sometimes even more than 1000 kPa at final closure. When such a high stress is applied on the GML/GCL/CCL liner system, consolidation of the GCL and CCL will occur and cause advective transport and variations in material properties both temporally and spatially, which can subsequently impact the process of contaminant transport through the GML/GCL/CCL liner system. Then, the question to be answered is to what extent will the consolidation of the liners affect the performance of the GML/GCL/CCL liner system?

This paper presents the simulation results of a numerical investigation of the effects of liner consolidation on the transport of benzene, a typical VOC in landfill leachates, through the GML/GCL/CCL composite liner system.

2 Methods

Numerical Model: The numerical simulations were performed using the CST3 model (see Pu and Fox 2016 on International Journal of Geomechanics, ASCE), which is a piecewise linear numerical model for coupled large-strain consolidation and solute transport in layered porous media and has been extensively validated using analytical and numerical solutions and laboratory experiments. The consolidation module of CST3 accounts for large strain, soil self-weight, general constitutive relationships, changing compressibility and hydraulic conductivity during consolidation, relative velocity of fluid and solid phases, unload/reload, time-dependent loading schedule and boundary conditions, externally-applied hydraulic gradient, depth-dependent preconsolidation stress profile, and multiple soil layers with different transport properties and material properties. The solute transport module of CST3 accounts for advection, diffusion, mechanical dispersion, linear and nonlinear sorption, equilibrium and nonequilibrium sorption, porosity-dependent effective diffusion coefficient, and first-order decay reactions.

Initial Configuration: In the simulations, the performed numerical simulations considered coupled consolidation and contaminant transport for a GML/GCL/CCL liner system. Figure 1 shows the initial configuration of the GML/GCL/CCL composite liner system, which is comprised of a leachate collection and removal system (LCRS), a GML, a GCL, a CCL, and a subgrade layer. The subgrade can represent an underlying leachate detection system or natural geological layer at atmospheric pore pressure, and is assumed to be freely drained. The GML is assumed to have no holes and is in excellent contact with the GCL, and thus there is no interface transmissivity between the GML and the GCL.

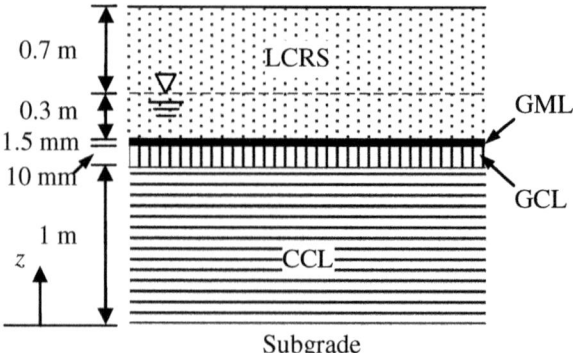

Fig. 1. Initial configuration for GML/GCL/CCL composite liner system.

3   Simulation Results

Figure 2(a) presents the trends for applied stress Δq and liner settlement S for the CC (i.e., "Consolidation Coupled") simulation of the GML/GCL/CCL system over the 40-year simulation period. Liner consolidation is finished at around $t = 10.7$ years (soon after the end of loading period of 10 years) and the induced final settlement is 98.15 mm, with 94.3 mm for CCL and 3.85 mm for GCL, corresponding to final average strain of 9.4% and 38.5%, respectively. The initial and final average void ratios, respectively, are 0.639 and 0.484 for the CCL. Corresponding values for the GCL are 4.824 and 2.577, respectively. Based on the changes in average clay void ratio due to consolidation, value of vertical hydraulic conductivity decreases by 83% for CCL and by 93% for GCL. The value of effective diffusion coefficient decreases by 16% for CCL and by 70% for GCL. Figure 2(b) presents the mass flux at the bottom of CCL per unit area F. For the non-sorbing CCL (i.e., $K_{dCCL} = 0$), Fig. 2(b) indicates the CC simulation cases result in benzene breakthrough at approximately 2 years, which is slightly earlier than the corresponding DO (i.e., "Diffusion Only") simulation cases. A small, temporary and sharp decrease in mass flux occurs between 10 and 10.5 years because during this short period advection suddenly decreases to zero due to the completion of consolidation (as shown in Fig. 2(a)). The corresponding DO simulations produce lower mass flux in the early stages but higher mass flux soon after the end of consolidation. At $t = 40$ years, benzene mass flux for DO simulation is 0.533 mg/m²/year, and is 29% higher than that for the CC simulation. For the sorbing CCL (i.e., $K_{dCCL} > 0$), breakthrough of benzene is significantly delayed and mass flux is reduced for all simulations due to the retardation effect. Interestingly, for the sorbing CCL, the DO simulations produce slightly earlier breakthrough relative to the CC simulations, which is in contrast to the phenomenon observed for the non-sorbing CCL.

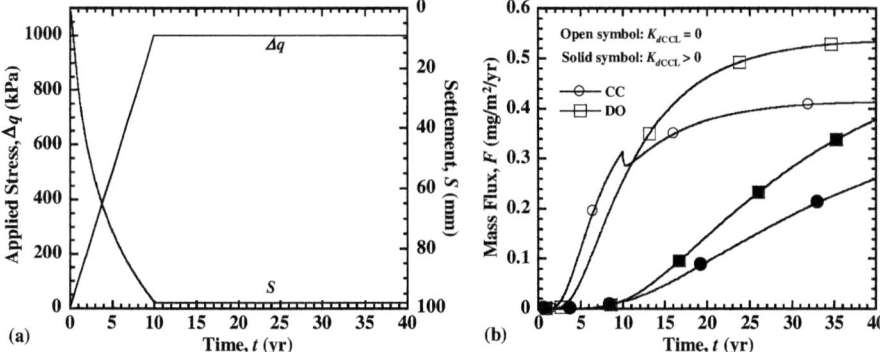

Fig. 2. Simulation results for GML/GCL/CCL composite liner: (a) applied stress and settlement; (b) benzene mass flux.

4 Conclusions

1. Depending on conditions, consolidation of the GCL and the CCL can have significant impact on the transport results of benzene, both during the consolidation process and long after consolidation has ended. The conventional approach for the assessment of liner performance neglects consolidation of the GCL and CCL and fails to consider the consolidation-induced transient advection and concurrent changes in material properties and, therefore, can lead to significantly different results.

2. The simulation results indicate that the impact of liner consolidation on benzene transport for a GML/GCL/CCL composite liner system is complex. On one hand, consolidation of GCL and CCL causes advection and a reduction in transport distance, which tends to accelerate contaminant transport. On the other hand, consolidation decreases the transport parameters (e.g., effective diffusion coefficient and porosity), which tends to slow down contaminant transport. Therefore, depending on the specific conditions and associated material properties, liner consolidation can either accelerate or delay contaminant transport

References

Pu, H., Fox, P.J.: Model for coupled large strain consolidation and solute transport in layered soils. Int. J. Geomech. **16**(2), 04015064 (2016)

Pu, H., Fox, P.J., Shackelford, C.D.: Assessment of consolidation-induced contaminant transport for compacted clay liner systems. J. Geotech. Geoenviron. Eng. **142**(3), 04015091 (2016a)

Pu, H., Shackelford, C.D., Fox, P.J.: Assessment of consolidation-induced VOC transport for a GML/GCL composite liner system. J. Geotech. Geoenviron. Eng. **142**(11), 04016053 (2016b)

New Method to Reduce Porosity of Rockfill Materials with Composite Slurry

Tao Wang, Sihong Liu$^{(\boxtimes)}$, and Yan Feng

College of Water Conservancy and Hydropower,
Hohai University, Nanjing 210098, China
sihongliu@hhu.edu.cn

Abstract. A new method of using composite slurry is proposed to increase the compression modulus and decrease creep deformation of rockfill materials for the construction of high rockfill dams. The composite slurry is a mixture of fly ash, cement and sand with the properties of easy flow and post-hardening. During the process of rolling compaction, the slurry admixture sprinkled on the rockfill surface will gradually infiltrate into the inter-granular voids of rockfill. Compression tests were carried out to investigate the compression characteristics of rockfill with composite slurry. Experimental results show the filling of composite slurry can effectively reduce the porosity of rockfill materials, especially for soft and gap-graded rockfill materials.

Keywords: Composite slurry · Rockfill materials · Porosity · Compressibility

1 Introduction

The rockfill dam has become the most widely used dam type, as it is reasonably adaptable to topography, geology and various kinds of rockfill materials [1, 2]. Currently, many high rockfill dams are under construction, such as Lianghekou (295 m), Shuangjiangkou (314 m) and Rumei (315 m) rockfill dams in western China. A key problem in high rockfill dam design and construction is controlling dam deformations. Large dam deformations may pose great threats to the performance and safety of rockfill dams [3, 4].

A significant number of researches have been done on the control of dam deformations. The proposed methods can be classified into two categories: one is the optimization of dam structure and construction, and the other is the reinforcement of rockfill materials. The former involves optimizing the dam zoning [5], setting pre-settlement periods [6], and raising compacting standards [7], etc. The latter involves using medium hard rock [8] and adding cementitious material to rockfill materials [9], etc. Dam deformations are closely related to the porosity of rockfill materials after rolling compaction. Previous studies show that raising compaction standards is usually adopted to get lower porosity of rockfill materials, and the porosity of rockfill materials should be controlled within 18% when they are used in 300 m-high rockfill dam [10]. However, such a low porosity is difficult to achieve only by raising compacting standards.

© Springer Nature Switzerland AG 2018
W. Wu and H.-S. Yu (Eds.): *Proceedings of China-Europe Conference on Geotechnical Engineering*, SSGG, pp. 1308–1311, 2018.
https://doi.org/10.1007/978-3-319-97115-5_89

Recently, a new method of using composite slurry is proposed to increase the compression modulus and decrease creep deformation of rockfill materials for the construction of high rockfill dams. The composite slurry is a mixture of fly ash, cement and sand with the properties of easy flow and post-hardening. As shown in Fig. 1, it is first sprinkled on the surface of rockfill materials and infiltrating into the rockfill materials under self-gravity. Then, in the process of rolling compaction, the composite slurry further fills the inter-granular voids of rockfill materials under exciting force, which is intended to reduce the porosity and increase the compression modulus of rockfill materials.

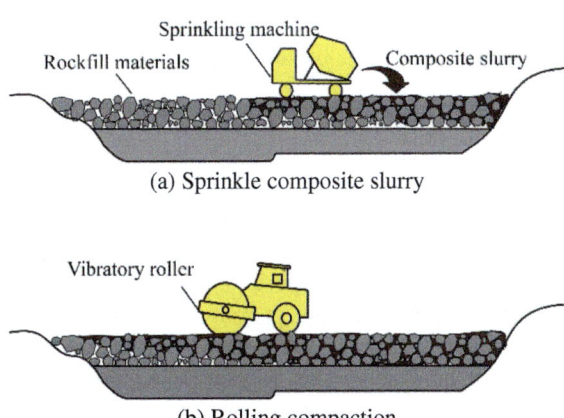

(a) Sprinkle composite slurry

(b) Rolling compaction

Fig. 1. Construction process using composite slurry.

In this paper, compression tests were carried out to verify the feasibility and effectiveness of the proposed method. The compression properties of rockfill with composite slurry were studied.

2 Compression Test

2.1 Materials and Method

Two different types of rockfill materials (denoted as, Rockfill A and Rockfill B) are used. Rockfill A is slightly-weathered dolomite, with a saturated uniaxial compression strength of 30 MPa–60 MPa (medium hard rock), and Rockfill B is strongly-weathered dioritic porphyrite, whose saturated uniaxial compression strength is 5 MPa–10 MPa (soft rock). Besides, two gradations are considered (shown in Fig. 2), gradation 1 is well graded (C_u = 30.2, C_c = 0.40.8), obtained on the basis of a dam's design gradation, while gradation 2 is gap-graded (C_u = 30.2, C_c = 0.48).

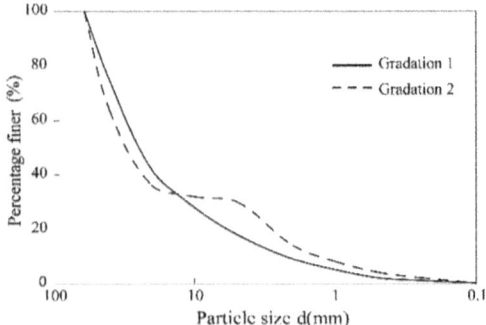

Fig. 2. Two gradation curves of rockfill materials.

2.2   Results and Discussions

The data of confined compression tests are plotted in Fig. 3 of void ratio, e versus the logarithm of vertical stress, logP. It can be seen that, under the same vibrating compaction conditions, the initial void ratio of rockfill with composite slurry is much smaller than that of rockfill without composite slurry, which means that the filling of composite slurry effectively reduces the porosity of rockfill. As the vertical stress increases, the curve transfers from a flat section to a steep section because of the compression and particle breakage of the rockfill. The value of vertical stress corresponding to the sudden decrease of void ratio is much larger for rockfill with composite slurry owing to the filling and cementation effects of composite slurry.

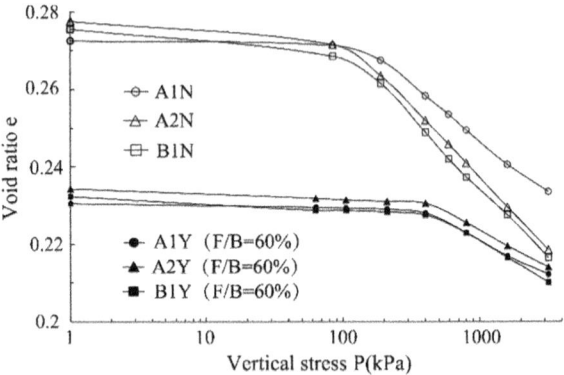

Fig. 3. Confined compression curves of rockfill with different types and gradations.

3 Conclusions

A new method of using composite slurry is proposed to increase the compression modulus of rockfill materials. The compression characteristics of rockfill materials with composite slurry were investigated. It is found that the filling of composite slurry can effectively reduce the porosity of rockfill materials, as well as increasing the compression modulus, especially for soft and gap-graded rockfill materials.

References

1. Alonso, E.E., Cardoso, R.: Behavior of materials for earth and rockfill dams: perspective from unsaturated soil mechanics. Front. Archit. Civ. Eng. China **4**(1), 1–39 (2010)
2. Zhang, B.Y., Zhang, J.H., Sun, G.L.: Deformation and shear strength of rockfill materials composed of soft siltstones subjected to stress, cyclical drying/wetting and temperature variations. Eng. Geol. **190**, 87–97 (2015)
3. Cetin, H., Laman, M., Ertunc, A.: Settlement and slaking problems in the world's fourth largest rock-fill dam, the Ataturk Dam in Turkey. Eng. Geol. **56**(3), 225–242 (2000)
4. Langroudi, M.F., Soroush, A., Shourijeh, P.T.: Stress transmission in internally unstable gap-graded soils using discrete element modeling. Powder Technol. **247**(10), 161–171 (2013)
5. Hunter, G., Fell, R.: Rockfill modulus and settlement of concrete face rockfill dams. J. Geotech. Geoenviron. Eng. **129**(129), 909–917 (2003)
6. Ma, H.Q., Cao, K.M.: Key technical problems of extra-high concrete faced rock-fill dam. Sci. China Ser. E: Technol. Sci. **50**(s1), 20–33 (2007)
7. Honkanadavar, N.P., Kumar, N., Ratnam, M.: Modeling the behaviour of alluvial and blasted quarried rockfill materials. Geotech. Geol. Eng. **32**(4), 1001–1015 (2014)
8. Cooke, J.B., Sherard, J.L.: Concrete-face rockfill dam: II Design. J. Geotech. Eng. **113**(10), 1113–1132 (1987)
9. Jin, F., An, X.H., Shi, J.J.: Study on rock-fill concrete dam. J. Hydraul. Eng. **36**(11), 1347–1352 (2005)
10. Ma, H., Chi, F.: Major technologies for safe construction of high earth-rockfill dams. Engineering **2**(4), 498–509 (2016)

Soil Nutrient Effects on Suction and Volumetric Water Content in Heavily Compacted Vegetated Soil

R. Tasnim[1(⊠)], J. L. Coo[1], C. W. W. Ng[1], and V. Capobianco[2]

[1] Hong Kong University of Science and Technology, Sai Kung, Hong Kong
rtasnim@connect.ust.hk
[2] University of Salerno, Fisciano, Italy

Abstract. Previous studies demonstrated that soil nutrients help plant growth and enhance the stability of bio-engineered slopes through plant-induced soil suction. Therefore, soil nutrient effects on the variations of suction and volumetric water content (VWC) in heavily compacted soil during evapotranspiration need to be studied. In this study, three replicates of *Schefflera heptaphylla* (Ivy tree) were grown for 6 months in nutrient poor and nutrient supplied compacted soil. Soil suction and VWC changes during drying of vegetated soils were measured after 3 and 6 months of plants growth. After the 3rd and 6th month of drying, suction increased by 5–10 kPa and 15–40 kPa, respectively in nutrient supplied vegetated soil compared to the nutrient poor soil due to increase in leaf area index (LAI) and root area index (RAI). In contrast, soil VWC was higher in all the vegetated soils after drying on the 6th month compared to the VWC after drying on the 3rd month. This might be due to soil-pore structure changes via bio-chemical root activities and organic matters.

Keywords: Vegetation · Nutrient · Suction

1 Brief Introduction and Methodology

Chemical composition of soil affects plant-growth and their effectiveness on infrastructures via hydrological reinforcement due to plant-induced soil suction changes [1]. NPK slow-release granular fertilizers are used for bio-engineering techniques in Hong Kong but lengthen establishment time of plants due to slow utilization of the fertilizers by plants, impairing their survival rate after transplantation [2, 3]. NPK fertilizer in a liquid form has shown to be a suitable substitute for a faster acclimation of plants in heavily compacted soil (used for man-made slopes) [7]. However, effects of NPK water soluble fertilizer on soil suction and volumetric water content (VWC) changes during drying of vegetated heavily compacted soil are not yet understood. This study investigates the effects of NPK (Nitrogen-Phosphorous-Potassium) water soluble fertilizer on the soil suction and VWC changes during drying of heavily compacted (Relative Compaction-95%) silty sand vegetated with Schefflera heptaphylla.

Six columns were constructed with an inner diameter of 200 mm and a height of 400 mm. Completely decomposed granite (CDG) soil was compacted up-to 390 mm

© Springer Nature Switzerland AG 2018
W. Wu and H.-S. Yu (Eds.): *Proceedings of China-Europe Conference on Geotechnical Engineering*, SSGG, pp. 1312–1315, 2018.
https://doi.org/10.1007/978-3-319-97115-5_90

depth at a RC of 95% (corresponding to a dry density of 1777 kg/m³ and 12% moisture content) [7]. An individual *Schefflera heptaphylla* (Ivy tree) was transplanted at the centre of the six columns with similar basal diameter (10 ± 2 mm) and root depths (125 ± 10 mm) for fair comparison. Four miniature tip tensiometers at 50, 130, 210 and 290 mm depth were installed at the centre of the columns. SM-300 probes were installed at 50 and 130 mm depth right next to the tensiometer to monitor VWC (See Fig. 1). All columns were placed in a temperature (25 ± 1 °C) and humidity (55 ± 5%) controlled room under a cool white fluorescent lamp for the whole testing period. Two test series were conducted; one series with nutrient supply in soil and another series without nutrient supply. For each series, 3 replicates of plants (total 6 plants) were grown for 6 months while continuous measurements of suction and VWC were monitored. Every 2 days, all columns were irrigated with similar amount of water and every 8 days 3 columns were instead irrigated with NPK (30-10-10) nutrient (2 gm) mixed with water (1 litre). At the end of the 3rd and 6th month, ponding head was applied on the surface of all soil columns until suctions at all four depths registered 0 kPa and percolation through the holes at the column base was observed. Then the columns were exposed to the environment to quantify the effects of evapotranspiration by the plant and soil on suction and VWC responses.

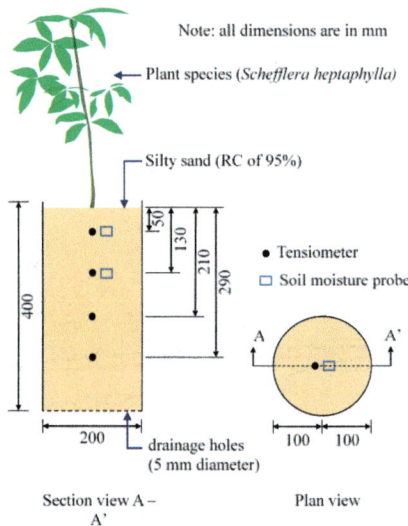

Fig. 1. Typical schematic setup and instrumentation of a tree-vegetated column in (a) Cross section view A–A' and (b) Plan view.

2   Interpretation of Test Results

Figures 2-(a) and (b) show the suction and VWC changes during 3 days of drying for vegetated soil with and without nutrient supply at 50 mm depth on the 3rd and 6th month. In Fig. 2-(a), bare soil suction was minimal (2–3 kPa) after drying for 3 days.

Vegetated soil suction increased by 15–75 kPa compared to bare soil due to water uptake by roots [5] and transpiration through stomata of leaves [4]. During drying on the 3rd month, 5–10 kPa more suction was observed in the nutrient supplied vegetated soil compared to the nutrient poor vegetated soil. This is due to the 50%–120% increase of LAI in nutrient supplied vegetated soil [7]. Both on the 3rd and 6th month, similar magnitude of suction was observed during drying in nutrient poor vegetated soil due to similar range (0.3–0.8) of LAI [7]. However, during drying on the 6th month, suction increased by 15–40 kPa in the nutrient supplied vegetated soil compared to the nutrient poor vegetated soil. This is because of the 160%–200% increase of LAI and 133% increase of RAI at 50 mm depth in nutrient supplied vegetated soil [7]. Larger root surface area can uptake more water [5] and larger leaf surface area has more stomata for water-transpiration in the atmosphere [4]. Moreover, 18–38 kPa suction increased on the 6th month in nutrient supplied vegetated soil compared to the suction measured on 3rd month due to 50%–70% increased LAI [7]. Increased soil suction can reduce soil permeability and increase soil shear strength [6].

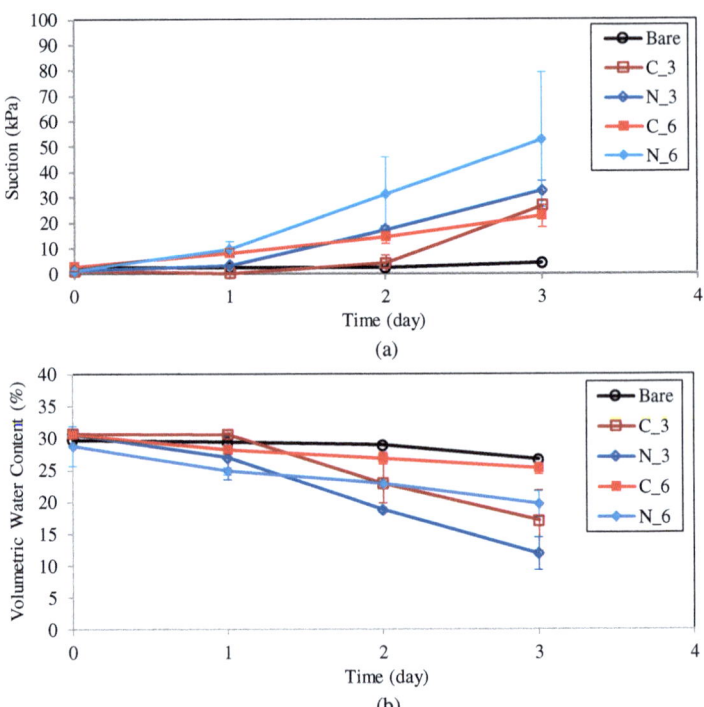

Fig. 2. Measured variations of (a) Suction and (b) VWC with time at 50 mm depth during drying of nutrient poor and nutrient supplied vegetated soil on the 3rd and 6th month of plant growth which are represented by 3 and 6, respectively. "C" represents the controlled tests without nutrient supply and "N" represents the tests with nutrient supply.

In Fig. 2-(b), during drying on the 3rd and 6th month, 3%–12% difference in VWC was observed between bare and vegetated soils and VWC decreased 6%–8% in nutrient supplied vegetated soil compared to the nutrient poor vegetated soil. Interestingly, VWC was higher in all vegetated soils during drying on the 6th month compared to the VWC on the 3rd month in contrast to the soil suction changes in Fig. 2-(a). This might be because of the biological factors such as the release of root exudates, organic matter or organic acid which takes place mainly within 2 mm from the root surface [8, 9]. These bio-chemical root activities and organic matters might have altered the soil pore structures which affected the VWC measurements in soil pores [9].

3   Conclusion

This study investigated the effects of NPK water soluble fertilizer on the soil suction and VWC changes during drying of heavily compacted (RC-95%) silty sand that were vegetated with *Schefflera heptaphylla* (Ivy tree) in a soil column.

During drying on the 6th month, suction increased by 15–40 kPa in the nutrient supplied vegetated soil compared to the nutrient poor vegetated soil due to increase of LAI and RAI. In contrast, soil VWC was higher in all vegetated soils during drying on the 6th month compared to the VWC during drying on the 3rd month due to changes in soil-pore structures through bio-chemical root activities and organic matters.

Acknowledgement. NSFC-51778166 and HKUST6/CRF/12R from the Research Grants Council of the Government of the HKSAR are acknowledged for this study.

References

1. Cazzuffi, D., Corneo, A., Crippa, E.: Slope stabilisation by perennial "gramineae" in southern Italy: plant growth and temporal performance. Geotech. Geol. Eng. **24**(3), 429–447 (2006)
2. GEO (Geotechnical Engineering Office): Technical guidelines on landscape treatment for slopes. Hong Kong, China (2011)
3. Hau, B.C., Corlett, R.T.: Factors affecting the early survival and growth of native tree seedlings planted on a degraded hillside grassland in Hong Kong, China. Restor. Ecol. **11**(4), 483–488 (2003)
4. Kelliher, F.M., Leuning, R., Raupach, M.R., Schulze, E.D.: Maximum conductance for evaporation from global vegetation types. Agric. For. Meteorol. **73**, 1–16 (1995)
5. McElrone, A.J., Choat, B., Gambetta, G.A., Brodersen, C.R.: Water uptake and transport in vascular plants. Nat. Educ. Knowl. **4**(5), 6 (2013)
6. Ng, C.W.W., Menzies, B.: Advanced Unsaturated Soil Mechanics and Engineering. Taylor and Francis, USA (2007)
7. Ng, C.W.W., Tasnim, R., Capobianco, V., Coo, J.L.: Influence of soil nutrients on plant characteristics and soil hydrological responses. Geotech. Lett. **8**(1) (2018, in press)
8. Sauer, D., Kuzyakov, Y., Stahr, K.: Spatial distribution of root exudates of five plant species as assessed by 14C labelling. J. Plant Nutr. Soil Sci. **169**(3), 360–362 (2006)
9. Traoré, O., Groleau-Renaud, V., Plantureux, S., Tubeileh, A., Boeuf-Tremblay, V.: Effect of root mucilage and modelled root exudates on soil structure. Eur. J. Soil Sci. **51**(4), 575–581 (2000)

Heat Transfer in a Geosynthetics Composite Liner System Containing Wrinkles

Mayu Tincopa[1(✉)], Abdelmalek Bouazza[1], and R. Kerry Rowe[2]

[1] Monash University, Melbourne, VIC 3168, Australia
mayu.tincopa@monash.edu
[2] Queen's University, Kingston, ON K7L 3N6, Canada

Abstract. Wrinkles commonly form in geomembrane-based liners system due to exposure to the fluctuations of solar radiation. Generally, air-gaps are present underneath the wrinkles in the geomembrane if devoid of defects. These air-gaps have low thermal conductivity, which may lead to a reduction of the temperature gradient across the liner including the subgrade soil. This paper presents a two-dimension numerical analysis of heat transfer for a geomembrane-geosynthetic clay liner composite liner system. The 2-D numerical analysis of heat transfer shows that the positive effect of the air-gap was minimised due to the lateral heat transfer. Two-dimensional finite element predictions have confirmed that the air-gaps may not significantly reduce the heat transfer towards the subgrade soil.

Keywords: Wrinkles · Geomembrane · Heat transfer

1 Introduction

Exposed geomembrane-based liners very often contain wrinkles due to the fluctuations of solar radiation [1–3]. Air-Gaps are present inside these wrinkles if the geomembrane (GMB) is devoid of defects [1]. Although wrinkles are unavoidable, GMB often is in an intimate contact with the material below it. As a result, the temperature along the composite liner and subsoil may vary at different locations within the liner system [2, 4]. However, very few studies have been conducted to quantify the influence of the presence of wrinkles on the temperature distribution along the subsoil in two-dimensional space numerically. This paper addresses this knowledge gap.

2 Method

A numerical calibration of a column test representing a GMB and geosynthetic clay liner (GCL) composite liner system was conducted based on the experimental test reported by [2]. Two numerical analyses were conducted. First, the laboratory condition was mimicked for different geometries to observe the effect of lateral heat migration. Second, the field condition reported by [1] was simulated based on the mean height and width of wrinkles. Three assumptions were made in the analysis: (1) Vapor and moisture transfer are neglected. (2) The thermal conductivity of the materials remains constant. (3) A perfect contact exists between GMB and GCL.

© Springer Nature Switzerland AG 2018
W. Wu and H.-S. Yu (Eds.): *Proceedings of China-Europe Conference on Geotechnical Engineering*, SSGG, pp. 1316–1319, 2018.
https://doi.org/10.1007/978-3-319-97115-5_91

2.1   Governing Equations

Comsol is a software which uses a finite element method and is based on solving differential equations was used to perform a non-linear analysis of heat transfer. The analysis was performed using the principle of conservation of energy and Fourier's law (Eq. 1).

$$\rho C_p \frac{\partial T}{\partial t} = \nabla \cdot (k \nabla T) \tag{1}$$

where ρ is the density [kg/m³], C_p is the specific heat capacity [J/(mK)], k is the thermal conductivity [W/(mK)] and T is the temperature [K]. This study examined only the steady state condition.

2.2   Conceptual Model

Figures 1a and b show the geometries and boundary conditions investigated as well as the discretisation adopted for modelling purposes. The air-gap was assumed to have a sinusoidal shape. The meshing size of the laboratory and field models had 17728 and 17248 triangular elements, and element area ratios equal to 0.011 and 0.0039, respectively.

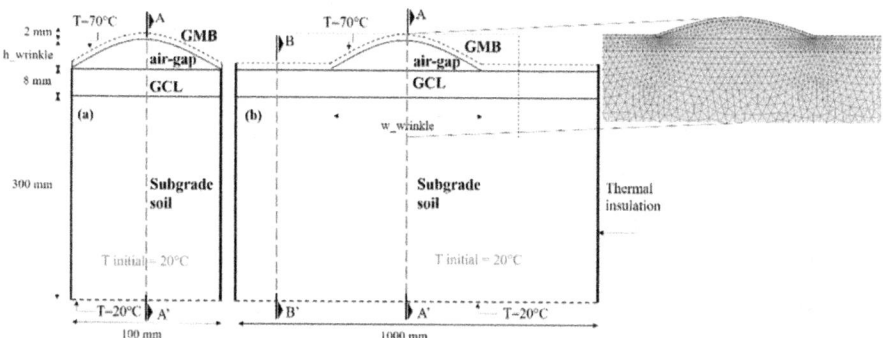

Fig. 1. Conceptual model of laboratory test (a) and field case (b).

Table 1. Material properties.

Properties	Units	GMB*	GCL*	Subgrade soil*	Air**
Thermal conductivity	w/(mK)	0.3	0.6	1.9	0.02
Specific heat capacity	J/(kgK)	1350	1515	1380	1005
Soil density	Kg/m³	800	1500	1620	1.6

*Reference [2] **Standard values

The conceptual model of the two cases consisted from top to bottom of four layers, namely: GMB, air, GCL and subgrade soil (Table 1). The temperature in each domain started at 20 °C as initial condition. The boundary conditions were imposed from data collected from the laboratory experiment reported by [1]. Two types of boundary conditions were considered, namely constant temperature and thermal insulation.

3   Results and Discussion

3.1   Calibration of the Model

Once the numerical model was built, based on the experimental conditions and the material properties, a parametric analysis was conducted to quantify the air-gap height numerically. A temperature profile at the end of the laboratory test was compared to the numerical analyses as shown in Fig. 2a. An air-gap height of 15 mm was identified as the optimum height following this comparison process.

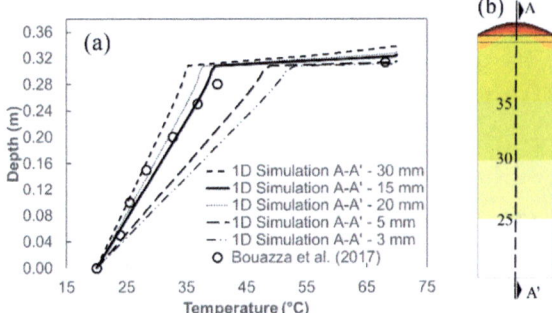

Fig. 2.   Cross section A-A' in the middle of the model (a) and a numerical model (b).

3.2   Effect of Wrinkle Size 1D

The attenuation of temperature due to the presence of an air-gap can be observed in Fig. 2b, which indicates that the air-gap under a wrinkle acted as an insulating layer. For example, a drop in temperature from 70 to 40 °C between the GMB and the subgrade occurred when the air-gap height reached 15 mm. Although the analysis was performed in 2-D, the closeness of the lateral boundary condition resulted in a simple 1D analysis.

3.3   Effect of Wrinkle Size 2D

As noted above, the 1D analysis led to an incomplete study of the effect of the air-gap due to heat transfer occurring in the horizontal direction. However, a numerical analysis can address this limitation of the laboratory test by examining alternative scenarios which are closer to field conditions. In this respect, two cases were examined. The first case consisted of modelling the presence of an air-gap height (15 mm) considering lateral heat effect (2D). The second case consisted of modelling the average height and width commonly reported in practice, i.e. 60 mm and 200 mm, respectively.

Figures 3b and c show that the temperature dropped significantly below the air-gap even though the effect of the lateral sides dissipated it. Figure 3a gives further evidence of the changes occurring across the liners. Air-gaps tend to increase the temperature gradient between the GMB and subgrade soil. However, the horizontal heat minimised such effect. For example, the laboratory case suggested that the temperature gradient dropped by 15 °C from 1D to 2D, while the field case revealed that the temperature gradient increased from 15 °C under 15 mm to 20 °C under 60 mm air-gap in 2D.

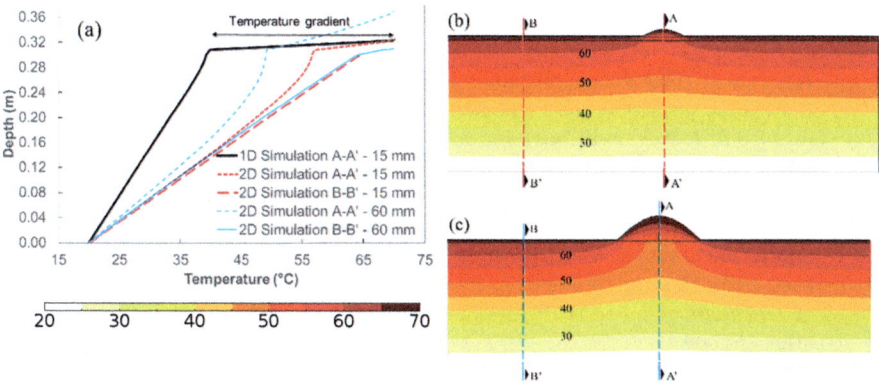

Fig. 3. (a) Isothermal temperature across the GMB, GCL and subgrade soil for (b) laboratory case 2D and (c) field case at 2D.

4   Conclusion

The modelling process captured the 2-D space effect of the wrinkle. Furthermore, the findings suggest that the air-gap acted as an insulating layer, which could reduce the temperature in the subgrade soil. However, horizontal heat flow minimised such favourable effect.

References

1. Rowe, R., Chappel, M., Brachman, R., Take, W.: Field study of wrinkles in a geomembrane at a composite liner test site. Can. Geotech. J. **49**(10), 1196–1211 (2012)
2. Bouazza, A., Ali, M., Rowe, R.K., Gates, W., El-Zein, A.: Heat mitigation in geosynthetic clay liners exposed to elevated temperatures. Geotext. Geomembr. **45**(5), 406–417 (2017)
3. Take, W.A., Watson, E., Brachman, R.W.I., Rowe, R.K.: Thermal expansion and contraction of geomembrane liners subjected to solar exposure and backfilling. J. Geotech. Geoenviron. Eng. **138**(11), 1387–1397 (2012)
4. Bouazza, A., Singh, R.M., Rowe, R.K., Gassner, F.: Heat and moisture migration in a geomembrane-GCL composite liner subjected to high temperatures and low vertical stresses. Geotext. Geomembr. **42**(5), 555–563 (2014)

Root Reinforcement to Stabilize Slopes

Rick Veenhof$^{(\boxtimes)}$ and Wei Wu

Institute of Geotechnical Engineering,
University of Natural Resources and Life Sciences, Vienna,
Feistmantelstraße 4, 1180 Vienna, Austria
{rick.veenhof,wei.wu}@boku.ac.at

Abstract. This study describes the effects of grass reinforced slopes. Centrifuge tests were performed on sandy slopes of 55° with a void ratio (e) of 0.9 (loose conditions) and a initial water content of 11.5%. Three different types of grasses were used as a root reinforcement, namely *Lolium perenne*, *Hordeum vulgare* and *Festuca arundinacea*. Centrifuge tests show that for slopes that are reinforced the period until failure is extended. The *Hordeum vulgare* reinforced slope performed best as no failure took place, likely this was due to the root system of *Hordeum vulgare* that developed more extensively compared to the other grasses.

1 Introduction

Mitigating shallow slope instability through vegetation has been proven to be an effective and cost decreasing method [2,4]. Early research on the effects of vegetation on the stability of slopes was done by [9,11]. They concluded that the vegetation stabilizes the upper layer of the slope through their roots, by contributing to the shear strength of the soil-root system. A prerequisite is that the roots should cross the slip surface to stabilize the slope [3]. This shallow depth is approximately 1–2 m deep [3,4]. Centrifuge modelling is an important method in revealing mechanisms of soil deformations and failure mechanisms, especially those which correspond to gravity and stress level [7]. Studies that tried to mimic the effects of real roots on slope stability in the centrifuge were [5] with 3D-printed plastic roots and [8] with polypropylene fibres. Whereas, for the real roots [1] planted Oat and [6] used three different species, namely Willow, Gorse and *Festulolium* grass to improve the stability. Both the artificial as the real roots studies state that improvement in stability is made due to the roots inclusion. Researchers have focused mostly on coarse roots to improve slope stability, less on the thinner roots that act as tensile elements within the soil matrix. Therefore, the aim of this study is to examine the stability of sandy slopes that are reinforced with different grass species on the basis of self weight-initiation in the centrifuge.

© Springer Nature Switzerland AG 2018
W. Wu and H.-S. Yu (Eds.): *Proceedings of China-Europe Conference on Geotechnical Engineering*, SSGG, pp. 1320–1323, 2018.
https://doi.org/10.1007/978-3-319-97115-5_92

2   Methods

The experiments are done using the beam-type centrifuge at the Institute of Geotechnical Engineering in Vienna. For a detailed background of the centrifuge, how the slope was created and the sand characteristics see [8]. To study the stability of root reinforced slopes in comparison to unreinforced slopes, 6 different models are tested. Slopes were built with a void ratio (e) of 0.9 (loose conditions), an initial water content of 11.5% and a slope angle of 55°, see Fig. 1. For each root reinforced slope a non-reinforced slope was built simultaneously and subjected to the same conditions. To reinforced the slopes, three different grasses have been used, namely *Lolium perenne*, *Hordeum vulgare* and *Festuca arundinacea*. In case of *Lolium perenne* the slope was planted with approximately 100 seeds (0.2 g). In case of *Hordeum vulgare* and *Festuca arundinacea* the seeds (both 70 g), were mixed with sand to create a reinforced layer of 20 mm along the slope, just like [8] did with the polypropylene fibres reinforcements. To water the seeds and prevent the top layer from drying out, 50 ml of water was uniformly sprayed on a day to day basis. A LED lamp with a timer of 6 h was installed above the slopes to mimic daylight. GeoPIV is used to analyse the digital images that were made with a digital camera close to the slope, to extract pre-failure slope deformation patterns [10].

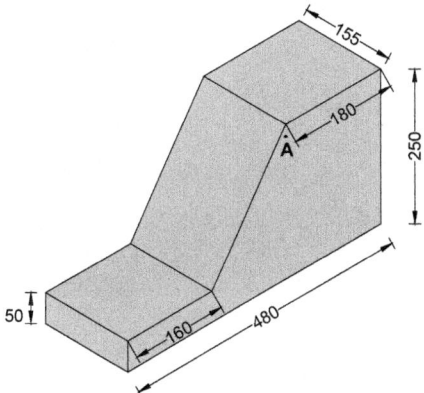

Fig. 1. Geometry of the slope [mm] and A is the point of total displacement

3   Results

Table 1 gives an overview of the centrifuge results. It shows the rpm (revolutions per minute), the g-level values and the total displacement at point A (Fig. 1) just before failure for each of the un-reinforced and reinforced slopes. Additionally, the length of the roots is recorded. The slopes that were reinforced with grasses, performed better. For all reinforced slopes an increase in g-level is observed, compared to their corresponding unreinforced situation. The with *Hordeum vulgare* reinforced slope did not fail. Figure 2 shows the failed *Lolium perenne* reinforced slope (Fig. 2a) and the displaced (but not failed) *Hordeum vulgare* reinforced slope (Fig. 2b).

Table 1. Root reinforced centrifuge tests

Grass species	Reinforced [−]	rpm [−]	g-level [−]	Total displ. [mm]	Root length [cm]
Lolium perenne	no	189.2	49.6	3.8	-
	yes	197.6	53.7	3.9	8.0
Hordeum vulgare	no	173.6	41.7	3.0	-
	yes	268.3	99.6	14.2	10.0
Festuca arundinacea	no	175.1	42.4	3.5	-
	yes	209.5	60.6	7.1	4.0

(a) *Lolium perenne*   (b) *Hordeum vulgare*

Fig. 2. Displacement of the root reinforced slopes

4   Conclusion

This study shows the effects of adding grasses as a root reinforcement on sandy slopes under loose conditions. Centrifuge tests were conducted to investigate the effect of grasses on slope failure. Reinforced slopes outperform unreinforced slopes, as the moment until failure is extended. The test which performed best was the slope reinforced with *Hordeum vulgare* as it did not fail. Further research is needed to determine the optimal design of grass reinforced slopes.

Acknowledgements. The research carried out in this paper is partly funded by the project 'GEORAMP' within the RISE programme of Horizon 2020 under grant number 645665. The first author would like to acknowledge the financial support from the Otto Pregl Foundation for Fundamental Geotechnical Research in Vienna.

References

1. Askarinejad, A., Springman, S.M.: Centrifuge modelling of the effects of vegetation on the responses of a silty sand slope subjected to rainfall. In: (IACMAG), Kyoto, Japan, pp. 1339–1344 (2015)
2. Chok, Y., Kaggwa, G., Jaksa, M., Griffiths, D.: Modelling the effects of vegetation on stability of slopes. To the ENZ of the Earth, vol. 1 (2004)
3. Genet, M., Stokes, A., Fourcaud, T., Norris, J.E.: The influence of plant diversity on slope stability in a moist evergreen deciduous forest. Ecol. Eng. **36**(3), 265–275 (2010)
4. Greenwood, J.R., Norris, J.E., Wint, J.: Assessing the contribution of vegetation to slope stability. Proceed. ICE-Geotech. Eng. **157**, 199–207 (2004)
5. Liang, T., Knappett, J.A.: Centrifuge modelling of the influence of slope height on the seismic performance of rooted slopes. Geotechnique **67**, 855–869 (2017)
6. Liang, T., Bengough, A.G., Knappett, J.A., MuirWood, D., Loades, K.W., Hallett, P.D., Boldrin, D., Leung, A.K., Meijer, G.J.: Scaling of the reinforcement of soil slopes by living plants in a geotechnical centrifuge. Ecol. Eng. **109**, 207–227 (2017)
7. Schofield, A.N.: Cambridge geotechnical centrifuge operations. Geotechnique **30**, 227–268 (1980)
8. Veenhof, R., Wu, W.: An experimental study on fibre reinforced slopes by centrifuge tests (2018)
9. Waldron, L.J., Dakessian, S.: Effect of grass, legume, and tree roots on soil shearing resistance. Soil Sci. Soc. Am. J. **46**, 894–899 (1982)
10. White, D.J., Take, W.A., Bolton, M.D.: Soil deformation measurement using particle image velocimetry (PIV) and photogrammetry. Geotechnique **53**, 619–631 (2003)
11. Ziemer, R.R.: The role of vegetation in the stability of forested slopes. In: Proc. First Union of For. Res. Org., Div. I, XVII World Congress, Kyoto, Japan (1982)

Engineering Properties and Mechanism of Solidifying Materials Treated Sludge

Zhenhua Wang[1], Wei Xiang[1,2(✉)], Qingbing Liu[2], and Xueting Wu[1]

[1] China University of Geosciences, Wuhan 430074, China
xiangwei@cug.edu.cn
[2] Three Gorges Research Center for Geo-Hazards, Ministry of Education,
China University of Geosciences, Wuhan 430074, China

Abstract. The treatment of organic-rich sludge is significant in geotechnical engineering, which contributes to the resource utilization of sludge and environmental protection. In this paper, several solidifying materials, namely cement, quick lime, potassium ferrate, and super absorbent polymers are selected to treat sludge. Experimental results show that the unconfined compressive strength of treated soil samples is significantly enhanced, and the heavy metal ions concentration measured from the exudation fluid of soil samples is reduced by treatments. Furthermore, the treatment mechanism is investigated by using X-ray diffraction and Scanning Electron Microscopy at the micro scale. The mixing of solidifying materials into sludge leads to the formation of cementitious compounds and the flocculation of clay particles.

Keywords: Sludge · Solidifying materials · Unconfined compressive strength
Heavy metal ions · Mechanism

1 Introduction

Subgrade construction in coastal areas often encounters organic-rich sludge that is difficult to utilize directly due to its low strength and high compressibility [1, 2]. In recent years, chemical stabilization with binders [1, 3, 4] such as cement, lime, and fly ash can be undertaken rapidly to improve the engineering properties of sludge. However, traditional binders are difficult to solve the problem of the contaminants in sludge effectively. In this study, cement, quick lime, potassium ferrate [5] and super absorbent polymers [6] are used as solidifying materials and mixed with sludge to increase the strength and immobilize possible contaminants. Thus, unconfined compressive strength test and leaching test have been performed to know the effect of solidifying materials. Further, X-ray diffraction and Scanning Electron Microscopy have been carried out to elucidate the mechanism of strength variation.

© Springer Nature Switzerland AG 2018
W. Wu and H.-S. Yu (Eds.): *Proceedings of China-Europe Conference on Geotechnical Engineering*, SSGG, pp. 1324–1327, 2018.
https://doi.org/10.1007/978-3-319-97115-5_93

2   Materials and Methods

2.1   Sludge and Samples

The sludge used in this study originated from Ningde in southeast China. The sludge was collected by open excavation from a depth of approximately 1 m below the natural ground level. The geotechnical properties of the sludge are presented in Table 1

Table 1. Geotechnical properties of sludge

Material	Symbol	ω (%)	γ (kN/m^3)	Gs	LL (%)	PL (%)	PI	SOC (%)	pH
Sludge	S-0	66.2	13.8	2.73	55.6	28.1	28.5	8.18	4.7

In this study, four different solidifying materials, cement, quick lime, potassium ferrate and super absorbent polymers were tested. Amounts of solidifying materials used to prepare mixtures of different combinations were taken on the basis of percentage by dry weight of the sludge. The samples were mixed with different solidifying materials singly or multiply in the order of cement 6% (C-6), lime 2% (CaO-2), potassium ferrate 0.5%, super absorbent polymers 2% (SAP-2) and all reagents in a certain proportion (CCPS).

2.2   Mechanical Test

In order to determine the strength characteristics of the samples, unconfined compression tests (UCS) were conducted in accordance with ASTM D2166. Before the test, the samples were sealed and cured for 7 days in the humidity room at a constant humidity and temperature condition.

2.3   Leaching Test

In the experiment, the samples were loaded into a pressure chamber permeated for 24 h under a certain seepage pressure. Then, the exudation fluid was filtered with a 0.2 m glass fiber filter. After filtration, the exudation fluid was analyzed by ICP-OES.

2.4   Analysis

The crystallographic composition of the samples was identified by X-ray diffraction (X'Pert PRO DY2198) with the steps of 0.01° at 4° min −1 in 2 ranging from 5° to 65°. The morphology of the samples was observed under a scanning electron microscopic (FE-SEM).

3 Results

3.1 Unconfined Compression Test

Figure 1 shows the relationships between the unconfined compression strength (UCS) values and additive content. The unconfined compressive strength of sludge increased significantly with addition of CCPS from 0.455 MPa to 4.098 MPa. Besides, the overall effectiveness of the treatment under the curing was in the order of CaO-2, C-6, SAP-2, and PF-0.5.

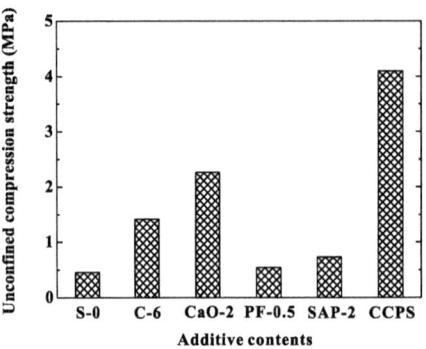 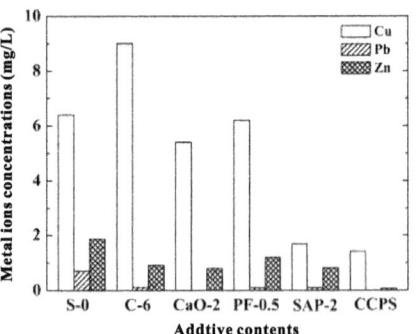

Fig. 1. Unconfined compressive strength of different samples

Fig. 2. Heavy metal ions concentrations in exudation fluid of different samples

3.2 Heavy Metals Analysis

Figure 2 shows heavy metal ions concentrations in exudation fluid of different samples. All solidifying materials decreased the contents of heavy metal ions Cu(II), Pb(II) and Zn(II) in varying degrees. The samples, SAP-2 and CCPS, mixed with super absorbent polymers had a great absorbability to the three heavy metal ions.

3.3 Mechanism Analysis

The XRD patterns of different samples are shown in Fig. 3. The sample, CCPS, revealed several new peaks clearly, which indicated the presence of ettringite and calcium silicate hydrate (C-S-H). However, other samples, only mixed with one material, had few changes in mineral composition.

Figure 4 presents SEM image of the sample CCPS. The SEM image revealed two significant reasons for the enhancement in strength of the sludge treated by all reagents. It was clearly observed ettringite and calcium silicate hydrate (C-S-H) filled in pores of the sludge. Besides, clay particles flocculated into larger size clusters.

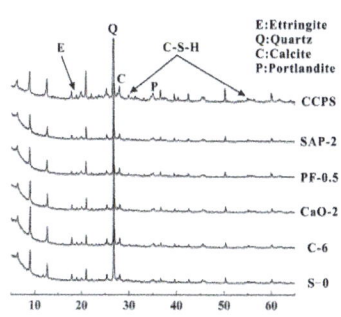

Fig. 3. X-ray diffraction patterns of different samples

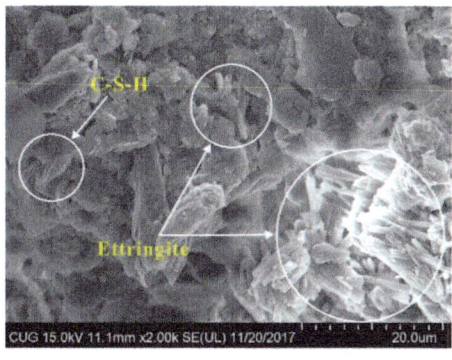

Fig. 4. SEM image of the sample CCPS

4   Conclusions

The foregoing discussion shows that the various facets of the engineering properties of sludge treated by different solidifying materials can be explained by the interplay of a few underlying mechanisms. Potassium ferrate and super absorbent polymers are propitious to hydration reaction and pozzolanic reaction. It promotes the formation of cementitious compounds. And clay particles flocculate into larger size clusters. Thus, the strength of sludge mixed with all reagents can reach 4.098 MPa. Besides, super absorbent polymers have an ability to absorb heavy metals. It can effectively reduce the concentrations of Cu(II), Pb(II) and Zn(II) in the exudation fluid of stabilized sludge.

References

1. Tastan, E.O., Edil, T.B., Benson, C.H., et al.: Stabilization of organic soils with fly ash. J. Geotech. Geoenviron. Eng. **137**(9), 819–833 (2011)
2. Tremblay, H., Duchesne, J., Locat, J., et al.: Influence of the nature of organic compounds on fine soil stabilization. Can. Geotech. J. **39**(3), 535–546 (2002)
3. Hampton, M.B., Edil, T.B.: Strength gain of organic ground with cement-type binders. Geotech. Spec. Publ. **81**, 135–148 (1998)
4. Bell, F.G.: Lime stabilization of clay minerals and soils. Eng. Geol. **42**(4), 223–237 (1996)
5. Jiang, J.Q.: Research progress in the use of ferrate(VI) for the environmental remediation. J. Hazard. Mater. **146**(3), 617–623 (2007)
6. Kesenci, K., Say, R., Denizli, A.: Removal of heavy metal ions from water by using poly beads. Eur. Polymer J. **38**(7), 1443–1448 (2002)

Experimental Study on Effects of NaCl Solutions on Soil-Water Characteristic Curves of Expansive Soil

Xiujuan Yang[1], Wojciech T. Sołowski[2(✉)], Henghui Fan[1], and Jinqian Dang[1]

[1] College of Water Resources and Architectural Engineering, Northwest A&F University, Yangling 712100, China
[2] Department of Civil and Environmental Engineering, Aalto University School of Engineering, 02150 Espoo, Finland
wojciech.solowski@aalto.fi

Abstract. Soil-water characteristic curve (SWCC) is an important characteristic of unsaturated soil. This paper investigates the total suction of the MX-80 bentonite, under different saturation and different concentrations of sodium chloride solutions. To create the soil-water characteristic curves for different amounts of NaCl, but same dry densities, the study used chilled-mirror dewpoint WP4 potentiometer. The experimental results show that the concentration of salt in pore water has significant influence on the bentonite. With the increase of the NaCl concentration, the samples total suction increased gradually and the matric suction decreased. When the soil saturation was greater than 50%, the pore solution's addition led to the rapid increase of the soil's total suction. The low concentration of salt has relatively little effect on the matric suction of the MX-80 bentonite; while when the ion concentration is high, the solutions' matric suction has a very significant impact on the soil matric suction.

Keywords: Expansive soil · Soil-water characteristic curve
Suction · Partially saturated soils with saline solutions
Chilled-mirror dew-point potentiometer

1 Introduction

Compacted expansive clays can create almost impermeable barriers, preventing transport of contaminants to the environment. The saturation process of those barriers is complex due to hydration involving water in both liquid and gas phases. The process is also influenced by the changing pore fluid salinity [1] due to water evaporation and transport in the gas phase, leading to possibly high concentration of salts in pore fluid. Earlier studies have shown that, among others, soil dry density, soil structure, pore water salt composition and concentration and exchangeable cations in the clay minerals all affect the soil-water retention curves (SWRC) e.g. [2, 3]. Number of investigated e.g. [4–6] investigated the effects of concentrations of the sodium chloride (NaCl), calcium chloride ($CaCl_2$) and potassium chloride (KCl). Those studies examined

© Springer Nature Switzerland AG 2018
W. Wu and H.-S. Yu (Eds.): *Proceedings of China-Europe Conference on Geotechnical Engineering*, SSGG, pp. 1328–1331, 2018.
https://doi.org/10.1007/978-3-319-97115-5_94

mainly low salt concentrations, whereas this study investigates higher salt concentrations.

2   Materials and Methods

This study concentrates on water retention of MX-80 sodium bentonite saturated with distilled water, as well as 1 mol/L, 2 mol/L, and 5 mol/L of NaCl solutions. In order to prepare the cylindrical samples (30 mm diameter and 5 mm height), the MX-80 bentonite was mixed with such amounts of the prepared NaCl solutions that after compaction, at the target dry density (1.60 g/cm3) the samples reached the target degree of saturation (10-100%). The samples dimensions and mass were measured after compaction in the mould. Based on the measurements, the volume, saturation and dry density of the samples was calculated. A chilled-mirror dew point hygrometer device (WP4C by Decagon Devices) measured the total suction of the prepared soil samples.

3   Results and Discussion

The powder and compacted samples were both measured using the WP4C chilled-mirror dew-point hygrometer technique (Fig. 1). The total measured suction fell within

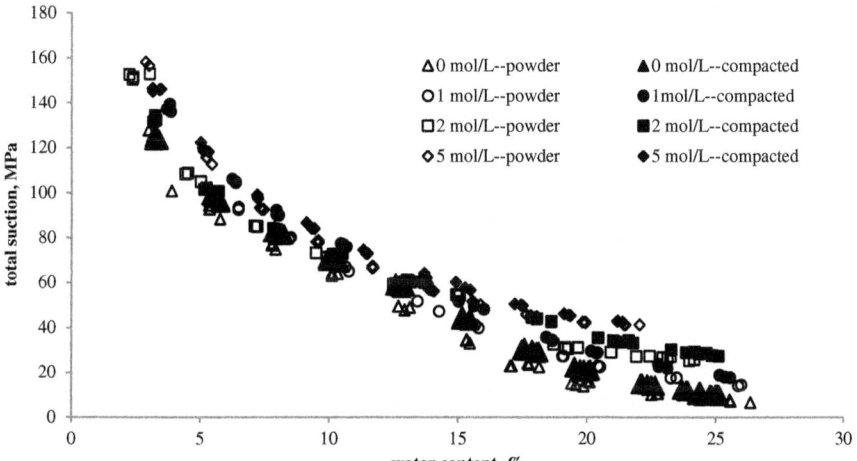

Fig. 1. The total suction of the powder and compacted bentonite samples.

range of 6 to 158 MPa. As expected, the bentonite powder total suction was only slightly smaller than the compacted samples, with respect to water content. As such, if the soil total suction must be roughly estimated on a very tight time schedule, the powder with the right water content may be used in place of the compacted samples.

Figure 1 shows the compacted specimens total suction, which varies from 10.9 to 145.7 MPa. In general, the trend shows that the total suction increases with the increase of the pore fluid salinity, similarly to findings of Thyagaraj and Salini [7] and He et al. [8]. However, somewhat unexpectedly, the increases in salt concentration in the saturating liquid in high range of suctions does not always lead to increase in suction – that is the case for 1 M and 2 M solutions.

The bentonite's matric suction can be theoretically calculated by the measured total suction and the solutions osmotic suction. The osmotic suctions were 4.66 MPa, 10.43

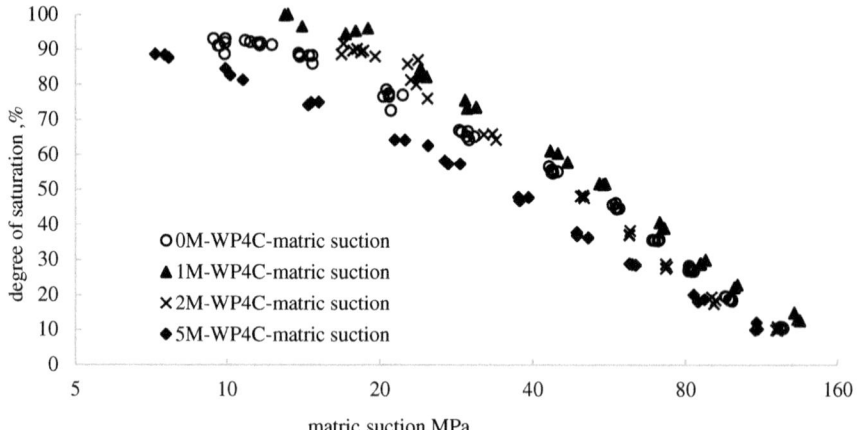

Fig. 2. The matric suction of the compacted bentonite in the WP4C.

MPa and 38.9 MPa for the 1M, 2M and 5M sodium chloride solutions, respectively. Under the rough approximation that the osmotic suction is the same as the initial osmotic suction of saturating liquid, the matric suction of the compacted bentonite can be recovered (Fig. 2).

The soil suction characteristic curves for bentonite seem to be possible to describe with very simple linear fitting. For example, assuming a linear logarithmic relationship between total suction and soil saturation:

$$s = a\ln(Sr) + b \tag{1}$$

where, Sr is the saturation of soil, %; s is the total suction of soil, MPa; a, b are fit-ting parameters, with very good correlation which is from 0.982 to 0.994.

4   Conclusions

Test results confirmed that soil water retention behaviour is influenced by the salt concentration in the pore fluid. In addition, the powdered bentonite and the compacted bentonite had the similar total suction with the same moisture content, though the suction of powder was always lower than the suction in compacted samples. The paper also proposes a very simple linear fitting of the water- soil characteristic curve which may be sometimes used instead of the more advanced formulations.

Acknowledgement. This research is financially supported by National Natural Science Foundation of China (51409217, 51579215), the China National Scholarship and KYT2018 programme via the THEBES project.

References

1. Villar, M.V.: Infiltration tests on a granite/bentonite mixture: influence of water salinity. Appl. Clay Sci. **31**, 96–109 (2006)
2. Vanapalli, S.K., Fredlund, D.G., Pufahl, D.E.: The influence of soil structure and stress history on soil water characteristics of a compacted till. Géotechnique **49**, 143–159 (1999)
3. Sun, D., Sun, W., Yam, W., Li, J.: Hydro-mechanical behaviors of highly compacted sand-bentonite mixture. Geotech. Geol. Eng. **2**, 79–85 (2010)
4. Leong, E.C., Tripathy, S., Rahardjo, H.: Total suction measurement of unsaturated soils with a device using the chilled mirror dew-point technique. Géotechnique **53**(2), 173–182 (2003)
5. Ravi, K., Rao, M.S.: Influence of infiltration of sodium chloride solutions on SWCC of compacted bentonite-sand specimens. Geotech. Geol. Eng. **31**, 1291–1303 (2013)
6. Ye, W.M., Zhang, F., Chen, B., Chen, Y.G., Wang, Q., Cui, Y.J.: Effects of salt solutions on the hydro-mechanical behavior of compacted GMZ01 bentonite. Environ. Earth Sci. **72**(7), 2621–2630 (2014)
7. Thyagaraj, T., Salini, U.: Effect of pore fluid osmotic suction on matric and total suctions of compacted clay. Géotechnique **65**(11), 952–960 (2015)
8. He, Y., Chen, Y.G., Ye, W.M., Chen, B., Cui, Y.J.: Influence of salt concentration on volume shrinkage and water retention characteristics of compacted GMZ bentonite. Environ. Earth Sci. **75**, 535 (2016)

Part X: Cold Regions Geotechnical Engineering

A Double-Yield-Surface Model for Frozen Saline Sandy Soil Incorporating Particle Crushing

Dan Chang[1,2(⊠)] and Yuanming Lai[1,3]

[1] State Key Laboratory of Frozen Soil Engineering,
Northwest Institute of Eco-Environment and Resources,
Chinese Academy of Sciences, Lanzhou 730000, Gansu, China
dchang@lzb.ac.cn
[2] School of Civil Engineering, Beijing Jiaotong University,
Beijing 100044, China
[3] University of Chinese Academy of Sciences, Beijing 100049, China

Abstract. In order to investigate the mechanical properties of frozen saline sandy soil, a series of triaxial compression tests were conducted on frozen saline sandy samples with Na_2SO_4 contents of 0%, 0.5%, 1.5%, 2.5% and 3.5% by weight at the temperature of $-6\,°C$. The strain softening/hardening, shear contraction and dilation properties as well as pressure melting of ice and particle crushing under high pressure are studied. The experimental results indicate that strain softening/hardening phenomena occur when the confining pressures are below and above 6 MPa, respectively, together with high dilation characteristics. The strength increases to a peak value and then decreases with the increasing confining pressure. The grain size distribution varies greatly after triaxial compression test compared with the original one, indicating that particle crushing occurs during the test. Thus, a double-yield-surface elastoplastic constitutive model is developed by employing the non-associated flow rule. The proposed model is verified by comparing the simulation results with the experimental data. It is found that the stress-strain relation and volumetric deformation can be predicted well under both low and high confining pressures.

Keywords: Frozen saline sandy soil · Constitutive model · Pressure melting
Particle crushing

1 Introduction

Frozen saline sandy soil is composed of solid mineral particles, ice inclusions, salt crystals, liquid water and gaseous inclusions. The mechanical properties of frozen saline sandy soil are very different from the general frozen soil without salt crystals. The constitutive model of general frozen soil has been frequently investigated in the literature. An increase in salinity considerably influences the mechanical properties of frozen saline soil. Besides, the particle crushing commonly occurs when granular materials undergo compression and shearing under high stresses. Thus, the purpose of this study is to develop an elastoplastic constitutive model considering the effect of particle crushing.

© Springer Nature Switzerland AG 2018
W. Wu and H.-S. Yu (Eds.): *Proceedings of China-Europe Conference on Geotechnical Engineering*, SSGG, pp. 1335–1339, 2018.
https://doi.org/10.1007/978-3-319-97115-5_95

2 Test Conditions and Results

2.1 Test Conditions

The sandy soil, as a kind of coarse sand, was adopted to make up the saline sandy samples with the salt contents of 0%, 0.5%, 1.5%, 2.5% and 3.5% by weight. The specimens were prepared as cylinders with the height of 125 mm and diameter of 61.8 mm. The confining pressure was maintained constant during the test and chosen from 0 MPa to 16 MPa, while the axial strain rate was controlled as 1.67×10^{-4} s^{-1} during shearing process. The grain size distribution of the specimens after triaxial compression test can be obtained by performing the sieve analysis.

2.2 Test Results

Figure 1 shows the results of stress-strain and volumetric strain curves from the tests. The samples exhibit strain softening when the confining pressure is lower than 6 MPa, while strain hardening occurs when the confining pressure is higher. Besides, the frozen saline sandy samples exhibit extremely dilation features during shearing process. Figure 2 shows the variation of relative breakage [1, 2] versus confining pressure at the salt content of 0%. The relative breakage shows obviously increasing tendency with the increase of confining pressure, indicating that considerable particle crushing is identified, and the growth rate slows down afterwards.

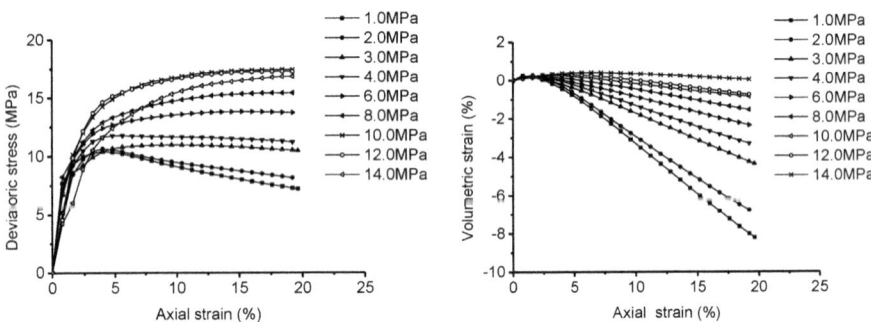

Fig. 1. Test results of samples with the salt content of 2.5%

3 Double-Yield-Surface Model for Frozen Saline Sandy Soil

3.1 Elastic Parameters

The elastic shear modulus is obtained from the loading-unloading-reloading triaxial compression test with the axial strain rate of 1.67×10^{-4} s^{-1}. The elastic volumetric

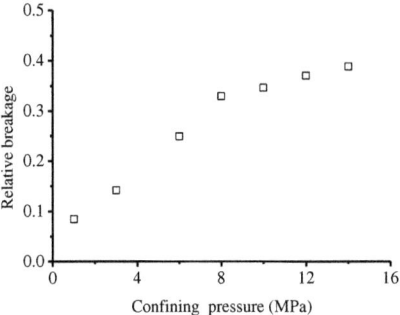

Fig. 2. Relation between relative breakage [1] and confining pressure of frozen saline sandy soil with the salt content of 0%

modulus can be obtained from the shear modulus and initial Poisson's ratio. From the test results, the shear modulus and volumetric modulus are expressed as:

$$G = (1 - \theta B)\left\{ G_0 + m_g[1 - \exp(-\frac{\sigma_c/p_a}{t_g})]\right\} \tag{1}$$

$$K = (1 - \theta B)m_k p_a \exp(n_k.\sigma_c/p_a) \tag{2}$$

where G_0, m_g, t_g, m_k and n_k are material parameters of frozen saline sandy soil related to salt content; p_a is atmospheric pressure and equal to 0.10133 MPa; B represents relative particle breakage; θ is related to the grain size distribution.

3.2   Plastic Mechanisms

The plastic compressive mechanism and plastic shear mechanism are considered to calculate the soil plastic deformation. Under plastic compressive mechanism, the yield function and potential function are supposed as:

$$f_v = \left[\frac{p - \vartheta\chi(p)p_0 h_v}{(1 - \vartheta)\chi(p).p_0 h_v}\right]^2 + \left[\frac{q}{\beta\chi(p).p_0 h_v}\right]^2 - 1 = 0 \tag{3}$$

$$g_v = \left[\frac{p - \vartheta\chi(p)p_0 h_v}{(1 - \vartheta)\chi(p).p_0 h_v}\right]^2 + \left[\frac{q}{\beta\chi(p).p_0 h_v + \alpha p}\right]^2 - 1 = 0 \tag{4}$$

where ϑ and β are two parameters which control the position and shape of the yield surface; $\chi(p)=\exp(-m/p_a<p-p_b>)$, p_b is the critical pressure of particle crushing; m is a constant describing the degree of particle crushing with increasing pressure; α is a material parameter.

Under plastic shear mechanism, the yield function and corresponding potential function are given as:

$$f_s = q^2 - [M_0 - \kappa(p)].p_a h_s.(p + A_0) = 0 \tag{5}$$

$$g_s = k_s q^2 - [M_0 - \kappa_s(p)].p_a h_s.(p + A_0) = 0 \tag{6}$$

where $\kappa(p) = m/p_a <p-p_b>$; M_0 is the initial slope of the critical state line in p-q plane; A_0 is the intercept of the critical line; k_s is a material parameter; $\kappa_s = m_s/p_a <p-p_b>$.

3.3   Comparison with Experimental Results

The comparisons between the simulation results and the experimental data for samples with the salt content of 0% are shown in Fig. 3. The proposed model can reflect the strain softening and high dilation properties under low confining pressures, as well as strain hardening and shrinkage characteristics under high confining pressures. The particle crushing has significant influence on stress-strain and volumetric strain properties.

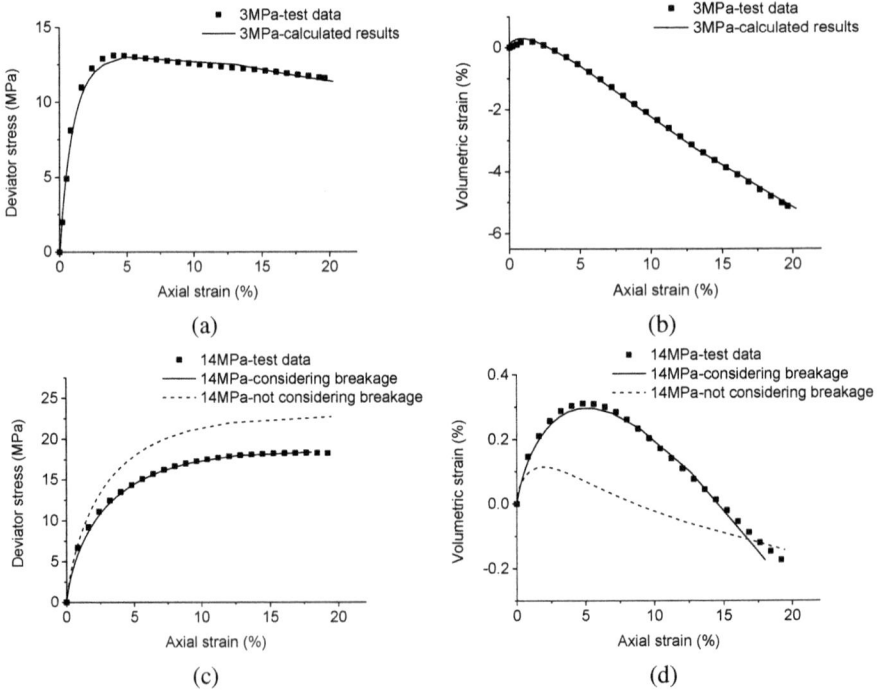

Fig. 3. Comparisons of stress-strain and volumetric strain relations between simulation results and experimental data

4 Conclusions

In this paper, a series of triaxial compression tests are conducted for frozen saline sandy soil. From the test results, plenty of particle crushing is identified. A double-yield-surface constitutive model is developed by employing the non-associated flow rule. Compared with the experimental data, the proposed model can simulate the deformation properties well.

References

1. Einav, I.: Breakage mechanics—part I: theory. J. Mech. Phys. Solids **55**(6), 1274–1297 (2007)
2. Einav, I.: Breakage mechanics—part II: modelling granular materials. J. Mech. Phys. Solids **55**(6), 1298–1320 (2007)

A New Strength Criterion for Frozen Clay Considering Temperature Effect

Dun Chen[1,2] (iD), Wei Ma[1(✉)] (iD), Yanhu Mu[1] (iD), Zhiwei Zhou[1] (iD),
Dayan Wang[1], and Lele Lei[1,2]

[1] State Key Laboratory of Frozen Soil Engineering, Cold and Arid Regions
Environmental and Engineering Research Institute, Chinese Academy of
Sciences, Lanzhou 730000, China
mawei@lzb.ac.cn
[2] University of Chinese Academy of Sciences, Beijing 100049, China

Abstract. To study the strength property of frozen soil under complex stress states, a series of directional shear tests on remoulded frozen clay were conducted under four mean principal stresses (p = 1, 3, 4.5 and 10 MPa) and four coefficients of intermediate principal stress (b = 0, 0.25, 0.5, and 0.75) at three temperatures (-6, -10, and -15 °C) using the hollow cylinder apparatus (HCA). The experimental results indicated that stress-strain curves of frozen clay all performs as strain hardening under directional shearing. In the p-q plane, the strength of frozen clay increases with increasing mean principal stress at first, and then decreases with a further increase of p. An elliptic strength criterion is proposed to describe this variation law. The test results also showed that the strength of frozen clay increases with decrease of temperature significantly. Then, temperature parameters are introduced into the elliptic function to consider the temperature effect.

Keywords: Strength criterion · Complex stress states · Temperature effect
Hollow cylinder apparatus

1 Introduction

Comparing with unfrozen soils, the mechanical properties of frozen soil are close related to the confining pressure and the temperature. Some researchers indicated that the Mohr-Coulomb criterion is suitable for frozen soil under a low confining pressure [1]. While under a high confining pressure, it was found that the strength of frozen soil changed nonlinearly with the increasing confining pressure. Based on triaxial compression results, Fish and Ma [1, 2] suggested a parabolic strength criterion of frozen soil. Qi and Lai [3, 4] modified the classical strength theory of unfrozen soils and established nonlinear strength criteria of frozen soil. In this paper, the objective of this study is to investigate the strength distribution law of frozen soil in the p-q plane under complex stress states, and then establish a new strength criterion to describe the multiaxial strength characteristics with consideration of the temperature effect.

© Springer Nature Switzerland AG 2018
W. Wu and H.-S. Yu (Eds.): *Proceedings of China-Europe Conference on Geotechnical Engineering*, SSGG, pp. 1340–1344, 2018.
https://doi.org/10.1007/978-3-319-97115-5_96

2   Test Condition

The test apparatus used in this study is HCA with negative temperature controlling. The soil used for the tests was clay from the Qinghai-Tibet plateau. The screened soils were mixed with 19.8% moisture content by weight. Then, they were into a split mold to make hollow cylindrical specimens (200 mm in height, 100 mm in outer diameter and 60 mm in inner diameter) with target densities by compaction. Then, the specimens were quickly frozen in the HCA under −30 °C for 48 h. Moreover, the coolants temperature was increased to the testing temperature and kept for 12 h. Finally, the directional shear tests were carried out under four coefficients of intermediate principal stress (b) and four mean principal stresses (p) at the three temperatures (T).

3   Results and Analyses

Figure 1 shows the stress-strain curves of test results with $b = 0.5$ and p in the range of 0.5 to 18 MPa at temperature of −6 °C. It can be seen that: (a) The generalized stress-strain curves all performs as strain-hardening characteristic; (b) the strength increases with increase of p when $p < 6.5$ MPa, but then decrease with a further increase of p when $p > 6.5$ MPa.

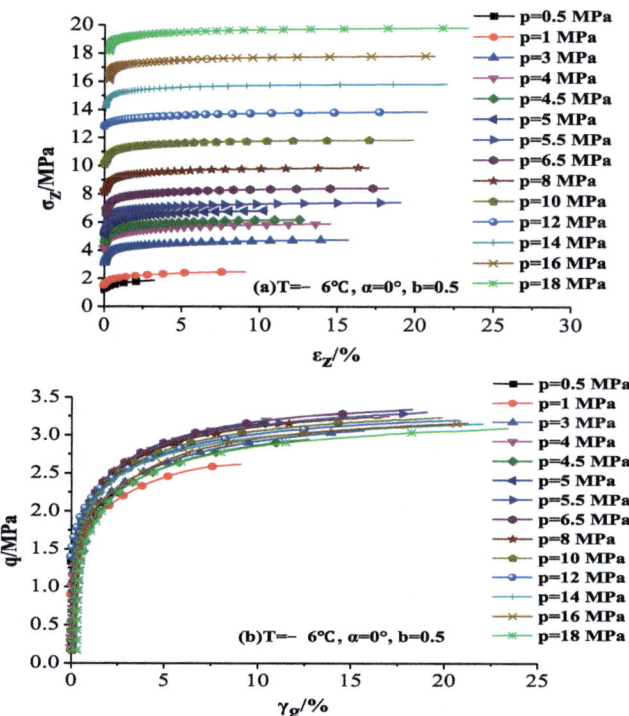

Fig. 1. Stress-strain curves of frozen clay under different mean principal stresses at −6 °C.

From Fig. 2, it can be found that the discrepancy of strength under b is great when the mean principal stress is low. With increase of the mean principal stress, this discrepancy decrease gradually. Under $b = 0.5$, the three-phase phenomenon is clearly observed within the range of the mean principal stress in this testing program. This phenomenon can be attributed to the so-called pressure melting and pressure crushing of the ice crystals in frozen soil [1, 2].

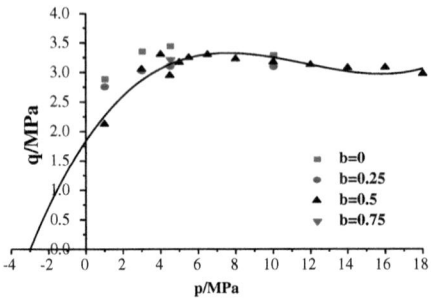

Fig. 2. The strength of frozen clay under different b. (left)

According to the law of the strength at the first and second phases, an elliptic function was proposed to describe the strength criterion of frozen clay in the p-q plane:

$$\left(\frac{q}{q_m}\right)^2 + \left(\frac{p - p_m}{f_{ttt} + p_m}\right)^2 = 1 \tag{1}$$

Where p_m denotes the maximum mean principal stress when the deviatoric stress reaches its maximum (q_m); f_{ttt} is the isotropic tensile stress of frozen clay.

In order to check the validity, we chose the different strength criteria for frozen soil to simulate test results. The previous strength criteria include the hyperbolic strength criterion [1, 2], modified Mohr-Coulomb strength criterion [3], and modified hydrostatic pressure strength criterion [4]. All the criteria can describe well the two phases of the strength variation with the mean principal stress increase (see Fig. 3).

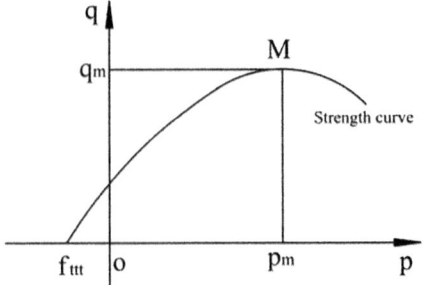

Fig. 3. Strength parameters in the p-q plane. (right)

4 Temperature Effect

With decrease of temperature, the strength of frozen clay increases significantly, as shown in Fig. 4. To consider the temperature effect, an impact factor is introduced into Eq. (1), and then the strength criterion can be expressed as:

$$q = q_m \left(1 - \left(\frac{p - p_m}{f_{ttt} + p_m}\right)^2\right)^{\frac{1}{2}} \left(\frac{\theta}{\theta_o}\right)^n \tag{2}$$

Where θ is test temperature; θ_o is a reference temperature valued of -1 °C; n is the material constant ($n = 0.01$ in experimental conditions of this study) (Fig. 5).

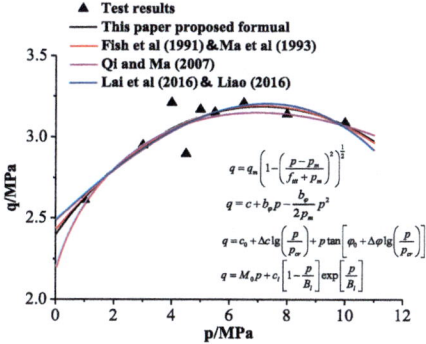

Fig. 4. The comparisons of strength criteria for frozen soils (left)

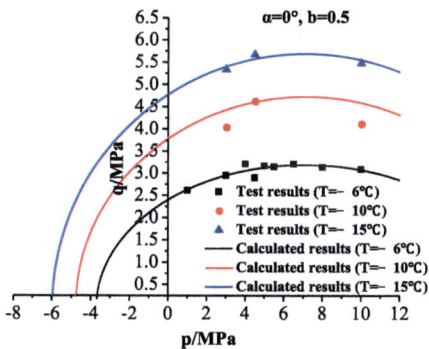

Fig. 5. Experimental results in the p-q plane (right)

5 Conclusions

Based on laboratory tests, some conclusions on strength of frozen clay under complex stress states are drawn as follows. The stress-strain curves of frozen clay under directional shearing all performed as strain-hardening. The strength of frozen clay first increases and then decreases with the increase of the mean principal stress in the p-q plane. According to the test data, an elliptic strength criterion was proposed for frozen clay under complex stress states with consideration of the temperature effect.

Acknowledgements. This work was supported by the National Natural Science Foundation of China (Nos. 41401077 and 41630636) and the State Key Development Program of Basic Research of China (No. 2012CB026106).

References

1. Fish, A.M.: Strength of frozen soil under a combined stress state. In: Proceedings of 6th International Symposium on Ground Freezing, vol. 1, pp. 135–145 (1991)
2. Ma, W., Wu, Z.W., Zhang, C.Q.: Strength and yield criteria of frozen soil. J. Glaciol. Geocryol. **15**(1), 129–133 (1993)
3. Qi, J.L., Ma, W.: A new criterion for strength of frozen sand under quick triaxial compression considering effect of confining pressure. Acta Geotech. **2**(3), 221–226 (2007)
4. Lai, Y.M., Liao, M.K., Hu, K.: A constitutive model of frozen saline sandy soil based on energy dissipation theory. Int. J. Plast. **78**, 84–113 (2016)

A Constitutive Model for Frozen Granular Soils

Roberto Cudmani$^{(\boxtimes)}$, Jian Sun, and Wei Yan

Technical University of Munich, Baumbachstr. 7, 81245 Munich, Germany
roberto.cudmani@tum.de

Abstract. In this contribution, an elastic-viscoplastic constitutive model of the Maxwell-Type to describe the time-dependent behavior of frozen soils is proposed based on results of unconfined and triaxial creep tests and strain rate-controlled compression tests from Orth (1985). The model is able to capture essential features of the behavior of frozen soils, as temperature-, rate- and mean-pressure-dependence for quasi-monotonic loading realistically, including the creep-failure-type observed in frozen soils under constant deviator stresses. The model has been validated by means of experimental data of a frozen sand and used to predict creep and rate-dependent behavior of frozen soils for monotonic loading.

Keywords: Frozen soils · Constitutive modelling

1 Introduction

Experimental results of laboratory tests show that under constant deviator stress, the strain rate of frozen soils initially decays over time until a minimum strain rate (turning point) is reached and then increases continuously until failure [1]. The time span between the begin of creep and the turning point, which decreases with increasing the deviator stress and increases with decreasing temperature, is called the "lifetime" of frozen soils, as beyond this time the specimen will inexorably fail under constant loading. In addition, when sheared at constant strain-rate, the shear strength increases with increasing strain-rate and decreasing temperature. These features of the mechanical behavior of frozen soils imply that, for example, a tunnel section made of frozen body will inexorably become unstable sometime after the begin of the tunnel excavation and the shear strength developed by the frozen soil will strongly depend on the excavation speed. Therefore, the numerical analysis of boundary-value problems including frozen soils requires the development of constitutive models with the ability of predicting the temperature-, rate- and time-dependent stress-strain- behavior, including a realistic prediction of the creep response, the lifetime and the strength observed in the laboratory. So far, the majority of constitutive models for frozen soils has been derived based on empirical and phenomenological theories [1, 2]. Based on a semi-empirical one-dimensional creep model developed by Orth (1985), Cudmani (2006) proposed an elastic-viscoplastic constitutive model for frozen soils in which the influence of the mean pressure was disregarded [2]. This model allows the simulation of the prediction of the material behavior observed in unconfined strain rate-controlled

© Springer Nature Switzerland AG 2018
W. Wu and H.-S. Yu (Eds.): *Proceedings of China-Europe Conference on Geotechnical Engineering*, SSGG, pp. 1345–1349, 2018.
https://doi.org/10.1007/978-3-319-97115-5_97

compression and creep tests. However, the model cannot predict the time- and temperature-dependent behavior of frozen soils for cylindrical stress paths other than one-dimensional compression realistically. In order to address these shortcomings, in this paper, the previous model is extended to account to the influence of the first invariant of the stress tensor.

2 Development of a Constitutive Model for Frozen Soils

2.1 Semi-empirical One-Dimensional Creep Model

Based on the results of unconfined creep tests, Orth (1985) proposed the following relationship between the creep rate and the elapsed time from the application of the vertical load on the frozen soil specimen

$$\dot{\varepsilon}_1 = \dot{\varepsilon}_m exp(-\beta)exp\left(\beta\frac{t}{t_m}\right)\left(\frac{t}{t_m}\right)^{-\beta} \tag{1}$$

Herein, β is a material constant. t_m is the creep time required to reach the minimal creep strain rate $\dot{\varepsilon}_m$ at the turning point. Orth found that t_m and $\dot{\varepsilon}_m$ are related through $\dot{\varepsilon}_m = c_1 t_m^{c_2}$, and c_1 and c_2 are material constants. In addition, the time- and stress-dependence of the strain rate $\dot{\varepsilon}_m$ is described by the following relationship:

$$\dot{\varepsilon}_m = \dot{\varepsilon}_\alpha \exp\left[\left(\frac{K_1}{\theta + 273.4} + \ln \dot{\varepsilon}_\alpha\right)\left(\frac{\sigma_1}{\sigma_\alpha(\theta)} - 1\right)\right] \tag{2}$$

σ_1 is the axial stress applied in the creep test. $\sigma_\alpha(\theta) = a_1(-\theta)^{a_2}$ is the reference axial stress causing the reference strain rate $\dot{\varepsilon}_\alpha$ in creep tests at the temperature θ [°C]. K_1, a_1 and a_2 are additional material constants.

2.2 An Elastic-Viscoplastic Constitutive Model

The proposed constitutive model is based on a rheological model of the Maxwell type. Thus, in tensorial form, the stress-strain relationship can be written as:

$$\dot{\sigma} = [D]\dot{\varepsilon}_{el} = [D](\dot{\varepsilon} - \dot{\varepsilon}_v) \tag{3}$$

Herein, $\dot{\sigma}$ is stress rate tensor, $\dot{\varepsilon}$, $\dot{\varepsilon}_{el}$ and $\dot{\varepsilon}_v$ represents the total, elastic and viscous strain rate tensors, respectively. [D] is the elastic stiffness. In [2], the viscous rate tensor was determined from Eq. (1) assuming that creep is only induced by deviator stresses, and $\dot{\varepsilon}_v$ is coaxial to the deviator stress tensor $\tilde{\sigma}$:

$$\dot{\varepsilon}_v = \sqrt{\frac{3}{2}}\frac{\tilde{\sigma}}{\|\tilde{\sigma}\|}\dot{\varepsilon}_m exp(-\beta)exp\left(\beta\frac{t}{t_m}\right)\left(\frac{t}{t_m}\right)^{\beta} \tag{4}$$

deviatoric stress tensor $\tilde{\sigma}$ is $\tilde{\sigma}_{ij} = \sigma_{ij} - \frac{1}{3}\sigma_{kk}\delta_{ij}$, $\|\tilde{\sigma}\| = \sqrt{\tilde{\sigma}_{ij}\tilde{\sigma}_{ij}}$ represent the Euclidean norm of $\tilde{\sigma}$. In order to account for the influence of the first invariant of σ, $\dot{\varepsilon}_m$ in Eq. (3) was modified. $\sigma_1 = q$ was replaced by $q = \sqrt{3J_2}$. In addition, a function $\varphi(I_1, J_2)$ as defined in Sect. 2.3 was introduced. The pressure-dependent strain-rate $\dot{\varepsilon}_m$ is given by

$$\dot{\varepsilon}_m = \dot{\varepsilon}_\alpha \exp\left[\left(\frac{K_1}{\theta + 273.4} + \ln \dot{\varepsilon}_\alpha\right)\left(\frac{\varphi(I_1, J_2)\sqrt{3J_2}}{\sigma_\alpha(\theta)} - 1\right)\right] \tag{5}$$

J_2 is the second invariant of the deviatoric stress tensor, $I_1 = tr(\sigma)$ is the first invariant of the stress tensor.

2.3   Determination of $\varphi(I_1, J_2)$

According to experimental results from Orth (1985), the relationship between the compression strength f_c and the axial strain rate $\dot{\varepsilon}_1$ can also be described by Eq. (2):

$$\dot{\varepsilon}_1 = \dot{\varepsilon}_\alpha \exp\left[\left(\frac{K_1}{\theta + 273.4} + \ln \dot{\varepsilon}_\alpha\right)\left(\frac{f_c}{\sigma_\alpha(\theta)} - 1\right)\right] \tag{6}$$

Analogously, we can use Eq. (5) to describe the behavior observed under constant strain rate-controlled triaxial compression tests:

$$\dot{\varepsilon}_1 = \dot{\varepsilon}_\alpha \exp\left[\left(\frac{K_1}{\theta + 273.4} + \ln \dot{\varepsilon}_\alpha\right)\left(\frac{(\varphi(I_1, J_2)\sqrt{3J_2})_f}{\sigma_\alpha(\theta)} - 1\right)\right] \tag{7}$$

By comparing the right-hand side of Eqs. (6) and (7), we realize that

$$\varphi(I_1, J_2) = \frac{f_c}{\sqrt{3J_2}} \tag{8}$$

To evaluate the parameter $\varphi(I_1, J_2)$, we adopt a four-parameter yield surface with smooth curved meridians proposed by Hsieh et al. (1982) as [3]

$$f(\hat{\rho}, \hat{p}, \vartheta) = A\hat{\rho}^2 + (B\cos\vartheta + C)\hat{\rho} + D\sqrt{3}\hat{\rho} - 1 = 0 \tag{9}$$

where $\hat{\rho} = \frac{\rho}{f_c}$, $\hat{p} = \frac{p}{f_c}$, $\rho = \sqrt{2J_2}$, $p = \frac{I_1}{3}$, $\cos 3\vartheta = \frac{3\sqrt{3}}{2}\frac{J_3}{J_2^{3/2}}$, $J_3 = \det \tilde{\sigma}$ is the third invariant of the deviatoric stress tensor, A, B, C and D are material constants. For a given stress state, the compressive strength f_c can be determined from Eq. (9).

2.4   Calibration of the Parameters

The procedure to determine the parameters c_1, c_2, K_1, a_1, and a_2 based on unconfined one-dimensional creep and compression tests are described in [1] and [2]. The four parameters A, B, C and D required to describe the function $f(\hat{\rho}, \hat{p}, \vartheta)$ and $\varphi(I_1, J_2)$ can be determined by fitting the yield surface to the experimental results of triaxial compression tests in the $\hat{p} - \hat{q}$ plane, where $\hat{q} = \frac{\sigma_1 - \sigma_3}{f_c}$. The elastic parameters E and v can be evaluated in unconfined compression tests. Exemplarily, the parameters for frozen Karlsruhe Sand are listed in Table 1.

Table 1. Parameters for the model of Karlsruhe frozen sand.

c_1	c_2	a_1	a_2	β	K_1	A	B	C	D
[-]	[-]	[MPa/°C]	[-]	[-]	[°C]	[-]	[-]	[-]	[-]
0.016	−0.193	3.050	0.591	0.692	3817	1.540	0.758	0.524	1.215

In Fig. 1, a comparison between experimental and numerical results of triaxial creep tests is presented. As it can be seen, the model can predict the time-dependent strain-rate during creep, including the lifetime of the samples realistically.

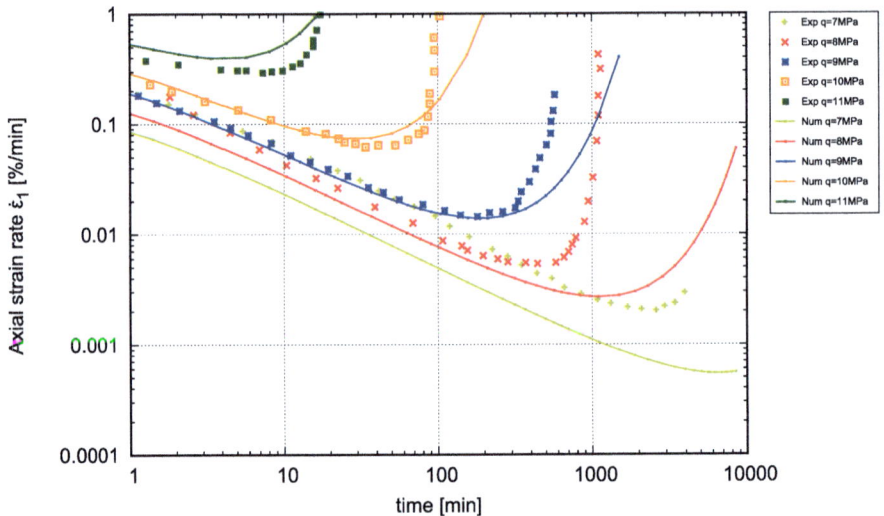

Fig. 1. Comparison between the results of experimental and numerical triaxial creep tests on Karlsruhe medium sand for constant mean normal stress p = −4 Mpa and different deviator stresses at −10 °C.

References

1. Orth, W.: Gefrorener Sand als Werkstoff. Veröffentlichung des Institutes für Bodenmechanik und Felsmechanik der Universität Fridericiana in Karlsruhe, Heft 100 (1986)
2. Cudmani, R.: An elastic-viscoplastic model for frozen soils. In: Proceedings of Numerical Modelling of Construction Processes in Geotechnical Engineering for Urban Environment, pp. 177–184. Balkema, Rotterdam, Bochum (2006)
3. Hsieh, S.S., Ting, E.C., Chen, W.F.: A plastic-fracture model for concrete. Int. J. Solids Struct. **18**(3), 181–197 (1982)

The Crystallization and Salt Expansion Characteristics of a Silty Clay

Jianhong Fang[1], Xu Li[1,2(✉)], Jiankun Liu[2], and Chenyinan Liu[2]

[1] Qinghai Research and Observation Base, Key Laboratory of Highway Construction and Maintenance Technology in Permafrost Regions, Xining, China
ceXuLi2012@gmail.com

[2] Key Laboratory of Urban Underground Engineering of Ministry of Education, Beijing Jiaotong University, Beijing, China

Abstract. To investigate mechanisms and principle for the salt expansion, theoretical analysis, micro-fabric observation and salt expansion tests are conducted to a silty clay containing sodium sulfate. Equations are provided to calculate the crystalized sodium sulfate through theoretical analysis. Long strip crystal bridges and bulky crystal are found in the micro scanning images of saline soil. When these crystals grow upon cooling, soil is supposed to expand. In salt expansion tests, the effects of compaction effort, water content, and salt content of soil are considered. Three compaction levels, i.e. 0.85, 0.90, and 0.95 are used. Three water contents, i.e. 13.2, 15.2 and 17.2% are used. Both pure Sodium sulfate soil and soil containing Sodium sulfate and Sodium chloride are used. The salt expansion data demonstrate that the volumetric content of crystal seems to have a unique relation with the volume strain of saline soil. A unified formula is established and can be used to predict the salt expansion strain for silty clay containing Sodium sulfate and Sodium chloride with any water content, salt content and dry density. Because the salt expansion tests used in this study use small sample and are without water supply, the salt and moisture migration in these tests is insignificant and the tests can be regarded as element tests. The unified formula can help to build constitute models for salt expansion.

Keywords: Saline soil · Salt expansion · Sodium sulfate

1 Introduction

In the northwest of China, Salt expansion is a typical disease existed in saline soil subgrade. When the temperature or moisture content decrease, soluble salt will crystallize. Sulfate salt crystallization can lead to the occurrence of expansion of roadbed, reduce or even destroy the road traffic conditions. The experimental study results indicate that the salt expansion of soil depends on the salt and water contents, the types of salt, temperature and cooling rate. The mechanism of saline expansion is very complicated and involves salt crystallization, moisture migration, and salt migration [1–3].

© Springer Nature Switzerland AG 2018
W. Wu and H.-S. Yu (Eds.): *Proceedings of China-Europe Conference on Geotechnical Engineering*, SSGG, pp. 1350–1354, 2018.
https://doi.org/10.1007/978-3-319-97115-5_98

As the solubility of sodium sulfate decrease fast with the decrease of temperature, crystallization of sodium sulfate will happen during the temperature reduction process. The chemical equation is as follows:

$$Na_2SO_4 + 10H_2O \leftrightarrow Na_2SO_4 \cdot 10H_2O \downarrow \tag{1}$$

The crystallization is controlled by the solubility product as,

$$C_{Na^+}^2 * C_{SO_4^{2-}} = k_{sp} \tag{2}$$

where C_{Na^+}, $C_{SO_4^{2-}}$ are the molar concentration of Na^+ and SO_4^{2-} respectively; k_{sp} is the solubility product of sodium sulfate. If NaCl is existed in the sulfate saline soil, the solubility of Na_2SO_4 will decrease. For a saturated Na_2SO_4 solution with a m molar concentration of NaCl, the molar concentration of Na_2SO_4 can be solved by,

$$n(2n+m)^2 = k_{sp}(T) \tag{3}$$

where n is the initial molar concentration of Na_2SO4 and m represents the molar concentration of NaCl; T is the final temperature after temperature decrease. If there is no NaCl existing in the solution, i.e. $m = 0$, the amount of crystalized sodium sulfate in unit of molar per liter can be calculated as,

$$x = \frac{n - n_S(T)}{1 - 0.18n_S(T)} \tag{4}$$

where $n_S(T)$ is saturated concentration of Na_2SO4 at temperature T.

If m is high, Na_2SO_4 cannot be fully dissolved in the water. In that case, the initial concentration of Na_2SO_4 may be irrational. Thus, the crystalized sodium sulfate will be overestimated. In such case, the concentration of NaCl m will be given and the initial concentration of Na_2SO_4 n can be solved by Eq. 3.

2 Experimental Study

In this study, the silty clay is taken from Xiangpi Mountain Region in Qinghai Province of China. The structure of saline soil samples is observed by a commercial available OLS4100 scanning microscope. Compacted saline soil samples are prepared at 25 °C, cooling to 0 °C, and then observed. As shown in Fig. 1, the crystalized sodium sulfate existed in soil pores. At the contact or throat among particles, the crystalized sodium sulfate is blocky and bonding the particles together. In larger pores, the crystalized sodium sulfate has long strip shapes. Such long strip crystals demand more space after the crystallization. If the crystalized saline bridge grows, the soil volume will subsequently expand.

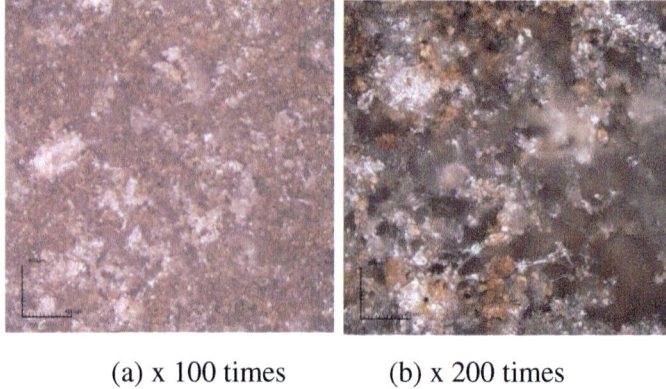

(a) x 100 times   (b) x 200 times

Fig. 1. Surface of saline soils with crystalized sodium sulfate (n_0 = 1.34 mol/l, $T \sim 0 \sim 4$ °C)

For this soil, about 36% of weight is composed by coarse soil (>0.075 mm) and it has a plastic limit and liquid limit of 20.1% and 30.7%, respectively. Thus, it can be classified as a silty clay. The result from Laboratory standard compaction test gives that the soil has a maximum dry density of 1.877 g/cm³ and an optimum water content of 13.2%. The specific gravity of solid is 2.72.

The indoor pure salt expansion tests are conducted to study the main factors and deformation behavior of silty clay containing sodium chloride and sodium sulfate. The effects of compaction effort, water content, and salt content of soil on salt expansion are considered in these tests. Three compaction levels, i.e. 0.85, 0.90, and 0.95 are used. Three water contents, i.e. 13.2, 15.2 and 17.2% are used. Both pure Sodium sulfate soil and soil containing Sodium sulfate and Sodium chloride are used.

The results of expansion test for soil samples only involving sodium sulfate are illustrated in Fig. 2a, where the relation between the increment of sample height h and the sample temperature T is reported. Referring to Fig. 2b, there are a unified relation between the volume strain ε_v and the reference expanding ratio χ, as:

$$\varepsilon_v = (1.12\chi - 0.03) \pm 0.01 \tag{5}$$

Equation 5 offers the 95% prediction band for salt expansion for soil containing sodium sulfate. the heave deformation of sample decreases with the increase of the concentration of NaCl m. That's to say, the existence of sodium chloride will inhibit the expansion of soil containing sodium sulfate. For all cases containing NaCl, the $\varepsilon_v - \chi$ relation is also consistent with Eq. 5 (Fig. 2d).

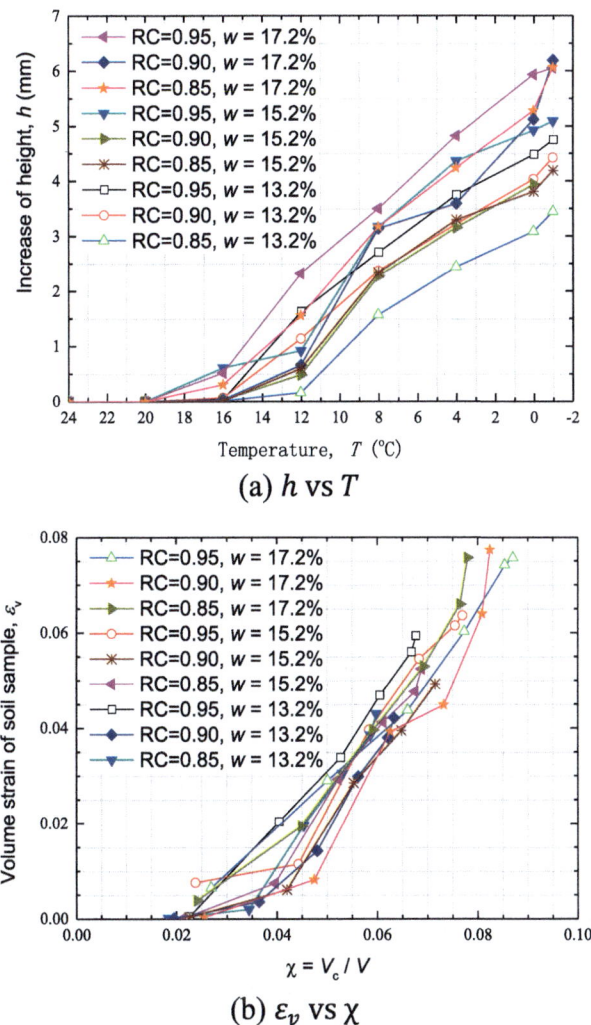

(a) h vs T

(b) ε_v vs χ

Fig. 2. Expansion of soil containing sodium sulfate

3   Conclusions

In this paper, theoretical analysis, micro fabric observation, and salt expansion tests are carried out to study the salt expansion phenomenon for a silty clay containing sodium chloride and sodium sulfate. It can be found that the volumetric saline crystal content can be used as a reference value for salt expansion. It seems that the volume strain of salt expansion has a unified relation with the volumetric saline crystal content. A unique formula is established to characterize the salt expansion of soil containing sodium chloride and sodium sulfate. That formula can be used to predict the salt expansion for soil with any dry density, water content and salt content.

Acknowledgments. This research is supported by the National Natural Science Foundation of China (Nos. 41471052, 51479001) and the Science and technology development plan of China Railway General Corporation (No. 2017G002-S).

References

1. Niu, X.R., Gao, J.P.: Expression for volume change of sulphate saline soil considering salt expansion and frost heave. Chin. J. Geotech. Eng. **37**(4), 755–760 (2015). (in Chinese)
2. Lai, Y., Wan, X., Zhang, M.: An experimental study on the influence of cooling rates on salt expansion in sodium sulfate soils. Cold Reg. Sci. Technol. **124**, 67–76 (2016)
3. Wan, X., You, Z., Wen, H., Crossley, W.: An experimental study of salt expansion in sodium saline soils under transient conditions. J. Arid Land **9**(6), 865–878 (2017)

Experimental Study on Temperature Threshold for Warm Frozen Sand in Terms of Mechanical Properties

Xueluan Guo, Junlin Zhao, Yansong Wang, and Jilin Qi[(⊠)]

Beijing University of Architecture and Civil Engineering, Beijing 100044, China
jilinqi@bucea.edu.cn

Abstract. Mechanical properties of warm frozen soils are similar to those of unfrozen soils. In recent years, the mechanical properties of warm frozen soils have attracted great attentions. However, there is no definition of warm frozen soil according to the mechanical behaviors of frozen soils so far. This paper attempts to define the temperature threshold for a warm saturated frozen sand in terms of mechanical properties. The Chinese ISO standard sand was taken as study object. Triaxial and confined compression tests were carried out on the saturated frozen samples. Mechanical properties such as cohesion, modulus and compression index were obtained and their changing tendency along with temperature were analyzed. It is found that the range of −1.0 °C to −0.5 °C seems to be the temperature when the mechanical properties change abruptly. Therefore, −1.0 °C can be defined as the temperature threshold for warm frozen soils with regard to the material tested in this program.

Keywords: Warm frozen soil · Mechanical properties · Temperature threshold

1 Introduction

When a soil is under a temperature close to the freezing temperature, the ice water phase changes severely, and its physico-mechanical properties also change drastically. The mechanical properties of this kind of material have attracted a great deal of attentions in the field of frozen soil mechanics and engineering, as it was found that the mechanical properties of warm frozen soil are close to that of unfrozen soils [1]. However, as a material, so far there is no clear temperature boundary to define the warm frozen soil. For geotechnical engineers, it seems necessary to have a temperature boundary from the perspective of mechanical properties.

The term "warm frozen soil" which refers to the frozen soil in the temperature range with severe phase changes, is often used to describe a frozen soil at a relatively high temperature. The mechanical properties of the so-called warm frozen soil have been extensively studied. For instance, Shields et al. [2] studied the creep characteristics of sand between −2.5 °C and −3.0 °C, and called the frozen soil in this temperature range a warm frozen soil; China's Temporary Regulations for Design of the Qinghai-Tibet Railway in Permafrost Regions [3] defines that a region with ground temperature higher than −1.0 °C is called the warm permafrost regions, and this is based on

© Springer Nature Switzerland AG 2018
W. Wu and H.-S. Yu (Eds.): *Proceedings of China-Europe Conference on Geotechnical Engineering*, SSGG, pp. 1355–1358, 2018.
https://doi.org/10.1007/978-3-319-97115-5_99

geographic perspectives. Tsytovich [4] defined temperature boundaries for stiff and plastic frozen soils for different soil classifications according to their failure characteristics, i.e., −0.3, −0.6, −1.0 and −1.5 °C for fine sand, silty sand, silty clay and clay, respectively. However, the test and mechanical indexes adopted for definition were very limited.

This paper takes the easily accessible Chinese ISO standard sand as the study object and uses both triaxial and confined compression tests to obtain cohesion, the secant modulus and the compression index at different temperatures so as to define a reasonable temperature threshold of warm frozen soil from the perspective of the mechanical properties.

2   Test Methodology

2.1   Description of the Test Program

The triaxial tests were carried out on a self-developed multifunctional material testing apparatus for frozen soils consisting of a loading system, a cooling system and a controlling system. It allows tests in both stress control and strain control modes. Temperature control ranges from −40 °C to +30 °C with an accuracy of ±0.1 °C.

The confined compression tests were carried out on a self-developed oedometer for frozen soils. It consists of a loading system, a temperature controlling system and a data collection system. The thermostat has a temperature control range from −50 °C to +90 °C with an accuracy of ±0.1 °C. The test material is the Chinese ISO standard sand in accordance with the Chinese code GB/T 17671-1999.

2.2   Experiment Method

For triaxial tests Multi-sieving pluviation was used to prepare samples so as to ensure the sample uniformity. Cylindrical samples were reconstituted in a steel tube, producing soil samples with a dry unit weight of 17.6 kN/m^3. The samples all had a diameter of 61.8 mm and a height of 125.0 mm, a water content of 17.2%. The prepared soil samples were fixed into a copper mold and saturated in a vacuum with access to distilled water, then frozen in a refrigerator. Afterwards, it was mounted on the triaxial apparatus for testing. The axial strain rate was $1.67 \times 10^{-4} \text{ s}^{-1}$, and the compression was stopped when the axial strain reached 20%. The four confining pressures of 0.5 MPa, 2 MPa, 5 MPa and 10 MPa, and the four temperatures of −0.5 °C, −1.0 °C, −2.0 °C and −5.0 °C, were used.

For the confined compression tests, the samples were prepared with a diameter of 61.8 mm and height of 80.0 mm. During the test, the axial pressure was applied at a rate of 0.01 MPa/s to a certain target load and maintained for one hour, then the next load was applied. Compression tests were carried out at temperatures of −0.5 °C, −1.0 °C, −2.0 °C, −3.0 °C and −5.0 °C, respectively, and four load steps of 1, 2, 5 and 10 MPa were used for each temperature.

3   Test Results and Discussions

3.1   The Relationship Between Cohesion and Temperature

The cohesion was obtained from the triaxial tests. For the strain softening failure, strength corresponds to the peak deviatoric stress σ_1–σ_3, while for strain hardening type, the peak deviatoric stress corresponding to 15% of the axial strain was taken as strength. According to the Mohr's circles of different confining pressures under a certain temperature, the shear strength envelope curve was drawn to obtain the cohesion c under this temperature. The change of cohesion in different temperatures are shown in Fig. 1. It can be found that the cohesion of frozen sand gradually decreases with the temperature rise, and the closer the temperature is to 0.0 °C, the faster it drops. The ratio of the cohesion difference to the absolute value of temperature difference is defined as k. When the temperature rises from −2.0 °C to −1.0 °C, $k_1 = 0.39$ while in the range of −1.0 °C to −0.5 °C, $k_2 = 1.16$. The drop rate of the cohesion increased rapidly, and there is a continuing trend of decline.

3.2   Relationship Between Secant Modulus and Temperature

The secant modulus E_{sec} is obtained from the beginning to a strain of 1% in each test, as is illustrated in Fig. 2. It can be seen that E_{sec} gradually decreases with rise of temperature, and the capacity of frozen soil to resist deformation decreases. It can also be noticed that the modulus increases with increasing confining pressure at the low temperature of −5.0 °C. However, at relatively high temperatures, say −0.5 °C, the modulus decrease tendency dominates along with increase in confining pressure. In other words, frozen soil at relatively high temperatures is very different from that at low temperatures. This is due to the fact that pressure-melting is easy to occur in warm frozen soils under certain pressure, which makes the material unstable in mechanical properties. The unstable tendency starts to appear at −1.0 °C in this test program.

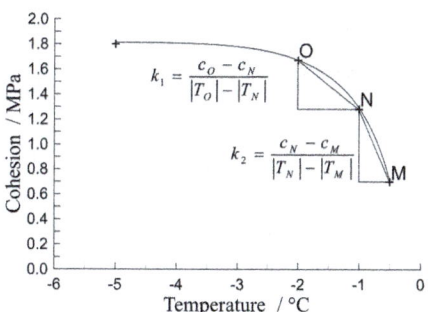
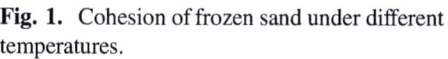

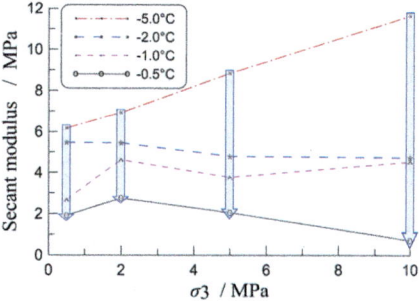

Fig. 1. Cohesion of frozen sand under different temperatures.

Fig. 2. E_{sec} of frozen sand under different temperatures.

3.3   Relationship Between Compression Index and Temperature

Confined compressive tests were performed using Chinese ISO standard sand under −0.5 °C, −1.0 °C, −2.0 °C, −3.0 °C, and −5.0 °C, respectively. Figure 3 illustrates the e-lgP curves at temperatures of 20 °C, −0.5 °C, −1.0 °C, and −2.0 °C. The compression index Cc is the change in porosity ratio when pressure increases from 1 MPa to 10 MPa, as this section on the e-lgP curve better reflects the overall compressibility of the material at this temperature. Cc at different temperatures is presented in Fig. 4. It can be seen that the compression index gradually increases with the rise of temperature with the most rapid change at −1.0 °C, and then the change slows down from −0.5 °C to 20.0 °C.

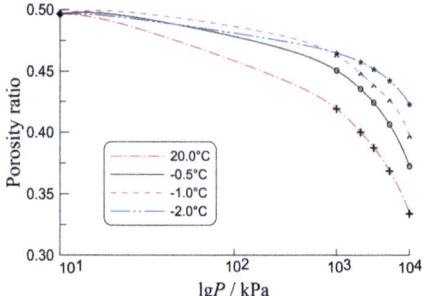
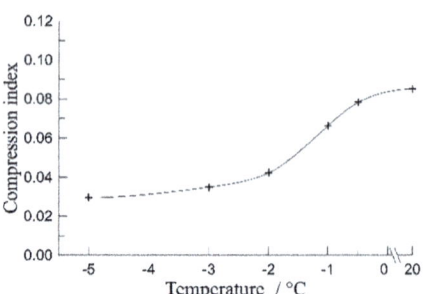

Fig. 3. e-lgP curves under 20 °C, −0.5 °C, −1.0 °C and −2.0 °C.

Fig. 4. Relationship between compression index and temperature

4   Conclusions

In this paper, triaxial and confined compression tests are carried out. Cohesion, secant modulus and compression index are obtained to define the temperature boundary of warm frozen soil. Their severe change occurs at −1.0 °C, which should be defined as the threshold for warm frozen soil in this test program.

Acknowledgements. This work was supported in part by the NSFC (Nos. 41572268), Joint Key Project of BNSFC and BMEC (No. KZ201810016020) and PCSIRT (No. IRT_17R06).

References

1. Ladanyi, B.: Creep behaviour of frozen and unfrozen soils. In: Proceedings of the 10th International Conference on Cold Regions Engineering - Putting Research into Practice, pp. 173–185. ASCE, Lincoln, NH, USA (1999)
2. Shields, D.H., Domaschuk, L., Man, C.S.: Deformation properties of warm permafrost. In: Strength Testing of Marine Sediments: Laboratory and In-Situ Measurements, pp. 473–486. American Society for Testing and Materials, PA, USA (1985)
3. Temporary Prescript for Engineering Surveying of Permafrost in the Qing-Tibet Railway. The First Survey and Design Institute of China Railway Construction Corporation, Lanzhou (2001)
4. Tshtovich, H.A.: Mechanics of Frozen Soil, pp. 11–13. Science Press, Beijing (1985)

Freeze-Thaw Processes of Soils in Active Layers in Northeast China

Ruixia He[1(✉)], Huijun Jin[1,2], and Xiaoli Chang[1,3]

[1] State Key Laboratory of Frozen Soil Engineering, Northwest Institute of Eco-Environment and Resources, Chinese Academy of Sciences, Lanzhou 730000, Gansu, China
hexuzhen.geo@gmail.com

[2] School of Civil Engineering, Harbin Institute of Technology, Harbin 150090, Heilongjiang, China

[3] Hunan University of Science and Technology, Xiangtan 411201, Hunan, China

Abstract. The processes of thawing and freezing and their associated complex hydrothermal coupling can significantly affect variations in mean annual temperatures and the formation of ground ice in permafrost regions. In this article, using soil temperature and moisture data in the permafrost region of the Nanweng'he River in the Da xing'anling Mountains, the freeze-thaw characteristics of the permafrost were studied. Variations in the soil temperature and the moisture were analyzed during each stage of the freeze-thaw process, and the effects of the soil moisture and ground vegetation on the freezing-thaw were discussed in this paper. The study results show that the thawing in the active layer is unidirectional, while the ground freezing is bidirectional (upward from the bottom of the active layer and downward from the ground surface). During the annual freeze-thaw cycle, the migration of soil moisture had different characteristics at different stages. In general, during of a freezing-thawing cycle, the soil water molecules will migrate downwards, i.e., soil moisture will transport from the entire active layer to the upper limit of permafrost. In the meantime, freeze-thaw in the active layer can be significantly affected by the soil moisture content and vegetation.

Keywords: Nanweng'he River National Natural Reserve · Active layer
Freezing-thaw processes · Moisture content · Vegetation · Effect

1 Introduction

The active layer is defined as the soil layer overlying the permafrost. This active layer thaws in the summer and freezes in the winter [1]. It is the most thermodynamically active rock or soil unit, and it have profound effects on many terrestrial processes [2]. The characteristics of the active layer include the freeze-thaw process, hydrothermal regimes, and the physical properties of the soil, which are key factors in mediating between the climate and the permafrost. Variations in temperature and thickness of active layer can greatly affect arctic ecosystems, geologic environments, and hydrologic processes [3–5]. Some local environmental factors, which can significantly affect the hydrothermal processes of soils in the active layer [6, 7]. There have been several studies

© Springer Nature Switzerland AG 2018
W. Wu and H.-S. Yu (Eds.): *Proceedings of China-Europe Conference on Geotechnical Engineering*, SSGG, pp. 1359–1363, 2018.
https://doi.org/10.1007/978-3-319-97115-5_100

of hydrothermal processes in and the characteristics of the active layer in China. A series of surveys and investigations of hydrothermal process in the active layer have been performed in the Tibetan Plateau [4–6, 8–10]. However, there is still an incomplete understanding of freeze-thaw processes in the active layer in Northeastern China. This scarcity of information is particularly relevant to Nanweng'he River National Natural Reserve, also, there was the diversity of vegetation form in the study area. There, we collected soil-temperature and water-content data in the active layer from 2011 to 2014 at two observation sites in Nanweng'he River National Natural Reserve set up by the State Key Laboratory of Frozen Soil Engineering of the Chinese Academy of Sciences. We used the data to study the freeze-thaw process and variations in soil temperature and water content in the active layer near the Nanweng'he River and analyzed the profound effects of water content and vegetation on the freeze-thaw processes.

2   Study Results: Freezing and Thawing Process of the Permafrost Active Layer

During the process of freeze-thaw cycle, the entire active layer undergoes cooling, initial freezing until complete, further cooling, warming, initial thawing until complete, and further warming. Providing 0 °C isotherm is the freezing point of soil water in study area, as shown in Figs. 1 and 2, the freezing process began to develop from the permafrost table upwards gradually in late September and early October from 2012 to 2014. The stable freezing began from ground surface within the following 10 days, and developed rapidly downwards. The bidirectional freezing processes then began to develop and lasted till middle October when the entire active layer became frozen. In late March of the following year, daily freezing and thawing appeared now and then, and gradually

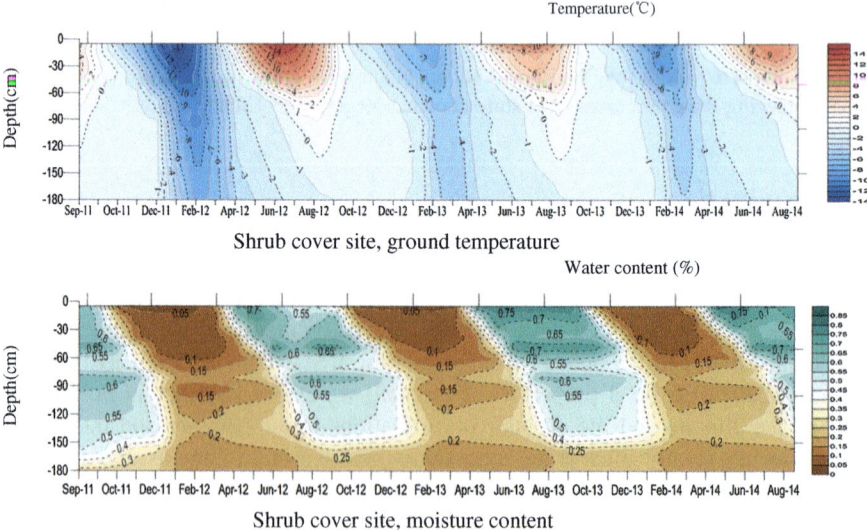

Fig. 1. Isotherms of active layer of the shrub cover site in the study area

changed to everyday one cycle from the beginning of April. Active layer began to thaw downwards in late April and early May until reaching a maximum depth in late August to early September (Figs. 1 and 2).

The study results show that the thawing in the active layer is unidirectional, while the ground freezing is bidirectional (upward from the bottom of the active layer and downward from the ground surface). During the annual freeze-thaw cycle, the migration of soil moisture had different characteristics at different stages. In general, during of a freezing-thawing cycle, the soil water molecules will migrate downwards, i.e., soil moisture will transport from the entire active layer to the upper limit of permafrost.

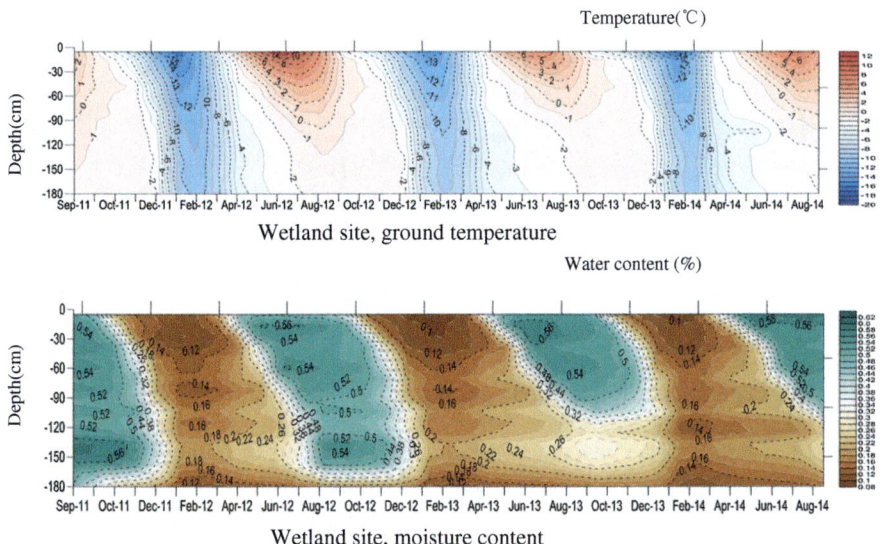

Wetland site, ground temperature

Wetland site, moisture content

Fig. 2. Isotherms of active layer of the wetland site in the study area

3 Discussions

Freeze-thaw in the active layer is controlled not only by the climate but also by the topography, soil texture, ground vegetation and hydrology. The study results show that freeze-thaw in the active layer can be significantly affected by the soil moisture content and vegetation.

3.1 Effect of Moisture Content on the Active Layer

The soil moisture content can slow the freeze-thaw process and substantially affect the heat distribution in the soil. As shown in Figs. 2 and 3, the soil temperature varied moderately, almost approaching the 0 °C isothermal line during the freezing stage, which became clear where the soil depth exceeded 140 cm. In the freeze-thaw cycle, the effect of the moisture content on the soil temperature variation was reflected in the slower variation of soil temperature relative to the moisture content variation in the freezing process (Figs. 1 and 2), which became more apparent with increasing moisture content.

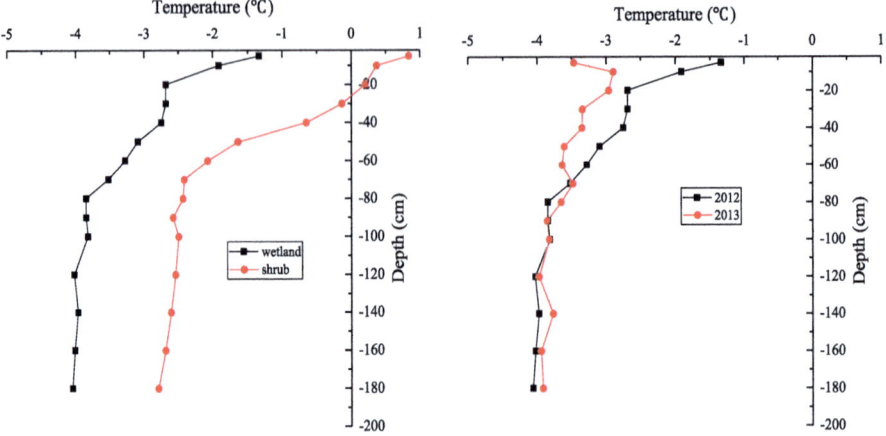

Fig. 3. Variation in mean annual temperature with depth at the two sites in 2012

Fig. 4. Variation in annual average temperature with depth at the wetland site in 2012 and 2013

3.2   Effect of Vegetation on the Active Layer

As discussed earlier, the vegetation differed between the two monitoring sites, which can further affect the ground temperature, the timing of the freeze-thaw period, the freezing time period, and the division of the freeze-thaw process (Figs. 3 and 4).

The study results show that the shrub cover site started to freeze in early October, whereas the corresponding date at the wetland site was mid-late September. There were similar differences in the date of thawing. The shrub cover site started to thaw in early May, whereas the corresponding date at the wetland site was mid-late May. These differences are due to the poor drainage conditions caused by the high vegetation coverage and thick organic humus layer in the wetland site, which generate a "thermal semiconductor" effect.

Figure 3 shows the mean annual temperature at various depths in the shrub cover site and wetland site in 2012. At any given depth, the annual average temperature of the wetland site was lower than that of the shrub cover site, indicating a negative correlation between the amount of vegetation coverage and the temperature, other geographic and topographic factors being equal. The temperature of the active layer was lower with increasing vegetation coverage. A similar pattern is evident in Fig. 4, the coverage in 2013 after a recovery of the vegetation was greater than in 2012, and the surface (5–10 cm) temperature was lower in 2013 than in 2012. However, at greater depths, the mean annual temperatures were generally identical, implying that the complex effects of vegetation differences on the temperature were concentrated primarily in the shallower soil. Because energy equilibrium can be achieved through heat exchange, such as thermal transfer between the ground and the atmosphere in the lower active layer, the upper portion of the active layer may undergo large temperature changes while the lower portion tends to remain at a uniform temperature.

Acknowledgements. This study supported by the National Natural Science Foundation of China (Grant No. 41401081) and the State Key Laboratory of Frozen Soils Engineering (Grant Nos. SKLFSE-ZT-41, SKLFSE-ZT-20 and SKLFSE-ZT-12).

References

1. Qin, D.H., Yao, T.D., Ding, Y.J.: Glossary of Cryosphere Science. China Meteorological Press, Beijing (2014)
2. Li, S.X., Nan, Z.T., Zhao, L.: Impact of soil freezing and thawing process on thermal exchange between atmosphere and ground surface. J. Glaciol. Geocryol. **24**, 506–511 (2002)
3. Liang, S.H., Wang, L., Li, Z.M.: The effect of permafrost on alpine vegetation in the source regions of the yellow river. J. Glaciol. Geocryol. **29**, 45–52 (2007)
4. Luo, D.L., Jin, H.J., Lü, L.Z.: Spatiotemporal characteristics of freezing and thawing of the active layer in the source areas of the Yellow River (SAYR). Chin. Sci. Bull. **59**, 3034–3045 (2014). https://doi.org/10.1007/s11434-014-0189-6
5. Luo, D.L., Jin, H.J., He, R.X.: Responses of surface vegetation on soil temperature and moisture of the active layer in the source area of the yellow river. Earth Sci. J. Chin. Univ. Geosci. **39**, 421–430 (2014)
6. Zhao, L., Cheng, G.D., Li, S.X.: Thawing and freezing processes of active layer in Wudaoliang region of Tibetan Plateau. Chin. Sci. Bull. **45**, 1205–1211 (2000)
7. Liu, G.S., Wang, G.X.: Influence of short-term experimental warming on heat-water processes of the active layer in a swamp meadow ecosystem of the Qinghai-Tibet Plateau. Sci. Cold Arid Reg. **8**(2), 125–134 (2016)
8. Wu, Q.B., Shen, Y.P., Shi, B.: Relationship between frozen soil together with its water heat process and ecological environment in the Tibetan Plateau. J. Glaciol. Geocryol. **25**, 250–255 (2003)
9. Pang, Q.Q., Li, S.X., Wu, T.H.: Simulated distribution of active layer depths in the frozen ground regions of Tibetan Plateau. J Glaci. Geocry. **28**, 390–395 (2006)
10. Wu, Q.B., Hou, Y.D., Yun, H.B.: Changes in active-layer thickness and near-surface permafrost between 2002 and 2012 in alpine ecosystems, Qinghai-Xizang (Tibet) Plateau. China. Glob. Planet. Change **124**, 149–155 (2014). https://doi.org/10.1016/j.gloplacha.2014.09.002

Effect of Thermokarst Lake on Foundation Under Embankment in Permafrost Regions

Xiaoying Hu[1,2(✉)], Yu Sheng[2], and Erxing Peng[2]

[1] Lanzhou University of Technology, Lanzhou 730000, Gansu, China
xiaoyinghu@lzb.ac.cn
[2] State Key Laboratory of Frozen Soil Engineering,
Northwest Institute of Eco-Environment and Resources, Chinese Academy of Sciences,
Lanzhou 730000, Gansu, China

Abstract. Thermokarst lakes can change the thermal regimes of foundation under the embankment and cause failures of embankment in permafrost regions. This paper analyzes the thermal impact of thermokarst lake on the permafrost based on the monitoring data. The results indicate that: (1) The thermokarst lake extend the time of warm season and shorten the time of cold season for permafrost; (2) The average ground temperature of permafrost is increased by the effect of thermokarst lake. The increment of temperature became larger with the decrease of distance from the thermokarst lake; (3) The thermokarst lake can inhibition the heat dissipation of permafrost in the cold season to make a thermal impact on the permafrost. Therefore, the embankment always absorbs heat from thermokarst lake, asymmetry thermal regimes of permafrost under the embankment is enhanced, which induces differential settlements and harms the safety and stability of embankment.

Keywords: Thermokarst lake · Permafrost embankment · Ground temperature
Thermal effect

1 Experiment Sites

Thermokarst lakes are widely distributed throughout the permafrost regions of the world as a result of thawing of ice-rich perma-frost [1]. Numerous studies have indicated that TLs have an obvious thermal effect on the underlying permafrost [2, 3]. The monitoring field (97.83 °E, 34.24 °N, 4725 m, Fig. 1) is located at the k577+210 and k577+310 along the Gonghe-Yushu Highway (Fig. 2). This experiment field is the Plateau landform with flat topography in the continuous ice-rich and ice-satiation permafrost where the swamping wetland and frost-heaving earth-hummocks are found to be developed well. The permafrost table is about −0.8~−2.5 m, the permafrost depth is about 60–100 m, and the ground temperature is generally −2.0~−1.5 °C.

© Springer Nature Switzerland AG 2018
W. Wu and H.-S. Yu (Eds.): *Proceedings of China-Europe Conference on Geotechnical Engineering*, SSGG, pp. 1364–1367, 2018.
https://doi.org/10.1007/978-3-319-97115-5_101

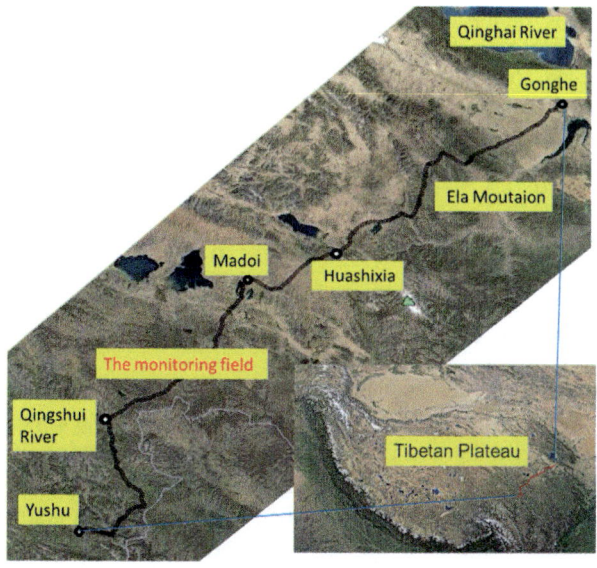

Fig. 1. The location of the study area

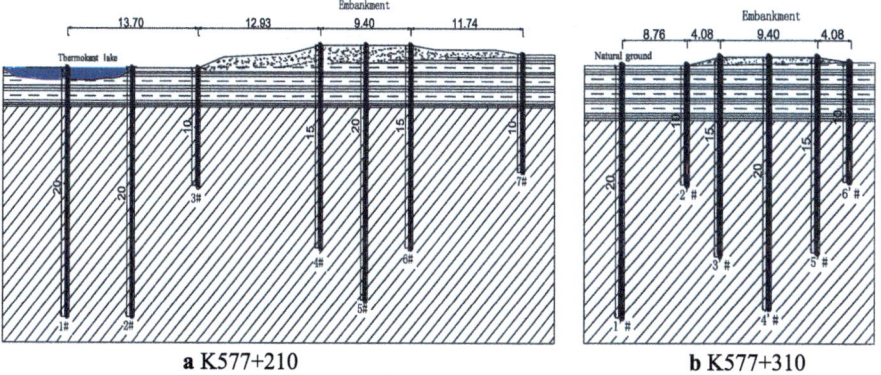

Fig. 2. Observation transverse sections of with embankment (unit: m)

2   Results and Analysis

2.1   Differences Between the Two Boundaries of Thermalkarst Lake and Natural Ground

The Table 1 shows that the results the frozen index and thawed index of boundaries within 1 m. The results indicate that the thermokarst lake can decline the frozen index of the underlying permafrost, and the frozen index of permafrost under the thermokarst lake is about 3% of permafrost under the natural ground. On the contrast, the thawed index of permafrost under the thermokarst lake is increased and is about 8.5 times of the permafrost under the natural ground.

Table 1. The contrast of the freezing and thawing index between the bottom of thermokarst lake and natural ground

Index/°C · day	Depth/m	Thermokarst lake	Natural ground
Frozen index	−1.0	−17.45	−684.12
Thawed index	−1.0	1715.10	202.04

2.2 Ground Temperatures Variations Under the Two Embankment

It can be seen that the degree of ground temperature increasing in March is greater than that in July from the Fig. 3a and b, which implies that the thermal impact from the high-water temperature at the bottom of thermokarst lake on the underlying permafrost in cold season is greater than that in warm season. This is because that the gradient of ground temperature in vertical under the embankment with thermokarst lake is changed from negative to positive in cold season, which makes the heat released of permafrost to be switched from the state of heat released to the state of heat absorbed. But in warm

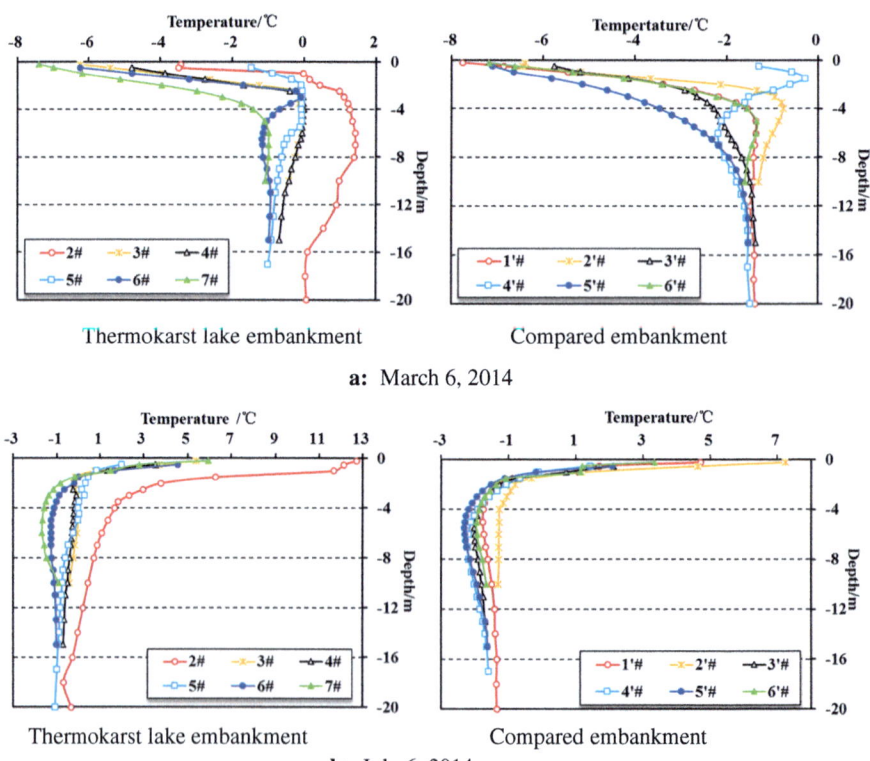

Fig. 3. Variations of ground temperature under the two embankment sections

season, the slope of positive gradient of ground temperature in vertical under the embankment with thermokarst lake is increased to strengthen the capacity of heat absorbed. So, it is key that the thermokarst lake can inhibition the heat released of permafrost under the embankment.

2.3   Heat Flux of Permafrost Under the Two Embankment

The heat flux at the depth of -5 m are shown as Fig. 4. The heat flux from the bore of 2# to the bore 3# is always positive, and the maximum heat flux is about 0.15 w/m^2. On the contrary, the heat flux from the bore of 3# to the bore of 4# is near to zero.

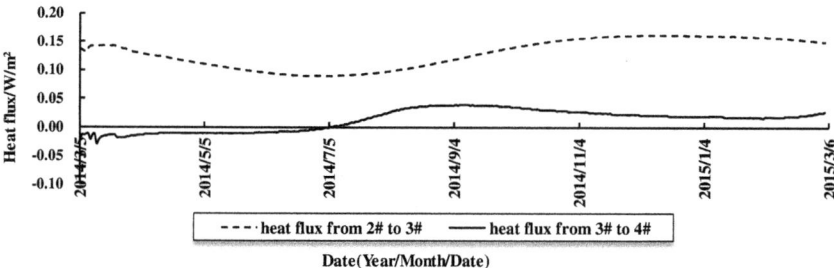

Fig. 4.   Heat flux at the depth of -3 m of permafrost under the two embankments

3   Conclusions

The thawed index and the permafrost frozen index at the bottom of thermokarst lake is increased 8.49 times and decreased 97% of them at the depth of -1 m of the ground. The negative ground geothermal gradient is changed to be positive under the embankment with thermokarst lake in cold season, and the slope of positive geothermal gradient is increased in warm season. The thermokarst lake can inhibition the heat released of permafrost under the embankment. Moreover, the thermokarst lake can transfer heat to permafrost around in two dimensional. So, for long time, the thermalkarst lake can enhance the ground temperature of permafrost under embankment and reduce the thermal stability of embankment in permafrost regions.

References

1. Qiu, G.Q., Liu, J.R., Liu, H.X.: Geocryological Glossary. Gansu Science and Technology Press, Lanzhou, China (1994)
2. Serreze, M.C., Walsh, J.E., Iii, F.S.C., et al.: Observational evidence of recent change in the northern high-latitude environment. Clim. Change **46**(1–2), 159–207 (2000)
3. Kokelj, S.V., Lantz, T.C., Kanigan, J., Coutts, R.: Origin and polycyclic behaviour of tundra thaw slumps, Mackenzie Delta region, Northwest Territories, Canada. Permafrost Periglac. Process. **20**(2), 173–184 (2009)

Experimental Study on Anti-frost Jacking of Belled Pile in Seasonally Frozen Ground Regions

Xubin Huang[1,2(✉)] and Yu Sheng[2]

[1] State Key Laboratory of Frozen Soil Engineering, Northwest Institute of Eco-Environment and Resources, Chinese Academy of Sciences, Lanzhou 730000, China
hxbxmty@lzb.ac.cn
[2] University of Chinese Academy of Sciences, Beijing 100049, China

Abstract. In seasonally frozen ground regions, the pile foundation embedded in frost-susceptible soils can be subjected to tangential frost-heave force, which is defined as an uplift force that acts along the pile-frozen soil interface. The belled pile in seasonally frozen ground regions was designed to produce a self-anchoring force so as to enhance bearing capacity against to frost-heave force. In order to reveal the function of anti-frost jacking of belled pile and the effect of different base angles belled pile embedded in seasonally frozen ground area, one-dimensional freezing test was conducted for three different base angel belled piles and a uniform-section pile. Freezing process, displacements of piles and soil surface, and soil pressure were monitored respectively. The results showed that the ultimate displacement of three belled piles were approximately 1/5 of uniform-section pile. The frost heave of ground surface beside the pile presented a funnel-shaped state, which illustrated that the pile restricted the free frost heave of the soil. The soil pressure beside the enlarged base presented different patterns with the change of frost depth and base angle. On this basis, the function of anti-frost jacking of belled pile embedded in frost-susceptible soils could be put forward ultimately.

Keywords: Seasonally frozen ground regions · Belled pile · Anti-frost jacking

1 Introduction

In seasonally frozen ground regions, the pile can be jacked out because of frost-heave force induced by frost heave of soil between pile and frozen soil. In order to reduce the effect of frost-heave force, the structure of belled pile is chosen to study. The belled pile (enlarged base pier) is a structure that the base diameter is larger than the pile shaft diameter, which can enlarge both bearing capacity and uplift capacity [1]. Many experimental studies have been conducted on the belled pile in unfrozen soil or sand [2–4]. Although there were lots of experiments of belled pile conducted in unfrozen soil to measure the ultimate uplift force or lord transfer of belled pile. However, there was few belled pile tests conducted in seasonally frozen ground. The paper showed the test of

© Springer Nature Switzerland AG 2018
W. Wu and H.-S. Yu (Eds.): *Proceedings of China-Europe Conference on Geotechnical Engineering*, SSGG, pp. 1368–1371, 2018.
https://doi.org/10.1007/978-3-319-97115-5_102

three different base angles belled pile and a uniform-section pile embedded in frost-susceptible soils, the freezing procedure of the experiment and the characteristics of anti-frost jacking of belled pile were discussed in the end.

2   Methods and Materials

One-dimensional frost heave experimental methodology was adopted. The arrangement of the piles and sensors presented in Fig. 1(a), the data collection frequency of each sensor was every five minutes. The procedure of the test was as follows: (1) The preparation of the experiment included the treatment of the soil and test the sensor; (2) Filled the soil layer by layer; (3) The wind was used as the method to lower the temperature of test box. (4) When the test of freezing was finished, the outer box was opened, and the frozen soil could melt at the room temperature.

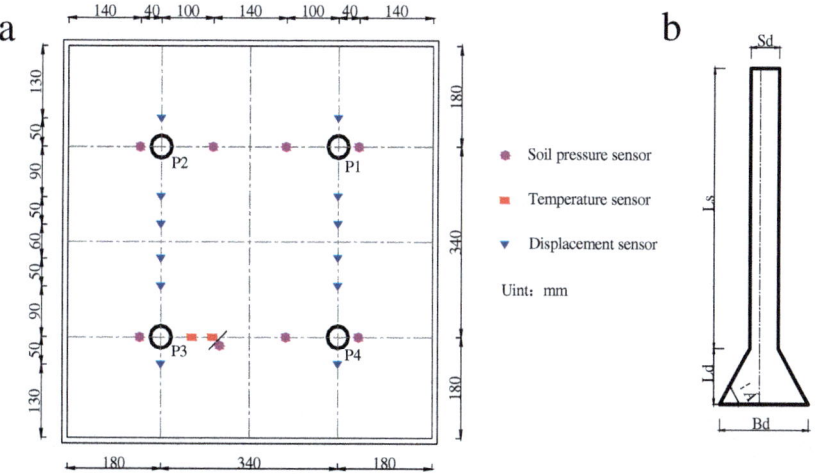

Fig. 1.   (a) Plane layout of sensors; (b) Schematic diagram of belled pile

The soil sample was chosen to be silty clay, which physical properties were as follows: the liquid limit of the experimental soil is 33.4% and the plastic limit is 17.9% and plasticity index is 15.5. The experimental soil water content was set as 34%, and the model pile was made up by concrete, the size of model belled pile was labeled in Fig. 1(b) and the geometrical parameters of each pile were presented in Table 1.

Table 1.   The geometrical parameters of each pile

No.	SD (mm)	BD (mm)	α (°)	Ls (mm)	Lb (mm)
P1	40	98	45	320	30
P2	40	98	60	300	50
P3	40	98	70	270	80
P4	40	40	90	350	–

3   Results

3.1   Temperature

The Fig. 2 showed the freezing depth during freezing process, we assumed that the freezing temperature was 0 degrees centigrade and the environmental temperature was set at −15 degrees centigrade. The test would be stopped once the temperature of 20 cm of embedded depth reached 0 degrees centigrade. The whole time from start to finish lasted 72 h.

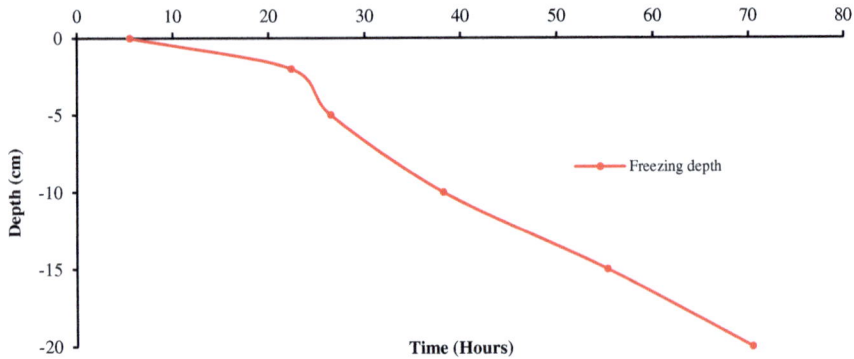

Fig. 2.   The freezing depth during freezing process

3.2   Displacement

The displacement of piles was presented in Fig. 3. With the freezing depth increased gradually, the uniform-section pile started to jack out at the beginning of soil freezing, however, the belled pile was not because of the exist of enlarged base. The Fig. 3 also illustrated that the ultimate displacement of belled pile was approximately 1/5 displacement of uniform-section pile, which proved that the effect of anti-frost jacking of belled pile was greatly apparent.

3.3   Soil Pressure

The Fig. 4 showed the change progress of soil pressure beside the enlarged base. When the freezing depth reached 10 cm, the soil pressure of P1, P3, P4 was almost the same. However, when freezing depth exceeded the 10 cm, the soil pressure was deeply developed. The Fig. 4 also presented that with the base angel increased, the ultimate soil pressure beside the enlarged base was different, P2 owned maximal soil pressure, the next were P3, P4 and P1. The belled pile will closer to uniform-section pile and the self-anchoring force of the enlarged base will decrease with the increase of base angel, that is why the belled pile with 45° base angle could receive more soil pressure than other belled pile with larger base angle and uniform-section pile.

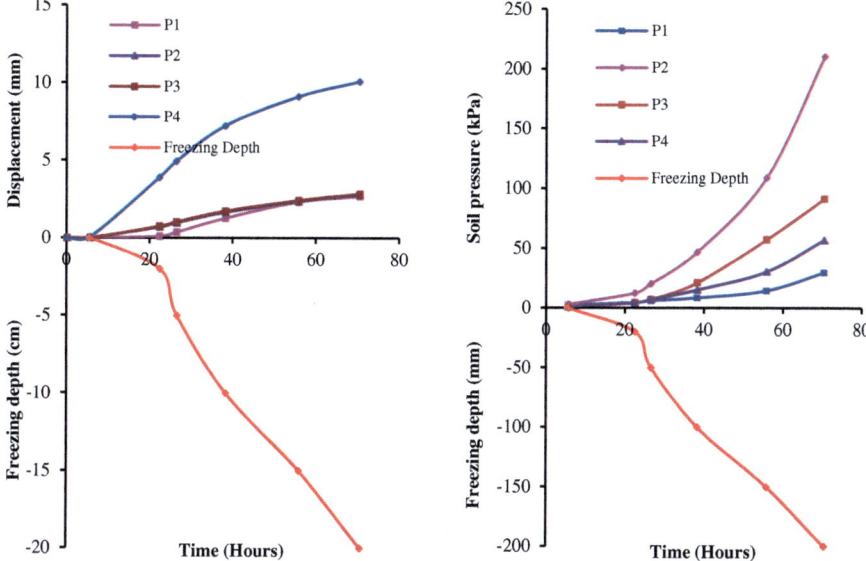

Fig. 3. Displacement of pile with the change of freezing depth

Fig. 4. Soil pressure beside the enlarged base with the change of freezing depth

4   Conclusions

The existence of belled pile in frozen soil restricted the free frost heave of soil beside the pile, then the tangential frost-heave force would appear at the interface of frozen soil and pile shaft conversely, that is the main reason of frost heave of the pile. So the displacement of frost heave of belled pile was much less than uniform-section pile and the soil pressure beside the enlarged base was larger than that pile with no enlarged base. On the basis of these results of one-dimensional frost heave experiment of belled pile, the characteristic of anti-frost jacking of belled pile was greatly apparent because of the existence of enlarged base.

References

1. Gao, G., Jian, J., Gu, B.: Comparative study on belled and equal-diameter piles. Chin. J. Rock Mech. Eng. **24**(03), 502–506 (2005)
2. Nazir, R., Moayedi, H., Pratikso, A., et al.: The uplift load capacity of an enlarged base pier embedded in dry sand. Arab. J. Geosci. **8**(9), 1–12 (2015)
3. Liu, W., Zhou, J., Mongke, T.: Uplift bearing tests and calculations of belled piles in Loess of Arid regions. In: International Conference on Case Histories in Geotechnical Engineering, New York (2004)
4. Chae, D., Cho, W., Na, H.Y.: Uplift capacity of belled pile in weathered sandstones. Int. J. Offshore Polar Eng. **22**(04), 297–305 (2012)

Experimental Study on the Interaction Between Pipe and Soil Under Frost Heave Condition

Long Huang[1,2(✉)] and Yu Sheng[1]

[1] State Key Laboratory of Frozen Soil Engineering, Northwest Institute of Eco-Environment
and Resources, Chinese Academy of Sciences, Lanzhou 730000, China
18189538646@163.com
[2] University of Chinese Academy of Sciences, Beijing 100049, China

Abstract. There are many large gas (or oil) pipelines whole or part to be built in permafrost areas all over the world, in these areas, the problem of pipeline engineering in low temperature environment is quite serious. For this reason, we carry out a mechanism experiment of the interaction between the pipe and the frozen soil in order to explore the deformation and stress coordination of the pipe-soil during the frost heave process. In this experiment, we can monitor the temperature, moisture content, soil pressure and pipeline strain of the whole model box. Finally, through analysis of experiment data, it turns out that with decreasing the temperature, unfrozen water content decrease, the frost heave increase, the soil pressure and pipeline strain increase gradually, at the same time also pipeline warp, but the amount of frost heaving is far less than free soil due to pipeline constraints, the pressure at the bottom of the pipe shows a tendency to be larger than the middle. With the increase of the number of freeze-thaw, the residual deformation of the pipe becomes more and more, and the pipeline strain and the pressure of pipe bottom gradually attenuate.

Keywords: Pipeline · Frost heave · Interaction · Mechanism experiment Monitor

1 Introduction

Since 1990s, there were many well-known foreign universities and research institutions have begun to carry out the research on the field observation, monitoring and testing of frost heave and thawing. At first, Nixon et al. Monitored and recorded the temperature changes along the Roman well oil pipelines, and simulated a simple mode of freezing and thawing around the pipeline, finally, they got an effective application model [1]. When pipeline passes through the uneven and sensitive permafrost regions, it will cause the stress gradient between frozen and thaw soil, and cause the vertical displacement of buried pipeline and additional stress of frozen pipe wall. Some Japanese scholars have done the pipe test under uneven subsidence, the results show that the pipeline presented an elastic reaction state in the nonsubsidence area, while a large deformation occurred in the subsidence area, and the deformation state and the stress state of the two areas

© Springer Nature Switzerland AG 2018
W. Wu and H.-S. Yu (Eds.): *Proceedings of China-Europe Conference
on Geotechnical Engineering*, SSGG, pp. 1372–1375, 2018.
https://doi.org/10.1007/978-3-319-97115-5_103

were different [2]. The University of Alaska and Hokkaido University in Japan had carried out the original model experiments. In order to predict the influence of vertical bending under frost heaving, ice planing, mobile slope, freezing and thawing to pipelines and facilities, Canadian researchers built an experimental equipment for frost heave in Calgary to study the law of dynamic phase transition [3]. In the early 21st Century, a geotechnical centrifuge model test machine was established to study the problem of frost heave and thawing for ground deformation and pipes, this technology not only simulated the operation mode of soil and pipeline, but also related to the plastic pipes considering the effect of fluid medium in the pipeline and evaluation of pressure in pipeline [4]. Since the beginning of the 70s of last century in China, a number of research institutions and scholars have carried out the scaling test of pipe-soil, and simulated the normal operation of pipelines, analyzed the problems of oil temperature and heat conduction in pipes, and used them to guide production. So far, there are a lot of theoretical analysis and results on the permafrost research by scholars at home and abroad, but there are few reports on buried pipelines in permafrost regions, some small-scale experiments are completed only on the basis of on-site monitoring and simple engineering, and basically rely on experience, the effect is not obvious, and it cannot be widely promoted, so it needs further development.

2 Principle and Material

The experimental model, based on the similarity principle and the dimensional analysis, will satisfy the similarity criteria as far as possible to simulate the cooperation between the pipe and the frozen soil. The experiment set some controllable parameters, to observe the law of frost heaving force and deformation by monitoring the temperature, moisture, frost heave displacement, soil pressure and pipeline strain, to explore the cooperation mechanism of pipeline and permafrost under different variables. The equipment involved in this experiment mainly includes a small environment model test box with a model of DRP305, Open thermal insulation box, Pipeline support, some data acquisition equipment, various types of sensors. In addition, the materials used in the experiment include: 201 stainless steel pipes, soil samples. According to the test needs, we carry out the initial and limit moisture content test, grain size test, thermal conductivity test, freezing temperature and unfrozen water content, etc. The basic test is based on the standard for soil test method (GB/T50123-1999) of China.

3 Frost Heave Experiment of Pipe-Soil

The experiment was carried out in the small environment model test box. Before the test started, preparation soil samples and the resistance strain gauges should be pasted on the corresponding position in the pipeline, Secondly, the pipe is fixed on the bracket according to the required constraint type, and then it was moved into the middle position of the test box. To fill in the saturated sand about 10 cm thick, and buried temperature, pressure, moisture sensors in the 7.5 cm plane, and lay seepage prevention in the 10 cm plane. Then to fill in the soil sample about 25 cm thick and set other sensors. The sensors were buried into four layers. The corresponding positions of sensors in each layer were: along the longitudinal section of the pipe bottom 1/2, 3/8, 1/4, 1/8. The experiment was divided into three stages: the first stage was the constant temperature stage at 5 °C, the second stage was the cooling stage at −15 °C, the third stage was the heating process with opening the box in the natural state. The structure chart of experiment is shown in Fig. 1 below.

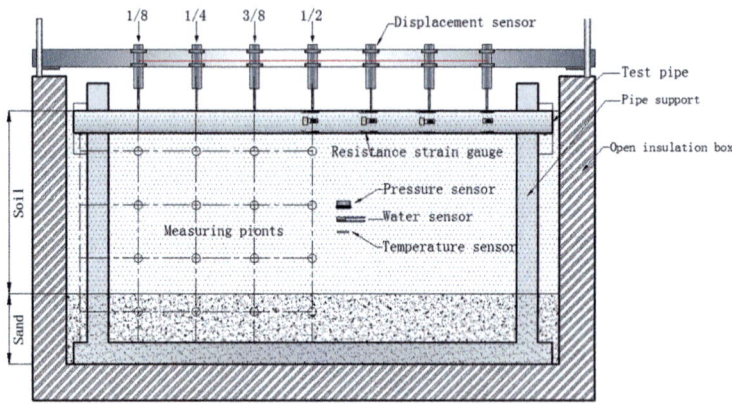

Fig. 1. The structure chart of experiment.

4   Experimental Results

1. During the freezing process of the experiment, when the water in the soil changes from the supercooled state to the frozen state, the content of unfrozen water has a sudden change. As the temperature continues to decrease, the pore water content will decrease rapidly, at this time, the temperature has great influence on the unfrozen water content in the soil. As the temperature continues to decrease, the content of unfrozen water in the soil tends to be stable (Fig. 2).

2. Along with the frost heaving at the bottom of pipe, the pipe was warped due to frost heave. As the pipe ends were constrained, the closer the pipe end was, the smaller the pipe deformation was, and the frost heave deformation of the pipe presented the maximum in the middle position (Fig. 3). At the same time, the changes of soil were also showing a similar law.

3. In the process of experiment, with the increase of frost heaving, the soil pressure at different positions also changed. We obtained the pressure variation curve in the frost heave process at each layer from different measuring points. It can be seen from the figure that with the gradual increase of frost heaving, the soil pressure was also increasing. The closer to the pipe end, the greater the pressure was, the pressure was released from the pipe deformation, Therefore, the pressure in the middle of the pipeline was smaller than that in the pipe end (Fig. 4).

4. In order to understand the change of the pipeline, after the test, we collected micro strain data and calculated the stress values of the vertical and horizontal layout points.

The stress values of 1/8, 1/4, 3/8 and 1/2 were analyzed to get the relationship that shown in Fig. 5 below. From the figure curve, with the pipe warpage increases, pipe stress increases gradually, and the stress value at middle section of the pipeline was greater than the pipe end, and the longitudinal stress and transverse stress also increases, it also indicated that the transverse deformation of the pipe has occurred.

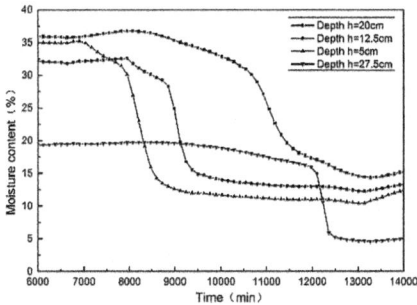

Fig. 2. Changes of moisture content

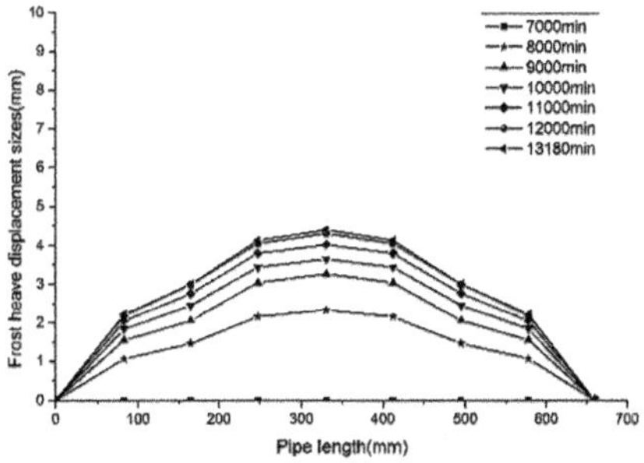

Fig. 3. Pipe displacement in frost heaving.

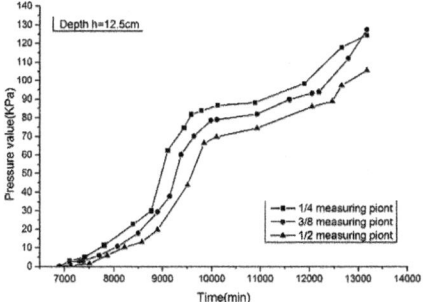

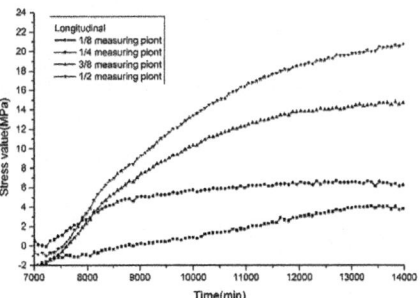

Fig. 4. Changes of soil pressure.

Fig. 5. Longitudinal stress changes of pipe.

References

1. Nixon, J.F., MacInnes, K.L.: Application of pipe temperature simulator for Norman Wells oil pipeline. Can. Geotech. J. 33(1), 140–149 (1996)
2. Gao, H.Y., Feng, Q.M.: Response analysis for buried pipelines through settlement zone. Earthquake Eng. Eng. Vibr. 17(1), 68–75 (1997)
3. Hu, Z.L., Wu, M., Chen, Y., et al.: The present situation and trend of pipeline safety study of frost heave. Oil Gas Storage Transp. 30(12), 881–882 (2011)
4. O'rourke, M., Gadicherla, V., Abdoun, T.: Centrifuge modeling of PGD response of buried pipe. Earthquake Eng. Eng. Vibr. 4(1), 69–73 (2005)

A Novel Preparation Device for Remolded Hollow Cylinder Specimen of Frozen Soil

Lele Lei[1,2], Dayan Wang[1(✉)], Yongtao Wang[3], Dun Chen[1], Yan Guo[1], and Wei Ma[1]

[1] State Key Laboratory of Frozen Soils Engineering, Northwest Institute of Eco-Environment
and Resources, CAS, Lanzhou 730000, Gansu, China
1064612453@qq.com
[2] University of Chinese Academy of Sciences, Beijing 100049, China
[3] Institute of Transportation, Inner Mongolia University,
Hohhot 010070, Inner Mongolia, China

Abstract. The frozen soil hollow cylinder apparatus is the main experiment instrument to study the mechanical behaviour of frozen soil under complex stress paths including stress rotation. However, it is difficult to prepare hollow cylinder specimen due to the thin wall. In this paper a novel preparation device for remolded hollow cylinder specimen of frozen soil is designed. The universal test machine is used to provide an axial compressive stress, for it can ensure accurate control in axial force and deformation along preparing process. Hence, it can improve the efficiency in specimen preparation. Two pedestals with 8 blades can directly shape grooves on the upper and bottom surfaces of specimen, which can reduce the disturbance while cutting the specimen. The applicability and reliability of specimen prepared by the novel preparation device in frozen soil hollow cylinder tests is verified.

Keywords: Frozen clay · Hollow cylinder specimen · Preparation device

1 Introduction

Specimen preparation is a key part in geotechnical tests and is also an important factor which would restrict the development of indoor tests [1]. In recent years, with the development of engineering construction in cold regions, the strength characteristics and related theories of frozen soil are experimentally investigated by frozen soil uniaxial tests and triaxial tests. However, the magnitude and directions of principal stress gradually change during loading process in nearly all geotechnical problems. In some cases, such as traffic and earthquake loading, these rotations are continuous and cyclic in nature. The existing test apparatus for frozen soil, such as uniaxial and triaxial instruments, cannot completely simulate these stress paths. Frozen soil hollow cylinder is an experimental device, which can control the rotation of the maximum-minimum principal stresses independently according to the axial load, torque, outer and inner cell pressures [2]. However, the preparation of thin-walled hollow cylinder specimens is a difficult problem in frozen soil laboratory tests.

© Springer Nature Switzerland AG 2018
W. Wu and H.-S. Yu (Eds.): *Proceedings of China-Europe Conference on Geotechnical Engineering*, SSGG, pp. 1376–1379, 2018.
https://doi.org/10.1007/978-3-319-97115-5_104

The research on the preparation method of remolded hollow cylindrical specimens mainly exists in non-frozen soil tests and has been rarely discussed in frozen soil tests. The main methods are the compaction method and the consolidation method. The self-made compaction device designed by Zhang et al. of Zhejiang University [3] can achieve the compaction for hollow cylindrical specimens, but the specimen size and target water content can be affected by compactive effort, stratification height and other factors. The consolidation method is originally proposed by Sheeran and Krizek [4], and it is then improved by Lin [5], Ji [6], and Shen [1] et al. This method is widely used for remolded hollow cylindrical specimen preparation, but the inevitable uniform distribution of soil particles will appear in the consolidation process, which may greatly influence the specimen's mechanical responses. To obtain reliable results in frozen soil hollow cylinder tests, it is necessary to develop an efficient preparation device for remolded hollow cylindrical specimens.

This paper presents a novel preparation device for remolded hollow cylindrical specimens. The constitutes of the novel preparation device is introduced. Then, the applicability and reliability of the novel device in the preparation of specimens of frozen soil hollow cylinder tests are verified.

2   The Novel Preparation Device

The novel preparation device consists of axial pressure bar, outer and inner steel cylinders, three tightening hoops, upper and bottom pedestals with 8 blades, as shown in Fig. 1. The outer steel cylinder is made up of three round steel plates, and the inner steel cylinders are made up of two round steel plates and a rectangular steel plate. The inner wall of outer cylinder and outer wall of inner cylinder are polished to make the removal of inner and outer cylinder easier and to reduce the disturbance to specimens when removing the modulus. The tightening hoops ensure that the modulus does not deform while preparing a specimen. Two pedestals with 8 ribs can directly shape grooves on the upper and bottom surfaces of specimens.

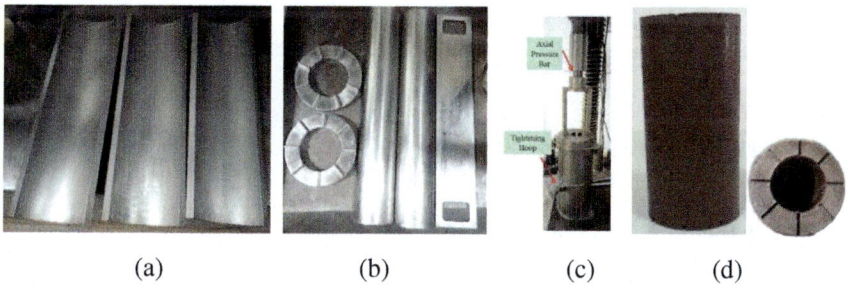

(a)	(b)	(c)	(d)

Fig. 1. Constitutes of the novel device. (a) Outer steel cylinder, (b) Inner steel cylinder (right) and upper and bottom pedestals (left), (c) Integral structure diagram, (d) The prepared specimen

The preparation device for hollow cylinder specimens has some advantages: (i) The design of three round plates for outer and inner steel cylinders can make the removal of inner and outer cylinder easier and reduce the disturbance to specimen when removing the modulus; (ii) The shaped grooves on the upper and bottom surfaces of specimens can avoid the disturbance to the specimen during the cutting process; (iii) The whole process of specimen preparation is conducted on the universal test machine, which can ensure the uniform force distribution in specimen and can improve the sample-making efficiency as well. These advantages provide strong guarantee for the stability and reliability of the test results.

3   Verification

3.1   Verification on Physical Characteristics

The stratified water content of specimens before and after freezing is measured by drying method to verify the physical characteristics of specimens prepared by the hollow cylindrical. First, a specimen is divided into 20 layers along the vertical direction with each layer height being 1 cm. The water content of each layer is averaged by three measuring points in each layer. The results are shown in Fig. 2 with the vertical line being the target water content. Figure 2 shows that the maximum error of the water content in the specimen is less than 0.65%. So, the water content of the specimen prepared by the novel preparation device is more uniform and meet the requirement of the frozen soil hollow cylinder tests.

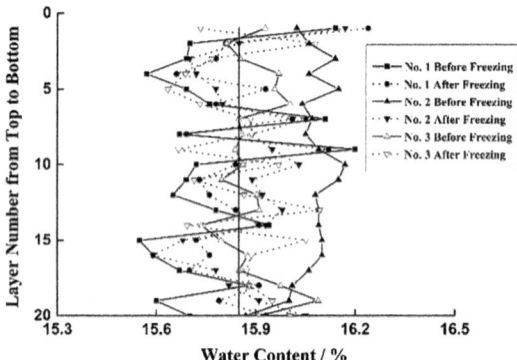

Fig. 2. Water content distribution in specimen before and after freezing

3.2   Verification on Mechanical Characteristics

Five groups repeated tests and two groups monotonic loading along fixed principals stress direction tests are carried out to verify the mechanical characteristics of prepared hollow cylindrical specimen. Each group of repeated tests includes two tests. From the test results, it can be seen that the stress-strain curves in two tests of each group are in great agreement with each other. The maximum strength difference is about 1.8%, and this can be verified that the prepared specimen by the novel device can achieve the repetition of the test. Each group of monotonic loading along fixed principals stress direction tests includes three tests, and the results are shown in Fig. 3. From Fig. 3, it is observed that the strength of frozen clay increases with the temperature decreasing, and the relationship between strength of frozen soil and temperature is nonlinear. The strength of frozen soil is also changed with b-value.

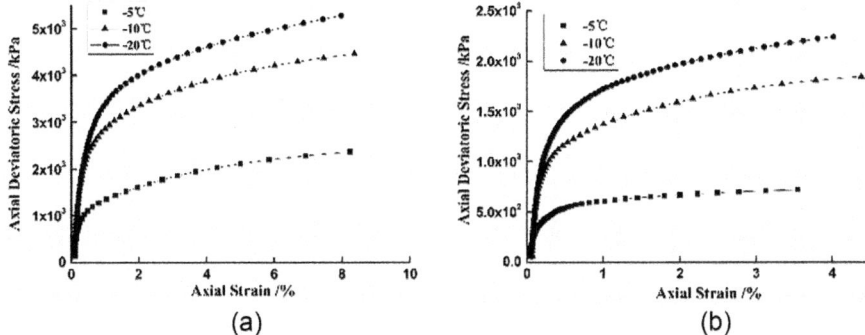

Fig. 3. The results in different temperature stress paths. (a) The result of first group (p = 4.5 MPa, b = 0, α = 30), (b) The result of second group (p = 4.5 MPa, b = 0.5, α = 30)

4   Conclusion

A novel preparation device for remolded hollow cylinder specimen of frozen soil and the constitutes of this device are introduced. From physical and mechanical properties, it is verified that the frozen soil specimen prepared by the novel preparation device meet the requirements of the frozen soil hollow cylinder tests.

References

1. Shen, Y., Zhou, J., Zheng, W.: Study and counter measure on side effects from principal stress axis rotation experiment with hollow cylinder apparatus. Water Resour. Hydropower Eng. **36**(11), 33–37 (2005)
2. Guo, Y., Wang, D.Y., Ma, W., et al.: Development and application of frozen hollow cylinder apparatus. J. Harbin Inst. Technol. **48**(12), 114–120 (2016)
3. Zhang, Q.F., Zhao, Y., Huang, B., et al.: Development of a compacting instrument for reconstituted samples in hollow cylinder apparatus. Res. Explor. Lab. **27**(5), 42–46 (2008)
4. Sheeran, D.E.: Preparation of homogenous soil samples by slurry consolidation. Civ. Environ. Eng. **6**(2), 356–373 (1971)
5. Lin, H., Penumadu, D.: Experimental investigation on principal stress rotation in kaolin clay. J. Geotech. Geoenviron. Eng. **131**(5), 633–642 (2005)
6. Cheng, J.Y., Wang, Y.S.: Technology and application of vacuum prepressing for soil sample. Port Waterw. Eng. **12**, 1–2 (1997)

Centrifuge Model Test on Performance of Thermosyphon Cooled Sandbags Supporting Warm Oil Pipeline Buried in Thawing Permafrost

Guoyu Li[1(✉)] 🆔, Hongyuan Jing[2] 🆔, Nikolay Volkov[3,4] 🆔, Wei Ma[1] 🆔, and Fei Wang[1] 🆔

[1] State Key Laboratory of Frozen Soil Engineering, Cold and Arid Regions Environmental and Engineering Research Institute, Chinese Academy of Sciences, Lanzhou 730000, China
guoyuli@lzb.ac.cn
[2] PetroChina Pipeline Company, Langfang 065000, China
[3] C-CORE, St. John's, NL, Canada
[4] GEOINGSERVICE LLP (Fugro Group), 29 Vernadskogo Avenue, Moscow, Russia

Abstract. The performance of thermosyphon cooled sandbags supporting the warm pipeline in permafrost was simulated in a centrifuge model test. Two pipes of an unsupported and a supported section were evaluated over a period simulating 20 years of operation with seasonal oil and air temperature cycles along the China-Russia Crude Oil Pipeline (CRCOP). The results showed that pipe settlement of the supported pipe was 45% of the settlement of the unsupported pipe. The final thaw bulbs extended about 3.6 and 1.6 times the pipe diameter below the unsupported and supported initial pipe elevations, respectively. The sandbag supports remained frozen throughout the test period due to cooling effect of thermosyphons. The maximum bending stress induced over the span length from bearing of the full cover is equivalent to 40% specified minimum yield strength. Potential buckling of the pipe should be considered as the ground thaws. Suggestions for design considerations for the field conditions along the CRCOP are made based on the model test observations.

Keywords: Oil pipeline · Permafrost thawing · Thermosyphon · Cooling effect
Thawing depth

1 Introduction

The CRCOP in Chinese territory is 953 km in length passing through a 441 km long discontinuous permafrost region in northeastern China [1]. This pipeline is being operated at positive oil temperature causing obvious thaw settlements [2]. A mitigative and control measure, thermosyphon cooled sandbags, was proposed and evaluated by centrifuge tests carried out in C-CORE in Canada [3]. This paper presents the design, results and interpretation of these physical model tests together with recommendations for the application of these results to the field.

© Springer Nature Switzerland AG 2018
W. Wu and H.-S. Yu (Eds.): *Proceedings of China-Europe Conference on Geotechnical Engineering*, SSGG, pp. 1380–1384, 2018.
https://doi.org/10.1007/978-3-319-97115-5_105

2   Testing Method and Process

Two oil pipes were tested in physical model tests at a 1/73rd scale, which was determined by two key dimensions of pipeline setup, namely the pipe span length, 20 m, and pipe diameter, 813 mm in prototype. One pipe was tested to investigate uniform thaw settlement and the second to investigate the effectiveness of thermosyphon cooled sandbags. Soil type and properties, pipeline characteristics, thermal insulation characteristics, oil and air temperatures, thermosyphon characteristics, characteristics of sandbag support in prototype and centrifuge model were determined and simplified partially according to filed conditions and scale law [3].

A standard pipe diameter of 11.1 mm with a wall thickness of 0.25 mm was selected to satisfy the requirement of g-level, and its span length was determined to be 270 mm. Mean monthly oil temperature in field was averaged for two seasons in one year to best control the thermal regime of the pipe. During the 43-minute "cold" season the oil temperature is 2 °C and during the 61-minute "warm" season the oil temperature is 7 °C. Mean monthly air temperature was determined according to the field meteorological data, ranging from a maximum of 19.7 °C and a minimum of −16.83 °C.

A substitute of model soil with a water content of 59% consisting of 50% silt and 50% Kaolin clay was selected based on the filed specific soil information and the plasticity index correlations developed by Nakase et al. [4] due to unavailability of in-situ soil. The inner insulated soil box with inner dimensions of 980 mm by 725 mm was fitted with components needed to fulfill the test requirements. A cooling base coil was installed at its bottom to keep the soil bed frozen during the test. A thin plate was installed along the center line of the inner box forming the two test beds for the two pipes. One layer of the slurry 15 mm thick was placed in the test box, leveled and frozen. After the freezing process, a new layer of slurry was added and the process was repeated. Nine layers of slurry were frozen to result in a soil block with a thickness of 133 mm. A trench 70 mm deep, 25 mm wide and 800 mm long was formed in each of two soil blocks for model pipe installment.

The temperature measurements under the pipes were made using Resistance Temperature Detectors (RTD). The RTD stack was placed in a drilled hole at the bottom of the trench. The locations of RTDs were confirmed with the designed dimensions. The Pore Pressure Transducer (PPT) was placed at the unprotected pipe invert. A linear variable differential transformer (LVDT) was positioned on a bearing plate on top of the clay to measure displacement of the soil block during the test. For the pipe settlement measurement, small plastic pipes were used to provide a conduit through the clay for the LVDT rods to rest on top of the pipe.

The model of sandbags was made of saturated frozen sand with a dimension 2.7 m wide, 1.4 m long and 2.7 m high. Sandbags were installed into the holes in the soil bed at the design locations. The test soil box was placed in an aluminum centrifuge strongbox with 50 mm insulation between the two boxes to minimize heat transfer. A lid containing an insulation and the refrigeration coil was positioned on top of the centrifuge strongbox. To enable cooling of experiments, the centrifuge is equipped with a refrigeration unit that circulates glycol from the refrigeration unit in the centrifuge motor pit through the centrifuge rotary joint and into the experiment box. The oil in the pipes was heated using a wire heater installed in the pipe along the entire length. A control RTD was also installed into the pipe so that it could measure the oil temperature directly. The heater is controlled by an electric supply by adjusting the electric current. The thermosyphons were modeled as required scale law using a 12.7 mm diameter brass rod attached to a copper pipe. Glycol refrigerant was pumped through the pipe to facilitate heat transfer up through the brass rod, out of the soil and into the flowing glycol. When the thermosyphons were scheduled to be turned off, the flow of glycol to the thermosyphon pipes was stopped. Two RTDs were attached to the thermosyphon so that the heat flow could be calculated. The heat flow was adjusted to be 1 W in winter for 35 min (at negative air temperatures) and 0 W in summer for 69 min (at positive air temperatures).

The data acquisition system provided instant monitoring of the test data, including the oil temperature, which allowed adjustments of the electric current to achieve the required oil temperature, the heat flow function, air and base soil temperatures, which allowed adjustments of temperature and/or volume flow rate of the liquid refrigerant to achieve the required temperatures. The centrifuge model test duration from reaching the target acceleration to spin-down was approximately 35.5 h. Warming of the oil pipes and cooling of thermosyphons began one hour 20 min after the test started and all the equipment was checked. The test included 20 thermal cycles of the oil pipeline and the surrounding thermosyphons. Each cycle lasted 104 min and modeled one-year operation of the pipeline.

3   Results and Discussion

3.1   Temperature

The temperature data with time for the RTDs attached to and under both pipes are collected and analyzed, respectively. They showed that the temperatures of the pipe wall varied between −0.5 °C and +4 °C in agreement with oil temperature cycles. The temperature at the soil-thermosyphon interface varied from −3 °C to −0.5 °C. The soil surface temperature stayed below 0 °C during the entire test. The thaw front penetration was calculated below the center of the pipes. For the unsupported pipe, the thaw front penetrated deeper corresponding to the absence of both thermosyphon cooling and sandbag supports. The thaw front advanced to about 20 to 40 mm below the initial pipe elevation for unsupported and supported pipes respectively.

3.2   Pore Pressure

The pore pressure data showed a general increase as the thaw front progressed and the thaw bulb grew. In turn, the cyclic drops of pore pressure are caused by suction as soil back-freezing occurs due to the temperature drop from +7 °C down to +2 °C.

3.3 Settlement

The LVDT measurements on both pipes were carried out. The thaw strain of unsupported pipe is more than that of supported pipe due to the sandbag support and cooling effect of thermosyphons. The initial delay in developing thaw settlement until thaw bulbs are 18 mm below the unsupported pipe and 8 mm below the supported pipe middle appeared to be associated with the initial growth of an axisymmetric thaw bulb. The response of the supported pipe was constrained by the sandbags.

The pipe settlement profile indicates a combination of global (rigid) and local (flexure) deformations. The unsupported pipe is settling with the thawed material at about 35% vertical strain, which is more than that of the supported pipe, 30% of the thaw bulb depth due to the influence of the supports.

4   Conclusions

This paper presents the results of a centrifuge model test to the CRCOP. The thaw settlement response of two pipes was evaluated. The sandbag supports under the pipe kept frozen by a pair of thermosyphons. The 4.5 mm pipe settlement of the supported pipe was 45% of the 10 mm settlement of the unsupported pipe over the 33 h period in model test. The settlement of the unsupported pipe was about 35% of the thaw bulb depth, which is more than that of the supported pipe, 30% of the thaw bulb depth. Finally, the thaw bulbs extended approximately 3.6 and 1.6 times the pipe diameter below the initial elevations of two pipes, respectively. The maximum bending stress induced over the span length from bearing of the full cover over the pipe is equivalent to about 40% specified minimum yield strength. Potential buckling of the pipe should be considered as the ground thaws.

A combination of schemes may be required for an optimal solution to thaw settlement. Two suggestions were made including oil chilling and foam injection with a low thermal conductivity material.

Acknowledgements. This work was funded by the National Natural Science Foundation of China (Grant Nos. 41672310, U1703244 and 41630636), the National Key Research and Development Program (2016YFC0802103) and STS Research Project of CAS (Grant No. CHHS-TSS-STS-1502). Authors also express kind thanks for Ryan Phillips, Gerry Piercey, Ken Chi, Karl Tuff, Derry Nicholl and Karl Kuehnemund in C-CORE in Canada for their efforts to carry out the test.

References

1. Li, G., Sheng, Y., Jin, H., et al.: Forecasting the oil temperatures along the proposed China-Russia crude oil pipeline using quasi 3-D transient heat conduction model. Cold Reg. Sci. Technol. **64**(3), 235–242 (2010)
2. Li, G., Ma, W., Wang, X., et al.: Frost hazards and mitigative measures following operation of China-Russia crude oil pipeline. Rock Soil Mech. **36**(10), 2963–2973 (2015)

3. C-CORE. Modelling of Thaw-Settlement Related to Mohe Daqing Crude Oil Pipeline. C-CORE Report R-12-080-928, Revision 1.0 (2012)
4. Nakase, A., Kamei, T., Kusakabe, O.: Constitutive parameters estimated by plasticity index. J. Geotech. Eng. **114**(7), 844–858 (1988)

Experimental Study on Creep Behavior for Thawed Saturated Clay

Bo Lin, Feng Zhang$^{(\boxtimes)}$, and Decheng Feng

School of Transportation Science and Engineering,
Harbin Institute of Technology, Harbin, China
zhangf@hit.edu.cn

Abstract. The shear properties and creep properties of thawed saturated clay were studied at different stress levels based on a series of undrained triaxial shear tests and creep tests. A hyperbolic creep model was employed to fit the obtained strain-time curves based on analyzing the evolution of the creep strain, the fitting parameters were also determined. The results show that the hyperbolic creep model can be used for predicting the evolution of creep strain for thawed saturated subgrade clay.

Keywords: Creep behavior · Hyperbolic creep model · Thawed saturated clay

1 Introduction

In seasonally frozen regions, subgrade soils are not only being subjected to large numbers of repeated traffic loading [1, 2], but also suffer from the effect of freeze-thaw cycles [3, 4], which could make the creep behavior of thawed subgrade soils as important as the dynamic behaviors.

2 Triaxial Creep Test

The saturated specimens were subjected to 7 cycles of freeze-thaw effect. The freezing process lasts for 14 h at a temperature of $-25\ °C$, the thawing process lasts for 10 h at a temperature of 25 °C. The shear properties were investigated first. Values of the failure strength under the confining pressures of 30 kPa, 60 kPa, 90 kPa, 120 kPa and 150 kPa are 49.1 kPa, 98.8 kPa, 128.4 kPa, 163.9 kPa and 198.3 kPa, respectively. The triaxial creep test program was showed in Table 1. Specimens TC-01 ∼ 05 were used to evaluate the effects of static deviatoric stress $\Delta\sigma$. A parameter D_r, which reflects the level of $\Delta\sigma$ was employed. Where $D_r = \Delta\sigma/(\sigma_1 - \sigma_3)_{failure}$, $(\sigma_1 - \sigma_3)_{failure}$ is the failure strength. Specimens TC-06 ∼ 08 were used to evaluate the effects of confining pressure σ_3. Each of the tests was lasted for 24 h. Specimens TC-09 was used to validate the empirical creep model, which was tested under five stages loading, each of the stage was lasted for 24 h.

© Springer Nature Switzerland AG 2018
W. Wu and H.-S. Yu (Eds.): *Proceedings of China-Europe Conference on Geotechnical Engineering*, SSGG, pp. 1385–1389, 2018.
https://doi.org/10.1007/978-3-319-97115-5_106

Table 1. Triaxial creep test program for thawed saturated clay.

Specimens	σ_3 /kPa	$\Delta\sigma$ /kPa	D_r	Specimens	σ_3 /kPa	$\Delta\sigma$ /kPa	D_r
TC-01	60	19.1	0.199	TC-05	60	70.0	0.731
TC-02	60	31.8	0.332	TC-06	90	70.0	0.545
TC-03	60	44.6	0.466	TC-07	120	70.0	0.427
TC-04	60	57.3	0.598	TC-08	150	70.0	0.353
TC-09	60	Loading sequence (kPa): $19.1 \rightarrow 31.8 \rightarrow 44.6 \rightarrow 57.3 \rightarrow 70.0$						

3   Test Results

It can be observed from Fig. 1 that all the axial strains present the same evolution law, that is, the axial strain increases sharply in the first 1 h, and gradually changes to a stable state when the creep duration exceed 4 h. Figure 1(a) presents that the higher the applied deviatoric stress $\Delta\sigma$, the larger instantaneous strain the specimens reach during the first 1 h. Meanwhile, Fig. 1(b) presents that, a larger confining pressure has a limit effect on evolution of the creep deformation. Figure 1(c) shows the evolution of axial strain under five kinds of deviatoric stress. Each of the creep curves experiences a same evolution law which is coincident with the evolution law presented in Fig. 1(a) and (b).

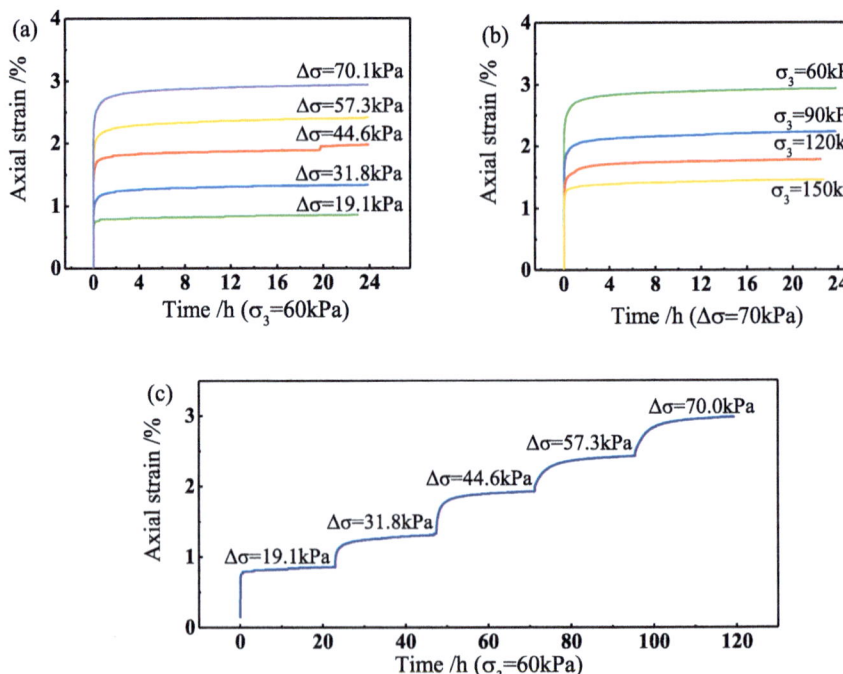

Fig. 1. Creep curves under different test conditions

Generally, the Singh-Mitchell model [5] and Mesri model [6] can be classified as the typical creep models. However, these two models could be imprecise in fitting the attenuate–stable creep behaviors, as the models describe the stress–time relationships by means of power functions. It is known that power function is incapable in fitting the test data when the creep behavior turns into the stable stage. Therefore, a hyperbolic function was introduced to fit the evolution of the creep strain for thawed saturated clay, which is written as,

$$\varepsilon = AD_r^n t/(t+B) \tag{1}$$

where ε is the axial strain. D_r reflects the level of applied deviatoric stress. A, n, B are the fitting parameters. t(h) is the creep duration. When t tends to be infinite, axial strain ε reach the maximum value ε_{max}, that is,

$$\varepsilon_{max} = \lim_{t\to\infty} AD_r^n t/(t+B) = AD_r^n \tag{2}$$

An equation which can describe relation between t/ε and t can be obtained by substituting Eq. (2) into Eq. (1):

$$t/\varepsilon = t/\varepsilon_{max} + B/\varepsilon_{max} \tag{3}$$

where ε_{max} and B are the constant. The t in the right side of Eq. (3) can be considered as an independent variable, and t/ε can be considered as a dependent variable. Values of ε_{max} and B were obtained by fitting the test data of specimens TC-01∼05 though Eq. (3). The fitting results are showed in Fig. 2. Values of ε_{max} for specimens TC-01∼05 are 0.862, 1.342, 1.964, 2.418 and 2.951, respectively. Values of B are 0.283, 0.251, 0.384, 0.219 and 0.168, respectively, but the ultimate value for B should be determined by examining the experimental data, because B controls the sharply increasing tendency of axial strain as mentioned in the discussion above. It was found that a smaller value for B can provide a better fitting results. The value of B was determined as 0.1 in consideration of the attenuate–stable creep behaviors presented in Fig. 1.

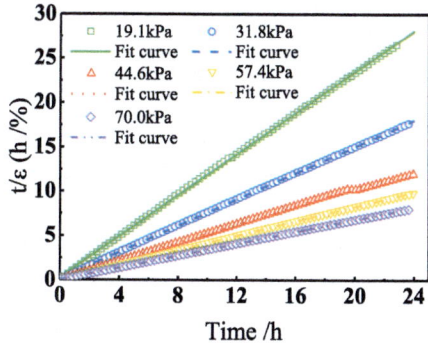

Fig. 2. Relations between t/ε and t

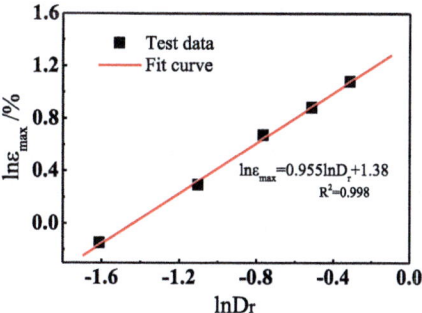

Fig. 3. Relation between ε_{max} and D_r

Taking the logarithm for Eq. (2) can form the following equation,

$$\ln \varepsilon_{\max} = n \ln D_r + \ln A \tag{4}$$

As values of $\varepsilon_{\max}$ and D_r were obtained above. Values of A and n can be determined as 3.976 and 0.955 by fitting the relations between $\ln \varepsilon_{\max}$ and $\ln D_r$ though Eq. (4). The fitting results are showed in Fig. 3. So far, values of A, n, B were determined based on the test data from specimens TC-01 $\sim$ 05. The expression for Eq. (1), i.e., empirical creep model for thawed saturated clay was determined accordingly,

$$\varepsilon = 3.976 D_r^{0.955} t / (t + 0.1) \tag{5}$$

It should be noticed that specimens TC-06 $\sim$ 08 can be used for validating the model between these specimens are not used for building the empirical model. Meanwhile, evolutions of the axial strain during each of the five-stage loading are uniform, the creep curves of the thawed saturated clay under five deviatoric stress states were obtained from the raw data by Mr. Chen's method. So specimen TC-09 can also be used for validating the empirical model. Comparisons between test data and calculated curves are presented in Fig. 4. It can be observed that, although the prediction of the axial strain appears to be somewhat inaccurate, such as the test deviatoric stress is 31.8 kPa presented in Fig. 4(a), the calculated curves can capture the increasing tendency of creep strain with the increasing of loading time, the overall agreement between the calculated curves and test data can be acceptable.

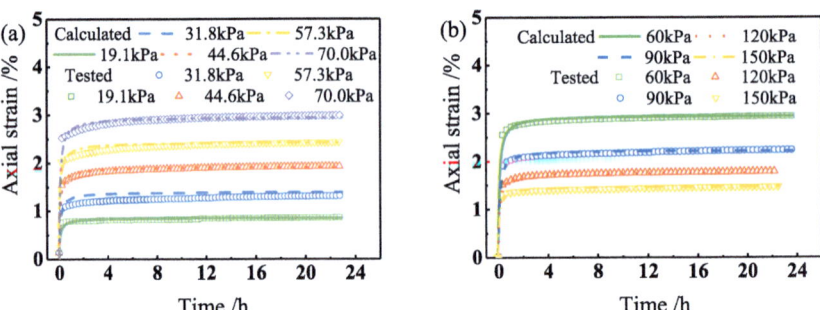

Fig. 4. Comparisons between test data and calculated results

References

1. Li, Q., Ling, X., Wang, L., et al.: Accumulative strain of clays in cold region under long-term low-level repeated cyclic loading: experimental evidence and accumulation model. Cold Reg. Sci. Technol. **94**, 45–52 (2013)
2. Lin, B., Zhang, F., Feng, D.: Long-term resilient behaviour of thawed saturated silty clay under repeated cyclic loading: experimental evidence and evolution model. Road Mater. Pavement Des. 1–15 (2017)

3. Qi, J., Vermeer, P.A., Cheng, G.: A review of the influence of freeze-thaw cycles on soil geotechnical properties. Permafr. Periglac. Process. **17**(3), 245–252 (2016)
4. Wang, D., Ma, W., Niu, Y., et al.: Effects of cyclic freezing and thawing on mechanical properties of Qinghai–Tibet clay. Cold Reg. Sci. Technol. **48**(1), 34–43 (2007)
5. Singh, A., Mitchell, J.K.: General stress-strain-time function for clay. J. Clay Mech. Found. Div. ASCE **94**(SM1), 21–46 (1968)
6. Mesri, G., Febres-Cordero, E., Shields, D.R., et al.: Shear stress-strain-time behaviour of clays. Géotechnique **31**(4), 537–552 (1981)

Compaction Behavior of Clay-Gravel Mixtures Under Normal and Low Temperature

Yang Lu[1,2(✉)], Sihong Liu[1], Meng Yang[1], and Yonggan Zhang[1]

[1] College of Water Conservancy and Hydropower, Hohai University,
Xikang Road, 1, Nanjing 210098, China
luy@hhu.edu.cn
[2] Department of Geotechnical Engineering and Geo-Sciences,
Universitat Politècnica de Catalunya, Calle Jordi Girona, 1-3, 08034 Barcelona, Spain

Abstract. Now in China, more and more clay-gravel mixtures in clod regions have to be used in a variety of geotechnical applications. Compaction state of compacted soil mixtures is closely linked to the properties of permeability, stiffness and strength. However, compaction behaviors under different temperature conditions are seldom studied. In this study, a series of dynamic compaction tests on clay-gravel mixtures under normal (room temperature) and cold temperature ($-10\ °C$) were conducted to investigate the effect of gravel content and temperature on the compaction behaviors. The experimental results show that the compaction behavior under cold temperature is significantly different from that under normal temperature. The finding has potential implications for the construction of earthworks in cold or seasonally frozen regions.

Keywords: Clay-gravel mixtures · Compaction behavior · Gravel content
Cold temperature · Frozen soil mechanics

1 Introduction

Clay-gravel mixtures have been used successfully in a variety of geotechnical applications, such as, homogeneous embankments, cores for dams, and compacted earth linings (Shelley and Daniel 1993; Li et al. 2013). Compaction has always been regarded as a process to 'fabricate' soils, and the compaction state of soil mixtures is generally closely linked to the properties of permeability, stiffness and strength (Tarantino and De Col 2008; Alonso et al. 2013). The compaction plane defined by Proctor (1933) is a convenient procedure to represent the compaction states of a given soil, which could reflect the fundamental relationship between achievable dry density and water content for a given compaction energy. Some previous attempts have been made to explore the compaction behaviors of clay-gravel mixtures (Donaghe and Torrey 1994; Chinkulkijniwat et al. 2010; Rücknagel et al. 2013). However, little attention has been paid on the compaction behaviors of soils under cold temperature conditions, especially for the clay-gravel

© Springer Nature Switzerland AG 2018
W. Wu and H.-S. Yu (Eds.): *Proceedings of China-Europe Conference
on Geotechnical Engineering*, SSGG, pp. 1390–1393, 2018.
https://doi.org/10.1007/978-3-319-97115-5_107

mixture. In this study, a series of dynamic compaction tests were performed on clay-gravel mixtures under normal and cold temperatures to investigate the effect of gravel content and temperature on the compaction behaviors.

2 Materials and Methods

The tested materials were taken from an earth-rockfill dam construction site located in the alpine-cold region of southwest China, where extreme low temperatures can reach −16 °C. Low temperature induced-freezing and freeze-thaw cycles will have potential effect on soil compaction during the construction of the dam core wall. The Atterberg limits of the brown clay were found to be a liquid limit of 28.0% and a plastic limit of 15.5%, and the specific gravity to be 2.71. The slate gravel with a maximum diameter of 20 mm was used in this study. The shape of the gravel was subangular. The grain size distribution curves for the tested materials were presented in Fig. 1(a). The heavy dynamic compaction test was used due to the maximum gravel particle size of 20 mm. The specimens measured 152 mm in diameter and 116 mm in height. Each was compacted in five equal layers with 56 blows per layer. With a rammer of 4.5 kg dropped from a height of 457 mm. The specific energy of the test can be calculated to be 2684.9 kJ/m^3. The compaction tests under cold temperature were conducted in the frozen soil laboratory at Hohai University. The testing environment and procedures were shown in Fig. 1(b)–(f). For the normal temperature, the dry density versus moisture content was tested for clayey soils with gravel contents of 0%, 10%, 30%, 50%, 70% and 80%. For the cold temperature, the compaction planes of the mixtures with gravel contents of 10%, 30% and 50% were obtained.

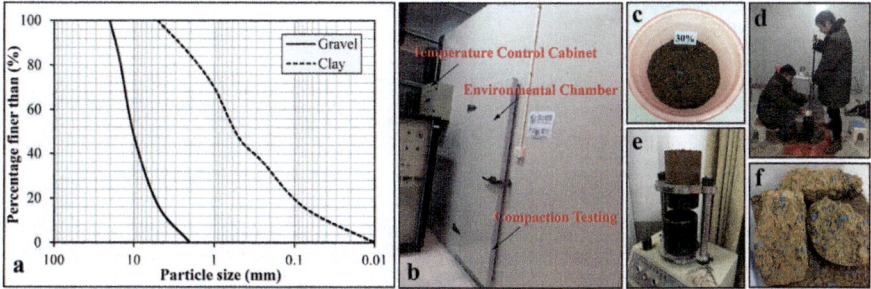

Fig. 1. (a) Grain size distribution of the tested materials; (b) Environmental chamber; (c) soil mixtures mixing; (d) dynamic compression testing under −10 °C; (e) Releasing sample mold; (f) Morphology of broken mixtures.

3 Results, Discussion and Concluding Remarks

The changes of the overall dry density versus moisture content for the tested clay-gravel mixtures were measured by a series of heavy dynamic compression tests as aforementioned. The results obtained at different gravel contents are plotted in Fig. 2(a). It can be seen that with the increasing of molding water content, the overall dry density grad-ually increased to a maximum value and then tended to decrease. The water content at the maximum dry density is the optimal moisture content for the tested soil mixtures. The compaction curves of soil mixtures for different gravel contents all presented a single peak shape, which is almost similar to that of pure clay soils. Therefore, the mixing of gravel into clay soils could not change the shape of compaction curves. Based on the experimental results in Fig. 2(a), the maximum dry density and corresponding optimum water content for each compaction curve of clay-gravel mixture can be determined, as shown in Fig. 2(b). It is apparent that the maximum dry density increased to a peak value of about 2.35 g/cm^3 until an abrupt reduction occurred with the increasing of gravel content at the threshold fraction (approximately 70%). However, the optimum water content presented a continuous decrease upon increasing of the gravel content.

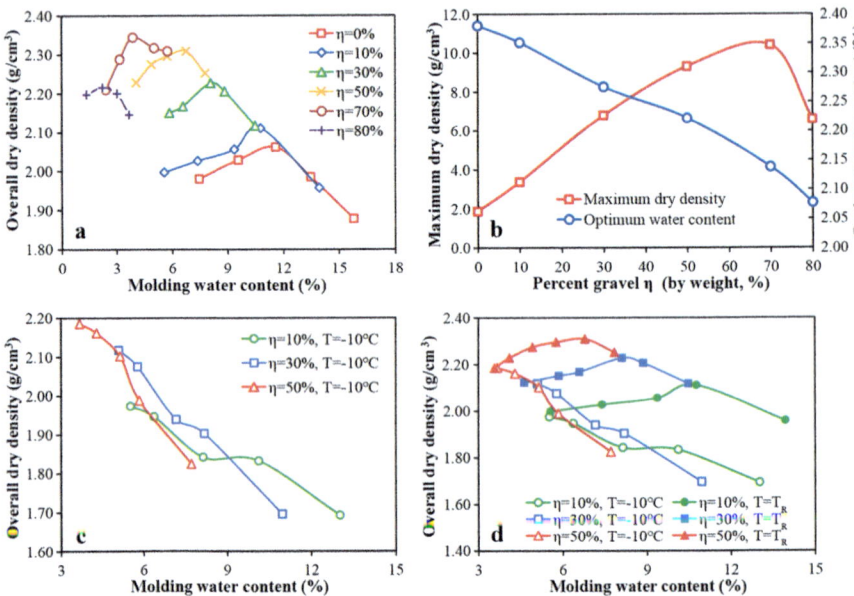

Fig. 2. (a) Dynamic compaction curves of clay-gravel mixtures with different gravel contents under normal temperature (at room temperature of approximately 20 °C); (b) Changes of maximum dry densities and optimum water contents with gravel contents; (c) Compaction curves of clay-gravel mixtures with different gravel contents under cold temperature (−10 °C); (d) Comparison of compression curves of clay-gravel mixtures under normal and cold temperatures.

Figure 2(c) shows the compaction curves of clay-gravel mixtures with three gravel contents under cold temperature of −10 °C. It is interesting that the compac-tion curves under cold temperatures are significantly different from those under normal temperatures. The overall dry densities of the three soil mixtures all decreased with the increasing of the molding water contents. No obvious optimum water content analogous to that under normal temperatures can be observed. This phenomenon was also found in the compaction tests of sandy soils under the temperature of $-10 \sim -15$ °C by Zhang et al. (2016). By plotting the compaction curves of soil mixtures with gravel contents of 10%, 30%, 50% at normal (room temperatures of about 20 °C) and cold temperatures (-10 °C), the comparison results are obtained in Fig. 2(d). It is found that for the mixtures with a certain gravel content value, the two compaction curves under normal and cold temperatures will intersect at one critical point. Herein, we define the maximum dry density and optimum water content at this critical point as critical maximum dry density, ρ_{cr}, and critical optimum water content, ω_{cr}, respectively. It is also indicated that clod temperature-induced freezing will have little effect on the compaction behaviors of clay-gravel mixtures when the molding water contents are small enough to the corresponding critical optimum water contents. As also can be seen, the clay-gravel mixture with higher gravel content will have greater ρ_{cr} and smaller ω_{cr}.

Acknowledgement. This work was supported by "The Joint Funds of the National Natural Science Foundation of China" (Grant No. U1765205) and "National Key R&D Program of China" (Grant No. 2017YFC0404800). It was also a part of work in the project Funded by the Priority Academic Program Development of Jiangsu Higher Education Institutions (PAPD) (Grant No. 3014-SYS1401). The first author also would like to appreciate the financial support of the China Scholarship Council (CSC) (Grant No. 201706710061).

References

Shelley, T.L., Daniel, D.E.: Effect of gravel on hydraulic conductivity of compacted soil liners. J. Geotech. Eng. **119**(1), 54–68 (1993)

Li, Y., Huang, R., Chan, L.S., Chen, J.: Effects of particle shape on shear strength of clay-gravel mixture. KSCE J. Civ. Eng. **17**(4), 712–717 (2013)

Tarantino, A., De Col, E.: Compaction behaviour of clay. Géotechnique **58**(3), 199–213 (2008)

Alonso, E.E., Pinyol, N.M., Gens, A.: Compacted soil behaviour: initial state, structure and constitutive modelling. Géotechnique **63**(6), 463 (2013)

Proctor, R.R.: Fundamental principles of soil compaction. Eng. News Rec. **111**(13), 372 (1933)

Donaghe, R.T., Torrey, V.H.: A compaction test method for soil-rock mixtures in which equipment size effects are minimized. Geotech. Test. J. **17**(3), 363–370 (1994)

Chinkulkijniwat, A., Man-Koksung, E., Uchaipichat, A., Horpibulsuk, S.: Compaction characteristics of non-gravel and gravelly soils using a small compaction apparatus. J. ASTM Int. **7**(7), 1–15 (2010)

Rücknagel, J., Götze, P., Hofmann, B., Christen, O., Marschall, K.: The influence of soil gravel content on compaction behaviour and pre-compression stress. Geoderma **209**, 226–232 (2013)

Zhang, S.J., Su, A.S., Li, Z.Y., Yan, J., Xu, L.L.: Study on winter construction technology of embankment engineering in cold regions. In: Proceedings of the 2016 Academic Annual Conference of Chinese Hydraulic Engineering Society, pp. 114–121. Hohai University Press, Nanjing (2016)

Experimental Study on the Strength Characteristics of Frozen Clay on π Plane

Yanhu Mu[1] , Wei Ma[1(✉)] , Dun Chen[1,2] , Zhiwei Zhou[1] ,
Dayan Wang[1], and Lele Lei[1,2]

[1] State Key Laboratory of Frozen Soil Engineering,
Cold and Arid Regions Environmental and Engineering Research Institute,
Chinese Academy of Sciences, Lanzhou 730000, China
mawei@lzb.ac.cn
[2] University of Chinese Academy of Sciences, Beijing 100049, China

Abstract. A set of directional shearing tests using hollow cylinder apparatus (HCA) was carried out on remolded frozen clay at −6 °C to study the strength properties of frozen soils in π plane. During the shearing, the stress Lode angles was fixed at five different values ($\theta\sigma$ = −30, −16.1, 0, 16.1 and 30°) under four mean principal stresses (p = 1, 3, 4.5 and 10 MPa). Through the tests, the strength locus of frozen soil in π plane with $\theta\sigma$ ranging from −30 to 30° were gained experimentally for the first time. The test results show that the strength envelope of frozen clay changes from a curve-sided triangle to a circle with increasing p. And a combination of the Spatially Mobilized Plane (SMP) criterion and Mises criterion, the Generalized Non-linear strength theory, can better describe this strength envelope evolution law in π plane.

Keywords: Strength of frozen soil · Π plane · Stress Lode angle
Hollow cylinder apparatus

1 Introduction

In actual engineering projects, frozen soil is generally subjected to multiaxial loads, and deformation and stress distribution are always non-uniform and complex. However, through triaxial compression test, the frozen soil strength can only be investigated in one *p-q* plane with stress Lode angle $\theta\sigma$ = −30°. To describe the strength property of frozen soil in principal stress space, some strength criteria of unfrozen soils were introduced into the strength criterion of frozen soil in π plane, such as the modified L-D strength criterion [1], the generalized nonlinear strength theory [2], and the M-N and L-D criteria [3].

In this paper, the hollow cylinder apparatus (HCA) was used to study the strength characteristics of frozen soil in π plane. A set of directional shearing tests along fixed stress Lode angle was carried out on remolded frozen clay at −6 °C. The strength locus of frozen clay in π plane were gained experimentally and simulated by four classical strength criteria of unfrozen soils.

© Springer Nature Switzerland AG 2018
W. Wu and H.-S. Yu (Eds.): *Proceedings of China-Europe Conference
on Geotechnical Engineering*, SSGG, pp. 1394–1398, 2018.
https://doi.org/10.1007/978-3-319-97115-5_108

2   Test Condition

The test apparatus used in this study is HCA with temperature controlling. Soil used in this study was clay, taken from the Qinghai-Tibet plateau, China. The screened soil was mixed with 19.8% moisture content by weight, and then put into a split mold to make hollow cylindrical specimens with target density by compaction. The compacted hollow cylinder specimens are 200 mm in height, 100 mm in outer diameter and 60 mm in inner diameter. After mounted into the FHCA-300, the specimens were quickly frozen by the coolants in the inner and outer cavities under −30 °C for 48 h to prevent the formation of ice lenses. Finally, the coolants temperature increased to the target testing temperature −6 °C and kept for 12 h before the test.

To study the strength characteristics of frozen clay in π plane, a set of directional shearing (DC) tests along five stress Lode angles ($\theta\sigma = -30, -16.1, 0, 16.1$ and $30°$) under four mean principal stresses ($p = 1, 3, 4.5,$ and 10 MPa) were carried out in this study. The corresponding values of the coefficient of intermediate principal stress (b) at the five $\theta\sigma$ are 0, 0.25, 0.5, 0.75 and 1.0, respectively. The test procedures were conducted as follows. Firstly, p was increased along the hydrostatic axis to the target value. Then, the shear stress q was increased with a rate of 30 kPa/min along a fixed stress Lode angle until the axial strain arrived to 25% or the specimen failed. During the shearing, p and the direction angle of major principal stress ($\alpha = 0$) were kept as constant.

3   Results and Analyses

Here, test results with the mean principal stress $p = 4.5$ MPa, as shown in Fig. 1, was taken as an example to analyze the stress-strain relationship of frozen clay under directional shearing. The stress-strain curves of frozen clay all performs as strain hardening. And the strength of frozen clay decreases considerably with increasing b. This means that impact of b on the strength of frozen clay should be taken into account.

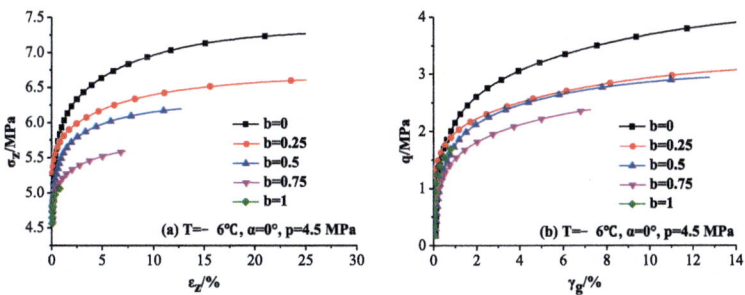

Fig. 1.   Stress-strain curves under different b of frozen clay.

The strength envelopes of frozen clay in π plane with different p are shown in Fig. 2. The shape of strength envelopes changes with increasing p. When $p = 1$, 3 MPa, the shape of the strength envelopes looks like a smooth curve-sided triangle. When $p = 4.5$ MPa, the strength envelope tends to change from a smooth curve-sided triangle to a circle. And when $p = 10$ MPa, the shape of the strength envelope tends to be a circle. Thus, the strength envelopes of frozen soils in π plane lie between a curve-sided triangle and a circle.

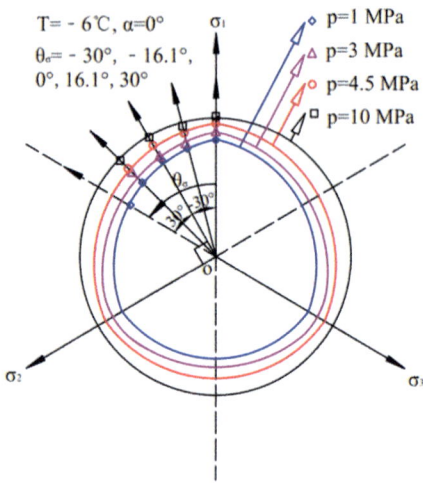

Fig. 2. Strength envelopes of frozen clay in π plane with different p.

Four classical strength criteria of geotechnical materials were used to simulate the test results above to validate their applicability for frozen clay (Fig. 3). The four strength criteria are Drucker-Prager strength criterion (D-P) [2], Lade-Duncan strength criterion (L-D) [1], Spatially Mobilized Plane strength criterion (SMP) [4] and Generalized non-linear strength theory (GNST) [3]. It can be seen that the GNST is more suitable for characterizing the strength envelopes of frozen clay in π plane with the tested range of p.

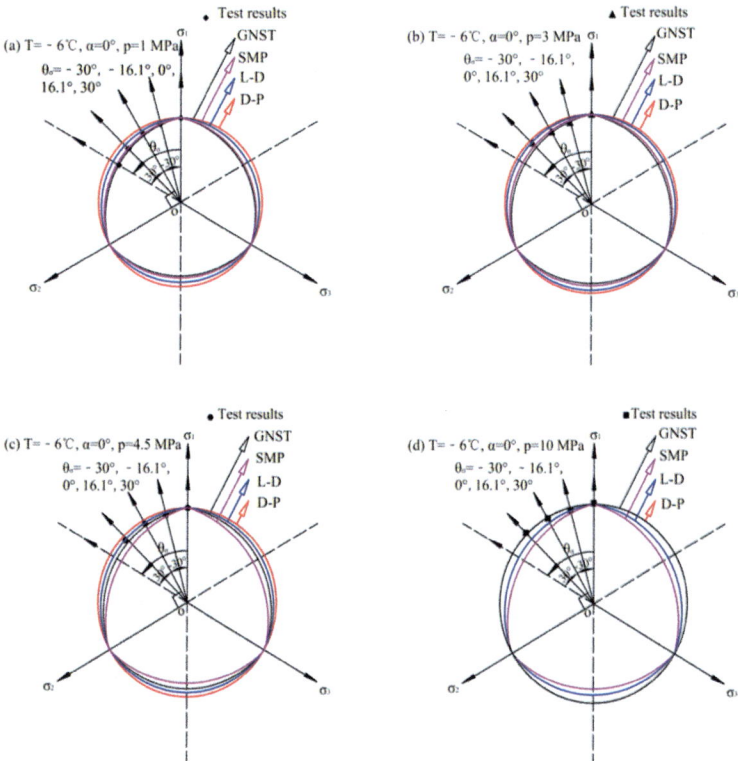

Fig. 3. Comparisons of strength envelopes between the test results and the simulated results of the four strength criteria (D-P, L-D, SMP, GNST).

4 Conclusions

A set of directional shear tests along five stress Lode angles ($\theta\sigma$ = −30, −16.1, 0, 16.1 and 30°) was carried out on remolded frozen clay at −6 °C using HCA. Through the directional tests, the strength locus of frozen clay in π plane with $\theta\sigma$ ranging from −30 to 30° were gained experimentally for the first time. The tested results show that the stress-strain curves of frozen clay all perform as strain hardening under directional shearing. Impact of b or $\theta\sigma$ on the strength of frozen clay is considerable and should be taken into account. But this impact decreases with increasing p. In π plane, the strength envelops of frozen clay changes from a curved triangle to a circle, and the GNST can better describe this evolution law of the strength envelops.

Acknowledgements. This work was supported by the National Natural Science Foundation of China (Nos. 41401077 and 41630636) and the State Key Development Program of Basic Research of China (No. 2012CB026106).

References

1. Lade, P.V., Duncan, J.M.: Elastoplastic stress-strain theory for cohesionless soil. J. Geotech. Eng. Div. **101**(10), 1037–1053 (1975)
2. Drucker, D.C., Prager, W.: Soil mechanics and plastic analysis or limit design. Q. Appl. Math. **10**(2), 157–165 (1952)
3. Yao, Y.P., Hu, J., Zhou, A., et al.: Unified strength criterion for soils, gravels, rocks, and concretes. Acta Geotech. **10**(6), 749–759 (2015)
4. Matsuoka, H., Nakai, T.: Stress-deformation and strength characteristics of soil under three different principal stresses. Proc. Jpn. Soc. Civ. Eng. **232**, 59–70 (1974)

Experimental Study of the Mechanical Properties of Coarse-Grained Soils from High Altitude and Cold Areas Under Freeze-Thaw Cycle

Yonglong Qu[1], Guoliang Chen[2], Fujun Niu[3], Wankui Ni[1], Yanhu Mu[3(✉)], and Tao Chen[4]

[1] College of Geology Engineering and Geomatics, Chang'an University, Xi'an 710054, China
[2] China Gold Group Tibet Tyrone Mining Development Co., Ltd., Tibet 850000, China
[3] State Key Laboratory of Frozen Soil Engineering, Northwest Institute of Eco-Environmental and Resources, Chinese Academy of Sciences, Lanzhou 730000, China
muyanhu@lzb.ac.cn
[4] School of Civil Engineering, Lanzhou University of Technology, Lanzhou 730050, China

Abstract. Experiments using GDS triaxial test apparatus were carried out on size-reduced coarse-grained soils with different moisture content (ω) (9%, 11.5%, 14%) and different dry density (ρ_d) (1.9 g/cm^3, 2.0 g/cm^3, 2.15 g/cm^3) to study the mechanical properties under different freeze-thaw cycles (C) (0, 1, 3, 5, 7, 9, 12, 15, 20 times). The results show that the uniaxial strength of coarse-grained soils decreases with the increase of cycle times, and tends to be steady around 9 times, the deformation modulus of coarse-grained soils was gradually decreasing. With the increase of water content, the strength of coarse-grained soil was constantly decreasing, the deformation modulus was reducing. The stress-strain curves of coarse-grained soils were gradually changed from strain softening to strain hardening with the increase of moisture content, but the high dry density was an exception. With the increase of dry density, the strength and deformation modulus of the coarse-grained soils were increasing, and the slope of post-peak curve was increasing, the residual strength was also aggrandizing, and the strain softening was more obvious. Meanwhile, the strain hardening characteristic of high moisture content also changed to softening with the increase of dry density.

Keywords: Coarse-grained soils · Freeze-thaw cycle · Moisture content Dry density · Mechanical property

1 Introduction

Generally, the coarse-grained soil was a mixture consist of diverse size and different property grains which less than 60 mm and larger than 0.075 mm. The scholars established a creep constitutive model for frozen soils with different contents of coarse-grains [1], put forward an elastic-visco-plastic double-yield-surface model for coarse-grained soils considering particle breakage [2], and also set up A strain hardening model for rock fills [3]. The influence of freeze-thaw action on hydraulic behavior of unsaturated

© Springer Nature Switzerland AG 2018
W. Wu and H.-S. Yu (Eds.): *Proceedings of China-Europe Conference on Geotechnical Engineering*, SSGG, pp. 1399–1402, 2018.
https://doi.org/10.1007/978-3-319-97115-5_109

volcanic coarse-grained soils was studied [4], and an experimental measurement and numerical simulation of frost heave unsaturated coarse-grained soil was carried out [5]. The influence of freezing-thawing cycle on the modulus of resilience of coarse-grained soil was studied [6], and the effect of fines content on frost heaving properties [7].

2   Test Conditions

This study was aiming at the mechanical properties of coarse-grained soils in high altitude and cold areas under freeze-thaw cycle weathering. The test temperature was −20 °C–20 °C, and one freeze-thaw cycle was 24 h, which including the freezing time of 12 h and melting of 12 h. By the compaction test results, the optimum moisture content was 9.10%, the maximum dry density was 2.368 g/cm^3, so the test dry densities were 1.9 g/cm^3, 2.0 g/cm^3 and 2.15 g/cm^3. The test moisture content was 9%, 11.5% and 14% according to the natural and saturated moisture content which were about 11.5% and 14.7%. The size of sample was Φ61.8 mm *125 mm and carried out the unconfined uniaxial compression test of coarse-grained soils under different freeze-thaw cycles (0, 1, 3, 5, 7, 9, 12, 15, 20 times) with different dry density and moisture content. The whole samples were 81 groups, and 3 samples per group.

3   Results and Analyses

According to the results, drew the total stress-strain curves of coarse-grained soils (see Fig. 1-a). It's easy to see that the coarse-grained soils were strain softening in this state. With the increase of freeze-thaw cycles, the uniaxial strength and deformation modulus were gradually decreasing, the slope of post-peak curve and the residual strength were also decreasing, and the strain softening properties showed more obvious, which illustrated the plastic deformation strengthening, the brittle weakening.

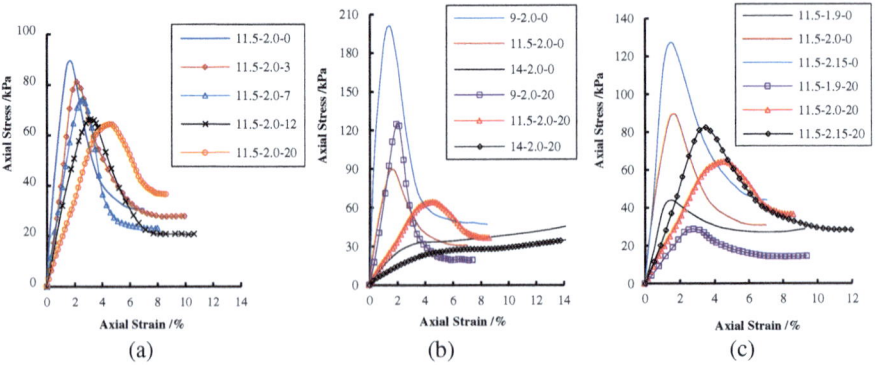

Fig. 1. The stress-strain curves with freeze-thaw cycle, dry density and moisture content, in the legend 'A-B-C': 'A' means moisture content, 'B' means dry density, 'C' means freeze-thaw cycles.

With the increase of moisture content, the uniaxial strength and deformation modulus were decreasing, the residual strength was also decreasing (see Fig. 1-b). The curves showed strain softening at low moisture content and change to strain hardening when near saturation state, but the high dry density was an exception.

With the increase of dry density, the uniaxial strength, deformation modulus and residual strength were all increasing, but the slope of post-peak curve was decreasing (see Fig. 1-c). It was strain-softening under low dry density and little moisture content, and softening was more obvious with the increasing of dry density. But it was strain-hardening under low dry density and high moisture content and may change from strain-hardening to strain-softening with the increase of dry density.

According to the test result of uniaxial strength (see Fig. 2-a), with the increase of freeze-thaw cycle, the uniaxial strength of coarse-grained soils was decreasing. At the same moisture content, the larger dry density, the faster damping of strength with freeze-thaw cycle, and the larger decrement. The fastest damping was about 6.46 kPa per times, and the largest decrement reached to 129.2 kPa. Meanwhile, the decreasing of strength tends to stabilize after 9times of freeze-thaw cycle. At the same dry density, the larger moisture content, the lower damping of strength with freeze-thaw cycle, and the lesser decrement. So, the decreasing strength of coarse-grained soils with the freeze- thaw cycle was closely related to moisture content and dry density.

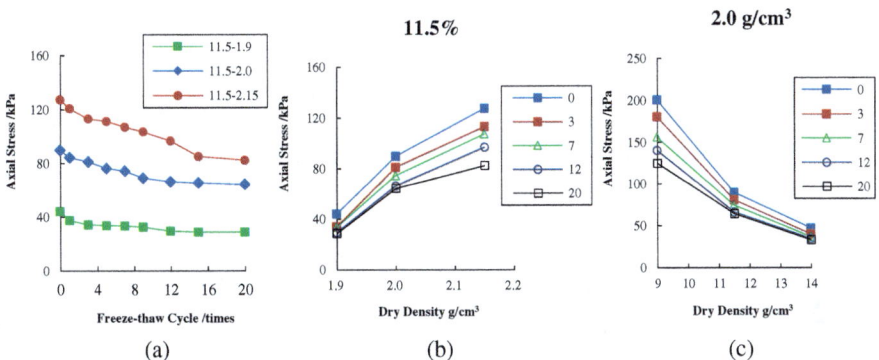

Fig. 2. The test results of uniaxial strength with freeze-thaw cycle, dry density and moisture content, in the legend 'A-B': 'A' means moisture content, 'B' means dry density. in the legend 'C': 'C' means freeze-thaw cycle.

The uniaxial strength of coarse-grained soils was increasing with the increase of dry density at the same moisture content and freeze-thaw cycle (see Fig. 2-c). The maximum increment was 220.33 kPa, the fastest growth was 881.33 kPa/(g/cm³), and the increasing speed slow down when near the max dry density, but near the saturation state was an exception.

The uniaxial strength was decreasing with the increase of moisture content at the same dry density and freeze-thaw cycle (see Fig. 2-c). The max decrement was 235.71 kPa, the fastest speed was 47.14 kPa/%. When near the optimum moisture content, the increment was max and the speed was the fastest. Meanwhile, with the increase of freeze-thaw cycle, the effect of moisture content on uniaxial strength was weakened.

4   Conclusions

The uniaxial strength of coarse-grained soils decreases with the increase of freeze-thaw cycle times, and tends to be steady around 9 times, the deformation modulus of coarse-grained soils was gradually decreasing. With the increase of water content, the strength of coarse-grained soil was constantly decreasing, the deformation modulus was reducing. The stress-strain curves of coarse-grained soils were gradually changed from strain softening to strain hardening with the increase of moisture content, but the high dry density was an exception. With the increase of dry density, the strength and deformation modulus of the coarse-grained soils were increasing, and the slope of post-peak curve was increasing, the residual strength was also aggrandizing, and the strain softening was more obvious. Meanwhile, the strain hardening characteristic of high moisture content also changed to softening with the increase of dry density.

Acknowledgements. This study is supported by the Key Scientific Research Project of China Gold Group (2016ZGHJ/XZHTL-YQSC-26) and Fundamental Research Foundation of the Central Universities (300102268716).

References

1. Hou, F., Lai, Y.M., Liu, E.L.: A creep constitutive model for frozen soils with different contents of coarse grains. Cold Reg. Sci. Technol. **145**(2018), 119–126 (2017)
2. Kong, Y.F., Xu, M., Song, E.X.: An elastic-visco plastic double-yield-surface model for coarse-grained soils considering particle breakage. Comput. Geotech. **85**(2017), 59–70 (2016)
3. Xu, M., Song, E.X.: A strain hardening model for rockfills. Rock Soil Mech. **31**(9), 2967–2972 (2010)
4. Ishikawa, T., Tokoro, T., Miura, S.: Influence of freeze-thaw action on hydraulic behavior of unsaturated volcanic coarse-grained soils. Soils Found. **56**(5), 790–804 (2016)
5. Li, A.Y., Niu, F.J., Zheng, H.: Experimental measurement and numerical simulation of frost heave unsaturated coarse-grained soil. Cold Reg. Sci. Technol. **137**(2017), 68–74 (2017)
6. Chen, Z.D., Chen, D.G., Chen, J.B.: Influence of freezing-thawing cycleon the modulus of resilience of coarse-grain fill with different water contents. J. Zhengzhou Univ. Eng. Sci. **25**(4), 9–13 (2014)
7. Wang, T.L., Yue, Z.R.: Influence of fines content on frost heaving properties of coarse-grained soils. Rock Soil Mech. **34**(2), 359–364 (2013)

Research on Bearing Capacity of Pile Foundation with Impact of Subpermafrost Water

Xiangyang Shi[1,2(✉)], Dongqing Li[1], and Ze Zhang[1]

[1] State Key Laboratory of Frozen Soil Engineering, Northwest Institute
of Eco-Environment and Resources, Chinese Academy of Sciences, Lanzhou 730000, China
shixiangyang2008@163.com
[2] University of Chinese Academy of Sciences, Beijing 100049, China

Abstract. Pile foundation has been widely applied in cold regions. In a special geological condition that subpermafrost water is developed, the shortage of bearing capacity is often encountered in these complex geological areas. In this paper, bearing capacity of pile foundation is studied by finite element method. Combined with geological and geothermal data of some bridge pile in permafrost region, temperature field is simulated by the finite element software ANSYS. Subpermafrost water at the depth of 20 m plays fixed boundary in the calculation model. Environmental temperature plays thermal boundary on the natural surface, including normal environmental temperature with rising amplitude 2.6 °C/50 a. According to the nodal temperature of pile-soil interface, the bearing capacity of pile foundation is calculated by interpolation method. The simulation results indicate that the subpermafrost water is especially significant to the geothermal field. Under impact of subpermafrost water and environment, pile-soil interface is warming, and permafrost base is rising. The bearing capacity of pile foundation decreases obviously with pile-soil interface warming.

Keywords: Permafrost · Pile foundation · Bearing capacity
Subpermafrost water

1 Introduction

Pile is the most common foundation form in permafrost regions, and it has been widely used. Pile foundation has little thermal disturbance to frozen soil, and it can be continuously constructed. Most piles are located below the active layer, so they are less affected by frost heave, so they can be applied to all types of geological conditions. With the effects of global warming and climate change, the environment of permafrost is changing. Studies have found that permafrost is warming, and active layer thickness is increasing in recent years [1]. Subpermafrost water is developed in some permafrost regions, which has great influence on permafrost degradation. Since the bearing capacity of pile foundation mainly depends on adfreeze strength, permafrost warming will directly decrease the bearing capacity of pile foundation. To the existing pile, large deformation or even a series of engineering disease would occur.

© Springer Nature Switzerland AG 2018
W. Wu and H.-S. Yu (Eds.): *Proceedings of China-Europe Conference on Geotechnical Engineering*, SSGG, pp. 1403–1406, 2018.
https://doi.org/10.1007/978-3-319-97115-5_110

2   An Engineering Case

2.1   General Information of Bridge Pile Site

The bridge is located in the north of Tanggula mountains, where is about 5000 m above sea level. The buried depth of confined subpermafrost water is about 19.9 m, and the field measured temperature of subpermafrost water is 0.6 °C. The bridge foundation is bored pile. The length of pile is 20.0 m, and its radius is 0.5 m (Fig. 1) (Table 1).

Fig. 1. Confined subpermafrost water leakage

Table 1. Geological conditions

Lithology	Depth (m)	Water content (%)
Fine breccia soil	0–3.0	15
Fine sand	3.0–4.0	24
Gravel soil	4.0–6.0	24
	6.0–8.0	23
Bedrock	8.0–10.0	15
	10.0–50.0	10

2.2   Engineering Disease

The pile foundation has been undergoing severe settlement deformation since 2009 [2], and the deformation is beyond the corresponding specification.

3   Numerical Results Based on Heat Conduction Theory

Considering the climate warming and confined subpermafrost water at 20.0 m depth, three-dimensional solid model is established, and the boundary conditions are applied. After the solution, extract the nodal temperature of pile-soil interface in the cold and warm seasons.

3.1 Some Basic Assumptions

Soil is regarded as a kind of homogeneous material in heat conduction. The buried depth of subpermafrost water is fixed. The temperature of confined subpermafrost water has no change with seasons, and the constant temperature is 0.6 °C.

3.2 Boundary Conditions

Combined with geological survey and the indoor test, parameters and boundary conditions are determined. The surface thermal boundary is the first boundary condition, which is represented by a sine function. The average surface temperature is −1.5 °C and the rising amplitude is 2.6 °C/50 a. The temperature of subpermafrost water plays a fixed boundary at 20.0 m depth. The initial geothermal temperature is the measured warm season data in 2014.

3.3 Numerical Results

Temperature field of pile foundation is simulated. The temperature of frozen layer has been increasing because of subpermafrost water. The permafrost table is almost stable, but the permafrost base is greatly rising.

Assume that the subpermafrost water is always confined. According to Chinese code for design of soil and foundation of building in frozen soil region, the bearing capacity is calculated on the basis of simulated interface temperature. The vertical bearing capacity of single pile can be obtained.

It can be seen that the bearing capacity of pile decreases with the degradation of permafrost. The characteristic value of bearing capacity is 2887.7 kN at first year but reduces to 1621.9 kN at the third decade. Bearing capacity drops with the most scope 43% or above. In summary, the subpermafrost water has a great influence on the pile bearing capacity (Fig. 2).

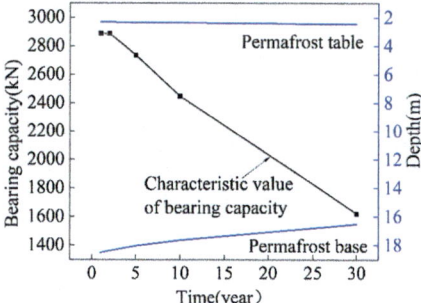

Fig. 2. Variation of permafrost thickness and characteristic value of bearing capacity

4   Bearing Behavior of Pile with Permafrost Base Change

This part is aimed to show the negative friction during permafrost base rising. Pile length is increased to 40.0 m, the rest is the same as the previous part. The axial force under the load of 1200 kN is calculated. During permafrost base rising, the pile resistance varied from conventional pile, and adfreeze strength would become load to the pile. Obviously, it is extremely unfavorable to the safety of pile foundation (Fig. 3).

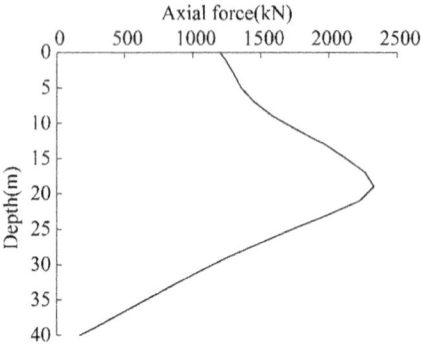

Fig. 3.   Axial force curve of pile

5   Engineering Advice

In summary, there's some uncertainty of pile bearing capacity in the areas of subpermafrost water developed. So, such geological areas should be avoided during the design phase. It's necessary to ascertain the permafrost table, permafrost base and subpermafrost water through geological survey and temperature measurement. If subpermafrost water is confined and far deeper than the pile, some engineering measures should be taken to reduce temperature of foundation. If permafrost water is not confined and causes permafrost degradation, negative friction should be eliminated or increase the length of pile to transfer the load to the deeper unfrozen layer.

Acknowledgement. This research was supported by the National Natural Science Foundation of China (No. 41271080 and No. 41771078), and the Key Research Program of Frontier Sciences of Chinese Academy of Sciences (No. QYZDY-SSW-DQC015).

References

1. Cai, H.C., Li, Y., Yang, Y.P., et al.: Variation of temperature and permafrost along Qinghai—Tibet railway. Chin. J. Rock Mech. Eng. **35**(07), 1434–1444 (2016)
2. You, Y.H., Wang, J.C., Wu, Q.B., et al.: Causes of pile foundation failure in permafrost regions: The case study of a dry bridge of the Qinghai-Tibet Railway. Eng. Geol. **230**, 95–103 (2017)

Experimental Study on Compressibility of Frozen Saturated Chinese Standard Sand

Xiaoyu Sun, Yansong Wang, Junlin Zhao, and Jilin Qi$^{(\boxtimes)}$

Beijing University of Civil Engineering and Architecture, Beijing 100044, China
jilinqi@bucea.edu.cn

Abstract. The compressibility of frozen soil must be taken into consideration when the deformation of highway and high-speed railway is strictly controlled in permafrost regions. In this study, the frozen saturated Chinese ISO standard sand was taken as the study object, and step load tests under different temperatures were carried out using a self-developed confined compression apparatus for frozen soils. Tests were conducted at the loads of 1, 2, 3, 5, 10 MPa and under temperatures of −0.5, −1.0, −2.0, −3.0, −5.0 °C. The coefficient of compressibility was obtained according to the e-σ_z curves for both unfrozen and frozen samples under different temperatures. The experimental results of temperatures ranging from room temperature to negative temperature were then obtained. The test results indicate that the compression curve of the frozen material are similar to that of the samples under room temperature; for warm frozen samples, the compressibility is considerable; the compressibility of frozen soil is closely related to temperature, i.e., the coefficient of compressibility increases with the increase of temperature in a form of exponential function.

Keywords: Frozen soil · ISO standard sand · Compressibility

1 Introduction

The compressibility of soil is a classical subject in geotechnical engineering. For frozen soils, the modulus is high at relatively low temperatures, and there is generally no obvious compressibility in industrial and civil infrastructures and is usually not considered. However, as the temperature rises the modulus of frozen soils decreases, and the mechanical properties of warm frozen soil are similar to that of unfrozen soil [1]. The compressibility should then be considered.

For cold regions engineering, the temperature change of the soil is a factor that must be taken into account in the engineering design, construction and maintenance. As early as 1956, Tsytovich found through laboratory tests that warm frozen soils (still negative temperature, near phase transition temperature) developed considerable compressive deformation under loads [2]. However, the study of the compressibility of frozen soil has not received enough attentions for a long time. Until 1982, Zhu and Zhang [3] studied the coefficient of compressibility of a frozen sand and a frozen clay, and analyzed the change law of compressibility of two different soil classifications at different

© Springer Nature Switzerland AG 2018
W. Wu and H.-S. Yu (Eds.): *Proceedings of China-Europe Conference on Geotechnical Engineering*, SSGG, pp. 1407–1411, 2018.
https://doi.org/10.1007/978-3-319-97115-5_111

temperatures and water contents. In recent years, Zhang and his team members [4–6] have carried out some studies on the compressive deformation of warm and ice-rich frozen soils. From the analysis of the above literatures, it can be seen that for the study of compressibility of frozen soil, there is currently no widely accepted conclusions. One of the important reasons is that different researchers used different study objects. On the other hand, there has been no systematic study on the compression characteristics of the step load under different temperatures. In order to make the results of peer researchers easy to compare, it is necessary to work on the same material which is also easily accessible for researchers, such as a standard sand. In recent years, much of the studies on mechanical properties of frozen soils are conducted by Chinese researchers, a Chinese ISO standard sand is then taken as study object in this program. The compressibility of the saturated frozen Chinese standard sand will be investigated.

2   The Confined Compression Test

2.1   Test Equipment

The equipment used in the test is a self-developed confined compression apparatus. The equipment consists of three parts: temperature control system, loading system and data acquisition system.

2.2   Test Material and Its Preparation

This work used the Chinese ISO standard sand as the study object. In order to obtain a more uniform cylindrical frozen soil sample, the multiple-sieving pluviation method was used for sample preparation [7], and the soil sample was saturated with vacuum. The saturated soil sample was placed in the refrigerator with the mold mounted to restrain possible frost induced swelling. The mold was removed after 12 h of freezing to make a cylindrical frozen saturated sand sample with an area of 30 cm^2, a height of 8 cm and a dry unit weight of 18.4 kN/m^3. The sample was mounted onto the apparatus and put under a target temperature for 6 h before loading.

2.3   Test Plan

Tests were conducted at the loads of 1, 2, 3, 5, 10 MPa and under temperatures of −0.5, −1.0, −2.0, −3.0, −5.0 °C as well as room temperature. During the entire loading program, the computer automatically records the axial stress and deformation of the specimen.

According to Chinese Standard for Soil Test Methods [8], when the deformation rate of the specimen after applying a load for 24 h is less than 0.01 mm/h, deformation can be considered as stable and the next step of load will be applied; otherwise, loading time should be extended. This stability criterion takes a very long time for a frozen soil with considerable creep characteristics. As this study focuses on the instantaneous deformation during step loading under different temperatures, creep deformation should not be included as far as possible. Therefore, in this test program compression deformation occurred in one hour was recorded under every loading step.

3   Analysis of Test Results

3.1   Curves of Strain with Time under Different Temperatures

This experiment used the 1-h stability criterion, which is different from the code. Here we will firstly present its reasonableness by analyzing the development of vertical strain ε_z along with time. Figure 1 illustrates the curve of ε_z with time at each load step. It can be noticed that at any temperature, the instantaneous ε_z developed from the sample at each load step is dominant. Taking a temperature of $-0.5\ ^\circ$C for example, as soon as the load reaches 1 MPa, strain develops very quickly, and the ε_z only increases by 0.07% in the following 1 h of constant loading, and 1.3% for $-5.0\ ^\circ$C. We therefore take the 1-h stability criterion in the test program.

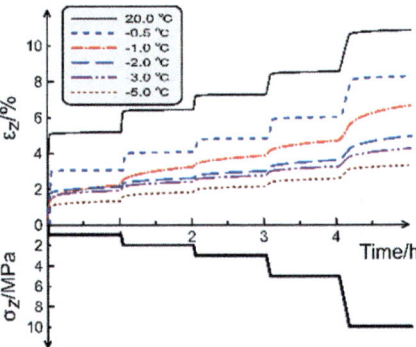

Fig. 1. The compression curves

3.2   Compressibility of the Material

The evaluation of the compressive properties of unfrozen soils is generally based on the compression curve. Following this methodology, we obtained the compression curves of the material under different temperature, as is shown in Fig. 2(a). The coefficient of compressibility versus temperature is presented in Fig. 2(b) and it can be seen from Fig. 2(b) that the coefficient of compressibility of the material in the pressure range of 0–1 MPa is obviously greater than that in the larger load ranges. However, regardless of load level, the coefficient of compressibility decreases with the temperature as an exponential function.

In addition, the e-σ_z curves of the frozen samples are similar to those for unfrozen soils, with one more influencing factor of temperature which is also dominant. Under lower temperatures, the compressibility of frozen sand is very small; however, under warmer temperatures, the compressibility is rather considerable. It is therefore significant to study the compressibility of frozen soils especially when warm frozen soil is encountered.

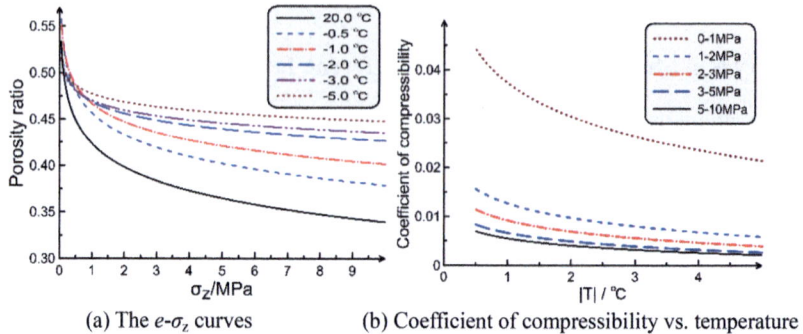

(a) The e-σ_z curves   (b) Coefficient of compressibility vs. temperature

Fig. 2. Compressibility for samples under different temperatures

4   Conclusions

In this paper, a frozen saturated Chinese standard sand was taken as study object. A self-developed confined compression apparatus was used to carry out the step-load compression test under different temperatures. The 1-h deformation stability criterion is firstly proposed and applied for obtaining compression strains. Major findings obtained from this test program are summarized as follows: (1) The e-σ_z curves of the frozen samples are similar to those of samples under room temperature. (2) The temperature has a great influence on the compressibility of the frozen standard sand samples. For warm frozen samples, the total amount of strain is considerable.

Acknowledgements. This work was supported in part by the NSFC (Nos. 41572268), Joint Key Project of BNSFC and BMEC (No. KZ201810016020) and PCSIRT (No. IRT_17R06).

References

1. Ladanyi, B.: Creep behavior of frozen and unfrozen soils. In: Proceedings of the 10th International Conference on Cold Regions Engineering-Putting Research into Practice, pp. 173–185. ASCE, Lincoln (1999)
2. Tsytovich, H.A.: Mechanics of Frozen Soil. Science Press, Beijing (1985). Translated by Changqing Z., Yuanlin Z.
3. Zhu, Y., Zhang, J.: Elastic and compressive deformation of frozen soils. J. Glaciol. Geocryol. **4**(3), 29–39 (1982)
4. Zheng, B., Zhang, J., Ma, X., Zhang, J.: Study on compression deformation of warm and ice-rich frozen soil. Chin. J. Rock Mech. Eng. **27**, 3064–3069 (2009)
5. Su, K., Zhang, J., Liu, S., Zhang, H., Ruan, G.: Compressibility of warm and ice-rich frozen soil. J. Glaciol. Geocryol. **35**, 369–375 (2013)

6. Ruan, G., Zhang, J.: Compression experimental research on warm and ice-rich permafrost on the Qinghai-Tibet Plateau. Hydrogeol. Eng. Geol. **41**(2), 50–56 (2014)
7. Ma, L., Qi, J., Yu, F., Yao, X.: Experimental study on variability in mechanical properties of a frozen sand as determined in triaxial compression tests. Acta Geotech. **11**, 61–70 (2016)
8. Ministry of Water Resources of the PRC: GB/T50123-1999 Standard for Soil Test Method. China Planning Press, Beijing (1999)

A Simple Equation for Predicting Freezing Point of Saline Soft Clay

Qinze Wang[1], Songhe Wang[1(✉)], Jilin Qi[2], Fengyin Liu[1], and Peng An[3]

[1] Institute of Geotechnical Engineering,
Xi'an University of Technology, Xi'an, China
wangsonghe@126.com
[2] College of Civil and Transportation Engineering, Beijing University of Civil Engineering and Architecture, Beijing 100044, China
[3] College of Geology Engineering and Geomatics,
Chang'an University, Xi'an, China

Abstract. Brine leakage due to pipe fracture is threatening the safety of frozen wall in artificial freezing facilitated engineering and the performance of the produced frozen soils will be deteriorated with salt inclusion. This paper takes the frequently soft clay in Wujiang District as the study object and five levels of calcium chloride solutions were supplemented to produce saline specimens. The soil freezing point for specimens at various water-salt combinations was measured by the thermocouple. Results indicate that the freezing point exhibits nonlinear increase at higher water contents and a critical value may exist close to the liquid limit, beyond which the gradient of freezing point to the water content lowers. An approximately linear decrease with the saline content was observed. By fitting experimental data, an empirical equation was proposed for the freezing point affected by both water and saline contents. The results may provide guidance in engineering design and numerical modeling in frozen ground engineering.

Keywords: Freezing point · Calcium chloride solution
Water-Salt combination

1 Introduction

Soil freezing point is a significant index in determining whether or not soils are in frozen state and is universally used in artificial freezing engineering, e.g., ascertaining the frost depth in natural ground and the thickness of frozen wall [1, 2]. Also, this index facilitates the temporary control of ground water in case of emergency during underground construction [3]. Kozlowski employed the method of differential scanning calorimetry and found the freezing point is a power function of both water content and plasticity index [2]. Bing and Ma noticed a decreasing trend in soil freezing point at higher salinity levels but growing with the water content for two types of silty clay such as the Lanzhou loess and Qinghai-Tibet silty clay [4]. Low et al. based on the principle of thermodynamics noted that the primary factor influencing the freezing point is the

© Springer Nature Switzerland AG 2018
W. Wu and H.-S. Yu (Eds.): *Proceedings of China-Europe Conference on Geotechnical Engineering*, SSGG, pp. 1412–1415, 2018.
https://doi.org/10.1007/978-3-319-97115-5_112

water content [5]. The inclusion of solute will diminish the unfrozen water in freezing soils [6]. Watanabe and Mizoguchi investigated the effect of solute in unfrozen water on the freezing point of silty sand and clay [7]. This paper aims to present a simple equation for estimating the freezing point of soft clay considering both the water content and salinity level.

2 Test Plan

The soft clay was taken from a foundation pit in Wujiang District, Jiangsu Province, with depth of 4.0–5.0 m. The natural water content is 49.3% and the dry density is 1.70 g/cm^3. The liquid and plastic limits are 41.8% and 21.6%, respectively. It mainly consists of three particle sizes, i.e., 7.5% (0.075–2.0 mm), 71.5% (0.005–0.075 mm) and 21.0% (<0.005 mm). Before testing, soil samples were leached in laboratory and then the calcium chloride solutions prepared by the 99% purity-calcium chloride powder were sprayed on the samples, and kept in a constant temperature and moisture environment for 24 h. Then, specimens were prepared by compressing the slurry, with diameter of 3.0 cm and height of 5.0 cm. The salinity level was controlled to be 0.0%, 0.2%, 0.5%, 1.0%, 2.0% and 5.0%. The water contents for specimens are 10%, 20%, 30%, 40%, 50% and 60%. The dry density for specimens is identical to that in natural state. The temperature of specimens was measured by a thermocouple with a high-precision temperature transducer placed into the center of the specimen and the live data can be collected and recorded by a data logger. The freezing point can be obtained from the equilibrium freezing phase based on the 'classic' cooling curve method [8].

3 Results and Analysis

3.1 Effect of Water and Salt on Soil Freezing Point

From Fig. 1a, the freezing point of saline soft clay tends to grow with the water content but still lower than the freezing point of conventional soft clay due to solute inclusion. As the water content exceeds 50.0%, it approaches to the freezing point of pure water and for specimens at various salinities, the ratio of freezing point to water content decreases and the difference among the six curves diminishes. This kind of change is not that obvious at higher salinity levels, e.g., a continuing increase can be noticed at salinities higher than 2.0%. The variation of freezing point with the salinity given in Fig. 1b shows an approximately linear decrease and at lower water contents, the salinity exhibits larger effect on the freezing point, as can be noticed from the larger slope of the curve. This may be related to the fact that the unfrozen water is in general absorbed by soil particles and this absorptive effect will diminish at higher contents due to the weak free energy in the ice-water interface and thus the freezing point approaches to that of pure water [9]. The inclusion of soluble salt enhances this kind of absorptive effect with lower soil water potential and thus the unfrozen water freezes as a lower temperature was reached, especially at higher salinity levels [10].

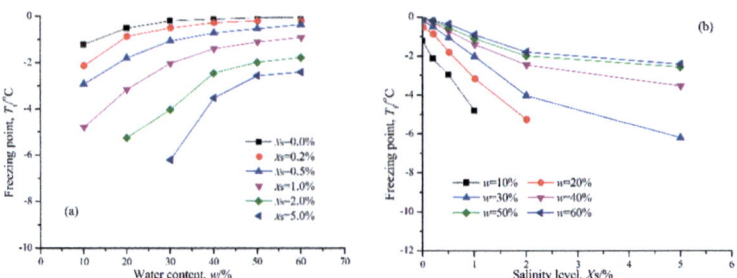

Fig. 1. Influencing factors for soil freezing point of soft clay: a. water content; b. salinity level.

3.2   An Empirical Equation for Predicting Soil Freezing Point

The empirical equation for predicting the freezing point of soft clay taking into account both the water content and salinity level can be described as

$$T_f = \alpha e^{\beta X_S} w^* \qquad (1)$$

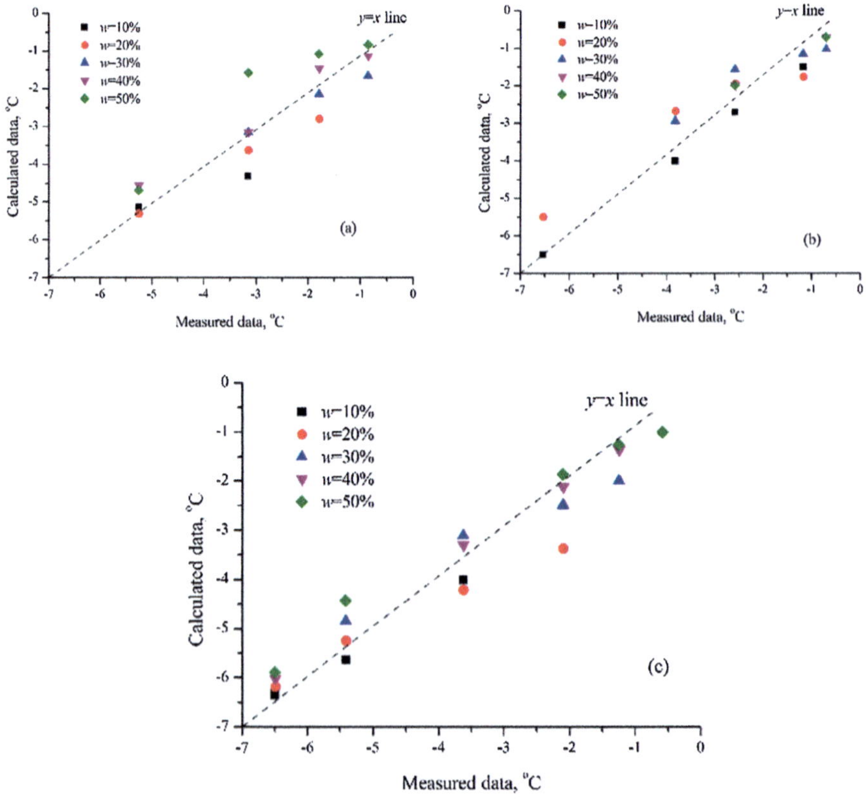

Fig. 2. Comparison of calculated and measured freezing point of soils

where, the parameters α and β can be fitted from test data. For soft clay used in this study, $\alpha = -37.15$, $\beta = 1.272$. The test data in literature [4] was also used to verify the rationality of the proposed equation. Figure 2 illustrates the comparison of calculated and measured data for three kinds of soils such as soft clay, silty clay in Qinghai-Tibet plateau and Lanzhou loess. It proves that the calculated freezing point for the three kinds of soils agrees well with the measured data.

4 Conclusions

This paper carried out experimental study on the freezing point of saline soft clay with calcium chloride solution and a simple equation for estimating the freezing point of saline soft clay was put forward taking into account both the water content and salinity level. Comparison of calculated and measured data shows good agreement.

Acknowledgements. This research was financially supported by the National Natural Science Foundation of China (No. 51778528, 41572268 and 51408486). These supports are greatly appreciated.

References

1. Lai, Y., Wu, D., Zhang, M.: Crystallization deformation of a saline soil during freezing and thawing processes. Appl. Therm. Eng. **120**, 463–473 (2017)
2. Kozlowski, T.: Some factors affecting supercooling and the equilibrium freezing point in soil–water systems. Cold Reg. Sci. Technol. **59**(1), 25–33 (2009)
3. Pierce, M.E., Detournay, C., Lagger, H.: Numerical modeling of ground freezing for subsurface construction. In: Alaska Rocks 2005, The 40th U.S. Symposium on Rock Mechanics (USRMS), 25–29 June, Anchorage, Alaska (2005)
4. Bing, H., Ma, W.: Laboratory investigation of the freezing point of saline soil. Cold Reg. Sci. Technol. **67**, 79–88 (2011)
5. Low, P.F., Anderson, D.M., Hoekstra, P.: Some thermodynamic relationships for soils at or below the freezing point: 1. freezing point depression and heat capacity. Water Resour. Res. **4**, 379–394 (1968)
6. Zhang, Y., Yang, Z., Liu, J., Fang, J.: Impact of cooling on shear strength of high salinity soils. Cold Reg. Sci. Technol. **141**, 122–130 (2017)
7. Watanabe, K., Mizoguchi, M.: Amount of unfrozen water in frozen porous media saturated with solution. Cold Reg. Sci. Technol. **34**(2), 103–110 (2002)
8. Grechishchev, S.E., Instanes, A., Sheshin, J.B., Pavlv, A.V., Grechishcheva, O.V.: Laboratory investigation of the freezing point of oil-polluted soils. Cold Reg. Sci. Technol. **32**(2–3), 183–189 (2001)
9. Lu, J., Zhang, M., Zhang, X., Yan, Z.: Experimental study on unfrozen water content and the freezing temperature during freezing and thawing processes. Chin. J. Rock Mech. Eng. **36**(7), 1803–1812 (2017). (in Chinese with English abstract)
10. Xu, X., Wang, J., Zhang, L.: Mechanism of Frost Heave and Salt Expansion of Soils. Science Press, Beijing (1995). (in Chinese)

Field Investigation on Thermo-Mechanical Behaviors of Pre-melting Lime Pile in Permafrost Regions

Jiliang Wang[(⊠)] and Chenxi Zhang

Heilongjiang Province Academy of Cold Area Building Research, Harbin, China
wangjiliang@sohu.com

Abstract. The lime soil pile was employed to melt the frozen soil and improve the bear capacity of thawed permafrost foundation. Firstly, the preliminary mixing proportion of lime soil was determined by indoor tests, and optimized by the insite temperature field monitoring. Secondly, a series of lime soil piles were constructed and the changes of temperature, displacement and bear capacity are monitored. Finally, the effects of lime soil pile on water content, uplift displacement and bear capacity pile and composite foundation were analyzed and discussed, respectively. The field experimental results could provide the experiences on the designation and construction of lime soil pile in warm permafrost regions.

1   Introduction

A mount of warm discontinues permafrost, temperature ranges from $0 \sim -0.5$ °C, covers in Northeast of China, and this frozen soil is very sensitive for environment temperature (Zhou et al. 2000). Construction of infrastructures on the warm permafrost site may change the temperature distribution of surrounding permafrost, and cause the frozen base move up, and some uneven settlement occurs and effects the stability of upper structure. Pre-melting frozen soil, such as using of steam or heating, is one effective procedure during construction the structures in warm discontinues permafrost regions (Tsytovich et al. 1975; Yershove 2004; Esch 2004). This research aims to investigate the engineering effects of pre-melting lime pile on the temperature, displacement and bear capacity of thawed permafrost site.

2   Materials and Results

The materials of pre-melting lime pile are composed of quick lime, clay soil, and cement, and the mass ratio is 60%, 35% and 5%, respectively.

- Quick lime: The content of CaO is greater than 70%, the heat release is greater than 960 kJ/kg, and the maximum diameter is smaller than 6 mm.
- Soil: The soil used in this research is clayed sand. The maximum diameter is smaller than 6 cm and the coarse content is greater than 60%.
- Cement: 325R Portland cement

© Springer Nature Switzerland AG 2018
W. Wu and H.-S. Yu (Eds.): *Proceedings of China-Europe Conference on Geotechnical Engineering*, SSGG, pp. 1416–1419, 2018.
https://doi.org/10.1007/978-3-319-97115-5_113

The field site is located in the Mohe, Northeast of China. According to the geology survey, the depth of permafrost ranges from 8.5 m to 16.0 m, and the temperature ranges from $0 \sim -0.1$ °C. The hole of pile is twist drilled with diameter 400 mm and length 18 m, the spacing distance is 2.5 and 3.0 times of pile diameter. For each of two test cases, 8 pre-melting piles were performed. To evaluate the melting effect, thermocouple temperature sensors with accuracy of 0.1 °C were installed in the permafrost surrounding the pile.

Figure 1 shows the changes of temperature after the quick lime pile was constructed. Note that when the pile is constructed, the permafrost begins to melt, the maximum temperature could reach to 37.3 °C after 102 h. Then, the temperature begins to decrease to 27.5 °C after 150 h.

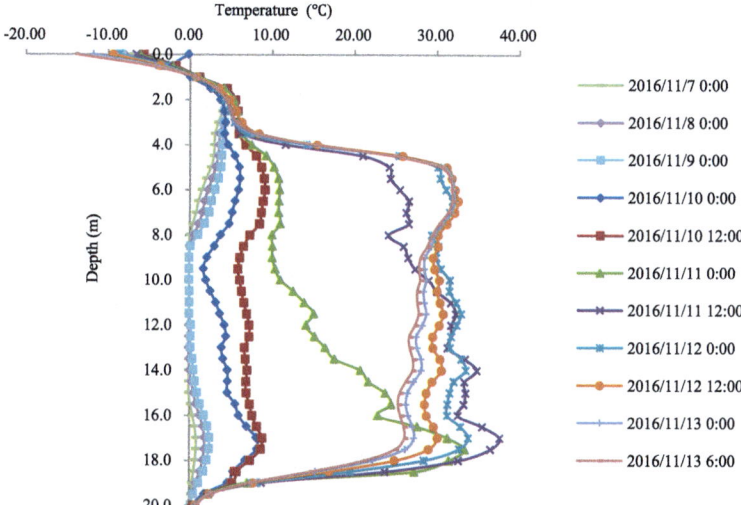

Fig. 1. Changes of temperature for pile-S3 in permafrost

Figure 2 shows the displacement of pile and soil, and Fig. 3 shows the layout of test point. Note that the uplift displacement of piles and soil increases with the increasing the time, especially during the first seven days, the increased uplift displacement shows a linear relationship. Then the uplift displacement presents a stably tendency. This test results shows the lime piles present a expand effects and the thawed permafrost soil was compacted.

Figure 4 shows the mean of uplift displacement for lime pile and soil. As is shown in this figure, the mean displacement of pile and soil is 2.17 cm and 2.63 cm for site 1 (2.5D) respectively, and is 2.67 cm and 3.00 cm for site 2(3.0D) respectively. The mean uplift displacement of soil is smaller than piles. However, it is strange that the uplift displacement in site 2 is greater than in site 1, the main reason is the ice content in site 2 is much greater than site 1.

Figure 5 shows the relationship curves of between loading and displacement. From the figure, the bear capacity of pile S1 is 70 kN, and the characteristic value is 35 kN.

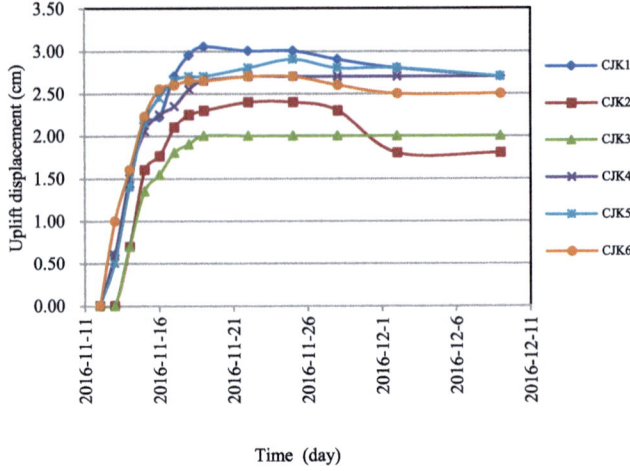

Fig. 2. The displacement of pile and soil

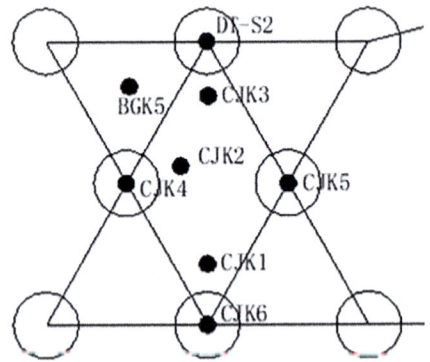

Fig. 3. Layout of test points

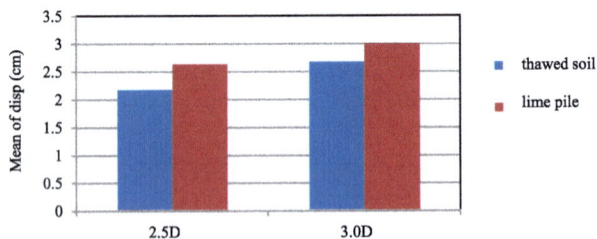

Fig. 4. Mean of uplift displacement for lime pile and soil

The bear capacity of soil T1 is 252 kPa, and the characteristic value is 126 kPa. The bear capacity of composite foundation D1 is 306 kPa, and the characteristic value is 153 kPa. The test data shows that the lime pile exhibits a notable reinforcement effects for the composite thawed permafrost foundation.

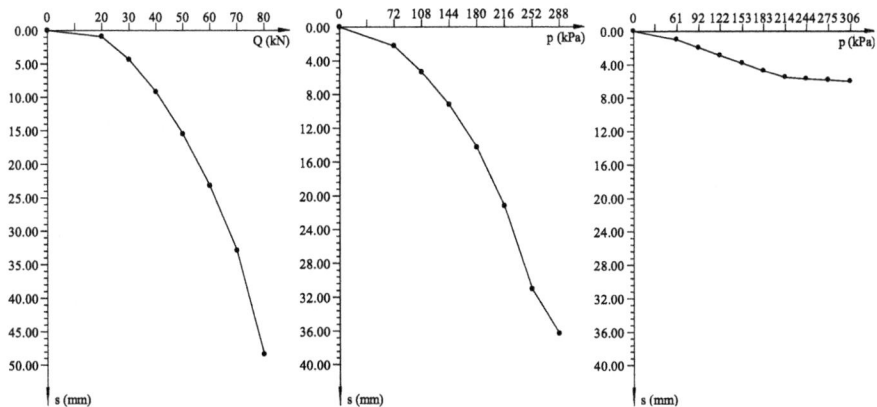

Fig. 5. Relationship curves of between loading and displacement (Left: pile; Middle: soil; Right: composite foundation)

References

Zhou, Y., Guo, D., Cheng, G.: Frozen Ground in China. Science Press, Beijing (2000)

Tsytovich, N.A., Swinzow, E., Tschebotarioff, G.: The Mechanics of Frozen Ground. McGraw-Hill, New York (1975)

Yershov, E.D.: General Geocryology. Cambridge University Press, Cambridge (2004)

Esch, D.C.: Thermal Analysis, Construction, and Monitoring Methods for Frozen Ground, vol. 13, pp. 57–77. Reston Va American Society of Civil Engineers (2004)

Heilongjiang Province Institute of Cold Area Building Research: Code for Design of Soil and Foundation of Building in Frozen Soil Region JGJ-98. China Architecture and Building Press, Beijing (2004)

Jin, H., Yu, Q., Lü, L.: Degradation of permafrost in the Xinganling Mountains, Northeastern China. Permafr. Periglac. Process. **18**(3), 245–258 (2007)

Study on Changes in Integrity Decay of Sandstone Subjected to Freeze-thaw Cycling

Liping Wang[1(✉)], Ning Li[1], Jilin Qi[2], Yanzhe Tian[3], and Shuanhai Xu[3]

[1] Xi'an University of Technology, Xi'an 710048, China
wangliping@xaut.edu.cn
[2] Beijing University of Civil Engineering and Architecture, Beijing 100044, China
[3] CCTEG XI'AN Research Institute, Xi'an 710077, China

Abstract. The integrity of the rock continues to decrease with the number of freeze-thaw cycles increases. This paper takes the coarse and fine sandstone collected from the Muli coal mine slope in Qinghai, northwest China, as the research object. Physical and mechanical properties of the rocks before and after freeze-thaw cycling are measured. Based on the decay equation suggested by the previous studies, the decay laws of rock integrity are analyzed by taking the uniaxial compressive strength (UCS) as the integrity index. It is found that the decay index λ is not a constant as generally recognized previously, but it keeps changing with the number of freeze-thaw cycles increases. At the same time, there are some differences in the changing laws of λ in different rocks. Finally, a relationship is established for the decay index and the velocity of longitudinal wave in the fine sandstones.

Keywords: Rock · Integrity · Freeze-thaw cycling · Decay equation
Decay index λ

1 Introduction

Field investigations in cold regions [1, 2] indicate that the long-term freeze-thaw (F-T) cycling can result in the rock weathering, a deep understanding of changes in integrity of rocks subjected to F-T cycles is key and necessary to deal with corresponding engineering issues.

Based on the results of 10 different rocks subjected to F-T cycling, Mutlutürk [3] found that if Shore hardness (SH) was the rock integrity index, there is a relationship between the rock integrity index and the number of F-T cycles.

$$I_N = I_0 e^{-\lambda N} \tag{1}$$

Where, λ is the decay index, I_0 is the rock initial integrity and I_N is the integrity after experiencing N freeze-thaw cycles. There have been the scholars [4–6] have adopted different physical and mechanical indexes of different rocks under the F-T cycles as the integrity to validate this equation. When determine λ, there is an implicit prerequisite,

© Springer Nature Switzerland AG 2018
W. Wu and H.-S. Yu (Eds.): *Proceedings of China-Europe Conference on Geotechnical Engineering*, SSGG, pp. 1420–1423, 2018.
https://doi.org/10.1007/978-3-319-97115-5_114

i.e., λ is considered as a constant in the whole process of F-T cycling, which does not conform to the observed phenomena in the experiments [7, 8]. It is reasonable to consider λ varying with the increases of F-T cycles for certain kind of rock.

This paper will investigate the changes in mechanical properties of two types of sandstone subjected to F-T cycling. The uniaxial compressive strength (UCS) will be taken as the integrity. The decay equation and the decay index of the rock integrity in terms of F-T cycling are studied.

2 Testing Methodology

The negative temperature in freeze-thaw test is set as −25 °C, the duration time is 8 h, the temperature for thaw is +25 °C, also for 8 h, i.e., one freeze-thaw cycle takes 16 h. UCS and P-wave velocity are measured individually after the rock specimens experienced 0, 30, 60, 90 and 120 freeze-thaw cycles.

3 Testing Results

The UCS of fine and coarse sandstone subjected to different cycles of freezing and thawing results of uniaxial compressive tests are listed in Table 1.

The average P-wave velocity of fine sandstone subjected to different numbers of F-T cycles is 3821 m/s, 3727 m/s, 3530 m/s, 3218 m/s, 2990 m/s, respectively.

Table 1. UCS and λ corresponding to different cycles of freezing and thawing

Rock classification	Numbers of F-T cycles	Measured UCS (MPa)	Decay index λ
Fine sandstone	0	114.8	
	30	98.8	5.00×10^{-3}
	60	86.7	4.68×10^{-3}
	90	79.6	4.07×10^{-3}
	120	72.3	3.85×10^{-3}
Coarse sandstone	0	104.1	
	30	90.1	4.81×10^{-3}
	60	83.9	3.60×10^{-3}
	90	74.2	3.76×10^{-3}
	120	64.4	4.00×10^{-3}

4 Rock Integrity Decay Laws Based on UCS

4.1 λ Corresponding to the Different Times of F-T Cycles

Taking the UCS as rock integrity, Eq. (1) can be used to obtain decay index λ:

$$\lambda = \frac{1}{N} ln \frac{I_0}{I_N} = \frac{1}{N} ln \frac{UCS_0}{UCS_N} \tag{2}$$

Where, UCS_0 is the UCS before F-T cycling; UCS_N is the UCS of the rock undergoing N cycles of freezing and thawing. The UCS of sandstone after different cycles of freezing and thawing and λ obtained by Eq. (2) are listed in Table 1.

As can be seen from Table 1, for fine sandstone, λ gradually decreases and finally tends to a constant, indicating that the integrity loss caused by single F-T cycles is rather serious in the initial stage, and then decreases until tends to a fixed value, for coarse sandstone, λ decreases firstly and then increases; While the λ of andesite [9] basically remains unchanged in the process of F-T cycles. The varying law can be different for different rocks, but it is also influenced by the factors such as structure components, grain shapes, pores and micro-cracks, etc.

4.2   Relationship of the Decay Index Versus Longitudinal Wave Velocity of the Fine Sandstone

The increment of the longitudinal wave velocity after N cycles of freezing and thawing are expressed as follows:

$$\Delta v = v_N - v_0 \tag{3}$$

Where, v_0 is the P-wave velocity before F-T cycling; v_N is the P-wave velocity undergoing N cycles of freezing and thawing, there exists a good fitting relationship between λ and the increment of longitudinal wave velocity ($R^2 = 0.985$), as shown in Eq. (4).

$$\lambda = 2E - 06\Delta v + 0.0051 \quad R^2 = 0.985 \tag{4}$$

Substituting Eq. (4) into decay equation, using UCS to express the rock integrity, we have the following:

$$UCS_N = UCS_0 e^{-(2E-06\Delta v + 0.0051)N} \tag{5}$$

As for the fine sandstone, the simple and easy measured change of velocity of longitudinal wave can be substituted into Eq. (5), the UCS after certain times of F-T cycles can be obtained.

5   Conclusion

In this paper, the F-T cycling tests are carried out on sandstone collected from the side slops of Muli coal mine in Qinghai Province under the saturated conditions. Physical and mechanical properties after undergoing different number of F-T cycles are investigated. Based on the decay equation suggested by Mutlutür [3] and taking UCS as the integrity, the rock integrity decay laws are analyzed. The following conclusions are obtained:

The decay index λ is not a constant value but changes with an increase in the number of F-T cycles. For the fine sandstone, λ gradually decrease with the increase in

F-T cycles and finally tends to become a certain constant; as to the coarse sandstones, λ decreases firstly and increases.

Taking UCS as the integrity, there exists a good fitting relationship between λ and velocity increment of longitudinal wave ($R^2 > 0.98$), λ can then be deduced by the velocity increment of longitudinal wave, further calculating the integrity indexes(UCS) subject to the number of different F-T cycles.

References

1. Sass, O.: Spatial patterns of rockfall intensity in the northern Alps. Zeitschrift für Geomorphologie Supp. 1(138), 51–65 (2005)
2. Stoffel, M., Lièvre, I., Conus, D., Grichting, M.A., Raetzo, H., Gärtner, H.W.: 400 years of debris-flow activity and triggering weather conditions: ritigraben, valais, Switzerland. Arctic Antarctic Alp. Res. 37(3), 387–395 (2005)
3. Mutlutürk, M., Altindag, R., Türk, G.: A decay function model for the integrity loss of rock when subjected to recurrent cycles of freezing–thawing and heating–cooling. Int. J. Rock Mech. Min. Sci. 41(2), 237–244 (2004)
4. Akin, M., Özsan, A.: Evaluation of the long-term durability of yellow travertine using accelerated weathering tests. Bull. Eng. Geol. Environ. 70(1), 101–114 (2011)
5. Jamshidi, A., Nikudel, M.R., Khamehchiyan, M.: Predicting the long-term durability of building stones against freeze–thaw using a decay function model. Cold Reg. Sci. Technol. 92 (92), 29–36 (2013)
6. Ghobadi, M.H., Babazadeh, R.: Experimental studies on the effects of cyclic freezing–thawing, salt crystallization, and thermal shock on the physical and mechanical characteristics of selected sandstones. Rock Mech. Rock Eng. 48(3), 1001–1016 (2015)
7. Hallet, B., Walder, J.S., Stubbs, C.W.: Weathering by segregation ice growth in microcracks at sustained subzero temperatures: verification from an experimental study using acoustic emissions. Permafrost Periglac. Process. 2(4), 283–300 (1991)
8. Xu, G., Liu, Q.: Analysis of mechanism of rock failure due to freeze-thaw cycling and mechanical testing study on frozen-thawed rocks. Chin. J. Rock Mech. Eng. 24(17), 3077–3082 (2005)
9. Yavuz, H.: Effect of freeze–thaw and thermal shock weathering on the physical and mechanical properties of an andesite stone. Bull. Eng. Geol. Environ. 70(2), 187–192 (2011)

Influence of Warm Oil Pipeline on Underlying Permafrost and Cooling Effect of Thermosyphon Based on Field Observations

Fei Wang[1,2] (ID), Guoyu Li[1(✉)] (ID), Wei Ma[1] (ID), Yanhu Mu[1] (ID),
Yuncheng Mao[1] (ID), and Bo Wang[3] (ID)

[1] State Key Laboratory of Frozen Soil Engineering, Cold and Arid Regions
Environmental and Engineering Research Institute, Chinese Academy
of Sciences, Lanzhou 730000, China
guoyuli@lzb.ac.cn
[2] University of Chinese Academy of Sciences, Beijing 100049, China
[3] Jiagedaqi Oil and Gas Transmission Sub-company, PetroChina Pipeline
Company, Jiagedaqi 165000, China

Abstract. Ground temperatures under a buried warm China-Russia oil pipeline were monitored to evaluate permafrost thawing and cooling performance of thermosyphons installed near the pipe in sporadic permafrost regions. Field observations demonstrated that warm oil thawed the underlying permafrost and increased the active layer thickness. Thermosyphons can cool the soils surrounding the pipe and effectively mitigate the permafrost thawing depending on their number and working duration. But now the heat dissipated from the warm pipeline and construction disturbance were not completely removed by two pairs of thermosyphons after two winters of operation. There was still a thawed layer beneath the thermosyphons even in winter. Further long-term monitoring of the cooling performance of thermosyphons is needed. This study provided basic data and analytic references for other similar cold region pipelines.

Keywords: Oil pipeline · Permafrost thawing · Thermosyphon
Cooling effect · Thawing depth

1 Introduction

Many man-made infrastructures have already been damaged by thaw settlement of permafrost in cold regions [1]. For example, the recorded ground surface subsidence was more than 2 m in the trench at a section of Norman Wells pipeline [2]. The China-Russia Crude Oil Pipeline (CRCOP) was built in winter of 2009–2010 using a conventional burial mode, which traverses 441 km discontinuous permafrost zone in China. Over seven years of operation, significant surface subsidence within the trench have occurred in warm and ice-rich permafrost regions [3]. Lots of investigations have shown that thermosyphon was an effective mitigation method to protect the permafrost from thawing [4, 5]. So thermosyphons were also installed along the CRCOP to avoid permafrost thawing. However, field observations of permafrost thawing and cooling performance of thermosyphons installed near the pipe is very limited. In this paper, the

© Springer Nature Switzerland AG 2018
W. Wu and H.-S. Yu (Eds.): *Proceedings of China-Europe Conference
on Geotechnical Engineering*, SSGG, pp. 1424–1428, 2018.
https://doi.org/10.1007/978-3-319-97115-5_115

evolution of ground temperatures on the right-of-way (ROW) of the CRCOP and cooling performance of thermosyphons were estimated.

2   Study Site and Monitoring Method

The study site is located in sporadic permafrost regions, near southern limit of permafrost in Northeast China. Four instrumented cross sections were established perpendicular to the pipeline at 20.0 m intervals. No.1 Cross Section was constructed in 2014 and two 20-m deep boreholes were drilled (see Fig. 1).

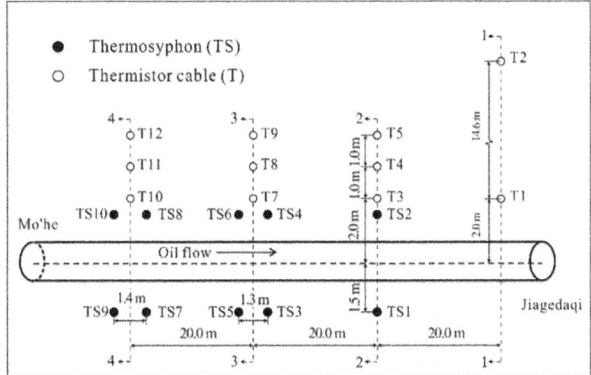

Fig. 1. Plane of the instrumented cross sections. Note: TS1-TS10 and T1-T12 indicate positions of thermosyphons and thermistor cables, respectively.

Due to a larger thaw bulb and pipe settlement, ten 9-m-long thermosyphons were installed vertically 1.5 m away from the pipe centerline with different numbers (one or two pairs of thermosyphons at different cross sections) and longitudinal spacings (1.3 or 1.4 m) in 2015. The lower 6 m of each thermosyphon was buried in soils and the upper 3-m section was exposed to the air. Correspondingly, three cross sections were instrumented in late June 2015. Nine boreholes 11 to 15 m deep were drilled 0.5, 1.5 and 2.5 m away from thermosyphons, respectively. The thermistor cables were placed accordingly to monitor the ground temperatures.

3   Results and Discussion

3.1   Evolution of Ground Temperatures

The oil temperature at this study site was not directly measured. We used the outlet oil temperature obtained from Jiadedaqi pump station (0.6 km away from our study site) as a substitute. Mean monthly oil temperatures ranged from 2.0 °C to 13.0 °C (May 2011 to March 2017). Figure 2 shows that ground temperatures on-ROW were substantially higher than those at an adjacent undisturbed site in Octobers of 2014 and

2017. Even for the deep permafrost (from −15 m to −20 m), the average temperature under the pipe was 0.2 °C higher than that at the natural site. Additionally, ground temperatures on-ROW increased significantly during the period of 2014–2017, particularly between −2 m and −8 m deep. The natural permafrost table was less changed (2.0 m) from 2014 to 2017, while the artificial permafrost table beneath the pipeline were much deeper, increasing by 2.7 m at a rate of 0.9 m/a over three years.

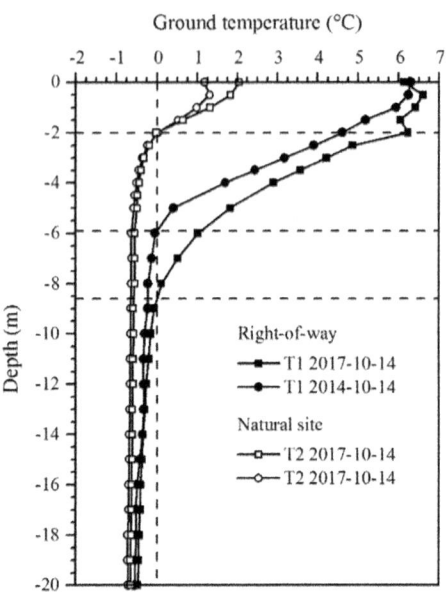

Fig. 2. Ground temperature profiles on-ROW (T1) and at the natural site (T2) in October of 2014 and 2017.

3.2   Cooling Performance of Thermosyphons

Figure 3 shows the ground temperature profiles in boreholes T3 and T7 (0.5 m away from the thermosyphon) in cold seasons. Ground temperatures in borehole T1 are also depicted together as the reference values to analyze thermosyphon performance. It can be seen in Fig. 3 that ground temperatures of the upper 6-m-thick soil layers near thermosyphons were much lower than those without thermosyphon in winter, indicating that the thermosyphons work better and have great cooling effect in winter. The best cooling effect was found at No. 3 cross section in borehole T7, which has two pairs of thermosyphons at a longitudinal spacing of 1.3 m.

Although the active layer thickness (ALT) from boreholes T3 and T7 kept increasing, different increasing rates were observed after installing thermosyphons. The increasing rate of the ALT was 0.9 m/a in borehole T1 (see Fig. 2), while it was 0.55 and 0.45 m/a in boreholes T3 and T7, respectively, during the monitoring period due to the cooling effect of thermosyphons. Descriptions above indicate that the

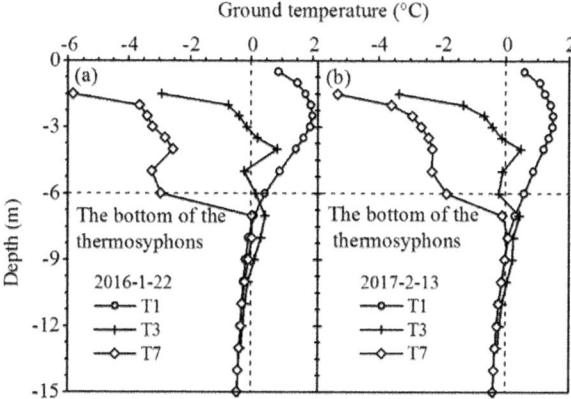

Fig. 3. Ground temperature profiles in boreholes T1, T3 and T7: (a) 2016-1-22; (b) 2017-2-13.

thermosyphons can alleviate the permafrost degradation by cooling the surrounding soils and reducing the increasing rate of ALT. But now they do not completely remove the heat dissipated from the warm oil pipeline and construction disturbance, resulting in a thaw bulb surrounding the pipe even in winter. That is, thermosyphons still need more time to cooling the soils surrounding the pipe. The long-term monitoring of the cooling performance of thermosyphon is also needed.

4 Conclusions

The CRCOP operated at year-round positive oil temperature and brought heat into the surrounding permafrost, resulting in the higher ground temperatures and the larger ALT, which dramatically accelerated the permafrost thawing on-ROW.

Thermosyphons can cool the surrounding soils and effectively mitigate the thawing of permafrost. Two pairs of thermosyphons are better to alleviate the permafrost degradation, but now they don't completely remove the heat dissipated from the warm oil pipeline and construction disturbance after two winters of operation.

Acknowledgements. This work was funded by the National Natural Science Foundation of China (Grant Nos. 41672310, U1703244 and 41630636), the National Key Research and Development Program (2016YFC0802103) and STS Research Project of CAS (Grant No. CHHS-TSS-STS-1502).

References

1. Schiermeier, Q.: Alpine thaw breaks ice over permafrost's role. Nature **424**, 712 (2003)
2. Smith, S.L., Burgess, M.M., Riseborough, D.W.: Ground temperature and thaw settlement in frozen peatlands along the Norman Wells pipeline corridor, NWT Canada: 22 years of monitoring. In: 9th International Proceedings on Permafrost, pp. 1665–1670. Fairbanks, Alaska (2008)

3. Li, G.Y., et al.: Frost hazards and mitigative measures following operation of China-Russia crude oil pipeline. Rock Soil Mech. **36**(10), 2963–2973 (2015)
4. Johnson, E.R., Hegdal, L.A.: Permafrost-related performance of the Trans-Alaska Oil Pipeline. In: 9th International Proceedings on Permafrost, pp. 857–864. Fairbanks, Alaska (2008)
5. Ma, W., Cheng, G.D., Wu, Q.B.: Construction on permafrost foundations: lessons learned from the Qinghai-Tibet railroad. Cold Reg. Sci. Technol. **59**(1), 3–11 (2009)

Study on the Stability of Spread-Footing Foundations on Permafrost Regions

Zhi Wen[⊠], Zhizhong Sun, and Shujuan Zhang

State Key Laboratory of Frozen Soil Engineering,
Northwest Institute of Eco-Environment and Resources,
Chinese Academy of Sciences, Lanzhou 730000, Gansu, China
wenzhi@lzb.ac.cn

Abstract. Spread-footing foundations embedded in frozen soil are subject to both the wind-induced uplift and frost heave forces. The stress and deformation of foundations in the Qinghai-Tibet Power Transmission Line (QTPTL) engineering were monitored to investigate the stress state of the tower foundations. The results showed that the stresses at the bases of tower foundations had a close relationship with air and ground temperatures. Seasonal variations in the contact stress depended on the seasonal freezing and thawing of foundation soil. The contact stress increased with the cooling of the underlying soils and decreased with the warming of the underlying soils. The numerical simulation results showed that the frost heave force induced by soil freezing potentially threatens the safety and normal operation of the QTPTL. Thaw settlement may lead to harmful deformation of tower foundations if global warming is considered in the numerical model. The remedial measure with thermosyphons only can reduce the settlement of the foundation and will increase the frost jacking risk of the foundation.

Keywords: Qinghai-Tibetan Plateau · Stress · Tower foundation
Frost heave force · Transmission line · Permafrost

1 Introduction

The Qinghai-Tibet Power Transmission Line (QTPTL) runs across 1,038 km of permafrost and seasonally frozen ground in the Interior of the Qinghai-Tibetan Plateau. The mean annual air temperature of permafrost and seasonally frozen ground along the QTPTL varies between −3 °C and −7 °C and the minimum air temperature is lower than −37 °C in short durations. The active layer is subjected to annual freeze-thaw cycles and its thickness varies between 2 and 3 m. Thus, substantial heave force is expected due to the existence of extensive frost-susceptible soils and cold climate. Moreover, the wind-induced uplift force is another load for the transmission towers and can reach more than 1,500 kN. Warm and ice-rich permafrost is widespread in permafrost regions of the Qinghai-Tibetan Plateau. The warming and subsequent thawing of ice-rich permafrost tends to result in serious engineering and environmental consequences. If not well designed or built, the tower foundations may be jacked up in the freezing periods and subside in the thawing periods, which would result in expensive maintenance costs and significantly threatens the safety and operation of the

© Springer Nature Switzerland AG 2018
W. Wu and H.-S. Yu (Eds.): *Proceedings of China-Europe Conference on Geotechnical Engineering*, SSGG, pp. 1429–1432, 2018.
https://doi.org/10.1007/978-3-319-97115-5_116

transmission lines. In arctic and sub-arctic regions, intense frost-induced heaving has considerably deteriorated the condition of numerous power network objects and significantly threatens the construction and operation of transmission lines [1–4]. Seasonal frost heave or thaw settlement deformation can impact the mechanical state of the foundations, while the stress state of tower foundations and surrounding soils has a close relationship with foundation failures [5–8]. Therefore, it is important to understand how the stress changes in order to reduce tower foundation damage.

2   On-Site Experimental Investigations

To investigate the contact stress characteristics of tower foundations, three thin-film pressure sensors were installed beneath each tower foundation. To monitor ground temperatures at the test foundation section, six boreholes were drilled in the vicinity of the tower foundation soils (Fig. 1). The depths of these boreholes varied from 10 to 23 m. In each borehole, thermistors were installed at depth intervals of 0.5 m from the ground surface. The thermistors were made by the State Key Laboratory of Frozen Soil Engineering, Cold and Arid Regions Environmental and Engineering Research Institute, Chinese Academy of Sciences, and with a calibrated accuracy of ±0.05 °C. The deformation of the tower foundations was measured by a water level instrument.

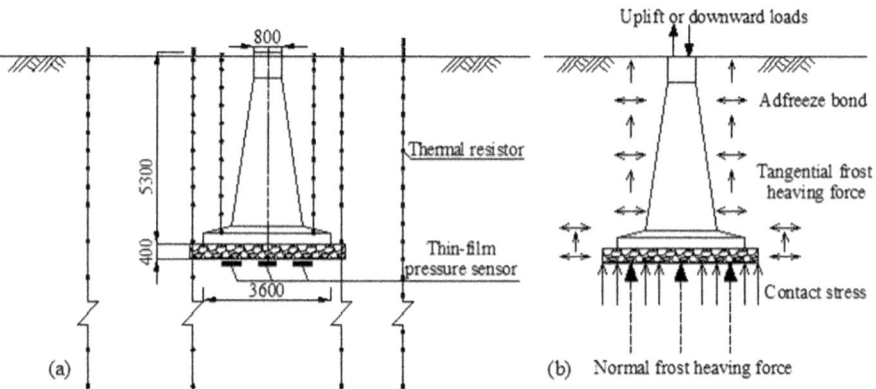

Fig. 1. Sensor installation design at the test tower foundation (a) and force analysis (b) (Unit: mm)

To study the cooling effect of two-phase closed thermosyphons, four thermosyphons were installed around the 492# tower foundation. The length of thermosyphons was 9 m; the evaporation part was 7 m-long and was embedded in the ground. The condensation part was 2 m-long and was exposed in the air. Figure 1 shows the sensor installation design at the test footing foundation and force analysis [9].

3   On-Site Experimental Investigations

Figure 2a shows the seasonal variation of contact stress under the 490# test tower foundation. The contact stresses of 490# showed significant variations. The absolute value of contact stress decreased gradually after October and reached approximately 0 kPa in January. Then the contact stress increased gradually to approximately 300 kPa in May, and then it remained constant during the summer.

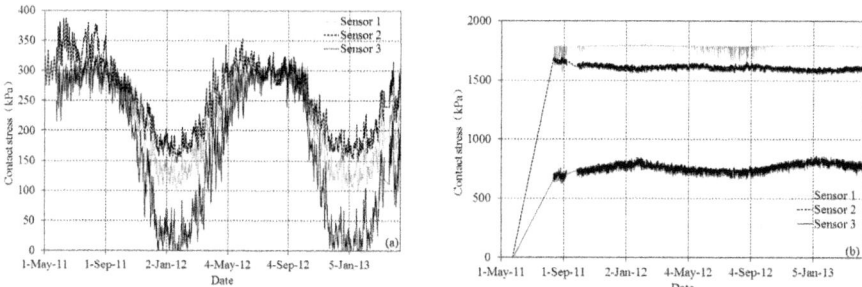

Fig. 2. Variations of contact stress at the 490# tower foundation (a), Variations of contact stress at the 492# tower foundation (b)

Figure 2(b) shows the seasonal variations in contact stress under the 492# test tower foundation. The value of contact stress at 492# is much higher than that at 490#. However, the seasonal variability of the contact stress at the tower 492# was much smaller than that at 490#. The contact stress increased linearly since May 2011 due to line hanging, and reached its peak in August 2011, after which it remained largely constant. In contrast to that at the tower 490#, the contact stress at 492# did not display significant seasonal variations and was free from the influence of air temperature and wind speed. Due to the cooling effect of thermosyphons, ground temperatures near the tower foundation clearly dropped since May 2011. The ground temperatures decreased from −0.5 °C in May 2011 to −2 °C in May 2013 (Fig. 3a). The monitoring of tower foundation deformation showed that a slight frost heave occurred at 492#, and reached 0.02 m in the past three years (Fig. 3b).

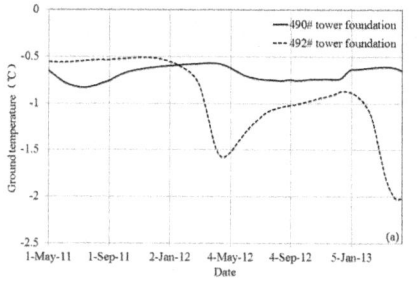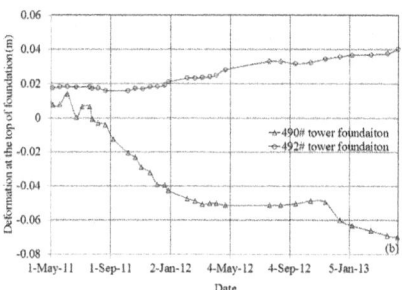

Fig. 3. Variations of ground temperature (a) and deformation (b) at the 490# and 492# tower foundations

4   Conclusions

The contact stresses under transmission tower foundations without thermosyphons showed significant seasonal variations, depending nearby air or ground temperatures. The contact stresses showed significant daily variation, which was closely related to variations in air temperature and ground temperature. The cooling effect of thermosyphons led to the occurrence of frost heave and high contact stress.

Acknowledgements. This research project was supported by the Natural Science Foundation of China (Grant No. 41771073, 41471061 and 41690144), the Research Project of the State Key Laboratory of Frozen Soils Engineering (SKLFSE-ZT-22) and the Major Program of Bureau of International Cooperation, the Chinese Academy of Sciences (131B62KYSB20170012).

References

1. Cheng, D.X., Zhang, J.M., Liu, H.J., Yu, Q.H., Liu, Z.W.: The influence factor analysis for site select of transmission line in frozen earth area. J. Eng. Geol. **17**(3), 329–333 (2009). (in Chinese)
2. Johnson, J.B., Esch, D.C.: Frost jacking forces on H and pipe pile embedded in Fairbanks silt. In: Proceedings of the 4th International Symposium on Ground Freezing, Sapporo, Japan, vol. 2, pp. 125–133 (2009)
3. Ladanyi, B., Foriero, A.: Evolution of frost heaving stresses acting on a pile. In: Lewkowicz A.G., Allard M. (eds.). Proceedings of the 7th International Conference on Permafrost, Yellowknife, N.W.T. Canada, pp. 623–633 (1998)
4. Lyazgin, A.L., Lyashenko, V.S., Ostroborodov, S.V., Ol'shanskii, V.G., Bayasan, R.M., Shevsov, K.P., Pustovoit, G.P.: Experience in the prevention of frost heave of pipe foundations of transmission towers under northern conditions. Power Technol. Eng. **38**(2), 124–126 (2004)
5. Penner, E.: Uplift forces on foundations in frost heaving soils. Can. Geotech. J. **11**, 323–338 (1974)
6. Perameswaran, V.R.: Adfreeze strength of frozen sand to model piles. Can. Geotech. J. **15**(4), 494–500 (1978)
7. Selvadurai, A., Hu, J.: Axial loading of foundations embedded in frozen soil. Int. J. Offshore Polar Eng. **6**(2), 650–653 (1996)
8. Tong, C., Guan, F.: Frost heaving and the prevention of freezing damage. China Water Power Press, Beijing (1985). (in Chinese)
9. Wen, Z., Sheng, Y., Jin, H., Li, S., Li, G., Niu, Y.: Thermal Elasto-plastic computation Model of a buried oil pipeline subject to frost heave and thaw settlement. Cold Reg. Sci. Technol. **64** (3), 248–255 (2010)

Nonlinear Numerical Analysis of Thaw Consolidation of Ice Rich Frozen Soil

Xiaoliang Yao[1(✉)], Boxiang Dang[2], and Jilin Qi[2]

[1] State Key Laboratory of Frozen Soil Engineering,
Northwest Institute of Eco-Environment and Resources,
Chinese Academy of Sciences, Lanzhou 730000, China
`yaoxl@lzb.ac.cn`
[2] School of Civil and Transportation Engineering,
Beijing University of Architecture and Civil Engineering, Beijing 100044, China

Abstract. A linear interpolation function fitting the nonlinear relationship between void ratio and compression modulus was proposed and corresponding strategy guaranteeing calculation accuracy and efficiency was developed for the nonlinear numerical analysis of 3-D large strain thaw consolidation of ice rich frozen soil. It was verified by a series of ice rich thaw consolidation tests that, with the proposed numerical implementation strategy, the calculated results match well with the tested results of pore water pressure and thaw displacement. Further analysis on the different stress-strain relationships shown that the prediction values and accuracies on pore water pressure and thaw displacement of nonlinear relationship is higher than that of linear relationship. This also leads to higher prediction accuracy on thaw consolidation degree and pore water pressure at thawing front of nonlinear relationship.

Keywords: Ice rich frozen soils · Thaw consolidation
Nonlinear stress-strain relationships

1 Governing Equations and Numerical Implementation

When ice rich frozen soil is involved, the large strain consolidation theory is usually employed to describe soil skeleton mechanical behavior and fluid flow in post-thawed domain [1]. Generally, the consolidation theory includes four parts (i.e., kinematic equation, constitutive equation, Darcy's law and fluid mass conservation equation), where the three dimensional linear constitutive theory us expressed as,

$$\dot{\sigma}_{ij} = \frac{E}{(1+v)}\dot{\varepsilon}_{ij} + \frac{vE}{(1-2v)(1+v)}\dot{\varepsilon}_{ij}\,\delta_{ij} - \delta_{ij}u \tag{1}$$

In which, σ_{ij} is total stress tensor, E is Young's modulus, v is Poisson's ratio, δ_{ij} is the Kronecker symbol, and $\dot{\varepsilon}_{ij}$ is symmetric deformation tensor.

© Springer Nature Switzerland AG 2018
W. Wu and H.-S. Yu (Eds.): *Proceedings of China-Europe Conference on Geotechnical Engineering*, SSGG, pp. 1433–1437, 2018.
https://doi.org/10.1007/978-3-319-97115-5_117

The thermal conductive equations are implemented to detect the post-thawed domain as following,

$$
\begin{cases}
- h_{i'i} + h_v = \rho c \dfrac{\partial T}{\partial t} \\
h_i = -\xi T'_i
\end{cases}
\tag{2}
$$

In Eq. (2), T is temperature (°C), h_v (W/m^3) is volumetric heat source intensity; c (J/kg·°C) and ξ (W/m·°C) are specific heat and thermal conductivity, respectively; ρ is the media density (kg/m^3). Both of the thermal parameters are temperature dependent, and the details can be referred to literatures [2, 3].

As it can be seen in Eq. (2), the linear stress-strain relationship was used in previous three dimensional analysis of frozen soil thaw consolidation [4, 5]. For ice rich frozen soil, the compressibility of which shows strong nonlinearity, and the nonlinear stress-strain relationship must be used to describe the soil skeleton mechanical behaviours. In the following, a nonlinear relationship between void ratio (e) and compression modulus (E_s) is used to modify the original linear thaw consolidation theory, i.e.,

$$
E_s = \frac{1 + e_0}{\lambda} \exp\left(\frac{e_0 - e}{\lambda}\right)
\tag{3}
$$

where, λ is the slope of K_0 compression e-log (stress) curve. The Young's modulus (E) can be further expressed as,

$$
E = \left(1 - \frac{2v^2}{1 - v}\right) E_s
\tag{4}
$$

In numerical analysis, a linear interpolation function Eq. (3) for fitting the nonlinear relationship between void ratio and compression modulus is proposed for guaranteeing calculation accuracy and efficiency as,

$$
E_{s(t+t\Delta)} = \left(1 - \frac{e_{t+t\Delta} - e_m}{e_{m+1} - e_m}\right) E_{s(m)} + \frac{e_{t+t\Delta} - e_m}{e_{m+1} - e_m} E_{s(m+1)} \left(e_m \leq e_{t+t\Delta} \leq e_{m+1}, \right.
$$
$$
m = 0, 2, 3, \ldots, 4)
\tag{5}
$$

where, $E_{s(m)}$ and e_{m-} are the data points obtained from the K_0 compression e-log (stress) curve.

2   Verification and Results Discussion

To verify the applicability of the proposed linear interpolation function for fitting the nonlinear relationship between void ratio and compression modulus, a series of 1-D thaw consolidation tests under different dry unit weight and surcharge loads were conducted, and the corresponding parameters for consolidation calculation are obtained in Table 1.

Table 1. Parameters for consolidation calculation.

Dry Unit Weight (kN/m^3)	λ	e_0	Surcharge Load (kPa)	k (m/s)	E_{s0} (kPa)	Poisson's Ratio
16.1	0.029	0.68	50	5.00E−08	271	0.25
			100	3.40E−08	508	
13.0	0.052	1.08	50	7.00E−08	189	0.30
			100	4.40E−08	346	
11.5	0.080	1.35	50	9.55E−08	137	0.30
			100	6.55E−08	255	

As the two main indexes representing the thaw consolidation behavior of frozen soil, thaw consolidation degree (*TD*) and the pore water pressure at thawing front (*PPTF*) are closely related thaw consolidation ratio (*TCR*) [2, 6], which is defined as,

$$TCR = \frac{\alpha}{2(c_v)^{1/2}} \qquad (6)$$

where, α is the thawing rate of soil sample, which is related to the thermal properties and boundaries; c_v is the consolidation coefficient and defined as,

$$c_v = \frac{kE_{s0}}{\rho_w} \qquad (7)$$

where, the secant compression modulus (E_{s0}) is used for ease of analyzing the difference between linear and nonlinear relationships.

Figures 1 and 2 show the relationships between thaw consolidation degree (*TD*), normalized pore water pressure at thawing front (N_{PPTF}) and *TCR*, where *TD* and N_{PPTF} are defined as,

$$TD = \frac{h(t)}{h_{\max}(t)} \qquad (8)$$

$$N_{PPTF} = \frac{u}{p_0} \qquad (9)$$

where, $h(t)$ is the thaw displacement, $h_{\max}(t) = \frac{E_{s0}}{P_0} x(t)$ and $x(t)$ is the thaw depth at time *t*. It can be seen that for the calculated results of both stress-strain relationships, N_{PPTF} is proportionally related to *TCR*, while it is opposite for *TD*. This indicates that with increase of *TCR*, more post-thawed pore water is generated, while the rate of drainage (c_v) is relatively decreased. Subsequently, the *TCD* decreases and N_{PPTF} increases. By comparing the calculated results of both relationships (linear and nonlinear), it can be found that for both of the indexes, the linear results are lower than that of nonlinear results, which is due to the calculated differences of both stress-strain relationships on thaw displacement and pore water pressure. In addition, the nonlinear stress-strain relationship shows a higher accuracy on of the *TCD* and N_{PPTF} (Figs. 1 and 2) than the

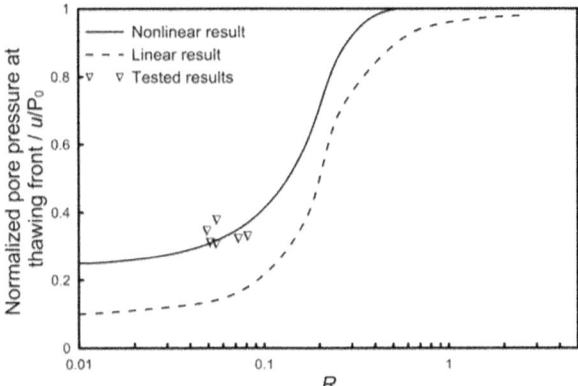

Fig. 1. Changes in N_{PPTF} vs. TCR (top)

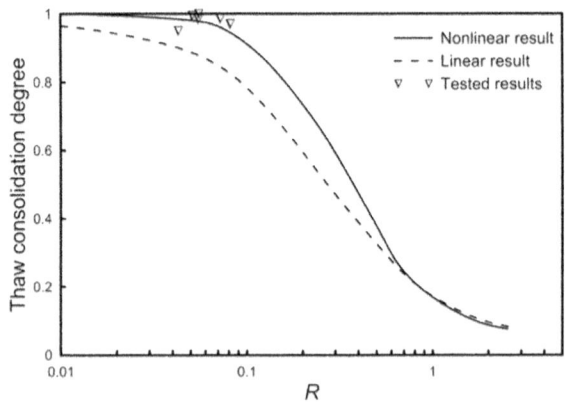

Fig. 2. Changes in TD with TCR (bottom).

linear relationship, which indicates the applicability of the proposed linear interpolation function in cold regions engineering when ice rich permafrost is involved.

Acknowledgements. This work was supported in part by the National Natural Science Foundation of China (No. 41671061) and the research Project of the State Key Laboratory of Frozen Soils Engineering granted to Dr. Xiaoliang Yao (SKLDSE-ZT-28).

References

1. Yao, X.L., Qi, J.L., Wu, W.: Three dimensional analysis of large strain thaw consolidation in permafrost. Acta Geotech. **7**(3), 193–202 (2012)
2. Qi, J.L., Yao, X.L., Yu, F.: Consolidation of thawing permafrost considering phase change. J. Civil Eng. KSCE **17**(6), 1293–1301 (2013)
3. Wang, S.H., Qi, J.L., Yu, F., Yao, X.L.: A novel method for estimating settlement of embankments in cold regions. Cold Reg. Sci. Technol. **88**(5), 50–88 (2013)
4. Sykes, J.F., Lennox, W.C., Charlwood, R.G.: Finite element permafrost thaw settlement model. J. Geotech. Eng. Div. **100**(11), 1185–1201 (1974)
5. Qi, J.L., Yao, X.L., Yu, F., Liu, Y.Z.: Study on thaw consolidation of permafrost under roadway embankment. Cold Reg. Sci. Technol. **81**(3), 48–54 (2012)
6. Morgenstern, N.R., Smith, L.B.: Thaw-consolidation tests on remoulded clays. Can. Geotech. J. **10**(1), 25–40 (1973)

Cooling Performance of a Composite Embankment for High-Grade Highways in Permafrost Regions

Mingyi Zhang[(⌧)], Yuanming Lai, Wansheng Pei, Qihao Yu, and Zhongrui Yan

State Key Laboratory of Frozen Soil Engineering,
Northwest Institute of Eco-Environment and Resources,
Chinese Academy of Sciences, Lanzhou 730000, China
myzhang@lzb.ac.cn

Abstract. In permafrost regions, high-grade highways with a wide and dark-colored asphalt pavement can cause the degradation of underlying permafrost. However, the embankments with single commonly cooling technique, e.g. two-phase closed thermosyphon (TPCT) embankment and crushed-rock embankment cannot effectively solve the problem. Therefore, a composite embankment for high-grade highways, combined with L-shaped TPCTs, crushed-rock revetments and insulation was designed. The L-shaped TPCTs were used to cool the core of the embankment, the crushed-rock revetments with different thicknesses was intended to cool the side slopes and to diminish the sunny-shady slope effect, and the insulation was designed to strengthen the cooling effect by increasing the thermal resistance of the embankment. Here, we experimentally and numerically evaluated the cooling performance of the composite embankment. The results indicate that the composite embankment can effectively raise the permafrost Table (0 °C isotherm) and cool the underlying permafrost under a separated high-grade highway with a wide and dark-colored asphalt pavement (double lanes each direction). Therefore, the composite embankment structure should be considered to be applied to the construction of high-grade highways in permafrost regions.

Keywords: Cooling performance · Composite embankment
High-grade highway · Permafrost region

1 Introduction

A wide and dark-colored asphalt pavement of high-grade highways will absorb much more solar energy than a narrow highway pavement, and causes the degradation of underlying permafrost. Although a series of embankments have been used to increase the stability of underlying permafrost, e.g. two-phase closed thermosyphon (TPCT) embankment and crushed-rock embankment [1–3], the cooling techniques cannot ensure alone the stability of high-grade highways. Therefore, a composite embankment, combined with L-shaped TPCTs, crushed-rock revetments and insulation was designed for high-grade highways. The cooling performance of the composite embankment was

© Springer Nature Switzerland AG 2018
W. Wu and H.-S. Yu (Eds.): *Proceedings of China-Europe Conference on Geotechnical Engineering*, SSGG, pp. 1438–1441, 2018.
https://doi.org/10.1007/978-3-319-97115-5_118

studied experimentally and numerically, used in a separated high-grade highway with a wide and dark-colored pavement (double lanes each direction).

2 Experimental Results and Analysis

The experimental site is located at Beiluhe in the Qinghai-Tibet Plateau, and is a typical warm and ice-rich permafrost region [4]. At the experimental site, we carried out a full-scale field experiment of a separated high-grade highway. The trend of the composite embankment is about 198°, and the sunny-shady slope effect is significant. The cross section of the composite embankment is shown in Fig. 1. The L-shaped TPCTs were inserted to the center of the embankment from the shoulders, the crushed-rock revetments with different thicknesses were paved on the side slopes, and the insulation was buried in the middle of the embankment.

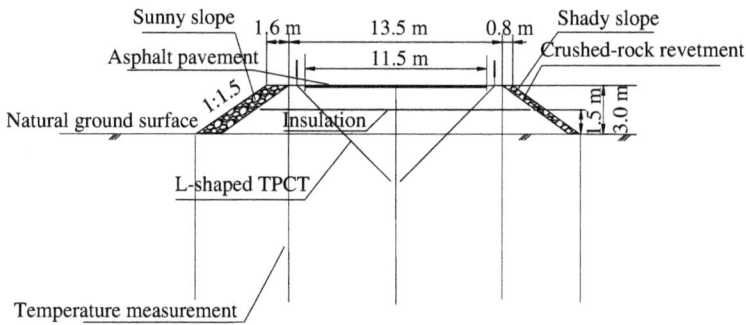

Fig. 1. Cross section of the composite embankment

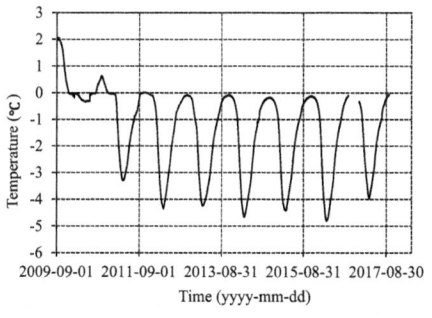

Fig. 2. Centerline temperature variation at the original ground surface (observed)

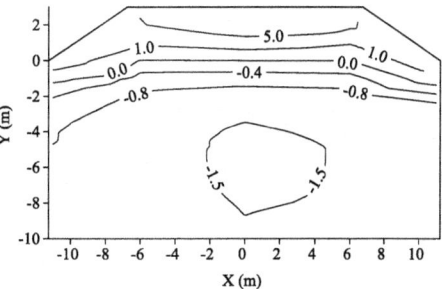

Fig. 3. Temperature distribution on October 15, 2017 (observed, Unit °C)

The centerline ground temperature of the composite embankment at the original natural ground surface is below 0 °C for the whole year since 2011, and the lowest temperatures are below −3 °C in cold seasons (Fig. 2). The permafrost Table (0 °C isotherm) rises and the −1.5 °C frozen zone exists under the embankment in 2017 (Fig. 3). This shows that the composite embankment is an effective method in protecting underlying permafrost. Furthermore, the temperature distributions in the composite embankment are symmetric (Fig. 3), illustrating that the crushed-rock revetments with different thicknesses can weaken the sunny-shady slope effect.

3   Numerical Results and Analysis

Based on the previous researches [3, 5–7], a three-dimensional coupled mathematical model was developed to simulate the cooling performance of the composite embankment (Fig. 1) by using the geography, geology and climate conditions at Beiluhe in the Qinghai-Tibet Plateau.

Similarly, the simulated centerline ground temperature of the composite embankment at the original natural ground surface is also below 0 °C for the whole year after it has been finished for 1 year (Fig. 4). Under the composite embankment, the permafrost table rises, and the temperature distributions are symmetric (Fig. 5); Furthermore, the −1.2 °C frozen zone still exists under the embankment in the 20th year after the construction under the climate warming (Fig. 5).

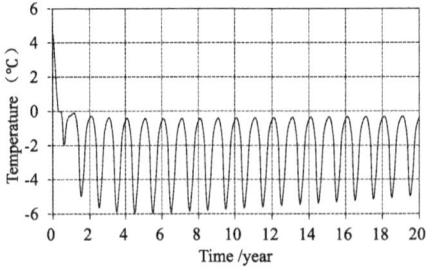
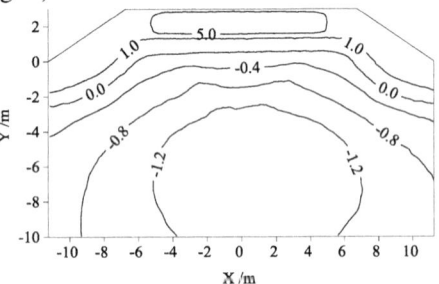

Fig. 4. Centerline temperature variation at the original ground surface (simulated)

Fig. 5. Temperature distribution on October 15, 20th year after the construction (Unit: °C)

4   Conclusions

The composite embankment with L-shaped TPCTs, crushed-rock revetments and insulation is an effective method in protecting the underlying permafrost and increasing the stability of the embankment in permafrost regions. It is also concluded that the cooling performance of the composite embankment is from the composite effects of the L-shaped TPCTs, crushed-rock revetments and insulation.

Acknowledgements. This research was supported by the National Natural Science Foundation of China (Grant Nos. 41471063, 41701070), the 100-Talent Program of the Chinese Academy of Sciences (Granted to Dr. Mingyi Zhang), and the Program of the State Key Laboratory of Frozen Soil Engineering (Grant No. SKLFSE-ZT-23).

References

1. Cheng, G.D.: A roadbed cooling approach for the construction of Qinghai-Tibet Railway. Cold Reg. Sci. Technol. **42**, 169–176 (2005)
2. Xu, J., Goering, D.J.: Experimental validation of passive permafrost cooling systems. Cold Reg. Sci. Technol. **53**, 283–297 (2008)
3. Lai, Y.M., Zhang, M.Y., Li, S.Y.: Theory and Application of Cold Regions Engineering. Science Press, Beijing (2009)
4. Gu, W., Yu, Q.H., Qian, J., Jin, H.J., Zhang, J.M.: Qinghai-Tibet Expressway experimental research. Sci. Cold Arid Reg. **2**(5), 396–404 (2010)
5. Zhuang, J., Zhang, H.: Heat Pipe Technology and Engineering Application. Chemical Industry Press, Beijing (2000)
6. Zhang, M.Y., Lai, Y.M., Zhang, J.M., Sun, Z.Z.: Numerical study on cooling characteristics of two-phase closed thermosyphon embankment in permafrost regions. Cold Reg. Sci. Technol. **65**, 203–210 (2011)
7. Zhang, M.Y., Lai, Y.M., Gao, Z.H., Yu, W.B.: Influence of boundary conditions on the cooling effect of crushed-rock embankment in permafrost regions of Qinghai-Tibetan Plateau. Cold Reg. Sci. Technol. **44**, 225–239 (2006)

Experimental Study of Elastic Properties of Saturated Clay Subjected to Freeze-Thaw Cycles

Feng Zhang[(⊠)], Bo Lin, Tao Li, and Decheng Feng

School of Transportation Science and Engineering,
Harbin Institute of Technology, Harbin, China
zhangf@hit.edu.cn

Abstract. This research aims to investigate shear wave velocity and small-strain shear modulus of saturated thawed clay. Various numbers of no-water supplied freeze-thaw cycling were experienced on the saturated clay firstly, then the wave velocity measurement setup was employed to carried out a series of non-destroyed wave velocity tests on each sample. During which, one pairs of BE were install on the top surface as transmitter and another on bottom surface as receiver, respectively, and another two pairs of BE were installed on horizontal direction. The effects of compaction degree and numbers of freeze-thaw cycles on elastic properties for vertical and horizontal direction were analyzed and discussed. The results indicate that initial density and freeze-thaw cycling exhibits a notable effect on elastic properties. The shear wave velocity and small strain shear modulus in vertical direction are greater than in horizontal direction, and both of them increase with increasing of initial density. However, they decrease after various numbers of freeze-thaw cycles and sharply drop at 1^{St} F-T.

Keywords: Saturated clay · Freeze-thaw cycles

1 Introduction

In seasonally frozen regions, the subgrade soil experienced repeated freezing and thawing cycling. During this process, the frozen water expands 9% in volume, and the pore between the soil particles was enlarged and may keep when the soil thawed. It will change the physical and mechanical behaviors (e.g. Simonsen et al. 2002; Wang et al. 2007; Qi et al. 2008).

Bender element is an electromechanical transducer to achieve converting between electrical energy and mechanical energy. The first bender element was introduced by Shirley and Hampton (1978), and then improved by numerous researches (Dyvik and Madshus, 1985; Lee and Santamarina, 2005). Wave velocity and elastic modulus of soil are important elastic parameters. The advantages of bender element have been demonstrated, and it also could easily get the wave velocity and initial tangent modulus even the strain as low as 10^{-6}. This research aims to investigate the effects of freeze-thaw cycling on elastic behaviors of clayed soil.

© Springer Nature Switzerland AG 2018
W. Wu and H.-S. Yu (Eds.): *Proceedings of China-Europe Conference on Geotechnical Engineering*, SSGG, pp. 1442–1446, 2018.
https://doi.org/10.1007/978-3-319-97115-5_119

The soil used in this study was obtained from Harbin, Northeast of China. According to guidance of the Test Methods of Soils for Highway Engineering (JTG E40-2007), the optimal water content and maximum dry density were determined as 17.4% and 1.74 g/cm^3 by the Modified Proctor Compaction Test. The specific gravity, liquid limit and plastic limit is 2.75, 36.98% and 25.19%, respectively. The soil is named as low liquid limit clay and classified as CL according to JTG E40-2007.

To prepare the samples, the air dried soil was sieved from a sieve with pores of 5 mm, and mixed with distilled water to achieve the optimal water content of 17.4%. After the soil was stored in a closed container for 12 h, the specimens were compacted at three compaction degrees of 90%, 94% and 98%, respectively. The size of samples is 62.8 mm in diameter and 125 mm in height. Saturating process was conducted by the suggestion of Standard for Soil Test Method (GB/T 50123-1999). Then the soil saturated samples were put into a freezer to begin the experience the freezing and thawing, the freezing temperature was determined as −25 °C and the duration is 12 h, and the thawing temperature was confirmed as 25 °C and duration is 6 h.

Figure 1 shows the wave velocity measurement setup and the layout the bender element in three directions. During the wave propagation tests, a step signal with peak-to-peak amplitude of 5 V was generated by Function Generator. The received signal was filtered by a bandpass Filter with frequency ranging from 500 Hz to 60 kHz, and the output signal was amplified to 20 dB. The filtered and amplified wave signal data was acquired by the Oscilloscope. Figure 2 shows the typical waveforms of saturated clay for un-FT and 9 FT.

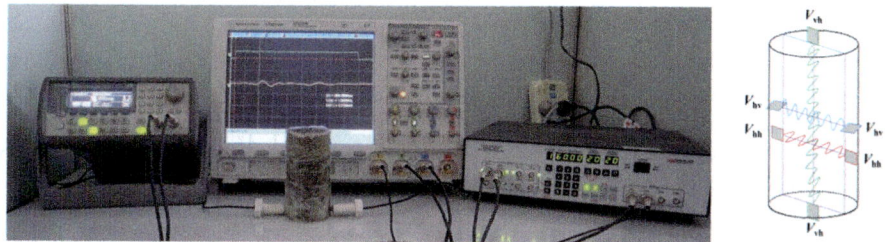

Fig. 1. Test setup of wave velocity and BE layout (Left: function generator; Middle: oscilloscope, sample and a pair of BE; Right: power amplifier with filter)

The shear wave velocity is calculated using the following equation:

$$V_s = L/\Delta t = L/(t - t_0) \tag{1}$$

Where, V_s is the shear wave velocity, L is the travel distance, Δt is the net travel time, t is the total travel time, t_0 is the system delay.

Based on the elastic wave propagation theory, the small-strain shear modulus can be evaluated by the following equations (Vinson 1978):

$$G = \rho V_s^2 \tag{2}$$

Where, ρ is the mass density.

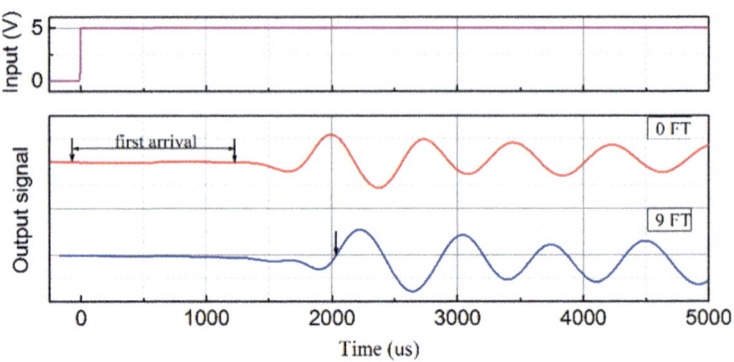

Fig. 2. Typical waveforms with various numbers of freeze-thaw cycling for saturated clay.

Figure 3 shows the shear wave velocity and small-strain shear modulus with different numbers of freeze-thaw cycling. As is shown in figures, the shear wave velocity in VH direction decreases with the increasing of numbers and the compaction degrees, especially experienced 1st freeze-thaw cycling, and the drop by 21%, 20% and 26%, respectively. Same tendency also could be found on the shear modulus, the decreases by 38%, 38% and 52%, respectively. However, wave velocity and modulus do not change much in HH and HV direction.

Figure 4 shows the anisotropy of elastic modulus caused by freeze-thaw cycling. Note that the G_{hv}/G_{vh} is 0.81 for un-experienced FT, and increases to 1.44 after deferent FT. The G_{hh}/G_{hv} is 1.17 for the sample no matter whether FT or not.

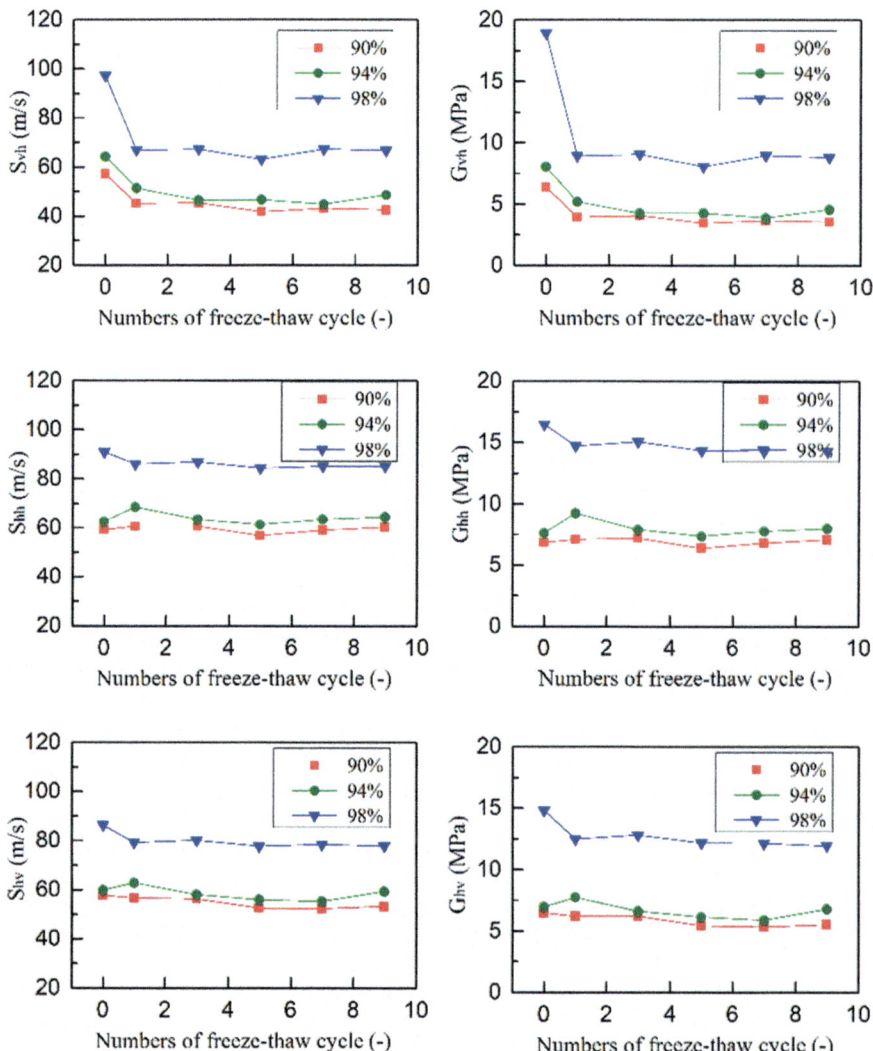

Fig. 3. Shear wave velocity and shear modulus with various numbers of FT cycling for vh-, hh- and hv- direction

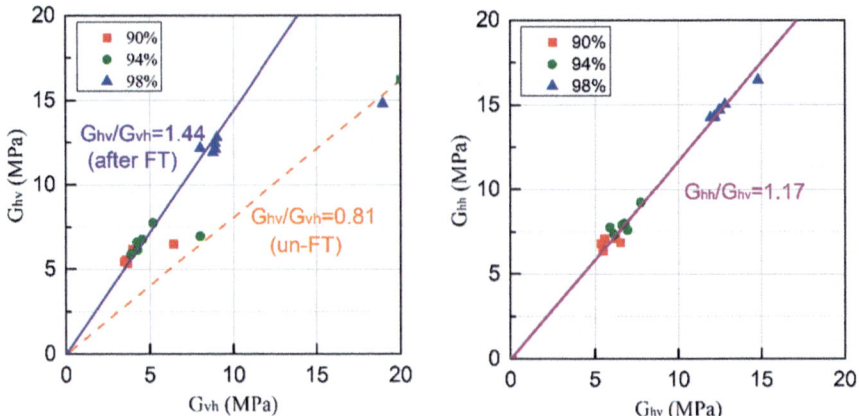

Fig. 4. Anisotropy of elastic modulus caused by freeze-thaw cycling.

References

Wang, D., Ma, W., Niu, Y., et al.: Effects of cyclic freezing and thawing on mechanical properties of Qinghai-Tibet clay. Cold Reg. Sci. Technol. **48**(1), 34–43 (2007)

Shirley, D.J., Hampton, L.D.: Shear-wave measurements in laboratory sediments. J. Acoust. Soc. Am. **63**, 607–613 (1978)

Simonsen, E., Janoo, V.C., Isacsson, U.: Resilient Properties of unbound road materials during seasonal frost conditions. J. Cold Reg. Eng. **16**(1), 28–50 (2002)

Qi, J.L., Ma, W., Song, C.X.: Influence of freeze-thaw on engineering properties of a silty soil. Cold Reg. Sci. Technol. **53**, 397–404 (2008)

Lee, J.S., Santamarina, J.C.: Bender elements: performance and signal interpretation. J. Geotech. Geoenvironmental Eng. **131**, 1063–1070 (2005). ASCE

Vinson, T.S.: Response of frozen ground to dynamic loading. In: Andersland, O.B., Anderson, D. M. Geotechnical engineering for cold regions, pp. 405–458. McGraw-Hill Book Company, New York (1978). Chap. 8

Dyvik, R., Madshus, C.: Lab measurements of Gmax using bender elements. In: Proceedings ASCE Annual Convention: Advances in the Art of Testing Soils Under Cyclic Conditions, pp. 186–197. Detroit, Michigan (1985)

Numerical Modeling of Rate-Dependence Behavior of Saturated Frozen Soil

Qiyin Zhu[1(✉)] [iD], Xiangyu Shang[1], Lianfei Kung[1], and Li Gang[2]

[1] State Key Laboratory for Geomechanics and Deep Underground Engineering,
China University of Mining and Technology, Xuzhou 221116, China
qiyin.zhu@gmail.com
[2] Jinan Rail Transit Group Co., Ltd., Jinan 250101, China

Abstract. The saturated frozen soil is composed of solid grains, ice and unfrozen water. The mechanical behavior of this engineering material is strongly associated with the amount of ice. Meanwhile, the amount of ice depends on the temperate and pressure and its existence is the main reason that induced the huge difference between the frozen and unfrozen soil. Furthermore, considering that ice is the highly rate-dependent material, the rate-dependency behavior of the frozen soil can be expected. A rate-dependent constitutive model based on the critical state theory is proposed for describing the rate-dependency of frozen soil. Being a variable parameter, the ice content is directly related to the temperature with can be correlated by the unfrozen water content curve. The tension strength and the pseudo-preconsolidation pressure can be expressed by the ice content. Meanwhile, the compression and recompression coefficient are related to the ice content and the values at the unfrozen condition. The rate-dependent behavior is considering by the over stress method. When the temperature is higher than $0°$, the ice content will be null. Then, the model will reduce to the conventional elastic-viscoplastic model. The predictive ability of the model is validated by simulating the 3D undrained triaxial compression test.

Keywords: Constitutive model · Rate dependency · Frozen soil
Ice content

1 Introduction

More and more engineering projects have been carried out on the frozen soils. For example, the method of freezing construction has also been widely used in the metro, tunnels and many other engineering activities. In these activities, the frozen soil was often subjected to the loading with different rates. A series of experiments have been conducted on the frozen soils under different rate conditions by the researches [1, 2]. The results show that the mechanical behaviors of frozen soil are rather depending the load rate. Some models have been widely used in the literature to describe the mechanical behavior of frozen soils [3–6]. However, few of them can simulate the rate-dependency behavior.

This paper attempts to propose an elastic viscoplastic model to address the rate-dependency behavior of saturate frozen soil.

© Springer Nature Switzerland AG 2018
W. Wu and H.-S. Yu (Eds.): *Proceedings of China-Europe Conference on Geotechnical Engineering*, SSGG, pp. 1447–1450, 2018.
https://doi.org/10.1007/978-3-319-97115-5_120

2 Constitutive Model

The constitutive model is referred to an isotropic and homogeneous material that is characterized by an elasto-viscoplastic behavior. The elastic part of the strain is defined through the relation between the shear modulus and the bulk modulus, similar to the Modified Clay Model (MCC). The viscoplastic strain is calculated based on the overstress theory.

2.1 Ice Content

The presence of ice is the chief cause for strengthening of freezing soils. Therefore, the volumetric ice content e_i is introduced as a key parameter in the description of the behavior of frozen soils

$$e_i = \frac{V_i}{V_s} \tag{1}$$

where V_i and V_s are the volume of ice and soil particle, respectively. e_i is obviously dependent on the temperature. The unfrozen water content w_u in the frozen soil can be calculated by the equation:

$$w_u = AT^B \tag{2}$$

where A and B are constants; and T is temperature. Thus, e_i can be obtained by

$$e_i = \frac{(w_0 - w_u) \cdot G_s}{\rho_i} \tag{3}$$

where w_0 is the initial water content; ρ_i is density of ice; G_s is specific gravity of soil particle.

2.2 Yielding Properties

A yield surface considering the tension behavior of frozen soil based on the MCC is adopted in

$$f = q^2 + M^2(p' - p_{0t})(p' - p_{0i}) = 0 \tag{4}$$

where p', q and M are the same as the definition in MCC model; p_{0t} and p_{0i} are the material properties representing yielding in isotropic tension and in isotropic compression. The expressions of the parameters relating to the frozen state can be obtained in Zhang et al. [6].

Figure 1 shows the illustration of the yielding surface of frozen soil at different temperatures from $0°$ to $-20°$. Thus, the model can simulate the behavior from a frozen state to an unfrozen state.

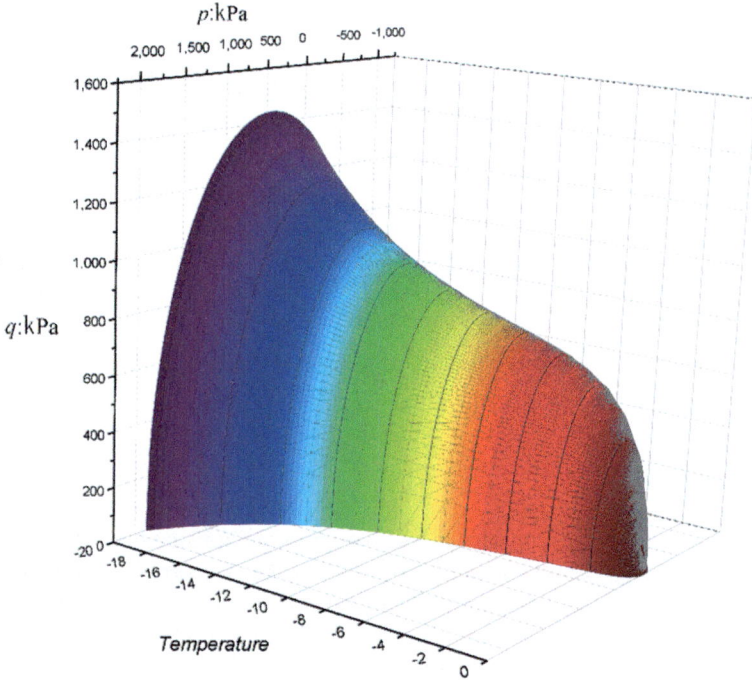

Fig. 1. Illustration of the evolution of yielding surface with temperatures

2.3 Viscoplastic Behavior

The viscoplastic part of the strain rate is calculated based on the overstress theory which is expressed as:

$$\dot{\varepsilon}^{vp} = \mu \left(\frac{F_d}{F_s} \right)^N \tag{5}$$

where, μ stands for the fluidity and N is the strain rate coefficient.

3 Predictive Ability

The predictive ability of the model is validated by simulating the undrained triaxial test under different strain rate and constant temperature. Figure 2a shows the relationship between deviatoric stress and primary stress under $-5°$ with three different strain rate。 Figure 2b shows the relationship between deviatoric stress and strain. As shown is the figure, the model can successfully follow the trend of the rate-dependency behavior of frozen soil.

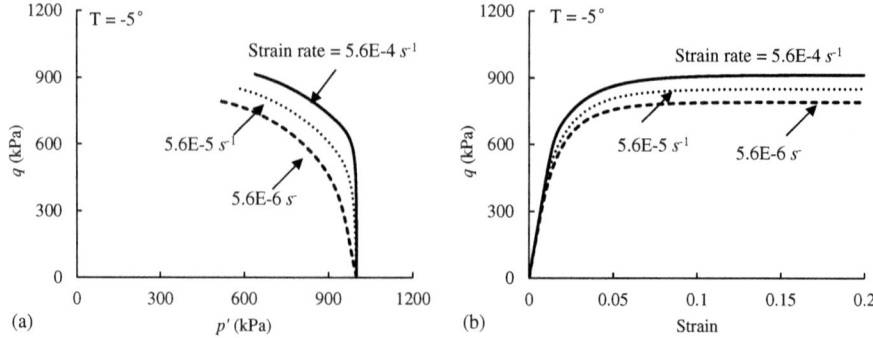

Fig. 2. The predictive results for triaxial tests under different strain rate: (a) q-p' relation and (b) stress –strain relation

4 Conclusion

In this paper, an elastic-viscoplastic model was proposed for simulating the behavior of frozen soils. The model was developed based on the overstress framework and the critical state theory. Ice content were selected as the key parameter relating to the temperature. Thus, the model can simulate the behavior from a frozen state to an unfrozen state with a single set of parameters. The predictive ability is validated by triaxial test under different strain rate.

Acknowledgment. This research project is financially supported by National Natural Science Foundation of China (41502271, 51504245).

References

1. Xu, G.: Hypoplastic constitutive models for frozen soil. Doctoral thesis. University of Natural Resources and Life Sciences, Vienna (2014)
2. Zhu, Y., Carbee, D.L.: Uniaxial compressive strength of frozen silt under constant deformation rates. J. Glaciol. Geocryol. **9**(1), 3–15 (1984)
3. Lai, Y., Liao, M., Hu, K.: A constitutive model of frozen saline sandy soil based on energy dissipation theory. Int. J. Plast. **78**, 84–113 (2016)
4. Ghoreishian Amiri, S.A., Grimstad, G., Kadivar, M., Nordal, S.: Constitutive model for rate-independent behavior of saturated frozen soils. Can. Geotech. J. **53**(10), 1646–1657 (2016)
5. Zhou, M.M., Meschke, G.: A three-phase thermo-hydro-mechanical finite element model for freezing soils. Int. J. Numer. Anal. Methods Geomech. **37**(18), 3173–3193 (2013)
6. Zhang, Y., Michalowski, R.L.: Thermal-hydro-mechanical analysis of frost heave and thaw settlement. J. Geotech. Geoenviron. Eng. **141**(7), 04015027 (2015)

Part XI: Geohazards-Risk Assessment, Mitigation and Prevention

Centrifuge Modelling of Slope Instability Due to Leakage of Buried Pipes

Kit Chan[1]([⊠]), Limin Zhang[1], Hong Zhu[1], and Te Xiao[2]

[1] Department of Civil and Environmental Engineering, The Hong Kong University of Science and Technology, Clear Water Bay, Kowloon, Hong Kong
kchanad@connect.ust.hk
[2] State Key Laboratory of Water Resources and Hydropower Engineering Science, Wuhan University, 8 Donghu South Road, Wuhan 430072, People's Republic of China

Abstract. Buried water mains, sewers and storm water pipes are critical infrastructures in the urban environment. In Hong Kong, a great number of pipes are buried in slopes, and catastrophic consequences due to landslides may happen upon pipe leakage. This study aims to investigate the infiltration process of leakage water with respect to the current Hong Kong mainlaying practice and explores its impacts on slope stability using geotechnical centrifuge modelling. Test results indicate that deep-seated slope failure, and surface erosion induced by concentrated flow of leaked water may occur when the pipes are subject to leakage.

Keywords: Buried pipe · Mainlaying · Slope stability · Leakage
Erosion

1 Introduction

Buried water mains, sewers and storm water pipes are critical infrastructures in Hong Kong. During their service, pipes may become defective and water may leak from the pipes. The leaked water will infiltrate into the surrounding soils. Since the pipe pressure for fresh water mains can be up to 400–600 kPa, the hydraulic gradients in the soil can be very high and hence internal erosion of the soil surrounding the pipe can happen.

After the 1994 Kwun Lung Lau landslide (Hui et al., 2007), the Code of Practice on Monitoring and Maintenance of Water-carrying Services Affecting Slopes (ETWB 2006) was published in response to the recommendations made by Morgenstern (2000). Water Supplies Department (2012) and the new code of practice (ETWB 2006) stipulate on pre-operation tests, inspections, and maintenance. Although attempt has been made to inspect and detect potential problems, pipe bursting still occurred frequently.

Only a limited research (Zhang and Li 2007) has been performed to study the infiltration process around a leaking pipe, the effects of leakage from pressurized BWCS on slope stability, and designs against slope failures caused by leaking BWCS. Therefore, it is necessary to investigate the slope stability upon leakage, as well as the infiltration process.

© Springer Nature Switzerland AG 2018
W. Wu and H.-S. Yu (Eds.): *Proceedings of China-Europe Conference on Geotechnical Engineering*, SSGG, pp. 1453–1457, 2018.
https://doi.org/10.1007/978-3-319-97115-5_121

2   Centrifuge Model Testing

2.1   Centrifuge Model Package

Three centrifuge model packages were developed to investigate the effects of trench width, fracture type and fracture orientation of the pipe on slope stability and infiltration process upon leakage of the buried pipe. All centrifuge model packages were successfully tested on the HKUST 400 g-ton geotechnical centrifuge. Figure 1 depicts a typical centrifuge model package with required monitoring instruments.

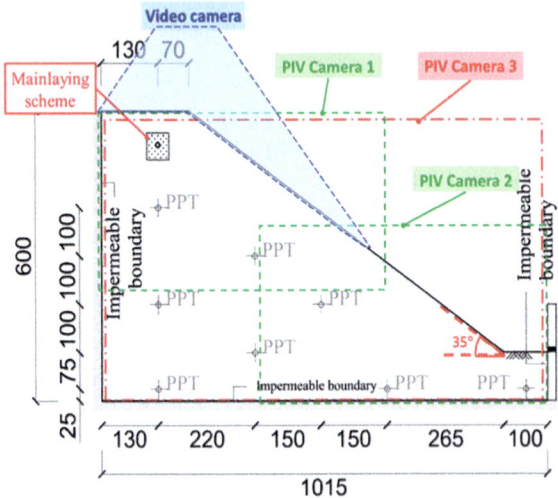

Fig. 1. Typical centrifuge model package with monitoring instruments. Dimensions in mm.

The dimension of the 1/30th-scale slope model was chosen to represent a 15 m high slope in prototype when it was tested at 30 g. The slope angle was 35°. Two trench widths, two types and two orientations of pipe fracture have been considered. The corresponding cover-to-main was 65 mm (1.95 m in prototype), which is measured from the ground surface to the crown of the buried pipe.

Completely decomposed granite (CDG) was adopted as the soil type in all centrifuge models. Particles larger than 2 mm were scalped to minimize the particle size effects. The soil was compacted to a relative compaction of 95% at a water content of 11.5% (dry density of 1757 kg/m³) using the under-compaction method. Each layer of soil was 25 mm. The CDG soil around the pipe and up to 10 mm (300 mm in prototype) above its crown was compacted to 85% relative compaction (RC) (dry density of 1572 kg/m³) in accordance with the Hong Kong Manual of Mainlaying Practice (WSD 2012). Beyond the 85% RC region, the CDG soil was compacted to 95% RC.

Figure 2 presents the details of the "mainlaying scheme" region for all model packages. Test 1 evaluated the effects of narrow trench with a slot-type fracture pointing upward; Test 2 considered a wide trench with a slot-type fracture pointing horizontally to the sloping surface; and Test 3 adopted a wide trench with a slot-type fracture pointing upward.

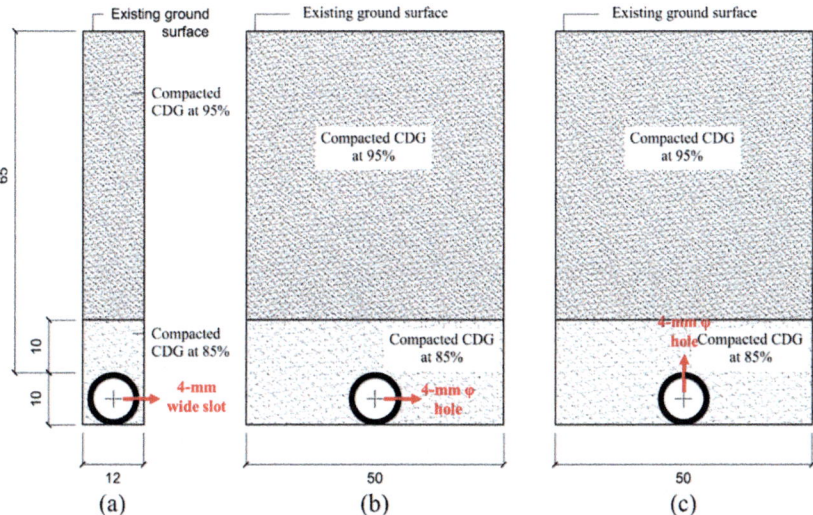

Fig. 2. Details of the "mainlaying scheme" regions for the model packages: (a) Test 1; (b) Test 2; (c) Test 3.

2.2 Test Procedures

In order to simulate leakage from a pressurized-buried water main, a water supply system is required. The water supply system utilizes generated hydrostatic pressure by controlling the water level difference. Water was continuously supplied to the water tank from outside the centrifuge. As a result, a relatively constant water pressure can be generated, and the pressurized water can be continuously supplied to the buried pipe to simulate the leakage process. The centrifuge tests were performed in five steps: swing up (rising g-level); generating sufficient pipe pressure; applying pipe pressure and simulating leakage; increasing pipe pressure; and swinging down the centrifuge and post-failure investigation.

3   Observations and Test Results

Leakage-induced surface erosion was observed in Tests 1 and 3, whereas deep-seated slope failure was observed in Test 2. In both Test 1 and 3, the fracture was oriented horizontally to the sloping surface. An erosion hole emerged at the ground surface and the leaking hole was exposed (Fig. 3(a)). After that, the majority of the leaked water was discharged through this erosion hole, gradually forming an erosion gully. At the end of the experiment, no significant signs of landslide and no deep-seated failure were observed. Nevertheless, in Test 2, the excavation trench was wider, and the fracture was oriented upward, and finally deep-seated failure occurred (Fig. 3(b)). Erosion holes were also observed during the test, but the surface discharge was not as much as that in Test 1 and 3. Most of the leaked water still infiltrated into the slope, causing water level rises. This was the main reason of the occurrence of deep-seated failure. Test 3 was conducted to differentiate the significance of trench width and fracture orientation effects. The only change was the fracture orientation, which was oriented horizontally to the sloping surface. No deep-seat failure was observed but leakage-induced surface erosion and concentrated surface flow.

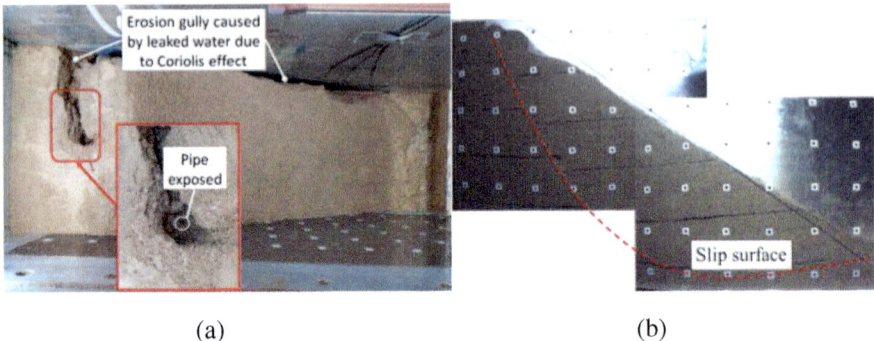

(a)   (b)

Fig. 3. Different modes of slope failure: (a) erosion failure; (b) deep-seated failure.

4   Conclusions

The current Hong Kong mainlaying practice with different trench widths, fracture types and fracture orientations were successfully studied through centrifuge modelling technique. The test results gave us some insight that the mode of slope failure due to leakage of BWCS may be dependent on the fracture orientation, fracture type and trench width.

Acknowledgement. This research would not have been possible without the support of the Construction Industry Council (CIC) (Project No. CIC15EG02), and technical assistance by AECOM Ltd., Geotechnical Engineering Office (GEO), and Hong Kong Institute of Engineers (HKIE). The contributions of these institutions are gratefully acknowledged.

References

Environment, Transport and Works Bureau (ETWB): Code of Practice on Monitoring and Maintenance of Water-carrying Services affecting Slopes, 2nd edn. ETWB, Hong Kong (2006)

Hui, T.H.H., Tam, S.M., Sun, H.W.: Review of Incidents Involving Slopes Affected by Leakage of Water-Carrying Services (GEO Report No. 203). Geotechnical Engineering Office, Hong Kong (2007)

Water Supplies Department (WSD): Manual of Mainlaying Practice. WSD, Hong Kong (2012)

Zhang, L.M., Li, J.H.: Determining the Safe Distance between Ductile Iron Gas Pipes and Sewage Main. The Hong Kong University of Science and Technology, Hong Kong (2007)

On Geological Hazards in Georgia

Diana Egiazarova[✉] and Zurab Tchkonia

Georgian Technical University, Kostava Street 77, 0175 Tbilisi, Georgia
diana.diana025@gmail.com

Abstract. In Georgia, landslide and debris flow/mudflow processes and water erosion are at the top of problems due to related eco-geological disaster risks and negative impact. By the scale of development of these events and their negative impact on the population and economy of the country, Georgia occupies leading position among high land countries of the world. Besides, the almost whole territory of the country is under the risk of earthquakes, the magnitude of which is 7–9. Impact of such earthquakes is directly stimulating and provoking gravitational landslide and debris flow processes. In Georgia major part of the population, agricultural lands, roads, oil and gas pipelines, hydro-technical – melioration facilities, power transmission lines and mountain tourism zones periodically are subjected to disastrous processes and the areas under the risk are expanding substantially. Around 70% of the territory of the country, about 3000 settlements (62%) are under the risk of geological disasters; 14,2% of agricultural lands were seriously damaged by geological processes and require conducting of cardinal protective measures; and 13,1% of agricultural lands are located within the high risk area. Consequently, more than 80% of the economic damage was caused to mountainous regions and the majority of eco-migrants are from highland areas. This causes vacation of villages.

Keywords: Geology · Hazards · Landslides · Mudflows

1 Study Area

1.1 Geomorphological, Geological, Climate General Conditions of Georgia

Georgia is a sovereign state in the Caucasus region of Eurasia. Situated at the juncture of Eastern Europe and Western Asia it is bounded to the west by the Black Sea, to the north by Russia, to the southwest by Turkey, to the south by Armenia, and to the southeast by Azerbaijan. Georgia covers a territory of 69,700 km^2 and its population is almost 4.5 million.

The Greater Caucasus Mountain Range is much higher in elevation than the Lesser Caucasus Mountains, with the highest peaks rising more than 5,000 m above sea level [1].

The modern relief of Georgia is comprised of forms with different hypsometric and morphographic features heavily dissected mountain slopes, deep erosive gorges, intermountain depressions, flat lowlands, plains, plateaus and uplands [5]. Rocks of different composition, age and stability, forming different geological structures, and

© Springer Nature Switzerland AG 2018
W. Wu and H.-S. Yu (Eds.): *Proceedings of China-Europe Conference on Geotechnical Engineering*, SSGG, pp. 1458–1461, 2018.
https://doi.org/10.1007/978-3-319-97115-5_122

determining the development of different hazardous geological processes are all found in the territory of Georgia. Rockslides and rock-falls are mainly observed on steep slopes comprising of hard rocks. Landslides and mudflows are, however, rare at these locations. On the other hand, both landslides and mudflows remain characteristic of the areas comprising of soft soil and easily dislodged rocks [5].

Climatic conditions also play an important role in triggering hazardous natural events within Georgia. There are often periods of heavy and prolonged rainfall, especially in the regions in the west of Georgia. In the mountain regions intensive and prolonged rains can lead to the development of mudflows and landslides. A rapid rise in the air temperature and the subsequent, rapid and intensive melting of the snowpack and/or the prolonged rains, occurring in late winter and early spring, further facilitate the rapid rise of the water levels in rivers which can cause floods and mudflows in low lying territories [5].

1.2 Key Factors Causing the Development of Natural Disasters

Recent activation of natural disasters and complication of geodynamic situation on the background of sensitive geological environment is preconditioned the following: Abnormal increase of negative meteorological events provoking geological processes on the background of global climate change; The intensity of earthquakes; Large scale technogenic impact on the environment and severe disturbance of balance; Low awareness of society on danger of geological disasters, insufficiently informed population and decision makers, lack of information on trends of activation of geological disasters, insufficient forecasts and assessment on possible development of these processes, unavailability of comprehensive disaster risk mitigation strategy and management tools [4].

2 Geodynamic Processes Statistic

2.1 Geodynamic Processes in Georgia

Natural disasters became even more topical at the beginning of 20th century; when the scale of disasters increased enormously on the background of human pressure on the environment and global climate change, as a result of which the risks, related to such disasters become unmanageable [4].

As of today in Georgia have been identified: about 53 thousand landslide-gravitational bodies and areas with high probability of their formation; about 3000 debris flow/mudflow transforming waterways; about 5000 locations of formation of avalanches; washout of river-banks and sea shores occurs on more than 1500 locations, with total length of more than 2000 km [4].

2/3 of landslides that occurred in the country have taken place in the mountain zones; such disastrous processes debris flow/mudflows, snow avalanches, rock avalanches, rock falls and glacial flows are typical for mountain zones [2–4].

2.2   Recorded Landslide, Mudflow and Rock Fall Events

The landscape, geology, geomorphology and climate of the territory of Georgia create favorable conditions for the development of active geological processes, such as landslides, mudflows and rock falls. A large number of settlements, agricultural lands, roads, oil and gas pipeline routes, high-tension power transmission towers, hydro-power and water treatment utilities, mining and tourist complexes have been periodi-cally affected by these processes. The negative social and economic, demographic and ecological situation caused by mudflows and landslide-gravitation processes developed within the stated zones cover all sectors of human activities.

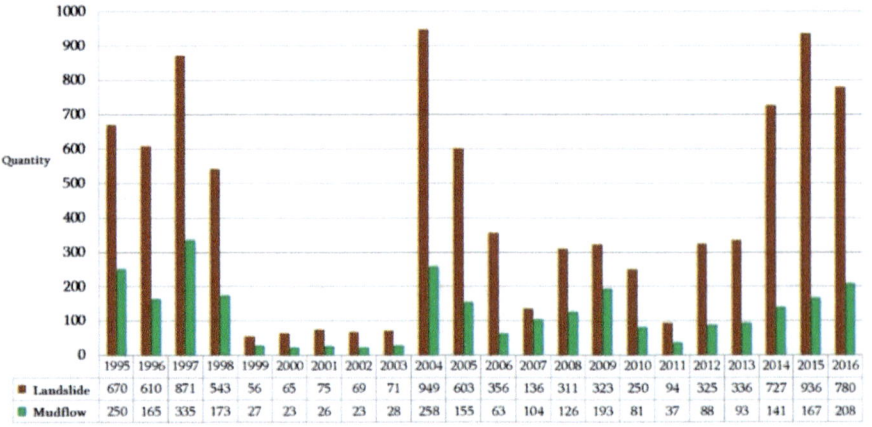

Fig. 1. Landslide and mudflow processes recorded in the territory of Georgia in 1995–2016 (Data from the National Environment Agency)

A particularly difficult situation has been created in the mountainous regions, where extreme activity of landslides and mudflows often requires evacuation of the popula-tion, and sometimes even their resettlement in other regions. The most alarming fact is that these events are frequently followed by human losses. From 1987 to present, more than 600 people died as a result of geological hazards in Georgia, and the number of victims exceeded 1,000 in the last 35 years. Around 60 thousand households have been resettled from damaged sites during the same period. The overall economic losses equal tens of millions of US dollars, and, including indirect losses, these exceed even hun-dreds of millions.

Until the end of 20th century in Georgia, extreme activity of landslide-mudflow processes was subject to a certain cycle and used to be repeated, on average, once every 3–5 and 8–11 years. This was due to the geological-climate conditions; however, since the 1990's the activity of these processes has increased and they now take place almost every year. As a result of this development, more and more new areas, settlements and engineering and technical installations have experienced the negative impacts of these processes (see Fig. 1) [5].

In Georgia, a national digital landslide inventory covering the entire country is not available. Landslide information is collecting by the Department of Geological of the National Environmental Agency [3].

Acknowledgements. The authors would like to thank National Environmental Agency of Ministry of Environment Protection and Agriculture of Georgia for providing various datasets used in this study. Shota Rustaveli National Science Foundation which financing project, Geodynamic processes hazard assessment and GIS analysis of Zugdidi-Jvari-Mestia-Lasdili motorway" (PhDF2016_44) which helped my PhD themes to research Svaneti region (Georgia).

References

1. Gaprindashvili, G.: Report on the 1st project of AES Geohazards Stream: Landslide hazard assessment in Georgia, ITC, Netherlands, p. 15 (2011)
2. Spanu, V., Gaprindashvili, G., Keith McCall, M.: Participatory methods in the Georgian Caucasus: understanding vulnerability and response to debrisflow hazards. Int. J. Geosci. **6**, 666–674 (2015)
3. National Environmental Agency, NEA: Information Bulletin on Outcomes of Geological Disasters in 2016 and Forecast for the 2017 Year in Georgia. Department of Geology, Ministry of Environmental Protection and Agriculture of Georgia, Tbilisi, p. 495 (2017)
4. Tsereteli, E., Gaprindashvili, M.: Geological Report envisaged by the Project on Development of climate resilient flood and flash flood and geological disaster management practices for Rioni river Basin, Tbilisi, p. 195 (2015)
5. Atlas of Natural Hazards and Risk in Georgia, CENN/ITC, Tbilisi, p. 124 (2012). http://drm.cenn.org/index.php/en/

Numerical Modelling of Main Shock and Aftershock Line of Chuya Earthquake 27.09.2003, Altay, Russia

Mikhail Eremin$^{(\boxtimes)}$ and Pavel Makarov

Institute of Strength Physics and Materials Science,
Siberian Branch of Russian Academy of Sciences,
Akademicheskii Av. 2/4, Tomsk, Russia
{eremin,pvm}@ispms.tsc.ru

Abstract. In this work the physical-mathematical model of stress-strain state evolution of loaded geomedium is developed with the purpose of numerical simulation of main shock and aftershock line of Chuya earthquake, occurred on the 27th of September 2003 in Gorny Altay, Russia. Structural model which includes main blocks and faults is constructed on the base of seismotectonic and paleoseismogeological investigations. It is shown that results of numerical simulation are in satisfactory agreement with field observations and simulated seismic process satisfies the laws of Guttenberg-Richter and Omori for aftershocks line.

1 Introduction

The history of instrumental observations of the last 300 years showed the several largescale earthquakes with magnitude M lying in the range of 6.9–8.3 occurred on the territory of Greater Altay which represents common seismotectonic province [1].

The Chuya earthquake with $M = 7.5$ on the 27th of September 2003, Gorny Altay, appeared to be one of the strongest earthquakes for the whole period of instrumental observations. Field studies carried out during 2003–2005 of the epicentral zone of main shock and aftershocks [1–5] gave a comprehensive information about the kinematics, stages, depths, focal mechanisms, reconstruction of natural stresses, etc. Seismotectonic and paleoseismogeological investigations carried out during fall of 2003 and summers of 2004, 2005 revealed mostly dextral strike-slip dislocations of main rupture represent the northern extension of faults of Mongolian and Gobi Altay [1].

As reported in article [2] a zone of seismotectonic fractures represent a fault zone in accordance with Riedel model with a maximum width of about 4 Km and is extended in North-West direction. It was also found that the main shock was located between the North Chuya and Chagan-Uzun blocks. It is also important that the right-lateral strike-slip fault occurred in Paleozoic basement and

© Springer Nature Switzerland AG 2018
W. Wu and H.-S. Yu (Eds.): *Proceedings of China-Europe Conference on Geotechnical Engineering*, SSGG, pp. 1462–1465, 2018.
https://doi.org/10.1007/978-3-319-97115-5_123

outcome to the surface of about 70 Km lineament of mostly R- and R' shears in a bed of Cenozoic sediments.

The field observations presented in aforementioned articles give a good and comprehensive insight into the process of seismoteconic deformations of Earth's crust in Gorny Altay. To get even a better insight into the process of main fault activation, seismotectonic deformations, redistribution of stresses, fracturing of Earth's crust elements an attempt to construct a physical-mathematical model of these processes was undertaken in particular work which represents the main goal.

In this work the physical-mathematical model, as well as simplified structural model of Chagan-Uzun block with adjacent territories is constructed.

As a result of numerical modelling, structure as well as phases of seismic process were studied numerically. The graph of seismic events recurrence was obtained. And the correspondence of simulated aftershock process to Omori law was also shown.

2   Physical-Mathematical Model of Deformation and Fracture of Earth's Crust Elements

In order to simulate the process of faults activation, seismotectonic deformations, redistribution of stresses, fracturing of Earth's crust elements and other features of stress-strain state evolution, we solved numerically the system of equations of solid mechanics by the means of finite-difference approach. The system of equations include the fundamental laws of mass, energy and momentum conservation and is closed with appropriate equations of state for elastic deformation (Hooke's law) and inelastic deformation (modified model of Drucker-Prager-Nikolaevskii). In detail an applied approach and physical meaning of all used parameters are described in [8].

3   Results of Simulation and Discussion

The simulation region was subjected to sub-meridional compression in accordance with existing geodynamic regime. The boundary conditions of uniaxial compression on upper and lower boundaries were set, on the left and right boundaries the symmetry conditions were set.

Presented results of numerical modelling show that constructed physical-mathematical model of stress-strain state evolution is capable of describing the seismic process in the epicentral zone of Chuya earthquake in accordance with recurrence graph (Fig. 1). If compared to the seismological data given in [2–5] a satisfactory agreement might be found. Total range of simulated seismic events was divided on 19 conditional classes. The location of main shock is in very good agreement with seismological data. In article [6] it was shown that real seismic process in epicentral zone follows the recurrence graph of Gutenberg.

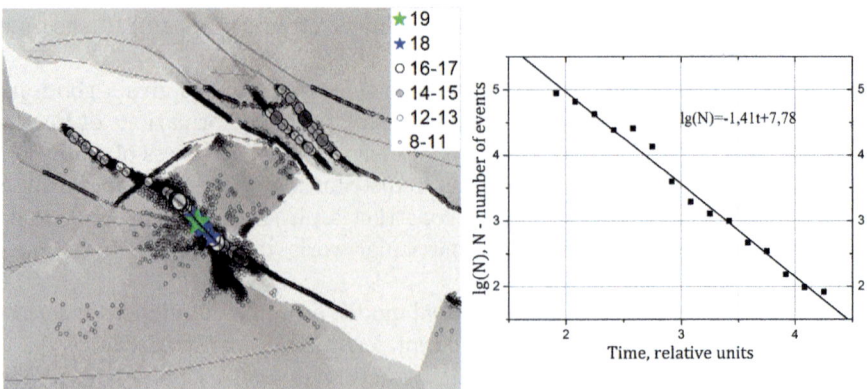

Fig. 1. Spatial structure at activation of main fault of epicentral zone of Chuya earthquake (left) and the graph of recurrence (right) of simulated seismic process.

For better understanding of temporal features of seismic process evolution a spatial-temporal structure was illustrated (Fig. 2). The results show that simulated seismic process also correspond to Omori law. The clustering of seismic events might be observed at activation of main fault of epicentral zone of Chuya earthquake, some fore-shock activity is also observed and then after the main

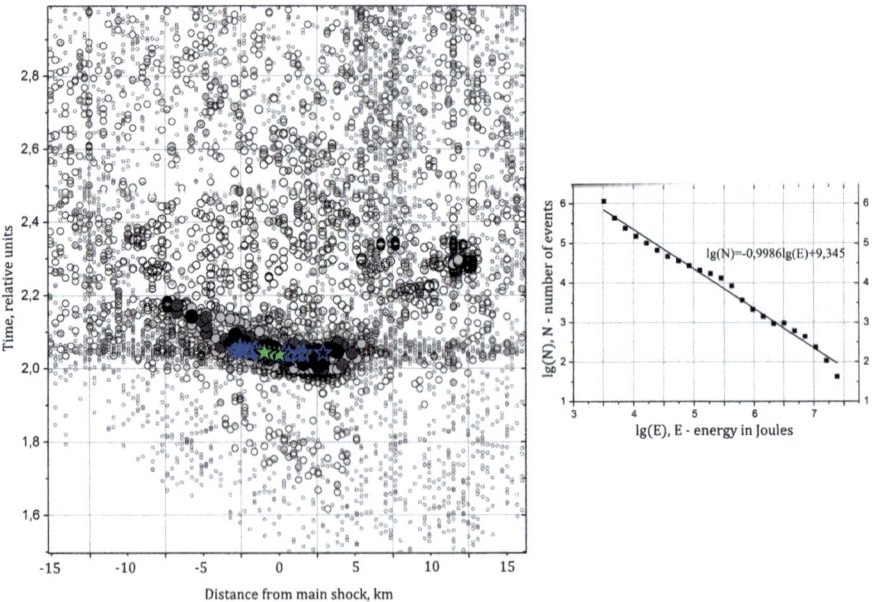

Fig. 2. Spatial-temporal structure at activation of main fault of epicentral zone of Chuya earthquake (left) and Omori law (right) of simulated seismic process.

shock the process follows the Omori law. For example, the aftershock process of Uregnuur earthquake which occurred in neighbouring region followed the Omori law [7].

4 Conclusions

We developed the physical-mathematical model of stress-strain state evolution of loaded geomedium capable of describing the features of seismic process. It is shown that simulated seismic process satisfies the laws of Guttenberg-Richter and Omori for aftershocks line and is in satisfactory agreement with field observations.

Acknowledgements. This work was granted by Russian foundation for basic research (grant No. 18-35-00224) and Plan of fundamental research of Russian state academies of science for 2013–2020 (project III.23).

References

1. Rogozhin, E.A., et al.: Tectonic setting and geological manifestations of the 2003 Altai earthquake. Geotectonics (2007). https://doi.org/10.1134/S001685210702001X
2. Lunina, O.V., et al.: Seismotectonic deformations and stress fields in the fault zone of the 2003 Chuya earthquake, Ms = 7.5, Gorny Altai. Geotectonics (2006). https://doi.org/10.1134/S0016852106030058
3. Lunina, O.V., et al.: Geometry of the fault zone of the 2003 Ms = 7.5 Chuya earthquake and associated stress fields, Gorny Altai. Tectonophysics (2008). https://doi.org/10.1016/j.tecto.2007.10.010
4. Novikov, I.S., et al.: The system of neotectonic faults in southeastern Altai: orientations and geometry of motion. Rus. Geol. Geop. (2008). https://doi.org/10.1016/j.rgg.2008.04.005
5. Leskova, E.V., Emanov, A.A.: Some properties of the hierarchical model reproducing the stress state of the epicentral area of the 2003 Chuya earthquake, Izvestiya. Phys. Sol. Earth (2014). https://doi.org/10.1134/S1069351314030057
6. Dorbath, C. et al.: Geological and Seismological Field Observations in the Epicentral Region of the 27 September 2003 Mw 7.2 Gorny Altay Earthquake, Bull. Seism. Soc. Amer. (2008). https://doi.org/10.1785/0120080166
7. Emanov, A.F. et al.: The Ms = 7.0 Uureg Nuur earthquake of 15.05.1970 (Mongolian Altai): the aftershock process and current seismicity in the epicentral area, Rus. Geol. Geop. (2012). https://doi.org/10.1016/j.rgg.2012.08.009
8. Makarov, P.V., Eremin, M.O.: Fracture model of brittle and quasibrittle materials and geomedia. Phys. Mesomech. (2013). https://doi.org/10.1134/S1029959913030041

Geotechnical Issues of Landslides in Ukraine: Simulation, Monitoring and Protection

Iurii Kaliukh$^{(\boxtimes)}$ ⓘ, Gennadiy Fareniuk ⓘ, and Iegor Fareniuk ⓘ

Research Institute of Building Constructions,
5/2 Preobrazhenska str., Kiev 03680, Ukraine
kalyukh2002@gmail.com

Abstract. More than 90% of the territory of Ukraine has complex ground conditions. As a result, Ukraine is becoming a region of integrated influence of regional flooding and seismic activity on formation of landslides. 4 case studies will be considered in the presentation: results of visual and instrumental inspections of the landslide protection structures (1); direct dynamic method for calculating and analysis of the stress-deformed state of landslide slopes under dynamic influences (2); technical and software means of the automated real time monitoring system for landslide of the Central Livadia Landslide System of the Autonomic Republic of Crimea of Ukraine (ARCU) (3); landslide stabilization in building practice: methodology and experimental investigation from ARCU (4). Only case study (4) is described in more detail because of the limited scope of the extended abstract. One of the most efficient solutions for landslide protection is the short underground piles (SUP). They are shortened bored piles, immersed below the slide surface on the calculated value and output above it at a distance providing overlapping of slipping soils. In the process of experimental work there appeared the need to choose the optimal testing mode. Selected results of experimental studies of the SUP are presented.

Keywords: Landslide protection structures · Experiments · Calculation

1 Introduction

Landslides are among the most serious and damaging geohazards worldwide Lacasse [1]. Wu et al. [2, 3], Yu et al. [4], Chen et al. [5], Trofymchuk et al. [6], Utili [7] studied landslide hazards and their origins, factor of safety in a partially saturated slope inferred from hydro-mechanical continuum modeling, use accelerogram of real earthquakes in the evaluation of the stress-strain state of landslide slopes, investigation by limit analysis on the stability of slopes with cracks and so on. One of the effective means of landslide danger mitigation is landslide protection structures that are widely used by the best practices [8]. One of the efficient decisions as for landslide protection structures is the use of SUP. They are shortened bored piles that dug into estimated value lower than slip surface and output higher than this surface on a distance providing covering of slipping soils (Fig. 1).

© Springer Nature Switzerland AG 2018
W. Wu and H.-S. Yu (Eds.): *Proceedings of China-Europe Conference on Geotechnical Engineering*, SSGG, pp. 1466–1469, 2018.
https://doi.org/10.1007/978-3-319-97115-5_124

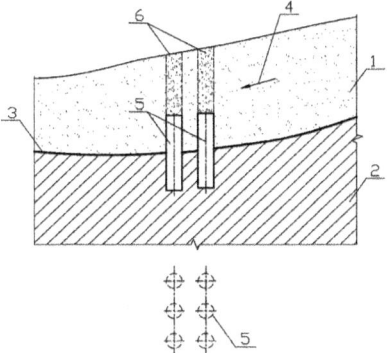

Fig. 1. The scheme of the retaining construction of key-piles: 1 – the landsliding body; 2 – firm soil; 3 – the line of slide; 4 – direction of landsliding dislocation; 5 – key-piles; 6 – backing of the wellhead.

2   Experimental Investigation of the SUP

2.1   Determination of π-Variables

Application π-theorem shows that for description of this task three π-variables are required. In case of square cross section of key-pile ($a = b$) number of π-variables decreases up to two (π_1 and π_2):

$$\pi_1 = \frac{a}{l}; \pi_2 = \frac{El}{Q}.$$

Here l – the key-pile length, a – typical dimensions of key-pile rectangular cross section, Q – intensity of side pressure along key-pile length, modeling intensity of landslide length pressure on the key-pile, E – modulus of elasticity. The resulting modeling scale was 1:10.

2.2   The Design of the *Volumetric Tray* and the Model

In order to create an analogue of the key-piles operation in landslide blocks, experimental unit was designed and constructed. The experimental unit consists of the following main parts (Fig. 2): steel fixed retaining structure 1; rolling bottomless tray 2 with the dimensions $0.57 \times 0.35 \times 0.33$ m, which can be moved on a wooden base 7; movement of bottomless tray 2 is measured by deflect to meter 6; loading is applied by the weight suspension 5. Strain gauge model 3 is installed in the 0.35×0.35 m slot in the base 7 and securely fixed upright by clamping device 8. Soil 4 is filled around the model layer by layer and is firmed by punning up to a predetermined value. In accordance with the purpose of laboratory investigations, 5 types of laboratory experiments with quadruple repetition were carried out: the shift of the tray without models and shifts with models of 0.30 m, 0.24 m, 0.15 m and 0.09 m lengths, which were determined on the basis of equality of π-variables.

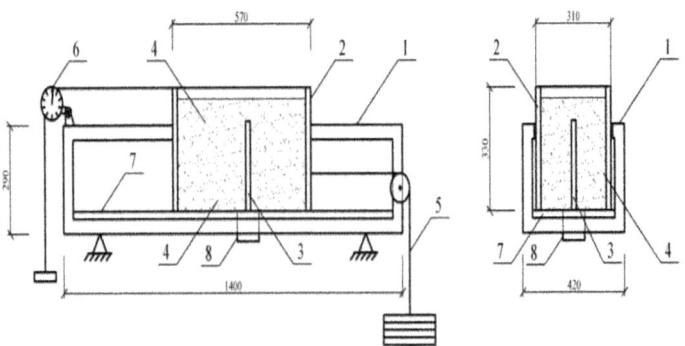

Fig. 2. Design of experimental unit

2.3   The Design of the *Volumetric Tray* and the Model

It was resolved that the load on pile E_i is the difference between the load during the tray shift with the model (P_i) and the load during the full tray shift without model (P_l) (№ 5 Fig. 2) on during different loading stages in the experiments: $E_i = P_i - P_l$. Table gives the loads on pile models, experimental M_* and ultimate M_i bending moments in fixing for different block steps as well as relative difference Δ between them.

Load stage, kg		E_i, кг	Model length					
			$l = 0.3$ m			$l = 0.15$ m		
P_i	P_l		M_*, kgm	M_i, kgm	$\Delta\%$	M_*, kgm	M_i, kgm	$\Delta\%$
40	32,5	7,5	81	75,0*	7,4	64	37,5*	41
				112,5	28		56,3	12
42		9,5	93	95,0*	2,1	71	47,5*	33
				142,5	35		71,3	0,1
44		11,5	107	115*	7,0	78	57,5*	26
				172,5	38		86,3	9,6
46		13,5	121	135*	10,0	87	67,5*	22
				202,5	40		101,3	14

*Note: there is an ultimate bending moment during the distribution pressure diagram on triangle (trapezoid) in numerator; in denominator in rectangle.

Obtained experimental material gives an opportunity to judge upon how bending moments got by theoretical way are correspondent to experimental values. It follows that experimental values of the moment are close to ultimate ones if the pressure diagrams are as following: triangle diagram Δ up to 10% - for the model of total length $l = H = 0.3$ m, the pressure center is on the height of $0.3 - 0.36$ l, that is comparable to the value $1/3$ H as for [9]; truncated triangle diagram Δ up to 6.7% - for the model of the length $l = 0.8$ H $= 0.24$ m, the pressure center is on the height of $0.35 - 0.4$ l; rectangular diagram Δ up to 14% - for the model of the length $l = 0.5$ H $= 0.15$ m, the pressure center is on the height of $0.43 - 0.57\ l$; rectangular diagram Δ up to 3.5% - for the model of the length $l = 0.3\ H = 0.09$ m, the pressure center is on the height of $0.49 - 0.52\ l$ (H – thickness of the tray filling). Construction of landslide protection structures using key-piles was offered to reinforce the slope in the area of the center "Yedinstvo" in Gurzuf town of the ARCU.

3 Experimental Investigation of the SUP

The integral error of unit experiment varied from to 8 to 14% depending on the length of the key-pile model, loading conditions and other factors. Experimental studies of interaction between key-pile model and soil mass in the tray showed the influence of the length of the key-pile in landslide depth on the form of pressure diagram. For the key-pile model modeling its full length $l = H$, the diagram of the pressure is mainly of triangular character that is not contrary to previously known results [9], which were based mainly on theoretical assumptions. With the reduction of the key-pile length from H to 0.5 H pressure diagram changes from triangular to trapezoidal shape. When the length of the pile $l = 0.5$ H and less the form of diagram transforms from rectangular to trapezoidal one.

References

1. Lacasse, S.: Terzaghi Oration: Protecting society from landslides – the role of the geotechnical engineer. In: Proceedings of the 18th International Conference on Soil Mechanics and Geotechnical Engineering, 2–6 Sept 2013, Paris, France, pp. 15–34 (2013)
2. Tensay, G.B., Wu, W.: Three dimensional analysis of seismic performance of an earthfill dam in Ethiopia. In: Proceedings of the 7th European Conference on Numerical Methods in Geotechnical Engineering, 2–4 June 2010, Trondheim, Norway, pp. 489–496 (2010)
3. Borja, R.I., White, J.A., Liu, X.Y., Wu, W.: Factor of safety in a partially saturated slope inferred from hydro-mechanical continuum modeling. Int. J. Numer. Anal. Meth. Geomech. **63**(2), 140–154 (2011)
4. Kim, J.M., Salgado, R., Yu, H.S.: Limit analysis of soil slopes subjected to pore pressures. J. Geotech. Geoenvironmental Eng. ASCE **125**(1), 49–58 (1999)
5. Chen, Zu-Yu., Morgenstern, N.R.: Extension to the generalized method of slices for stability analysis. Can. Geotech. J. **20**(1), 104–119 (2011)
6. Trofymchuk, O., Kaliukh, I., Silchenko, K., Polevetskiy, V., Berchun, V., Kalyukh, T.: Use accelerogram of real earthquakes in the evaluation of the stress-strain state of landslide slopes in seismically active regions of Ukraine. In: Lollino, G., et al. (eds.) Engineering Geology for Society and Territory, vol. 2, pp. 1343–1346, Springer, Cham (2015)
7. Utili, S.: Investigation by limit analysis on the stability of slopes with cracks. Geotechnique **63**(2), 140–154 (2013)
8. Kaliukh, I., Senatorov, V., Marienkov, N., Trofymchuk, O., Silchenko, K., Kalyukh, T.: Arrangement of deep foundation pit in restricted conditions of city build-up in landslide territory with considering of seismic loads of 8 points. In: Proceedings of the XVI ECSMGE, 13–17 Sept 2015, Edinburgh, Great Britain, pp. 535–540 (2015)
9. Ginzburg, L.: Landslide protection retaining constructions. Stoiizdat, Moscow, USSR, 80 p. (1979)

Back Analysis Algorithm Based
on Particle Swarm Optimization

Abdoulie Fatty[(⊠)] and A. J. Li

National Taiwan University of Science and Technology,
Taipei City 106, Taiwan (R.O.C.)
afatty13@gmail.com

Abstract. In geotechnical engineering, back analyses are generally used to investigate uncertain parameters. Back analyses can be done by considering known conditions, such as failure surfaces, displacement, and structure performance. In fact, a complex nonlinear optimization function is typically required for most geotechnical problems in order to gain better understandings of these uncertainties. Therefore, Particle Swarm Optimization is utilized in this study to assist in back analyses based on finite element upper bound limit analysis method. This technique is an evolutionary computational approach, which is suitable for solving nonlinear global optimization problems. By using Particle Swarm Optimization with numerical upper bound method, a simple case study showed that the obtained results are reasonable and reliable. The usability of the proposed method is demonstrated. It could be seen as a new potential approach for geotechnical back calculations.

Keywords: Particle Swarm Optimization · Back analysis

1 Introduction

The determination of geotechnical parameters is carried out either at the site or in the laboratory. Due to complex field conditions and the limitation of testing conditions it is sometimes difficult to obtain accurate results using testing parameters. By means of an iterative adjustment, a parameter back calculation can be used to obtain an accurate result. This approach uses the result obtained from a simulation and compares it to a real measurement. By iteratively adjusting the geotechnical parameters, the simulation result is successively fitted to the field measurement. It has been substantiated that back analysis can provide significant information regarding geotechnical parameters. Particle Swarm Optimization (PSO) is applied in this paper to improve the back analysis approach and obtain better results.

PSO is a population based stochastic optimization algorithm inspired by the social behaviors of animals such as fish schooling and flying birds [1]. The implementation of PSO is quite convenient since it requires simple mathematical operators. It is found to be very successful in solving complex optimization problems [2].

In this study, the finite element upper bound limit analysis method [3, 4] is used in conjunction with the Hoek Brown failure criterion. The latest version of the Hoek-Brown failure criterion [5] is employed.

© Springer Nature Switzerland AG 2018
W. Wu and H.-S. Yu (Eds.): *Proceedings of China-Europe Conference on Geotechnical Engineering*, SSGG, pp. 1470–1473, 2018.
https://doi.org/10.1007/978-3-319-97115-5_125

$$\sigma_1' = \sigma_3' + \sigma_{ci}(m_b \frac{\sigma_3'}{\sigma_{ci}} + s)^\alpha; \tag{1}$$

Where

$$m_b = m_i \, exp\left(\frac{GSI - 100}{28 - 14D}\right); \tag{2}$$

$$s = exp(\frac{GSI - 100}{9 - 3D}); \tag{3}$$

$$\alpha = \frac{1}{2} + \frac{1}{6}\left(e^{-GSI/15} - e^{-20/3}\right). \tag{4}$$

GSI ranges from 10, for extremely poor rock masses, to 100 for intact rock. The parameter *D* depends the degree of disturbance. The value of *D* ranges from 0 for undisturbed in situ rock mases and 1 for disturbed rock mass properties.

2 Particle Swarm Optimization

The algorithm is called particle swarm optimization, as it simulates the behavior of a flock of flying birds or fish schooling. Every single bird in the flock is considered as a particle. The algorithm is first initialized by generating a random number of particles and a fitness function is used to evaluate the fitness of each particle. Each particle has a velocity, which is used to direct its flying pattern. The flying of each particle is altered in accordance with its own flying experience and those of its companions' flying experience this scenario is a good representation of a possible solution to a problem. The best solution obtained by a particle so far is known as its personal best (p_b) and the best solution obtained by any of the particles in the swarm so far is known as the global best (g_b). The velocity and position of each particle are updated using its p_b and gbest g_b according to the equations below.

- Velocity

$$v_k(t+1) = w * v_k(t) + c_1 r_1 (p_b - x_k(t)) + c_2 r_2 (g_b - x_k(t)) \tag{5}$$

Where, $v_k(t)$: velocity of the particle k at iteration t
$c_1 = c_2 = 2.05$: constants
$w = 0.7$: inertia weight
$r_1 = r_2$: random numbers ranging for [0,1]
p_b: personal best
g_b: global best
$x_k(t)$: current position of particle k at iteration t

- Position

$$x_k(t+1) = x_k(t) + v(t+1) \qquad (6)$$

Equation (5) is used to determine the new velocity of particle k. The inertia weight (w) determines the extent to which the particle remains along its original course unaffected by the pulls of p_b and g_b [6]. c_1 and c_2 imitate the weights whether returning to the location of the personal best or exploring the location toward the global best and the values can be determined through empirical study. The purpose of r_1 and r_2 is to mimic the unpredictable behavior of nature swarms. Particle k flies towards a new position according to Eq. (6). A fitness function, which is usually the objective function under consideration, is used to measure the performance of each particle.

3 Back Analysis of Geotechnical Parameters of a Rock Slope

3.1 The Objective Function

The efficiency of the proposed PSO algorithm was tested based on back analysis of geotechnical parameters for a rock slope. The objective function is to minimize the difference between the targeted unit weight (γ_m) and the predicted unit weight (γ_p) from the finite element upper bound limit analysis method by successively fitting the simulation result to the targeted unit weight.

$$min\delta = \sqrt{(\gamma_m - \gamma_p)^2} \qquad (7)$$

δ: the objective function

3.2 Introduction of the Rock Slope

In this study, a simple slope with a slope angle (β) of $45°$ is considered for back analyses. The geotechnical parameters of the rock slope are $GSI = 10$, $D = 0.7$, $m_i = 20$, $\sigma_{ci} = 4$ Mpa. Using the finite element upper bound limit analysis method, the predicted unit weight (γ_p) is 16.42 KN/m^3. However, the targeted unit weight (γ_m) of this particular slope is assumed to be 18.2 KN/m^3. Therefore, a model optimization approach is required to find the reasonable inputs.

3.3 Back Analysis Results and Discussion

Among the four geotechnical parameters GSI, D, m_i, and σ_{ci}, only D is considered for back calculations. Because D is a relatively sensitive parameter. The given range of D is from 0.5 to 1 for the back calculations.

The unit weight obtained from the finite element upper bound limit analysis method is quite different from the targeted unit weight. With the aid of back analysis by iteratively adjusting the value of D the predicted unit weight is successively fitted to the targeted unit weight as shown in the Table 1 below.

Table 1. Parameter comparison before and after back calculation

Initial parameters	Back-calculated parameters
$D = 0.7$	$D = 0.68$
$\gamma_p = 16.42$ KN/m^3	$\gamma_p = 18.2$ KN/m^3

4 Conclusion

The result of the simulation of the case study show that the proposed method performed well in reaching an optimal solution with a satisfying degree of accuracy. It is possible to apply PSO to solving more complicated geotechnical back calculation problems.

References

1. Kennedy, J., Eberhart, R.C.: Particle swarm optimization. In: IEEE International Conference on Neural Networks, vol. 4, pp. 1942–1948. IEEE Press (1995)
2. Kennedy, J., Eberhart, R.C.: Swarm Intelligence. Morgan Kaufmann Publishers (2001)
3. Lyamin, A.V., Sloan, S.W.: Upper bound limit analysis using linear finite elements and nonlinear programming. Int. J. Numer. Anal. Methods Geomech. **26**, 181–216 (2002)
4. Krabbenhoft, K., Lyamin, A.V., Hjiaj, M., Sloan, S.W.: A new discontinuous upper bound limit analysis formulation. Int. J. Numer. Anal. Meth. Geomech. **63**, 1069–1088 (2005)
5. Hoek, E., Carranza-Torres, C., Corkum, B.: Hoek-Brown Failure criterion-2002 edition. In: Proceedings of the North American Rock Mechanics Symposium Toronto (2002)
6. Shi, Y., Eberhart, R.C.: Empirical study of particle swarm optimization. In: Proceedings of the Congress on Evolutionary computation. Piscataway, NJ, IEEE Service Centerm, 1999, 1945–1956 (1999)

Seismic Performance of Geosynthetic-Reinforced Earth Retaining Walls Subjected to Strong Ground Motions

Domenico Gaudio$^{(\boxtimes)}$ ⓘ, Luca Masini ⓘ, and Sebastiano Rampello ⓘ

Sapienza University of Rome, Via Eudossiana 18, 00184 Rome, Italy
{domenico.gaudio, luca.masini,
sebastiano.rampello}@uniroma1.it

Abstract. There is general evidence that the good performance of geosynthetic-reinforced earth retaining walls (GRWs) observed after strong seismic events can be attributed to their capacity to redistribute seismic-induced deformations within the reinforced zone, provided that the reinforcements are characterised by adequate extensional ductility. Therefore, it is desirable to promote the activation of internal (or local) plastic mechanisms, involving the reinforcement strength, since the design phase. In this study, the seismic performance of two earth retaining walls is compared. Specifically, a pseudo-static approach is adopted to conceive a GRW with a seismic resistance, expressed by the critical seismic coefficient k_c, that involves activation of an internal plastic mechanisms, and a conventional retaining wall, in which the same critical seismic coefficient k_c is attained for an external (or global) plastic mechanism. Finite difference dynamic analyses are carried out to evaluate the seismic performance of the two walls. In the analyses, a real acceleration time history is applied at the base of the models. The results of the dynamic analyses show that, compared to the wall in which external plastic mechanisms develop, the wall designed to activate internal plastic mechanisms exhibits a better seismic performance, with lower permanent displacements computed at the end of the seismic event.

Keywords: Reinforced soils · Pseudo-static approach · Dynamic analyses

1 Problem Definition

Seismic performance of geosynthetic-reinforced earth retaining walls (GRWs) subjected to strong ground motions is usually deemed satisfactory compared to conventional retaining walls. Lower permanent displacements, resulting from the temporary activation of plastic mechanisms, are typically attained, as observed from shaking-table experiments [1], centrifuge tests [2] and numerical analyses [3]. The good seismic performance can be attributed to the capability of the reinforcement layers to redistribute seismic-induced deformations into the reinforced zone, provided that they possess an adequate extensional ductility and that are involved in the plastic mechanisms. Then, GRWs can profitably be designed to activate internal plastic mechanisms during strong seismic events. To this aim, Gaudio *et al.* [4] have recently proposed a procedure to conceive GRWs with an internal seismic resistance lower than the external one.

© Springer Nature Switzerland AG 2018
W. Wu and H.-S. Yu (Eds.): *Proceedings of China-Europe Conference on Geotechnical Engineering*, SSGG, pp. 1474–1478, 2018.
https://doi.org/10.1007/978-3-319-97115-5_126

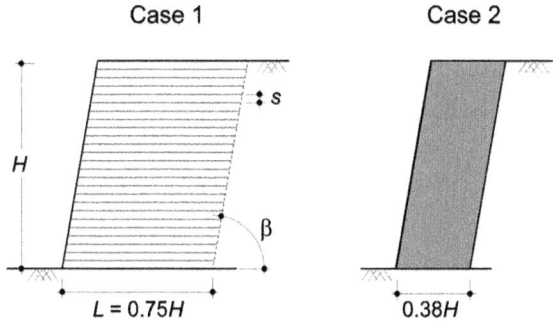

Fig. 1. Schematic layout of the two considered retaining walls.

In this paper, the seismic performance of two different idealised walls, retaining the same backfill, founded on the same soil deposit but activating different plastic mechanisms, is compared. The backfill is characterised by a height $H = 15$ m and a slope $\beta = 80°$ (Fig. 1). The first wall (case 1) was designed, using the procedure proposed in [4], with a critical seismic coefficient related to internal mechanisms, k_c^{int}, lower than the one related to external (or global) mechanisms, k_c^{ext}. It is characterised by 25 uniformly spaced geogrid layers ($s = 0.6$ m), a length-to-height ratio $L/H = 0.75$ ($L = 11.75$ m), and a tensile strength of the reinforcements $T_T = 25$ kN/m, with infinite extensional ductility. Values of the internal and external critical seismic coefficient computed through the kinematic theorem of limit analysis are $k_c^{int} = 0.101$ and $k_c^{ext} = 0.196$, respectively. The second wall (case 2) is representative of a conventional gravity wall; it is endowed with the same seismic resistance of the GRW, though associated to external plastic mechanisms ($k_c = k_c^{ext}$). Specifically, the length L of the second wall was selected to obtain the same k_c value of case 1 performing plane-strain pseudo-static analyses with the finite difference code FLAC 5 [5]. In these analyses the horizontal component of inertial forces, expressed as a fraction k_h of gravity g, was progressively increased until convergence, evidenced by a steady reduction of the unbalanced forces, became no longer possible. Under this circumstance, the numerical model exhibited a well-defined plastic mechanism and the seismic coefficient k_h that activates the mechanism is the critical seismic coefficient k_c [3, 6]. Soil behaviour was described by using a linear elastic-perfectly plastic model with Mohr-Coulomb strength criterion and zero dilatancy. Geogrid levels were modelled with FLAC *strip* elements, capable of reacting to axial deformation only. The fill is made of a coarse-grained soil with an angle of shearing resistance $\varphi' = 35°$, whereas foundation soils are charac-terized by a cohesion $c' = 10$ kPa and an angle $\varphi' = 28°$. Fill-reinforcement contact resistance is assigned a Mohr-Coulomb strength criterion with an angle $\varphi'_{s/GSY} = \varphi' = 35°$. All materials have a unit weight $\gamma = 20$ kN/m^3. For case 2, the wall was modelled as a linear elastic material in order to inhibit any internal plastic mechanism. Further details of the numerical model can be found in Masini *et al.* [3].

A critical seismic coefficient $k_c = 0.060$ was computed for both the walls, although an internal plastic mechanism is activated for case 1 in which the reinforcements are intersected by the sliding surfaces, while an external mechanism develops for case 2, in which both the foundation soils and the backfill are involved. It is worth mentioning that for the GRW, the internal critical seismic coefficient k_c^{int} provided by the numerical analysis is slightly lower than the one obtained from the kinematic theorem of limit analysis; this because the latter provides upper-bound values of the system's resistance, and because of the different assumptions about the flow rule, associated in the limit analysis and non-associated in the numerical analyses.

2 Dynamic Numerical Analyses

Seismic performance of the walls was evaluated by performing time domain dynamic analyses, using the same grid adopted for the pseudo-static analyses but replacing the static fixities with FLAC free-field boundary conditions. Both models were subjected to the scaled horizontal acceleration time history recorded in Monte Cavallo during the October 2016 central Italy earthquake (peak acceleration a_{max} = 0.37 g, Arias intensity I_A = 0.75 m/s, mean period T_m = 0.20 s, significant duration T_D = 4.23 s). Soil cyclic behaviour was described through the hysteretic damping model available in FLAC.

In Fig. 2 the magnified deformed grids obtained at the end of the dynamic calculation are depicted. The deformation patterns agree with the plastic mechanisms of the pseudo-static analysis; for case 1 (GRW) two concurrent shear bands are seen to develop: an internal surface, within the reinforced zone, which is compatible with a log-spiral and a two-block mechanism extending to the upper portion of the backfill, both starting from the toe of the wall. On the contrary, the conventional gravity wall of case 2 exhibits a well-defined shear band, involving both the backfill and the foundation soils.

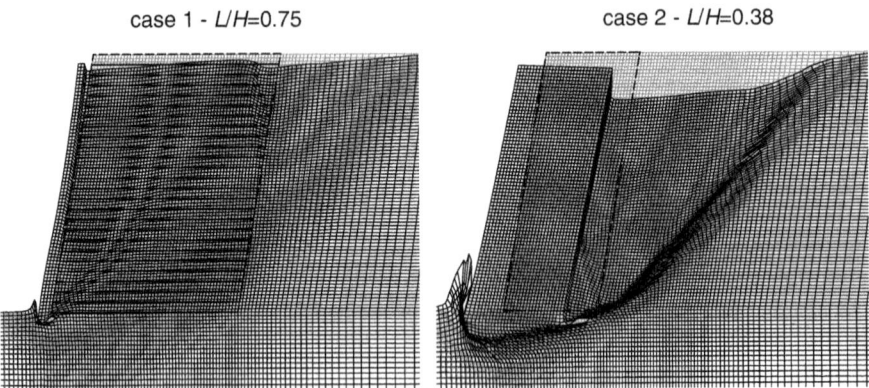

Fig. 2. Deformed mesh (magnification factor = 3).

The different deformation patterns result in a quite different seismic performance of the two walls. Figure 3 shows the time histories of horizontal displacements u, relative to free-field boundary, computed at the top of wall façade. Displacements u = 31 cm

($u/H \approx 2.0\%$) and $u = 56$ cm ($u/H \approx 3.7\%$) were computed for the GRW and the conventional retaining wall, respectively. Huang *et al.* [7] indicated values u/H 2.0% as permissible horizontal displacements: then, the seismic performance of the GRW can be deemed acceptable, while it is not for the conventional wall.

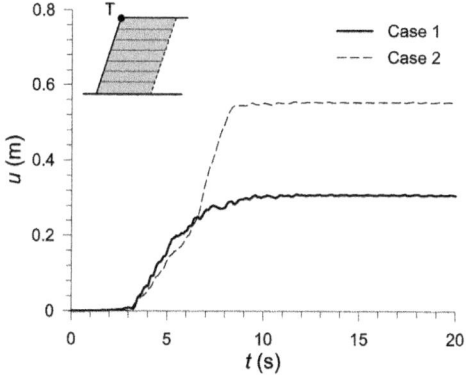

Fig. 3. Time histories of horizontal displacement relative to free-field.

3   Conclusions

Geosynthetic-reinforced earth retaining walls (GRWs) exhibit good performance under severe seismic loadings, thanks to their capability of redistributing seismic-induced deformations within the reinforced zone. Accordingly, GRWs should be conceived to promote the activation of internal plastic mechanisms. The procedure proposed in Gaudio *et al.* [4] was followed to design a GRW with an internal seismic resistance lower than the external one. A conventional gravity wall with the same seismic resistance but activating an external plastic mechanism was also designed for comparison. Plain strain dynamic analyses were then performed, in which the two walls were subjected to the same seismic input. The permanent displacements computed for the GRW at the end of the ground motion were seen to be substantially lower than the ones computed for the conventional wall, as a result of the activation of temporary plastic mechanisms within the reinforced zone.

References

1. Yazdandoust, M.: Seismic performance of soil-nailed walls using a 1 g shaking table. Can. Geotech. J. **55**, 1–18 (2018)
2. Kramer, S.L., Paulsen, S.B.: Seismic performance evaluation of reinforced slopes. Geosynth. Int. **11**(6), 429–438 (2004)
3. Masini, L., Callisto, L., Rampello, S.: An interpretation of the seismic behaviour of reinforced-earth retaining structures. Géotechnique **65**(6), 349–358 (2015)

4. Gaudio, D., Masini, L., Rampello, S.: A performance-based approach to design reinforced-earth retaining walls. Geotext. Geomembr. 46(4), 470–485 (2018). https://doi.org/10.1016/j.geotexmem.2018.04.003
5. Itasca: FLAC fast Lagrangian analysis of Continua v.5.0. User's manual. Itasca Consulting Group, Minneapolis, USA (2005)
6. Masini, L., Rampello, S., Callisto, L. Seismic behaviour of large earth dams: from site investigations to numerical modelling. In: Proceedings of the 1st IMEKO TC-4 International Workshop on Metrology for Geotechnics, Benevento, Italy, pp. 288–294 (2016)
7. Huang, C.C., Wu, S.H., Wu, H.J.: Seismic displacement criterion for soil retaining walls based on soil strength mobilization. J. Geotech. Geoenv. Eng. ASCE 135(1), 74–83 (2009)

Effect of Air Compression and Counterflow on Shallow Landslides Under Intense Rainfall

Tongchun Han[(⊠)], Shiguo Ma, and Riqing Xu

Research Center of Coastal and Urban Geotechnical Engineering,
Zhejiang University, Hangzhou, China
htc@zju.edu.cn

Abstract. Under heavy rainfall, the surface soil is saturated quickly. Rainwater percolates down as a wetting front, displacing air in soil. Rainwater infiltration into the unsaturated zone is potentially affected by air compression ahead of the wetting front. For a large and shallow slope area, air compression ahead of the wetting front will lead to a fluctuant air pressure in excess of atmospheric pressure. In this study, the process of air compression and counterflow ahead of wetting front was simply analyzed. An extended Green Ampt model is introduced considering air compression and counterflow. Based on the model, the infiltration of a shallow slope subjected to an intense rainfall was studied. The limit equilibrium method and the Mohr-Coulomb failure criterion were readily applied to calculate the safety factor of slope. Considering the effects of the air pressure fluctuation produced by air compression and counterflow ahead of the wetting front, a stability analysis model was presented, which can be readily used for estimating the likelihood of a shallow landslide triggered by intense rainfall. It is found that the gauge air pressure ahead of the wetting front, on the one hand, leads to the formation of static water pressure that serves to reduce the normal stress of the interface at wetting front in saturated zone, hence has an adverse effect on the stability of slope, on the other hand, delayed the downward movement of the wetting front, accordingly, decrease the likelihood of landslide.

Keywords: Wetting front · Intense rainfall · Infiltration · Landslide
Air compression and counterflow

1 Introduction

Many landslides are caused by long periods of intense rainfall around the world, especially in the subtropical areas. Research has brought about the realization that most landslides are caused by the infiltration of rainwater into slopes. But few of them have considered the effects of the air pressure produced by air compression ahead of the wetting front under an intense rainfall. Sun et al. (2015) used a simulator to analyze the features of water–air two-phase flow in a soil slope under rain conditions. Results showed that pore air pressure is unfavourable for slope stability for a given slip surface. Cho (2016) analyzed two-phase flow in a slope by numerical modelling method.

A better understanding of the process of rain infiltration contributes to the evaluation of slope stability. In the paper, the process of air being compressed with rainfall is

© Springer Nature Switzerland AG 2018
W. Wu and H.-S. Yu (Eds.): *Proceedings of China-Europe Conference
on Geotechnical Engineering*, SSGG, pp. 1479–1482, 2018.
https://doi.org/10.1007/978-3-319-97115-5_127

analyzed first. Then, an approximate method is proposed to evaluate the likelihood of this special slope failure. Finally, an example of an infinite slope is used to address the problem.

2 Infiltration Model

2.1 Air Pressure Ahead of the Wetting Front

When intense rainfall infiltrates through the soil surface over a large area, the soil surface will be saturated instantly. Initially, air pressure in the soil is equal to atmospheric pressure u_a, then an air pressure head h_{af} (in excess of atmospheric pressure head) will form and increase constantly with the wetting front moving downward. The air pressure head h_{af} can be calculated from Boyle's law for a perfect gas.

With the downward advancement of the wetting front, the air below the wetting front is compressed gradually, hence air pressure increase until eventually reaches to a point where the infiltration rate $i_w = 0$, and the corresponding time is denoted by t_0. Wang et al. (1997) determined two extreme air pressures, the maximum air pressure head H_b occuring at the time when air erupts and the minimum air pressure head H_c occuring immediately after air escapes. After air eruption, the air pressure head h_{af} will fluctuate dynamically between H_c and H_b after t_0.

2.2 Extension of Green-Ampt Model

Before t_0. h_{af} is less than the air-breaking value, H_c. During this time period, the capillary pressure at the wetting front is always $h_{cf} = h_{wb}$. The infiltration rate is

$$i_w = K_e \frac{z_f + h_0 + h_{wb} - h_{af}}{z_f} \qquad (1)$$

By integrating Eq. (1), the $t \sim z$ relation is given

$$t = \frac{1}{K_e} \left[z - \left(h_0 + h_{wb} - h_{af} \right) \log \left(1 + \frac{z}{h_0 + h_{wb} - h_{af}} \right) \right] \qquad (2)$$

Where K_e is the effective conductivity.

After t_0. The air pressure below the wetting front begins to fluctuate between H_c to H_b. The infiltration rate after t_0 can be averaged as

$$i_w = \frac{i_{min} + i_{max}}{2} = \frac{K_c}{2} \frac{h_{ab} - h_{wb}}{z_f} \qquad (3)$$

The infiltration depth of the wetting front may be calculated by

$$z_f = \left[z_0^2 + K_e(h_{ab} - h_{wb})(t - t_0)\right]^{1/2} \tag{4}$$

Where z_0 is the depth of wetting front where the air pressure equals the air-breaking value H_b for the first time.

3 Stability Analysis

3.1 Stability Calculation

The Safety Factor of Slope at the Wetting Front with GA Infiltration Model. The air pressure in soil is always the atmospheric pressure. Therefore, for an infinite slope, the safety factor of slope at the wetting front can be expressed as

$$F_s = \frac{c' + \left(\gamma_t z_f + h_0 \gamma_w\right) \cos^2 \alpha \tan \phi'}{\left(\gamma_t z_f + h_0 \gamma_w\right) \cos \alpha \sin \alpha} \tag{5}$$

The Safety Factor of Slope at the Wetting Front with AE Infiltration Model. The air is confined below the wetting front, an excess atmospheric pressure that serves to reduce the normal stress of the interface at wetting front in the saturated zone occurs. There is a period of time, in which the air pressure below wetting front is gradually increased until it reaches the closing pressure. In this period, the corresponding safety factor is given by

$$F_s = \frac{c' + \left(\gamma_t z_f + h_0 \gamma_w - \gamma_w h_{af}\right) \cos^2 \alpha \tan \phi'}{\left(\gamma_t z_f + h_0 \gamma_w\right) \cos \alpha \sin \alpha} \tag{6}$$

After air eruption, the matrix suction at the wetting front begins to fluctuate between h_{ab} and h_{wb}. In comparison, $h_{cf} = h_{ab}$ is more adverse for slope stability, therefore, is used to calculate safety factor. Thereby, the minimum safety factor of slope with air pressure can be expressed as

$$F_s = \frac{c' + \left[(\gamma_t - \gamma_w)z_f - \gamma_w h_{ab}\right] \cos^2 \alpha \tan \phi'}{\left(\gamma_t z_f + h_0 \gamma_w\right) \cos \alpha \sin \alpha} \tag{7}$$

For any given time t, slope stability can be judged by Eq. (7), and the corresponding depth of wetting front is given by Eq. (4).

3.2 Application of the Infiltration Model to Slope Stability

An idealized homogeneous and isotropic infinite soil slope is assumed. The groundwater table is identical to the bedrock plane. The parameters for soil are taken from

HYDRUS for silt loam. The top is subjected to rain infiltration. Slope surface has no ponded water because of surface runoff, thus the ponded water head on slope surface is taken as zero during the intense rainfall, i.e. $h_0 = 0$.

Equations (5)–(7) are used to calculate the safety factor F_s of slope at the wetting front, which are plotted against time as shown in Fig. 1.

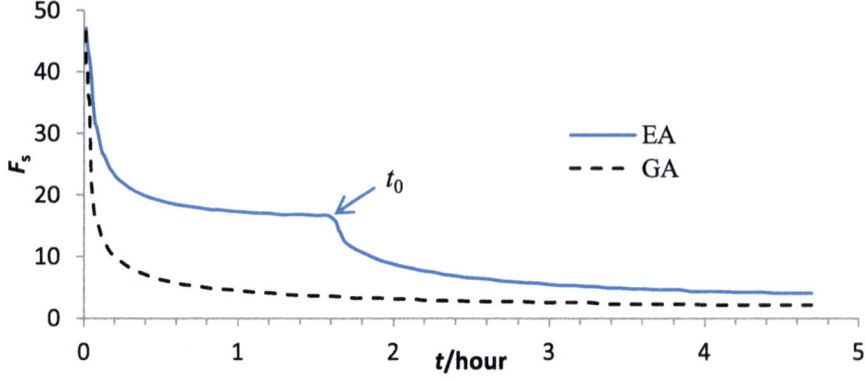

Fig. 1. Safety factors vs. time

4   Conclusions

Infiltration of rainfall over a large and shallow slope area involves water inflow and air outflow. This paper studied the infiltration and stability of slope under an intense rainfall by the extended Green and Ampt model. The results indicate that the delayed effect of air pressure on landslides is obvious. With the wetting front moving slowly downward, a temporary perched water table occurs at the wetting front. The normal stress at the potential slide surface in the saturated zone is reduced owning to the perched water table. Consequently, it is necessary to consider the effect of air pressure on shallow landslides triggered by intense rainfall.

References

Cho, S.E.: Stability analysis of unsaturated soil slopes considering water-air flow caused by rainfall infiltration. Eng. Geol. **211**, 184–197 (2016)

Sun, D.M., Zhang, Y.G., Semprich, S.: Effects of airflow induced by rainfall infiltration on unsaturated soil slope stability. Transp. Porous Media **107**(3), 821–841 (2015)

Wang, Z., Feyen, J., Nielsen, D.R., van Genuchten, M.T.: Two-phase flow infiltration equations accounting for air entrapment effects. Water Resour. Res. **33**(12), 2759–2767 (1997)

Upper Bound Stability Analysis of MSW Slope Layered by Fill Age Considering Particle Compressibility

Maosong Huang[1]([✉]), Xinping Fan[1], Haoran Wang[2], and Xilin Lu[1]

[1] Department of Geotechnical Engineering,
Tongji University, Shanghai 20092, China
mshuang@tongji.edu.cn
[2] Urban Construction Design and Research Institute, Shanghai, China

Abstract. The shear strength of the municipal solid waste (MSW) varies during the service years of the landfill and it is also affected by the particle compressibility. First, a pore-water pressure reduction coefficient, which attributes to the compressibility of a particle and the solid matrix, is introduced to the effective stress formulation to modify the Terzaghi's principle. This coefficient is also dependent of the landfill age. Then, based on the upper bound limit analysis theorem, the rotational-translational mechanism previously introduced by Huang [1] is generalized to analyze the multilayer municipal solid waste landfill slope. Finally, the stability of a landfill slope case is analyzed considering the influence of the particle compressibility. The computed safety factors of the landfill slopes decrease when the particle compressibility is considered.

Keywords: Landfill · Fill age · Particle compressibility · Upper bound

1 Rotational-Translational Mechanism

Because of the liner system in the landfill, the limit analysis presented here is based on the rotational-translational mechanism rather than the conventional rotational failure mechanism for homogeneous slopes. As show in Fig. 1, the wedge a is the rotating block, and the wedge b is the translating block. It is assumed that the failure surface (the BN part) is a logarithmic spiral rotating around the point O. The wedge b translates along the liner system.

In the calculative process, it is worth noting that the frictional angle should be consistent with the corresponding layer that the gap contains. Thus, the kinematically admissible velocity field can be constructed as shown in Fig. 1.

2 Fill Age and Particle Compressibility

When the Mohr-Coulomb criterion is adopted to describe the shear strength of MSW, the strength parameters obtained from drained triaxial tests and undrained triaxial tests are different. Based on the method of Skempton [2], the strength criterion of MSW can be described as

© Springer Nature Switzerland AG 2018
W. Wu and H.-S. Yu (Eds.): *Proceedings of China-Europe Conference on Geotechnical Engineering*, SSGG, pp. 1483–1486, 2018.
https://doi.org/10.1007/978-3-319-97115-5_128

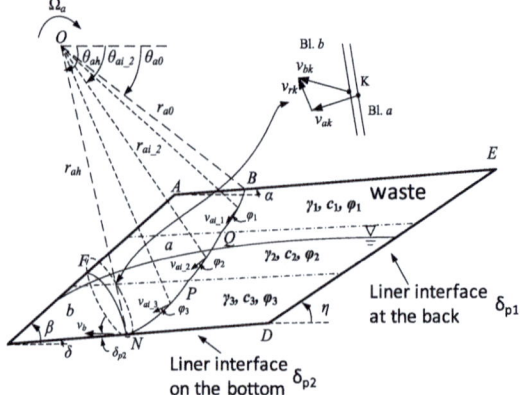

Fig. 1. Rotational-translational mechanism and velocity hodograph

$$\tau = c' + (\sigma - Au_w)\tan \varphi' \tag{1}$$

Lian [3] conducted the undrained triaxial tests of MSW at different fill ages and the shear strength parameters were obtained without considering particle compressibility. In this paper, the shear strength parameters are matched by Eq. (1). For example, the results of MSW for fill age at 6–9.5 year are presented in Fig. 2.

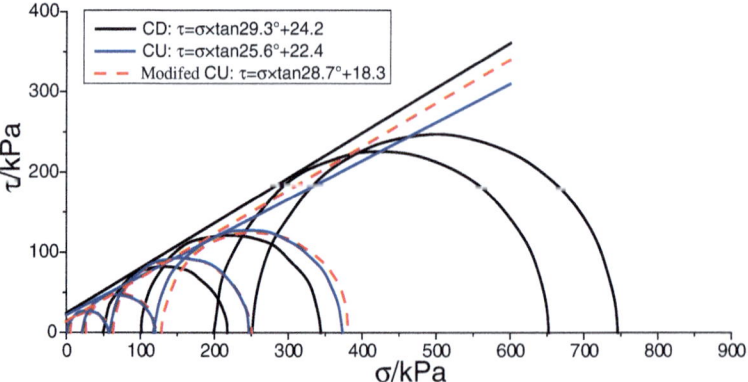

Fig. 2. Strength envelope line of MSW (fill age: 6–9.5 year)

3   Case Study

A landfill slope layered by aging is presented in Fig. 3. Parameters of the waste are shown in Table 1, which are matched by Eq. (1) from the undrained triaxial tests of MSW at different fill ages by Lian (2011). At first, the homogeneous landfill slopes in

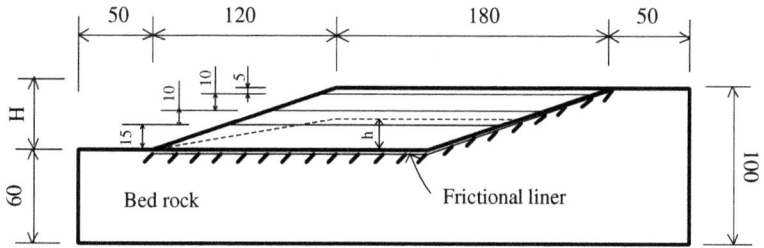

Fig. 3. Geometry of the landfill slope considering particle compressibility

Table 1. Parameters of the waste

Age (y)			0–3.5	3.5–6	6–9.5	9.5–13
Reduction coefficient of pore-water pressure			0.76	0.85	0.89	0.95
CU	(not considering particle compressibility)	c' (kPa)	15.4	24	22.4	21.8
		φ' (°)	22.5	29.4	25.6	41.3
	(considering particle compressibility)	c' (kPa)	22	20	18.3	13.7
		φ' (°)	18.5	30.1	28.7	39.4
CD		c' (kPa)	45.2	17.3	24.2	15.6
		φ' (°)	19.3	30.3	29.3	40

four different fill ages at a water table level of $h/H = 0.4$ are analyzed with and without considering particle compressibility, of which the safety factors are given in Fig. 4.

Then, for different water table levels, the stability of layered landfill slope is analyzed considering compressibility. The results are shown in Fig. 5. From Figs. 4 and 5, it can be seen that the larger the fill age is and the influence of the particle compressibility is less. The safety factor of the landfill slope also decreases with the increase of the water height.

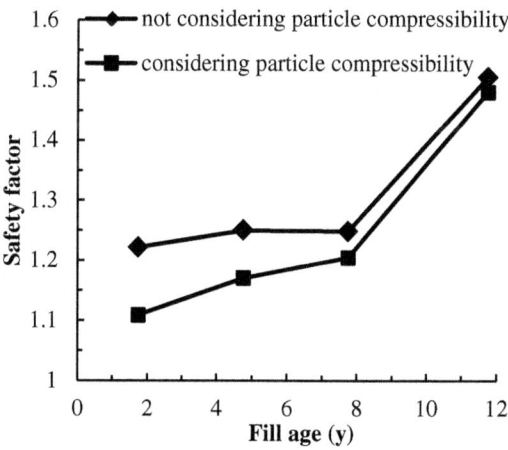

Fig. 4. Safety factors of the landfill slopes in different fill ages

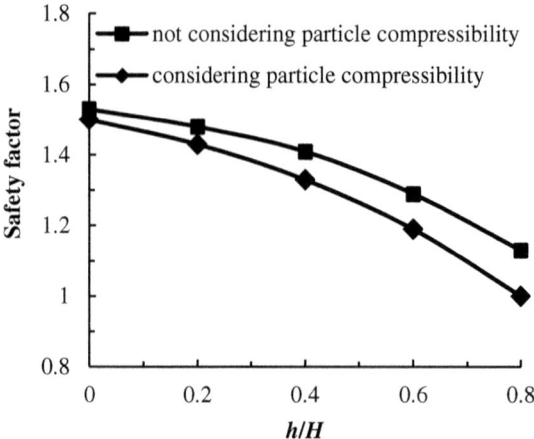

Fig. 5. Safety factors of the layered landfill slope in different water height levels

Acknowledgement. This work was financially supported by the National Basic Research Program of China (973 Program) (Grant No. 2012CB719803). This support is gratefully acknowledged.

References

1. Huang, M., Wang, H., Sheng, D., et al.: Rotational-translational mechanism for the upper bound stability analysis of slopes with weak interlayer. Comput. Geotech. **53**, 133–141 (2013)
2. Skempton, A.: Effective stress in soils, concrete and rocks. In: Pore Pressure and Suction in Soils, pp. 4–16. Butterworths, London (1961)
3. Lian, B.Q.: Experimental study on triaxial test of municipal solid waste. Ph.D. thesis, Zhejiang University, Hangzhou (2011)

Modeling of Landslides and Assessment of Their Impact on Infrastructure Objects in Ukraine

Olena Ivanik[(✉)]

Taras Shevchenko National University of Kyiv,
90 Vasylkivska Street, Kiev 03022, Ukraine
om.ivanik@gmail.com

Abstract. The area of Ukraine is characterized by active development of landslide processes within different tectonic and landscape-climatic zones. For assessing and predicting landslide processes two main approaches have been applied. The first approach is based on a comprehensive analysis of the factors of formation of landslide processes using statistical methods, GIS analysis and cartographic modeling. The second approach assumes field work, monitoring and numerical modeling of landslide processes with the aim of local prediction of landslide hazard. Determination of stress-strain state of rocks within a particular landslide-prone slope has been made taking into account a number of geological and geomorphological factors. The basic summary factors taken into account in the construction of physical and mathematical models and calculations are water saturation; effect of the gravitational field of the Earth; type of rocks with its inherent thermomechanical properties, linear expansion ratio, density; boundary conditions on the boundary of the proposed mass (by displacement or stress); geometric characteristics of the selected mass (size, slope angle). The studies determined the composition and structure of the most dangerous landslides. There are predominantly structural landslides, landslides formed in inhomogeneous, anisotropic environment.

Keywords: Landslide · Modeling · Natural hazards

1 Introduction

The area of Ukraine is characterized by active development of natural hazards processes within different structural and tectonic and landscape-climatic zones. The greatest development of landslides was observed on the coast of the Black and Azov Seas, as well as in Cherkasy, Kharkiv, Lviv, Chernivtsi and Transcarpathian regions. The total number of landslides was 22,948 units, and the area of distribution of landslides only in 200 settlements covers more than 44.0 km^2. The largest landslide hazard is observed in the Transcarpathian region, where the number of landslides was 3279 units, with a total area of 385.1044 km^2.

For the protection of ecological environment, we should be able to control or minimise catastrophic impact of natural hazards processes. Detecting landslides, as

© Springer Nature Switzerland AG 2018
W. Wu and H.-S. Yu (Eds.): *Proceedings of China-Europe Conference on Geotechnical Engineering*, SSGG, pp. 1487–1490, 2018.
https://doi.org/10.1007/978-3-319-97115-5_129

well as monitoring their activity using integrated techniques is carried out in order to provide maps, risk assessment and forecasting of geohazardous events in Ukraine.

World experience in the field of research of hazardous geological processes underlines the effectiveness of modern GIS-methods and modeling in preventing and predicting the negative impact of landslide processes, mudflows, etc. (Foster et al. 2012). Simulation methods and GIS are used for predictive mapping, creating models for multifactorial spatial estimation of landslides. Thus, the risks of landslide, formation and possible areas of flooding are estimated on the basis of statistical analysis, remote sensing data and field observations. As a result, methods for assessing the state of man-made systems under the influence of hazardous geological processes on the basis of modeling geological processes and structures, as well as using GIS have been developed (Wu et al. 2015; Jaboyedoff et al. 2012).

The proposed methodology for modeling of hazardous geological processes and quantitative assessment of their impact on infrastructure objects includes: consistent analysis of geological processes and geological situations within the areas of occurrence of hazardous geological processes and model polygons; determination of input parameters of modeling on the basis of theoretical, empirical, experimental data and GIS analysis; development of physical, geological and mathematical models of the geological environment and processes; prediction of the impact of natural hazards on man-made objects and calculations of their stress-strain state; development of preventive measures to minimize the negative impact of hazardous geological processes on critical infrastructure objects.

2   Mechanisms and Factors of Formation of Landslide Processes

Mechanisms and factors of formation of landslide processes are examined in terms of model objects in the *Carpathian, Crimean and Middle Dnieper regions* having complex geological structure.

Natural hazards cause substantial damage in the *Carpathian Mountains.* In the Ukraine this area has well-developed infrastructure that includes roads, bridges, railways, pipelines, commercial and residential structures. This is why any geohazardious event disrupts sustainable development of this area and causes many accidents. Based on the use of the above-mentioned effective methods for assessing and predicting landslide processes within the Carpathian model area, two main approaches were applied. The first approach is based on a comprehensive analysis of the factors of formation of landslide processes using statistical methods, GIS analysis and cartographic modeling. Statistic technique also includes weight factors analysis. Five main categories of these methods are identified by: geomorphological mapping; analysis of landslides inventories; statistical methods and process based conceptual models. A spatial database of 1478 landslides was analyzed using GIS to map landslide susceptibility in Zakarpataska region. The second approach assumes field work, monitoring and numerical modeling of landslide processes with the aim of local prediction of landslide hazard. The special field studies determined the composition and structure of the most dangerous landslides. Thus, in the Carpathian region there are predominantly structural landslides, landslides formed in inhomogeneous, anisotropic environment. Most landslides were found in Oligocene flysch sediments affected by destructive zones with varying orientation relating to slopes. The researches confirmed the special role of destructive zones characterized by intense fracturing, brecciation and stratification of flysch deposits. Special studies are also being carried out to assess the physical parameters of rocks (Young's modulus, Poisson's ratio, Lame constant, etc.) within the landslide slopes, which are used in assessing slope stability and calculating the stress-strain state of rock complexes in landslide hazard zones on the basis of the theory of linear elasticity.

Gravitational processes of the *Crimean region* are subjected to effects of seismic, structural, lithological, geomorphological and hydrogeological factors. Development of landslide processes is observed in the Jurassic and Cretaceous rock complexes.

Middle Dnieper area is comprised of two regions: Kyiv and Kaniv, which differ in geological structure and neotectonic regime; while processes of gravitational nature vary widely. The presence of Neogene and Quaternary sands and clays creates a prerequisite for the formation of structural landslides in the heterogeneous environment.

The in-depth geological analysis confirmed that in terms of rheological behavior of rocks landslides can be divided into the following types: elastic, elastic-plastic, viscous and visco-plastic (Ivanik 2015). They can be formed in geological environments with different orientations of the geological structure in relation to the slope (see Fig. 1). This structure of landslide slopes defines the approaches and methods of landslide hazard modeling.

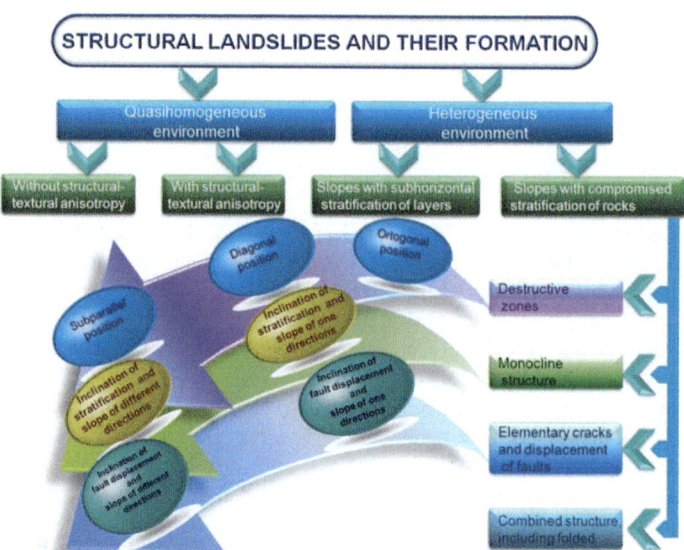

Fig. 1. Classification of structural landslides and there formation.

3   Calculation of the Stress-Strain State of the Landslide-Prone Slopes

Determination of stress-strain state of rocks within a particular landslide-prone slope has been made taking into account a number of geological and geomorphological factors. The basic summary factors taken into account in the construction of physical and mathematical models and calculations based on an algorithmic scheme of the mathematical model are water saturation; effect of the gravitational field of the Earth; type of rocks with its inherent thermomechanical properties, linear expansion ratio; density; boundary conditions on the boundary of the proposed mass (by displacement or stress); geometric characteristics of the selected mass (size, slope angle). Under the proposed mathematical model, the examined phenomenon is described as thermoelastic-plastic equilibrium of the isotropic matrix under effect of applied mass (gravitational field of the Earth) and surface efforts, inhomogeneous stationary temperature field.

4   Results

Two approaches have been used for the assessment and forecasting of geohazards. The first approach is based on GIS and statistical analysis of geo-environmental factors related to the occurrence of hazardous processes. Based on the results of the analysis of the reference landslide area, landslide susceptibility maps have been prepared. The second approach is based on field work, monitoring and modeling of landslides for the local forecasting. The proposed classification of structural is the basis of, physical and mathematical models of landslide slopes, and subsequent modeling of the landslide hazard based on the determination of the stress-strain state of slopes.

References

Foster, C., Pennington, C.V.L., Culshaw, M.G., Lawrie, K.: The national landslide database of great Britain: development, evolution and applications. Environ. Earth Sci. **663**, 941–953 (2012). https://doi.org/10.1007/s12665-011-1304-5

Ivanik, O.M.: Classification of the structural landslides for the natural hazard assessment. In: 2015 Abstract of the 77th EAGE Conference and Exhibition 2015, Madrid, Spain, 1–4 June 2015. https://doi.org/10.3997/2214-4609.201413442

Jaboyedoff, M., Oppikofer, T., Abellan, A., Derron, M.H., Loye, A., Metzger, R., Pedrazzini, A.: Use of LIDAR in landslide investigations: a review. Nat. Hazards **61**, 5–28 (2012). https://doi.org/10.1007/s11069-010-9634-2

Wu, W.: Recent Advances in Modelling Landslides and Debris Flows. Springer International Publishing, Switzerland (2015). https://doi.org/10.1007/978-3-319-11053-0

The Ultimate Lateral Soil Pressure on Stabilizing Piles in Slopes

Guoping Lei$^{(\boxtimes)}$ and Wei Wu

Institute of Geotechnical Engineering,
University of Natural Resources and Life Sciences, Vienna, Austria
guoping.lei@students.boku.ac.at, wei.wu@boku.ac.at

Abstract. In this study, the ultimate lateral soil pressure p_u on stabilizing piles is discussed and the empirical equations and analytical models to predict p_u are reviewed. The plastic deformation theory is modified by introducing a new parameter K_f, which is back calculated by a numerical calculation. The parameter is then compared with the corresponding soil stress in the numerical model. The modified theory is able to evaluate the p_u in a larger range of pile spacing.

Keywords: Ultimate lateral pressure · Stabilizing piles

1 Introduction

Rows of piles are widely used to increase the stability of slopes. For the ultimate lateral soil pressure p_u on the pile, various theories or empirical equations were proposed for clay or sandy soils. For sandy soil, the simplest approach is to use the suggestion of Broms (1964) [1] in which $p_u = aK_p\sigma'_{vo}$, where $K_p = \text{Tan}^2(45 + \frac{\phi}{2})$ is the Rankine passive pressure coefficient; ϕ is the internal friction angle of soil; σ'_{vo} is the effective overburden pressure and a is a coefficient ranging between 3 and 5. The equation was initially proposed for the laterally loaded piles in stable soil. For the piles in the sliding soil, $\frac{a}{2}$ can be taken into account (Poulos 1995 [2]). Barton (1982) [3] found from lateral pile tests that the simple variation $p_u = K_p^2\sigma'_{vo}$ could be used.

Reese et al. (1974) [4] proposed a wedge failure mechanism in front of the pile, which leads to a variation of p_u, with depth that is initially proportional to K_p, but at greater depths becomes proportional to k_p^3 (Fleming et al. (2008) [5]). Norris (1986) [6] proposed a strain wedge model to predict the response of a flexible pile under lateral loading. The model was later applied in the analysis of stabilizing piles (Ashour and Ardalan (2012) [7]). The basic of the theory is assuming a conceptualized three-dimensional wedge, in which the soil is at critical state at the wedge surface.

© Springer Nature Switzerland AG 2018
W. Wu and H.-S. Yu (Eds.): *Proceedings of China-Europe Conference on Geotechnical Engineering*, SSGG, pp. 1491–1494, 2018.
https://doi.org/10.1007/978-3-319-97115-5_130

Considering the arching effect between adjacent piles, Ito and Matsui (1975) [8] have developed a plastic deformation theory for the flow of soil through a row of piles. Their equations show that the limiting pressure developed on a pile by flowing soil depends on the strength properties of the soil, overburden pressure, and the spacing between the piles relative to their diameter. Their equations are meant to apply for the portion of the piles in the unstable or moving soil. However, the equations are only valid over a limited range of spacings, since, at large spacings or at very close spacings, the mechanism of flow through the piles adopted by Ito and Matsui (1975) [8] is not the critical mode (Poulos (1995) [2]). Nevertheless, this theory has provided a physical meaningful alternative to the empirical equations. Numerical study from Lei (2018) [9] has shown that, when the pile spacing increases to a certain critical value ($s/d = 3$), the soil pressure on the pile is fully mobilised and the arching effect is formed. As the pile spacing increases further, the soil flows through the piles and more pressure is transferred to the down-slope soil, which in return strengthens the arching effect to enable more pressure on the pile. This brings up a question to an assumption in the plastic deformation theory, which is that the normal stress at the pile line (AA' in Fig. 1) is represented by the active earth pressure:

$$\sigma_x = \gamma z K_p^{-1} - 2cK_p^{-\frac{1}{2}} \tag{1}$$

where, z is the soil depth, γ is the unit weight of soil, and c is cohesion. This assumption serves as a boundary condition to solve the differential equation. However, it does not hold for very small or big pile spacings. In this study, the plastic deformation theory is improved by introducing a new parameter.

2   The Modified Plastic Deformation Theory

A new parameter K_f is introduced into the plastic deformation theory to evaluate the normal soil stress at the pile line as:

$$\sigma_x = K_f \gamma z \tag{2}$$

In the plastic deformation theory, two sliding surfaces, which occur along the lines AEB and $A'E'B'$ (Fig. 1), are assumed. The soil layer becomes plastic only in the soil zone $AEBB'E'A'$, in which the Mohr-Coulomb's yield criterion is applied. As the soil stress on the plane AA' is given, the soil stress on the plane BB' can be derived using the framework of the plastic deformation theory as follows:

$$\begin{aligned} |\sigma_x|_{BB'} &= (\frac{D_1}{D_2})^{N_A}(K_f\gamma z + c\frac{N_D}{N_C})\exp(N_C\frac{D_1 - D_2}{D_2}\tan(\frac{\pi}{8} + \frac{\phi}{4})) \\ &+ (\frac{D_1}{D_2})^{N_A}c(\frac{N_B}{N_A} - \frac{N_D}{N_C}) - c\frac{N_B}{N_A} \\ &= K_1 * |\sigma_x|_{AA'} + K_2 c \end{aligned} \tag{3}$$

where

$$N_A = N_\phi^{\frac{1}{2}} \tan\phi + N_\phi - 1; \quad N_B = 2\tan\phi + 2N_\phi^{\frac{1}{2}} + N_\phi^{-\frac{1}{2}};$$

$$N_C = N_\phi \tan\phi; \quad N_D = 2N_\phi^{\frac{1}{2}} \tan\phi + 1; \quad N_\phi = \tan^2\left(\frac{\pi}{4} + \frac{\phi}{2}\right).$$

The lateral force acting on a pile per unit thickness of layer is the difference between lateral forces acting on the plane BB' and AA', as:

$$p_u = p_{BB'} - p_{AA'} = D_1|\sigma_x|_{BB'} - D_2|\sigma_x|_{AA'} \tag{4}$$

When the lateral force on pile p is measured at field, from model tests or calculated by numerical simulation, the coefficient K_f can be back calculated by:

$$K_f = \frac{p_u - K_2 D_1 c}{(K_1 D_1 - D_2)\gamma z} \tag{5}$$

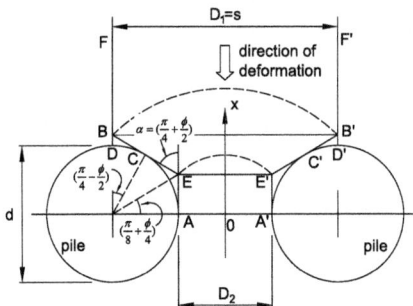

Fig. 1. State of plastic deformation in the ground just around piles (D_1, D_2 are the pile axis distance and the pile clear distance, respectively)

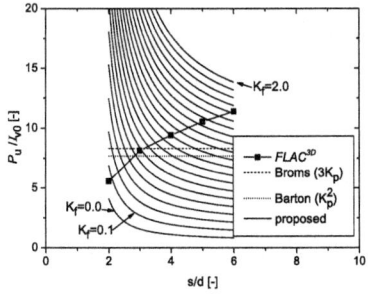

Fig. 2. The normalised P_u

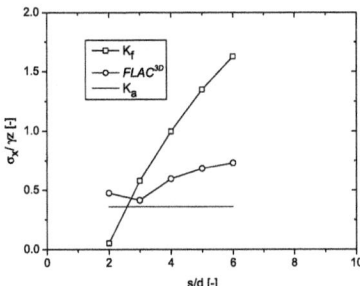

Fig. 3. Back-calculated K_f

The total ultimate force on the pile P_u can be obtained by integrating p_u along the pile depth. For better comparison, P_u is normalised by $I_{vo} = \frac{1}{2}\gamma h^2$, where h is the pile length in the sliding soil. Figure 2 shows the normalised P_u varying with s/d calculated by the proposed method adopting various K_f values. The objective is a sliding layer of loose silty sand with a thickness of 4 m, a density of 1900 kg/m^3, a friction angle of 28° and a cohesion of 3 kPa. The pile diameter $d = 1$ m. Some widely used empirical predictions and the result from $FLAC^{3D}$ are also plotted in the figure. It is seen that the empirical predictions intersect with the curve from $FLAC^{3D}$ with a big angle, which implies that the empirical predictions are only precise in a small range of s/d. The curves using the proposed theory with various K_f can intersect with the curve from $FLAC^{3D}$ in the whole range of s/d. The K_f value is back calculated using the numerical results and is shown in Fig. 3. The σ_x at AA' in the numerical model is normalised by γz, which is equivalent to K_f, and its average value is also plotted in the figure. With the increase of s/d, the stress at AA' has experienced a decrease to $s/d = 3$ and a increase afterwards. The back-calculated K_f has shown an unrealistic variation with even a value larger than 1 when the spacing is bigger than 4. This has also reflected the limitation of the plastic deformation theory. However, comparing with K_a, which is the active soil pressure coefficient and is adopted in the original plastic deformation theory, the proposed modification makes the theory more flexible to be applied in a larger range of pile spacing.

References

1. Broms, B.B.: Lateral resistance of piles in cohesionless soils. J. Soil Mech. Found. Div. ASCE **90**, 123–158 (1964)
2. Poulos, H.G.: Design of reinforcing piles to increase slope stability. Can. Geotech. J. **32**, 808–818 (1995)
3. Barton, Y.: Laterally loaded model piles in sand: centrifuge tests and finite element analyses. University of Cambridge (1982)
4. Reese, L.C., Cox, W.R., Koop, F.D.: Analysis of laterally loaded piles in sand. Offshore Technology in Civil Engineering: Hall of Fame Papers from the Early Years, pp. 95–105 (1974)
5. Fleming, K., Weltman, A., Randolph, M., et al.: Piling Engineering. CRC Press, Boca Raton (2008)
6. Norris, G.: Theoretically based BEF laterally loaded pile analysis. In: Proceedings of the 3rd International Conference on Numerical Methods in Offshore Piling, pp. 361–386 (1986)
7. Ashour, M., Ardalan, H.: Analysis of pile stabilized slopes based on soil-pile interaction. Comput. Geotech. **39**, 85–97 (2012)
8. Ito, T., Matsui, T.: Methods to estimate lateral force acting on stabilizing piles. Soils Found. **15**, 43–59 (1975)
9. Lei, G.: Slope stabilised by piles: centrifuge tests, numerical analysis and analytical method. University of Natural Resources and Life Sciences, Vienna (2018)

Influence of Bulk Density and Slope on Debris Flows Deposit Morphology: Physical Modelling

Shuai Li, Xiaoqing Chen[✉], Gongdan Zhou, Dongri Song,
and Jiangang Chen

Key Laboratory of Mountain Hazards and Earth Surface Process,
Institute of Mountain Hazards and Environment,
Chinese Academy of Sciences (CAS), Chengdu, China
xqchen@imde.ac.cn

Abstract. Prediction of the runout of a debris flow is of great significance for hazard mitigation. Laboratory tests were carried out to simulate debris flows under different bulk densities and slopes. A linear negative relationship between bulk density and runout-ratio-with and volume was found. Results show that the morphology of the deposited sediment is strongly influenced by the inclination of the channel. Moreover, a new empirical formula has been developed to estimation of inundated volume and runout-ratio-width.

Keywords: Debris flow · Deposit morphology · Physical modelling

1 Introduction

Debris flows have a great deal of mobility and strong destructive power. It can destroy the local resident and fatalities [1]. Therefore, accurate prediction of their runout distances is critical to protect downstream infrastructure and towns [2].

Debris flow runout distance prediction is a complicate issue because it was governed by the properties of sediment, water content, topographical factors, volume of mass, and so on [3]. Although extensive studies regarding debris flow runout have been conducted, how the deposition shape and size (length, width and height) are affected by various influencing factors is still poorly understood. Here, the effects of bulk density and flume slope on the debris-flows runout distance were investigated through the laboratory experiments on small-scale flume in this study. Moreover, a new empirical formula has been developed to calculate the inundated volume and runout-ratio-width of debris flow.

2 Experimental Method

The Experimental model has a rectangular channel with a width and a depth of 0.30 m and 0.35 m, respectively. The two channel sections inclined at different angles are herein referred to as the upstream channel (UC) and the downstream channel (DC). Two flume configurations were used in this study. In configurationI, the UC is 3.0 m in length and is inclined at 45°, followed by the DC, which is 4.0 m in length and

© Springer Nature Switzerland AG 2018
W. Wu and H.-S. Yu (Eds.): *Proceedings of China-Europe Conference on Geotechnical Engineering*, SSGG, pp. 1495–1499, 2018.
https://doi.org/10.1007/978-3-319-97115-5_131

horizontal, and transitions into an un-channelized horizontal outflow plane. In configurationII, the UC is also 3.0 m in length and inclined at 30°, followed by the DC, which is 2.0 m in length and inclined at 7.6°.

The sediments used for the debris flow mixtures are from the natural deposition fan of the Jiangjia Gully in Yunnan Province, China. The bulk densities of the debris flows were varied for each test, from 1.912×10^3 kg/m^3 to 2.183×10^3 kg/m^3.

3   Interpretation of Test Result

3.1   Effects of Bulk Density

Debris flows with bulk density large than 2.044×10^3 kg/m^3 cannot be reach the end of the DC. Decreasing bulk density enhances runout and deposit area, but the maximum deposit height is reduced (Fig. 1). Debris flows spread to both sides for small resistance in the lateral direction, meanwhile, the runout distance is reduced whereas the maximum deposit height is increased. Additionally, a part of large bulk density debris flows stopped at the (DC), also reduce the volume of debris on the outflow plain. An unambiguous negative linear correlation between the bulk density and runout-ratio-width and deposit volume was observed (Fig. 2).

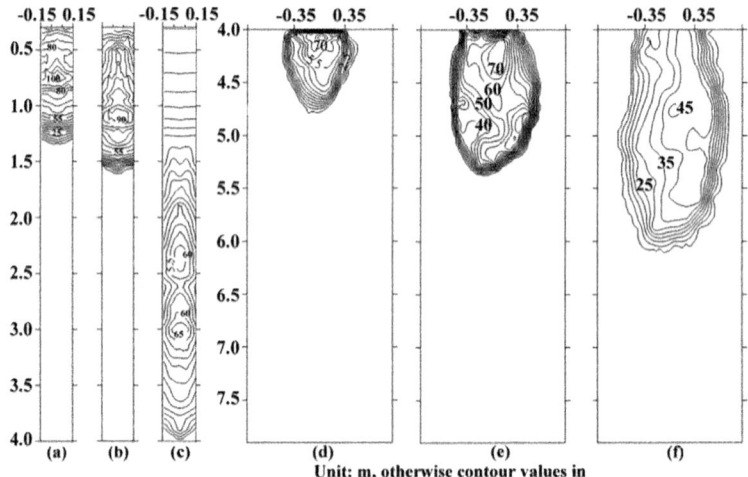

Fig. 1. Influence of bulk density on deposition morphological. (a) $\rho_m = 2.183$; (b) $\rho_m = 2.128$; (c) $\rho_m = 2.044$; (d) $\rho_m = 2.026$; (e) $\rho_m = 1.965$; and (f) $\rho_m = 1.912$ ($\times 10^3$ kg/m^3).

3.2   Effects of Flume Slope

The results show that the morphology of configurationI exhibits elliptic, while the morphology of configurationII presents a strip-like shape (Fig. 3). The runout distance and deposit area increased, whereas the maximum deposit height decreased in configurationII compared to that of configurationI. Although a steeper UC slope (45°) can

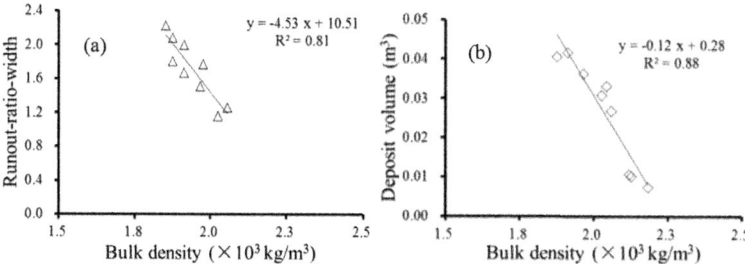

Fig. 2. Relation between debris flow bulk density and deposit morphology. (a) Runout-ratio-width; (b) Deposit volume.

increase gravitational potential energy, the abrupt change in slope to 0.0° (a flat DC slope) correspondingly increases the vertical component, leading to intensive collisions between the debris and bottom floor, thus decreasing debris flow kinematics and effectively limiting the runout distance. In contrast, the gentler UC slope (30°) in configuration II followed by a non-horizontal DC slope (7.6°) provides a smoother transition for the flowing debris, promoting a more efficient downstream motion which effectively increases runout distance. Beyond the break in slope, the attenuation in flow velocity induces deposition of debris and partial transfer of momentum into the generation of turbulence [4]. Especially, the local curvature at the break in slope controls the separate of velocity components perpendicular and parallel to the flow, which determines the flow motion and velocity attenuation.

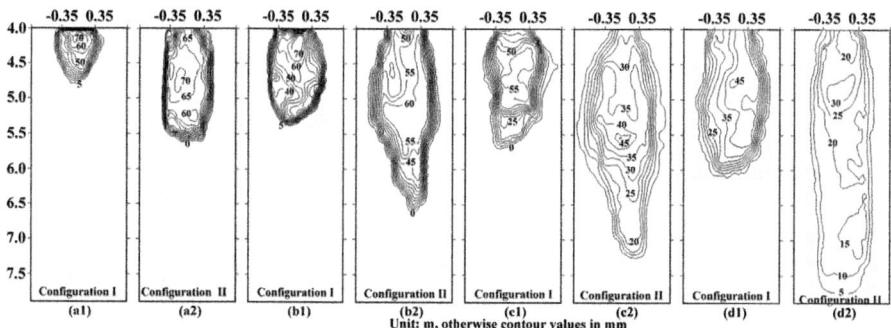

Fig. 3. Influence of slope on morphology of deposited debris. (a1) and (a2): $\rho_m = 2.026 \times 10^3$ kg/m³; (b1) and (b2): $\rho_m = 1.965 \times 10^3$ kg/m³; (c1) and (c2): $\rho_m = 1.912 \times 10^3$ kg/m³; (d1) and (d2): $\rho_m = 1.875 \times 10^3$ kg/m³.

Based on the present experimental results, a regression equation is developed to predict the debris flow deposition on the outflow plain in terms of the slopes and lengths of the approaching UC and DC, and the bulk density. The relationship can be given as follows:

$$V = 0.01i_1l_1 + 0.008i_2l_2 - 0.13\rho_m + 0.26 \tag{1}$$

$$L/W = 10.1012i_1l_1 + 1.1616i_2l_2 - 1.65\rho_m \tag{2}$$

where, V is deposit volume, m^3; i_1, i_2 are the approaching UC and DC slopes, respectively; l_1, l_2 are referred to as the approaching UC and DC lengths, respectively, m; ρ_m is bulk density, 10^3 kg/m^3; L/W is runout-ratio-width, which L is the maximum runout distance and W is the maximum deposit width.

Equations (1) and (2) are also tested against the experimental data (Fig. 4). It can be seen that calculation results agree well the experimental data.

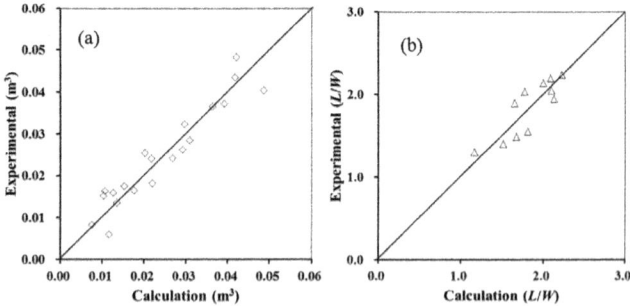

Fig. 4. The calculation and experimental data. (a) deposit volume; (b) runout-ratio-width.

4   Conclusions

This study attempts to provide some improvements to the understanding of the runout of the deposited debris by incorporating the effects of bulk density and slope. The deposits from low bulk density flows are longer and thinner than those from high bulk density flows. Runout distance and deposit area are enhanced with an increase in channel slope. Gentler slope transitions normally lead to longer runout distances. Furthermore, a simple empirical method for preliminary estimates of debris-flow runout-ratio-with and volume was developed. The results may aid in future explorations and evaluations of debris flow hazards.

Acknowledgment. The authors acknowledge the Science and Technology Service Network Initiative (grant KFJ-STS-ZDTP-015), the National Natural Science Foundation of China (grant 11672318), the CAS "Light of West China" Program (grant Y6R2220220), the China Post-doctoral Science Foundation (grant 2016M602716), and the Chinese Academy of Sciences (CAS) Pioneer Hundred Talents Program.

References

1. Iverson, R.M.: The physics of debris flows. Rev. Geophys. **35**(3), 245–296 (1997)
2. Hürlimann, M., McArdell, B.W., Rickli, C.: Field and laboratory analysis of the runout characteristics of hillslope debris flows in Switzerland. Geomorphology **232**, 20–32 (2015)
3. Haas, T., Braat, L., Leuven, J.R., Lokhorst, I.R., Kleinhans, M.G.: Effects of debris flow composition on runout, depositional mechanisms, and deposit morphology in laboratory experiments. J. Geophys. Res. Earth Surf. **120**, 1949–1972 (2015)
4. Kim, Y., Paik, J.: Depositional characteristics of debris flows in a rectangular channel with an abrupt change in slope. J. Hydro-environ. Res. **9**(3), 420–428 (2015)

Simulation of Building Failure by Landslide Impact

Hongyu Luo[(⊠)] and Limin Zhang

Department of Civil and Environmental Engineering,
The Hong Kong University of Science and Technology,
Clear Water Bay, Kowloon, Hong Kong
hluoae@connect.ust.hk

Abstract. A significant research gap in quantitative risk assessment for land-slides is the quantification of physical vulnerability of buildings subject to landslide impacts. In this paper, an explicit time integration program LS-DYNA is utilized to analyze the response of RC framed buildings. The analysis gives an insight of the failure process of buildings impacted by a landslide and provides a viable approach to assessing the vulnerability of buildings to landslides.

Keywords: Landslides · Building · Vulnerability · Impact analysis
LS-DYNA

1 Introduction

As urban development expands into sloping terrains, more landslide hazards can be induced, which can lead to loss of lives, damage to properties and the environment. It is generally accepted that quantitative risk assessment (QRA) for landslides is preferred over qualitative analysis [1]. However, the quantification of vulnerability is a main obstacle for the development of this method. An extensive review of these methods has been presented by Totschnig and Fuchs [2]. Among them, the empirical method is the most frequently used in the literature. The accuracy of this method is highly data-dependent and some important data about landslide intensity and structure damage are hard to retrieve from past events and then missing in the vulnerability model. In earthquake engineering, the structure behavior and vulnerability have been studied using analytical methods for decades. However, the corresponding methods for flow-type landslides are very poor and the dynamic response of buildings to landslide/debris is not clear.

2 Finite Element Model

2.1 Model Description

An explicit dynamic analysis of progressive failure of a three-dimensional RC framed structure was carried out using LS-DYNA. This progressive failure analysis was conducted on a 2-bay (4.8 m), 3-span (3.6 m) and 3-storey (3.6 m) RC framed structure. Detail information about the numerical model is shown in Fig. 1.

© Springer Nature Switzerland AG 2018
W. Wu and H.-S. Yu (Eds.): *Proceedings of China-Europe Conference on Geotechnical Engineering*, SSGG, pp. 1500–1503, 2018.
https://doi.org/10.1007/978-3-319-97115-5_132

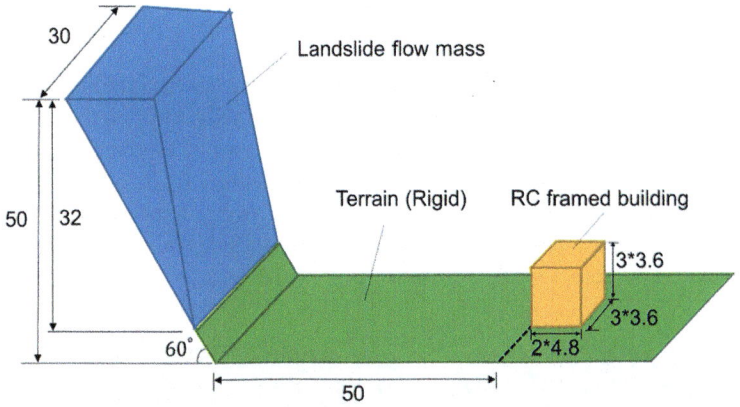

Fig. 1. Finite element model.

2.2 Material Model and Failure Criterion

For landslide flow mass, a single-phase fractional soil material (MAT_SOIL_AND_FOAM), which is assumed to be elasto-plastic following the Drucker-Prager yield criteria, was applied. This material has been studied and proved to be able to simulate landslides dynamics [3]. In this paper, the undrain strength index has been used by assuming an undrain condition during the flow process. The constitutive model parameters are summarized in Table 1.

Table 1. Constitutive parameters.

Landslide [4] (Yin et al. 2016)

Constitutive model	Density (kg/m³)	Elastic modulus (MPa)	Poisson's ratio	Internal friction angle, f_u (°)	Apparent cohesion, c_u (kPa)
MAT005	2000	20	0.3	20	10

Concrete (beam/column/slab/wall)

Constitutive model	Density (kg/m³)	Elastic modulus (MPa)	Poisson's ratio	Compressive strength (MPa)	Failure strain
MAT159	2400	23	0.15	20.1 (Grade30)	0.20

Reinforcement/Stirrup

Constitutive model	Density (kg/m³)	Elastic modulus (MPa)	Poisson's ratio	Yield stress (MPa)	Failure strain
MAT003	7850	200	0.3	400	0.10

3 Impact Process

The impact process is shown in Fig. 2. The landslide reaches the building with a frontal velocity of 8.8 m/s. After the initial impact, diagonal cracks occur on the external walls and the front walls are destroyed suddenly due to the low out-of-plane flexural capacity (Fig. 2(a)). Then, the side walls behind the corner columns are heavily damaged under large high compressive pressures. After that, the two corner columns are totally destroyed, and the side walls start to crush. Finally, the side walls are destroyed but the residual structural components still stand (Fig. 2(b)).

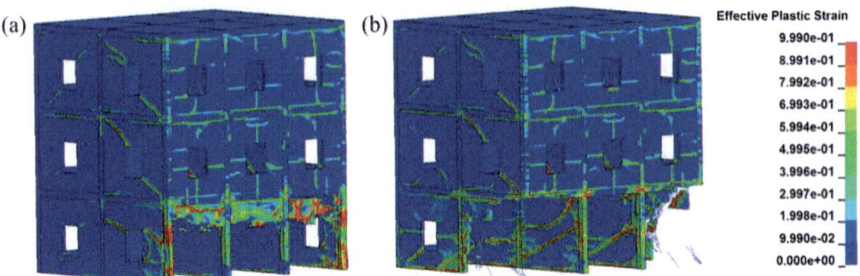

Fig. 2. Snapshots of global structure response.

It is worth noting that an obvious torsional failure occurs only on the two corner columns. A torque is generated only on the corner columns due to the pressure deference acted on the two sides of corner columns (Fig. 3). This pressure deference mainly comes from the velocity deference. Because of the high compressive strength and low tensile and shear-resisting capacity, the corner columns can be twisted easily due to the loss of lateral constrains caused by the damage of the side walls.

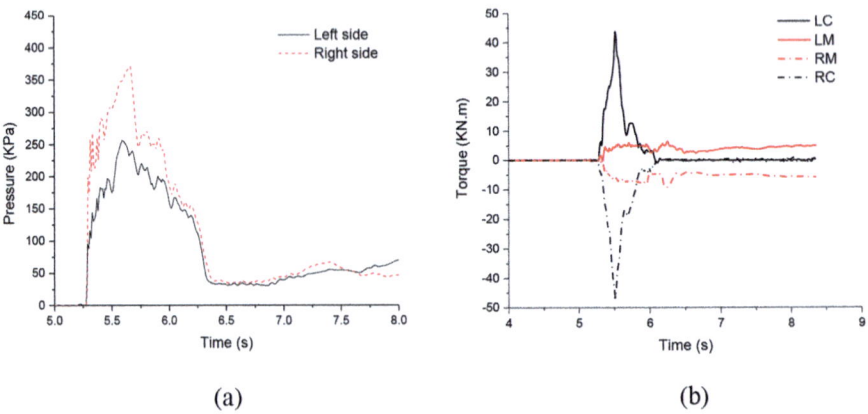

Fig. 3. (a) Time history of the pressure acted on the two sides of the corner column; (b) Torque on the bottom cross section of the two corner columns (L: left side; R: right side; C: corner column; M: middle column)

Impact pressure is commonly used to describe the landslide intensity and assess the vulnerability of buildings. For RC buildings, the vulnerability value reaches 1.0 at 37 kPa based on numerical simulations [5]. In this impact analysis case, the maximum pressure is 350 kPa and the building was partially damaged. This large pressure may be caused by the fact that the side walls have a high compressive strength, which can be regarded as a concrete shear wall. It has high lateral stiffness and can resist higher impact loading by the flow material. It is also worth noting that the damage state of building depends on many factors. A single index of landslide intensity cannot fully describe this complicated interaction process. This may be the main reason that discrepancy occurs on the limit pressures for different damage states in the literature.

4   Summary and Conclusions

This paper presents an explicit analysis of RC framed structure impacted by a landslide. In this idealized impact case, a torsional failure mechanism of corner column duo to pressure differences is observed and the damage of building continues to develop until the landslide flow stops. However, because of the presence of shear wall, the progressive collapse is not observed. Discrepancy occurs on the limit pressures for different damage states. Further investigation based on the proposed method needs to be conducted to establish more reliable vulnerability models.

Acknowledgements. This research is supported by the Research Grant Council of the Hong Kong Special Administrative Region (Project Nos. T22-603/15N and C6012-15G).

References

1. Uzielli, M., Nadim, F., Lacasse, S., Kaynia, A.M.: A conceptual framework for quantitative estimation of physical vulnerability to landslides. Eng. Geol. **102**(3–4), 251–256 (2008)
2. Totschnig, R., Fuchs, S.: Mountain torrents: quantifying vulnerability and assessing uncertainties. Eng. Geol. **155**, 31–44 (2013)
3. Kwan, J.S.H., Koo, R.C.H., Ng, C.W.W.: Landslide mobility analysis for design of multiple debris-resisting barriers. Can. Geotech. J. **52**(9), 1345–1359 (2013)
4. Yin, Y., Li, B., Wang, W., Zhan, L., Xue, Q., Gao, Y., Zhang, N., Chen, H., Liu, T., Li, A.: Mechanism of the December 2015 catastrophic landslide at the Shenzhen landfill and controlling geotechnical risks of urbanization. Engineering **2**(2), 230–249 (2016)
5. Quan Luna, B., Blahut, J., Van Westen, C.J., Sterlacchini, S., van Asch, T.W., Akbas, S.O.: The application of numerical debris flow modelling for the generation of physical vulnerability curves. Nat. Hazards Earth Syst. Sci. **11**(7), 2047–2060 (2011)

Characteristics and Application of Micropiles in Slope Engineering

Li Ma$^{(\boxtimes)}$, Yifu Hu, Desheng Gu, and Chong Jiang

Central South University, Changsha 410083, Hunan, China
mary_moli@foxmail.com

Abstract. As the application of micropiles being extended from foundation engineering to landslide engineering, some transforms of the mechanism and behaviors of micropiles have happened. A scaled slope was modeled in the lab to simulate three slope failure modes respectively, including sole slope, slope reinforced with single piles and composited piles. Besides, theoretical analysis about influencing factors was researched which was aimed to compare the slope failure and resistance and anti-sliding characteristics of micropiles. Taking the displacement, lateral loading and moments along micropiles shaft as main targets, the results illustrate that micropiles have provided higher resistance to slope compared to sole sliding slope. The reinforcement effect of composited piles performs better than single piles. It also turns about to present several deformation stages covering compaction, elastic deformation and plastic deformation and ultimate failure stages. In addition, a close insight into the influence of soil load, pile diameter and length, coefficient of subgrade on micropiles' performance has been given respectively. The application condition and circumstance of micropiles can be concluded consequently. It is expected that micropiles are considerably suitable for shallow landslide reinforcement while slope height less than 10 m, or sliding force under 300 kN/m, or coefficient of subgrade greater than 10000.

Keywords: Micropiles · Scaled slope model · Impact factors
Micropile's application

1 Introduction

Some research on micropiles, or mini-piles have been conducted in order to investigate the role they play, such like micropiles footings resting on dense sand [1], Underpinning an existing foundation [2], or application on earth embankment landslides [3]. Comparing to large diameter piles like manual excavation piles, micro plies have a fast construction speed, high safety, and flexible access technology. These advantages promote its application in landslides projects and especially those facing limitation of construction site and time. However, the mechanism and behaviors have changed as micropiles have been utilized in different circumstance [2, 4, 5].

In most cases, micropiles reinforcing landslides have been regarded as lateral loading structures [6], while a number of micropiles made of metal materials have been researched in sands [7]. As the usage of reinforced concrete micropiles in more

© Springer Nature Switzerland AG 2018
W. Wu and H.-S. Yu (Eds.): *Proceedings of China-Europe Conference on Geotechnical Engineering*, SSGG, pp. 1504–1507, 2018.
https://doi.org/10.1007/978-3-319-97115-5_133

complicated cases increases, more explorations are implemented. Therefore, the implementation of modeling experiments in our research will focus on the characteristics and application range of micro anti-sliding pile as well as ensuring the safety of slope.

2 Slope Model

The anti-sliding characteristics of micro piles were studied mainly through scaled slope which were designed and modeled and shown in Fig. 1. The prototype came from a weak rock embankment slope of highways. The key scale factor of length of 10 was selected for the physical modeling. Other variables will obtain and applied under the laws of similitude. The scaled slope model was 2.24 m in length, 1.0 m in height and 1.2 m in width, which was made of mixture of sand, gypsum and cement to simulate weak rock. Single piles and composited piles were casted with proportioned and mixed concrete. Herein, composited piles refer to two single piles being connected on the top with a reinforced concrete beam. Modeled pile has a diameter of 0.03 m considering a prototype diameter of 0.3 m and scaling law. The main method used to simulate slope failure was to apply incremental loading to slipping mass until obvious deformation appeared. The experiments included the following model tests, namely sole slope model, slope model with two rows of single piles and slope model with composited piles used for reinforcement.

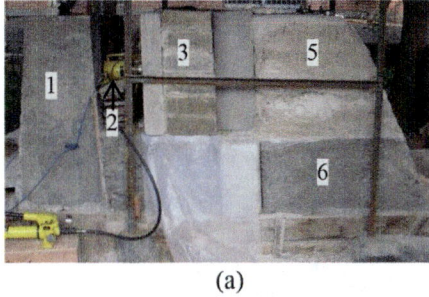 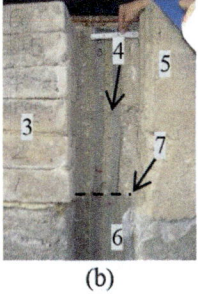

(a)   (b)

Fig. 1. Scaled slope model with micropiles (Note: 1-Reaction wall; 2-loading system; 3-load-transmitted mass; 4-micropiles; 5-slipping mass; 6-bed stratum; 7-Shearing surface)

3 Results and Discussion

3.1 Loading-Displace Curve of Model Experiments

It is obvious that the force-displacement curve under B status of all three experiments shows highly similarity. The first stage is compaction in which loading increased without any displacement changes on the top of slope. The second stage between A and B is elastic stage when loading keeps increasing and slipping mass began to move slowly. Micropiles make a big contribution to resistance compared to the force tested in sole slope. As the curve developed to C, slope appeared faster deformation rate (Fig. 2).

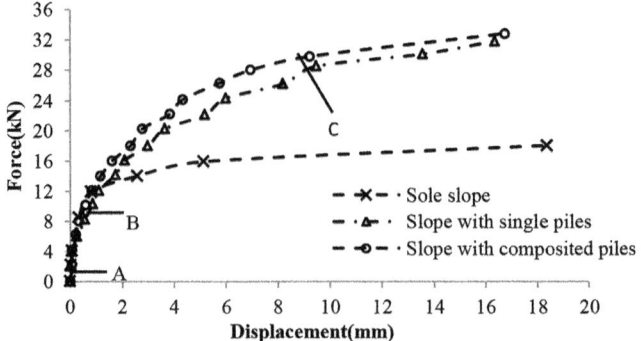

Fig. 2. Loading-displacement curves of model experiments.

3.2 Influence of Loading on Piles' Behavior

The change of length and diameter of piles, coefficient of subgrade play important roles in reinforcement effect of micropiles as well. The pile displacement is inversely proportional to the pile diameter. The smaller the pile diameter is, the larger the displacement is. On the other hand, bending stiffness increases while displacement decreases which might mean the resistance around piles surfaces weakens. Micropiles with smaller diameters seem to be highly likely to make the best of elastic characteristics of piles and soil resistance.

Considering pile diameter of 0.3 m and coefficient of subgrade of 10000 kN/m^3, the response of pile shaft to lateral loading is shown in Figs. 3 and 4.

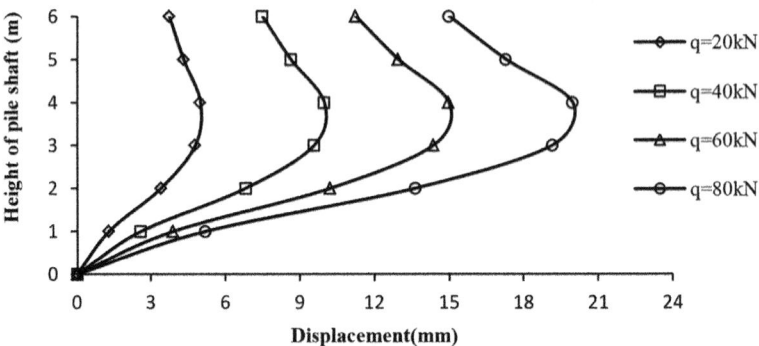

Fig. 3. Loading-displacement curves of model experiments.

Last but not the least, effect of coefficient of subgrade was analyzed under the condition that pile diameter of 300 mm, lateral load of 60 kN/m are adopted.

The result indicates displacement and bending moment are inversely proportional to the coefficient. Higher coefficient of 10000 is suggested for micropiles' application.

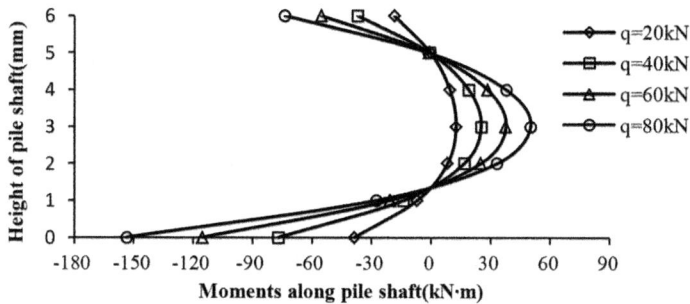

Fig. 4. Loading-displacement curves of model experiments.

4 Conclusion

Based on modeling experiments and theoretical analysis results, the applicable environment of micro anti-slide piles could be concluded as the followings: Micropiles can provide relatively bigger sliding resistance and decrease deformation rates. Besides, the reinforcement effect of composited piles performs better than single piles. It is expected that micropiles are suitable for shallow slope with slipping depth less than 10 m, or sliding force under 300 kN/m, or higher coefficient of subgrade.

References

1. Elzoghby, A., Elsaied, F.: Performance of footing with single side micro-piles adjacent to slopes. Alex. Eng. J. **53**, 903–910 (2014)
2. JAMP: Design and execution manual for seismic retrofitting of existing pile found-anon with high capacity micropiles. Foundation Engineering Research Team, Structures Research Group, Public Works Research Institute, Japan (2002)
3. Sun, S., Wang, J., Bian, X.: Design of micropiles to increase earth slopes stability. Cent. South Univ. **20**, 1361 (2013)
4. Benslimane, A.: Dynamic behavior of micropiles systems subjected to sinusoidal ground motion. Doctor Dissertation of Polytechnic University (2000)
5. Pearlman, S.L., Wolosick, J.R.: Pin piles for bridge foundations. In: 9th Annual International Bridge Conference, Pittsburgh, Pennsylvania, 15–17 June 1992, 8 p. (1992)
6. Binu, S.: A model study of micropiles subjected to lateral loading and oblique pull. Indian Geotech. J. **41**(4), 196–205 (2011)
7. Kershaw, K., Luna, R.: Full-scale field testing of micropiles in stiff clay subjected to combined axial and lateral loads. J. Geotech. Geoenviron. Eng. **140**(1), 255–261 (2014)

Influence of Wetting-Drying Cycle in Road Cut Slope in Loess in Northwest China

Yuncheng Mao[1] , Guoyu Li[1(⊠)] , Wei Ma[1] , Yanhu Mu[1] ,
and Fei Wang[1,2]

[1] State Key Laboratory of Frozen Soil Engineering,
Cold and Arid Regions Environmental and Engineering Research Institute,
Chinese Academy of Sciences, Lanzhou 730000, China
guoyuli@lzb.ac.cn
[2] University of Chinese Academy of Sciences, Beijing 100049, China

Abstract. Wetting-drying cycle is a strong weathering process which considerably changes the engineering properties of metastable loess. It has been considered in slope stability evaluation and slope surface protection. In this study, rainfall intensity, volumetric water contents and ground temperatures in loess road cut slopes were monitored at two study sites, Dingxi and Pingliang, northwest China, to determine infiltration depth of rainfall and frost depth. The monitored data indicated that water contents from the ground surface to −20 cm deep changed significantly, less changed below the depth of −40 cm. The surficial loess of road cut slope had experienced strong wetting-drying cycles in the rainy season. Based on the monitored data, the wetting-drying and direct shear tests were carried out in laboratory to estimate the influence of wetting-drying cycle on dry density and shear strength of undisturbed loess samples. The mean dry density and shear strength of Dingxi loess samples decreased from 1.60 g/cm^3 and 30.3 kPa to 1.30 g/cm^3 and 17.8 kPa after 30 cycles, respectively. Concerning the Pingliang soil samples, they decreased from 1.66 g/cm^3 and 31.7 kPa to 1.40 g/cm^3 and 15.7 kPa, respectively. Wetting–drying cycle significantly reduced the dry density and strength of loess samples and became a key factor leading to the failure of loess road cut slope.

Keywords: Wetting-drying cycle · Loess road cut slope
Volumetric water content · Frost depth · Dry density · Shear strength

1 Introduction

Slope surface peeling is one of the common problems on loess cut slopes in northwest China, which leads to serious soil erosion, and eventually affects the secure operation of highway [1]. The literatures reported that the factors resulting in slope surface peeling included the contents of clay, soluble salts, the wetting-drying and freezing-thawing cycles [2, 3]. Owing to periodic wetting-drying cycles (WDC), the water contents in loess slope surface changes, which changed the engineering properties of loess, very sensitive to water. Long-term WDC will inevitably affect the strength and deformation characteristics of loess due to changes in the structure [4]. However, field

© Springer Nature Switzerland AG 2018
W. Wu and H.-S. Yu (Eds.): *Proceedings of China-Europe Conference
on Geotechnical Engineering*, SSGG, pp. 1508–1511, 2018.
https://doi.org/10.1007/978-3-319-97115-5_134

monitored data are very limited about the water contents and ground temperatures in loess road cut slope in rainy season in arid and semiarid areas, and deterioration mechanisms of loess road slope due to WDC were not studied. In this study, the volumetric water contents and ground temperatures were observed in field to determine their relationship and the experimental conditions in laboratory. Then the wetting-drying and direct shear strength tests for undisturbed loess samples taken from road slopes were carried out to investigate the variations in dry density and shear strength due to WDC, revealing deterioration mechanisms of loess road slope in loess regions.

2   Field Monitoring

2.1   Study Site and Monitoring Method

Two automatic meteorological stations were installed at Pingliang and Dingxi, northwestern China, respectively. The ground temperatures and volumetric water contents in loess road cut slopes at different depths (−5 cm, −20 cm, −40 cm, −80 cm) were monitored and measured in 2007. The schematic figure of the road cut slope and instrumentations are shown in Fig. 1.

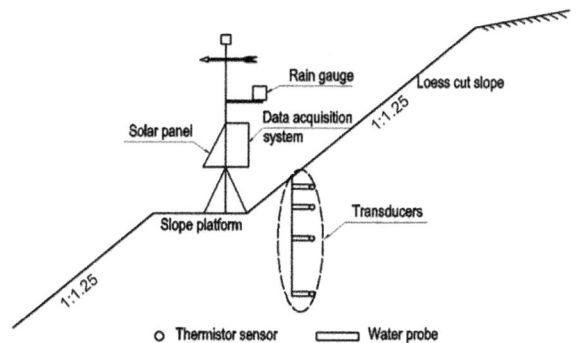

Fig. 1. Cross section of the monitored slope.

2.2   Monitored Results

The monitored results are depicted in Figs. 2 and 3. They showed that: (1) after a rainfall of 28 mm, the volumetric water content at the depths of −5 cm and −20 cm at Dingxi site increased from 15.9% and 20.5% to 31.8% and 25.7%, respectively. At Pingliang site, they increased from 12% and 13.8% to 20% and 14.9% after a rainfall of 23 mm; (2) the daily minimum ground temperatures at the depths of −5 cm and −20 cm at Dingxi site decreased by 2.9 °C and 2.0 °C after a rainfall of 28 mm, respectively. The daily maximum ones decreased by 6.6 °C and 2.8 °C, respectively. At Pingliang site, the daily minimum ground temperatures decreased by 0.9 °C and 2.3 °C at depth of −5 cm and −20 cm after a rainfall of 23 mm, respectively. The maximum ones decreased by 5.2 °C and 2.2 °C respectively; (3) During the observation period, the water contents and ground temperatures in loess road cut slope below the depth of −40 cm at two sites were less changed.

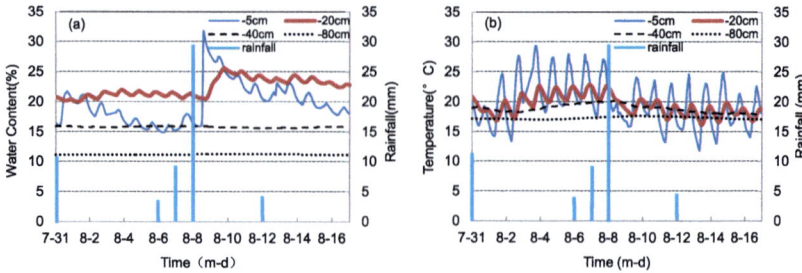

Fig. 2. Variations in volumetric water content and ground temperature versus rainfall at different depths in a road cut slope at Dingxi.

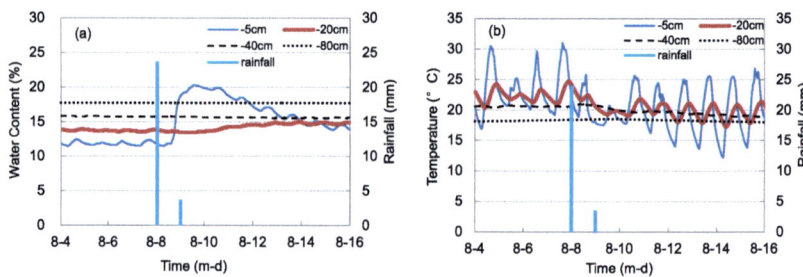

Fig. 3. Variations in volumetric water content and ground temperature versus rainfall at different depths in a road slope at Pingliang.

3   Laboratory Tests

Some laboratory experiments were carried out to study the influence of WDCs on the engineering properties of undisturbed loess samples taken from road cut slopes. The laboratory testing results are plotted in Fig. 4.

They showed that: (1) the mean dry densities of Dingxi and Pingliang samples decreased with increasing number of WDC. For example, they decreased from 1.60 g/cm³ and 1.66 g/cm³ to 1.30 g/cm³ and 1.40 g/cm³ after 30 WDCs, respectively; (2) the mean shear strengths of Dingxi and Pingliang samples also decreased with increasing number of WDC. They decreased from 30.3 kPa and 31.7 kPa to 17.8 kPa and 15.7 kPa after 30 WDCs, respectively.

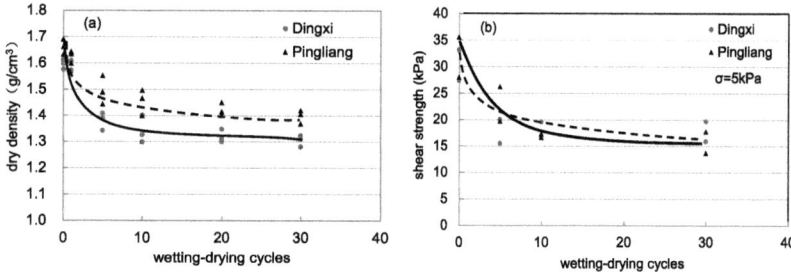

Fig. 4. Variations in dry density and shear strength with number of WDC for Dingxi and Pingliang undisturbed loess samples.

4 Conclusions

The monitored results indicated that the loess road cut slope had experienced strong WDCs in the rainy season, and the water contents from the ground surface to −20 cm deep changed significantly, but it was less changed below the depth of −40 cm at Dingxi and Pingliang sites after a rainfall of 23 mm or 28 mm. These rainfalls led to a significant decrease in the ground temperature above −20 cm. The daily minimum and maximum ground temperatures at depths of −20 cm at Dingxi site decreased by 2.0 °C and 2.8 °C, respectively. At Pingliang site, the daily minimum and maximum ground temperatures at depths of −20 cm decreased by 2.3 °C and 2.2 °C, respectively. The laboratory experimental results indicated that both dry densities and shear strengths of undisturbed loess samples at Dingxi and Pingliang sites decreased with increasing number of WDC. The mean dry densities of Dingxi and Pingliang samples decreased by 18.2% and 15.9%, and the mean shear strength decreased by 41.2% and 50.6%.

Acknowledgements. This work was funded by the National Natural Science Foundation of China (Grant Nos. 41702333, 41672310, U1703244 and 41630636), the Research and Development Program of Gansu Province (Grant No. 1207TCYA009), Research project of the State Key Laboratory Frozen Soil Engineering of CAS (Grant No. SKLFSE-ZY-16), the National Key Research and Development Program of China (Grant No. 2016YFC0802103), the Science and Technology Major Project of Gansu Province (Grant No. 143GKDA007) and STS Research Project of Chinese Academy of Sciences (Grant No. CHHS-TSS-STS-1502).

References

1. Feng, L.C., Zheng, Y.W.: China Collapsible Loess, pp. 188–189. China Railway Publishing House, Beijing (1982)
2. Fan, H.Y.: Research for Mechanism and Treatment Technique of Spalling in Loess Slope, Doctor, Chang'an University, Xi'an, China (2011)
3. Fang, P.: Study on Punishment Technology of Flake in Loess Side Slope, Master, Chang'an University, Xi'an, China (2007)
4. Yuan, Z.H., Ni, W.K., Tang, C., Hu, S.M., Guan, J.J.: Experimental study of structure strength and strength attenuation of loess under wetting-drying cycle. Rock Soil Mech. **38**(7), 1894–1902 (2017)

Numerical Investigation of Drilled Shafts Near an Embankment Slope Under Combined Torque-Lateral Load Scenario

Aigul Mussabayeva[1] ⓘ, Jong Kim[1] ⓘ, Deuckhang Lee[1] ⓘ,
Taeseo Ku[2] ⓘ, and Sung-Woo Moon[1(✉)] ⓘ

[1] Department of Civil and Environmental Engineering,
Nazarbayev University, Astana, Kazakhstan
sung.moon@nu.edu.kz
[2] National University of Singapore,
1 Engineering Dr. 2, Singapore 117576, Singapore

Abstract. Generally, cantilevered structure-foundation systems supporting highway signs, signals, and luminaires in the areas exposed to severe wind loadings (e.g., hurricane) have been designed under coupled torsion and lateral load scenario. Especially, mast arm cantilevered structures constructed near or on an embankment slope may have more concerns on the torsional and lateral resistance of the foundation. However, most research works have merely considered drilled shaft foundations under torsion-lateral load case with an embankment in proximity. In this study, a numerical study is performed with different soil layers to: (1) understand the combined torque-lateral load behavior of drilled shafts near an embankment slope; (2) examine the effect of both the torsion and lateral resistances in the proximity of an embankment slope. It was found that torsional stiffness decreases with increase in slope angles. Finally, design criteria (e.g., minimum allowable distance from the embankment, maximum allowable point load near the embankment) of the mast arm assembly and loads are provided.

Keywords: Drilled shafts · Torsion · Lateral load · Embankment

1 Introduction

It is common practice to install overhead cantilevered structures supporting signs and signals at intersections, which are typically supported by drilled shaft foundations. Currently, there is no standard methodology adopted to determine design parameters (e.g., minimum allowable distance from the embankment, maximum allowable point load near the embankment) of mast arm assembly and drilled shafts [1]. Especially, there is high necessity for hurricane prone areas to address the issue of lateral-torsional resistance of drilled shafts to avoid failures of the mast arm systems. Previous research works assumed either lateral load or torsion being applied separately, while [2–5] have proven by full scale field and/or centrifuge tests that overhead cantilevered structures should be designed for coupled torque-lateral load scenario. [6] investigated combined torque-lateral load behavior of drilled shafts near an embankment slope though

© Springer Nature Switzerland AG 2018
W. Wu and H.-S. Yu (Eds.): *Proceedings of China-Europe Conference on Geotechnical Engineering*, SSGG, pp. 1512–1515, 2018.
https://doi.org/10.1007/978-3-319-97115-5_135

numerical investigation. The study showed that both the torsion and lateral resistances are influenced by the proximity of an embankment slope. However, the model by [6] was simulated with one simple soil profile (sand only), which might be a reason for insignificant reduction in torque resistance. The objective of this study is to evaluate the influence of the embankment with multiple soil layers on the structure stability and define safe distance and point load parameters for the structure.

2 Numerical Simulation

In order to simulate a combined torque-lateral load scenario, a three dimensional (3D) model was created using PLAXIS finite element 3D code (FEM software), as shown in Fig. 1(a). The model was validated by the full-scale field test results [5]. Table 1 illustrates the parameters for Mohr-Coulomb model to represent soil behavior. Lateral load and torsion have been simulated by the application of the concentrated load (89 kN) at an eccentric distance of 10.67 m from the pole. Figure 1(b) compares the simulated torque-rotation response with the field measurement. The result of numerical simulation is in good agreement with the field test result, which shows applicability of the model. The maximum range of the rotation was chosen based on the failure occurrence of the field test.

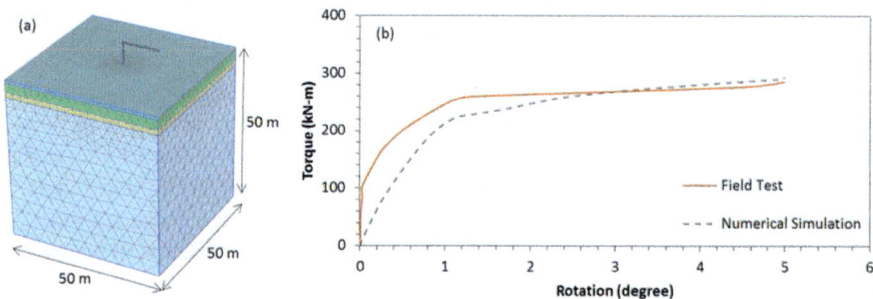

Fig. 1. (a) 3D model of the validation test, and (b) comparison of torque-rotation responses.

Table 1. Properties of materials and soils at the test site

	Drilled shaft	Mast arm	Clay	Sand with silt		
				1	2	3
Diameter (m) × Length (m)	1.22 × 5.49	–	–	–	–	–
Pole height (m) × Arm length (m)	–	6.1 × 9.1	–	–	–	–
Depth (m)	–	–	0.75	3.8	5.5	7.6
Unit weight, γ (kN/m^3)	25	75	17.9	18.1	19.0	19.6
Undrained shear strength, c_u (kPa)	–	–	–	–	–	–
Friction angle, φ (°)	–	–	–	31	34	36
Modulus of elasticity (kPa)	1.7×10^7	2.0×10^8	–	–	–	–

Figure 2(a) presents a schematic view of the mast arm drilled shaft near an embankment slope for numerical simulation. Based on the validated model, 3D model was developed, as shown in Fig. 2(b). The slope angles (θ) are varied between 0° and 33.7°, while the typical point load of 50 kN according to [6] is applied at an eccentric distance.

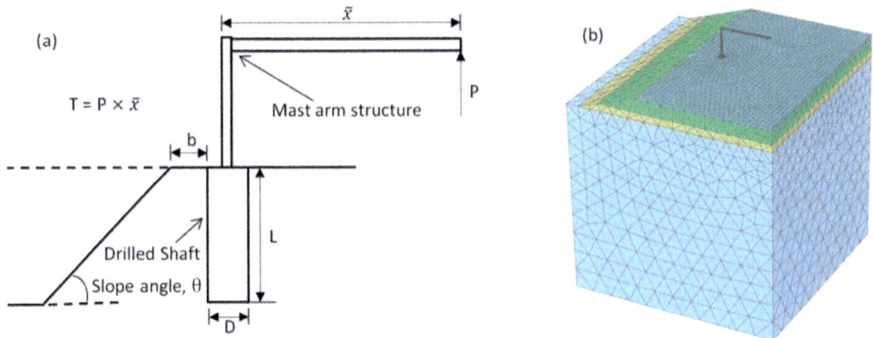

Fig. 2. (a) A schematic profile of the embankment [5], and (b) 3D model.

3   Results and Discussion

Based on the numerical modelling, the following results were obtained. Stability analysis for mast arm assembly is conducted by varying the distance and point load (Table 2). Maximum allowable loads are determined where the structure is exactly near the embankment slope for different slope angles, respectively. The results lead to conclusion that the tolerable point load increases with decrease in the slope angle. In addition, slope angle influences the stability of the structure at the distance from 12 to 18 m. Minimum allowable distance from the embankment decreases with decrease in the slope angle.

Table 2. Allowable loads and distances corresponding to different slope angles

Slope angle, θ° (slope)	Maximum allowable P (kN) at b = 0	Minimum allowable b (m) corresponding to P = 50 kN
33.7° (1:1.5)	9	18
26.6° (1:2)	13	14
21.8° (1:2.5)	15	13
18.4° (1:3)	17	12

Note, P = applied load, b = the closest distance from embankment to drilled shaft

The maximum allowable b for all cases was used to understand the variation of torque with rotation. Figure 3 shows the comparison of torsional resistance of the embankment cases and flat ground case with typical load of 50 kN. The torsional

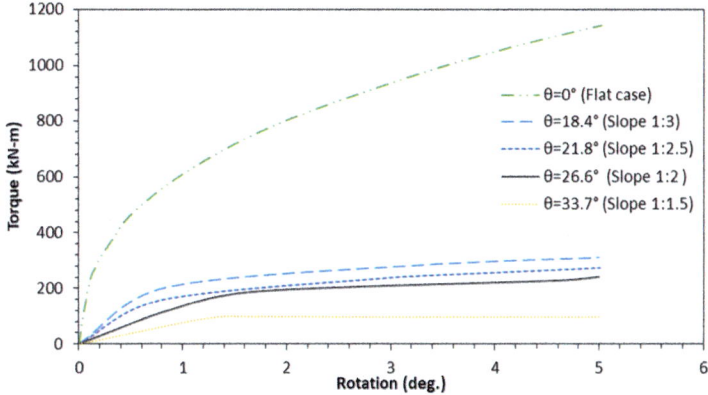

Fig. 3. Variation of torque with rotation.

resistance corresponding to 5° of rotation was used consistently for the comparison in Fig. 3. It is found that torsional resistance decreases with an increase in slope angle.

4 Summary

In conclusion, the developed model was successfully validated and used for the investigation of combined torque-lateral load behavior of drilled shafts near an embankment slope. Parameters such as allowable distance of the structure from the embankment and allowable point load of the structure when it is near the embankment were determined. The results showed that reduction in torsional resistance is influenced by different slope angles of the embankment. In the future study, the reduction ratio should be examined and compared to the case of homogeneous soil. Moreover, more cases with varied drilled shaft parameters (diameter, length) should be studied.

References

1. Li, Q., Asce, S.M., Stuedlein, A.W., Asce, M., Barbosa, A.R., Asce, A.M.: Torsional load transfer of drilled shaft foundations. J. Geotech. Geoenviron. Eng. **143**(8), 1–13 (2017)
2. Hu, Z., McVay, M., Bloomquist, D., Herrera, R., Lai, P.: Influence of torque on lateral capacity of drilled shafts in sands. J. Geotech. Geoenviron. Eng. **132**(4), 456–464 (2006)
3. Hu, Z.: Determining the optimum depth of drilled shafts subject to combined torsion and lateral loads in saturated sand from centrifuge testing (2003)
4. Thiyyakkandi, S., McVay, M., Lai, P., Herrera, R.: Full-scale coupled torsion and lateral response of mast arm drilled shaft foundations. Can. Geotech. J. **53**(2), 1928–1938 (2016)
5. Irsainova, A., Mussabayeva, A., Kkand, S.T., Kim, J.: Numerical analysis of torsionally loaded drilled shafts near an embankment slope in cohesionless soils. J. Eng. Appl. Sci. **12**(5), 3834–3838 (2017)
6. AASHTO: Standard Specifications for Structural Supports for Highway Signs, Luminaries and Traffic Signals, 4th edn. (2001)

Evaluation of Residual Shear Strength of Landslide Reactivated Soil

V. Senthilkumar and S. S. Chandrasekaran[✉]

Department of Structural and Geotechnical Engineering, School of Civil Engineering, Vellore Institute of Technology, Vellore, Tamil Nadu, India
chandrasekaran.ss@vit.ac.in

Abstract. Reactivated landslides are generally associated with residual shear strength. Residual shear strength mobilized in such landslides are important parameters in slope stability and landslide simulation analysis. The present study involves estimation of residual shear strength of landslide reactivated soil at Marappalam area of Nilgiris district, Tamil Nadu state, India where landslides being reactivated frequently due to rainfall infiltration. A series of torsional ring shear tests performed under different normal loads to study the shear behaviour and estimate the residual shear strength parameters of soil. Since the residual soil at Marappalam area formed due to weathering, an effort made to study the weathering effects along with clay fractions and clay minerals on residual shear strength of soil. The test conducted on soil samples collected from various depths which is having different clay fractions with varying plasticity characteristics and fines content. The mineralogical and micro fabric analysis shows the presence of kaolinite clay mineral formed due to weathering. As the depth increases, the gradation of soil increases that decreases the clay content and degree of weathering. Hence, it is observed that the residual shear strength of the soil decreases with increase in clay content in presence of kaolinite clay mineral.

Keywords: Residual shear strength · Clay minerals · Weathering
Landslides

1 Introduction

Nilgiris is a district of Tamil Nadu state in southern India. The district is a part of Western Ghats mountain ranges located at the tri-junction of Tamil Nadu, Karnataka and Kerala states. The study area Marappalam is a place situated in Nilgiris district at latitude of $11°\ 20'\ 12.80''$ N and longitude of $76°\ 49'\ 25.28''$ E. Landslide is one of the major natural hazards occur frequently in Nilgiris district. Marappalam location experienced several major landslides in the past. Landslide occurred in the year 1993 and 2009 at Marappalam area were among the major landslides in Nilgiris landslide history [2, 4]. The rainfall induced landslide made significant damages to infrastructures and environment at Marappalam location [2]. As the Marappalam slope has gentle topography with thick soil cover and past history of landslide occurrence and receiving heavy rainfall in all monsoon seasons, landslide reactivation being a major problem. The aim of the present study is to evaluate the residual shear strength of landslide

© Springer Nature Switzerland AG 2018
W. Wu and H.-S. Yu (Eds.): *Proceedings of China-Europe Conference on Geotechnical Engineering*, SSGG, pp. 1516–1520, 2018.
https://doi.org/10.1007/978-3-319-97115-5_136

reactivated soil at Marappalam area of Nilgiris district and study the effect of weathering and influence of clay content and clay minerals on residual shear strength of soil.

2 Sample Collection and Laboratory Investigation

Geotechnical characterization of soil and rock at Marappalam location has been carried out by authors and reported in Senthilkumar et al. [4]. The study consists of detailed borehole investigation, geophysical investigation and numerical simulation of landslides. Four boreholes were drilled at Marappalam 2009 landslide location. The samples were collected using split spoon sampler by conducting standard penetration test (SPT) at regular intervals. For the present study, the representative soil samples collected from a borehole at three different depths (0.30 m, 3.00 m and 6.00) have been used for evaluation of residual shear strength. Detailed laboratory tests were performed on soil samples collected from borehole based on relevant ASTM standards. Scanning Electron Microscopic (SEM) and Energy Dispersive X-ray (EDX) diffraction analysis were performed to study the micro fabric and mineralogical characteristics of soil. A series of ring shear tests performed with an air-dried representative soil sample passing through U.S. Standard No. 200 sieve [1]. The water content equal to liquid limit was added to soil sample and batch-mixed to ensure uniformity and minimize the entrapped air [1]. The processed soil sample was placed in the sample container and the samples of each depth were consolidated with three normal stresses (98 kPa, 147 kPa, 196 kPa) until the sample reaches complete consolidation. The samples were then sheared under a lower shearing speed of 0.1 mm/min to avoid pore water pressure generation during shearing.

3 Results and Discussion

3.1 Sample Characterization

The soil samples collected from borehole at three different depths have been characterized based on laboratory investigations. The grain size distribution curve is shown in Fig. 1. As shown in figure, the soil at 0.30 m depth having 81% of fines content which includes clay content of 18%. Figure also shows that the percentage of fines and clay content decreases as the depth increases. According to plasticity characteristics, the plasticity of soil decreases with depth due to decrease in clay content. The soil at 0.30 m depth having high plasticity and it reduced to low plastic at 3.00 m and 6.00 m depth. Based on the laboratory investigation results the soil at 0.30, 3.00 and 6.00 m depths are classified as Fat clay with sand (CH), Sandy lean clay (CL) and Sandy Silt (ML) respectively. Figure 2 shows the SEM images of soil at 0.30 m, 6.00 m depths and EDX spectrum of soil at 0.30 m depth respectively. The SEM image shown in Fig. 2a depicts the presence of kaolinite clay mineral with highly porous structure at 0.30 m depth. Decrease in the pore volume and kaolinite clay mineral proportion with depth is clearly visible from Fig. 2b. The presence of kaolinite clay mineral is confirmed by EDX spectrum shown in Fig. 2c.

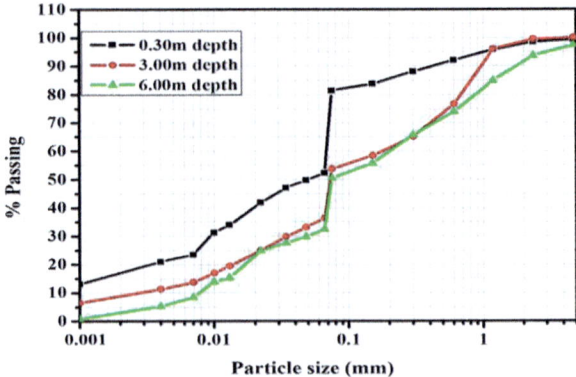

Fig. 1. Grain size distribution curve

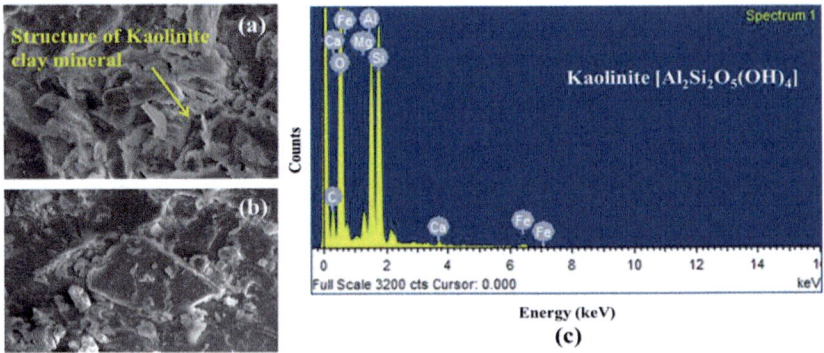

Fig. 2. SEM & EDX results, (a) & (b) SEM image at 0.3 and 6.00 m (c) EDX spectrum at 0.3 m

3.2   Residual Shear Strength

As the samples were sheared under three different normal stresses, the shear stress and shear displacement plot and failure envelop plot under each normal stress for three types of soil are shown in Fig. 3. As shown in Fig. 3 the residual shear resistance of fat clay with sand under the normal stresses of 98, 147 and 196 kPa are 50, 69 and 91 kPa respectively (Fig. 3a). Accordingly the residual shear resistance of sandy lean clay and sandy silt under three normal stresses are 58, 79, 97 kPa and 59, 80, 100 kPa respectively (Fig. 3c and e). The residual friction angle has been estimated from failure envelope plotted between normal stress and shear stress. The residual friction angle (φ_r) of fat clay with sand, Sandy lean clay and Sandy silt are 25.7°, 27.6° and 28.1° respectively (Fig. 3b, d and f).

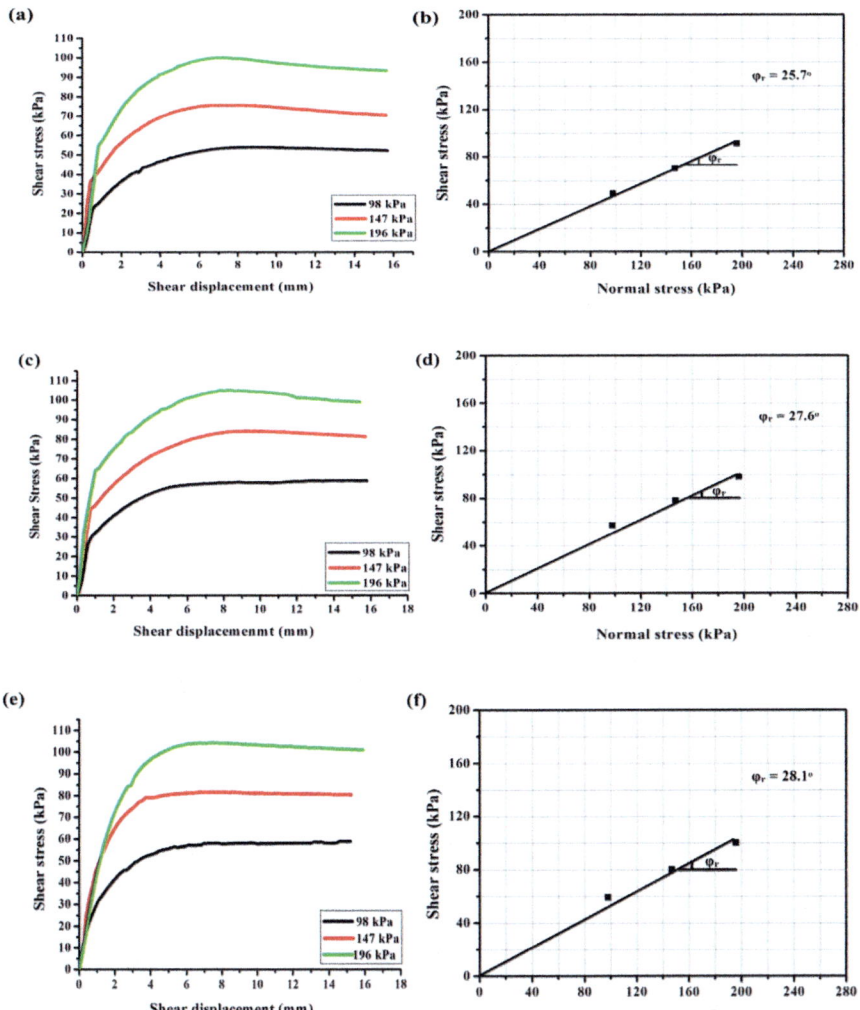

Fig. 3. Shear stress (vs) shear displacement and failure envelope plots of soil at, (a & b) 0.30 m depth, (c & d) 3.00 m depth, (e & f) 6.00 m depth

3.3   Effect of Weathering, Clay Content and Clay Minerals on Residual Shear Strength

According to Little [3] six grade weathering classification, the weathering degree of Marappalam slope has been identified based on field observations, laboratory investigations and micro fabric and mineralogical analysis [4]. The upper portion of slope is totally weathered and transformed to Grade VI residual soil (Fat clay with sand) upto 2.00 m depth followed by completely weathered sandy lean clay layer (Grade V) upto 4.50 m depth [3, 4]. A highly weathered rock layer (Grade IV) presents upto 9.00 m depth at which more than 80% of rock disintegrated to soil. The soil of this layer contains sandy silt type soil [4]. The weathering profile shows that the weathering degree decreases as depth increases which attributes to decrease in clay content and clay mineral proportions with depth (Figs. 1 and 2). Hence, the residual shear strength of fat clay with sand is less than that of sandy lean clay and sandy silt due to presence of higher fines and clay content with high proportion of kaolinite clay mineral formed due to high degree of weathering. Though the proportions of clay content and kaolinite clay minerals varies between sandy lean clay and sandy silt, the residual shear strength between two soils varies marginally due to less variation on grain size distribution and fines content (Fig. 2).

4   Conclusions

Based on the results and preceding discussions from the present study, following conclusions are drawn.

- The soil at 0.30 m, 3.00 m and 6.00 m depths were classified as fat clay with sand, sandy lean clay and sandy silt respectively. The soil at top surface having higher fines content due to high degree of weathering. The fines content decrease with depth as the weathering degree decreases.
- The top soil layer contains higher clay content and kaolinite is the dominant clay mineral formed due to high degree of weathering. The clay content and proportions of kaolinite clay mineral decreases due to decrease in fines content and degree of weathering
- As the residual shear strength is the function of clay content and clay minerals, the shear strength parameters and residual shear resistance value decreases due to increase in clay content and proportions of kaolinite clay mineral with respect to degree of weathering.

References

1. ASTM, D 6467: Standard Test Method for Torsional Ring Shear Test to Determine Drained Residual Shear Strength of Cohesive Soils, West Conshohocken, PA (2013)
2. Chandrasekaran, S.S., Sayed Owaise, R., Ashwin, S., Jain, R.M., Prasanth, S., Venugopalan, R.B.: Investigation on infrastructural damages by rainfall-induced landslides during November 2009 in Nilgiris India. Nat. Hazards 65(3), 1535–1557 (2013)
3. Little, A.L.: The engineering classification of residual soils. In: Seventh International Conference on Soil Mechanics and Foundation Engineering, pp. 1–10. ISSMFE, Mexico (1969)
4. Senthilkumar, V., Chandrasekaran, S.S., Maji, V.B.: Geotechnical characterization and analysis of rainfall-induced 2009 landslide at Marappalam area of Nilgiris district, Tamil Nadu, India. Landslides 14, 1803–1814 (2017)

Investigation of Rockfall Impact Against Gravel Cushion via a Discrete Element Approach

Weigang Shen, Tao Zhao[(✉)], Feng Dai, Jiawen Zhou, and Nuwen Xu

State Key Laboratory of Hydraulics and Mountain River Engineering,
College of Water Resource and Hydropower,
Sichuan University, Chengdu 610065, China
zhaotao@scu.edu.cn

Abstract. This study investigated the impact of rockfall impact against a granular cushion layer via 3-D discrete element method (DEM). In the numerical model, the rock boulder was modeled as a single sphere with an incident velocity, and the granular layer was modeled as an assembly of poly-dispersed spherical particles. The numerical model was calibrated and validated by comparing the numerical results with experimental data reported in the literature. It was further used to investigate the effect of rock boulder mass on the maximum impact force, impact duration and penetration depth. The obtained numerical results are in good agreement with available experimental observations.

Keywords: Rockfall · Gravel cushion · Discrete element method
Impact force · Impact duration · Penetration depth

1 Introduction

Concrete rock sheds are widely used to protect highways along steep slopes from rockfall or rock avalanches in mountainous areas. This structure is generally composed of a reinforced concrete roof slab and a cushion layer. The cushion layers play a crucial role in absorbing energy and reducing the maximum impulsive forces applied on the rock shed, and they commonly consist of gravels or soils. The design of such protection structure requires a proper estimation of the dynamic forces resulting from the impact and the penetration depths of rock boulders into cushion layer. Up to now, several empirical methods have been developed to determine the impact force in engineering practice, such as the Japanese and Swiss design codes [1, 2]. Even though these methods are very simple to use from a practical point of view, none of them have been acknowledged as an universal formula, notably because these methods were obtained in specific impact and boundary conditions [3]. Thus, the impacting of rockfall against granular cushion covering rock shed is always of concern.

The macro and micro mechanical response of impacted granular layer can be addressed by the discrete element method (DEM). When carefully calibrate on the basis of some well-established experimental data, numerical modeling of the impacts allows researchers to analyze some quantities which can be hardly obtained from experiments.

© Springer Nature Switzerland AG 2018
W. Wu and H.-S. Yu (Eds.): *Proceedings of China-Europe Conference
on Geotechnical Engineering*, SSGG, pp. 1521–1525, 2018.
https://doi.org/10.1007/978-3-319-97115-5_137

Roethlin et al. [4] used a three-dimensional DEM to study the distribution of stresses on the concrete slab. The results indicate that the stresses are determined mainly by the shape and size of rock boulder. Calvetti et al. [5] studied the impact of a boulder on a shelter covered by a cushion layer of granular soils using the commercial three-dimensional DEM software PFC3D. The kinetic energy of falling boulder was set up to 5000 kJ. Such high impact energy is generally not considered in experimental tests in order to avoid the risk of damaging the measurement devices (e.g., loading cells).

In the present study, a numerical model of rockfall impact against a granular cushion layer is established using the open source DEM code ESyS-Particle. The objective is to investigate the response of granular cushion under various rock boulder masses.

2 Numerical Model Configuration

The model configurations (see Fig. 1) in this study are similar with the experimental and numerical model commonly used in the literature [3, 6]. In our model, the concrete slab is represented by a layer of fixed particles with radii of 0.1 m. The granular soil is modeled as an assembly of poly-dispersed rigid spherical particles with the same radii of 0.2 m. The boundaries of granular soil are defined by four lateral walls and the fixed particle layer. The dimensions of granular cushion layer are 20 m in length and 20 m in width, respectively. The falling rock boulder is modeled as a rigid sphere. The radius of rock boulder (R_b) is a varying parameter whose effect on the response of impacted granular layer is presented in Sect. 3. During simulations, gravitational forces are applied to all soil particles and a calculation procedure is run until the total granular kinetic energy reduces to almost zero. After gravity deposition, the rock boulder is positioned just above the center surface of the soil layer with an initial vertical velocity (V_0) which is computed as a function of the falling height H ($V_0 = \sqrt{2gH}$). The input parameters of the DEM are listed in Table 1.

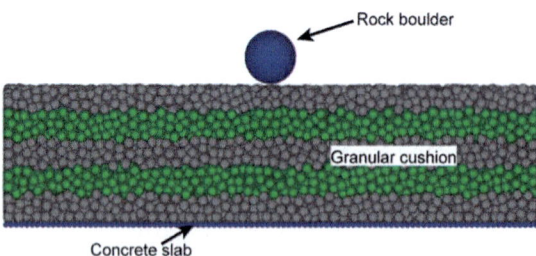

Fig. 1. Numerical model configuration. The rock boulder is modeled as a spherical particle, and the cushion layer is modeled as an assembly of spherical particles.

Table 1. Input parameters used in the simulations

DEM parameters	Value	DEM parameters	Value
Soil particle radius, R_m (m)	0.2	Viscous damping coefficient, β	0.05
Slab particle radius (m)	0.1	Particle friction coefficient, μ_1	0.577
Particle density, ρ (kg/m^3)	2650	Gravitational acceleration, g (m/s^2)	9.81
Particle Young's modulus, E (MPa)	1×10^1	Time step size, Δt (s)	1×10^{-6}
Particle Poisson's ratio, υ	0.25		

3 Results

In order to calibrate the numerical model, experimental results of Calvetti et al. [5] were compared with the results of simulations under similar conditions. In the experimental study, the cushion layer was composed of coarse granular soils placed in a reinforced concrete pit. The rock boulder is represented by an 850 kg concrete sphere with diameter of 0.9 m. The set of falling height (H) selected is 18.45 m, 13.7 m and 10.0 m. The experimental results for H = 18.45 m is reproduced to calibrate the particle Young's modulus. The remaining tests of H = 13.7 m and H = 10.0 m were employed to validate the numerical model. The comparison between the experimental results and the numerical results is shown in Fig. 2. With regard to the time histories of impact force ($F_{boulder}$), the model is able to reproduce the main qualitative features if the experimental results. In general, after t_0 = 0.0 s, impact forces firstly increase to the maximum values, and then decrease to zero at t_b. From a quantitative point of view, impact forces are slightly overestimated. Defining the impact duration (T_{imp}) as the time difference between t_b and t_0, it is clear that the model can reproduce the impact duration. All in all, the performance of the model is satisfactory.

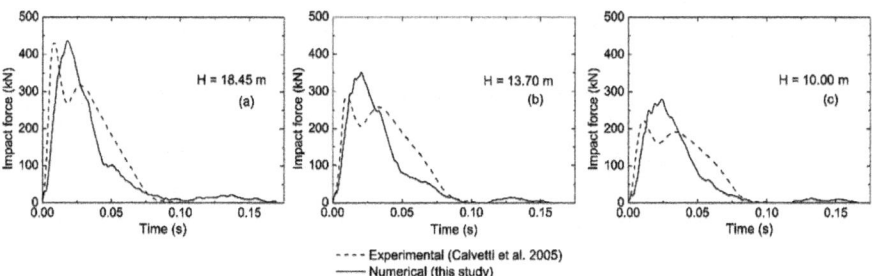

Fig. 2. Time histories of the impact force for the tests of (a) H = 18.45 m, (b) H = 13.7 m and (c) H = 10.0 m.

Through experimental tests and dimensional analysis, Pichler et al. [6] related the maximum impact force, impact duration and penetration depth to the boulder mass, falling height and soil indentation resistance. The numerical results under conditions of various boulder masses are compared with results from the literature [6] (see Fig. 3). The falling heights (H) of the boulders in all cases are 20 m. In terms of the impact duration and penetration depth, the DEM results are close to the results reported in Pichler et al. [6]. This is because the particle friction was conserved during gravity deposition, which led to a loose soil layer. In terms of the peak value of $F_{boulder}$, the DEM results are slightly greater than the results of Pichler et al. [6] corresponding to the 5% quantile of strength. This is possibly because a spherical boulder is used in the DEM simulations instead of cubic boulders. Some studies in the literature indicate that compared to the spherical boulders, angular boulders have lower peak values of impact force. This might be due to the fact that, on the same granular layer, spherical boulders encounter greater resistance than angular boulders.

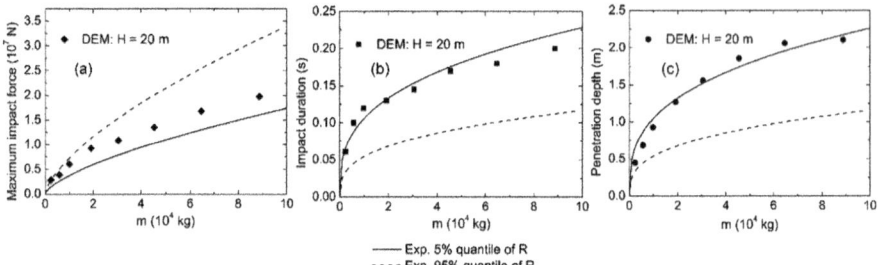

Fig. 3. Comparison between the DEM results and the results from Pichler et al. (2005): evolution of (a) maximum impact force on the boulder, (b) impact duration of the boulder and (c) penetration depth in the soil layer depending on the boulder mass.

4   Conclusions

A numerical model of rockfall impact against a granular cushion layer is developed in this study using the open-source code ESyS-Particle. It was calibrated by comparing the numerical results with the available experimental data. The comparison shows that the numerical model can qualitatively reproduce the impact response of rockfall, even though the maximum impact force is overestimated. In addition, the dependence of maximum impact force, impact duration and penetration depth on the rock boulder mass is agreement with the theoretical results from Pichler et al. (2005). The agreement confirms that the numerical model can be used to investigate the impact of rockfall against granular bedding layers.

References

1. ASTRA: Actions de chutes de pierres sur les galeries de protection, Office fédéral des routes OFROU (2008)
2. Japan Road Association: Manual for anti-impact structures against falling rocks, Japan (2000)
3. Zhang, L.R.: Local field modeling of interaction between a soil body and a falling boulder. Université Grenoble Alpes (2015)

4. Roethlin, C., Calvetti, F., Yamaguchi, S., Vogel, T.: Numerical simulation of rockfall impact on a rigid reinforced concrete slab with a cushion layer. In: Fourth International Workshop on Performance, Protection and Strengthening of Structures, Mysore, India (2013)
5. Calvetti, F., di Prisco, C., Vecchiotti, M.: Experimental and numerical study of rock-fall impacts on granular soils. Rivista Italiana di Geotecnica **4**, 95–109 (2005)
6. Pichler, B., Hellmich, C., Mang, H.A.: Impact of rocks onto gravel design and evaluation of experiments. Int. J. Impact Eng. **31**(5), 559–578 (2005)

Impact of Precipitation on Dissipation of Pore Pressure in Colluvium of the Carpathian Flysch Landslide

J. Stanisz$^{(\boxtimes)}$ (iD), P. Krokoszyński, and R. Kaczmarczyk

AGH University of Science and Technology,
Mickiewicza 30 av., 30059 Cracow, Poland
jstanisz@agh.edu.pl

Abstract. Measurement of pore water pressure is helpful to locate the position of the zone of weakness, and its changes may indicate the formation of slip surface at hazard of mass movements. It is shown in the paper, based on some studies, that the pore water pressure measurements have helped to determine the area where the increase of deviator stress and this way the effective stress reduction and the development of the slip surface. In the process of destroying the geological structure of the center the apparent increase of pore pressure follows, and then it declines shortly after the end of the process. The changes are particularly evident after periods of heavy rainfall and are related to the permeability of ground. The paper discusses the results of pore pressure tests in dry and wet conditions.

Keywords: Pore pressure · Dissipation test · Landslide

1 Introduction

The measurement of pore pressure is important in determining the state of stress and strength parameters of geotechnical layers for slope stability analysis. On precipitation the stress state changes together with the pore pressure. This condition leads to the destruction of the solid skeleton leading to the growth of the deviator. The suction pressure decrease. Filtration and its cause which is the increase in pore pressure affect the reduction of effective stress. This condition causes the development of volumetric and plastic deformations. This process intensifies in a cracked or disturbed medium. The permeability of the layer and the rate of dissipation of the pore pressure influence the deformation development. One of the commonly used methods for its determination is CPTU static sounding.

© Springer Nature Switzerland AG 2018
W. Wu and H.-S. Yu (Eds.): *Proceedings of China-Europe Conference on Geotechnical Engineering*, SSGG, pp. 1526–1529, 2018.
https://doi.org/10.1007/978-3-319-97115-5_138

2   Research Area and Methodology

Research work was carried out on a part of the landslide, located in Tęgoborze-Just about 100 km to the south from Cracow (Little Poland) (Fig. 1A). From the geographical point of view, the research area is located on the border of two mesoregions: the Beskid Wyspowy Range and Rożnowskie Foothills, which are the part of Outer Western Carpathians. Colluvial deposits cover residual soil and discordant flysch bedrock - Magura sandstones, shale and shale-sandstone complexes of Sub-Magura

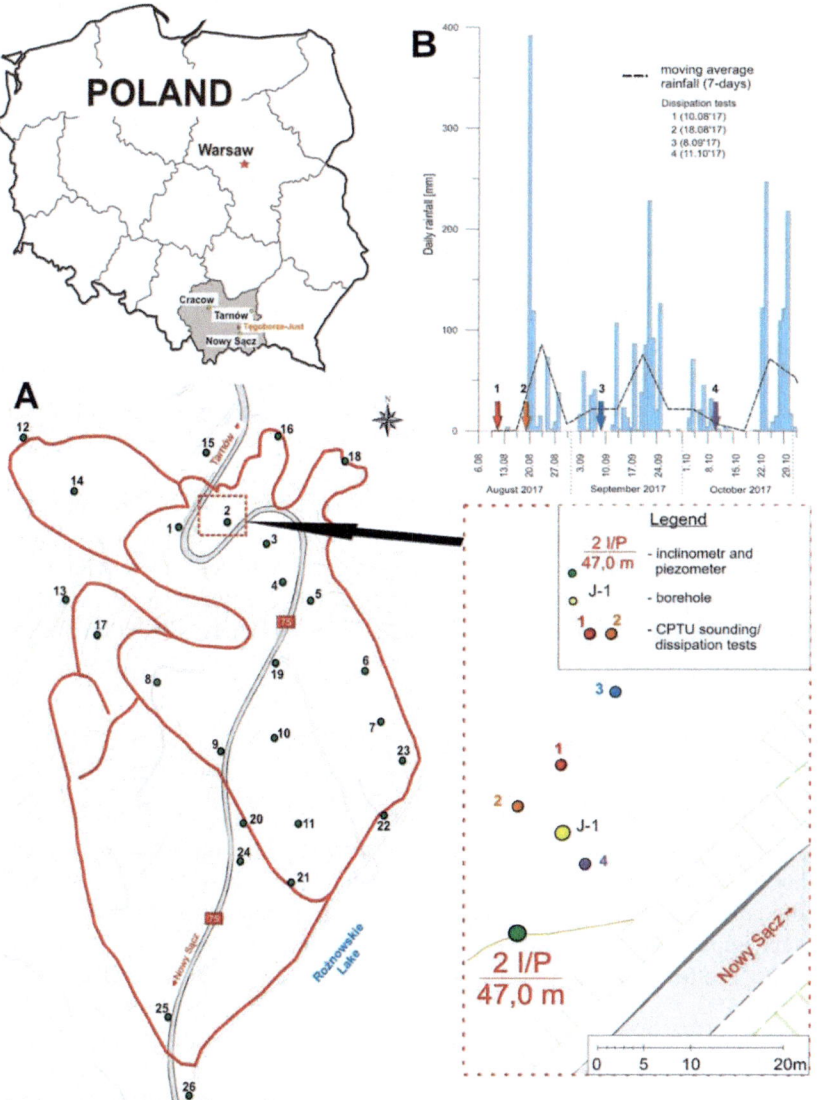

Fig. 1. Documentation map (A) with precipitation graph (B)

and hieroglyphic beds (Tertiary) and deluvial material (clayey silt). The azimuth of its movement is 130°, and its average slope angle is 11.8°. Landslide slopes are of convex-concave type and the terrain relief is strongly diversified. There are steep slopes and flattenings on the study area. Slopes are cut by small coomb valleys, dells and gullies. Elevation varies from about 430 m a.s.l. to about 267,5 m a.s.l. by the Rożnowskie lake. The landslide covered the area of 28 ha and is classified as periodically active and locally continuously active.

The aim of the research work is to determine the rate of dissipation of the pore pressure of the colluvial material in dry and wet conditions. The results of field and laboratory works will be used to calibrate the numerical model of landslide.

2.1 Piezocone Dissipation Tests

The research was conducted from August to October 2017. Geotechnical recognition (chart a map, borehole drilling) was carried out. Samples for laboratory tests were taken from the borehole. The physical and mechanical parameters were measured, including the filtration coefficient. In the geotechnical profile clayey silt (clSi) were distinguished in plastic consistency condition and shale-sandstone (grsaCl). The holes were recognized to outflow water at depths of 0.8; 1.5; 2.3 and 2.7 m.

Dissipation of pore pressure were carried out using a CPTU NOVA System. The tests were performed using a 10 cm^2 electronic cone with interchangeable tip. The probe was performed at the standard 2 $cm \cdot s^{-1}$ rate. A separate filter made of sintered brass was used for each sounding. Every filter was factory-vented and filled with glycerine. Pore pressure (u_2) measurements were obtained for the filter placed after the cone. Each of the dissipation lasted about 2–3 h.

Measurements were carried out in dry conditions (no. 1 and 2) and in wet conditions (no. 3 and 4). Measurements 1 and 2 were held on 10 and 18 August. During this period, there were no precipitation. Measurement 3 was performed on 8 September after precipitation of 35 mm (6.09) and 41 mm (7.09). Measurement 4 was made on 11 October during a slight precipitation (2 mm), 10.10 was a day without no precipitation (Fig. 1B).

3   Analysis of Results

Movement of the probe cone in the soil causes a suction pressure. This is seen as the negative pressure and is associated with the tension of the medium. According to the authors, the negative pressure can lead to the colmatization of the filter - especially in plastic soils and containing clay particles (e.g. clSi). As a consequence, it has a significant impact on the measurement results. During the study, an increase in pore pressure to the pressure specific to measured layer or hydrostatic pressure was observed.

The time of pressure stabilization was varied. For relatively homogeneous layers (clSi) the rate of pressure dissipation is similar (Figs. 2A and C). The increase in pore pressure takes place in the range of 40–80 kPa. In residual shale-sandstone complex, the gradient of the dissipation curve varies and depends on the size of the particles and grains (Figs. 2B and D). In the precipitation, there is an increase in pore water pressure (75–160 kPa) and an increase in dissipation rate. The dissipation rate is affected by soil and water conditions, filtration coefficient, grain size and pore filter silting-up.

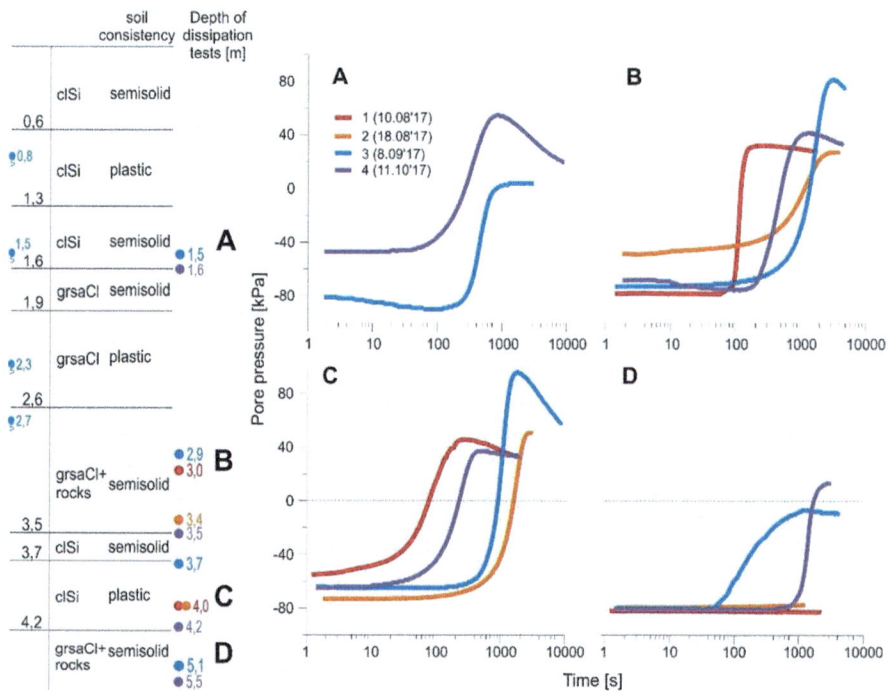

Fig. 2. Dissipation of pore pressure for layers of clayey silt (A and C) and residual shale-sandstone complex (B and D)

Impact of Root System on Soil Strength in Shallow Landslide

J. Stanisz[1]([⊠]) [ID], Ł. Kaczmarek[2] [ID], T. Zydroń[3], A. Gruchot[3] [ID],
T. Wejrzanowski[2], and P. Popielski[2] [ID]

[1] AGH University of Science and Technology,
Mickiewicza 30 av., 30059 Cracow, Poland
jstanisz@agh.edu.pl
[2] Warsaw University of Technology,
Politechnika 1 sq., 00661 Warsaw, Poland
[3] University of Agriculture in Kraków,
Mickiewicza 21/24 av., 33332 Cracow, Poland

Abstract. The main aim of this study was determination of tree root architecture and its influence on soil reinforcement on the case of shallow landslide activated in May of 2010 in Polish Flysch Carpathians. The research was divided into few stages: recognition of geotechnical conditions of analyzed site, determination of root area ratio (RAR) of tress using trench wall method, laboratory tensile tests of root samples, shear strength tests of non-reinforced and root reinforced samples, characterization of soil microstructure by high resolution X-ray computed microtomography (mXCT). The site investigations revealed surficial soil layer, which is composed of silt and silt-loam soil. RAR measurements showed the significant decrease of root number with the depth. Maximum value of RAR was equal to 0.25%. The direct shear apparatuses were used for shear strength tests of rooted samples. The roots cause slight increase of internal friction angle and cohesion values. The mXCT method provide evaluation of the density and orientation of the roots in the sample. This data enabled for more accurate assessment of the root effect on the development of shear stress within soil.

Keywords: Root system · Shallow landslide
X-ray computed microtomography (mXCT)

1 Introduction

The stability of forested and Bushed slopes is an important issue related to soil characteristics and its evolution caused by vegetation. Tree roots may perform a strengthening or weakening (e.g. influence wind; [2]) function within geological medium. Understanding of this effect is of key importance for shaping properties of landslide by sensible planting of trees with specific root system. Thus, the main aim of this study was determination of tree root architecture and its influence on soil reinforcement by using different characterization techniques.

© Springer Nature Switzerland AG 2018
W. Wu and H.-S. Yu (Eds.): *Proceedings of China-Europe Conference on Geotechnical Engineering*, SSGG, pp. 1530–1534, 2018.
https://doi.org/10.1007/978-3-319-97115-5_139

The strength of the soil is mainly affected by the dimension and type of the roots system as well as their thickness and number. The most effective solution is the presence of several species. Thick and medium roots limit the development of the failure process. Type of roots depends on the location of the potential slip surface in shallow landslides.

1.1   Characteristic of the Research Area and Methodology

Within these studies the field recognition work were carried out on a landslide located in Winiary, at a distance of 30 km southwards form Cracow (southern Poland, Little Poland province). This geographical area is part of the Wieliczka Foothills. In terms of geology, the area is built of silt, silty-clay and cretaceous sandstone-shale complex. The landslide area is mainly covered by common birch, hornbeam, black locust, elm. Landslide movement was initiated on May 2010, what cause taking off silty material. The research plan consisted of the following stages:

- I stage: drilling boreholes and recognition of soil-water conditions,
- II stage: execution of research excavations and input soil samples with an intact structure — without and with a root system. In the wall of excavations, were profiled along with the roots' intake for research,
- III stage (laboratory tests): determination of physical and mechanical properties of soil with roots,
- IV stage: characterization of soil microstructure by high resolution X-ray computed microtomography (mXCT).

2   Analysis of Results

2.1   Physical and Mechanical Soil Properties

Relevant techniques enabled determination of physical parameters of soils, such as: natural moisture, bulk density, Atterberg limits (Casagrande cup) and granulometric composition (aerometer test). Mechanical parameters (i.e. angle of internal friction and cohesion) were calculated based on the direct shear test. Samples were sheared for loads 50,100,150 and 200 kPa.

The silty samples had average natural moisture content of 9–14% and a density of 2.09 $g \cdot cm^{-3}$. The plastic limit was 14–16% and the liquid limit was 22–25%. The soil was in a very stiff and stiff consistency. Strength tests were carried out for samples with and without root system. In the first case, the angle of internal friction was 24.9–27.0° and the cohesion was 22.4–29.6 kPa. The angle of internal friction of rooted-soil was 24.7–29.5° and its cohesion was 28.3–34.6 kPa.

2.2   Roots' Properties

Root area ratio (RAR) values of selected trees were measured by profile trench method. Trenches were situated 1.0 m from trees' stem. The width and depth of trenches were equal to 1.0 m. The measurements of root were made at 10 cm thick layers of soil, which were marked on the vertical profile walls using pins and string. Counting RAR all roots with a diameter between 1 and 10 mm were considered (Fig. 1). Roots smaller than 1 mm were not considered in the calculation of RAR due to its great uncertainty in identification and mapping. Roots greater than 10 mm were excluded from analysis of RAR in order to fulfill requirements of Wu-Waldron model [1].

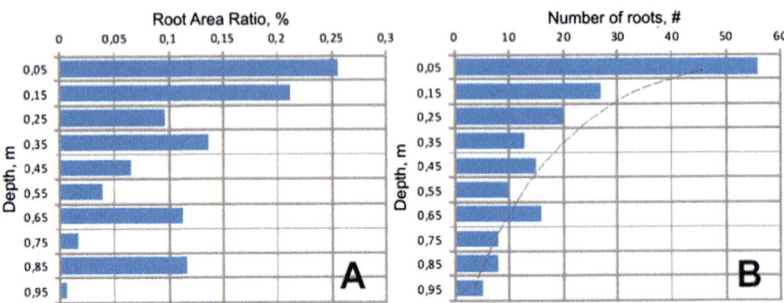

Fig. 1. Exampary of Root Area Ratio - RAR (A) and roots number (B) measurement results.

Live roots collected from trenches were used for tensile strength tests. Prior to measurements the root samples were saturated with water for at least 24 h. Tensile rate was 10 mm·min−1 and the initial sample length was equal to 10 cm. Diameter of each root was measured before the test in three different points of the sample. Thanks to tensile tests correlations between maximum tensile force, elasticity modulus and root diameter were determined (Fig. 2).

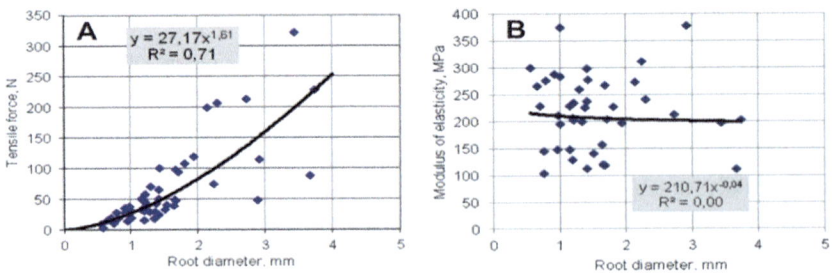

Fig. 2. Maximum tensile force (A) and modulus of elasticity (B) versus root diameter.

2.3   X-Ray Computed Microtomography

Within these studies, novel, non-invasive and non-destructive method, namely high-resolution X-ray computed microtomography (mXCT) was applied to investigate 3D structure of roots interpenetrated soils. The root system, pores and fractures space and soil itself have huge water absorption coefficient range. 3D images obtained by mXCT were used for systematic simulation of fluid flow throughout porous soils and their permeability. Calculations have been performed by Finite Volume Method.

The microstructure of analyzed soil samples is heterogeneous, where some regions are prone to failure initiation. On the other side, identified dozens of roots elements make the soil more cohesive. Average porosity of the analyzed material is 13% and the dominant equivalent diameter of voids is ~ 0.19 mm. The range of voids equivalent diameter starts from several dozens of microns and can reach even few millimeters. The average tortuosity of the flow path is ~ 2.2. The presence of roots system causes greater straightness of the flow path (thus, the tortuosity of flow decreases). That is directly related to the straightness of the root pile system. The relative permeability of the tested soil is increased due to soil disintegration by the root system, accompanied by increased porosity. The dominant pathway for the fluid flow is located near the roots (Fig. 3).

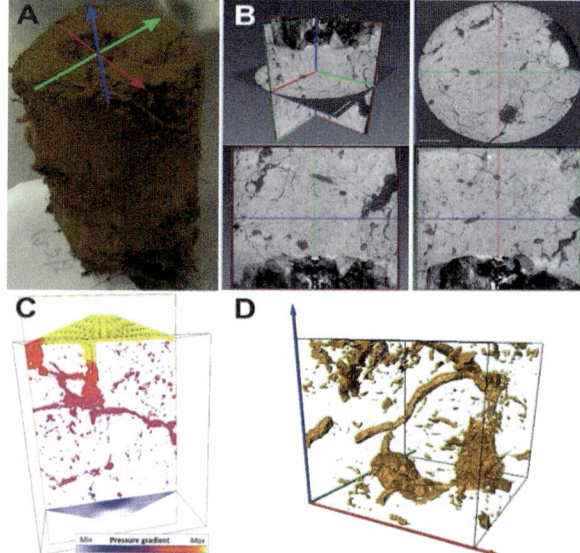

Fig. 3. Soil sample (A), mXCT images (B) and stress field of fluid flow through sample (C). Root spatial distribution in sample (D).

In the present studies, a significant influence of the root system on the physical and mechanical parameters of the soil is visible. The use of non-invasive mXCT tests provide evaluation of the density and orientation of the roots in the sample. This data enabled for more accurate assessment of the root effect on the strength characteristics of soil. Future research will be focused on extension of successive populations of soils type and their characterization towards better understanding of the root system effect on the slope stability of forested areas.

References

1. Bischetti, G.B., Chiaradia, E.A., Epis, T., Morlotti, E.: Root cohesion of forest species in the Italian Alps. Plant Soil **324**, 71–89 (2009)
2. Greenwood, J.R., Norris, J.E., Wint, J.: Assessing the contribution of vegetation to slope stability. Geotech. Eng. **157**(4), 199–207 (2004)

A Simple Method for Evaluating Progressive Failure Process of Rainfall-Induced Shallow Landslide

Yang Tang[1], Kunlong Yin[2(✉)], and Wei Wu[3]

[1] Institute of Geological Survey, China University of Geosciences,
Wuhan 430074, Hubei, China
[2] Faculty of Engineering, China University of Geosciences,
Wuhan 430074, Hubei, China
yinkl@126.com
[3] Institute of Geotechnical Engineering, University of Natural Resources
and Life Sciences, 1180 Vienna, Austria

Abstract. Rainfall-induced shallow landslide is a significant issue in the south and southwest of China. It generally exhibits the behavior of progressive failure. However, the overall stability of coefficients can't reflect the development of the local failure. In this article, a simple method, which is utilizing the progressive failure combined with the improved Green-Ampt model, is proposed to solve this problem during the rainfall. Taking the Guzhang shallow landslide as an example, it shows that the initial failure area of the shallow landslide is revealed in the front and middle of the landslide during the rainfall.

Keywords: Shallow landslide · Green-Ampt model · Progressive failure

1 Introduction

Rainfall is one of the important environment factors that is affecting the stability of landslides [1, 2]. The rainfall-induced landslides are widely distributed in mountains areas of the south and southwest of China, most of these landslides are shallow landslides [3]. The rainfall-induced shallow landslides are always occurred suddenly, which can cause severe losses during a very short time. Therefore, it is very important to figure out the failure mechanism of rainfall-induced shallow landslide, which can help us to prevent the hazard [4].

In this article, a simple method is proposed to evaluate the progressive failure process of rainfall-induced shallow landslide. In this method, Improved Green-Ampt model is used to evaluate the rainfall infiltration process, the progressive failure analysis method is used to evaluate the failure process during the rainfall infiltration. The Guzhang shallow landslide, which is located in Xiangxi state, Hunan province, China, is used as an example.

© Springer Nature Switzerland AG 2018
W. Wu and H.-S. Yu (Eds.): *Proceedings of China-Europe Conference on Geotechnical Engineering*, SSGG, pp. 1535–1538, 2018.
https://doi.org/10.1007/978-3-319-97115-5_140

2 Method

2.1 Improved Green-Ampt Infiltration Model

The wetting front will be formed in each slice, due to the rainfall infiltration, and it will extend to the bottom of the soil. Moreover, the water from the saturated zone will supply to another slice, which will be affected by the geometrical and yield the seepage force. According to the Darcy' law, principle of water balance and Green-Ampt model, the improved Green-Ampt infiltration model can be used to calculate the depth of the wetting front during the process which is mentioned above (Fig. 1).

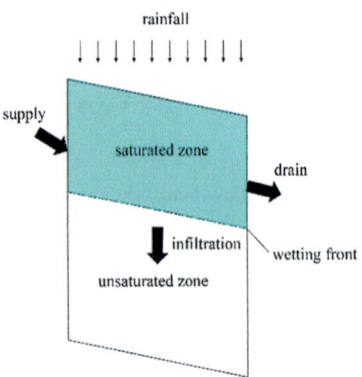

Fig. 1. Rainfall infiltration sketch.

2.2 Progressive Failure Analysis Method

The progressive failure calculation of the landslide is based on the security redundant, which is defined as the difference between the sliding force and the resistant force. The force of each slice is decomposed into two components, one is parallel to the sliding surface and the other is perpendicular (Fig. 2).

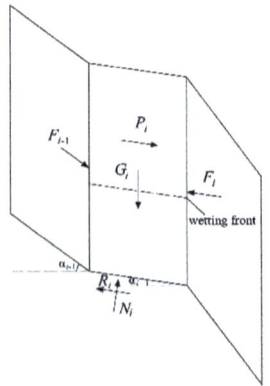

Fig. 2. The stress of each slice.

3   Landslide Event

In July 17th, 2016, due to the rainfall infiltration for nearly 5 h, the Guzhang shallow landslide hazard occurred in Xiangxi state, Hunan Province, China. The landslide destroyed 5 houses and caused the train to stop for several hours. Fortunately, no injuries are reported in this hazard.

The geological cross-section is shown in Fig. 3. According to the shape of the geological cross-section, we generalize the shape as shown in Fig. 4, which is assumed that the sliding surface is parallel to the slope surface.

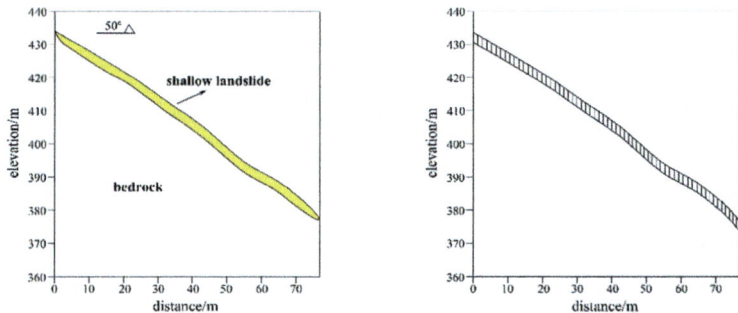

Fig. 3. Geological cross-section of the landslide (left) and generalized model (right).

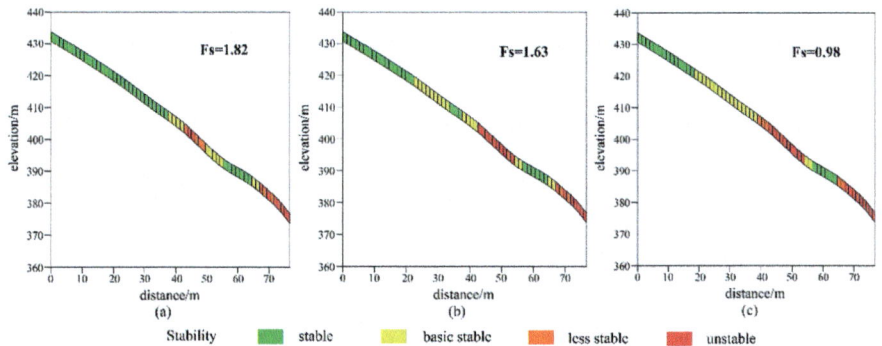

Fig. 4. (a) Landslide stability after rainfall infiltration of 3.2 h. (b) Landslide stability after rainfall infiltration of 3.5 h. (c) Landslide stability after rainfall infiltration of 4 h.

4   Results

According the Fig. 4, the initial failure area of the shallow landslide is in the front and middle of the landslide. And it appears after the 3.2 h of the rainfall. Then, the area of the failure is developing from the first failure area to any other adjacent area.

5 Conclusions

Rainfall-induced shallow landslides are widely distributed in the south and southwest of China. To analyze the failure process of the shallow landslide during the rainfall, the improved Green-Ampt model combined progressive failure calculation method is proposed. The method here proposed efficiently modeled real landslide, which is called Guzhang shallow landslide, in Xiangxi state, Hunan province, China.

According to the result of the calculation, the initial failure area of the shallow landslide is in the front and middle of the landslide. Then, the landslide will develop from the local failure to the general failure.

Method for analysis of landslide based on local failure can be more comprehensive, more practical understanding the landslide deformation and failure of the most prone to initial position. In the monitoring of rainfall-induced shallow landslide, the prediction of potential damage initial position can instruct us to lay crucial monitoring points in the landslide deformation. In addition, it is very helpful to the work of the monitoring and the protection of shallow landslide.

References

1. Tang, Y., Yin, K.L., Liu, L., Zhang, L., Fu, X.L.: Dynamic assessment of rainfall-induced shallow landslide hazard. J. Mt. Sci. **14**(7), 1292–1301 (2017)
2. Wang, F.D.: Preliminarily study on features of shallow accumulation landslide and relationship between it and precipitation. Hydrogeol. Eng. Geol. **1**, 20–23 (1995)
3. Guo, X., Zhao, C.G., Yu, W.W.: Stability analysis of unsaturated soil slope and its progress. China Saf. Sci. J. **15**(1), 14–18 (2005)
4. Khan, Y.A., Lateh, H.: Failure mechanism of a shallow landslide at Tun-Sardon road cut section of Penang Island, Malaysia. Geotech. Geol. Eng. **29**(6), 1063 (2011)

Dimensionless Stability Charts for *c-φ* Slopes with Tension Cracks Subject to Seismic Action

S. Utili[1(✉)] 🆔 and A. Abd[2]

[1] Newcastle University, Newcastle upon Tyne NE1 7RU, UK
stefano.utili@newcastle.ac.uk
[2] University of Tikrit, Tikrit, Iraq

Abstract. A set of analytical solutions achieved by the upper bound theorem of limit analysis and the pseudo-static approach is presented for the assessment of the stability of homogeneous c, φ slopes manifesting vertical cracks and subject to seismic action. Rotational failure mechanisms are considered for slopes with cracks of known or unknown depth and location. Charts providing the stability factor for fissured slopes subject to both horizontal and vertical accelerations for any combination of c, φ and slope inclination are provided. Yield seismic coefficients are also provided. Finally, Newmark's method [1] was employed to assess the effect of cracks on earthquake induced displacements.

Keywords: Limit analysis · Fissure · Earthquake · Stability number
Landslide

1 Introduction

Plenty of experimental evidence shows that the presence of cracks reduces slope stability [2]. Here, an analytical method based on the upper bound theorem of limit analysis and on the so-called pseudo-static approach [3] is presented for the assessment of the stability of uniform c, φ slopes manifesting vertical cracks and subject to seismic action.

Three situations are considered in the Paper:

(i) the most unfavourable scenario of cracks present in the slope (such a scenario may be assumed by practitioners in the absence of reliable information on the presence of cracks);
(ii) slopes subject to cracks of known depth;
(iii) slopes subject to cracks of known location.

With regard to the first problem, (*i*), the assumption of the most unfavourable scenario reflects the fact that often neither the position nor the depth of a crack is known. Assuming the terminology of [4, 5], the "stability factor" for a slope at impending failure is defined as $N = \gamma H/c$, with γ being the ground unit weight, H the slope height (see Fig. 1) and c the ground cohesion.

However, if reliable information on the cracks existing in the slope is available, the conservative assumption of the most unfavourable scenario is no longer justified. In this eventuality, either the depths of the cracks (problem ii) or their locations (problem

© Springer Nature Switzerland AG 2018
W. Wu and H.-S. Yu (Eds.): *Proceedings of China-Europe Conference on Geotechnical Engineering*, SSGG, pp. 1539–1545, 2018.
https://doi.org/10.1007/978-3-319-97115-5_141

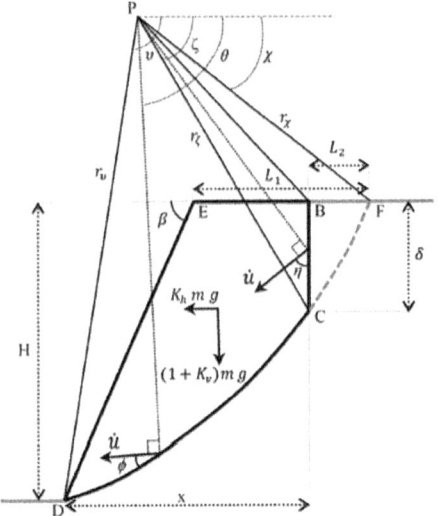

Fig. 1. Failure mechanism. Note that $\eta \neq \phi$.

iii) can be prescribed reducing the number of potential failure mechanisms to be considered in the search for the least upper bound.

2   Theory Leading to the Analytical Solution

The failure mechanisms assumed in our analysis are 2D single wedge rigid rotational mechanisms (see Fig. 1). The failing wedge E-D-C-B rigidly rotates around point P with the ground lying on the right of the log-spiral D-C and of the vertical crack C-B remaining at rest. In the adopted formulation cracks are treated as no-tension non-cohesive perfectly smooth (no friction) interfaces, therefore no energy is ever dissipated along a crack and the angle η is $0° < \eta < 180°$ [6, 7].

The upper bound is derived by imposing energy balance for the failing wedge E-D-C-B:

$$\dot{W}_d = \dot{W}_{ext} \tag{1}$$

where $\dot{W}_d$ and $\dot{W}_{ext}$ are the rate of dissipated energy and of external work respectively. The detailed calculation of $\dot{W}_d$ and $\dot{W}_{ext}$ is reported in [6]. The stability factor, $N = \gamma H/c$, is obtained as:

$$N = \frac{\gamma H}{c} = f(\chi, \upsilon, \zeta, \phi, \beta, K_h, \lambda) \tag{2}$$

with $\lambda = K_v/K_h$ (consistently with Fig. 1, the + sign indicates vertical downward acceleration, whereas the − sign indicates vertical upward acceleration). The global minimum of $f(\chi, \upsilon, \zeta, \phi, \beta, K_h, \lambda)$ over the three geometrical variables χ, υ, ζ provides the least (best) upper bound on the stability factor having assumed that the most unfavourable crack for the slope is present.

By solving Eq. (1) with respect to K_h instead, the upper bound on the yield seismic coefficient, K_y, is obtained:

$$K_y = f_y(\chi, \upsilon, \zeta, \phi, \beta, c/\gamma H, \lambda) \tag{3}$$

The global minimum of $f_y(\chi, \upsilon, \xi, \phi, \beta, c/\gamma H, \lambda)$ over the three geometrical variables χ, υ, ζ provides the least upper bound on K_y.

3 Stability Factor

The global unconstrained minimization of $f(\chi, \upsilon, \xi, \phi, \beta, K_h, \lambda)$ in Eq. (2) provides the stability factor when the most unfavourable crack is present. In Fig. 2 the obtained stability factors are plotted against the inclination of the slope face, β, for $\phi = 20°$, $30°$ and $40°$ for $K_h = 0.2$ and 0.4, and for λ ranging from -1 to $+1$. The charts of Fig. 2 are useful to practitioners in order to get an immediate estimate, erring on the safe side, of the stability of a fissured slope subject to seismic excitation when no data on either the depth or the position of the existing cracks are known.

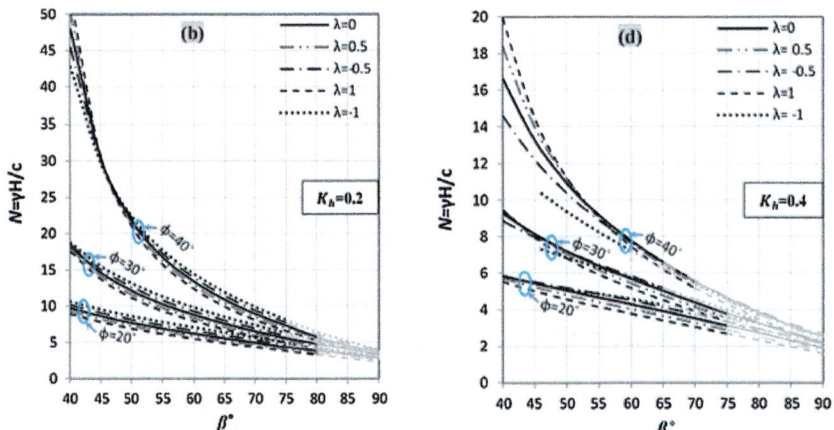

Fig. 2. Stability factor against slope inclination for the most unfavourable crack scenario with $\lambda = K_v/K_h$. (a) $K_h = 0.2$; (b) $K_h = 0.4$. Grey lines indicate the cases where the log-spiral failure surface (rotational failure mechanism) degenerates into a plane (translational failure mechanism).

4   Yield Seismic Coefficient

The yield horizontal acceleration, gK_y, is a key parameter informing practitioners of the level of seismic acceleration for which a given slope, stable under static conditions, becomes unstable. The global minimum of $f_y(\chi, \upsilon, \zeta, \phi, \beta, c/\gamma H, \lambda)$ over the three geometrical variables χ, υ, ζ (see Eq. 3) provides the least upper bound on the yield seismic coefficient, K_y, assuming that the most unfavourable crack for the slope is present. In Fig. 3 the difference in percent between the obtained yield seismic coefficients and the corresponding coefficients for a slope of the same characteristics but intact is plotted. It can be seen that the presence of cracks causes substantial reduction of the yield seismic coefficient, especially for steep slopes of low ϕ. Looking at the charts for $\beta = 75°$ (see Fig. 3b) it can be noticed that the reduction of Ky due to the presence of cracks becomes less significant for c increasing. However, in case of gentle slopes (see Fig. 3a), there is an inversion of the trend at $\phi = 30°$: for slopes with $\phi > 30°$ the reduction in Ky due to the presence of cracks becomes more significant for c increasing.

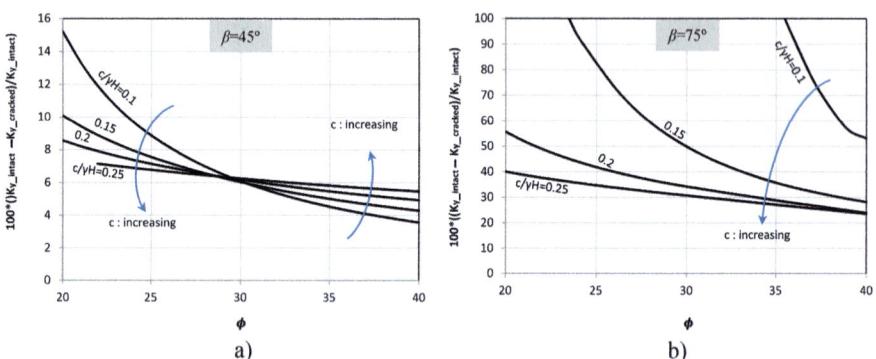

Fig. 3. Percentage of reduction in the yield acceleration due to the presence of the most unfavourable crack for the stability of the slope with $\lambda = 0$. (a) $\beta = 45°$, (b) $\beta = 75°$.

5   Zones Unaffected by Cracks

In [7, 8], it is shown that the presence of cracks reduces the stability of a slope only if they are located in a region inside the slope. The effect of seismic acceleration on the extension of this zone was investigated. The location of the crack needs to be pre-scribed by imposing a crack location into the minimisation of $f(\chi, \upsilon, \zeta, \phi, \beta, K_h, \lambda)$ in Eq. (2) (problem iii in 'Introduction'). It was found that for a sufficiently high value of K_h, the extension of the zone where the presence of cracks affects slope stability is no longer a function of the slope inclination, but of ϕ and K_v solely. With regard to ϕ, when friction is low (so cohesion tends to contribute more to the shear resistance against sliding) the zone where the presence of cracks affects slope stability is larger than when friction is high (so friction tends to contribute more to shear resistance against sliding).

6   Influence of Cracks on Earthquake-Induced Displacements

The presence of cracks makes the geometry of the failing wedge rotating away substantially different (see Fig. 1) and, as a consequence, makes the analytical expression needed to calculate the induced displacements different too [6]. To assess the influence of the presence of cracks on seismic induced displacements the records of two well monitored earthquakes, the Northridge earthquake in 1994 (California, USA) and the Loma Prieta earthquake in 1989 (California, USA), whose features are provided in Table 1, were applied to a slope with $\phi = 20°$, $c/\gamma H = 0.1$, $\beta = 55°$ and $\lambda = 0$.

Table 1. Main characteristics of the earthquakes considered in the example cases.

Earthquake	Northridge	Loma Prieta
Date	17/1/1974	9/2/1989
Station	24283 Moorpark-Fire Sta.	67476 Gilroy-Historic Bldg.
Magnitude	6.7	6.9
Direction	180°	180°
Peak accel. (g)	0.292	0.241
Eqicentre distance (km)	23	28.1

Newmark's pseudo static approach was employed to calculate the seismic induced displacements. The horizontal displacement of the slope toe accumulating over time is plotted in Fig. 4a whilst the final accumulated displacement is plotted against ϕ values in Fig. 4b for both cases of intact slope and slope subjected to the most unfavourable crack. By comparing the two curves for the same given earthquake, it turns out that the presence of cracks increases the amount of displacement significantly: for instance, in the case of the Northridge earthquake, cracks make the total accumulated displacement 5 times larger than the displacement occurring if the slope is un-fissured.

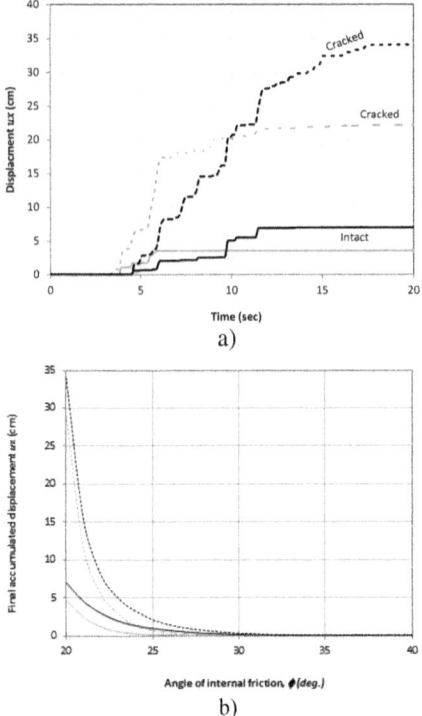

Fig. 4. (a) Horizontal displacement of the slope toe versus time ($\phi = 20°$, $\beta = 55°$, $\lambda = 0$, and c/γH = 0.1). (b) final accumulated displacement versus angle of internal friction ($\beta = 55°$, $\lambda = 0$, and c/γH = 0.1). Black and grey lines represent the displacements induced by the Northridge Loma Prieta earthquake respectively.

7   Conclusions

It was found that fissures may substantially reduce slope stability, i.e. lower both stability factor and yield acceleration up to 30% in comparison with the case of intact slope, with the amount of reduction depending on both the geometrical characteristics of the slope and the ground strength parameters: the reduction is higher for steep slopes of low friction angle subject to high accelerations, whereas for gentle slopes of high ϕ subject to moderate earthquakes it is negligible. Horizontal yield accelerations were calculated for any combination of β, ϕ, and K_h of engineering interest, having assumed the most unfavourable crack for the stability of the slope to be present.

Then, Newmark's approach was employed to calculate seismic induced displacements. The relationship between crack depth and final accumulated displacements was investigated for an example slope subjected to the accelerograms of two past earthquakes. It emerges that the displacements induced for a fissured slope can be significantly larger, up to five times, than the case of intact slope depending on the slope characteristics.

Acknowledgment. The Higher Committee for Education Development in Iraq (HCED) and H2020 Marie Skłodowska-Curie Actions RISE 2014 'Geo-ramp' grant number 645665 are acknowledged.

References

1. Newmark, N.M.: Effect of earthquakes on dams and embankments. Geotechnique **15**, 139–159 (1965)
2. Utili, S., Castellanza, R., Galli, A., Sentenac, P.: Novel approach for health monitoring of earthen embankments. J. Geotech. Geoenv. Eng. ASCE **141**(3), 04014111 (2015)
3. Terzaghi, K.: Mechanisms of landslides. In: Engineering Geology (Berkeley), pp. 83–123. Geological Society of America, Boulder, USA (1950)
4. Taylor, D.W.: Fundamentals of Soil Mechanics. Wiley, New York (1948)
5. Chen, W.F.: Limit Analysis and Soil Plasticity. Elsevier, Amsterdam (1975)
6. Utili, S., Abd, A.: On the stability of fissured slopes subject to seismic action. Int. J. Numer. Anal. Meth. Geomech. **40**(5), 785–806 (2016)
7. Abd, A., Utili, S.: Design of geosynthetic - reinforced slopes in cohesive backfills. Geotext. Geomembr. **45**, 627–641 (2017)
8. Utili, S.: Investigation by limit analysis on the stability of slopes with cracks. Geotechnique **63**, 140–154 (2013)

Introduction to the Badong Field Test Site for Landslide Research in the Three Gorges Reservoir Area of China

Jinge Wang(✉), Wei Xiang, Aijun Su, and Chengren Xiong

Three Gorges Research Center for Geohazards, China University of Geosciences,
Wuhan 430074, Hubei, China
wangjinge@cug.edu.cn

Abstract. The initial impoundment and periodic water level variation may change the balance of geological environment of the reservoir area and bring different kinds of geohazards. Reservoir induced landslides typify such hazards that present challenges to the long-time operation of the Three Gorges Water Conservancy and Hydropower Project of China. The Huangtupo landslide, with multiple sliding stages and masses, is one of the largest and most destructive landslides still deforming in the Three Gorges Reservoir area. From 2008, we take the Huangtupo landslide as a typical case, and build a large field test site to provide an unprecedented opportunity for the research on the evaluation and prevention of reservoir induced landslides. The test site is composed with a tunnel group with a total length of 1.1 km and a series of monitoring system on both earth surface and underground. Through the tunnel group and multiple monitoring systems, visitors can directly enter the Huangtupo No.1 riverside sliding mass to closely observe the bedrock, sliding zone, and sliding mass, and also operate related scientific experiments and deep monitoring, such as large scale in-situ mechanical tests, underground hydrological tests, deep stress, deformation and environment monitoring. Now, the test site becomes an important site for international research, education, and academic communication about geohazards.

Keywords: Three Gorges Reservoir · Landslide · Field test site
Monitoring

1 Background

The Yangtze Three Gorges, composed with Qutang Gorge, Wu Gorge and Xilin Gorge, is a well-known section of Yangtze River that attracts the research interests from all over the word. For the flood management, power generation, and shipping, the Three Gorges Water Conservancy and Hydropower Project (Three Gorges Project for short), one of the largest civil engineering projects in human history, dams the Yangtze River to form a 660 km long reservoir that impounds about 39.3 km^3 of water. From 2003 to 2008, the water level of the Three Gorges Reservoir has raised more than a hundred meters and fluctuates annually between 145 m and 175 m above sea level. The initial impoundment and periodic water level variation may change the balance of geological environment of

© Springer Nature Switzerland AG 2018
W. Wu and H.-S. Yu (Eds.): *Proceedings of China-Europe Conference on Geotechnical Engineering*, SSGG, pp. 1546–1550, 2018.
https://doi.org/10.1007/978-3-319-97115-5_142

the reservoir area and bring different kinds of geohazards. Reservoir induced landslides typify such hazards which present challenges to the long-time operation of the project. According to statistics, there are more than 2,000 landslides with different sizes in the Three Gorges Reservoir area. And among them, at least 250 are identified as large-scale landslides with volumes larger than a million cubic meters [1].

Badong County of Hubei Province is located on the south bank of Yangtze River between Wu Gorge and Xiling Gorge in the Three Gorges Reservoir area. The original Badong downtown is one of the largest urban areas wholly below the highest water level of the reservoir. Because of the impoundment, downtown Badong start to be rebuilt in the Huangtupo area from 1982, and the new community was mostly completed in 1991. However, the Huangtupo Slope was found to be a revived ancient landslide, and the main evidences are two sliding events near the toe area occurred in 1995. The following investigation indicate that the Huangtupo landslide is a huge and complex landslide group with multiple sliding stages and masses, which contains four parts, named No.1 riverside sliding mass (Northwest), No.2 riverside sliding mass (Northeast), transformer substation landside (Southeast), and garden spot landslide (Southwest). The total volume is up to 6.9×10^7 m^3, makes it one of the largest and most harmful landslides in the Three Gorges Reservoir area. Since 1992, the Badong government officially started the second resettlement of Huangtupo residents. Until now, although the resettlement is finished, the threats of such a huge landslide to the safety of nearby residents and the Yangtze main channel still exist (Fig. 1).

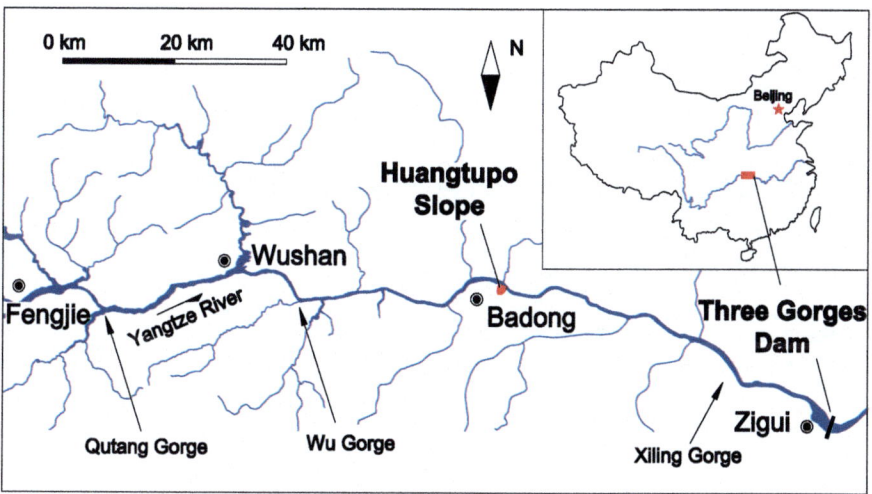

Fig. 1. Location map of study area

2   Introduction of the Badong Field Test Site

From 2008, the Three Gorges Research Center for Geohazards (TGRC) of China University of Geosciences (CUG) takes the Huangtupo landslide as a typical case and builds a large field test site for research on the evaluation and prevention of reservoir induced landslides. Now, it becomes an important site for research, education, and academic communication about geohazards. The test site is composed with a tunnel group and a series of monitoring system. The tunnel group is built in the Huangtupo No.1 riverside sliding mass, includes a main tunnel (908 m long and 5 m diameter), 5 branch tunnels (5 m to 145 m long and 3.5 m diameter), 2 test tunnels and several observation windows. The test tunnels are entirely constructed in the sliding zones of the landslide and connect with the main tunnel by No.3 and No.5 branch tunnels. Through the tunnel group and multiple monitoring systems, researchers can directly enter the sliding mass to closely observe the bed rock, sliding zone, and sliding mass, and also operate related scientific experiments and deep monitoring, such as large scale in-situ mechanical tests, underground hydrological tests, deep stress, de-formation and environment monitoring (Fig. 2).

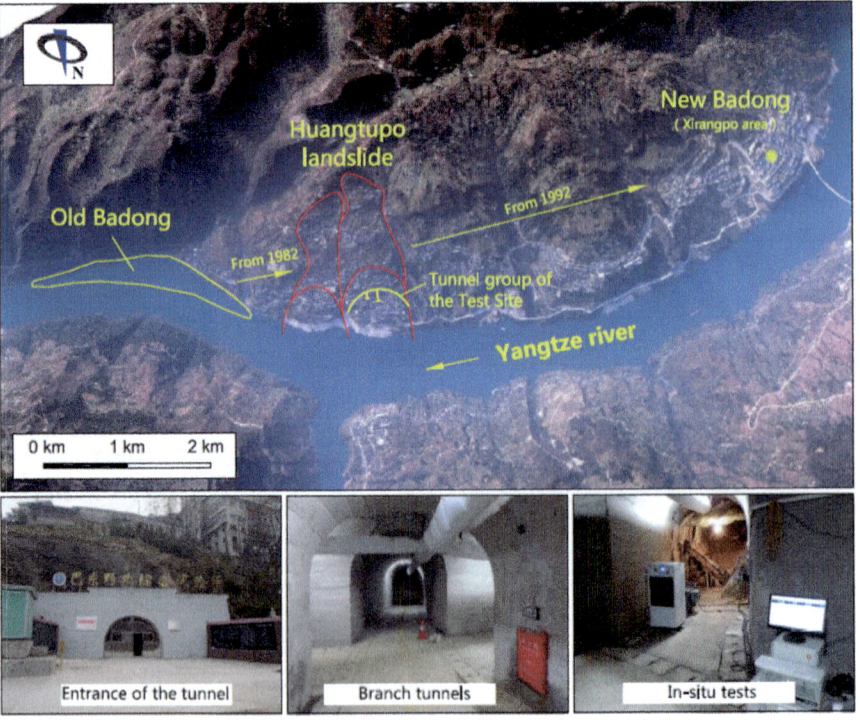

Fig. 2. Image and photos of Badong field test site

The main research works operated on the test site include geological investigation, in-situ monitoring, and scientific experiment. The methods of geological investigation involve earth surface exploration, mapping, borehole drilling, tunneling and geophysical prospecting. Until now, a multi fields and real time monitoring system has been established in the test site on both earth surface and underground. The environmental monitoring involves the precipitation and reservoir water level. The earth surface deformation monitoring devices involve the Global Navigation Satellite System (GNSS) such as GPS and BDS, the Interferometric Synthetic Aperture Radar (InSAR), and the three-dimensional laser scanning system. The deep deformation monitoring devices involve the borehole inclinometers, and the crack meter and distributed optical fiber sensing system in the underground tunnel group. The groundwater monitoring system collects the data of ground water level, flow rate, chemical composition and temperature. For the scientific experiment, the large size tunnel group is not only convenient for deep sampling for laboratory tests, but also provides the necessary site for the underground large scale in-situ tests. In the rupture zone exposed area of the tests tunnel, an underground laboratory, 65 m beneath the earth surface, has been excavated along the strike direction of the sliding surface. In-situ direct shear test, direct shear creeping test and triaxial creeping test were conducted to obtain mechanical parameters of sliding zone soils.

3   Some Research Results

A general conceptual model for landsliding triggered by reservoir operation: The monitoring data of Huangtupo landslide indicates that the deformation appears to be highly correlated to the reservoir water level. A hydro-mechanical numerical model is constructed to investigate the quantitative links among the episodic movements and subsurface water content, pore water pressure, total stress, and effective stress variations. We find that the variations in pore water pressure, suction stress, hydrostatic reservoir water loading, and slope self-weight induced by the fluctuating water levels have different effects on stress conditions and slope stability [2].

The shallow and deep deformation characteristics of Huangtupo No.1 sliding mass during seasonal water level fluctuation: The monitoring data indicate that the earth surface and deep deformation of this landslide have different characteristics. During increases in the water level, the earth surface deformation velocity decreases, and then increases obviously in the subsequent water level decreasing stage. But on the contrary, the deformation velocity of the deep sliding mass accelerates obviously during the water level increasing periods. The distinction of surface and deep deformation regulations indicates that the effects of water level fluctuation on the stability of the shallow and deep regions of reservoir wading landslides are different [3].

References

1. Li, L.R.: Geological hazards and prevention in the Three Gorges Reservoir Area. Land Resour. **42**(2), 4–7 (2002)
2. Wang, J.E., Xiang, W., Lu, N.: Landsliding triggered by reservoir operation: a general conceptual model with a case study at Three Gorges Reservoir. Acta Geotech. **9**(5), 771–788 (2014)
3. Wang, J.E., Su, A.J., Xiang, W., et al.: New data and interpretations of the shallow and deep deformation of Huangtupo No. 1 riverside sliding mass during seasonal rainfall and water level fluctuation. Landslides **13**(4), 795–804 (2016)

Factors Influencing Landslide Deformation from Observations in the Three Gorges Reservoir

Beibei Yang[1,2(✉)], Suzanne Lacasse[2], Kunlong Yin[1],
and Zhongqiang Liu[2]

[1] Faculty of Engineering, China University of Geosciences,
Wuhan 430074, China
cugyangbeibei@163.com
[2] Norwegian Geotechnical Institute, Oslo, Norway

Abstract. Since the impoundment of the Three Gorges Reservoir in mid-2003, slope displacements have occurred and existing landslides have been reactivated. The movement are a threat to the community and the environment. High precipitation in the area increases landslide susceptibility. The relationships among landslide displacement, rainfall and reservoir drawdown were analyzed for the Baijiabao landslide with the Grey relational analysis approach and using eight years of monitored displacement. The results suggest that rainfall triggered the landslide deformation in the earlier years. Then reservoir drawdown and the combined effect of rainfall and drawdown became preponderant. The identification of the triggers is important for selecting mitigation measures and evaluating the risk due to slope movement.

1 Introduction

The Three Gorges Reservoir in China is characterized by long and narrow valleys, where more than 500 km of unstable reservoir banks and 5300 landslides have been recorded in the reservoir area (Yi et al. 2011; Wu et al. 2017). The long-term fluctuation of the reservoir water level influences bank stability by changing the hydraulical and mechanical loading (Cojean and Cai 2011). Rainfall can also trigger landslide deformations and slope failure (Coe et al. 2003; He et al. 2010). The mean annual rainfall in the Three Gorges area from 1960 to 2010 was 1113.2 mm. The combined effect of reservoir water level and rainfall makes the triggering of landslide movement more complex and can increase the risk of landslide (Li et al. 2010; Tan et al. 2018). Researchers have studied deformations, failure modes and slope stability under reservoir water level variations and rainfall (Yang et al. 2017). The key influence factors and deformation characteristics probably keep changing as the landslide develops (Brunsden 2001; Tang et al. 2018). In the present study, the quantitative relationships between landslide deformation and triggering factors were established.

© Springer Nature Switzerland AG 2018
W. Wu and H.-S. Yu (Eds.): *Proceedings of China-Europe Conference on Geotechnical Engineering*, SSGG, pp. 1551–1555, 2018.
https://doi.org/10.1007/978-3-319-97115-5_143

2   Observed Displacements for the Baijiabao Landslide

The Baijiabao landslide is a deep colluvial landslide, which occurred in the Zigui town on the west side of the Xiangxi River, a major tributary of the Yangtze River. Four GPS real time monitoring stations were installed on the ground surface of the landslide at the end of 2006. As the observed movements were uniform, Station ZG326 was taken as a good indicator of the movement. The relationships among displacement, reservoir water level and rainfall intensity were analyzed individually and combined. Figure 1 shows the reservoir level, monthly accumulated rainfall and displacement velocity over eight years at the location of the Baijiabao landslide.

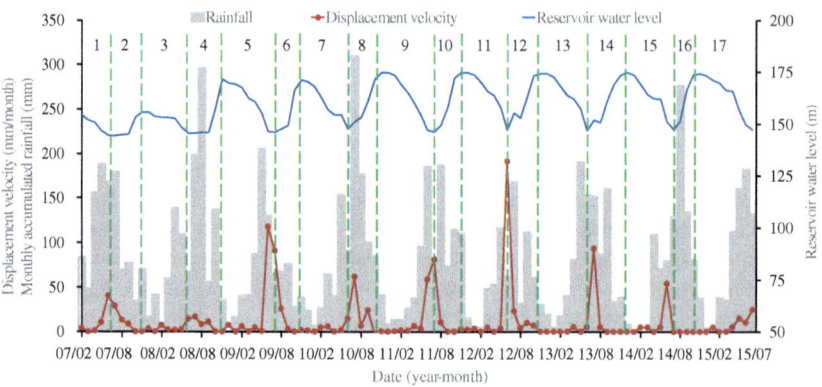

Fig. 1. Observed displacement velocity, reservoir water level and monthly accumulated rainfall at the Baijiabao landslide location, Station ZG326.

2.1   Reservoir Water Level

Figure 1 indicates that the displacement velocity has a positive correlation with reservoir drawdown (increased with drawdown and decreases with rise of water), peaking usually towards the end of reservoir drawdown except in 2008. The effect of reservoir drawdown on landslide deformation, however, varied with time. The monthly displacement in 2009 had a sharp rise of 118 mm when the water level decreased from 175 to 145 m, but was much smaller in 2010. This variation may be due to the geological conditions with the foundation "adjusting" itself as the first scheduled full filing and drawdown occurred in 2008/2009.

Little displacement, if any, was recorded during the rise of the water in the reservoir. With a full reservoir, the slope is under hydrostatic pressure, with water force orthogonal to and in the direction of the slope, stabilizing the slope. As cracks appear on the slope, reservoir rise results in increased forces and softened soil masses. As the water level drops, the pore pressure decrease even if the water level in the slope decreases more slowly than in the reservoir, a hydraulic gradient can be formed, and landslide deformations increase.

2.2 Rainfall

The period of May to September is the rainy season in the area. Figure 1 shows that the monthly peak displacement velocity occurred during the rainy season. Many of the other months had insignificant deformation. The displacement rate had a positive correlation with rainfall intensity. The data show, however, that the displacement velocity from May to June 2012 was much larger than in the rainy season in 2010 and 2014, although higher rainfall was recorded in 2010 and 2014.

2.3 Combined Effect of Rainfall and Reservoir Water Level

From January to March 2014 (Fig. 1), the deformation was virtually negligible, but from May to June 2014, the displacement velocity showed a significant increase with high intensity rainfall and reservoir drawdown. In August 2008 and 2014, the displacement rate is only 7.8 mm and 0 mm although the monthly rainfall is up to 296.4 mm and 276.7 mm. But then, the reservoir water level is rising.

3 Relationships Between Landslide Deformation and Landslide Triggers

Reservoir water level and rainfall are the major triggers of the Baijiabao deformations. The degree of influence however varied with time. To quantify the relationship, the observations were divided into 17 periods following the reservoir level rise and drawdown (Fig. 1). The grey relational analysis approach was used.

3.1 Grey Relational Analysis Approach

Grey relational analysis is a model to measure the degree of correlation among factors (Liu et al. 2017). Landslide displacement velocity was selected as primary sequence (X0) and three causal factors, i.e., reservoir water level (X1), rainfall (X2) and the combined rainfall and rate of reservoir level change (X3) as sub-sequences. With X = [X0, X1, X2, X3], the sequence becomes:

$$X_k(i)' = X_k(i) \Big/ \frac{1}{n} \sum_{i=0}^{n} X_k(i) \tag{1}$$

where $i = 0,1,\ldots, n$; $k = 0,1,\ldots, m$; n and m are the number of data points and influence factors, respectively. The correlation coefficients are calculated from:

$$\xi\big(x_0(i)', x_k(i)'\big) = \frac{\min\limits_{k} \min\limits_{i}\big(x_k(i)' - x_0(i)'\big) + \rho \max\limits_{k} ax\big(x_k(i)' - x_0(i)'\big)}{\big|X_k(i)' - X_0(i)'\big| + \rho \max\limits_{k} \max\limits_{i}\big(x_k(i)' - x_0(i)'\big)} \tag{2}$$

where ρ is the resolution coefficient and normally set as 0.5. The grey relational grade (GRG), used to evaluate the correlation of the variables, is obtained by Eq. (3).

The GRG ranges from 0 to 1 and a value of 0.6 or higher designates strong correlation: the higher the grade, the higher the correlation.

$$r(x_0, x_i) = \frac{1}{n} \sum_{k=1}^{n} \xi\left(x_0(i)', x_k(i)'\right) \tag{3}$$

3.2   Results for Station ZG 326, Baijiabao Landslide

The grey relational grade at Station ZG326 are shown in Fig. 2. Before 2008, the landslide displacement showed a much stronger correlation with rainfall. The Baijiabao landslide lies at elevation 125 to 275 m. Most of the landslide body was not submerged, and rainfall was the preponderant trigger. In June 2008 (period 3 in Fig. 2), the landslide deformation became correlated with all three factors analyzed, as the reservoir was lowered. During reservoir rise (e.g. period 4), rainfall is the sole strongest trigger. In periods 5, 7 and 8 (Fig. 2), the GRG for the combined effect was larger than the two single triggers, and was therefore the more unfavorable condition. Similar trends are seen in later periods, with GRG-values greater than 0.6.

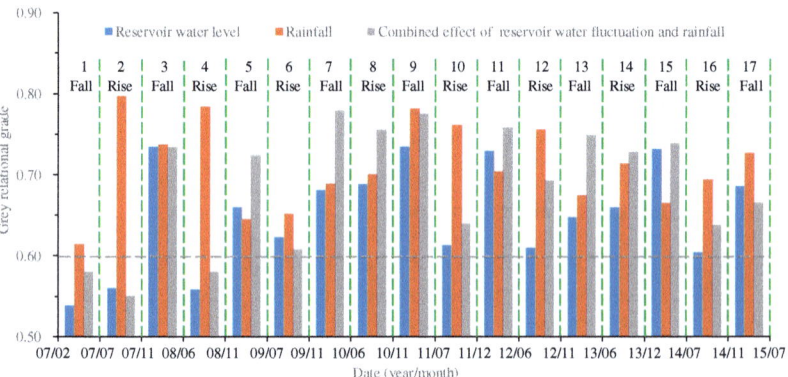

Fig. 2. Grey relational grades for each factor with time, Baijiabao landslide, Station ZG326.

4   Conclusion

The grey relational analysis of the Baijiabao landslide quantified the relative influence of rainfall, reservoir drawdown and their combined effect on the deformation of the Baijiabao landslide. The landslide deformation was more strongly influenced by rainfall before full impoundment. Reservoir drawdown and the combined rainfall and drawdown became key triggers in later years. Such estimates are simple to do and are important for selecting of mitigation measures and evaluating risk. In the analysis, it was important to carefully select the reference period for the calculations. A further refinement would be to do the analysis with each of the monthly measurements rather than subdividing the eight years into 17 periods. (The authors appreciate the help for data sources from the Headquarter of Geo-hazard Prevention of Three Gorges Reservoir.)

References

Brunsden, D.: A critical assessment of the sensitivity concept in geomorphology. Catena **42**, 99–123 (2001)

Coe, J.A., Ellis, W.L., Godt, J.W., et al.: Seasonal movement of the Slumgullion landslide determined from global positioning system surveys and field instrumentation, July 1998–March 2002. Eng. Geol. **68**, 67–101 (2003)

Cojean, R., Cai, Y.J.: Analysis and modeling of slope stability in the Three-Gorges Dam reservoir (China) - The case of Huangtupo landslide. J. Mt. Sci. **8**(2), 166–175 (2011)

He, K.Q., Wang, S.Q., Du, W., et al.: Dynamic features and effects of rainfall on landslides in the Three Gorges Reservoir region, China: using the Xintan landslide and the large Huangya landslide as the examples. Environ. Earth Sci. **59**(6), 1267–1274 (2010)

Li, D.Y., Yin, K.L., Leo, C.: Analysis of Baishuihe landslide influenced by the effects of reservoir water and rainfall. Environ. Earth Sci. **60**(4), 677–687 (2010)

Liu, SF., Yang, Y., Forrest, J.: Grey Data Analysis. Springer, Singapore (2017)

Tan, F.L., Hu, X.L., He, C.C., et al.: Identifying the main control factors for different deformation stages of landslide. Geotech. Geol. Eng. **36**, 469–482 (2018)

Tang, H.M., Liu, X., Wen, T., et al.: Probabilistic forecasting of landslide displacement accounting for epistemic uncertainty: a case study in the Three Gorges Reservoir. Landslides (2018). https://doi.org/10.1007/s10346-017-0941-5

Wu, Y.P., Miao, F.S., Li, L.W., Xie, Y.H., Chang, B.: Time-varying reliability analysis of Huangtupo Riverside No.2 Landslide in the Three Gorges Reservoir based on water-soil coupling. Eng. Geol. **226**, 267–276 (2017)

Yang, B.B., Yin, K.L., Xiao, T., et al.: Annual variation of landslide stability under the effect of water level fluctuation and rainfall in the Three Gorges Reservoir. China. Environ. Earth Sci. **76**, 564–580 (2017)

Yi, W., Meng, Z.P., Yi, Q.L.: Theory and Method of Landslide Stability Prediction in the Three Gorges Reservoir Area. Science Press, Beijing (2011)

Modelling of Castaño Viejo Tailings Flow Case History

Francisco Zabala[✉], Gustavo Navarta, and Luciano A. Oldecop

Earthquake Engineering Research Institute, Universidad Nacional de San Juan,
San Juan, Argentina
fzabala@unsj.edu.ar

Abstract. The Castaño Viejo mine, located in the Andes Mountains of San Juan Province, Argentina, was in operation between 1956 and 1964. A series of tailings dams were built during operation. One of these dams collapsed and the tailings flowed downstream for 3 km causing 3 causalities. This paper describes the Material Point Method modelling of this case history. An ASTER GDEM digital elevation terrain model and in-situ GPS measurements were used to generate the boundary condition for the MPM three-dimensional model. The tailings material was modelled as an equivalent fluid using Bingham constitutive equation. Results of the model runout are compared with GPS measurements of flow relicts and estimation of tailings deposits depth.

Keywords: MPM · Tailings · Flow

1 Introduction

The Castaño Viejo mine, located in the Andes Mountains of San Juan Province, Argentina, was in operation between 1956 and 1964. The mine produced lead, zinc, and gold. A series of tailings dams were built during its operation along a creek downstream of the processing plant using wood logs for slope reinforcing. These dams were evaluated by a team of the National University of San Juan, Argentina and some geotechnical and geophysical studies were accomplished more than 50 years since the dams' abandonment. The tailings are fine-grained soils classified as low plasticity clays (CL), silts and sands with a high degree of saturation nowadays (80%) with the exception of a 1–2 m thick layer near the surface. This fact is interesting because the area is a desert with high hydric deficit and the tailings in the dam's body remain almost saturated. Low values of SPT blows (N = 2–4) were measured in one of the larger dams and also low values of shear velocities.

One of the dams collapsed during operation. The reason of the failure is unknown but there is some evidence suggesting it was caused by water injection, due the rupture of a decant pipe, followed by erosion of the external slope and tailings liquefaction. The tailings flowed downstream causing 3 causalities without reaching the Castaño River. A detailed GPS survey of the tailings relicts in the stream margins was carried out. Figure 1 shows one of the relicts at the downstream farthest location which has a thickness of approximately 30 cm. Figure 2 is a plan view of the dam deposit (black)

© Springer Nature Switzerland AG 2018
W. Wu and H.-S. Yu (Eds.): *Proceedings of China-Europe Conference on Geotechnical Engineering*, SSGG, pp. 1556–1560, 2018.
https://doi.org/10.1007/978-3-319-97115-5_144

and the relicts position (R01 to R10). Also, colored in blue, the estimated run out of the tailings derived from field observations, is displayed.

This paper describes the Material Point Method three dimensional modelling of this case history. MPM [4] is a lagrangian "particle-mesh" numerical method previously used in modeling dynamic problems with large displacements. With MPM, a body is discretized into a collection of lagrangian particles, which carry all the data needed to define the body's state. The interaction between particles takes place in a background fixed mesh, similar to those used in the finite element method.

Fig. 1. Tailings relicts at the R10 location.

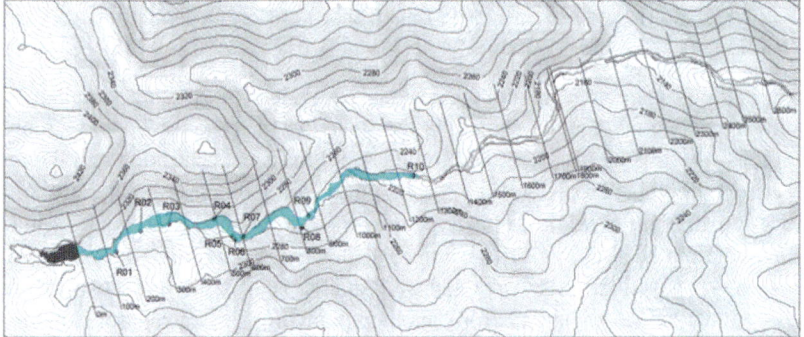

Fig. 2. Plan view of the dam deposit (black) and relicts position (R01 to R10)

2 MPM Model

An ASTER GDEM digital elevation terrain model and in-situ GPS measurements were used to generate the terrain boundary condition for the MPM three-dimensional model. A region of 268 hectares with very high average slopes was modelled with an interpolating grid of 5 m per 5 m. Figure 3 shows the MPM mesh, with a dimension of 5 m × 5 m in plan, and 1 m in elevation, which covers the model extent, and an image of the terrain surface superposed. The mesh has approximately 40 million cells and the fluidized portion of the dam deposit was modeled with 4832 particles (8 per cell).

Fig. 3. Three dimensional MPM mesh superposed to the terrain surface.

The tailings material was modelled as an equivalent fluid using Bingham plastic constitutive equation. A main subject is the selection of reasonable yield strength for the tailings which is highly dependent on the solid mass concentration and the mineral type as it is obtained in tests [1]. As a guide for calibration, the ratio of post-liquefaction strength to vertical effective stress S_u/σ'_v obtained from liquefaction cases [3] can be estimated as 0.05 to 0.10 using SPT blow count obtained for one of the dams (N = 2 to 4). Considering an average soil thickness of 2 m, the post-liquefaction strength could be 1.5 a 3 kPa. From another point of view, a measure of the yield strength can be obtained from equilibrium of a Bingham material layer resting on an inclined plane [2], at zero velocity, using the thickness of the farthest relict (R10) and the average slope of the stream at that location.

$$\tau_y = \gamma_t h \sin i \tag{1}$$

τ_y: yield strength, γ_t: fluidized tailings specific weight, h: material height, i: slope. This value is estimated as 15 kN × 0.30 m × 0.05 = 0.225 kPa

3   Simulation Results

Several simulation results were obtained varying the yield strength with a constant viscosity of 10 Pa s using GEOPART code [5]. Figure 4 displays the run out simulations 10 min after the instant of dam breach for different values of the yield strength (identified with different colors). For a yield strength of 8 kPa, the flow barely moves 150 m (grey) from the dam toe and using 0.10 kPa the tailings approximately reach the farthest relict location (1200 m from the dam). This last yield strength is in the range of the tests reported in [1] for liquid sand tailings with a solid mass fraction of 50–60%. These results also depend on the adopted discretization which is relatively coarse in the horizontal dimensions. Analyses with finer MPM meshes are in progress, even though are very time-consuming.

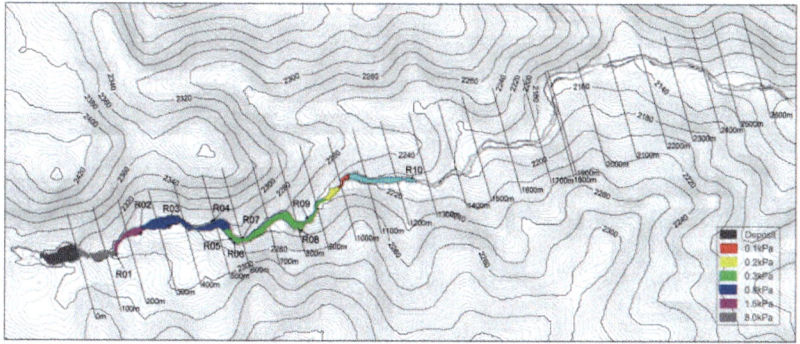

Fig. 4. Plan view of different simulations run out with different yield strength.

4   Conclusions

Results of the model run out for Castaño Viejo case history are compared with GPS measurements of the location of flow relicts and field estimation of tailings deposits depth. The simulations output is quite realistic using values for the tailings liquefied strength in the range of rheological test reported in the literature.

References

1. Boger, D.: Rheology and the resource industries. Chem. Eng. Sci. **64**, 4525–4536 (2009)
2. Coussot, P.: Yield stress fluid flows: a review of experimental data. J. Non-Newton. Fluid Mech. **211**, 31–49 (2014)
3. Olson, S.M.Y., Stark, T.D.: Liquefied strength ratio from liquefaction flow failure case histories. Can. Geotech. J. **39**, 629–647 (2002)
4. Sulsky, D., Schreyer, H.L., Zhou, S-J.: Application of a particle-in-cell method to solid mechanics. Comput. Phys. Commun. **87**, 236–252 (1995)
5. Zabala, F., Rodari, R.: GEOPART. MPM 3D code. Earthquake Engineering Research Institute. UNSJ (2002–2018)

Soil - Structure Interaction at the Bogatići Landslide in Bosnia and Herzegovina

Sabid Zekan[1(✉)], Mato Uljarević[2], Majda Mešić[3], and Alen Baraković[4]

[1] University of Tuzla, Univerzitetska 2, 75000 Tuzla, Bosnia and Herzegovina
sabid.zekan@untz.ba
[2] University of Banja Luka, Stepe Stepanovića 77/3,
78000 Banja Luka, Bosnia and Herzegovina
[3] NNM Inženjering, Bosne Srebrene 56, 75000 Tuzla, Bosnia and Herzegovina
[4] Ministry of Mining, Industry and Energetics FBiH, Alekse Šantića bb,
88104 Mostar, Bosnia and Herzegovina

Abstract. Movement of sliding mass, also known as the mechanism of sliding, is very complex. It is possible to observe the products of movement, i.e. displacements and deformations in the landslide body. In addition, displacements and deformations have an impact on artificial structures. The soil-structure interaction in landslide body has to be understood, whether we decide to save the existing structure or to design a new one.

In this paper, the observation of displacements, deformations and soil-structure interaction at the landslide Bogatići is presented. The structure of hydro-power dam was exposed to the passive pressure of sliding mass, while the retaining structures have been overturned as a result. A geological fault was the barrier to the sliding. The Bogatići landslide is about 1400 m long and from 80 to 100 m wide. It spreads up to 300 m in the foot. Sliding surface is about 10 to 15 m deep in the upper part and up to 30 m in the foot. Activation of the landslide occurred in May 2010 after a period of heavy rainfall. Reactivation of the landslide occurred in June 2011 due to heavy rainfall again.

Transport of sliding mass has been characterized by "push-pull" system. Reduction of strength in the foot and the secondary overload produce secondary sliding. The paper analyses different types of soil-structure interactions. Research results are to improve methodology of structure remediation.

Keywords: Landslide · Bogatići · Soil-structure interaction · Passive pressure

1 Introduction

The landslide Bogatići is located 15 km south from Sarajevo, Bosnia and Herzegovina, at the left bank of the Željeznica river. The first triggering of the landslide occurred in June, 2010, and the second one in June, 2011. Both triggering occurred due to heavy raining. The landslide has discontinued character in regard to time, because there has been no sliding without heavy rains in June. It is estimated that there are two fossil parts of the landslide, marked as "F_1" and "F_2", and the active part of the landslide marked as "A" [1] (Fig. 1).

© Springer Nature Switzerland AG 2018
W. Wu and H.-S. Yu (Eds.): *Proceedings of China-Europe Conference on Geotechnical Engineering*, SSGG, pp. 1561–1564, 2018.
https://doi.org/10.1007/978-3-319-97115-5_145

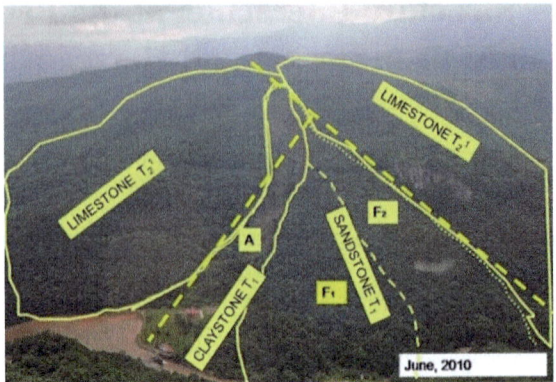

Fig. 1. Panorama of the Bogatići landslide [1]

Genetically, this is an old multiphase consequent landslide with delapsive type of development, likely to be reactivated frequently. By shape, a very elongated, so-called "glacial" type, landslide was formed with the typical fan-like ending.

2 Deformations and Displacements in the Landslide Body

The main causes of triggering mechanism of Bogatići landslide are: high water level and organic silt material in the landslide body. The mechanism of triggering is timely discontinued and occurs in June during heavy raining period. The landslide is divided in two parts: "A_g" –the upper and "A_d" – the lower part. The "A_g" upper part of the landslide moves as translation, while the "A_d" part moves as semi rotation [1] (Fig. 2).

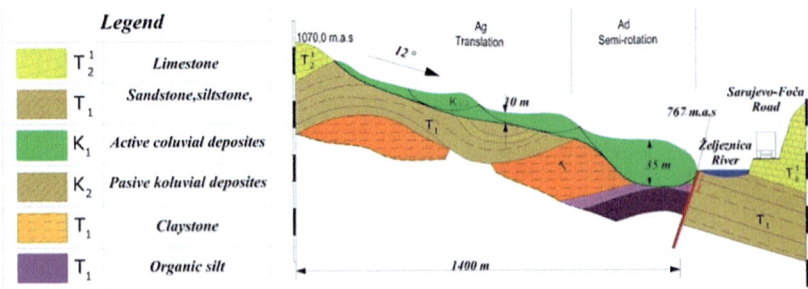

Fig. 2. Crossection of the active (A) landslide

The investigation has proved that there is a difference in lithological composition of the lower and upper part of the landslide. The lower part of the landslide is basically made up of claystones, while the upper part of the landslide built of mica sandstone and siltstone. Both materials are subject to decay as poorly petrified rocks. Landslides foot has been moved depending on the balance established after each sliding. The lower levels of the terrain, on the left bank of Željeznica river, were the place of the deposition of mass, transported by gravity down the slope. Željeznica river has partly eroded away some of the mass deposited. Due to accumulation mass greater than erosion, the current morphology of the terrain has a form of "fan-type reservoir belly".

Investigation has proved that there is a fault along the left bank of Željeznica river. There are more solid rocks in the east part of fault, on the right bank of the river. The rocks are gray and brown sandstones, hard shales partly, and above them is the limestone also visible at the terrain's surface. The west wing, on the left bank of Željeznica river, is composed of weak materials, such as shales and organic soil [1, 2].

3 Soil-Structure Interaction in the Landslide Body

During sliding period, the mass comes into contact with fault and hard rock, resulting in fracture, according to "The states of plastic equilibrium" by Rankine. In this process, the extreme passive pressures occurred in the area directly in front of the fault zone, due to increased horizontal stress versus vertical stress. The soil fracture is the result, based on the passive pressures increase. This fact indicates that there is a large number of cracks in the landslide foot, as well, soil "extrusion" at the slope toe (Figs. 3, 4 and 5).

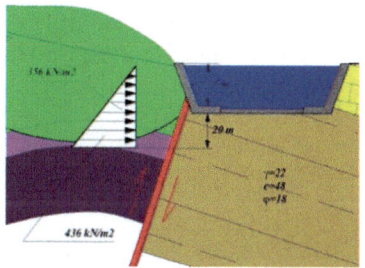

Fig. 3. Passive pressure at the natural fault and retaining structure

Fig. 4. Retaining structure overturned by passive pressure [2]

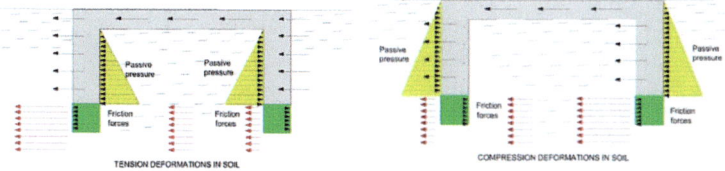

Fig. 5. Passive pressures and friction forces depending of deformations

Two houses have been exposed to the movement in the landslide. One of the houses were exposed to compression and the other one to tension deformations. Both of the houses were built on the rigid concrete foundations. Since rigid structures do not display deformation behavior as soil, the differential displacement occurs on the place of the "soil-structure" interaction. Passive pressures and friction forces react on the outer surface of structure in compression zone. In tension zone they react towards the inner surface of structure.

Passive pressures, by Rankine, increase with depth, and friction forces increase with weight of the building. Therefore, shallow foundation on larger depths and the buildings with larger weights are not acceptable in the semistable slopes.

4   Conclusion

The soil-structure interaction in landslide body has to be understood, whether we decide to preserve the existing structure or to design a new one. The observation of displacement, deformation and soil-structure interaction at the landslide Bogatići is presented in this paper. Investigation has proved that there is a fault along the left bank of the Željeznica river. Due to increased horizontal versus vertical stress, the extreme passive pressures occurred in the area directly in front of the fault zone. The soil fracture is the result of the passive pressures increase. This indicates that there is a large number of cracks in the landslide foot, as well, soil "extrusion" at the slope toe. Since rigid structures do not display deformation behavior as soil, the differential displacement occurs on the place of the "soil-structure" interaction. While, in compression zone, the passive pressures and friction forces react on the outer surface of structure, they react towards the inner surface in tension zone.

References

1. Zekan, S., Suljić, N.: Sliding causes and triggering mechanisms at the Bogatići landslide. In: Landslide and Flood Hazard Assessment, 1st Regional Symposium on Landslides in the Adriatic-Balkan Region, Zagreb, Croatia, pp. 141–146 (2014)
2. Zekan, S., Avdić, M., Hodzic, M.: Causes and mechanism of the Bogatići landslide. In: Proceedings GEO-EXPO, Tuzla 2013, Bosnia and Herzegovina, pp. 283–291 (2013)

Evaluation of Stability Analysis Methods of Embankments on Soft Clays

Akzhunis Zhamanbay⬦, Jong Kim⬦, and Sung-Woo Moon(✉)⬦

Department of Civil and Environmental Engineering,
Nazarbayev University, Astana 010000, Kazakhstan
sung.moon@nu.edu.kz

Abstract. Construction on soft clay deposits is assumed to be significant concern in geotechnical engineering field. Soft clays are characterized by low bearing capacity, high ductility and low permeability, which lead to certain constraints in embankment design. Therefore, to ensure safety of structures on soft grounds, it is necessary to define the capacity that foundation can bear before construction process. In addition, the behavior of structure has to be predicted to avoid failures or other unfavorable circumstances that could take place in the future. A number of prediction methods have been proposed, but the predictions could suffer from a lack of accuracy, resulting in a lack of confidence in practice. In this study, numerical simulation of embankments on soft soils using finite element method (FEM) is performed to evaluate existing methods for predicting performance of embankments on soft clays and to propose the most accurate stability analysis approach among reviewed methods.

Keywords: Embankment · Stability analysis · Soft clays

1 Introduction

Stability of embankments can be evaluated before or after construction. In this study, the prediction is conducted after embankment construction. Figure 1 presents indicators used for the prediction of embankment stability [1]: (1) height (H); (2) settlement under embankment center (Δs) (Point A); (3) lateral deformation (Δy) at embankment toe (Point B) which is assumed to be related to occurrence of embankment failure [3]; (4) vertical deformation at toe (Δz) (Point C) and at point beyond toe (Δz_x) (Point D) of the structure; (5) excess pore water pressure (Δu) under the center and toe of the embankment [3].

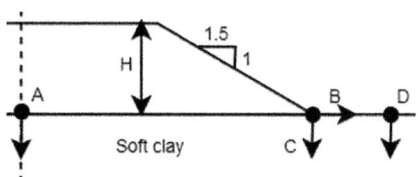

Fig. 1. Embankment parameters

© Springer Nature Switzerland AG 2018
W. Wu and H.-S. Yu (Eds.): *Proceedings of China-Europe Conference on Geotechnical Engineering*, SSGG, pp. 1565–1569, 2018.
https://doi.org/10.1007/978-3-319-97115-5_146

Literature review revealed numerous embankment stability evaluation methods. In this study, three analysis methods for evaluating embankment stability are used, as indicated in Table 1.

Table 1. Summary of three existing methods of stability analysis

#	Failure indicators	Authors
1	Plot of maximum embankment settlement (s_{max}) versus the ratio of incremental changes of maximum lateral displacement to settlement ($\Delta y_{max}/\Delta s_{max}$)	Matsuo and Kawamura [2]
2	Plot of maximum embankment settlement against maximum lateral displacement at the toe of embankment (slope between 0.7–0.9)	Tavenas and Leroueil [4]
3	Rate of lateral displacement at the toe of embankment is considerably increased with approach of failure condition, when height is between 70–90% of failure height	Hunter and Fell [1]

2   Numerical Simulation

Numerical simulation (Plaxis 2D) is carried out using soil properties from a test site given in Table 2 with Mohr-Coulomb soil constitutive model [5]. The embankment is constructed with base of 15 m and slope of 1.5:1.

Table 2. Soil properties at the test site

Parameters		Fill	Very soft to soft silty clay (-4 m)	Dark greenish grey silty clay (-13 m)	Dark grey silty clay (-22 m)	Whitish grey and firm silty clay (-24 m)
Unit weight (kN/m^3)	γ_{dry}	16.6	12.5	13.5	11.6	14.4
	γ_{sat}	18.0	15.8	16.5	14.6	17.3
Permeability (m/day)	k_x	4.00E-02	1.43E-03	1.63E-03	1.47E-03	1.01E-03
	k_y	4.00E-02	7.15E-04	8.13E-04	7.35E-04	5.05E-04
Young's modulus, E (kN/m^2)		2000	1286	1724	1088	1465
Poisson's ratio, ν		0.3	0.35	0.35	0.35	0.35
Cohesion, c (kN/m^2)		10	15	17	37	20
Friction angle, φ (°)		24	2	1	5	10

In order to evaluate three existing methods, 5 cases are simulated with different rest periods (i.e., 1, 2, 3, 5, and 10 days) between construction stages. Moreover, embankment height is increased with increments of 0.5 m corresponding to each construction stage until it reaches the failure (Fig. 2 (a)). For example, Fig. 2

(a) presents the results for case 3 with the rest period of 3 days. A rapid increase is occurred in settlement and lateral displacement close to failure height at 35 days, obtained from the deformed model for the chosen case (Fig. 2 (b)). Figure 3 plots stability analyses determined from each method. The failure from method 1 occurs at the intersection with the contour line, while the failure from method 2 is observed when the slope (Δy/Δs) suddenly increases and becomes more than 0.9. From method 3, the point when the height reaches 70% of the failure height (0.7) is referred as the failure.

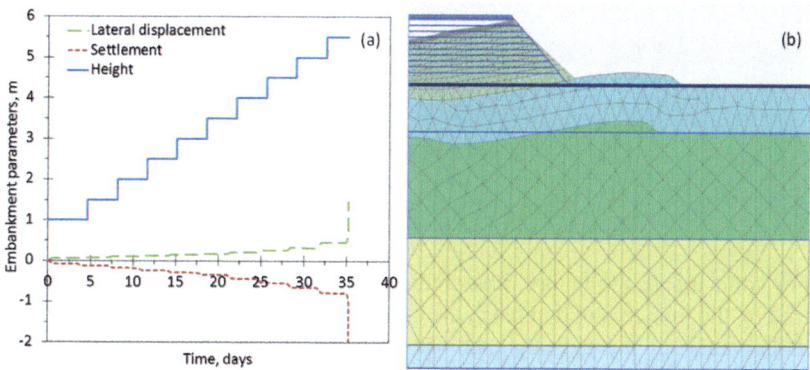

Fig. 2. (a) Change in settlement and height of the embankment with time and (b) numerical simulation output (case 3)

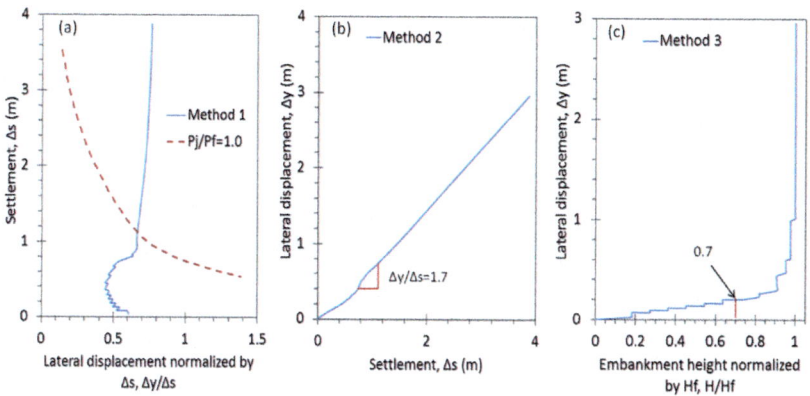

Fig. 3. Stability analysis plots by (a) method 1, (b) method 2, and (c) method 3 (case 3)

Figure 4 compares the failure height estimated from this study with that determined from three existing methods. Method 1 leads to the same failure height as this study. Although two methods (2 and 3) produce more conservative values in comparison with simulation, the predicted and simulated failure heights have similar trends with the rest time.

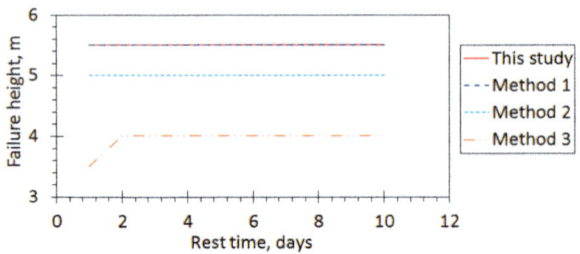

Fig. 4. Comparison of failure heights

For comparison, the failure time predicted from the existing methods can be normalized by the time required to reach the failure from the simulation. Table 3 compares the failure time of the existing methods normalized by failure time of simulation. Method 3 gives the most moderate results with high standard deviation (SD). Method 1 gives the closest period ratios of 0.96 with low SD.

Table 3. Comparison of predicted failure time normalized by failure time of simulation

#	Normalized time, days			
	This study	Method 1	Method 2	Method 3
1	1.0	1.0	0.9	0.6
2	1.0	1.0	0.9	0.7
3	1.0	1.0	0.9	0.6
4	1.0	1.0	0.9	0.7
5	1.0	1.0	0.9	0.7
Average (SD)	1.00 (0)	0.96 (0.003)	0.88 (0.013)	0.65 (0.017)

3   Conclusion

Conclusions of this study are as follows:

- The result of method 1 (Matsuo and Kawamura) is similar to that obtained from the numerical simulation, and method 3 (Hunter and Fell) gives the most conservative result among these three methods.
- The failure periods determined from numerical simulation and those estimated from three existing methods are comparable with each other, based on the average value of the normalized failure time.

Because there was no failure during construction at the test site, the three existing methods as well as other reviewed methods would be verified by performing numerical simulations with real cases with the failure during construction for future study.

References

1. Hunter, G., Fell, R.: Prediction of impending failure of embankments on soft ground. Can. Geotech. J. **40**(1), 209–220 (2003)
2. Matsuo, M., Kawamura, K.: Diagram for construction of embankment on soft ground. Soils Found. **17**(3), 37–52 (1977)
3. Leroueil, S.: Embankments on soft clays. In: CIGMAT, Houston, Texas (2006)
4. Tavenas, F., Leroueil, S.: The behaviour of embankments on clay foundations. Can. Geotech. J. **17**(2), 236–260 (1980)
5. Aziz, B.A.: Stability and Deformation Analysis of Embankment on Soft Clay, Undergraduate, Universiti Teknologi Malaysia (2010)

Centrifuge Model Test on Excavation-Induced Failure of Soil Slopes Overlying Bedrock

Yiying Zhao[(⊠)] and Ga Zhang

State Key Laboratory of Hydroscience and Engineering, Tsinghua, China
zhaoyiying13@163.com

Abstract. A great proportion of slopes are located on the bedrock, and the excavation at the slope toe often induces the slope failure. In this paper, centrifuge model tests were performed to investigate the excavation-induced failure of soil slopes overlying the bedrocks. The slope failure mode could be categorized to two types: sliding along the bedrock or failing inside the soil. The slopes underwent significant progressive failure process, and the failure mechanism could be described with a significant coupling of deformation localization and failure. The bedrock shape influenced the deformation localization and induced different failure morphology.

Keywords: Soil slope · Bedrock · Excavation centrifuge model test
Failure

1 Introduction

A great proportion of slopes are located on the bedrock. The bedrock has been proven to affect the deformation and failure of the slope significantly, however, the influential mechanism has not been clarified though numerical and physical simulations were conducted. Centrifuge modeling has been widely used to studies on the failure behavior of slopes under different loading conditions [1–4]. The toe excavation often triggers landslide [5]. In this paper, centrifuge model tests were performed to investigate the excavation-induced failure of soil slopes overlying the bedrocks of different shapes. The effect of bedrock on deformation and failure of soil slopes was analyzed on the basis of test observations.

2 Methodology

The centrifuge model tests were conducted using the centrifuge at Tsinghua University. The soil used in the tests was a type of silty clay with a specific gravity of 2.7. The dry density of the soil was 1.55 g/cm^3, and the water content was 17%. The cohesion of the soil and the internal frictional angle were 24 kPa and 27°, respectively. Two cambered bedrocks of different shapes were used in the two tests respectively (Fig. 1). The bedrock was simulated by the organic glass.

© Springer Nature Switzerland AG 2018
W. Wu and H.-S. Yu (Eds.): *Proceedings of China-Europe Conference on Geotechnical Engineering*, SSGG, pp. 1570–1573, 2018.
https://doi.org/10.1007/978-3-319-97115-5_147

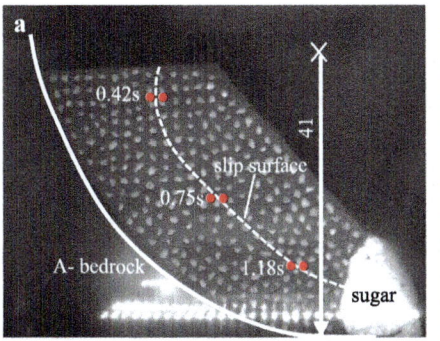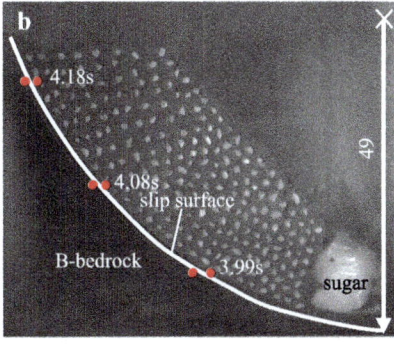

Fig. 1. Photos of slope model and failure morphology (unit: cm). **a** A-bedrock, **b** B-bedrock.

The slope model was 28 cm high with a gradient of 1.2:1. The bedrock was placed behind the slope. When preparing the soil slope model, a kind of soft sugar was used to simulate the soil to be excavated at the slope toe (Fig. 1). The excavation height was 9.6 cm.

During the centrifuge model tests, the centrifugal acceleration increased from 1 g gradually to 50 g, and then was kept unchanged for a few minutes to stabilize the deformation of the slopes. After the deformation of the slopes was stable at 50 g level, water was drawn off to dissolve the sugar to simulate the process of toe excavation. An image record system was used to capture a series of images of the slope during the excavation through the transparent organic glass of the model container [6]. The displacement of the slope was analyzed according to the photograph series.

3 Key Results

3.1 Failure Mode

The slippage displacement of the slopes along the bedrock gradually increased during the excavation (Fig. 2). For the slope overlying A-bedrock, the slippage displacement at different elevations all increased slowly and remained small when the slip surface inside the soil appeared. This result indicated that the sliding failure did not occur along the bedrock and appeared in the soil. For the slope overlying B-bedrock, the slippage displacement exhibited a significant increase at approximately 4.2 s from the beginning of excavation. The inflection points on the history curves indicated that there were sliding failure along the bedrock. There were not any slip surfaces inside the soil at that time. Therefore, the two soil slopes exhibited different failure modes. The slope overlying B-bedrock failed at the interface by sliding along the bedrock, and the slope overlying A-bedrock failed inside the soil. The shape of the bedrock changed the failure mode of the slope and affected the failure morphology of the slope induced by excavation.

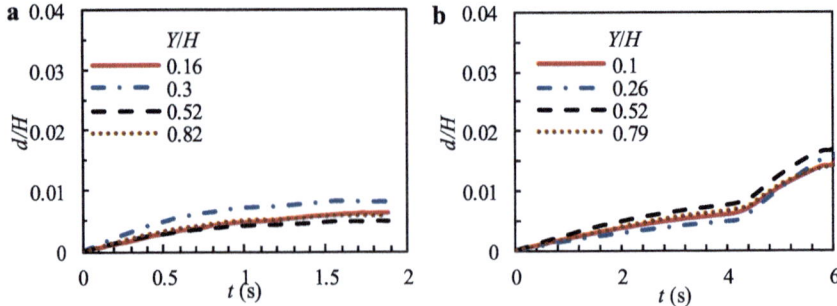

Fig. 2. Displacement histories along the bedrock. *d*, displacement along bedrock; *H*, slope height; *t*, time; *Y*, vertical distance from slope toe. **a** slope with A-bedrock, **b** slope with B-bedrock.

3.2 Failure Mechanism

To determine the development process of the slip surface, three point couples at different elevations were selected on opposite sides of the slip surface of the two slopes (Fig. 1). The relative displacements of all the point couples increased over time and had a remarkable inflection after which the relative displacement began to increase significantly. The inflection indicated the appearance of the slip surfaces of the slopes (Fig. 1). The two slopes exhibited different failure orders: for the slope overlying A-bedrock, the local failure first appeared at the lower part of the slope and it gradually developed upwards to the slope top; however, the case is contrary for the slope overlying B-bedrock. This result indicated that both slopes underwent significant progressive failure process, however, bedrocks of different shapes led to different failure orders of the slopes.

Figure 3 shows the horizontal gradient of displacement of the two slopes along horizontal profiles. The horizontal gradient was small with a uniform distribution at the beginning of the excavation. Then, the horizontal gradient increased rapidly within a certain area with an evident peak. This demonstrated that the deformation localization appeared in the slope during the toe excavation. The peak displacement gradient grew, indicating that the deformation localization became stronger. The local failure turned up inside the deformation localization zone near the position with the peak. After the slip surfaces appeared, the deformation localization near the slip surface continued to increase significantly. Therefore, the failure mechanism could be described with a significant coupling of deformation localization and failure. In other words, the deformation localization developed and caused local failure of the slope, and the local failure induced new deformation localization. The deformation localization position of the two slopes were quite different. The deformation localization of the slope overlying A-bedrock appeared inside the soil, however that of the slope overlying B-bedrock was near the bedrock. It manifested that the bedrock shape influenced the deformation localization and induced different failure morphology.

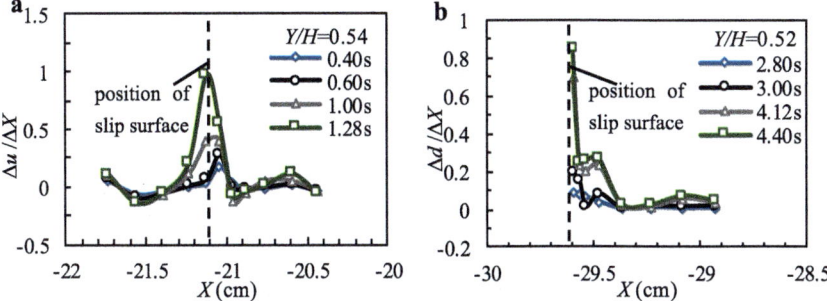

Fig. 3. Horizontal gradient of displacement of two slopes during excavation. Δd, relative displacements along bedrock; Δu, relative horizontal displacements; H, slope height; X, Y, horizontal and vertical distances from slope toe. **a** slope with A-bedrock, **b** slope with B-bedrock.

4 Conclusions

1. The failure mode of the slopes overlying the bedrock was categorized to two types: sliding along the bedrock or failing inside the soil under excavation.
2. The slopes underwent significant progressive failure process. The bedrock shape could alter the failure sequence of the slopes.
3. The failure mechanism of the excavation-induced failure of soil slopes overlying the bedrock could be described with a significant coupling of deformation localization and failure. The bedrock shape influenced the deformation localization and induced different failure morphologies.

References

1. Viswanadham, B.V.S., Rajesh, S.: Centrifuge model tests on clay based engineered barriers subjected to differential settlements. Apply Clay Sci. **42**(3–4), 460–472 (2009)
2. Zhang, G., Hu, Y., Wang, L.P.: Behaviour and mechanism of failure process of soil slopes. Environ. Earth Sci. **73**(4), 1701–1713 (2015)
3. Zornberg, J.G., Sitar, N., Mitchell, J.K.: Performance of geosynthetic reinforced slopes at failure. J. Geotech. Geoenviron. Eng. **124**(8), 670–683 (1998)
4. Wang, L.P., Zhang, G.: Progressive failure behavior of pile-reinforced clay slopes under surface load conditions. Environ. Earth Sci. **71**(12), 5007–5016 (2014)
5. Stark, T.D., Arellano, W.D., Hillman, R.P., Hughes, R.M., Joyal, N., Hillebrandt, D.: Effect of toe excavation on a deep bedrock landslide. J. Perform. Constructed Facil. **19**(3), 244–255 (2005)
6. Zhang, G., Hu, Y., Zhang, J.M.: New image analysis-based displacement-measurement system for geotechnical centrifuge modeling tests. Measurement **42**(1), 87–96 (2009)

Discrete Element Analyses
of Earthquake-Induced Landslide

Tao Zhao[1]($\boxtimes$) (iD), Giovanni B. Crosta[2] (iD), and Nuwen Xu[1]

[1] State Key Laboratory of Hydraulics and Mountain River Engineering,
College of Water Resource and Hydropower, Sichuan University,
Chengdu 610065, China
zhaotao@scu.edu.cn
[2] Department of Earth and Environmental Sciences,
Università degli Studi di Milano Bicocca,
Piazza della Scienza 4, 20126 Milan, Italy

Abstract. The discrete element method (DEM) has been employed to analyze the Tangjiashan landslide induced by the 2008 Ms 8.0 Wenchuan earthquake. In the DEM model, the layered structure of the slope mass can be obtained by generating each sub-layer separately, with assigned different strength and stiffness properties. The bottom intact rock layer was initially bonded together with the basal failure plane to represent initially intact slope, and the site-recorded seismic shaking wave components are used as base excitations. The numerical results show that the slope mass can maintain a stable state under relatively low seismic shaking motions. As the seismic shaking intensity increases, the fractures/cracks occur and propagate gradually along the basal failure plane. The numerical simulations illustrate clearly the progressive failure and subsequent valley damming of the Tangjiashan landslide, from which some mechanisms of slope motion and deformation, as controlled by the complex layering geometries are presented.

Keywords: Discrete element method · Landslides · Layered structure
Seismic shaking · Fracture

1 Introduction

Earthquake induced landslides can be catastrophic as they are always associated with almost instantaneous slope collapse and spreading, posing significant hazards to human lives and lifeline facilities worldwide. These events are widely observed in the main earthquake fault and on the hanging-wall side [1]. In addition, landslides occurred near the deeply incised river valleys with steep bank slopes could potentially create landslide dams blocking the river channel. These landslide dams may frequently fail catastrophically, leading to serious downstream inundation and flooding, often with large social and economic consequences. For instance, more than 15,000 rockslides, landslides and debris flows were induced by the 2008 Wenchuan earthquake, leading to the formation of more than 250 landslide dams, posing high risk of secondary hazards (e.g. landslides and floods) to the downstream areas [2]. These phenomena are under

W. Wu and H.-S. Yu (Eds.): *Proceedings of China-Europe Conference on Geotechnical Engineering*, SSGG, pp. 1574–1578, 2018.
https://doi.org/10.1007/978-3-319-97115-5_148

intensive research due to their significant destructive power as well as the still unexplained initiation and propagation mechanisms of slope failure.

The 2008 Ms 8.0 Wenchuan earthquake triggered lots of catastrophic landslides, rock falls and debris flows, distributed along the Longmenshan seismic fault zone within a 300 km long and 10 km wide region. Among these landslides, the Tangjiashan landslide is one of the largest and most dangerous one, with a total displaced mass of approximately 2.04×10^7 m^3 [2]. This landslide occurred immediately after the earthquake and it moved atop of the fragmented bedrock scouring the bank of Jianjiang River for nearly 2400 m and subsequently forming a large landslide dam [3] (see Fig. 1). The landslide dam, as described by Xu et al. (2009), in a first approximation, consists of mainly three layers of the grayish black siltstone of the Qingping Formation of Cambrian age, with a slight increase in grain size with depth. The current research has focused on modeling the progressive slope failure under seismic shakings via discrete element method (DEM) [4], aiming to provide new insights into the detailed micro- and macro- responses of rock mass during the slope failure and landslide propagation.

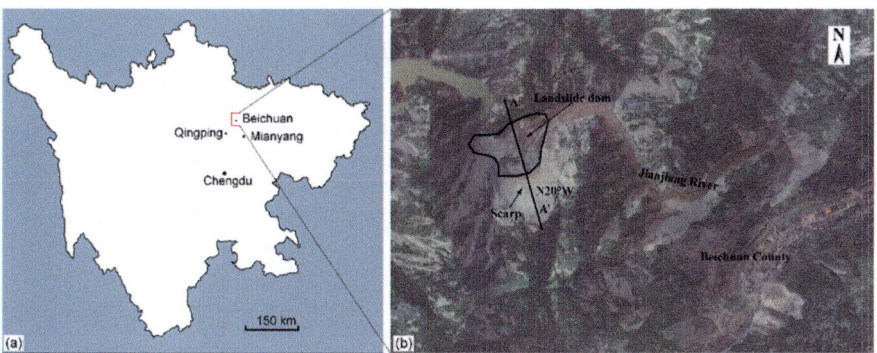

Fig. 1. Location and aerial view of the Tangjiashan landslide dam (A–A' is the cross section studied in this paper)

2 DEM Model Configurations

In the current study, the open source discrete element method (DEM) code ESyS-Particle was employed to run simulations presented herein. The basic DEM theory and rigorous calibrations can be found in Zhao *et al.* [5], and will not be repeated here. In short, the brittle slope rock mass was simulated as an assembly of particles cemented together via the so-called parallel bond model. The motion of each single particle is governed by the Newton's second law of motion, and the interaction between bonded particles can be calculated by the well-defined linear-elastic model.

According to the layered deposit structure, the initial slope mass is assumed to be composed of three layers aligning approximately parallel to the slope failure plane in the DEM simulations. The performed simulations of Tangjiashan landslide have a plane strain boundary condition in which the size of the out of plane direction model is set as 20 m. In this framework, a number of frictionless particles were used to fill up this unit slope domain which can be regarded as one fraction of the real Tangjiashan slope. The failure plane is represented by a collection of triangular meshes, onto which a layer of uniformly graded grains (r = 1500–2000 mm) were glued to simulate a non-erodible and rough base. The generated DEM model is shown in Fig. 2, and the total number of particles in layer I, II and III are 15,787, 34,911 and 46,712, respectively. The granular assembly is bonded together to represent the initially intact slope mass, with very high strength for the non-completely disintegrated rock strata (I), normal strength for the boulders and blocks (II) and low strength for the fragmented rocks and soils (III). The seismic acceleration records from Qingping seismic station in Mianzhu county are used as the representative seismic wave data for the Tangjiashan landslide.

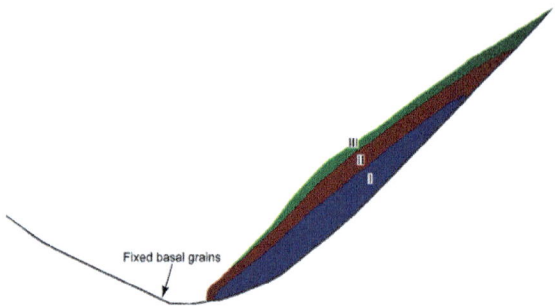

Fig. 2. DEM model of the Tangjiashan slope mass consisting of three layers: I: non-completely disintegrated rocks; II: boulders and blocks; III: fragmented rocks with soil.

3   Results

The distribution of slope damages can be visualized by plotting red dots within the slope profile at locations where bonds break, as illustrated in Fig. 3. According to the figure, it is clear that before 20 s (b), the slope mass remains almost intact, and only very few bond breakages occur near the basal failure plane. Then, as the ground vibration intensity increases, many bond breakages accumulate gradually near the base to form a connected fragmented debris layer beneath the slope mass (see (c)), which facilitates the subsequent landslide motion. At 22 s (c), several transversal cracks can be observed at the bottom, middle and upper rear regions of the slope, as indicated by the arrows. The subsequent landslide propagation and deposition would cause more damage of the slope structure, especially in the sliding front (A), middle (B, C) and rear (D) regions, as reflected by the enlargement of the cracks. In these regions, the sliding front (A) has experienced very intensive interaction with the bed rock as the landslide collides onto the other bank of the valley, while the middle slope suffers strong compressions from the lower and upper slope regions. The upper rear region contains mainly eluvial debris with very weak rock structure which is prone to disintegration during landslide. During the deposition stage (g–i), the transversal cracks can grow to completion, together with the gradual increase of their sizes due to intensive compression and shearing. In particular, the region just behind the sliding front (II and III) suffers very strong compression from the upper slope, resulting in very high and concentrated rock damage there (see the regions enclosed by dashed circles). The final debris deposition in (i) shows clearly five large intact rock blocks as delimited by the major cracks.

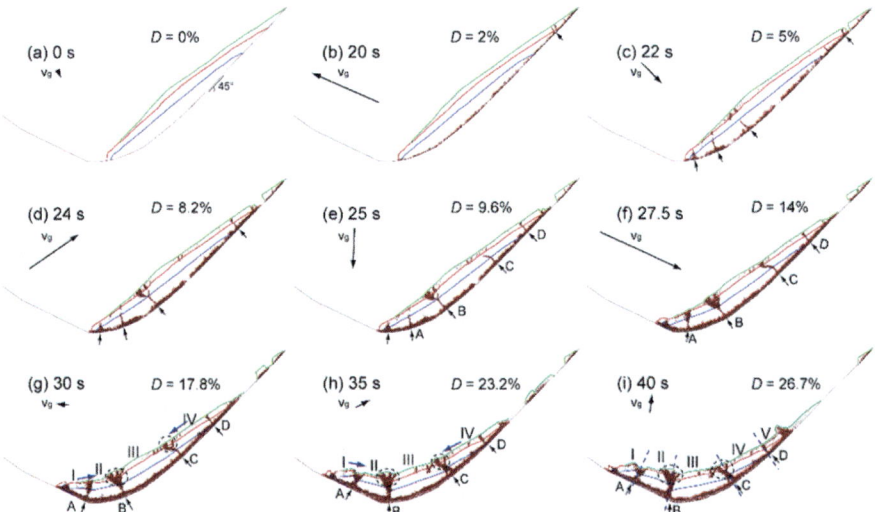

Fig. 3. Distribution of accumulative slope rock damage during the landslide. Small arrows indicate the major crack developed region, and the dashed circles in (g–i) show the compression induced concentrated damage zone. The ground seismic motion (v_g) is plotted as a vector on each figure. The percentage of internal rock damage (D) is also labeled on each plot.

4   Conclusions

During the co-seismic landslide, the rock damages concentrate mainly near the slope failure plane, and the landslide motion causes more internal rock damages, which finally divides the deposit into five large intact rock blocks. The rock fragmentation also increases the overall mobility of landslide spreading.

References

1. Huang, R.Q., Li, W.L.: Analysis of the geo-hazards triggered by the 12 May 2008 Wenchuan Earthquake, China. Bull. Eng. Geol. Environ. **68**(3), 363–371 (2009)
2. Xu, Q.: Landslide dams triggered by the Wenchuan Earthquake, Sichuan Province, south west China. Bull. Eng. Geol. Environ. **68**(3), 373–386 (2009)
3. Yin, Y., Wang, F., Sun, P.: Landslide hazards triggered by the 2008 Wenchuan earthquake, Sichuan, China. Landslides **6**(2), 139–152 (2009)

4. Cundall, P.A., Strack, O.D.L.: A discrete numerical model for granular assemblies. Géotechnique **29**(1), 47–65 (1979)
5. Zhao, T.: Investigation of rock fragmentation during rockfalls and rock avalanches via 3-D discrete element analyses. J. Geophys. Res. Earth Surf. **122**(3), 678–695 (2017)

Deformation and Instability Mechanism of Reservoir Landslide: A Case Study

Changjun Zhao$^{(\boxtimes)}$ and Minghui Xu

Changjiang Survey, Planning and Design Research Co., Ltd., Wuhan, China
genghiskhanchina-m@163.com

Abstract. The reservoir landslide is of sudden, uncertain and destructive characteristics, and the deformation stage of the classic landslide is divided into three stages: initial deformation, constant velocity deformation and acceleration deformation (see Fig. 1). In the case of a landslide with accelerated deformation indication, it is of great significance to accurately judge which stage of acceleration deformation it is in for forecasting impending slide and risk and disaster assessment. Taking the 5 million m^3 slide mass high-speed reservoir entry event of Muzhuping landslide as an example, in this paper, the failure mechanism and instability characteristics of high-speed deformation landslide are analyzed. Combined with the successful forcasting experience of two high-speed reservoir entry landslide cases and a wide range of geological survey and related theoretical analysis, this study provides some useful experience and enlightenment for landslide hazard and forecasting impending slide.

Keywords: Reservoir landslide · Muzhuping landslide
High speed reservoir entry · Critical sliding prediction

1 Deformation Process of Muzhuping Landslide

Muzhuping landslide on the Modao River as the tributary of Qingjiang River had the high-speed entry event of 5 million m^3 sliding mass (surge height of 48 m, see Fig. 2) during the initial impoundment period. The landslide is located in the quaternary huge deposit slope area, and the substances focus on the collapse deposit broken stones and heavy stones. The slide mass is about 57–110 m thick and has an area of 1.065 km^2 and total volume of 78 million m^3 (see Fig. 2) and it is divided into 3 areas: Area A, Area B, Area C. The slope gradient of the Modao River reservoir bank slope is 26°–35°, and the normal impoundment level of the reservoir is 400 m, with the main flood control season between May and August every year. From May 2 to May 11, 2007, for the front edge of landslide from the occurrence of ground crack, small-scale band caving to large-scale collapse and slump and transient barrier lake, the time cycle for actually constituting the large-scale landslide event had three times, and the deformation pattern could be summarized as "creep - acceleration -destruction - balance" (see Fig. 3).

© Springer Nature Switzerland AG 2018
W. Wu and H.-S. Yu (Eds.): *Proceedings of China-Europe Conference on Geotechnical Engineering*, SSGG, pp. 1579–1582, 2018.
https://doi.org/10.1007/978-3-319-97115-5_149

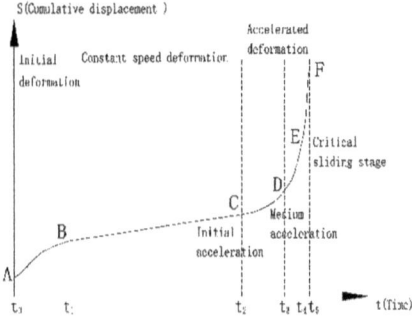

Fig. 1. Typical landslide displacement (S) – time (t) curve and division of deformation stages

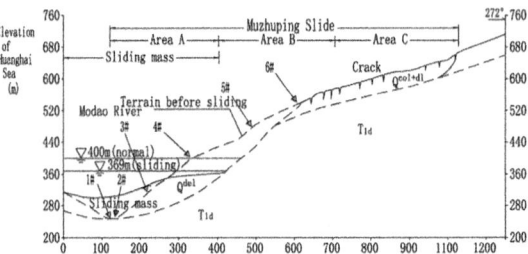

Fig. 2. B-B' longitudinal profile of Muzhuping landslide sliding area

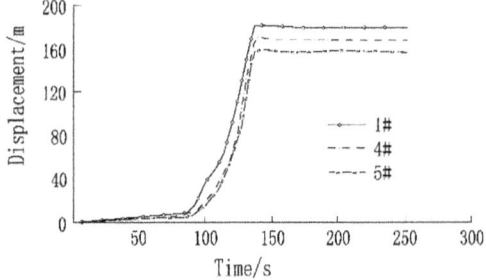

Fig. 3. Curve graph of sliding displacement process of sliding mass

2 Landslide Monitoring and Deformation Status

2.1 Landslide Monitoring

The engineer has continuously conducted the professional deformation monitoring since May 2007, lasting for 10 years. The monitoring result reveals that, the sliding area basically tends to balance now. After experiencing the severe deformation – constant speed deformation, the maximum horizontal displacement of surface in Area A is about 889.6 mm, as the area with the largest and fastest deformation at present. Area B has larger deformation, with the general deformation of about 727.3–1,068.5 mm and the maximum deformation of about 1,084.7 mm. Area C has the slowest rate with the maximum deformation of about 20.3 mm. The analysis suggests that, the overall landslide deformation now is at the stage of constant speed deformation (see Fig. 4). However, influenced by the hundred-year flood of reservoir on July 19, 2016, some accelerated deformation signs occurred, and it was very difficult to predict the future deformation trend and disaster risk accurately.

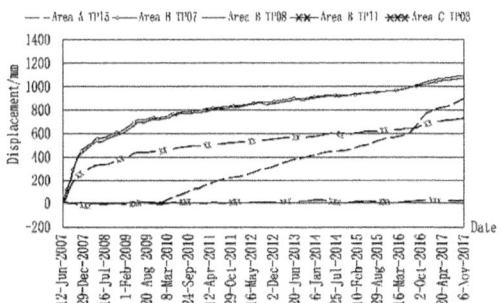

Fig. 4. Curve graph for horizontal cumulative displacement changing process of monitoring points

2.2   Trend Prediction and Risk Assessment of Landslide Deformation

After slide mass high-speed reservoir entry event of sliding area, the geological engineer made the correct disaster risk assessment and deformation trend prediction of three stages in 2007, i.e. 24-h warning and monitoring at the first stage, continuous monitoring and warning and removal of some residents at the second stage, and removal of all residents and designation of disaster affected area at the third stage. Meanwhile, the engineer deemed that, the landslide deformation would continue for a long term, but two possibilities existed, i.e. landslide deformation possibly tending to balance (the first prediction conclusion) or accelerated slippery risk (the second prediction conclusion). The latest comprehensive assessment considers that, Muzhuping landslide is wholly at the stage of constant speed deformation now, but some factors involving rainfall and flood leads to the increase in the risks of local sliding mass and instability reservoir entry again, which conforms to the previous judgment in 2007 and the second prediction conclusion.

2.3   Mechanism Analysis of Landslide Deformation

While summarizing the high-speed reservoir entry mechanism of Muzhuping landslide, it can be simply expressed as "initial impoundment and soaking inducing the small-scale critical reservoir collapse → fast rise of reservoir water level increasing the buoyancy and sliding mass weight → heavy rainfall increasing the pore water pressure and seepage pressure → sliding mass forming the circular sliding surface yet sliding at high speed". The exterior key controlling factors are mainly manifested as the atmospheric rainfall, rising rate and frequency of reservoir water (flood), and seismic activity intensity, etc.

3 The Intrinsic Inducement Mechanism of High - Speed Deformation Landslide

By applying the above deformation analysis mode and prediction theory, the geological engineer had successfully predicted two high speed instability reservoir entry events and to avoid the heavy casualties of 165 people and great property losses.

For the landslide with high speed slide risks, it always has "creep - drastic deformation (jerk) - destruction" deformation pattern and generally has the following characteristics. (1) The whole landslide has relatively steep terrain and large height difference. The bank slope angle is more than 25°. (2) It mostly presents the loose or broken bank slope geologic structure. (3) The deformation speed mainly depends on the size of corresponding sliding force. (4) The (approximately) forward bank slope structure is easier to induce the landslide with high speed sliding risks.

4 Experience and Enlightenment

- The landslide prediction theories and methods are various, and the diversity, uncertainty and fuzziness criteria must be abandoned, while the uniqueness and certainty criteria should be regarded as the critical sliding prediction criteria.
- The accelerated deformation stage of landslide displacement – time curve can be divided into the initial acceleration, medium acceleration, and critical sliding stages. The three stages have large essential difference before and after motion states.

References

1. Zhou, J.J., Zhang, Y., Dong, Z.H., Tang, Y.C., Huang, X.G.: Numerical simulation analysis on deformation failure process of Muzhuping landslide. Yellow River **31**(8) (2009)
2. Xu, Q., Tang, M.G., Xu, K.X., Huang, X.B: Research on landslide spatial-temporal evolution rule and warning and prediction. Chin. J. Rock Mech. Eng. **27**(6) (2008)
3. Zeng, Y.P.: Research on prediction and forecast of heavy sudden landslide disaster. Doctoral dissertation, Chengdu University of Technology (2009)

Effect of Slope Angle on Stabilizing Piles in C-φ Soil

Mingxing Zhu[1][($\boxtimes$)] (iD), Hongqian Lu[1], Weiming Gong[2],
and Zhihui Wan[2]

[1] China Energy Engineering Group Jiangsu Power Design Institute,
Nanjing 211102, China
phd_mxingzhu19856l@vip.163.com
[2] Southeast University, Nanjing 210096, China

Abstract. To evaluate the influence of slope angle on lateral force acting on anti-slide row piles in c-φ soil landslide, this work firstly proposes the numerical solution to determine nonlinear sliding surface of sliding wedge of soil. Accordingly, recursive solution for lateral pressure acting on active side of stabilizing piles is derived by flat element method. Furthermore, analytical solution for distribution of soil-pile pressure acting on anti-slide piles is obtained by utilizing the improved plastic deformation theory which takes account of soil-arching effect. The subsequent comparison shows that the results from predictions match well with which from tests and implies that the proposed approach can be employed to evaluate lateral force on stabilizing piles effectively. Finally, parametric study is performed and results illustrate that landslide thrust (i.e., passive load) on stabilizing pile will increase with increasing soil inner friction angle (φ), inclined slope surface angle (α) and with decreasing cohesion (c) and the ratio of pile spacing to pile diameter (D_1/d).

Keywords: Stabilizing piles · Landslide thrust · C-φ soils
Flat element method

1 Introduction

As an effective reinforcement technique, anti-slide row piles are widely used in slope projects to prevent landslide. During the design procedure of stabilizing piles, determination of the landslide thrust acting on pile shaft has drawn much attention. Ashour *et al.* [1] obtained analytical solutions of passive load acting on pile by strain wedge method. Ito *et al.* [4] has proposed plastic deformation theory considering soil-arcing effect to derive ultimate landslide thrust. Based on the double-soil-arcing effect theory, He *et al.* and Zhu *et al.* [2, 7] have done the same work for piles embedded in sandy slope.

To effectively examine the influence of inclined slope surface angle on lateral force (i.e., passive load) acting on anti-slide row piles in c-φ soil landslide, this work firstly establishes recursive solution for lateral pressure acting on active side of stabilizing piles. Furthermore, analytical expression of landslide thrust is derived and verified by

© Springer Nature Switzerland AG 2018
W. Wu and H.-S. Yu (Eds.): *Proceedings of China-Europe Conference
on Geotechnical Engineering*, SSGG, pp. 1583–1587, 2018.
https://doi.org/10.1007/978-3-319-97115-5_150

case study. Finally, parametric studies are performed to reveal its influence factors and corresponding effects.

2   Solution for Landslide Thrust Acting on Pile Shaft

As shown in Fig. 1, vertically row piles with diameter d, pile spacing D_1 and net distance D_2 ($= D_1-d$) are embedded in c-φ slope soil of surface slope α, unit weight γ. The internal friction angle and cohesion of interface between soil and pile are δ and c_w, respectively. The length of row piles in sliding zone is h which is divided into n identical parts. Thus, based on the equilibrium relationship of lateral forces, vertical forces and moments, the lateral pressure acting on pile shaft can be calculated by

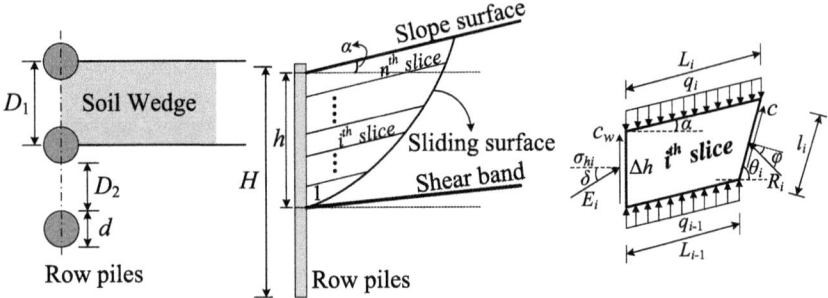

Fig. 1. Analytical models of stabilizing row piles in inclined slope

$$\sigma_{hi} = (q_iL_i - q_{i-1}L_{i-1} + \gamma S_i - c_w\Delta h - cl_i\cos\varphi/\sin(\theta_i - \varphi)) \times \frac{\sin(\theta_i - \varphi)\cos\delta}{\Delta h\cos(\delta + \varphi - \theta_i)} \tag{1}$$

where $L_i = \Delta h \sum_{j=1}^{i} \cos\theta_i/\sin(\theta_i - \varphi)$, $\Delta h = h/n$; $l_i = \Delta h\cos\alpha/\sin(\theta_i - \alpha)$; $S_i = \Delta h \cos\alpha l_{mi}$, $l_{mi} = (L_i + L_{i-1})/2$; $\theta_i = 0.5[\varphi + \alpha + \arccos(J\sin\alpha/(c\cos\varphi + J\sin\varphi))]$ can be obtained by method of [7] and parameter J can be given by method of [3]:

$$J = \left\{ \frac{(\gamma z\cos^2\alpha + c\cos\varphi\sin\varphi) -}{\sqrt{\gamma^2 z^2\cos^2\alpha(\cos^2\alpha - \cos^2\varphi) + c^2\cos^2\varphi + 2c\gamma z\cos^2\alpha\cos\varphi\sin\varphi}} \right\}/\cos^2\varphi \tag{2}$$

in which $z = i \times \Delta h$, and relation of q_{i-1} and q_i can be described as

$$q_{i-1} = q_iB_0 + \gamma\Delta hB_1 + c_wB_2 - cB_3 \tag{3}$$

in   which $B_0 = T_0/T_1, B_1 = l_{mi}^2\cos\alpha(2\sin(\delta - \alpha)\sin(\theta_i - \varphi) - \cos\alpha\cos(\delta + \varphi - \theta_i))/T_1$; $B_2 = 2l_{mi}\Delta h \times (\cos\alpha\cos(\delta + \varphi - \theta_i) - \sin(\delta - \alpha)\sin(\theta_i - \varphi))/T_1$; $B_3 = 2l_il_{mi}\sin(\delta - \alpha)\cos\varphi/T_1$; $T_0 = 2l_{mi}L_i\sin(\delta - \alpha) \times \sin(\theta_i - \varphi) - L_i\cos(\delta + \varphi - \theta_i)$

$(L_i\cos\alpha - l_i\cos\theta_i)$, $T_1 = 2l_{mi}L_{i-1}\sin(\delta - \alpha)\sin(\theta_i - \varphi) - L_{i-1}\cos(\delta + \varphi - \theta_i)$ $(L_{i-1}\cos\alpha - l_i\cos\theta_i)$ Combined lateral pressure along pile shaft of Eq. 1 and the improved plastic deformation theory [5], landslide thrust acting on pile side can be described as

$$p_{hi} = \sigma_{hi}f(\varphi) + g(c, \varphi) \qquad (4)$$

in which p_{hi}/(kN/m) is the landslide thrust (i.e., passive load); σ_{hi}/kPa is lateral pressure calculated by Eq. 1; functions $f(\varphi)$ and $g(c, \varphi)$ are as followings:

$$\begin{cases} f(\varphi) = \beta_\varphi N_\varphi + \lambda_\varphi N_\varphi \left[1 + \beta_\varphi N_\varphi / D_2\right] \\ g(c, \varphi) = 2c\beta_\varphi \sqrt{N_\varphi} + c\left[((D_1 - D_2)/D_2)N_\varphi\left(1 + 2\sqrt{N_\varphi}\right)\tan(\pi/8 + \varphi/4) + 2\sqrt{N_\varphi}\right] \times \\ \quad [(D_1 - D_2)(1 + \tan\varphi)\cot(\pi/4 + \varphi/2)] + c(D_1 - D_2)\cot(\pi/4 + \varphi/2) \end{cases}$$

$$(5)$$

where for sandy soil, $g(c, \varphi) = 0$; $N_\varphi = \tan^2(45° + \varphi/2)$; $\beta_\varphi = (D_1 - D_2)\tan\varphi \times \tan(\pi/8 + \varphi/4)$; $\lambda_\varphi = (D_1 - D_2)[1 + \tan\varphi\cot(\pi/4 + \varphi/2)]$.

3 Cases and Parameters Study

[6] has reported a research on landslide thrust on stabilizing piles as shown in Fig. 2a, in which piles with diameter $d = 0.4$ m, pile spacing $D_1 = 0.9$ m and net distance $D_2 = 0.5$ m are embedded in slope of surface slope $\alpha = 11°$, $\varphi = 28°$, unit weight $\gamma = 19$kN/m³; $\tan\delta = 0.7\tan\varphi$ and $c_w = 0.7c$ ($c = 0$). The comparison results are illustrated in Fig. 2b which implies that compared with other theoretical methods, results from proposed approach match well with measured and Flac³D results.

Based on the case studied in Fig. 2b, influence of cohesion c, inner friction angle φ, inclined surface slope α and ratio of pile spacing to diameter D_1/d on landslide thrust acting on pile shaft has been performed and corresponding results are shown in Fig. 3,

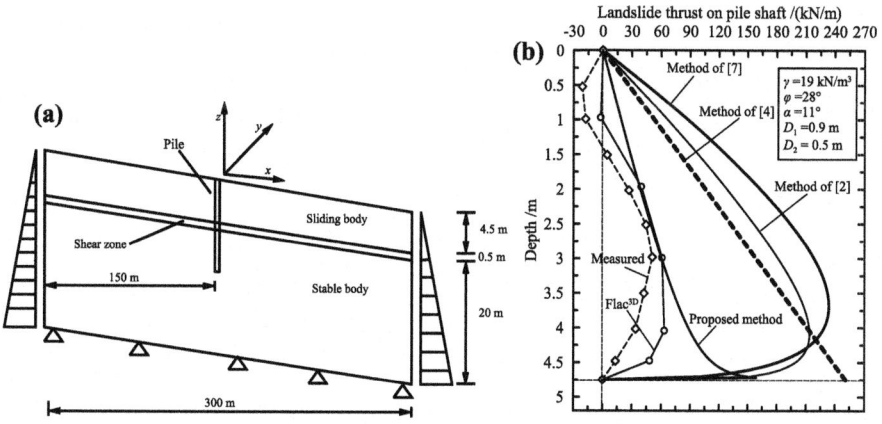

Fig. 2. a Sketch of stabilizing piles in slope, **b** experimental, numerical and theoretical results

which implies that landslide thrust on stabilizing pile will increase with increasing soil inner friction angle (φ), inclined slope surface angle (α) and with decreasing cohesion (c) and the pile spacing ratio (D_1/d).

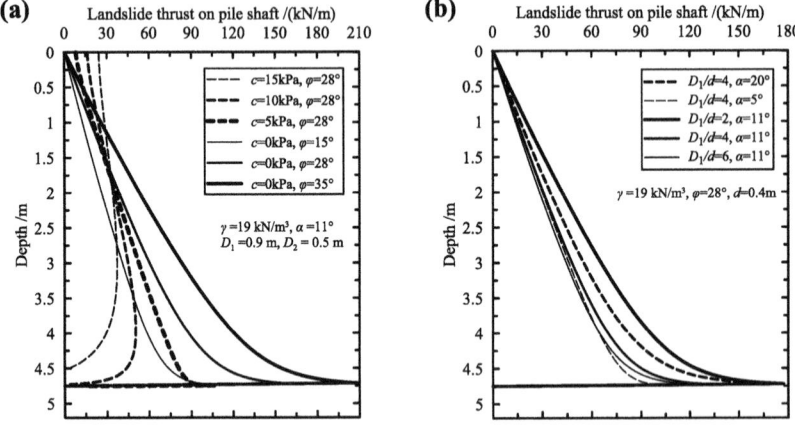

Fig. 3. **a** Influence of c and φ, **b** influence of α and D_1/d ($c = 0$)

4   Conclusions

Combined with Rankine theory and Mohr's pole theory, this work proposed method to calculate nonlinear sliding surface of soil wedge. Then, by means of improved plastic deformation theory, this paper deduced recursive solution for landslide thrust acting on active side of stabilizing piles, which is verified by case study. Finally, parametric studies indicate that passive load will increase with increasing φ, α and with decreasing cohesion c and the pile spacing ratio D_1/d.

Acknowledgements. The authors would like to acknowledge financial support from China Postdoctoral Science Foundation (2017M611955), Jiangsu Province Postdoctoral Science Foundation (1701028B) and Science & Technology Project of JSPDI (32-JK-2016-003).

References

1. Ashour, M., Ardalan, H.: Analysis of pile stabilized slopes based on soil-pile interaction. Comput. Geotech. **39**, 85–97 (2012)
2. He, Y., Hazarika, H., Yasufuku, N.: Evaluating the effect of slope angle on the distribution of the soil-pile pressure acting on stabilizing piles in sandy slopes. Comput. Geotech. **69**, 153–165 (2015)
3. Iskander, M., Chen, Z., Omidvar, M.: Active static and seismic earth pressure for c–φ soils. Soils Found. **53**(5), 639–652 (2013)

4. Ito, T., Matsui, T.: Methods to estimate lateral force acting on stabilizing piles. Soils Found. **15**(4), 43–59 (1975)
5. Kumar, S., Hall, M.L.: An approximate method to determine lateral force on piles or piers installed to support a structure through sliding soil mass. Geotech. Geol. Eng. **24**(3), 551–564 (2006)
6. Lirer, S.: Landslide stabilizing piles: experimental evidences and numerical interpretation. Eng. Geol. **149**, 70–77 (2012)
7. Zhu, M.X., Zhang, Y., Gong, W.M.: Discussion on "Evaluating the effect of slope angle on the distribution of the soil-pile pressure acting on stabilizing piles in sandy slopes". Comput. Geotech. **79**, 176–181 (2016)

Part XII: Unsaturated Soils and Energy Geotechnics

Energy Utilisation and Ground Temperature Distribution of a Field Scale Energy Pile Under Monotonic and Cyclic Temperature Changes

Mohammed Faizal$^{(\boxtimes)}$ and Abdelmalek Bouazza

Monash University, Melbourne, VIC 3800, Australia
mohammed.faizal@monash.edu

Abstract. The operating modes of energy piles depend on the thermal energy requirements of the built structures. The various operating modes of the energy piles lead to variations in the energy utilisation and ground temperature distribution. This paper presents the results obtained from a field-scale energy pile subjected to different modes of operations. In particular, it is found that cyclic temperatures provide better energy utilisation compared to monotonic temperatures. Also, the ground temperatures reduce with increasing radial distance from the pile for all operating modes.

Keywords: Energy pile · Field tests · Energy utilization · Ground temperatures

1 Introduction

Energy piles are foundation piles that act as underground heat exchangers when coupled with ground source heat pumps (GSHPs) for heating and cooling built structures [1]. The operating principle of energy piles is that the ground at a constant temperature below ground surface is utilised as a heat source or sink to operate a heat pump, which helps maintain comfortable temperatures in built structures. Depending on the usage requirements, energy piles undergo various temperatures including monotonic heating and cooling and daily cyclic temperature changes from intermittent operations of the GSHP [2]. The ground temperatures in the intermittent operating modes during the non-operating times of the GSHP can recover naturally or can be forcefully recharged using solar energy or cooling towers to improve geothermal energy utilisation and to help maintain a balance of ground temperatures [4]. The effects of monotonic and cyclic temperature changes have been widely studied for borehole heat exchangers, whereas very limited studies have been conducted on field scale energy piles. Investigations of energy utilisation and ground temperature distribution are required from field scale tests for better evaluation of the effects of temperature variations under real boundary conditions. The present study discusses results from tests conducted on a field scale energy pile for monotonic heating, monotonic cooling, intermittent operation with natural ground thermal recovery, and intermittent operation with forced ground thermal recovery.

© Springer Nature Switzerland AG 2018
W. Wu and H.-S. Yu (Eds.): *Proceedings of China-Europe Conference on Geotechnical Engineering*, SSGG, pp. 1591–1594, 2018.
https://doi.org/10.1007/978-3-319-97115-5_151

2 Energy Pile Details and Experimental Procedures

The experiments were conducted on a 0.6 m diameter, 16.1 m long energy pile installed in mostly dense sand. A schematic of the energy pile is shown in Fig. 1. There were three U-loops formed from high-density polyethylene pipes which extended to a depth of 14.2 m in the pile. The fluid inlet and outlet temperatures were recorded at the pile head. The ground temperatures were monitored at radial distances of 0.5 m and 2 m from the pile edge, and up to a depth of 18 m at 2 m intervals. There were no head restraints on the energy pile, and the pile head and the ground surface were exposed to the atmosphere. More details of the energy pile are reported elsewhere [6]. There were four sets of experiments conducted in this study, the details of which are listed in Table 1. Only 20 days of data are considered herein for assessing the effects of monotonic and cyclic temperature change on the energy extracted/injected and on the ground temperatures. The thermal responses of the energy pile are presented elsewhere [3].

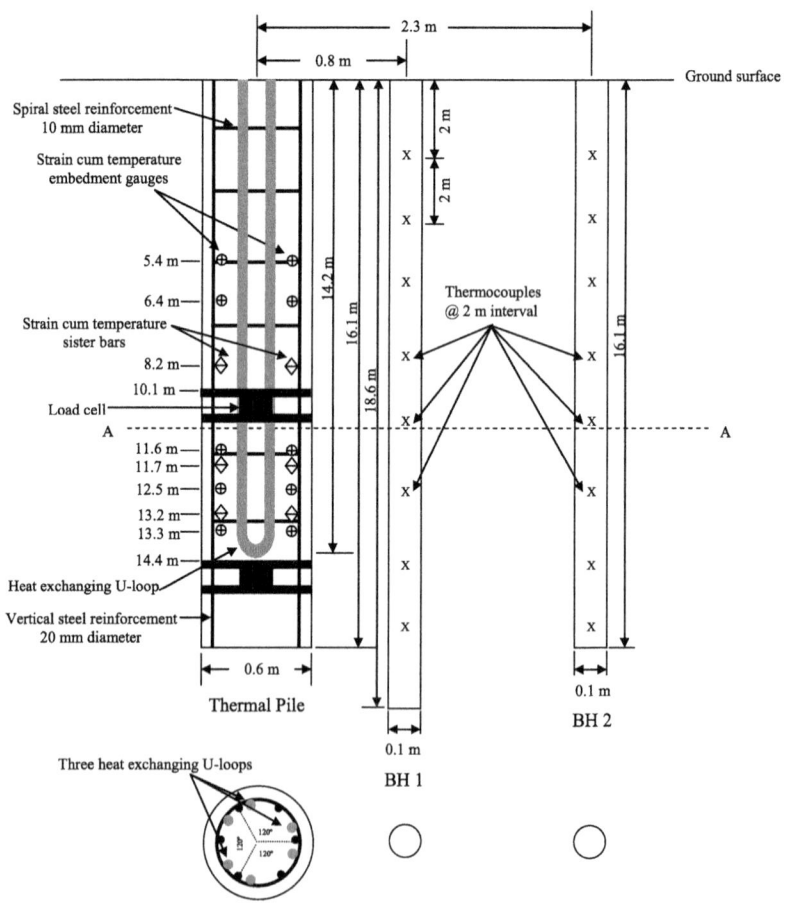

Fig. 1. Schematic diagram of the energy pile showing the sensor locations [5].

Table 1. Summary of experiments [3].

Operating mode	Description	Inlet water temperatures	Inlet water flowrates	Experiment duration
24H	24 h heating, daily	45 °C	10 LPM	52 days
24C	24 h cooling, daily	5 °C	15 LPM	24 days
16N	16 h cooling and 8 h rest, daily	5 °C	15 LPM	25 days
16F	16 h cooling and 8 h heating, daily	7 °C to 16 °C in the cooling cycle	15 LPM in the cooling cycle	24 days
		30 °C to 55 °C in the heating cycle	13.5 LPM in the heating cycle	

3   Results and Discussions

The average energy extracted/injected from/to the ground is shown in Fig. 2a, where the positive magnitudes are energy extracted from the ground and negative magnitudes are energy injected into the ground. The energy extracted in the 24C and 16N modes are similar due to similar changes in fluid temperatures between the inlet and outlet of the pile. Energy extraction from the ground is improved in the 16F mode compared to the 24C and the 16N modes. This is due to a higher recovery in ground temperatures in the 16F mode, shown in Fig. 2b. The energy injected into the ground in the 16F mode is also larger than the 24H mode, as the cold water in the cooling cycle of the 16F mode enhances the temperature difference between the ground and the fluid inlet temperature, hence larger heat transfer occurs. The results indicate that ground recharging during stop-run operations will be beneficial in geothermal energy usage, hence, improved performance of the ground source heat pump.

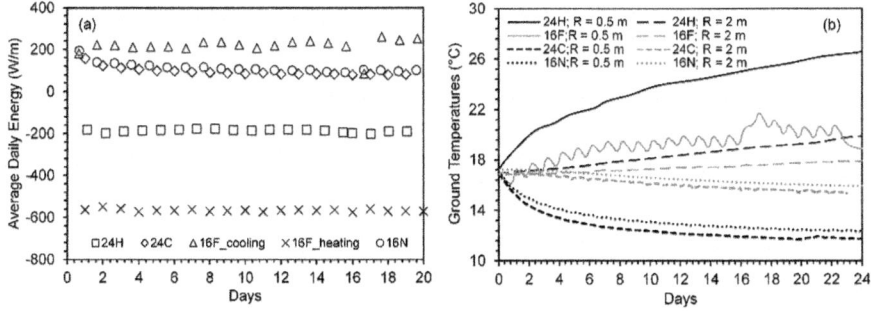

Fig. 2. Time series of energy utilisation and ground temperatures for all the experiments (a) average daily geothermal energy utilisation (b) ground temperature changes at radial distances of 0.5 m and 2 m from the pile edge.

The ground temperatures at a depth of 12 m for all the experiments are shown in Fig. 2b. The ground temperatures were monitored at radial distances of 0.5 m and 2 m from the pile edge. It is seen that the highest effect of pile temperature variations on ground temperatures is at a radial distance of 0.5 m, and becomes low at a radial distance of 2 m. The results indicate that the ground temperature changes of a single energy pile operation will not be affected beyond 2 m radial distances for similar soil conditions. Furthermore, it is seen that forced ground recovery in the 16F mode greatly recovers the ground temperatures compared to the natural recovery in the 16N mode. The ground temperatures in the 16F mode are also closest to the initial ground temperatures, indicating that forced ground thermal recharging overall induces lower ground temperature changes with optimal geothermal usage.

4   Summary

This study investigated the effects of monotonic and cyclic temperature changes on the energy utilisation and ground temperature distribution of a field scale energy pile installed in dense sand. The results indicated that forced ground thermal recharging during the non-operating times of the heat pump greatly enhanced the energy extracted/injected and maintained the ground temperatures closer to initial conditions. Furthermore, the results of the present study indicated that the effect of a single pile operation on the ground temperature changes reduced with increasing radial distances and became almost negligible at a radial distance of 2 m from the pile edge.

References

1. Brandl, H.: Energy foundations and other thermo-active ground structures. Géotechnique **56** (2), 81–122 (2006)
2. Faizal, M., Bouazza, A., Singh, R.M.: An experimental investigation of the influence of intermittent and continuous operating modes on the thermal behaviour of a full scale geothermal energy pile. Geomech. Energy Environ. **8**, 8–29 (2016)
3. Faizal, M., Bouazza, A., Haberfield, C., McCartney, J.S.: Axial and radial thermal responses of a field scale energy pile under monotonic and cyclic temperature changes. J. Geotech. Geoenviron. Eng. (2018, accepted)
4. Yi, M., Hongxing, Y., Zhaohong, F.: Study on hybrid ground-coupled heat pump systems. Energy Build. **40**(11), 2028–2036 (2008)
5. Singh, R., Bouazza, A., Wang, B.: Near-field ground thermal response to heating of a geothermal energy pile: observations from a field test. Soils Found. **55**(6), 1412–1426 (2015)
6. Wang, B., Bouazza, A., Singh, R., Haberfield, C., Barry-Macaulay, D., Baycan, S.: Posttemperature effects on shaft capacity of a full-scale geothermal energy pile. J. Geotech. Geoenviron. Eng. **141**(4), 04014125-1-12 (2015). https://doi.org/10.1061/(ASCE)GT.1943-5606.0001266

Thermo-Mechanical Behavior of Reinforced Concretes for Energy Piles

Wei Huang[1], Wei Xiang[2(✉)], and Jin Luo[1]

[1] Faculty of Engineering, China University of Geosciences (Wuhan),
Wuhan, China
{huangwei,jinluo}@cug.edu.cn
[2] Three Gorges Research Center for Geo-hazards,
China University of Geosciences (Wuhan), Wuhan, China
xiangwei@cug.edu.cn

Abstract. Energy pile is a ground source heat pump technology that couples foundation pile with ground heat exchangers for geothermal energy exploitation. However, the thermally induced contraction and expansion of the concrete affect the heat transfer performance and sustainable operation of the energy piles. In order to minimize these effects, a set of concrete samples was tested by deploying heating-recovery-cooling-recovery cycle. Three different concrete samples including plain concrete, polypropylene reinforced and steel fiber reinforced concretes are prepared. Temperature and stain development of the concrete samples were continuously recorded during the testing process. The results show that heating expansion and cooling contraction are both reduced for steel fiber reinforced concrete. For polypropylene fiber reinforced concretes, only the heating expansion is reduced but the contraction is larger than the plain concrete. Finally, the content of steel fiber of 1.3% is optimized to minimize the thermally induced mechanical effects on steel fiber reinforced concrete for energy piles.

Keywords: Energy pile · Thermo-mechanical behavior
Fiber reinforced concrete · Strain

1 Introduction

The energy pile is a renewable geothermal energy coupled pile foundations as ground heat exchangers [1]. Compering with the traditional borehole heat exchangers, the energy piles are known to be cost economically. Energy piles put heat exchanges into the foundations and save the cost of drilling and area to install borehole. Pahud put the u-type heat exchanger into the concrete piles and applied the concrete piles as foundation of Munich airport building [2]. Since 21 century, energy piles installed widely in the developed countries and were being an important research topic [3–5]. It was the first time that energy piles were used in Tianjin China in 2004.

With reference of experiences on projects at home and abroad, the operating of energy piles is a complicated thermal-mechanical process. the expansion and shrinkage of the plies material affects long-term running and the safety of buildings due to energy

© Springer Nature Switzerland AG 2018
W. Wu and H.-S. Yu (Eds.): *Proceedings of China-Europe Conference on Geotechnical Engineering*, SSGG, pp. 1595–1599, 2018.
https://doi.org/10.1007/978-3-319-97115-5_152

piles operate in cooling and heating thermal load for different seasons, Laloui tested the energy piles in the Swiss federal Institute of Technology in Lausanne, the length of energy piles was set for 15 m, the diameter was set for 117 cm, when temperature difference between the energy piles and ground approached to 15 °C, additional temperature stress was measured to be 2 MPa and the deformation on the top of piles was 4 mm [6]. P Bourne-Webb, BL Amatya researched the energy piles at the Lambeth Academy in London, the length and diameter of energy were set to 23 m and 55 cm, when energy piles heated to 28 °C, thermal load was measured to 500 kN, what's more, the expansion of energy piles was about 2 mm [7, 8]. AD Donna developed a 3-dimensional finite model and simulated the deformation of energy piles after seasonal thermal storage operation, the energy piles heated to 40 °C, the expansion was 4 mm [9].

In conclusion, additional temperature stress affects long-term running and the safety of buildings during seasonal thermal storage operation [10, 11]. To deal with problem above, this paper aims to investigate thermal-mechanical behavior of grouting concretes, polypropylene and steel fiber reinforced concretes in energy piles, and find the optimal proportion of energy piles.

2 Material and Test Program

2.1 Materials

Samples are made by ordinary Portland cement, the maximum size of coarse aggregate size is about 5 mm. the length of steel fiber is 10 mm, cross sectional area is measured to be 1 mm × 0.8 mm, respectively, the length of polypropylene is 8 mm. Concrete mix proportion listed in Table 1 is obtained by The Design Regulations Of Mix Ratio Of Ordinary Concrete (JGJ55-2011), the sample is Cylindrical and the diameter and height of sample are 50 mm and 100 mm, respectively. According to thermal conductivity experiment, the maximum thermal conductivity is measured to be 2.44 W/m · K in the concrete with content of the steel fiber of 1.3%, the optimal content of steel fiber and polypropylene is set to be 1.3% and 0.7% respectively. All samples named C1 C2 C3.

Table 1. Mix proportion of pile material sample

Sample number	Water cement ratio	Cement (kg/m³)	Water (kg/m³)	Sand (kg/m³)	Gravel (kg/m³)	Content of steel fiber (%)	Content of polypropylene fiber (%)
C1	0.55	381	210	668	1240	1.3	0
C2	0.55	381	210	668	1240	0	0.7
C3	0.55	381	210	668	1240	0	0

2.2 Test Program

To investigate the thermal-mechanical characteristics of reinforced grouting concretes in energy piles. As shown in Figs. 1, 2, 3 and 4, we have developed 815 rock mechanic test system, heating panel, cooling system. The test steps divide into heating, recovery,

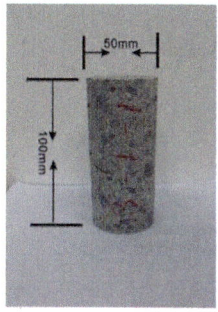

Fig. 1. Cylindrical sample

Fig. 2. 815 rock mechanic test system

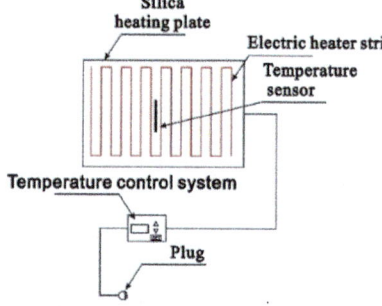

Fig. 3. Heating panel

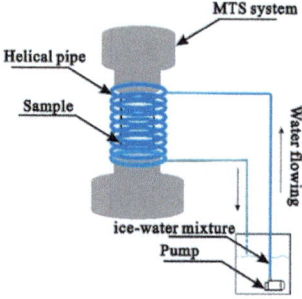

Fig. 4. Cooling system

cooling, recovery stages and simulate energy piles operation during seasonal thermal storage. The samples were set into MTS system with the heating panel, the MTS system provided the axial stress 28 kN and kept stable for 20 min, firstly, in the heating stage, the heating panel was set to 55 °C, and kept stable for 30 min. Then, stopped heating and recovered to the indoor temperature for 60 min, cooling system kept the temperature to 0 °C at next cooling stage. Finally, recovered to the indoor temperature for 60 min. In terms of monitor, MTS system kept the axial stress stable and stored the stain of each samples per second throughout the experiment, meanwhile. The stain of samples developed with time was obtained.

3   Results

As shown in Fig. 5, comparing with the plain concrete C3, steel fiber reinforced concrete samples C1 reduced the strain of the heating expansion and cooling shrinkage, and polypropylene fiber reinforced concrete samples C2 reduced the strain of the heating expansion but increased the strain of cooling shrinkage. Respectively, that means steel fiber added into plain concrete can improve the both compressive strength

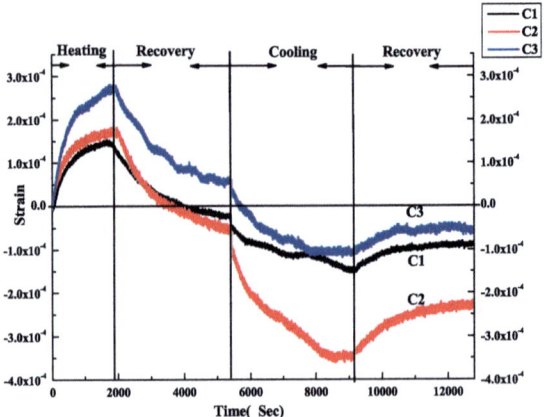

Fig. 5. Strain development of three different reinforced samples with four stages.

and tensile strength of concrete, however, polypropylene fiber added into plain concrete can only increase the tensile strength but reduce the compressive strength of concrete.

Table 2 shows that strain of the both steel fiber and polypropylene fiber reinforced concrete was reduced by adapting heating loads, comparing with plain concrete C3, steel fiber reinforced concrete sample C1 reduced about 49.6% and polypropylene fiber reinforced concrete sample C2 reduced about 38.1% in heating stage, meanwhile, in the cooling stage, the strain of steel fiber reinforced concrete sample C1 was reduced about 28.8% but polypropylene fiber reinforced concrete sample C2 increased about 81.9%.

Table 2. Strains development of fiber reinforced sample with four stages

Specimen number	Stain			
	Heating	Recovery	Cooling	Recovery
C1	1.44×10^4	1.71×10^4	-1.14×10^{-4}	0.65×10^{-4}
C2	1.75×10^4	-2.05×10^4	-2.91×10^4	1.17×10^{-4}
C3	2.78×10^4	-2.22×10^4	-1.60×10^4	0.50×10^{-4}

4   Conclusions

The results show that the maximum thermal conductivity is measured to be 2.44 W/m · K in the concrete with content of the steel fiber of 1.3%. The strain of the both steel fiber and polypropylene fiber reinforced concrete is reduced by adapting heating loads, steel fiber reinforced concrete reduces about 49.6% and polypropylene fiber reinforced concrete reduces about 38.1%. Furthermore, heating expansion and cooling shrinkage strain are both reduced for steel fiber reinforced concrete during the heating and cooling load cycles testing, but polypropylene fiber reinforced concretes increase cooling shrinkage strain. Comparison of the three different pile concretes, steel fiber reinforced concrete is suggested to be a suitable material for the grouting of energy piles and the optimal proportion of steel fiber is 1.3%.

References

1. Hamada, Y., Saitoh, H., Nakamura, M.: Field performance of an energy pile system for space heating. Energy Build. **39**, 517 (2007)
2. Pahud, D., Fromentin, A., Hadorn, J.C.: The Duct Ground Heat Storage Model (DST) for TRNSYS Used for the Simulation of Heat Exchanger Piles. User Manual, December 1996 Version (1996)
3. Sekine, K., Ooka, R., Hwang, S., Nam, Y., Shiba, Y., Eng, M.: Development of a ground-source heat pump system with ground heat exchanger utilizing the cast-in-place concrete pile foundations of buildings. ASHRAE Trans. **113**, 558–566 (2007)
4. Shiba, Y., Ooka, R., Sekine, K.: Development of a high-performance water-to-water heat pump for ground-source application. ASHRAE Trans. **113**, 261–270 (2007)
5. Omer, A.M.: Ground-source heat pumps systems and applications. Renew. Sustain. Energy Rev. **12**, 344 (2008)
6. Laloui, L., Nuth, M., Vulliet, L.: Experimental and numerical investigations of the behaviour of a heat exchanger pile. Int. J. Numer. Anal. Methods Geomech. **30**, 763 (2006)
7. Bourne-Webb, P., Amatya, B., Soga, K., Amis, T., Davidson, C., Payne, P.: Energy pile test at Lambeth College, London: geotechnical and thermodynamic aspects of pile response to heat cycles. Géotechnique **59**, 237 (2009)
8. Amatya, B.L., Soga, K., Bourne-Webb, P.J., Amis, T., Laloui, L.: Thermo-mechanical behaviour of energy piles. Géotechnique **62**, 503 (2012)
9. Donna, A.D., Loria, A.F.R., Laloui, L.: Numerical study of the response of a group of energy piles under different combinations of thermo-mechanical loads. Comput. Geotech. **72**, 126 (2015)
10. Loria, A.F.R., Gunawan, A., Shi, C., Laloui, L., Ng, C.W.W.: Numerical modelling of energy piles in saturated sand subjected to thermo-mechanical loads. Geomech. Energy Environ. **1**, 1 (2015)
11. Jin, L., Rohn, J., Wei, X., Bertermann, D., Blum, P.: A review of ground investigations for ground source heat pump (GSHP) systems. Energy Build. **117**, 160 (2016)

Flow Behaviour of Fractured Geothermal Reservoir Rocks Under In-Situ Stress and Temperature Conditions

W. G. P. Kumari[(⊠)] and P. G. Ranjith

Department of Civil Engineering, Monash University,
Building 60, Melbourne, VIC 3800, Australia
pabasara.wanniarachchige@monash.edu

Abstract. Enhanced Geothermal Systems (EGS) is a potential carbon-neutral form of renewable energy whereby fractured granite at depths of around 2–4 km. Such geothermal reservoirs are high-temperature rock formations located at deep underground and therefore with ultra-low permeable characteristics. Therefore, natural and artificially created rock fractures provide major flow pathways for the reservoir fluid circulation process. Evolution of permeability through such rock fractures under potentially existing extreme geothermal conditions is quite important for an effective application of EGS systems. This study therefore discusses the experimental results of a series of flow experiments conducted on artificially fractured Australian Strathbogie granite under wide range of pressure (30 MPa confining pressure, 5 MPa to 25 MPa injection pressure) and temperature (from room temperature to 250 °C) conditions using a newly developed high temperature-high pressure rock triaxial test apparatus with capability of simulating environment of EGS reservoirs. The steady state flow rates through fractures found to linearly increase with increasing injection pressure under the considered injection pressures and temperatures. Permeability along rock fractures was therefore calculated by employing cubic law and relevant temperature and pressure dependent fluid properties. According to the experimental results, both stress level and the temperature have significant influences on fracture flow characteristics of the tested granite. Increasing of temperature caused a significant non-linear increment in flow rate and permeability through fractures due to the thermally induced micro crack generation.

Keywords: Enhanced geothermal reservoirs · Fracture permeability
High pressure · High temperature

1 Introduction

Provision of sufficient and sustainable energy to full fill the demands of the growing population has been identified as a primary concern of the world. Recent attempts have been made to explore unconventional geothermal resources which are located at deep underground associated with high-temperature rock formations which are known as enhanced/Engineered Geothermal Systems (EGS). EGS is created by drilling a well into the target hot rock formation and then creating a reservoir by injecting high-pressure

© Springer Nature Switzerland AG 2018
W. Wu and H.-S. Yu (Eds.): *Proceedings of China-Europe Conference on Geotechnical Engineering*, SSGG, pp. 1600–1603, 2018.
https://doi.org/10.1007/978-3-319-97115-5_153

water to open up the existing fracture/joints or to create new fractures. Once the injected water is gradually heated with the thermal energy of the rock, it is extracted to the surface through a production well which is drilled in the reservoir zone. During the fluid circulation process, natural or artificially created fractures provide major pathways for fluid flow and therefore, it is essential to understand how fluids are able to migrate through those fractures. Since EGS systems are located at high temperature (generally 100–300 °C) and high pressure (1–3 km depths) formations [1], flow performance under these extreme conditions needs to be evaluated.

Flow performance of fractured rock is significantly influenced by elevated stress levels and temperatures, where fractures can be closed by application of high normal stresses, conversely can be extended by a decrease of effective stresses due to the increase of injection pressures [2]. Further, high temperatures can result in alteration the microstructure of the rock through thermally induced volumetric expansion, mineralogical changes, development of new micro-cracks and extending of the existing micro-cracks [3]. In addition, when the fracture is subjected to a temperature field thermal over closure [4] can occur resulting reduction of permeability. Further, and alteration of physical properties (density and viscosity) of the circulation fluid including the phase of the fluid can significantly influence the flow performance along the fractures however this aspect has not sufficiently understood to date. Therefore, in this paper, we aim to understand the flow performance of artificially fractured granite rock under a wide range of stress and temperature conditions, making an important contribution to deep geothermal engineering applications.

2 Methodology

Granite samples which were collected from Strathbogie, Victoria, Australia were employed for the flow through experiments. It is comparatively a fine grain granite type with grains ranging from 0–150 μm. Based on the XRD analysis it was identified that the selected granite mainly consists of quartz, K-feldspar, plagioclase, and biotite. Cylindrical granite specimens of 22.5 mm diameter and 45 mm height were prepared for the flow through experiments. The diametrical fracture was induced along the cylindrical axis of the specimen employing the Brazilian test. Next, drained permeability tests were conducted utilizing the newly developed high-pressure high-temperature triaxial testing apparatus of Deep Earth Energy Laboratory at Monash University [3]. This triaxial rig is capable of simulating the extreme geothermal environments with a thermal capacity of up to 300 °C, confining pressure of 137 MPa and injection pressure of 165 MPa. Distilled water at room temperature was injected along the fluid flow line and it was assumed water is entering to the sample at the target temperature since the fluid flow line acts as a heat exchanger. Using a sensitive electronic balance mass of downstream fluid mass was recorded. Assuming fracture flow occurs through two idealized parallel plates, the cubic law was employed to calculate the fracture permeability employing relevant temperature and pressure dependent fluid properties.

3   Results and Discussions

3.1   The Effect of Temperature on Permeability of Fractured Granite

Figure 1(a) illustrates the variation of permeability with temperature under different injection pressures. Generally, two tendencies were observed in all the injection pressures; firstly, up to 100 °C, reduction of permeability with increasing temperature and secondly increasing of permeability in the higher temperatures. Further, in all the temperatures, it was found that with increasing of injection pressure permeability is increased. Permeability is the parameter which represents the change of the rock matrix/flow path irrespective of the fluid properties. Therefore, considering the variation of permeability, it can be identified any changes of the fracture flow path. For example, the reduction of permeability implies shrinkage or blockage of the flow path while increment of permeability indicates enhancement or widening of the flow path. Therefore, it can be identified that up to 100 °C, shrinkage of fracture path occurs while at latter temperatures, widening or improving the fracture. The calculated hydraulic aperture further confirmed this hypothesis such that reduction of fracture aperture up to 100 °C and then increments of fracture aperture at latter temperatures. As a result of the thermally induced volumetric expansion, it can be expected to the closure of the induced fracture resulting in the reduction of permeability through the sample. Next, at latter temperatures, the increment of permeability, suggests that either the induced fracture has been widening or new fractures have been created due to the thermally induced damage [3].

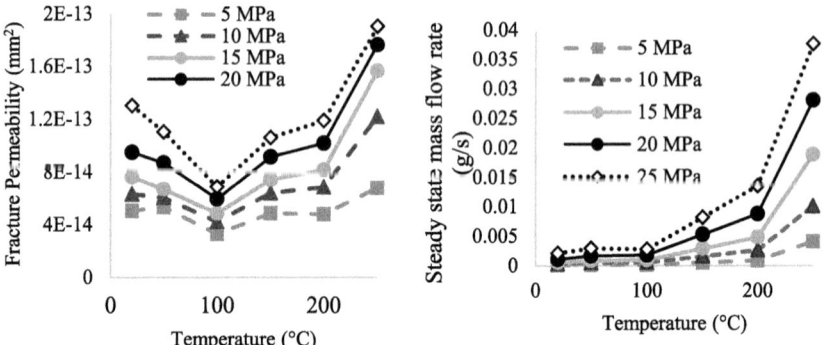

Fig. 1. (a) Permeability vs. temperature at different injection pressures (b) Steady-state mass flow rates vs. temperature at different injection pressures

3.2   The Effect of Temperature on Permeability of Fractured Granite

Next, the flow performance of the fractured granite was evaluated in terms of the steady state mass flow rate because, in geothermal energy production, the mass flow rate directly correlates with the efficiency of the system. Figure 1(b) illustrates the variation of steady state mass flow rate with increasing temperature. Interestingly, it was observed that a nonlinear increment of steady state mass flow rate with increasing temperature. Up to 100 °C, the increment of steady state mass flow rate was less evident, however, at higher temperatures a significant increment as observed. Incorporating the temperature dependent fluid properties, and the changes of fracture aperture, these variations can be clearly explained. Both viscosity and density are decreasing with increasing temperature however compared to density reduction, viscosity reduction is significant. This significant viscosity reduction results in less frictional resistance at high temperatures hence enhancement of the flow performance. However, at low tempera-ture, although the flow properties are enhanced, the flow path has been shrunk resulting less improvement of the steady state mass flow rate below 100 °C. At latter tempera-tures, both fracture aperture and viscosity are improving resulting significant enhancement of mass flow rate at higher temperatures.

4 Conclusions

Major findings of the study can be summarised as follows;

- With the increment temperature from room temperature to 100 °C, fracture per-meability decreases due to the thermal overclosure and further increase of tem-perature causes them to be considerably increased due to the thermally induced damage.
- Increasing of temperature results non-linear increment of steady state mass flow rate due to the enhanced viscosity of the injection fluid and thermally induced damage at higher temperatures.
- Therefore, it can be expected, reservoir temperature and pressure are critical parameters in evaluating the flow performance of the reservoir because the circu-lation fluid properties significantly depend on them.

References

1. Tester, J.W., et al.: The future of geothermal energy: impact of enhanced geothermal systems (EGS) on the United States in the 21st century. Massachusetts Institute of Technology, p. 209 (2006)
2. Bandis, S.C., Lumsden, A.C., Barton, N.R.: Fundamentals of rock joint deformation. Int. J. Rock Mech. Min. Sci. Geomech. Abstracts 20(6), 249–268 (1983)
3. Kumari, W., Ranjith, P., Perera, M., Shao, S., Chen, B., Lashin, A., Al Arifi, N., Rathnaweera, T.: Mechanical behaviour of Australian Strathbogie granite under in-situ stress and temperature conditions: an application to geothermal energy extraction. Geothermics 65, 44–59 (2017)
4. Barton, N., Makurat, A.: Hydro-thermo-mechanical over-closure of joints and rock masses and potential effects on the long term performance of nuclear waste repositories. In: Proceedings of Eurock2006 Multiphysics Coupling and Long Term Behavior in Rock Mechanics, Liege, pp. 445–450 (2006)

Calculation of Osmotic Suctions for Bentonite in Saline Solutions

Xiaoyue Li, X. J. Zheng, and Yongfu Xu$^{(\boxtimes)}$

Department of Civil Engineering,
Shanghai Jiao Tong University, Shanghai 200240, China
yongfuxu@sjtu.edu.cn

Abstract. Bentonites are usually selected as the engineered barrier material of repositories for radioactive waste. The saline solution from surrounding rock fissures can affect the mechanical behaviour of bentonite for the reason that the osmotic suction in pore water can act as an additional total stress component on bentonite. The osmotic coefficient, as the key to calculate the osmotic suction, is usually obtained by measuring the vapor pressure of a solution and that of the pure solvent with a differential manometer in experimental method. Considering that the vapor pressure is affected by many factors such as solute type, concentration, and temperature, it is very complicated to obtain the osmotic coefficient by experimental method. In this paper, the osmotic coefficient is calculated according to the modified Debye-Hückel equations and the calculated results are validated by comparing with the experimental data in other literature. In this way, the osmotic suction for different solutions under different temperatures can be obtained by calculation.

Keywords: Bentonite · Saline solution · Osmotic suction

1 Introduction

Compacted bentonite is widely used as a barrier material for HLW repositories mainly because it has the property of swelling in water to fill the surrounding rock fissures and form an impermeable layer to prevent the emission of nuclear waste into the surrounding environment. The mechanical behaviour may be strongly influenced by physicochemical effects when saline concentrated pore fluids are introduced to clays [1]. The increasing of pore water concentration will weaken the swelling ability of clay [2]. The composition and concentration of the pore water solution can significantly affect the shear strength of bentonite [3]. The influence of pore water on the mechanical behavior of bentonite is mainly due to the reason that osmotic suction from pore water acts as an additional total stress component that favours the reduction in swell potential and the increase in shear strength of the compacted clay specimens [4]. Thus, the quantitative research of the osmotic suctions of different solutions is sig-

© Springer Nature Switzerland AG 2018
W. Wu and H.-S. Yu (Eds.): *Proceedings of China-Europe Conference on Geotechnical Engineering*, SSGG, pp. 1604–1608, 2018.
https://doi.org/10.1007/978-3-319-97115-5_154

nificant to study the mechanical behavior of bentonite in saline solution. Van't Hoff gives the osmotic suction π related to solution concentration as follows:

$$\pi = \zeta RTm\phi \tag{1}$$

where ζ is the number of ions that the solute can dissociate in solution (e.g., NaCl = 2), R the universal gas constant (8.31 J/mol/K), T the absolute temperature in Kelvin, m the molality of solute (mol/kg) and ϕ the osmotic coefficient given as [5]:

$$\phi = \frac{\rho_w}{\zeta m M_w} \ln\left(\frac{P_w}{P_0}\right) \tag{2}$$

where the M_w is the molar mass of water (18.016 g/mol), ρ_w the unit weight of water, P_w the saturated vapor pressure on the surface of the pore water solution and P_0 the saturated vapor pressure on the surface of the pure water. The osmotic coefficient is usually obtained in experimental method by measuring the vapor pressure of a solution and that of the pure solvent with a differential manometer [6]. Considering that saturated vapor pressure is affected by factors such as solute type, concentration and temperature [5], the experimental method of measuring the saturated vapor pressure to get the osmotic coefficients of different solutions is very complicated and inconvenient for engineering application. In addition, according to the van't Hoff equation, only the osmotic coefficient of a highly diluted solution can be considered as 1. Thus, the difficulty in obtaining osmotic coefficients hinders the calculation of the osmotic suction π. In this paper, a calculation method based on the modified Debye-Hückel equations is found to obtain the osmotic coefficients of different saline solutions, and the calculated results are verified by experimental results from other literature.

2   Calculation of Osmotic Suctions

The osmotic coefficient can be derived by differentiation from the excess free energy theory, which is complicated and inconvenient to calculate [7]. Based on Debye-Hückel equations, Kenneth and Guillermo [8, 9] simplified the determination method of the osmotic coefficients, which is empirically superior to the conventional form. The equation to calculate the osmotic coefficients of single electrolytes MX is as follows:

$$\phi = |Z_M Z_X| f^\phi + m\left(\frac{2V_M V_X}{V}\right) B^\phi_{MX} + m^2 \frac{2(V_M V_X)^{3/2}}{V} C^\phi_{MX} + 1 \tag{3}$$

where V_M and V_X are the number of M cation and X anion respectively in the chemical formula, $V = V_M + V_X$. Z_M and Z_X are respective charges, and m is molality. $B_{MX} = \beta^{(0)}_{MX} + \beta^{(1)}_{MX} e^{-\alpha I^{1/2}}$, hereinto, $\beta^{(0)}_{MX}$, $\beta^{(1)}_{MX}$ and C^ϕ_{MX} are empirical parameters related to the solute type obtained by looking up the table given by Kenneth and Guillermo [9], and

$\alpha = 2$ is found to satisfy most of the common solutes. The f^ϕ is given as:

$$f^\phi = -A^\phi \frac{I^{1/2}}{1 + bI^{1/2}} \tag{4}$$

where I, the ionic strength, equals to $\sum (m_i z_i^2)/2$, b as an empirical parameter is selected 1.2 here and A^ϕ, the Debye-Hückel coefficient for the osmotic function has the value of 0.392 for water at 298 K [8]. The osmotic coefficients of $NaNO_3$ and $NaCl$ solute under different concentrations in water are calculated according to the equations above and the results are compared with experimental data from other literature [6, 10] (see Fig. 1). The calculated results are consistent with the experimental results.

For the 2–2 electrolytes solutions, adding a term to B_{MX} would create a better

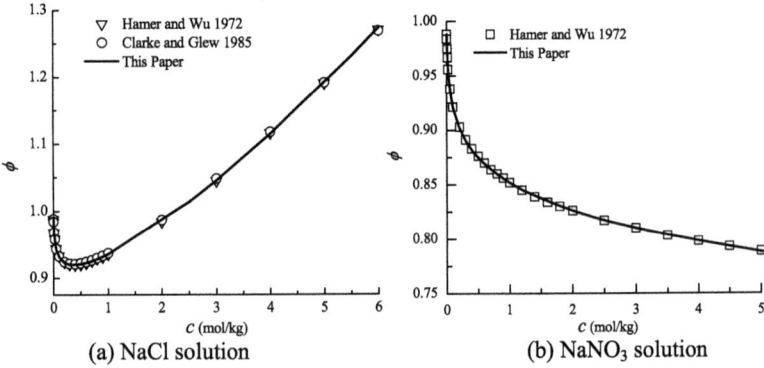

(a) NaCl solution   (b) $NaNO_3$ solution

Fig. 1. Relationship between osmotic coefficient and concentration in water (298 K).

agreement with the experimental results [9]. In this case, B_{MX} has the following form:

$$B_{MX} = \beta_{MX}^{(0)} + \beta_{MX}^{(1)} e^{-\alpha_1 I^{1/2}} + \beta_{MX}^{(2)} e^{-\alpha_2 I^{1/2}} \tag{5}$$

where $\beta_{MX}^{(2)}$ is an empirical parameter related to the solute type, $\alpha_1 = 2$, $\alpha_2 = 1.4$ and other parameters are the same as they are defined in Eq. (3) previously.

The osmotic coefficients of $CaCl_2$ and $Ca(NO_3)_2$ in ethanol at 298 K have also been calculated by using the same method in this paper. Under the same calculation method, the osmotic efficient in ethanol is different from that in the water due to the differences of some parameter values (e.g., $A^\phi = 2.006$ for ethanol). The comparison between calculated results and the experimental results from other literature [11, 12] in Fig. 2 indicates that the osmotic coefficients of $CaCl_2$ and $Ca (NO_3)_2$ in ethanol at 298 K calculated from the mentioned method in this paper are in agreement with the experimental ones. Therefore, the calculation method is proved reliable and the osmotic coefficients of other kinds of single electrolytes solutions or double electrolytes in

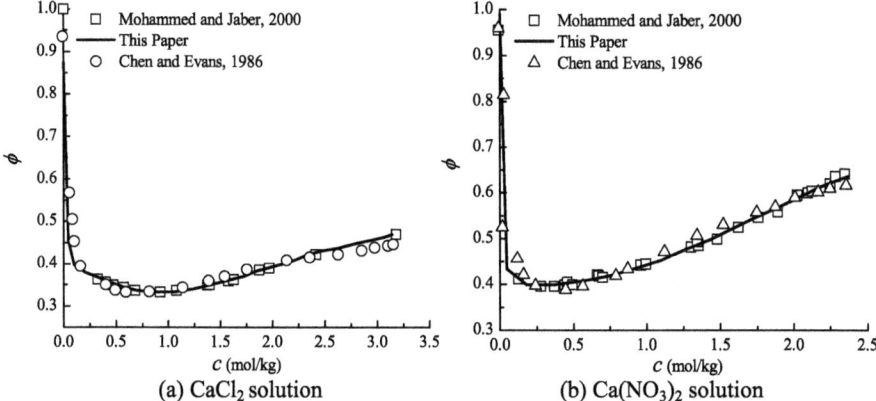

Fig. 2. Relationship between osmotic coefficient and concentration in ethanol (298 K).

different kinds of solutions at different temperatures can be calculated by the modified Debye-Hückel equations.

3 Conclusions

According to the results of this study, the osmotic coefficients that determine the osmotic suction of different solutions are calculated by the modified Debye-Hückel equation, and the correctness of the calculation results has been validated by other literature. Further research about the osmotic suction of mixed saline solution is needed. In addition, some tests, such as obtaining swelling pressure and shear strength for bentonite in saline solution, are still required in order to obtain further calculating of the mechanical behaviour of bentonite in saline solution.

References

1. Ye, W.M., Zhang, F., Chen, B.: Effects of salt solutions on the hydro-mechanical behavior of compacted GMZ01 bentonite. Environ. Earth Sci. **72**, 2621–2630 (2014)
2. Xu, Y.F., Xiang, G.S., Jiang, H.: Role of osmotic suction in volume change of clays in salt solution. Appl. Clay Sci. **101**, 354–361 (2014)
3. Zhang, L., Sun, D.A., Jia, D.: Shear strength of GMZ07 bentonite and its mixture with sand saturated with saline solution. Appl. Clay Sci. **132–133**, 24–32 (2016)
4. Rao, S.M., Thyagaraj, T.: Swell-compression behavior of compacted clays under chemical gradients. Can. Geotech. J. **44**, 520–532 (2007)
5. Apelblat, A., Dov, M., Wisniak, J., Zabicky, J.: The vapour pressure of water over saturated aqueous solutions of malic, tartaric, and citric acids, at temperatures from 288 K to 323 K. J. Chem. Thermodyn. **27**(1), 35–41 (1995)
6. Hamer, W.J., Wu, Y.C.: Osmotic coefficient and mean activity coefficients of uni-univalent electrolytes in water at 25 °C. J. Phys. Chem. Ref. Data **1**(4), 1047–1099 (1972)

7. Scatchard, G.: Excess free energy and related properties of solutions containing electrolytes. J. Am. Chem. Soc. **90**(12), 3124–3127 (1968)

8. Kenneth, S.P., Guillermo, M.: Thermodynamics of electrolytes. II. activity and osmotic coefficients for strong electrolytes with one or both ions univalent. J. Phys. Chem. **77**(19), 2300–2308 (1973)

9. Kenneth, S.P., Guillermo, M.: Thermodynamics of electrolytes. III. activity and osmotic coefficients for 2–2 electrolytes. J. Solution Chem. **3**(7), 539–546 (1974)

10. Clarke, E.C.W., Glew, D.N.: Evaluation of thermodynamic function for aqueous sodium chloride from equilibrium and calorimetric measurements below 154 °C. J. Phys. Chem. Ref. Data **14**(2), 489–610 (1985)

11. Mohammed, T.Z.M., Jaber, J.S.: Measurement and correlation of osmotic coefficients and evaluation of vapor pressure for solutions of $CaCl_2$ and $Ca(NO_3)_2$ in ethanol at 298K. Fluid Phase Equilib. **172**, 221–235 (2000)

12. Chen, C.C.: Evans LBA.: a local composition model for the excess Gibbs energy of aqueous electrolyte systems. AIChE J. **32**, 444–454 (1986)

Principle of Effective Stress is an Approximated Method

Chenggang Zhao, Zhaoyang Song, Jian Li$^{(\boxtimes)}$, Guoqing Cai,
and Weihua Li

Key Laboratory of Urban Underground Engineering of Ministry of Education,
Beijing Jiaotong University, Beijing 100044, China
jianli@bjtu.edu.cn

Abstract. Soil mechanics is a powerful theoretical analysis tool in geotechnical engineering. The principle of effective stress is a basis of classic soil mechanics. However, it in fact is an approximated method. This is attributed to three factors. Firstly, the effective stress is an equilibrium equation rather than an independent state variable. Secondly, the effective stress is a variable controlled the deformation behavior and the shear strength of soil, but not the only one. Thirdly, the principle of effective stress is an equivalent method, but not an accurate one. The above viewpoints are justified in this paper as follows. Firstly, it is demonstrated that the soil is a variable and sensitive material. Thus, it is difficult to quantitatively describe soil behaviors by only one variable. Secondly, the definition and the comprehension of the effective stress are discussed. Finally, to deal with complex problems most frequently encountered in geotechnical engineering, two approaches are advised, i.e., choosing additional variables except for the effective stress and using the multi-factor theory.

Keywords: Effective stress · Predicted uncertainty · Multi-factor theory

1 Introduction

Due to the proposing of the effective stress principle for saturated soils by Terzaghi [1, 2], soil mechanics is established as an independent discipline distinct separated from solid mechanics, which is successful applied in geotechnical engineering in the past. The principle of effective stress is a basic component of the classical soil mechanics and is applied in the theory of consolidation and the critical state soil mechanics. However, limitations of the effective stress principle are recognized by dealing with complex problems most frequently encountered in geotechnical engineering at present, e.g., collapse or losing the stability of unsaturated soil zone due to wetting, landslide due to climate change, subgrade settlement during freezing and thawing cycles, long-term settlement of high-speed railway subgrade under traffic loading, radioactive waste disposal and landfill waste disposal, etc. Soil mechanics is a subject contained half theory and half experience, which doesn't meet basic requirements of science, e.g., rigorous, consistent and without contradiction. To deal with the above complex problems, it requires the soil mechanics theory developing and reforming. There are two things should be considered, which are how to establish a rigorous soil mechanics

© Springer Nature Switzerland AG 2018
W. Wu and H.-S. Yu (Eds.): *Proceedings of China-Europe Conference on Geotechnical Engineering*, SSGG, pp. 1609–1612, 2018.
https://doi.org/10.1007/978-3-319-97115-5_155

theory, and how to reduce uncertainty of the theoretical prediction by soil mechanics. Some preliminary explorations have been done for the former one by the authors [3]. In this paper, the latter one and limitations of the effective stress principle are discussed.

2 Predicted Uncertainty by Soil Mechanics Theory

Soil is a deposit of particles and pores are filled by the pore water or air. The cohesion between particles is relatively weak compared to other materials, hence the soil is a kind of friction material. However, the magnitude of interaction forces, i.e., cohesion, between adjacent particles caused by physicochemical effects cannot be ignored as the gravity of soil particles is small. The physicochemical effects include capillarity of meniscus, crystallization of salt in pore water, electrostatic force, Van der Waals force, hydration force and so on [4]. These interaction forces are not constants which are varied with many factors, e.g., temperature, humidity, salinity of pore fluid, etc. It makes the soil be a variable and sensitive material [5]. In other words, engineering properties of soil are sensitive to these factors.

The deformation behavior and the shear strength of soil are the most frequent problems in geotechnical engineering, and their influence factor isn't single. Whitman [6] pointed out that the shear strength τ_f depends on not only the effective stress but also other factors, and its function could be expressed by:

$$\tau_f = f(\sigma'_{ii}, e, T, H, S, E, S_r, F, C, \sigma'_{2f}) \tag{1}$$

where σ'_{ii} is effective stress on the failure plane, e is void ratio at failure, T is time, H is stress history; S is structure, e.g., flocculated or dispersed; E is environment conditions; S_r is degree of saturation; F is formation conditions of soil; C is capillary tension; σ'_{2f} is effective stress on the plane with the greatest shear strain. These factors also affect the deformation behavior because that the strength is usually determined by a special state of the whole process of deformation during shearing. The influences of effective stress on the deformation behavior and the shear strength are the most important when external factors, e.g., temperature, humidity, are constant. Then simple expressions of strength and deformation are obtained:

$$\tau_f = f(\sigma'_{ij}), \quad \varepsilon_{ij} = f'(\sigma'_{ij}) \tag{2}$$

The specific expressions of Eq. 2 are proposed by previous studies, and Eqs. 1 and 2 could be degenerated to the Mohr-coulomb criterion when a linear function is adopted. When variations of other variables in Eq. 1 except for the effective stress are not obvious, the influences of these variables could be reflected by values of model parameters. Otherwise, it will lead to serious errors of predicted results by Eq. 2. We realize that soil mechanics is a simplified theory neglected lots of factors and established on the basis of many unreal pre-hypothesis, which leads to a discrepancy between the theoretical model and the reality situation.

3 Principle of Effective Stress

The principle of effective stress consists of two parts, i.e., the definition of effective stress and the comprehension of effective stress. The authors shall discuss three fundamental topics of the above two parts. (1) Is effective stress an independent state variable? (2) Is it the unique variable controlled the deformation behavior and the shear strength of soil? (3) Is it an accurate method?

The total stress in saturated soil is undertaken by liquid pore water and soil skeleton. The part undertaken by the soil skeleton is called effective stress [2]. The pressure in the water produces neither a measurable compression nor a measurable increase of the shearing resistance, which is called the neutral stress. The effective stress is characterized as a difference between total stress σ_{ij} and pore water pressure u_w. Based on the thermodynamics theory, an independent state variable should be path independent, that is, they have no memory of their own history. Generally, the total stress and the pore water pressure are considered directly controllable and measurable. It is true for sand. However, the value of pore water pressure is dependent on the structure of soil and the surface potential associated with Van der Waals force, electrostatic force and so on for clays, saline soils and soils under non-isothermal condition [4, 7]. For example, the pore water pressure gets larger as the decreasing of void ratio when the total potential of pore water is constant. It implies that the pore water pressure is not a controllable variable, although it can be indirectly measured. In fact, the pore water pressure is an internal variable rather than an independent state variable. Therefore, the effective stress should be treated as an equilibrium equation dependent on stress paths.

Typically, there are two viewpoints on the comprehension of effective stress. (1) Effective stress is the only variable controlled deformation and strength of soil. (2) It is a stress variable controlled deformation and strength of soil, but not the only one. Most of the researchers correctly recognize that the effective stress is not the only variable controlled the deformation behavior and the shear strength as shown in Eq. 1. Regrettably, lots of constitutive relationships of deformation and strength are established by the single independent variable, i.e., effective stress, although researchers established these models do not agree with the first viewpoint subjectively.

Substantially, the saturated soil constituted of two phases is equivalent as a single-phase material by the principle of effective stress. However, the effect of seepage on the mechanical behavior could not totally represented by the value of pore water especially under non-equilibrium conditions. This is not to say that there are any mistakes for this method, but merely to illustrate its limitations. Overall, based on the above, the authors think that the principle of effective stress is an approximated method.

4 Additional Variables and Multi-factor Theory

To deal with complex problems in geotechnical engineers, the common method is choosing additional variables except for the effective stress. Cam-Clay model was established for normal consolidation clays by the effective stress and the void ratio [8]. However, it could not depict a shear dilatancy for over-consolidated clays and dense sands. In order to overcome the shortcoming, an addition state parameter $\psi = e - e_c$ was adopted by Li et al. [9], where e and e_c are the current void ratio and the critical void ratio on the critical-state line in the $e - p'$ space corresponding to the current mean effective stress p'. The state parameter ψ can be considered a measure of how far the material state is from the critical state in terms of density. In another instance, it is generally agreed by researchers that at least two constitutive variables are required to depict the mechanical behaviors of unsaturated soil. Usually, the first variable represents an overall stress state of soil, which can be expressed as the net stress or the Bishop's stress. The second variable represents an effect of suction changes on the structure of soil, which can be expressed as a function of suction and degree of saturation.

At present, most of discussions on the effective stress principle are carried out under the framework of classical soil mechanics. However, it is advised to adopt multi-factor theory to deal with complex problems in geotechnical engineering, in which the influence factors except for the effective stress are considered as independent variables and the coupling effects between multi-factors are taken into account. Obviously, it is a more complete and scientific method to study the deformation behavior and the shear strength of soil. Maybe someone would ask, the more variables the better? The answer is no. It will greatly increase complexity of constitutive relationships. Therefore, the less the better when the precision is met. Maybe someone considers that it is complicated to depict the deformation behavior and the shear strength of soil by the multi-factor theory. The authors realize that scientific knowledge always develops from simple to complex, and so it is of the soil mechanics theory.

References

1. Terzaghi, K.: Erdbaumechanik auf bodenphysikalischer Grundlage. Deuticke, Vienna (1925). (in German)
2. Terzaghi, V.K.: Theoretical Soil Mechanics. Wiley, New York (1943)
3. Liu, Y., Zhao, C.G., Cai, G.Q.: Rational Soil Mechanics and Thermodynamics. Science Press, Beijing (2016). (in Chinese)
4. Gonçalvès, J., Roussea-Gueutin, P., de Marsily, G., et al.: What is the significance of pore pressure in a saturated shale layer? Water Resour. Res. 46(4), W04514 (2010)
5. Zhao, C.G., Bai, B.: Fundamentals of Soil Mechanics, 2nd edn. Tsinghua University Press and Beijing Jiaotong University Press, Beijing (2017). (in Chinese)
6. Whitman, R.V.: Some considerations and data regarding the shear strength of clays. In: Research Conference on Shear Strength of Cohesive Soil, pp. 581–614. ASCE, Soil Mechanics and Foundations Division, Colorado of USA (1960)
7. Nitao, J.J., Bear, J.: Potentials and their role in transport in porous media. Water Resour. Res. 32(2), 225–250 (1996)
8. Roscoe, K.H., Schofield, A.N., Thurairajah, A.H.: Yielding of soils in states wetter than critical. Géotechnique 13(3), 211–240 (1963)
9. Li, X.S., Dafalias, Y.F., Wang, Z.L.: State-dependent dilatancy in critical-state constitutive modeling of sand. Can. Geotech. J. 36(4), 599–611 (1999)

Thermal Conductivity of Unsaturated Soil: Equivalent Microstructure Approach

Dariusz Łydżba, Adrian Różański[(✉)], and Damian Stefaniuk

Faculty of Civil Engineering, Wrocław University of Science and Technology,
Wybrzeże Wyspiańskiego 27, 50-370 Wrocław, Poland
adrian.rozanski@pwr.edu.pl

Abstract. The aim of the paper is to adopt the equivalent microstructure approach, originally formulated for saturated soil with respect to thermal conductivity [1], to the case of unsaturated one (3-phase medium). For that purpose, we propose two methodologically different approaches within the framework of Mori-Tanaka homogenization scheme. The first one is based on the replacement of the 3-phase medium with a 2-phase one. In this case, thermal conductivity of the mixture of water and air is estimated using the Hashin-Shtrikman lower and upper bounds. The latter approach treats the soil as the 3-phase medium. This requires a deep reformulation of the equivalent microstructure approach introduced in [1]. The second approach is recognized as the proper one, providing the remarkable agreement between the measurements and predictions of thermal conductivities at whole range of saturation degrees.

Keywords: Porous media · Unsaturated soil · Homogenization

1 Introduction

Analytical homogenization schemes based on the solution of single inclusion problem, e.g., the Mori-Tanaka (M-T) approach, are computationally attractive tools for estimating the homogenized properties of porous media. The main disadvantage of these methods is the choice of proper simplified microstructure; in case of M-T scheme, which assumes the 'matrix-ellipsoidal inclusion' morphology, one has to prescribe inclusion families, ellipsoids differentiated by the semi-axis aspect ratio θ.

In the earlier Authors' paper [1], the problem of simplified microstructure identification, with respect to the overall thermal conductivity of saturated soils, was considered. The idea was to replace a real microstructure with a virtual one, i.e., the original mixture of pores of diverse shapes is replaced by the set of oblate spheroids of randomly distributed aspect ratios, characterized by probability density function M^{eq}, being the so called equivalent microstructure [1]. The identified microstructure can be then successfully used in the approximation homogenization schemes (e.g., the M-T method) for the efficient prediction of soil thermal conductivity.

The aim of this work is to extend the equivalent microstructure approach for the case of 3-phase heterogeneous medium, namely the unsaturated soil (where three distinct phases coexist: solid, water and air). For that purpose, the approach originally proposed in [1] is deeply reformulated. The details are provided below.

© Springer Nature Switzerland AG 2018
W. Wu and H.-S. Yu (Eds.): *Proceedings of China-Europe Conference on Geotechnical Engineering*, SSGG, pp. 1613–1617, 2018.
https://doi.org/10.1007/978-3-319-97115-5_156

2 Preliminary Analysis

As a first trial we propose to replace the 3-phase medium (unsaturated soil) with a 2-phase one and then follow the procedure proposed in [1]. This replacement consists in the evaluation of the thermal conductivity of the mixture of water and air (λ_f^{hom}) and then filling the pore space of the soil with the fluid characterized by λ_f^{hom}. Therefore, a problem considered here is a fully saturated porous medium characterized by the conductivities of the solid phase and fluid, λ_s and λ_f^{hom}, respectively. We use two distinct approaches for the λ_f^{hom} evaluation, i.e., the Hashin-Shtrikman (H-S) lower and upper bounds. In Fig. 1, computed values of λ_f^{hom} are presented as a function of water content, S_r.

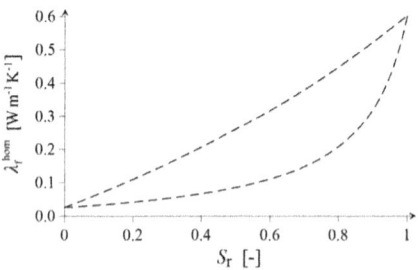

Fig. 1. H-S bounds for a mixture of air and water as a function of water content.

Next, the inverse problem for the identification of M^{eq} is solved using the approach proposed in [1]. The only given data within the inverse problem is the set of laboratory measurements of thermal conductivities at different saturation degrees S_r, porosity (data from [2]) and soil solids conductivity (estimated from Johansen approach [3]). During the M^{eq} identification process, the soil porosity (ϕ) remains undistorted, so the equivalent microstructure has the same volume fraction of pores as the real material. The equivalent microstructure is identified within the framework of the M-T approach which, as it was shown in [1], can be expressed as the linear Fredholm equation of the first kind, i.e.:

$$\int_0^1 P_m\left(\lambda_f^{hom}, \theta\right) M^{eq}(\theta) d\theta = \frac{(1-\phi)}{\phi} \frac{\lambda_s - \lambda^{hom}}{\lambda^{hom} - \lambda_f^{hom}}, \tag{1}$$

where P_m is the so called localization operator (for more details see, e.g., [4]) and λ^{hom} is the measured value of the soil thermal conductivity. Hence, the inverse problem that is to be solved is as follows: given the kernel function P_m as well as the right side of Eq. (1), identify the probability density function, $M^{eq}(\theta)$.

The inverse problem is solved separately for two cases: λ_f^{hom} evaluated by the lower and upper H-S bounds. Identified M^{eq} functions are then used for the prediction of soil thermal conductivity using M-T scheme. In Fig. 2 the predictions and laboratory measurements are presented as a function of λ_f^{hom}. Evaluated probability density functions M^{eq} for both cases are also shown in bottom right part of plots.

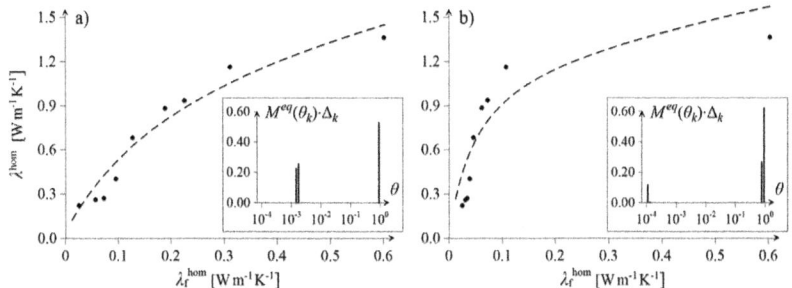

Fig. 2. Prediction of soil thermal conductivity as a function of λ_f^{hom} obtained by: (a) upper and (b) lower H-S bounds.

It can be seen, that the simplified approach, based on the replacement of 3-phase medium with the 2-phase one, does not "a good job". Even though the M^{eq} is identified by the solution of the inverse problem, the agreement between the predictions and measurements is not satisfactory. Therefore in the next Section we propose another reformulation of the equivalent microstructure approach for unsaturated soil.

3   Equivalent Microstructure Approach for Unsaturated Soil

The unsaturated soil is now treated as a 3-phase medium, so in the M-T approach all phases are included and Eq. (1) is properly reformulated. We assume that the pore space (in general, being filled by water and air at the same time) is still described by one, common pdf, M^{eq}. However, the novel in this approach is that for a given saturation degree (S_r) the corresponding ellipsoid aspect ratio θ_{Sr} is identified according to the relation below:

$$S_r = \int\limits_0^{\theta_{Sr}} M^{eq}(\theta)\, d\theta. \tag{2}$$

Therefore, for a given S_r a portion of M^{eq} function (corresponding to the interval 0 to θ_{Sr}) is prescribed to the water phase while the remaining part represents the air phase. It is graphically presented in Fig. 3, where the dashed area represents the ellipsoidal families of inclusions 'filled' with water. Hence, the increase of S_r governs the kinetic of 'filling' of consecutive inclusion families. At the extreme cases, i.e., when $S_r = 0$ M^{eq} corresponds to the air phase only whereas for $S_r = 1$ – to water phase. In order to apply this novel approach, the original inverse problem proposed in [1] is deeply reformulated. Due to the limits of extended abstract, the details cannot be presented.

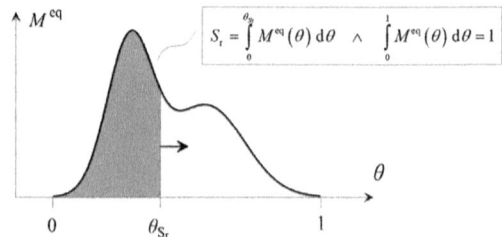

Fig. 3. The novel concept of equivalent microstructure approach for unsaturated soil.

The solution of the inverse problem provides the equivalent microstructure for unsaturated soil under consideration. Once again, the identified M^{eq} is substituted into the M-T scheme. Obtained predictions together with laboratory measurements are presented in Fig. 4. The figure is supplemented with M^{eq} function. What is remarkable, the novel approach provides a very well agreement between predicted and measured conductivities.

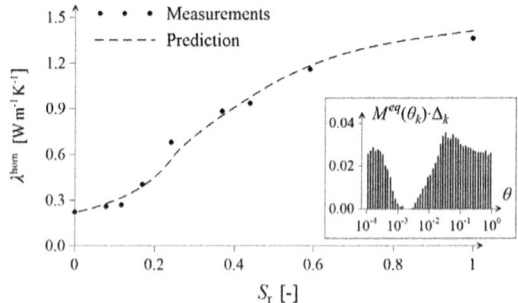

Fig. 4. Prediction of soil thermal conductivity as a function of saturation degree.

4 Final Remarks

Based on the conducted analyses, the following conclusions can be drawn:

- the equivalent microstructure approach, originally formulated for fully saturated soil (2-phase medium) [1], can be adopted for the case of unsaturated soil (3-phase medium),
- nevertheless, it was shown that the simplified approach based on the replacement of 3-phase medium with the 2-phase one does not give satisfactory results,
- the novel approach, where S_r governs the kinetic of 'filling' of consecutive inclusion families, provides the satisfactory results for the whole range of S_r.

References

1. Łydżba, D., Różański, A., Stefaniuk, D.: Equivalent microstructure problem: mathematical formulation and numerical solution. Int. J. Eng. Sci. **123**, 20–35 (2018)
2. Lu, S., Ren, T., Gong, Y., Horton, R.: An improved model for predicting soil thermal conductivity from water content at room temperature. Soil Sci. Soc. Am. J. **71**(1), 8–14 (2007)
3. Johansen, O.: Thermal conductivity of soils (No. CRREL-TL-637). Cold Regions Research and Engineering Lab Hanover NH (1977)
4. Torquato, S.: Motivation and overview. In: Torquato, S. (ed.) Random Heterogeneous Materials, pp. 1–19. Springer, New York (2002)

Modelling Hydro-Chemo-Mechanical Behaviour of Active Clays Through the Fabric Boundary Surface

Mario Manassero$^{(\boxtimes)}$, Andrea Dominijanni, and Nicolò Guarena

Politecnico di Torino, c.so Duca degli Abruzzi 24, 10129 Turin, Italy
mario.manassero@polito.it

Abstract. The osmotic, hydraulic and self-healing efficiency of bentonite based barriers (e.g. geosynthetic clay liners) for containment of polluting solutes are governed both by the physico-chemical intrinsic parameters of the bentonite, i.e. the solid density (ρ_{sk}), the total specific surface (S), and the total fixed negative electric surface charge (σ), and by the state and fabric parameters able to quantify the soil density and microstructure, i.e. the total (e) and nano (e_n) void ratio, the average number of platelets per tactoid ($N_{l,AV}$), the effective electric fixed-charge concentration ($\bar{c}_{sk,0}$), and the Stern fraction (f_{Stern}). In turn, the fabric parameters seem to be controlled by the effective stress history, ionic valence and related exposure sequence of salt concentrations in the pore solution. A theoretical framework able to describe chemical, hydraulic and mechanical behaviours of bentonites has been set up. In particular, the relationships, linking the aforementioned intrinsic, state and fabric parameters of a given bentonite with its hydraulic conductivity (k), effective diffusion coefficient (D_s^*), osmotic coefficient (ω) and swelling pressure (u_{sw}) under different stress-histories and solute concentration sequences, are presented. The proposed theoretical framework has been validated by comparison of its predictions with some of the available experimental results on bentonites.

Keywords: Active clays · Bentonite microstructure · Fabric Boundary Surface
Landfill barriers · Hydro-Chemo-Mechanical behaviour

1 Bentonite Structure

Bentonite is a clay soil that usually contains a significant percentage (e.g., $\geq 70\%$) of the three-layered (2:1) clay mineral montmorillonite. Isomorphic substitution in montmorillonite usually results in the replacement of a portion of tetravalent silicon (Si^{4+}) and trivalent aluminium (Al^{3+}) in the crystalline structure with a divalent metal, such as magnesium (Mg^{2+}), and this leads to a negative surface charge, σ, equal to 0.114 $C \cdot m^{-2}$. Montmorillonite crystals consist of parallel-aligned, alumino-silicate lamellae, which are approximately 1 nm thick and 100–200 nm in the lateral extent. The total specific surface of a single platelet, S, available for water adsorption is approximately equal to 760 $m^2 \cdot g^{-1}$, assuming a solid density, ρ_{sk}, of 2.65 $Mg \cdot m^{-3}$ or a specific gravity, G_s, of 2.65 (-).

© Springer Nature Switzerland AG 2018
W. Wu and H.-S. Yu (Eds.): *Proceedings of China-Europe Conference on Geotechnical Engineering*, SSGG, pp. 1618–1626, 2018.
https://doi.org/10.1007/978-3-319-97115-5_157

Montmorillonite particles can be represented as infinitely extended platy particles, also called platelets or lamellae. The half distance, b, between the montmorillonite particles can be estimated from the total void ratio, e. Norrish [1] showed that bentonite can have a dispersed structure or fabric in which clay particles are present as well separated units, or an aggregated structure that consists of packets of particles, or tactoids, within which several clay platelets are in a parallel array, with a characteristic interparticle distance of 0.9 nm.

The formation of tactoids has the net result of reducing the surface area of the montmorillonite, which then behaves like a much larger particle with the diffuse double layer only fully manifesting itself on the outside surfaces. The formation of tactoids is due to internal flocculation of the clay platelets, and mainly depends on the concentration and the valence of the ions in the soil solution and, to a lesser extent, on the effective isotropic stress history or, in turn, on the total void ratio, e. The average number of clay platelets or lamellae forming tactoids, $N_{l,AV}$, increases with an increase in the ion concentration and valence of cations in the soil solution, whereas there is not apparent unique trend in $N_{l,AV}$ versus the micro void ratio, e_m, for a given concentration. Unfortunately, the average number of platelets per tactoid is a very difficult parameter to be experimentally assessed by both direct (e.g. Nuclear Magnetic Resonance, X-ray Diffraction, Small Angle X-ray Spectrometry, Transmission Electron Microscopy) and indirect methods (e.g. Hydraulic Conductivity, Swelling, Osmosis, Anion Available Pore Volume); therefore, the possibility to link, for a given bentonite, the latter parameter to the total void ratio, e, and to the salt concentration of the equilibrium solution, c_s, can be considered a really important step forward to enhance the bentonite barriers design, as shown in the following of this paper.

The average half distance, b, in perfectly dispersed clays may be estimated, assuming a uniform distribution of the clay platelets in a parallel orientation [2] from the relation:

$$b = \frac{e}{\rho_{sk}S} \tag{1}$$

If the clay has an aggregated structure, only the external surface of the tactoids is in contact with the mobile fluid, therefore the void space within the platelets in the tactoids should be subtracted from the total void space to obtain the micro-void space, e_m, with reference to the conducting pores [2]. If $N_{l,AV}$ is the average number of platelets per tactoid, the external or effective specific surface, S_{eff}, and the internal specific surface, S_n, are given by:

$$S_{eff} = \frac{S}{N_{l,AV}} \tag{2a}$$

$$S_n = S - S_{eff} = \frac{(N_{l,AV} - 1)}{N_{l,AV}} S \tag{2b}$$

The average half distance between the platelets in the tactoid, b_n, as determined by means of X-ray measurements, can vary between 0.2 nm and 0.5 nm [3]. The total void ratio, e, of the bentonite is given by the sum of the void ratio inside the tactoid representing the nano or non-conductive pores (nm size), e_n, and the void ratio representing the micro or conductive pores (μm size), e_m. The water inside the tactoids together with the water of the Stern Layer can be considered part of the solid particles and is excluded from the transport mechanisms.

Therefore, if d_d is the thickness of the Stern Layer consisting of hydrated cations wrapping the external surface of the tactoid, d_{Stern}, divided by the average half distance between the platelets in the tactoid, b_n, the void ratio associated with the internal surfaces of the tactoid, e_n, can be estimated as follows:

$$e_n = b_n \rho_{sk} (S_n + \frac{S \cdot d_d}{N_{l,AV}})$$ (3)

where ρ_{sk} = density of the solid particles. The corrected half distance, b_m, between the tactoids, in the case of an aggregate microstructure of bentonite, can be estimated from an equation similar to Eq. 1, or:

$$b_m = \frac{e_m}{\rho_{sk} S_{eff}}$$ (4)

where $e_m = e - e_n$ = void ratio representing the void space between the tactoids.

2   Fabric and State Parameters

Referring to the conceptual scheme and the possible evolutions of the bentonite structure previously described, other fabric parameters can be derived from the basic average number of platelets per tactoid, $N_{l,AV}$. Equation 2a gives the effective specific surface, S_{eff}, based on the single platelet specific surface, S ($= 760$ m$^2 \cdot$g^{-1}), and the basic fabric parameter $N_{l,AV}$. Moreover, another useful fabric parameter can be directly derived from the latter through the following equation (see also [4]):

$$\bar{c}_{sk,0} = \frac{(1 - f_{Stern}) \cdot \sigma}{F} \cdot \rho_{sk} \cdot S_{eff}$$ (5)

where: $\bar{c}_{sk,0}$ = effective fixed charge concentration of the solid skeleton relative to the solid volume (mol$\cdot$m^{-3}); f_{Stern} = fraction of electric charge compensated by the cations specifically adsorbed in the Stern Layer ranging between 0.70 and 0.95 (-), and F = Faraday's constant ($= 96.485$ C$\cdot$mol^{-1}). The proposed fabric parameters can be influenced primarily by the concentration of ions in the pore solution and by the void ratio, related in turn to the effective isotropic component of the stress history of the considered bentonite.

Once the family of fabric and intrinsic parameters describing the bentonite structure and the physico-chemical properties at the nano and micro scale have been defined, a parallel development can be drawn with some aspects of the well known elasto-plastic-work-hardening models within the traditional soil mechanics (e.g. the Cam Clay

Table 1. Intrinsic, fabric and state parameters for mechanical and chemical models.

Fields	Actions	Intrinsic parameters	Fabric and state parameters
Mechanical	Shear stress: τ	ρ_{sk}, φ_{cv}	e, p', ψ
Chemical	Ion concentration: c_s	S, σ	$N_{l,AV}$, S_{eff}, $\bar{c}_{sk,0}$, f_{Stern}, d_{Stern}

Notes: p': isotropic effective stress component, ψ: dilatancy angle, φ_{cv}: friction angle at the critical state.

model) that are capable of simulating the mechanical behaviour of soils on the basis of a series of state and intrinsic parameters as reported in Table 1.

As shown in [2, 5], the framework that includes the aforementioned fabric parameters is able to link the coupled transport phenomena of water and ions by imposing the chemical equilibrium between the bulk electrolyte solutions and the internal micro-pore solution at the macroscopic scale level, through the Donnan, Navier-Stokes and Nernst-Planck equations. Moreover, also some specific aspects of the mechanical behaviour can be modelled and coupled with the chemical and transport behaviour by taking into account the different types of intergranular actions, beyond the solid contact stress and the bulk pore pressure, such as electro-magnetic attraction/repulsion and osmotic swelling/suction forces.

In principle, and similar to the evolution of the basic Cam Clay model by Alonso et al. [6] through the UPC model for unsaturated soil, the intrinsic and fabric parameters listed in Table 1 and the related framework can extend the basic theoretical approaches to include the mechanical behaviour that is related to the structural features of the active fine grained soils under fully saturated conditions, taking into account not only actions such as the stress history and the related void ratio, but also the ion species and the concentration changes in the pore fluids.

3 Validation of the Solute Transport and Swelling Model

The theoretical framework proposed by Dominijanni and Manassero [2, 5] was found to provide a satisfactory interpretation of the experimental results of Malusis and Shackelford [7, 8], Malusis et al. [9] and Dominijanni et al. [4] in terms of osmotic efficiency, ω, of two bentonites (see Fig. 1) tested with solutions characterized by salt (KCl, NaCl) concentrations up to a maximum of $c_s = 100$ mM. The experimental data are in good agreement with the linear relationship, predicted by the theoretical model, relating the restrictive tortuosity factor, τ_r, to the osmotic efficiency coefficient, ω, as follows:

$$\tau_r = 1 - \omega \tag{6}$$

where τ_r represents the ratio between the osmotic effective diffusion coefficient, D_ω^*, that is measured during an osmotic test [4, 8] and the effective salt diffusion coefficient, D_s^*, that is obtained by extrapolating the value of D_ω^* at $\omega = 0$.

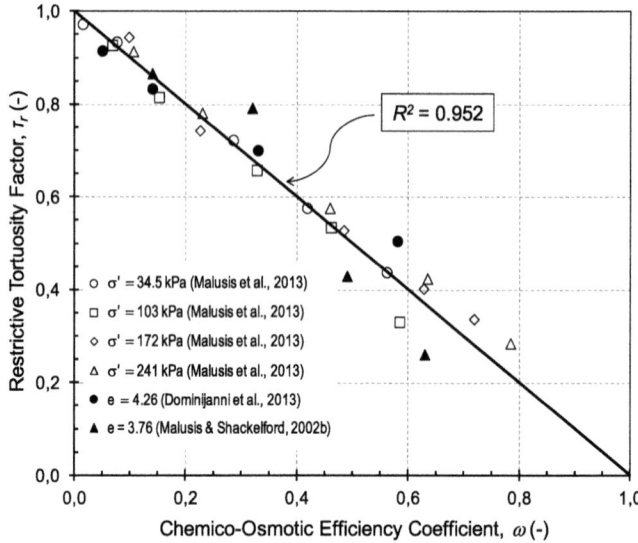

Fig. 1. Restrictive tortuosity factor versus chemico-osmotic efficiency coefficient with the theoretical linear relation.

Also, the evaluation of the osmotic swelling pressure theoretically calculated on these samples by the use of the same input parameters, referring in particular to $N_{l,AV}$, S_{eff}, $\bar{c}_{sk,0}$, was in very good agreement with the related experimental results (see [10]). However, the range of salt concentrations, c_s, investigated within the experimental tests previously noted, was not sufficient to induce any significant variation in bentonite structure and, therefore, the values of the defined fabric parameters ($N_{l,AV}$, S_{eff} and $\bar{c}_{sk,0}$).

For this reason, a more general and reliable validation of the proposed theoretical framework for modelling the bentonite hydro-chemico-mechanical behaviour requires consideration of other experimental results, such as both the hydraulic conductivity and the swelling pressure, using ion concentrations higher than 200 mM and, possibly, ≥ 1000 mM. In the case of the higher ion concentrations, the fabric of the bentonites undergoes major changes due to flocculation phenomenon under low confining stress (high void ratio), resulting in a significant increase in the average number of platelets per tactoid and a correspondent decrease in the effective specific surface and fixed charge concentration of the solid skeleton.

A series of hydraulic conductivity, swelling and oedometer tests performed on different bentonites by different authors (see [11]) have been analysed to validate the proposed general framework. The comparison of experimental and theoretical results consists of the following steps:

- Referring to a first series of hydraulic conductivity tests on different bentonites and permeant solutions, an assessment of the effective specific surface, S_{eff}, has been performed using the following equation [4]:

$$k = \frac{\tau_m}{3} \frac{e_m^3}{(1+e_m)\cdot(\mu_w+\mu_{ev})} \frac{\gamma_w}{(\rho_{sk}S_{eff})^2} \tag{7}$$

Neglecting as a first approximation the electro-viscosity coefficient, μ_{ev}, the definitions of the remaining undefined terms are: τ_m = matrix or steric tortuosity factor (≤ 1) that takes into account the tortuous nature of the actual permeant pathways through the porous medium due to the geometry of the interconnected pores, μ_w = water viscosity coefficient, and γ_w = water unit weight.

- The evaluation of the average number of platelets per tactoid, $N_{l,AV}$, and $\bar{c}_{sk,0}$ has been carried out via Eqs. 2a and 5, respectively, and the assessment of the theoretical results in terms of swelling pressure, u_{sw}, has been obtained as follows (from [4]):

$$u_{sw} = 2RTc_s \left[\sqrt{\left(\frac{\bar{c}_{sk,0}}{2\cdot e_m \cdot c_s}\right)^2 + 1} - 1 \right] \tag{8}$$

where R is the ideal gas constant and T is the absolute temperature (K).

Finally, all the available bentonite samples, in contact with both deionized water and different chemical solutions, have been considered for the assessment of the basic fabric parameter, $N_{l,AV}$, throughout the experimental hydraulic conductivity and swelling pressure measurements. This parameter is uniquely linked with the other two electro-fabric parameters (i.e. S_{eff}, $\bar{c}_{sk,0}$) and, moreover, provides, as previously illustrated, a satisfactory prediction of chemico-osmotic, hydraulic and swelling/shrinking behaviours of bentonites. The obtained $N_{l,AV}$ values have been plotted versus the micro-void ratio, e_m, in Fig. 2.

In order to relate $N_{l,AV}$ to e_m and c_s, the following phenomenological equation has been proposed:

$$N_{l,AV} = N_{l,AV0} + \frac{\alpha}{e_m} \cdot \left(\frac{c_s}{c_0}+1\right) + \beta \cdot e_m \cdot \left[1 - \exp\left(-\frac{c_s}{c_0}\right)\right] \tag{9}$$

where c_0 represents the reference concentration (=1 M), $N_{l,AV0}$ is the ideal average minimum number of lamellae per tactoid when $c_s = 0$ and $e_m \to \infty$; $\alpha = e_m \cdot (N_{l,AV} - N_{l,AV0})$ for $c_s = 0$ is a coefficient relating $N_{l,AV}$ and e_m when $c_s = 0$ and β is a constriction degree coefficient of the platelets. The parameters $N_{l,AV0}$ and β are both depending on bentonite type, pre-treatments (e.g. removal of soluble salts, consolidation), hydration and chemicals exposure sequence.

The micro-void ratio e_m in Eq. 9 can be derived from the total void ratio through the following equation:

$$e_m = \frac{e \cdot N_{l,AV} - S\rho_{sk}b_n(N_{l,AV}+d_d-1)}{N_{l,AV}} \tag{10}$$

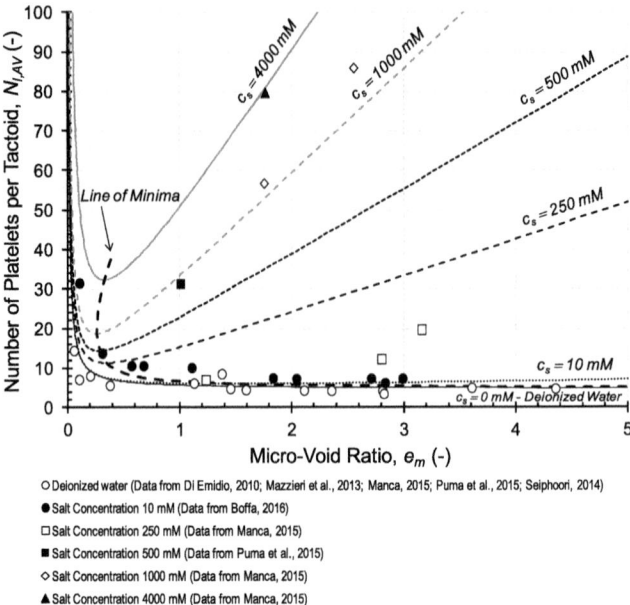

Fig. 2. Average number of platelets per tactoid ($N_{l,AV}$) versus micro void ratio (e_m) based on values from the interpretation of hydraulic conductivity, osmotic and swelling tests (Fitting parameters: $N_{l,AVO} = 4.79$, $\alpha = 0.91$, $\beta = 42.45$; coefficient of determination, $R^2 = 0.89$).

Inserting Eq. 10 into Eq. 9 the number of lamellae per tactoid is related to the total void ratio and the salt concentration through a cubic equation, which can be solved analytically or numerically for given values of the parameters $N_{l,AVO}$, α, β, S, ρ_{sk}, b_n and d_d.

All the available experimental data, previously mentioned, were regressed by imposing $N_{l,AVO} = 4.79$, $\alpha = 0.91$ and $\beta = 42.45$ (coefficient of determination, $R^2 = 0.89$).

It is now interesting to observe the regression results, in terms of average number of lamellae per tactoid, obtained from the interpretation of the aforementioned hydraulic conductivity and swelling tests. The average number of lamellae per tactoid plotted versus the micro void ratio (i.e. the inter-tactoids void ratio) in particular shows, as expected, a very interesting trend. In fact, for any given electrolyte concentration (apart from the unique case of deionized water), an initial decrease in $N_{l,AV}$ to a minimum value is followed by a continuous increasing trend with corresponding increase in the micro-void ratio.

More specifically, the ideal line, which represents the minimum loci of the afore-mentioned function at different ion concentrations, c_s, of the solutions in contact with the bentonite versus e_m may represent a separation locus between flocculating and dispersive behaviour of the considered bentonites, in a similar way to the case of unsaturated soils where swelling and shrinking behaviours are dependent on the degree of saturation (or suction) versus the confining stress and related void ratio [6].

Some 3D views of the proposed Fabric Boundary Surface (FBS) in the domain defined by the number of lamellae per tactoid, $N_{l,AV}$, the micro-void ratio, e_m, and the electrolyte concentration, c_s, are reported in Fig. 3, together with the profiles defined by planar sections orthogonal to the main axes.

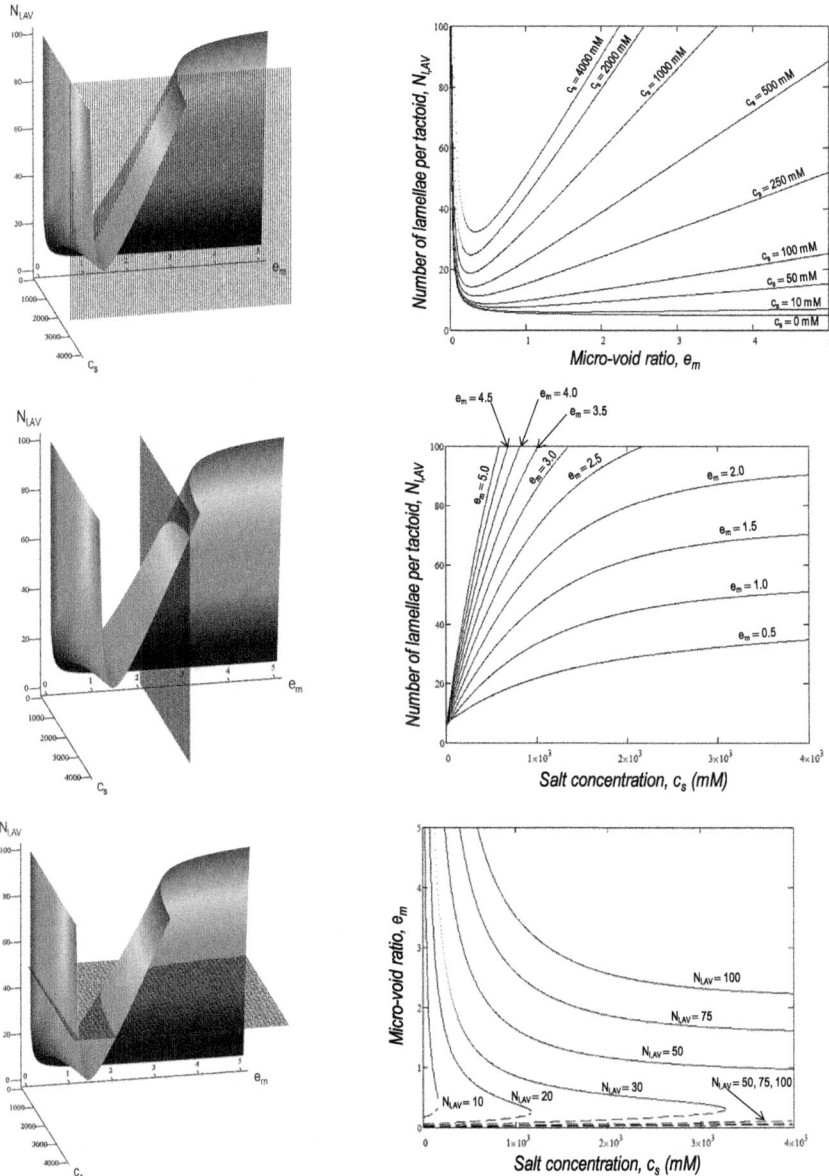

Fig. 3. (a) 3D image of the FBS and (b) planar section traces of the FBS.

References

1. Norrish, K.: The swelling of montmorillonite. Discuss. Faraday Soc. **18**, 120–134 (1954)
2. Dominijanni, A., Manassero, M.: Modelling the swelling and osmotic properties of clay soils. Part II: the physical approach. Int. J. Eng. Sci. **51**, 51–73 (2012b)
3. Shainberg, I., Bresler, E., Klausner, Y.: Studies on Na/Ca montmorillonite systems. 1. The swelling pressure. Soil Sci. **111**(4), 214–219 (1971)
4. Dominijanni, A., Manassero, M., Puma, S.: Coupled chemical-hydraulic-mechanical behavior of bentonites. Géotechnique **63**(3), 191–205 (2013)
5. Dominijanni, A., Manassero, M.: Modelling the swelling and osmotic properties of clay soils. Part I: the phenomenological approach. Int. J. Eng. Sci. **51**, 32–50 (2012a)
6. Alonso, E., Gens, A., Josa, A.: A constitutive model for partially saturated soils. Géotechnique **40**(3), 405–430 (1990)
7. Malusis, M.A., Shackelford, C.D.: Chemico-osmotic efficiency of a geosynthetic clay liner. J. Geotech. Geoenviron. Eng. **128**(2), 97–106 (2002a)
8. Malusis, M.A., Shackelford, C.D.: Coupling effects during steady-state solute diffusion through a semipermeable clay membrane. Environ. Sci. Technol. **36**(6), 1312–1319 (2002b)
9. Malusis, M., Kang, J., Shackelford, C.D.: Influence of membrane behavior on solute diffusion through GCLs. In: Proceedings of the International Symposium on Coupled Phenomena in Environmental Geotechnics (CPEG), ISSMGE TC 215, Torino, Italy, 1–3 July 2013, pp. 267–274. CRC Press Taylor & Francis Group, London (2013)
10. Manassero, M., Dominijanni, A., Musso, G., Puma, S.: Coupled phenomena in contaminant transport. Theme Lecture in: Proceedings of the 7th International Congress on Environmental Geotechnics. Lessons, Learnings & Challenges, Engineers Australia (EA) (AUS), 7th International Congress on Environmental Geotechnics, Melbourne, Australia, 10–14 November 2014, pp. 144–169 (2014)
11. Manassero, M.: On the fabric and state parameters of active clays for contaminant control. In: Lee, W., Lee, J.-S., Kim, H.-K., Kim, D.-S. (eds.) Proceedings of the 19th International Conference of Soil Mechanics and Geotechnical Engineering, 17th–22nd September 2017, Seoul, Korea, pp. 167–189 (2017)

Soil Texture Based Approach for Thermal Conductivity Evaluation

Adrian Różański[(✉)]

Faculty of Civil Engineering, Wrocław University of Science and Technology,
Wybrzeże Wyspiańskiego 27, 50-370 Wrocław, Poland
adrian.rozanski@pwr.edu.pl

Abstract. The proper estimation of soil solids conductivity, λ_s, is of primary importance for evaluation of soil conductivity at various degrees of saturation. This property, i.e. λ_s, cannot be measured directly. Its value is usually estimated by various semi-empirical relations which have been reported in the literature to be improper in many of engineering applications. One of the possible reason is the fact that these models do not take into account the influence of organic matter. The aim of this paper is to create the '*Feret-like* triangles' where λ_s values are plotted as a function of soil texture. For that purpose, a series of numerical simulations is carried out. The micromechanics approach, formulated in [1], is used. The nomograms provide information on the solids conductivities as a function of the soil texture, namely the percentage of individual separates. Different cases, with respect to the organic matter content, are also analyzed.

Keywords: Micromechanics · Soil solids conductivity · '*Feret-like*' triangle

1 Introduction

It was shown in earlier Author's paper [1] that a proper estimation of thermal conductivity of soil solids, λ_s, is of primary importance in prediction of overall soil conductivity, λ. In particular, accurate evaluation of λ at a given degree of saturation is mostly affected by how accurately the conductivities in the dry, λ^{dry}, and saturated states, λ^{sat}, are estimated. The magnitude of heat transfer in dry soil is related mainly to soil porosity, so the value of λ^{dry} is usually well estimated with the use of various semi-empirical relations. On the contrary, thermal conductivity in the saturated state, which is usually estimated by the so called geometric mean equation, requires a good prediction of λ_s.

In [1], a micromechanical model for evaluation of soil solids conductivity was proposed. It was based on the assumption that the individual soil fractions, i.e., sand (Sa), silt (Si) and clay (Cl) can be associated with the mineral composition of the soil skeleton. In other words, the complex structure of the soil at the microscale scale was modelled by considering the spatial variation in thermal conductivity in relation to the soil texture. It was carried out by postulating that thermal conductivity coefficient of each soil separate can be characterized by the proper probability density function (pdf) $f(\lambda)$. Using the back analysis algorithm, based on both computational micromechanics and simulated annealing approach, probability density functions, $f(\lambda)$, were identified in [2].

© Springer Nature Switzerland AG 2018
W. Wu and H.-S. Yu (Eds.): *Proceedings of China-Europe Conference
on Geotechnical Engineering*, SSGG, pp. 1627–1631, 2018.
https://doi.org/10.1007/978-3-319-97115-5_158

The aim of this work is to create the nomograms for different types of soils, which could be then used in engineering applications for fast and efficient evaluation of soil solids conductivity. Differentiation of soils is due to the content of organic matter, whose presence in the soil skeleton is very important from the point of view of its thermal conductivity. It is proposed to perform a series of numerical simulations (using approach proposed in [1] and pdfs identified in [2]) in order to estimate λ_s values for different soil textures and organic matter contents. The nomograms are expected to be in a form of 'Feret-like triangles'; So, as a result, the solids conductivity is expected to be easily recognized providing the soil texture is available. The details are provided below.

2 Framework of the Computational Process

Following the procedure proposed in [1], within the micromechanics approach, the soil skeleton (being a homogenization domain) is generated via a specific stochastic process, i.e., a soil fraction, either sand, silt or clay, is assigned to each voxel and is associated with a conductivity value drawn randomly from the pdf [2]:

$$f(\lambda) = \phi_{Sa} f_{Sa}(\lambda) + \phi_{Si} f_{Si}(\lambda) + \phi_{Cl} f_{Cl}(\lambda),$$ (1)

where ϕ and f are, respectively, the volume fraction and the thermal conductivity's pdf of individual soil fraction (the subscripts Sa, Si or Cl indicate the soil separate). Since, the homogenization domain is also supplemented with voxels that correspond to organic matter, so the following relation holds true:

$$\phi_{Sa} + \phi_{Si} + \phi_{Cl} = 1 - \phi_{OM},$$ (2)

where ϕ_{OM} is the content of organic matter. The organic matter is described by the deterministic (non-random) value of thermal conductivity, $\lambda_{OM} = 0.25 \ \mathrm{Wm^{-1} \ K^{-1}}$.

The exemplary random realization of soil skeleton is presented in Fig. 1. Black voxels correspond to organic matter. Remaining voxels are the "cells" with thermal conductivities prescribed as random numbers generated from appropriate pdfs, as was identified in [2].

Generation of random numbers (random thermal conductivities) is performed as a so-called inverse method (Fig. 2). In general, this problem is divided into two parts. First, a simple generator is used to generate uniformly distributed numbers, which in a second step are transformed to follow the required distribution $f(\lambda)$.

Within the generation process, the analytical pdfs identified in [2] in the form of triangular distributions, are used. For particular soil separates, the formulas are as follows:

$$f_{Sa}(\lambda) = \begin{cases} 1.136 \cdot 10^{-2} \cdot \lambda + 8.727 \cdot 10^{-2}, & 2 \le \lambda \le 8.8 \\ 0, & \text{otherwise} \end{cases}$$

$$f_{Si}(\lambda) = \begin{cases} -4.325 \cdot 10^{-2} \cdot \lambda + 3.806 \cdot 10^{-1}, & 2 \le \lambda \le 8.8 \\ 0, & \text{otherwise} \end{cases}$$

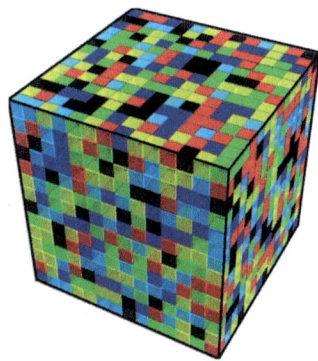

Fig. 1. Homogenization domain, i.e., a soil skeleton, used in micromechanics based approach.

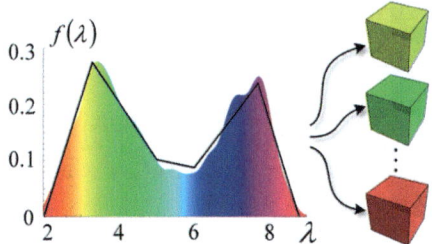

Fig. 2. Graphical presentation of the process of random numbers generation.

$$f_{Cl}(\lambda) = \begin{cases} -5.945 \cdot 10^{-2} \cdot \lambda + 4.637 \cdot 10^{-1}, & 2 \leq \lambda \leq 7.8 \\ 0, & \text{otherwise} \end{cases} \tag{3}$$

Obviously, the final pdf, for given soil texture (percentages of individual soil separates), is created using Eqs. (1) and (3).

3   Results

Using the micromechanical approach (for details see, e.g., [1]) the values of soil solids conductivities, for various soil textures as well as organic matter contents, are evaluated. The results are then used to create the '*Feret-like triangles*'. In Fig. 3 the '*Feret-like triangle*', obtained with a series of numerical simulations (the computations are carried out in the framework of the Finite Volume Method), for the case of no organic matter content is displayed. It represents the values of soil solids conductivities (the unit is W m^{-1}K^{-1}) as a function of soil texture, which is expressed in terms of percentages of individual soil separates, i.e., sand, silt and clay.

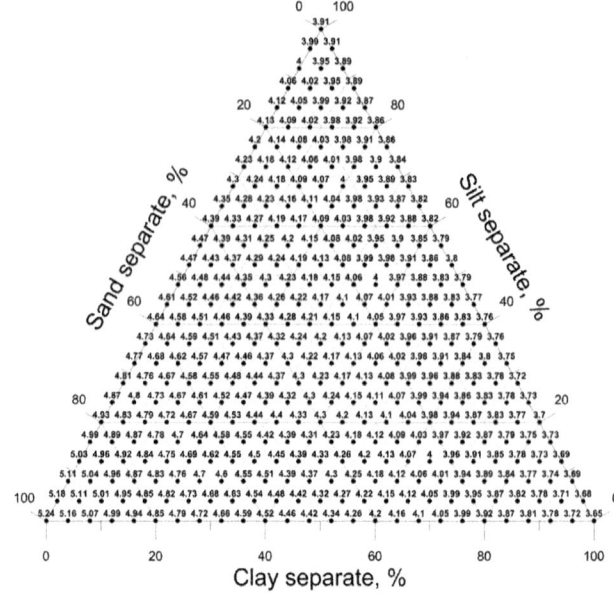

Fig. 3. Soil solids conductivity plotted as a function of percentages of soil separates; the case of no organic matter content ($\phi_{OM} = 0\%$).

The extreme cases considered as a 100% of individual soil separate, i.e., sand, silt and clay provide following values of solids conductivity (results are arranged in the same way as the fractions): 5.24, 3.91 and 3.65 W m^{-1} K^{-1}. These order of results is consistent with those reported in the literature, e.g. [3].

In the considerations different contents of organic matter contents are also analyzed. These results, due to the extended abstract limit, are not presented here.

4   Final Remarks

The aim of this paper was to create the '*Feret-like triangles*' where λ_s values are plotted as a function of soil texture. For that purpose, a series of numerical simulations was carried out. The micromechanics approach, formulated in [1], was used. The nomograms provide information on the solids conductivities providing the soil texture and organic matter content are the only information available. These graphs can be very useful in engineering applications, where fast and efficient estimation of soil conductivity is extremely important.

References

1. Różański, A., Stefaniuk, D.: Prediction of soil solid thermal conductivity from soil separates and organic matter content: computational micromechanics approach. Eur. J. Soil Sci. **67**(5), 551–563 (2016)
2. Stefaniuk, D., Różański, A., Łydżba, D.: Recovery of microstructure properties: random variability of soil solid thermal conductivity. Stud. Geotech. Mech. **38**(1), 99–107 (2016)
3. Tian, Z., Lu, Y., Horton, R., Ren, T.: A simplified de Vries-based model to estimate thermal conductivity of unfrozen and frozen soil. Eur. J. Soil Sci. **67**(5), 564–572 (2016)

Experimental Study on Highly Compacted Bentonite Aggregates Subjected to Wetting and Drying

Haiquan Sun[(⊠)] [ID], David Mašín [ID], and Jan Najser [ID]

Hydrogeology, Engineering Geology and Applied Geophysics,
Faculty of Science, Charles University,
Albertov 6, 128 43 Prague, Czech Republic
haiquan.sun@natur.cuni.cz

Abstract. Microstructure of compacted bentonite is an important aspect which controls its hydro-mechanical behavior, also is vital in constitutive modelling. In order to understand this mechanism, the wetting and drying behavior of compacted bentonite with initial dry densities of 1.9 g/cm^3 was investigated by environmental scanning electron microscopy (ESEM). The attempt of quantitative interpretation was done by digital image analysis at aggregates level (in micro meters). The macro pores between aggregates were clearly sensitive to change of suction, while there were only small volume changes of the aggregates. Volumetric strain of the measured aggregates varied. After one wetting-drying cycle, some new macro pores were formed and the hysteretic behavior of inter-aggregate pores was observed.

Keywords: Aggregates · Bentonite · ESEM · Volume strain

1 Introduction

Compacted bentonites are popular as buffer and backfill material in high level nuclear waste repositories. Compacted bentonites are usually placed between canister and surrounding host rocks. In the repository, bentonite is supposed to be hydrated by adsorption of water from the side of host rocks, while it may dehydrate near canister side due to high temperature. In the final state, the whole bentonite buffer is expected to fully homogenize, but in the initial state, it is obvious that the bentonite must undergo significant change of water content. These drying and wetting cycles will modify the macro and micro structure of bentonite. The changes in bentonite structure are important also for the constitutive modelling of expansive soils [1, 2].

Environmental scanning electron microscopy (ESEM) is a more advanced technique compared to SEM with the ability to test samples with different degree of saturation. The temperature and pressure inside ESEM chamber can be controlled and modified at variety of ranges [3]. In this paper, the wetting and drying paths of compacted bentonite were analyzed in the ESEM chamber. The volume strain of aggregates was calculated by digital image analysis.

© Springer Nature Switzerland AG 2018
W. Wu and H.-S. Yu (Eds.): *Proceedings of China-Europe Conference on Geotechnical Engineering*, SSGG, pp. 1632–1635, 2018.
https://doi.org/10.1007/978-3-319-97115-5_159

2 Materials and Methods

Compacted Czech B75 bentonite extracted from the Cerny vrch deposit (north-western region of the Czech Republic), was used in this study. It contented a montmorillonite content around 60% and initial water content about 10%. The plastic limit, liquid limit and specific density of solid particles were 65%, 229%, and 2.87, respectively. For more detailed information about the Czech bentonite B75, see Sun et al. [4].

The samples equilibrated at the suction of 286.7 MPa by vapor equilibrium method were used for ESEM observations. The suction value of 286.7 MPa is equivalent to relative humidity of 10% by Kelvin's equations [5]. The water vapour pressure of 93 Pa (relative humidity of 10%, suction of 290.75 MPa) was determined as optimal initial state for the experiment. Then the vapour pressure was gradually increased up to 850 Pa (the relative humidity 97%, suction of 3.85 MPa). After the maximum value of the relative humidity was reached, the relative humidity was gradually decreased again down to 10%. The interval between vapour pressure changes was 15 min.

3 Results and Discussion

The micrographs of compacted bentonite upon wetting and drying for the initial dry density of 1.90 g/cm³ after the compaction were presented in Fig. 1. The double arrow is indicating the distance between selected aggregates. The distinctive macro-structural changes with changing suction can clearly be observed. The left aggregate moved up,

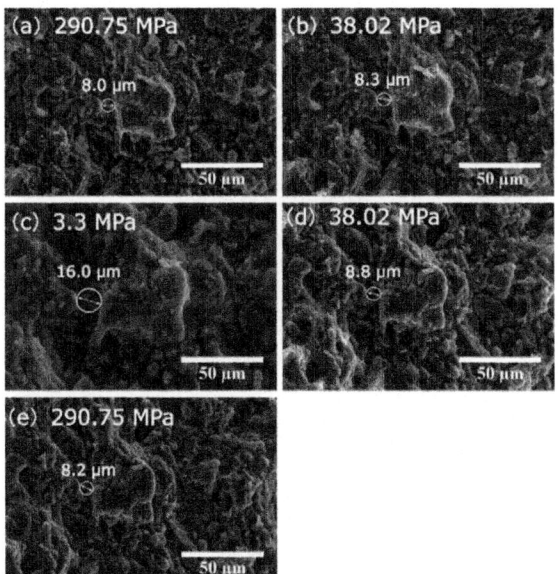

Fig. 1. Selected ESEM micrographs of compacted bentonite with a dry density of 1.90 g/cm³ under the wetting-drying path. (a), (b), (c)-wetting path; (c), (d), (e)-drying path.

while the right aggregate moved downwards enlarge the macropore diameter at the same time. It is clear that the inter aggregate distance increases upon wetting and decreases upon drying. We could also observe hysteretic phenomenon occuring after one wetting-drying cycle. The aggregate distance increase as a result of wetting-drying cycle (see Fig. 1).

In order to quantitatively analyse the aggregate volume change upon wetting and drying, the digital image analysis technique was used [4]. The original ESEM photo represents a plan view in two-dimensions. The boundaries of aggregates were carefully discovered and processed. The volume deformation of the aggregates can be obtained by each ESEM photos with the assumption of spheric shape of the aggregates. Thus, the volume strain (Eq. 1) was used to quantitative analyse the surface volume change upon wetting and drying.

$$\varepsilon_v = \frac{v_i - v_0}{v_i} \times 100\% \tag{1}$$

where ε_v is volume strain, v_i is the volume at stage i, v_0 is the initial volume of the aggregates.

In the paper, the first observation was setted as starting point (suction of 290.75 MPa). Then the volume strain of each stage followed by wetting and drying can be calculated. We selected four different aggregates to analyse its volume strain changing with suction. The surface strain upon wetting and drying path was shown in Fig. 2. The volume strain increased with the decrease of suction. However, this increase was relatively minor up to the low suction of 3.9 MPa. The water entering the macropores and immersing the aggregates at low suction range may cause the aggregate boundaries to be less clearly defined at the micrograph. Above the suction of 38.02 MPa the aggregate size appeared to be insensitive to the hydraulic path and only very small effect of hydraulic hysteresis was measured: aggregate volumetric response thus appeared to be reversible. It must be noted here that the aggregates boundaries

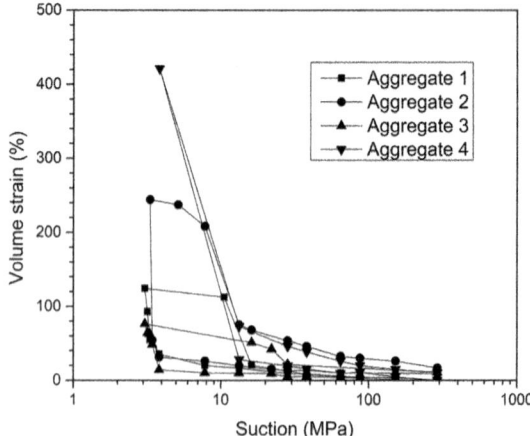

Fig. 2. Volume strain versus suction for compacted bentonite along wetting and drying path.

were hard to recognize at the suctions lower than approximately 10 MPa. The clay aggregates were covered by water which had already entered the macropores at this suction level, which may cause overestimation the surface area of the aggregates.

4 Conclusions

Wetting and drying cycles were performed in an ESEM chamber to quantitatively analyse the deformation of the inter-aggregate pores. Observations and measurements show that swelling of aggregates occurs under wetting and there was a hysteretic behavior observed after one wetting-drying cycle. The deformation of aggregates was low under high suction, while there was a sudden increase of volume strain at suctions below 10 MPa. The volume strain of four selected aggregates was not unique.

Acknowledgements. This project receives funding from the Euratom research and training programme 2014-2018 under grant agreement No 745942. The first author acknowledges support by the grant No. 846216 of the Charles University Grant Agency. Institutional support by Center for Geosphere Dynamics (UNCE/SCI/006) is greatly appreciated.

References

1. Mašín, D.: Double structure hydromechanical coupling formalism and a model for unsaturated expansive clays. Eng. Geol. **165**, 73–88 (2013)
2. Mašín, D.: Coupled thermohydromechanical double-structure model for expansive soils. ASCE J. Eng. Mech. **143**(9), 04017067 (2017)
3. Jenkins, L.M., Donald, A.M.: Observing fibers swelling in water with an environmental scanning electron microscope. Text. Res. J. **70**, 269–276 (2000)
4. Sun, H., Mašín, D., Najser, J., Neděla, V., Navratilova, E.: Bentonite microstructure and saturation evolution in wetting-drying cycles evaluated using ESEM, MIP and WRC measurements, Géotechnique (2018, under review)
5. Sun, H., Mašín, D., Najser, J.: Thermal Water Retention Characteristics of Compacted Bentonite. Geoshanghai 2018. Shanghai, China (2018)

Interpretation of Plate Load Tests on Unsaturated Sand

Yi Tang[(⊠)] and Zhengyin Cai

Department of Geotechnical Engineering, Nanjing Hydraulic Research Institute,
Nanjing, China
hhtmjdty@sina.com

Abstract. This study presents the results of plate load tests conducted on an unsaturated silty sand. The soil samples were subjected to wetting and drying, by raising and lowering of the water table, to mimic conditions which exist in the field. The suction values at different depths were measured by the vibrating wire piezometers. The effects of hydraulic loading histories, soil densities and suctions on the ultimate bearing capacity were studied. An equation was developed to interpret the experimental results based on the relative density dependent peak friction angles and suction profiles. The estimated bearing capacities agree well with the values measured in the plate load tests and the differences can be partly attributed to the superposition of the bearing capacity factors.

Keywords: Plate load tests · Effective stress principle · Unsaturated soil

1 Introduction

The plate load test (PLT) is a widely used field test to determine a soil's ultimate bearing capacity and settlement under a given load. Unsaturated soils are widely distributed throughout the world, accounting for almost 40% of earth's land surface where about 60% of the world's population lives (Khalili et al. 2000). There is a need to understand the response of plate load tests on unsaturated soils to be able to evaluate or back-calculate the strength parameters of the ground properly so shallow foundations can be designed with confidence. The conventional interpretation of the results of plate load tests performed on unsaturated soil would lead to unrealistically large strength and stiffness parameters if the effect of suction is not considered. The way suction may influence PLT results is not straight forward. This is because there is a non-unique relationship between suction (s), strength and degree of saturation (Sr) due to a phenomenon known as hydraulic hysteresis. For a given s profile, PLT results will differ depending on whether the soil has undergone a drying or wetting process prior to achieving that profile.

This paper presents the results of laboratory controlled PLTs performed on unsaturated silty sand samples. The samples are subjected to wetting and drying, by raising and lowering of the water table. The suction profiles are measured using vibrating wire piezometers. The PLT results are interpreted using an equation based on the effective stress principle.

© Springer Nature Switzerland AG 2018
W. Wu and H.-S. Yu (Eds.): *Proceedings of China-Europe Conference on Geotechnical Engineering*, SSGG, pp. 1636–1639, 2018.
https://doi.org/10.1007/978-3-319-97115-5_160

2 Plate Load Tests on Unsaturated Silty Sand

2.1 Soil Classification and Model Test Apparatus

The tested soil is decomposed granite sourced from the catchment area of Lyell dam, NSW, Australia. It is classified to be well-graded silty sand (SM) according to the Unified Soil Classification System. A critical state friction angle of $\phi'_{cs} = 35.7°$ for Lyell silty sand was determined by triaxial compression tests for both saturated and unsaturated conditions. Also at the critical state it was observed that $c' = 0$ kPa. The testing apparatus consists mainly of a steel-framed testing rig, loading devices and a data acquisition system. The testing rig is 0.69 m wide by 2.07 m long by 1.14 m deep. The details of the testing rig are shown in Fig. 1(a). The PLT equipment comprises a rigid circular steel plate, a load cell, a vertical displacement transducer and a worm gear actuator. The circular steel plate has a diameter of 0.15 m. The load cell and the displacement transducer are connected to the lab-view data acquisition system. The details of the loading devices are shown in Fig. 1(b).

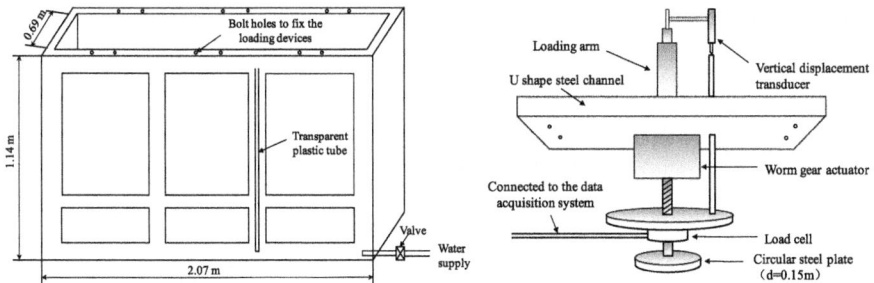

Fig. 1. (a) The details of the testing rig; (b) The details of the loading devices.

2.2 Testing Program and Tests Results

Large dry soil samples with different densities were prepared using a modified method combining pluviation and vibration. Soil samples with heights of 0.44 m and 0.43 m were prepared, corresponding to void ratios of 0.34 and 0.29 and relative densities of 0.82 and 0.92. No obvious changes in void ratio with depth were observed, indicating the soil samples are uniform. After preparing the dry samples water is added through the valve to the soil sample from the bottom of the testing rig. The samples are subjected to wetting and drying, by raising and lowering of the water table. Different suction profiles were developed in unsaturated soil samples which underwent drying and wetting processes. Four piezometers are used in each test to measure the suction values and are inserted at three different depths. Two piezometers are inserted at the depth of 0.07 m (about half of the diameter of the plate) and the other two piezometers are placed at the depths of 220 mm and 370 mm (1.5 times and 2.5 times the plate diameter). The measured suction values at different depths are shown in Table 1.

Table 1. Suctions measured at different depths of each test.

Test number	Suction at 70 mm (kPa)	Suction at 220 mm (kPa)	Suction at 370 mm (kPa)
USW082-1 and USW082-2	5, 4.8	4.6	1
USD092-1 and USD092-2	22.9, 23.4	21.1	5.6

Four plate load tests (PLTs) were conducted on unsaturated soil samples. Two PLTs were performed at different locations on the surface of each sample to investigate the repeatability. The plate was loaded at a rate of 0.03 mm/s. The test was ended when the failure load was reached.

Figure 2 shows the load-displacement curves for the PLTs performed on unsaturated samples having undergone wetting (USW082-1 and USW082-2) and drying (USW092-1 and USW092-2). The initial void ratio is 0.34 and 0.29, respectively. The peak loads were taken to be the failure loads (P_{max}) and were used to determine the ultimate bearing capacity.

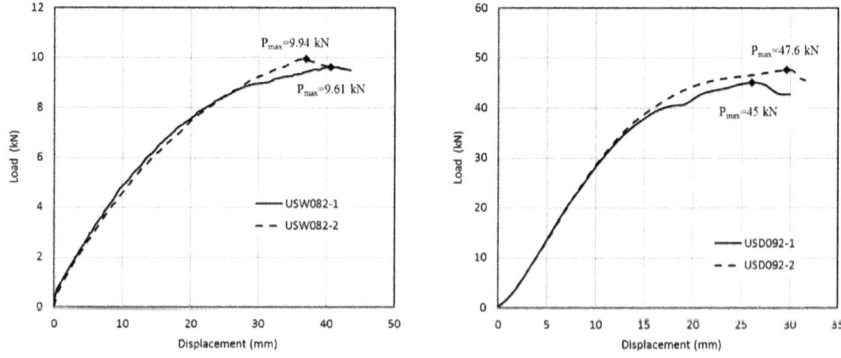

Fig. 2. Plate load tests results for unsaturated silty sand having undergone (a) wetting with test void ratio of 0.34; (b) drying with test void ratio of 0.29.

3   Interpretation of the Plate Load Tests Results

Linear approximations of χs (is the effective stress parameter) profiles could be assumed for both drying and wetting as:

$$\chi s = \chi s_0 - \rho z \qquad (1)$$

where χs_0 is the value of χs at the base of the footing, z is the depth and ρ is a constant defining the variation of χs with depth. For the PLTs performed in this study, values of χ corresponding to the measured suction values are calculated for different initial locations of hydraulic states on the void ratio dependent SWCC. For the sample with an initial void ratio of 0.34 and 0.29, the χs profiles are defined based on the measured suction values as $\chi s = 1.9 - 0.36z$ kPa and $\chi s = 12.8 - 5.6z$ kPa, respectively.

The cohesion (c') and χs have similar and independent effects on the shear strength and the bearing capacity of unsaturated soil. The effect of the linear variation of χs with depth can be incorporated into the Terzaghi's bearing capacity equation similar to a linear cohesion profile in saturated soil. Based on the effective stress principle and the shear strength of unsaturated soils, a bearing capacity equation for unsaturated soil may be obtained by extending Terzaghi's bearing capacity equation:

$$q_u = (c' + \chi s_0 \tan\phi')N_c + qN_q + 0.5B(\gamma - \rho)N_\gamma \qquad (2)$$

where ϕ' is the friction angle, N_c, N_q and N_γ are the bearing capacity factors, q is the overburden pressure, γ is the unit weight of the soil, and B is the footing width.

Rigorous values of the bearing capacity factors, N_c, N_q and N_γ, for rough circular footings, presented by Martin (2004), are used. According to the unique critical state friction angle of $\phi'_{cs} = 35.7°$, the peak friction angles are assumed to be 39° and 40° for soil samples with initial relative densities of 0.82 and 0.92, respectively. Figure 3 compares the bearing capacity calculated using Eq. (2) and the measured values. It is shown that generally the estimated bearing capacity agrees well with those obtained by the plate load tests and the errors are less than 20%. The differences may be due to the superposition used in the bearing capacity equation. The bearing capacity factors are evaluated independently, as it is the case for all conventional bearing capacity equations. It has long been recognized that the interaction between the bearing capacity factors makes the bearing capacity equation conservative.

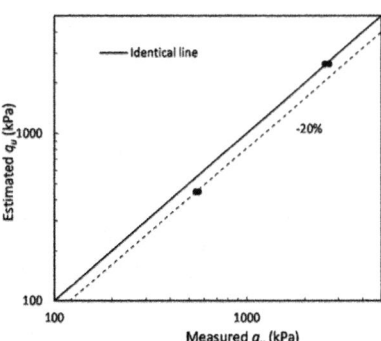

Fig. 3. Comparison between the measured and estimated bearing capacities.

References

Khalili, N., Khabbaz, M.H., Valliappan, S.: An effective stress based numerical model for hydro-mechanical analysis in unsaturated porous media. Comput. Mech. **26**, 174–184 (2000)

Martin, C.M.: User Guide for ABC-Analysis of Bearing Capacity. University of Oxford, Oxford (2004)

Thermo-Mechanical Behaviour of Inventory Materials in a Packed Bed Thermal Energy Storage

Xuetao Wang$^{(\boxtimes)}$ and Christoph Niklasch

Tunnel Engineering, Zentrale Technik, Ed. Züblin AG, Stuttgart, Germany
{xuetao.wang, Christoph.niklasch}@zueblin.de

Abstract. The paper presents a study of thermo-mechanical behaviour of heat energy storage materials of a packed bed made of ceramic pebbles undergoing thermal charging and discharging cycles. Both experimental and numerical results show that the pressure on the storage wall increased over the charging and discharging cycles, but tended asymptotically to a stable threshold value, which can be derived as design values for the packed bed pressure.

Keywords: Thermal energy storage · Thermo-mechanical behaviour
Packed bed

1 Background

One of the most attractive large-scale storage option for electricity is still pumped hydro storages. Considering the need of clean energy, however, the further expansion of pumped hydro storages is constrained by the geographic requirements for these storage systems and also the environmental impact and local acceptance. As an alternative to pumped hydro storages, adiabatic compressed air energy storages are being developed. The concept of compressed air energy storage is to store electrical energy in compressed air. If a surplus of electrical energy is in the grid, air is compressed and stored for example in large salt caverns. The thermal energy generated by compressing the air will be stored in a thermal energy storage in adiabatic compressed air energy storages. During generating electrical energy, the stored thermal energy is able to pre-heat the compressed air before expanding it, saving fossil fuel. In order to store and deliver heat, the materials used in the thermal energy storage as inventory material should meet the design requirement of chemical stability, mechanical stability and thermal stability. Using packed beds of ceramic pebbles or packed beds of natural stones combines large surfaces available for heat exchange with economic construction. The paper focuses on the study of the thermo-mechanical stability of inventory materials in an energy storage using numerical methods. Comparisons with experimental results are done for verification of the numerical models.

© Springer Nature Switzerland AG 2018
W. Wu and H.-S. Yu (Eds.): *Proceedings of China-Europe Conference
on Geotechnical Engineering*, SSGG, pp. 1640–1643, 2018.
https://doi.org/10.1007/978-3-319-97115-5_161

2 Numerical Study

The verification experiments for studying the thermo-mechanical behaviour of the heat storage materials of a packed bed made of ceramic pebbles undergoing charging and discharge cycles were performed by the German Aerospace Centre (DLR). The laboratory model of heat storage is a cuboid with dimension of 0.58 m (L) × 0.875 m (D) × 1.8 m (H). Figure 1 shows the dimension of the storage and setup of measurement sensors. There are total 8 sensors for measuring the forces on the storage wall and 31 sensors for measuring the temperature inside of the packed bed. The force sensors are plates with diameter of 13 cm. The heat storage material is an assembly of man-made ceramic balls with diameter ranges from 14–17 mm.

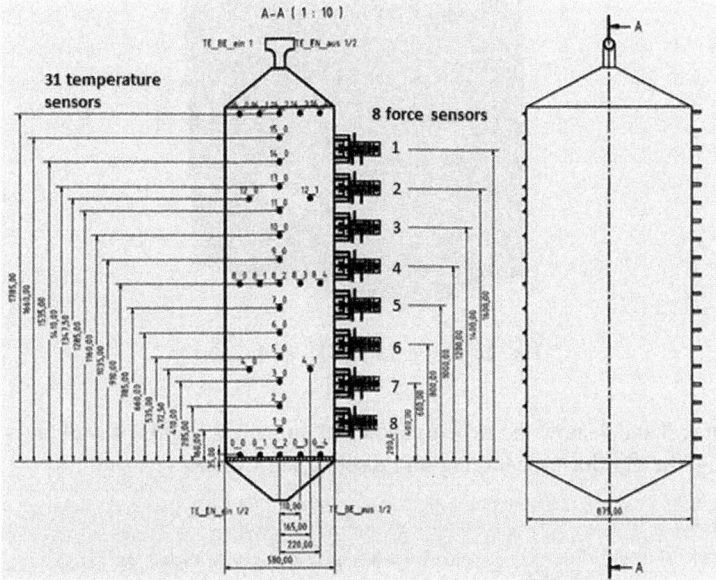

Fig. 1. Laboratory model plan of heat energy storage of DLR.

In the experiments, more than 10 cycles of heat charging (from the bottom of the cuboid) and discharging (from the top of the cuboid) with highest temperatures of about 580 °C have been performed. During these charging and discharging cycles, the forces on the storage walls and the temperatures in the storage materials have been measured.

In numerical study, the experiments are simulated by using Discrete Element Method [1]. In this study, an Open Source Discrete Element Method Particle Simulation Software – LIGGGHTS [2] is used. The material parameters for ceramic balls have been studied [3] before the experiments and numerical simulations of the thermo-mechanical behaviour were performed. The numerical model uses the mechanical parameters determined by the mechanical material tests.

As temperature input for the simulation of the thermo-mechanical test, measured temperatures were used to simulate the charging and discharging cycles. The numerical model of storage is shown in Fig. 2.

Fig. 2. Numerical model of the storage.

Figures 3 and 4 show the experimental and numerical results of wall pressure along the storage height during charging and discharging cycles.

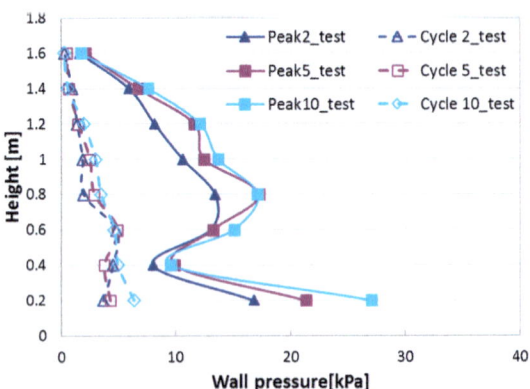

Fig. 3. Experimental results from the experiment 1.

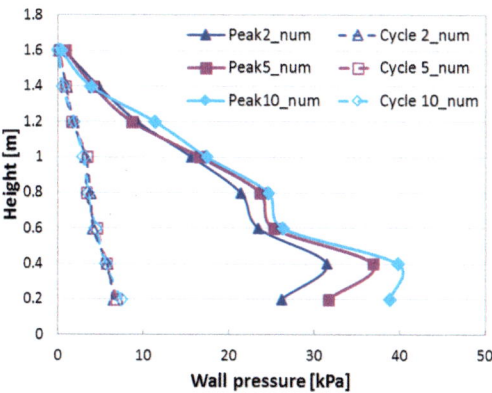

Fig. 4. Numerical results for the experiment 1.

3 Discussions

The numerical results of the wall pressure in Fig. 4 show a good agreement with the measured wall pressures in Fig. 3 except for the pressure in the lower part of the thermal energy storage from 0.4 m–0.6 m. It can be seen in the experimental results that the measurements from force sensor at 0.4 m and 0.6 m height of the storage (force sensors Kraft 6 und Kraft 7) are abnormally smaller than other measurements. The possible reason for this abnormality could be that the ceramic balls contacted with the force sensor cells were too few. As shown in Fig. 1, the diameter of force sensor cells is about 13 cm which is just about 10 times of the diameter of ceramic balls (14–17 mm). If there are not enough ceramic balls in contact with the force sensors, the measurements of the sensors could be less accurate.

Both the numerical results and measurements show that the pressure on the storage wall increases over the charging and discharging cycles, though the differences after 5 cycles of the charging and discharging are minor. It implies that the pressure on the storage wall would tend asymptotically to a stable threshold value. Based on this, design values for the packed bed pressure can be derived for the design of the pressure vessel of the thermal energy storage.

References

1. Cundall, P.A., Strack, O.D.L.: A discrete numerical model for granular assemblies. Geotechnique **29**(4765), 47–65 (1979)
2. Kloss, C., Goniva, C., Hager, A., Amberger, S., Pirker, S.: Models, algorithms and validation for opensource DEM and CFD-DEM. Prog. Comput. Fluid Dyn. Int. J. **12**(2/3), 140–152 (2012)
3. Wang, X.T., Niklasch, Ch., Mayer, P-M.: Comparison of shear strength of particulate materials determined by the direct shear test and DEM simulation. In: IV International Conference on Particle-Based Methods, PARTICLES 2015, 28–30 September, Barcelona (2015)

A New Setup to Measure Hydraulic Properties of Unsaturated Soils

Tiande Wen, Longtan Shao$^{(\boxtimes)}$, and Xiaoxia Guo

State Key Laboratory of Structural Analysis for Industrial Equipment,
Department of Engineering Mechanics, Dalian University of Technology,
Dalian 116024, China
shaolt@hotmail.com

Abstract. The measurement of the hydraulic properties of an unsaturated soil is time consuming and complex, and the water evaporation loss may cause error to the measurement of hydraulic conductivity in multi-step outflow experiment. In this paper, an improved pressure plate instrument is introduced, which can overcome the deficiency of the impact of the water evaporation loss on the measurement of the hydraulic properties. Then, the hydraulic properties of silicon micro-powder (SMP) and Guangxi Guiping clay (GGC) were conducted under five cycles of drying and wetting by this instrument, respectively. We found that the hysteresis effect was apparent less with the number of drying and wetting cycles increases, and tended to exhibit a zero hysteresis under fifth cycle. Additionally, the transient water content curves (TWCCs) of these two soils can be obtained, and the hydraulic conductivity for these two soils can be calculated by Shao et al. (2017) method based on TWCC and SWRC, which can get more data points to present the variation of hydraulic conductivity during each step of the outflow process.

Keywords: Unsaturated soil · Hydraulic properties
Multi-step outflow experiments · Soil water retention curve

1 Introduction

The multi-step outflow method is a transient measurement method for estimating the hydraulic conductivity in the laboratory. The major advantage of this method is that the matric suction can be precisely controlled, and the hydraulic conductivity and SWRC can be derived simultaneously. This method has become one of the most promising and practical methods for estimating the hydraulic conductivity of unsaturated soil [2–5]. However, the duration of the test increases as the water content in the soil decreases during the multi-step outflow experiment. The hydraulic conductivity of unsaturated soil is generally measured without considering the water evaporation loss during this experiment. The main sources of measurement error in multi-step outflow experiment is the water evaporation from (i) the soil surface and (ii) the outflow vessel.

In this paper, multi-step outflow experiments on SMP and GGC were conducted in an improved pressure plate instrument with a water evaporation compensation system. It can be ignored the evaporation from the unsaturated soil surface in this pressure plate

© Springer Nature Switzerland AG 2018
W. Wu and H.-S. Yu (Eds.): *Proceedings of China-Europe Conference on Geotechnical Engineering*, SSGG, pp. 1644–1648, 2018.
https://doi.org/10.1007/978-3-319-97115-5_162

instrument, and the major evaporation is from water outflow vessel. Then, the SWRC of SMP and GGC could be conducted under five drying and wetting cycles by this instrument. Additionally, Shao et al. (2017) method was introduced, which can be calculated hydraulic conductivity.

2 Apparatus

The improved pressure plate instrument is composed of a pressure chamber, a ring cutter, a gas pressure controlling system, an exhaust gas flushing system, a water outflow measurement system, an evaporated compensation system and a computer (see Fig. 1). The increased gas pressure leads to elevated gas pressure while maintaining the pore water pressure constant; the real-time volume of cumulative outflow water with high-precision balance is then measured; the evaporation compensation is measured by the same type of balance at the same time.

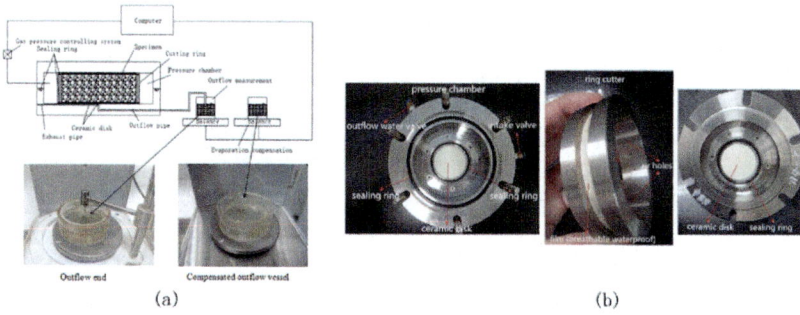

(a)   (b)

Fig. 1. (a) A diagram of experimental setup; (b) Pressure chamber

In order to avoid the water evaporation from the unsaturated soil surface in this setup, the top and the bottom of the soil specimen in the ring cutter are tightly attached to the two saturated ceramic disks during the experiments. Air holes around the ring cutter guarantees that the gas pressure is applied around the specimen and the water is trapped in the specimen all the time with a film (breathable waterproof), therefore the evaporation from the soil surface could be ignored (see Fig. 1). The outflow vessel that is attached to the setup is covered only with a tiny hole to let needle insert. In spite of this, evaporation losses from the outflow end would happen. The evaporation is from the outflow end during an unsaturated hydraulic conductivity test, which was compensated for by taking mass measurements on an identical outflow vessel not connected to the test setup.

3   Materials and Experimental Procedures

The compacted silicon micro-powder (SMP) and Guangxi Guiping clay (GGC) specimens were used in this study (see Shao et al. 2017). The drying path of SMP is [0 kPa,
10 kPa, 20 kPa, 40 kPa, 60 kPa, 100 kPa, 160 kPa, 300 kPa], and the wetting path is
[160 kPa, 100 kPa, 60 kPa, 40 kPa, 20 kPa, 10 kPa, 0 kPa]. Similarly, the drying path
of GGC is [0 kPa, 10 kPa, 20 kPa, 40 kPa, 80 kPa, 110 kPa, 170 kPa, 250 kPa,
350 kPa, 450 kPa], and the wetting path is [110 kPa, 40 kPa, 10 kPa, 0 kPa].
   The cumulative outflow volume can be obtained by measuring the amount of
outflow and evaporation loss. It can be obtained the relationship between the cumulative outflow volume and time in the entire process of the experiment. Then, it can be
applied in the calculation of the relationship between the volumetric water content and
time. Consequently, the SWRC can be derived (Fig. 2).

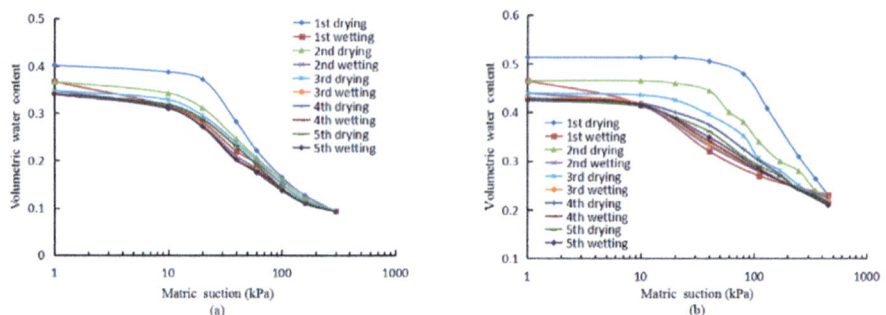

Fig. 2. Measured soil water retention curves for (a) SMP; (b) GGC

   The hysteresis of SWRCs under the first drying and wetting cycles is more significant than that on the subsequent cycles due to the hysteresis for these two soils. The
wetting SWRCs always below the drying SWRCs. Only minor hysteresis is noted in
the SWRC under five cycles of drying and wetting (see Fig. 2).

4   Measuring the Hydraulic Conductivity

Shao et al. (2017) presented a measurement method to describe the change of hydraulic
conductivity during each step outflow experiment, which shows better agreement with
test data comparing with the Gardner outflow analysis method and the van Genuchten-
Mualem prediction model (see Fig. 3). In this method, the transient water content curve
(TWCC) and SWRC of the soil could be obtained in multi-step outflow experiments.
As a gas pressure is exerted, the difference of the matric potential between the TWCC
and the SWRC at the same water content could be calculated. This difference is
considered to have the potential to drive the flow of pore water. Therefore, the
hydraulic conductivity is calculated by using Darcy's law.

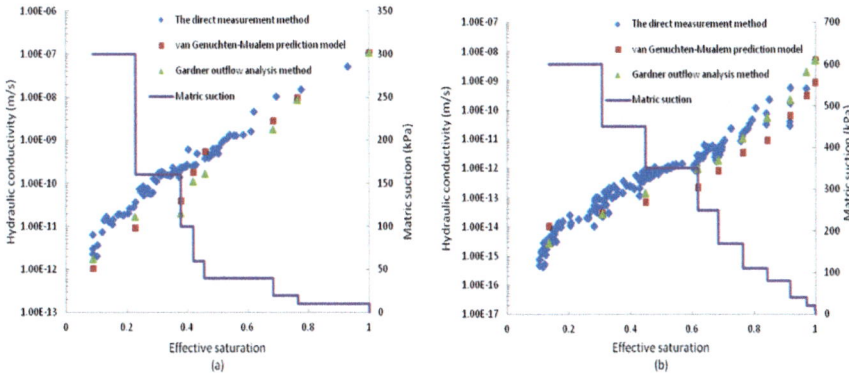

Fig. 3. Hydraulic conductivity versus effective saturation curves and matric suction versus effective saturation curves for (a) SMP; (b) GGC.

$$k_u = -\frac{\Delta V}{A\Delta t}\left(1 + \frac{\Delta u_a^1}{\gamma_w \Delta z}\right)^{-1} \tag{1}$$

where ΔV is the cumulative outflow volume at the time interval Δt, A is the cross-sectional area of the specimen, Δz is the thickness of the specimen, and γ_w is the unit weight of water.

5 Summary

In the multi-step outflow experiments, the evaporation losses of SMP and GGC are remarkable and will lead to a measurement error in the SWRC without compensating for evaporation. The measurement value of the water content will be overestimated. It also reported the SWRC of two soils with a focus on the SWRCs under five drying-wetting cycles. And the main finding is that the effects of hydraulic hysteresis on the SWRC become less and less with the increase of the number of drying and wetting cycles. Otherwise, the Shao et al. (2017) method is feasible comparing with the Gardner outflow analysis method and the van Genuchten-Mualem prediction model.

References

1. Gardner, W.R.: Calculation of capillary conductivity from pressure plate outflow data. Soil Sci. Soc. Am. J. **20**(3), 317–320 (1956)
2. Chen, P., Wei, C., Yi, P., Ma, T.: Determination of hydraulic properties of unsaturated soils based on non-equilibrium multistep outflow experiments. J. Geotech. Geoenviron. Eng. **143**(1), 0416087 (2016)
3. Shao, L.T., Wen, T.D., Guo, X.X., Sun, X.: A method for directly measuring the hydraulic conductivity of unsaturated soil. Geotech. Test. J. **40**(6), 20160197 (2017)

4. Sadeghi, M., Tuller, M., Gohardoust, M.R., Jones, S.B.: Column-scale unsaturated hydraulic conductivity estimates in coarse-textured homogeneous and layered soils derived under steady-state evaporation from a water table. J. Hydrol. **519**, 1238–1248 (2014)
5. Wayllace, A., Lu, N.: A transient water release and imbibitions method for rapidly measuring wetting and drying soil water retention and hydraulic conductivity functions. Geotech. Test. J. **35**(1), 103–117 (2012)

Part XIII: Geotechnics in Transportation, Structural and Hydraulic Engineering

Analytical Study on the Effect of Moving Surface Load on Underground Tunnel

Zhigang Cao[1(✉)], Si Sun[1], Zonghao Yuan[2], and Yuanqiang Cai[2]

[1] Zhejiang University, Hangzhou 310027, China
Caozhigang2011@zju.edu.cn
[2] Zhejiang University of Technology, Hangzhou 310027, China

Abstract. To investigate the influence of the surface moving load on the underground tunnel, an analytical solution for calculating vibrations from a circular tunnel buried in a half-space due to a surface moving load was firstly given in this paper. The surface load was represented by a moving harmonic point load, and the half-space with a circular tunnel was visco-elastic with a cylindrical hole. The analytical solution of the half-space foundation and tunnel were obtained in the frequency domain based on the fundamental solutions of governing equation of elastic ground in Cartesian and cylindrical coordinate systems. Also, the transformations between the plane wave functions and the cylindrical wave functions and the surface boundary conditions should be used. Then response in the time domain was obtained by inverse Fourier transform. The influence of surface moving loads on the vibration of underground tunnel can be investigated by using the analytical model. The displacement and acceleration of the tunnel surface under different load velocities and tunnel buried depth are analyzed in this paper. The results show that both the dynamic stress and the acceleration responses above the vault of the tunnel increases significantly as the moving speed of the load increases. The dynamic stresses decay rapidly as the buried depth of the tunnel increases, while the acceleration responses decay relatively slowly.

Keywords: Surface loads · Wave form transformation
Environmental vibration assessment · Analytical solution

1 Introduction

With the increasing development of the urban surface space, the demand for the development of underground space is becoming more and more intense. A large number of underground shopping malls, subway and pipe culverts are built in main cities, and underground buildings are always close to ground buildings. The resulting interaction influence between the surface environment and underground buildings is attracting widespread attention. Some researchers [1–3] had established several three-dimensional models to study the dynamic response of a homogeneous elastic half-space due to the moving load applied on the ground surface. Furthermore, an increasing number of researchers [4–6] focused on the ground vibration subjected to underground traffic load. However, the research on dynamic response of the existing underground

© Springer Nature Switzerland AG 2018
W. Wu and H.-S. Yu (Eds.): *Proceedings of China-Europe Conference on Geotechnical Engineering*, SSGG, pp. 1651–1655, 2018.
https://doi.org/10.1007/978-3-319-97115-5_163

tunnel to ground traffic load has not been seen. It is urgent to carry out the research on the evaluation of the impact of the ground traffic load on the existing underground tunnel.

In the present paper, the infinite cylindrical cavity model in three-dimensional elastic half-space under surface load is established and the response of underground tunnel induced by surface moving load is calculated. The influence of load speed, tunnel depth and lining thickness on the displacement, acceleration and dynamic stress of the inner surface of the cylindrical cavity is analyzed.

2 Governing Equations and Solutions

A detailed three-dimensional model is shown in Fig. 1 and Cartesian and cylindrical coordinate systems used in this paper is presented.

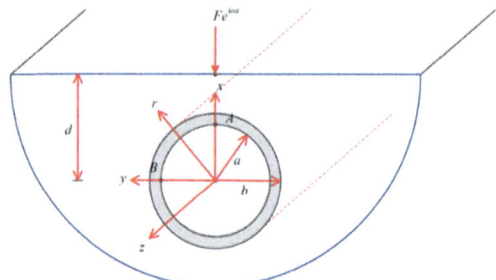

Fig. 1. Model of an underground tunnel in the half-space

The Lamé equations of the half-space are given as:

$$(\lambda_e + 2\mu_e)\nabla(\nabla \cdot \mathbf{u}_e) - \mu_e\nabla \times (\nabla \times \mathbf{u}_e) = \rho_e\ddot{\mathbf{u}}_e \tag{1}$$

Constitutive relations of the elastic soil are given as:
Cartesian coordinate system:

$$t^{(e_x)}(\mathbf{u}) = \mathbf{e}_x\lambda\nabla \cdot \mathbf{u} + \mu\partial_x\mathbf{u} + \mu\nabla u_x \tag{2}$$

cylindrical coordinate system:

$$t^{(\mathbf{e}_r)}(\mathbf{u}) = \mathbf{e}_r\lambda\nabla \cdot \mathbf{u} + 2\mu\partial_r\mathbf{u} + \mu\mathbf{e}_r \times (\nabla \times \mathbf{u}) \tag{3}$$

The fundamental solutions of governing equation of elastic ground in Cartesian and cylindrical coordinate systems:

$$\phi_1^- = \nabla \times (e_z e^{i(qz-h_s x)} \sin py) \qquad \chi_{1m}^+ = \nabla \times (e_z H_m^{(1)}(g_s r) \sin m\varphi e^{iqz})$$
$$\phi_2^- = \nabla \times \nabla \times (e_z e^{i(qz-h_s x)} \cos py) \quad \chi_{2m}^+ = \nabla \times \nabla \times (e_z H_m^{(1)}(g_s r) \cos m\varphi e^{iqz}) \qquad (4)$$
$$\phi_3^- = \nabla (e^{i(qz-h_p x)} \cos py) \qquad \chi_{3m}^+ = \nabla (H_m^{(1)}(g_p r) \cos m\varphi e^{iqz})$$

The displacement of the circular tunnel is the sum of the displacement caused by the down-going waves and the displacement caused by the out-going waves. Thus, the expression of the displacement of the circular tunnel can be obtained:

$$\mathbf{u} = \mathbf{u}_1 + \mathbf{u}_2$$
$$= \int_{-\infty}^{\infty} dq \int_0^{\infty} dp \sum_{j=1}^{3} A_j(q,p)\phi_j^-(q,p,\mathbf{x}) + \int_{-\infty}^{\infty} dq \sum_{j=1}^{3} \sum_{m=0}^{\infty} B_{jm}(q)\chi_{jm}^+(q,\mathbf{r}) \qquad (5)$$

The displacement of the lining structure can be expressed as:

$$\mathbf{u}_1(\mathbf{r}) = \int_{-\infty}^{\infty} dq \sum_{j=1}^{3} \sum_{m=0}^{\infty} (C_{jm}(q)\chi_{jm}^{10}(q,\mathbf{r}) + D_{jm}(q)\chi_{jm}^{1+}(q,\mathbf{r})) \qquad (6)$$

Stress boundary conditions on ground surface and inner surface of lining:

$$\tau(x=d) = \sigma \qquad \tau_1(r=a) = \sigma_1 \qquad (7)$$

Boundary conditions of the contact surface of lining and soil:

$$\tau(r=b) = \tau_1(r=b) \qquad \mathbf{u}(r=b) = \mathbf{u}_1(r=b) \qquad (8)$$

3 Results and Analysis

In Fig. 2, the time history of the vertical displacement of the observation point A to a moving harmonic load is presented when the tunnel depth is 10 m and three load velocities are considered: $c = 20$ m/s, $c = 60$ m/s and $c = 100$ m/s. In Fig. 2, when the moving load moves to position of the observation point ($t = 0$), the vertical displacement of the tunnel reaches the peaks and the peak values under the three cases are 1×10^{-9} m, 2×10^{-9} m and 3.5×10^{-9} m, respectively. It can be seen that the displacement response of the tunnel obviously increases as load velocity increases. It can be seen that the frequency components of the three cases mainly distribute around 5 Hz, as shown in the corresponding frequency spectrum which is presented in Fig. 2(b). At the speed of 20 m/s, the frequency components of the vertical displacement distribute mainly in the range between 3–7 Hz. As the load velocity increases to 60 m/s and 100 m/s, the corresponding frequency components mainly distribute in the range of

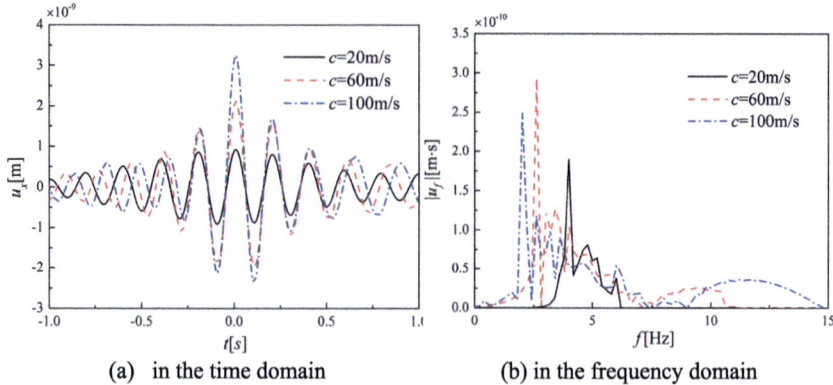

(a) in the time domain   (b) in the frequency domain

Fig. 2. The vertical displacement of the observation point A

1–10 Hz and 0–15 Hz. This is due to the well-known Doppler effects, and the frequency range can be calculated by $f_{cr} = f_0/2\pi (1 \pm V/c_R)$, and f_{cr} is the upper and lower limit of the frequency range.

The influence of the lining thickness and tunnel depth on the acceleration response of the observation point A is investigated in Fig. 3. At the lining thickness of 0.5 m, the maximum acceleration is 7×10^{-7} m/s^2. When lining thickness decreases to 0.25 m and 0.125 m, the maximum acceleration response increases by 20% and 60%, respectively. In Fig. 3(b), the maximum acceleration response at the case of $d/r = 3$ is 2.2×10^{-4} m/s^2. When $d/r = 4$ and $d/r = 5$, the maximum acceleration response is 1.0×10^{-4} m/s^2 and 0.5×10^{-4} m/s^2. It can be obviously seen that the lining thickness and tunnel depth have important influence on the acceleration response.

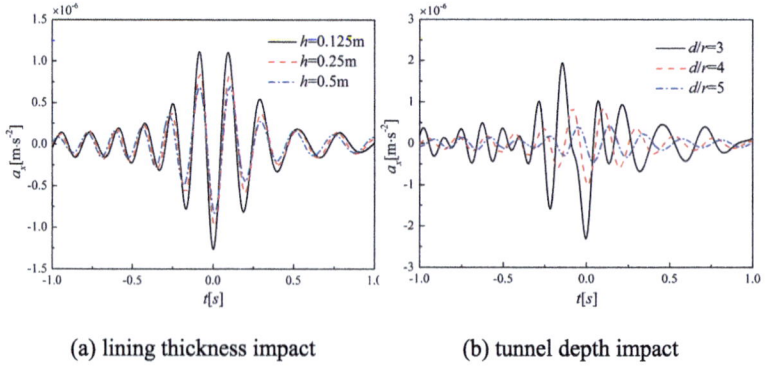

(a) lining thickness impact   (b) tunnel depth impact

Fig. 3. The influence of tunnel depth and lining thickness on the acceleration response

4 Conclusion

An analytical solution for calculating vibrations from a circular tunnel buried in a half-space due to a surface moving load was firstly given in this paper. The influence of load velocity, tunnel depth and lining thickness on dynamic response of the tunnel is investigated. The results show that the dynamic responses above the vault of the tunnel increases significantly as the moving speed of the load increases. The acceleration response of tunnel increases significantly when the depth of tunnel is increased or the thickness of lining is increased. It shows that increasing the depth of tunnel and the thickness of lining can be used as an effective way to reduce the surface vibration.

References

1. Sneddon, I.N.: The stress produced by a pulse of pressure moving along the surface of a semi-infinite solid. Rendiconti del Circolo Matematico di Palermo 2, 57–62 (1952)
2. Bierer, T., Bode, C.: A semi-analytical model in time domain for moving loads. Soil Dyn. Earthq. Eng. 27, 1073–1081 (2007)
3. Xu, B., Lu, J.F., Wang, J.H.: Dynamic response of a layered water-saturated half space to a moving load. Comput. Geotech. 35, 1–10 (2008)
4. Metrikine, A.V., Vrouwenvelder, A.: Surface ground vibration due to a moving train in a tunnel: two-dimensional model. J. Sound Vib. 234(1), 43–66 (2000)
5. Lu, J.F., Jeng, D.S., Lee, T.L.: Dynamic response of a piecewise circular tunnel embedded in a poroelastic medium. Soil Dyn. Earthq. Eng. 27(9), 875–891 (2007)
6. Sheng, X., Jones, C.J.C., Thompson, D.J.: Prediction of ground vibration from trains using the wavenumber finite and boundary element methods. J. Sound Vib. 293(3), 575–586 (2006)

Influence of Matric Suction on the Long-Term Behavior of Fouled Road Base Materials Under Traffic Loading

Jingyu Chen[1(✉)], Yuanqiang Cai[1], Zhigang Cao[1], and Chuan Gu[2]

[1] Zhejiang University, Hangzhou 310027, People's Republic of China
chenjingyucsu@163.com
[2] Wenzhou University, Wenzhou 325035, People's Republic of China

Abstract. The road base and subbase materials are normally in unsaturated condition, and fouled by fines due to subgrade pumping. These unsaturated fouled materials usually present different properties due to combined effects of matric suction and fine content. To study the long-term behaviors of these materials, a series of large-scale cyclic triaxial tests were conducted. Crushed tuff aggregates, incorporated with corresponding kaolin contents, were selected to be the testing materials. Cyclic loadings with two amplitudes were applied on the mixture specimens under four initial matric suctions and two fine contents. The matric suction was controlled by axis-translation method. The accumulated axial strain and resilient modulus were focally analyzed. The testing results show that for materials with higher fine content, the hysteresis phenomenon during the drying and wetting path of the soil-water characteristic curve will be more obvious. Under cyclic loadings, the resilient modulus increases with the increase of matric suction, and the increase rate decreases with the increase of matric suction. The accumulated axial strain decreases with the increase of the matric suction, and the decrease rate decreases with the increase of matric suction. Larger cyclic loading amplitude and higher fine content will amplify the influence of matric suction.

Keywords: Matric suction · Large-scale triaxial test · Fine content
Traffic loading · Long-term cyclic characteristics

1 Introduction

The base and subbase layers are important structures of the road pavements. They play a significant role to bear the traffic loadings and transfer them downward to the subgrade layers. During their service periods, traffic loadings can lead to the accumulation of plastic strain and differential settlement, which may cause the pavement cracks, affect traffic safety and increase the maintenance cost.

Coarse granular materials are widely used as road base and subbase materials due to their good mechanical properties and easy exploitation. A series of experimental studies have been conducted on deformation characteristics of coarse granular materials in saturated condition [1–4]. While the actually base and subbase layers were built above the ground water level, and are normally in unsaturated condition. So, some

© Springer Nature Switzerland AG 2018
W. Wu and H.-S. Yu (Eds.): *Proceedings of China-Europe Conference on Geotechnical Engineering*, SSGG, pp. 1656–1659, 2018.
https://doi.org/10.1007/978-3-319-97115-5_164

researchers have studied the cyclic characteristics of road base and subbase materials under unsaturated conditions [5, 6]. These studies concluded that the resilient modulus and accumulated deformation are dependent on both moisture content and stress state. However, during these studies, only moisture contents were controlled, and the matric suction remained unknown. Though during the traditional unsaturated tri-axial tests, it is realized to control and measure matric suction of the specimen accurately [7, 8]. Due to size effects, the traditional unsaturated tri-axial apparatus cannot be applied to test the characteristics of unsaturated coarse granular materials.

Craciun et al. [9] and Ishikawa et al. [10] improved the original large-scale tri-axial apparatus with unsaturated system, which realized the control and measure of matric suction during tests on unsaturated coarse granular materials. The laboratory results indicated that the influence of matric suction on long-term behavior of road base and subbase layers cannot be neglected. However, these studies mainly focus on the improvement of testing apparatus, and detailed studies on the long-term characteristics of unsaturated road base and subbase materials are needed. In addition, seldom researches have been designed specially to investigate the engineering characteristics of unsaturated crushed aggregates derived from tuff, which are widely distributed in the southeastern coast area in China.

2   Testing Programs

Large-scale triaxial tests were conducted on crushed tuff aggregates with two Kaolin contents ($F = 3\%$ and 6%) under four initial matric suctions ($s = 0$ kPa; 30 kPa; 60 kPa and 90 kPa) subjecting to two cyclic loading amplitudes ($q^{ampl} = 60$ kPa and 100 kPa). A representative low confining pressure $\sigma_{net} = 40$ kPa was applied to all the tests. The cyclic loadings were applied in a load-controlled fashion with compression wave form. The loading cycles N was set to 50,000, and the loading frequency was 1 Hz. During the cyclic loading process, data storage occurred for every 10th cycle, and in every storage cycle, the data were recorded at 50 points.

3   Results and Analysis

3.1   Accumulated Axial Strain

Figure 1 shows the curves of accumulated axial strain, ε_1^{acc} at $N = 50,000$ versus matric suctions s for two cyclic stress magnitudes ($q^{ampl} = 60$ kPa and 100 kPa) at two fine contents ($F = 3\%$ and 6%), respectively. For each test, ε_1^{acc} increases at a high rate during the beginning of tests as compaction occurs, and then tends to increase slowly at almost constant small rate for all the tests. The increase of s will decrease ε_1^{acc} due to larger effective stress for specimens under higher matric suction. In addition, the ratios of ε_1^{acc} at $s = 90$ kPa to that at $s = 0$ kPa will be 70.7% and 54.6% for $q^{ampl} = 60$ kPa and 100 kPa at fine content of 3%, and 56.6% and 48.2% for $q^{ampl} = 60$ kPa and 100 kPa at fine content of 6%, respectively. It can be concluded that larger cyclic loading amplitude and higher fine content will amplify the influence of matric suction on the accumulated axial strain of coarse granular materials.

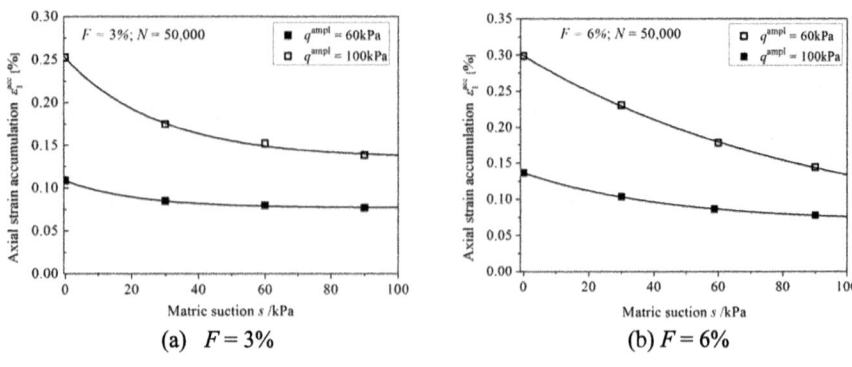

(a) $F = 3\%$ (b) $F = 6\%$

Fig. 1. The relationship between accumulated axial strain (ε_1^{acc}) versus matric suctions (s).

3.2 Resilient Modulus

Figure 2 gives the relationship between resilient modulus (M_r) and matric suction (s) for two cyclic stress magnitudes (q_{ampl} = 60 kPa and 100 kPa) at two fine contents (F = 3% and 6%), respectively. It is can be seen that M_r increases with the increase of s at decreasing rate, which is also due to larger effective stress for specimens under higher matric suction. In addition, the ratios of M_r at s = 90 kPa to that at s = 0 kPa will be 108.2% and 117.7% for q^{ampl} = 60 kPa and 100 kPa at fine content of 3%, and 110.2% and 119.2% for q^{ampl} = 60 kPa and 100 kPa at fine content of 6%, respectively. It can be concluded that larger cyclic loading amplitude and higher fine content will amplify the influence of matric suction on the resilient modulus of coarse granular materials.

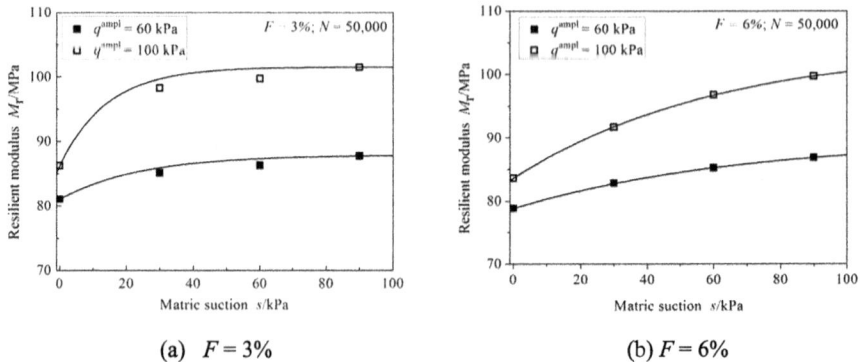

(a) $F = 3\%$ (b) $F = 6\%$

Fig. 2. The relationship between resilient modulus (M_r) versus matric suctions (s).

3.3 Conclusions

The hysteresis phenomenon during the drying and wetting path of the soil-water characteristic curve will be more obvious for materials at higher fine content than that at lower fine content.

The increase of matric suction will increase the accumulated axial strain at decreasing rate and decrease the resilient modulus of testing materials at decreasing rate, respectively.

Larger cyclic loading amplitude and higher fine content will amplify the influence of matric suction on both accumulated axial strain and resilient modulus of coarse granular materials.

References

1. Suiker, A.S.J., Selig, E.T., Frenkel, R.: Static and cyclic triaxial testing of ballast and subballast. J. Geotech. Geoenviron. Eng. **131**(6), 771–782 (2005)
2. Thakur, P.K., Vinod, J.S., Indraratna, B.: Effect of confining pressure and frequency on the deformation of ballast. Géotechnique **63**(9), 786–790 (2013)
3. Wichtmann, T., Rondón, H.A., Niemunis, A.: Prediction of permanent deformations in pavements using a high-cycle accumulation model. J. Geotech. Geoenviron. Eng. **136**(5), 728–740 (2009)
4. Cai, Y., Chen, J., Cao, Z., Gu, C., Wang, J.: Influence of grain gradation on permanent strain of unbound granular materials under low confining pressure and high-cycle loading. Int. J. Geomech. **18**(3), 04017156 (2017)
5. Cerni, G., Cardone, F., Virgili, A.: Characterisation of permanent deformation behaviour of unbound granular materials under repeated triaxial loading. Constr. Build. Mater. **28**(1), 79–87 (2012)
6. Cao, Z., Chen, J., Cai, Y., Gu, C., Wang, J.: Effects of moisture content on the cyclic behavior of crushed tuff aggregates by large-scale tri-axial test. Soil Dyn. Earthq. Eng. **95**, 1–8 (2017)
7. Yang, S., Lin, H., Kung, J.H.S., Huang, W.: Suction-controlled laboratory test on resilient modulus of unsaturated compacted subgrade soils. J. Geotech. Geoenviron. Eng. **134**(9), 1375–1384 (2008)
8. Blatz, J.A., Graham, J.: Elastic-plastic modelling of unsaturated soil using results from a new triaxial test with controlled suction. Géotechnique **53**(1), 113–122 (2003)
9. Craciun, O., Lo, S.C.R.: Matric suction measurement in stress path cyclic triaxial testing of unbound granular base materials. Geotech. Test. J. **33**(1), 33–44 (2009)
10. Ishikawa, T., Zhang, Y., Tokoro, T., Miura, S.: Medium-size triaxial apparatus for unsaturated granular subbase course materials. Soils Found. **54**(1), 67–80 (2014)

Land Subsidence in Shanghai and Its Influence on Transportation Infrastructure

Weiwei Cao, Mingguang Li$^{(\boxtimes)}$, Yujin Shi, Jinjian Chen, and Jianhua Wang

Shanghai Jiaotong University, 800 Dongchuan Road, Shanghai 200240, China
lmg20066028@sjtu.edu.cn

Abstract. Shanghai is located in the coastal region of the Yangtze River Delta which has a phreatic aquifer and five confined aquifers labeled from Confined Aquifer I (CA I) to Confined Aquifer V (CA V). During nearly 100 years of groundwater exploitation, the main exploited aquifers have been switched from CA II and CA III to CA IV and CA V. The deep soil layers went through great compaction because of groundwater withdrawal and became relatively stable after adoption of groundwater recharging measures. However, the shallow soil layers are still deforming as a result of construction activities i.e. underground space exploitation, which causes different responses of transportation infrastructures buried in different depths. Of the three investigated transportation infrastructures, the elevated roads suffer the least from land subsidence and the subsidence of the ground roads is the largest while the subsidence of metro tunnels is between the other two infrastructures.

Keywords: Groundwater exploitation and recharge · Land subsidence
Transportation infrastructures

1 Introduction

Shanghai is located in the coastal region of the Yangtze River Delta which has a phreatic aquifer and five confined aquifers, with confined aquifers labeled from Confined Aquifer I (CA I) to Confined Aquifer V (CA V) shown in Fig. 1. Land subsidence has been a severe problem in Shanghai due to groundwater withdrawal since 1920s. [1] During nearly 100 years of groundwater exploitation, the main exploited aquifers have been switched from CA II and CA III to CA IV and CA V. The deep soil layers (below the roof of CA I) went through great compaction after groundwater pumping. Nowadays, with both exploitation of groundwater resources and adoption of water recharging measures, the deformation of deep soil layers has been relatively stable. However, due to engineering construction especially underground space exploitation, the shallow soil layers (above the roof of CA I) are still deforming to varying degrees, which causes different responses of transportation infrastructures.

The investigated transportation infrastructures mainly consist of elevated roads, ground roads and metro tunnels. The elevated roads are supported by pile groups buried in sandy clays and silty clays while ground roads are constructed on the ground, which is greatly affected by land subsidence [3]. Through research by Shen [4], it is

© Springer Nature Switzerland AG 2018
W. Wu and H.-S. Yu (Eds.): *Proceedings of China-Europe Conference on Geotechnical Engineering*, SSGG, pp. 1660–1663, 2018.
https://doi.org/10.1007/978-3-319-97115-5_165

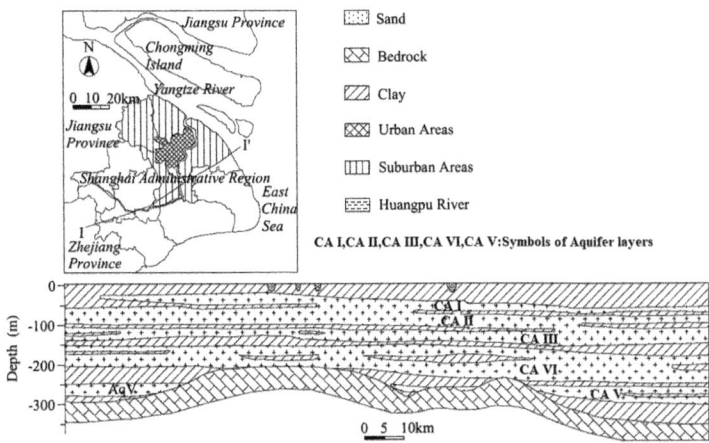

Fig. 1. Distribution of aquifers in Shanghai (cross-Sect. 1-1') (modified from Xu [2]).

acknowledged that the metro tunnels in Shanghai are generally constructed at a depth of 9–15 m in a very soft clay layer and his research shows that metro tunnel settlement is correlated to sublayer settlement rather than ground surface settlement. The deformation of deep soils caused by groundwater exploitation and recharging measures has been investigated and research shows that the settlement has been well controlled, some areas even presenting rebound phenomenon, while the differential settlement is still a great concern in Shanghai. Besides, the acceleration of urbanization such as the exploitation of underground space also has a great impact on land subsidence, which causes the deformation of shallow soil layers. However, the comprehensive analysis of the two soil layers has not been well conducted and its impact on different transportation infrastructures are still not clear. This paper aims to investigate the coupling deformation patterns of both shallow soil layers and deep soil layers and figure out different responses of transportation infrastructures.

2   Method

This paper fuses field data based on a long-term comprehensive monitoring program in Shanghai to reveal the deformation patterns of the confined aquifers and responses of different transportation infrastructures due to soil layer compaction. Firstly, by analyzing field data, the volumes of the extracted and recharged groundwater in main extracted aquifers is presented in Fig. 2. Then by processing monitoring data of layerwise marks, the compression of the deep soils (below the roof of CA I) is investigated and the average annual settlement rate is presented in Fig. 3. We compare the two figures to analyze the relations between the groundwater pump-recharging pattern and compression of the deep soil layers and investigate the impact which groundwater pumping-recharging pattern has on deep soil layers. Secondly, we analyze construction data of foundation and tunnel excavation and compare the results of the consequent settlement with those of the deep soil layers to investigate the comprehensive impact on land subsidence. Finally, the responses of transportation infrastructures (ground roads, elevated roads and metro tunnels) of different depths and structures are analyzed using construction and monitoring data to determine the compression impact which they suffer from the two kinds of soil layers.

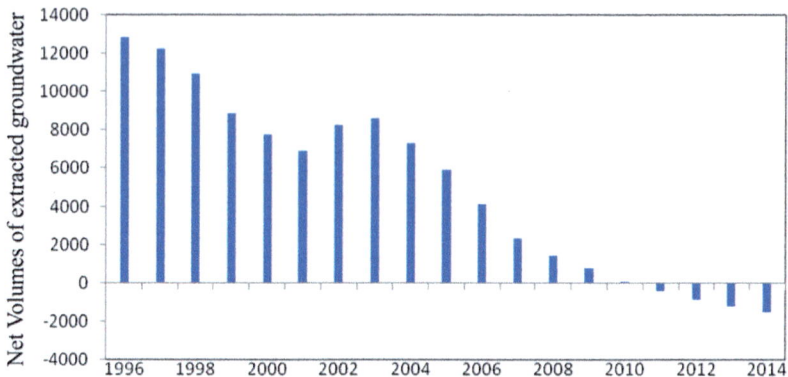

Fig. 2. Annual extracted and recharged volumes of groundwater.

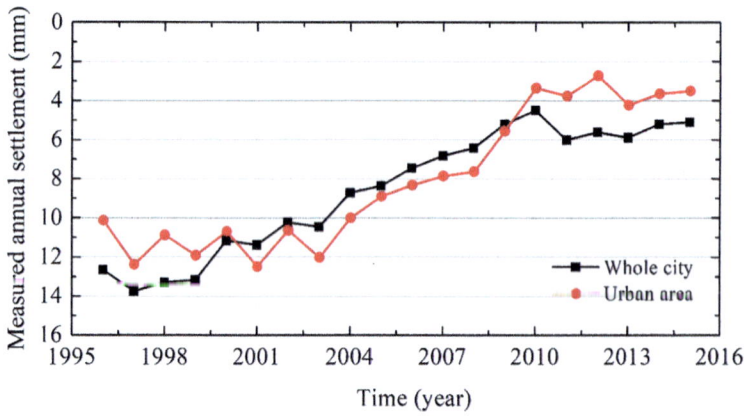

Fig. 3. Settlement rate of the whole city and central city in Shanghai.

3   Results and Discussion

The analysis results show that the settlement rate of Shanghai city has a negative correlation with the exploitation of groundwater and an approximative positive correlation with the recharge of the groundwater, which indicates that the compaction of deep soil layers has been effectively mitigated and presents stable rebound due to groundwater recharging measures. However, shallow soil layers are still in deformation as a result of engineering construction such as tunnel excavation, foundation pit excavation and subgrade construction, which causes settlement of the transportation infrastructures. The responses of different infrastructures are diverse because they are constructed in different depths and their supporting layers are also different. Their deformation increases with the increasing distance from the ground surface. The results show that of the three investigated transportation infrastructures, the elevated roads suffer the least from land subsidence for its supporting piles are buried relatively deep in the stable soil, which is less affected by construction activities compared with the other two. The subsidence of the ground roads is the largest because they are built on the ground and affected by compression of both deep soil layers and shallow soil layers, and its subsidence coincides with the ground settlement. The subsidence of metro tunnels is between the other two infrastructures because they are built in the middle depth of shallow soil layers which are also affected by construction activities but not that severe.

4 Conclusions

The compaction of deep soil layers has been effectively mitigated and presents stable rebound due to groundwater recharging measures. However, shallow soil layers are still in deformation as a result of engineering construction such as tunnel excavation, foundation pit excavation and subgrade construction, which causes settlement of the transportation infrastructures. the elevated roads suffer the least from land subsidence for its supporting piles are buried relatively deep in the stable soil. The subsidence of the ground roads is the largest because they are built on the ground and affected by compression of both deep soil layers and shallow soil layers. The subsidence of metro tunnels is between the other two infrastructures.

References

1. Chai, J.C., Shen, S.L., Zhu, H.H., Zhang, X.L.: Land subsidence due to groundwater drawdown in Shanghai. Géotechnique **54**, 143–147 (2004)
2. Xu, Y.-S., Shen, S.-L., Ren, D.-J., Wu, H.-N.: Analysis of factors in land subsidence in Shanghai: a view based on a strategic environmental assessment. Sustainability **8**(12), 573 (2016)
3. Shi, Y.J., Li, M.G., Chen, J.J., Wang, J.H.: Long-term settlement behaviour of a highway in land subsidence area. J. Perform. Constr. Facil. (2018)
4. Shen, S.-L., Wu, H.-N., Cui, Y.-J., Yin, Z.-Y.: Long-term settlement behaviour of metro tunnels in the soft deposits of Shanghai. Tunn. Undergr. Space Technol. **40**, 309–323 (2014)

Geotechnical Properties of Phosphogypsum and Its Use in Road Engineering

Barbara Cichy[1], Cezary Kraszewski[2($\boxtimes$)], and Leszek Rafalski[2]

[1] Inorganic Chemistry Division IChN, New Chemical Syntheses Institute,
Gliwice, Poland
[2] Road and Bridge Research Institute "IBDiM", Warsaw, Poland
`ckraszewski@ibdim.edu.pl`

Abstract. Phosphogypsum is an industrial waste generated during the production of phosphoric acid. Its amounts in the world are significant and still increase. This waste is mainly stored and its storage is a problem for the natural environment. Road construction industry allows for the use of large amounts of industrial waste, such as fly ash, coal mining waste and metallurgy industry by-products. However, due to its properties, phosphogypsum is a waste with high limitations for use. The research of technical and chemical properties of Phosphogypsum is still being carried out. According to previous studies, the use of unprocessed phosphogypsum in building materials causes deterioration of their properties, therefore processing is required. The paper presents the results of testing phosphogypsum-fly ash mixture and this mixture processed into a granular form with sulphur and various additions of a modifier. The solidification of phosphogypsum into granules was aimed at improving geotechnical properties and reducing the leaching of heavy metals. The paper discusses the geotechnical properties of the mixture before and after granulating it in terms of the possibility of use its in road engineering as the geotechnical structures. The presented research was carried out as part of the RID-6 (Road Innovation Development) research project "Use of recycled materials".

Keywords: Waste · Phosphogypsum · Earth works · Road engineering

1 Introduction

In the wet phosphoric acid process, the phosphate rock, containing mainly minerals from the apatite group, is treated with sulphuric acid. As a result of the chemical reaction, two products are formed, i.e. phosphoric acid and gypsum, which, due to the specific contamination, is known as phosphogypsum. The phosphoric acid is mainly used for production of mineral fertilizers. Nowadays 90% of the amount of phosphoric acid is produced using wet method [1, 2]. Phosphogypsum is dominant by-product of phosphoric acid production process. It is estimated that 100–280 million tons of waste phosphogypsum is produced globally each year [1, 3]. Lopez et al. even give the value of 200–280 million tons per year [4]. The composition of phosphogypsum is determined by characteristics of the raw material from which it is derived. In all cases, the main component of phosphogypsum is gypsum. It also contains many impurities such as phosphates, sulphates and fluoride compounds that originate directly from the output

© Springer Nature Switzerland AG 2018
W. Wu and H.-S. Yu (Eds.): *Proceedings of China-Europe Conference
on Geotechnical Engineering*, SSGG, pp. 1664–1667, 2018.
https://doi.org/10.1007/978-3-319-97115-5_166

phosphate rock. Phosphates, sulphates and fluoride compounds occur in the form of free acids and as salts of metals, including heavy metals. In addition to the impurities, phosphogypsum contains several radioactive isotopes from the decay chain of uranium and thorium [5]. The content of impurities such as toxic metals and radionuclides poses a threat to the environment, which limits the range of methods for using the waste [3]. Because of environmental restrictions, different in different countries, only 15% of the phosphogypsum produced is consumed in agriculture, construction or cement industry. A well-known method of disposing of solid wastes is to use them as materials for construction of roads. Research on the use of phosphogypsum in road engineering works was carried out in Florida [6], France [7], Finland [8] and Poland [9, 10].

The paper presents the research on granulated phosphogypsum as a material for road construction.

2   Materials and Testing

Phosphogypsum and fly ash from combustion of hard coal were used for the preparation of phosphogypsum-fly ash granulate (PG/FA) through simultaneous mixing and granulation of phosphogypsum (PG) 70% m/m with fly ash (FA) 30% m/m in a laboratory mixer, the granulator. During selection of the composition of the granulation mixture, attention was paid to neutralize the acid phosphogypsum with alkaline fly ash to obtain the pH of aqueous granulate extract from 6.5 to 9.0. The phosphogypsum-fly ash granulates (PG/FA) was used to produce the sulphur phosphogypsum fly ash composite. For this purpose, the granulate was stirred with melted sulphur (S) (above 140 °C) and a modifying agent (M = dicyclopentadiene, stabilized with 100–200 ppm of 4-tert-Butylcatechol). The method of preparation of both materials is presented in Fig. 1.

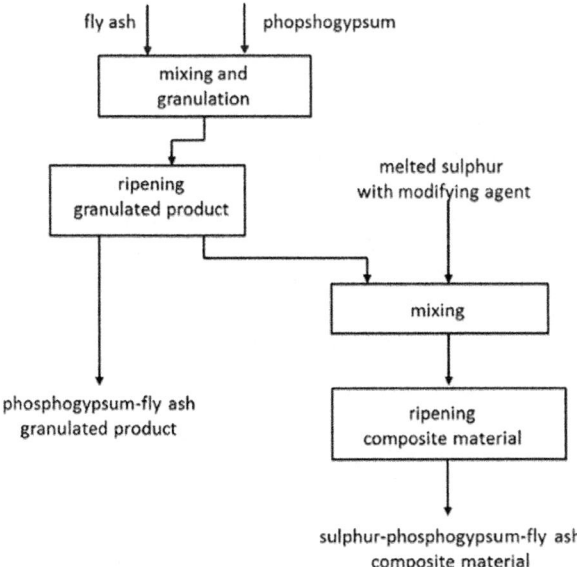

Fig. 1. The flow sheet for sulphur phosphogypsum fly ash composite

PG/FA+S without modify.   PG/FA+S 1% modify.   PG/FA+S 5% modify.

Fig. 2. Illustration depicting the produced PG/FA+S granulate.

3 Results of Geotechnical Tests

See Table 1.

Table 1. Geotechnical properties of PG/FA and PG/FA+S.

Tested feature	PG/FA	PG/FA+S 5% modification
Sand content (0.063 mm–2.0 mm), %	28.7	83.4
Silt content (0.002 mm–0.0063 mm), %	68.5	14.8
Clay content (<0.002 mm), %	2.8	1.8
Sand equivalent SE_4, -	-	45
Maximum dry density, g/cm^3	1.070	1.530
Optimum moisture content, %	47.3	6.3
CBR (immediately after molding), %	36	61
CBR (after 4 days soaking in water), %	23	76
LS linear swelling, %	0	0.85
Angle of internal friction ϕ, °	35	32
Cohesion c, kPa	30	24

4 Conclusion and Suggestions

The test results show that the mixture of phosphogypsum with hard coal fly ashes (PG/FA) has significant technological constraints in terms of direct use in earth roadworks. In terms of grain size, its geotechnical classification corresponds to silt. Testing of the granulate (PG/FA+S) confirms that it is possible to obtain from the silty form (PG/FA) a particulate material with grain size similar to sands. It is noted that the degree of modification of the sulphur has an impact on the grain size distribution of the granulate. The larger the addition of the modifying agent to the sulphur, the finer granules are obtained, and vice versa (Fig. 3). After granulation with the use of sulphur, the PG/FA mixture has much more favorable geotechnical properties for use in earth roadworks, even in capping and subgrade improvement layers. The issue of swelling of the granulate requires a wider research using different quantities of sulphur and additives of the modifying agent to minimize or eliminate this phenomenon in the

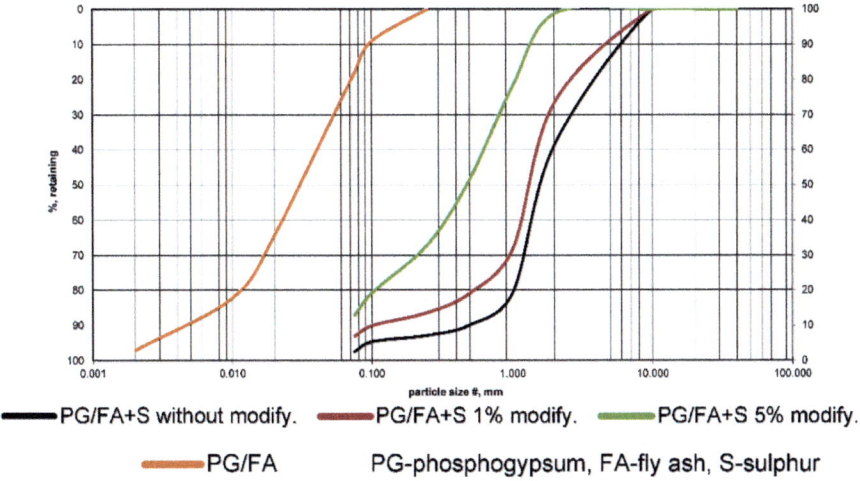

Fig. 3. Particle size distribution of tested materials.

production of granulate. An interesting phenomenon is the growth of CBR value of the granulate, even after swelling. In the long term, the development of the bearing capacity of compacted granulate should be necessarily investigated.

References

1. Tayibi, H., Choura, M., López, F.A., Alguacil, F.J., López-Delgado, A.: Environmental impact and management of phosphogypsum. J. Environ. Manage. **90**(8), 2377–2386 (2009)
2. Cichy, B., Jaroszek, H.: Phosphogypsum management. world and polish practice. Przem. Chem. **92**(7), 1336–1340 (2013)
3. Macías, F., et al.: An anomalous metal-rich phosphogypsum: characterization and classification according to international regulations. J. Hazard. Mater. **331**, 99–108 (2017)
4. López, F.A., Gázquez, M., Alguacil, F.J., Bolívar, J.P., García-Díaz, I., López-Coto, I.J.: Microencapsulation of phosphogypsum into a sulphur polymer matrix: physico-chemical and radiological characterization. J. Hazard. Mater. **192**(1), 234–245 (2011)
5. El-Didamony, H., Gado, H.S., Awwad, N.S., Fawzy, M.M., Attallah, M.F.J.: Treatment of phosphogypsum waste produced from phosphate ore processing. J. Hazard. Mater. **244–245**, 596–602 (2013)
6. Chang, W.F., Chin, D.A., Ho, R.: Phosphogypsum for Secondary Road Construction. FIPR Publication #01-033-077 (1989)
7. Mangin, S.: Le Phosphogypse: utilisation d'un sous- roduit industriel en techni ue routiére. Laboratoire central des ponts et chaussées, Paris (1978)
8. Disposal Management System for Utilization of Industrial Phosphogypsum and Fly Ash. Project No.: LIFE98ENV/FIN/000566. Design Guide: Road Construction Based on Phosphogypsum and Fly Ash (2002)
9. Folek, S., Walawska, B., Wilczek, B., Miśkiewicz, J.: Use of phosphogypsum in road construction. Pol. J. Chem. Technol. **13**(2) (2011)
10. Patent PL224994

Mechanically Stabilized Earth Walls and Uneven Reinforcement Lengths (Trapezoidal Walls) – Design Development and Challenges

Ching Dai[1(✉)] and James Livingston[2]

[1] Chartered Principal with Coffey Services (NZ) Limited,
Auckland, New Zealand
ching.dai@coffey.com
[2] Geotechnical Engineer with Coffey Services (NZ) Limited,
Auckland, New Zealand

Abstract. A composite trapezoidal MSE-embankment system was designed and constructed to support the embankment widening project at a Highway Bridge in New Zealand. The widening necessitated the construction of retaining walls on the existing abutment slopes and under the existing bridge. To suit the constructability and achieve a customized cost effective solution, the use of traditional rectangular MSE reinforcement lengths, which would require vertical excavation into the existing abutment slope, was replaced by the use of Trapezoidal-shaped reinforcement lengths. This paper assesses the consideration of a designed and constructed Trapezoidal Wall for a highway Bridge abutment, and variations on the conventional assessment of failure mechanisms to ensure the wall achieved the design intent. A few options are discussed and compared. Numerical analysis is carried out to establish the achievement of wall stability and serviceability requirements. The successful application of this wall is demonstrated by observed stability and the monitoring during and post-construction.

Keywords: MSE wall · Trapezoidal · Uneven reinforcement

1 Introduction

Coffey was engaged by New Zealand Transport Agency (NZTA) to prepare a MSE wall bridge abutment design, including both temporary construction and permanent wall design, for the widening of the highway overbridge abutments as part of overall highway improvements. The project proposes to widen the existing northbound and southbound carriageways from two lanes and a hard shoulder to three lanes plus a bus lane. A shared cycleway and pedestrian path will be provided adjacent to the northbound carriageway.

This paper presents the challenges faced in the design of the proposed widening, the design development and methodologies, assessment outcomes, and the construction performance of the final embankment widening solution, which comprises trapezoidal shaped reinforced soil walls.

© Springer Nature Switzerland AG 2018
W. Wu and H.-S. Yu (Eds.): *Proceedings of China-Europe Conference on Geotechnical Engineering*, SSGG, pp. 1668–1671, 2018.
https://doi.org/10.1007/978-3-319-97115-5_167

2   Retaining Wall – Challenges and Design Development

The embankment widening comprised cut heights up to 8 m above natural ground level. The MSE wall design underwent two stages of development, starting with a conventionally designed rectangular block by others, and concluding in a trapezoidal-shaped block (uneven reinforcement lengths) buttressing the existing embankment. A trapezoidal MSE wall was adopted in the final redesign. There were two options prepared for the construction phase, refer to Fig. 1. Option 1 – soil nail, and option 2 – inner soldier (pin) pile. Option 2 was selected for temporary construction support after cost comparison. This approach resolved the constructability issue in the previous design, where the whole abutment cut can be completed in one operation with the vertical rectangular cut requiring sheet piling and soil nail support being avoided. The discussion below is concentrated on the permanent trapezoidal wall.

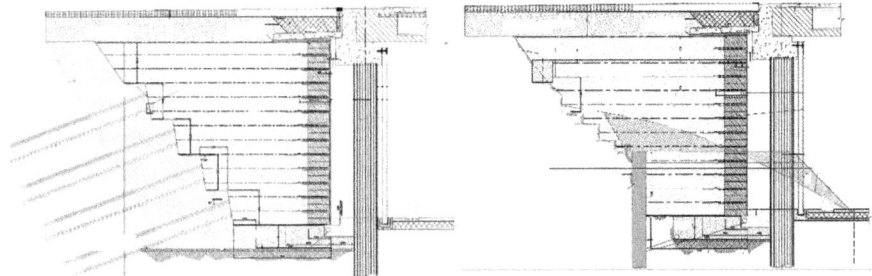

Fig. 1. Options for temporary construction phase Left: Option 1–soil nail; Right: option 2–inner pin pile

In the absence of design standards specifically for the proposed trapezoidal-MSE solution, the Federal Highways Administration (FHWA 2006) document on "Shored Mechanically Stabilised Earth (SMSE) Wall Systems Design Guidelines" was used as a guiding document. Design refinements were made based on FHWA's (2006) recommendations, resulting in the final design solution, whereby the "base" of the RSW was shortened to 4 m. The design necessitated approximately a 1V:1H temporary excavation into the existing embankment to accommodate the widened base. Reference can be made to Fig. 1 (right) for an illustrative sketch summarising the relative dimensions and design features adopted during at the Final Design phase for the Road Abutments.

The updated final redesign challenged the original design mechanisms. The following sections provide some detailed discussion of the issues.

3   Design Analysis – Geotechnical Capacity and Discussion

The trapezoidal MSE walls have been analysed and designed for internal stability and global stability using appropriate computer software.

All design loads, geotechnical parameters and criteria are summarized in the Design Report (Coffey 2017). Global stability was assessed using the software SLIDE version 7.0. We have modelled the maximum above ground wall height of 8 m for the western abutment and 7 m for the eastern abutment, the trapezoidal MSE wall has been superimposed on to the existing slope surface which is approximately 35°.

The bearing capacity was checked through our spreadsheet-based program. On the basis of this work, a 1 m thick No Fines Concrete foundation is required to support the MSE wall due to surface traffic loads.

Static sliding and over turning analysis were ignored in the analyses based on the discussion presented in Sect. 3. However, the potential seismic sliding mechanism was analysed as discussed below (cases C1 and C2).

The following Global Stability Cases have been analysed for the longitudinal wall and outputs are presented in the Design Report (Coffey 2017).

- Case A (required FoS>1.5): Long-term static case, using drained parameters and anticipated long-term groundwater profiles;
- Case B (required FoS>1.2): Short Term Static Case – using the elevated groundwater profile which is approximately 2 m higher than the long-term groundwater profile;
- Case C1 (required FoS>1.2): Seismic Case – Ultimate Limit State (ULS) Peak Ground Acceleration (PGA) with total stress parameters; and
- Case C2 (required FoS>1.0 or displacement <50 mm): Maximum Credible Earthquake (MCE) Seismic Case – MCE Peak Ground Acceleration (PGA) with total stress parameters. An MCE event has been considered as 1.5 times the ULS PGA. Where a factor of safety is below 1.0 under the MCE Seismic Case a displacement analysis has been calculated in accordance with Jibson, 2007

Deformation analysis was undertaken for the proposed MSE wall using finite element software Plaxis 2D under the long-term groundwater case. The forecasted lateral movement under SLS conditions is less than 18 mm, and the majority of this was expected to happen during the construction phase.

4   Construction Performance

The monitoring plan was prepared, and the observation prisms were installed during and post construction to confirm the performance of the trapezoidal MSE wall as per the design predictions.

The corresponding trigger levels were derived from the design analysis and referred to bridge structural acceptance or capacity from the structural team. The trigger levels that were used are: Alert Level of 5 mm, alarm level of 10 mm and action level of greater than 15 mm. The relevant contingency plan was summarized in the IFC drawings and The Design Report (Coffey 2017).

5 Conclusions

A composite trapezoidal MSE-embankment system was adopted to support an embankment widening project for a highway overbridge abutment. The concept behind this system exploits the self-supporting characteristics of the existing embankment and consequent elimination of lateral earth pressures imposed on the buttressing MSE wall. The mechanics behind this system deviates from conventional MSEW design theory, such as those followed by typical design standards, AS4678 and R57 (2012). Adoption of design guidelines provided in FHWA (2006) "SMSE Wall Systems Design Guidelines" was utilised in the development of the Final Design, which rendered a trapezoidal-MSE wall with narrow a 4 m base width for an 8 m high wall sitting on a stepped 1V:1H interface with the underlying embankment. Finite element analyses were undertaken to assess stability and formulate deformation predictions. Instrumentation points on the existing bridge and MSE wall registered deformations within the expected magnitudes at the time of writing, thereby confirming the observed stability and successful performance both during and post-construction.

Based on the detailed comparison with the conventional design and construction, it is demonstrated that the Coffey trapezoidal MSE wall design and construction was a most cost-effective and constructible solution for this project.

References

Elias, V., Christopher, B.R., Berg, R.R.: Mechanically stabilized earth walls and reinforced soil slopes, design & construction guidelines. Report No. FHWA-NHI-00-043. Federal Highway Administration, p. 394, March 2001

National Concrete Masonry Association (NCMA): Design Manual for Segmental Retaining Walls. In: Collin, J.G. (ed.), Second Edition, p. 289 (2002)

British Standards Institution (BSI), Amended: Code of Practice for Strengthened/Reinforced Soils and Other Fills. BS8006 (2010)

Geotechnical Engineering Office (GEO): Guide to Reinforced Fill Structure and Slope Design, Geoguide 6. Civil Engineering Department, Government of the Hong Kong Special Administrative Region, p. 240 (2002)

Federal Highway Administration: Shored Mechanically Stabilised Earth (SMSE) Wall Systems Design Guidelines, Publication No. FHWA-CFL/TD-06-001, Federal Highway Administration, Lakewood, CO 80228 (2006)

Australian Standard: Earth-retaining structures, AS4678 (2002)

Coffey: Design Report 197759, Auckland, NZ (2017)

Field Test Study of Long-Term Displacement of Bridge Foundation Subjected to Lateral Loads

Guangming Yu[1,2], Weiming Gong[1,2(✉)], and Guoliang Dai[1,2]

[1] School of Civil Engineering, Southeast University,
Nanjing 210096, Jiangsu, China
101004924@seu.edu.cn
[2] Key Laboratory of C&PC Structures, Ministry of Education,
Nanjing 210096, Jiangsu, China

Abstract. To investigate the effect of creep characteristics of soft soils on time-dependent deformation of root-caisson foundation, long-term lateral loading tests were performed about 120 days in Wangdong Yangzi River Highway Bridge. The displacement of caisson head, the deformation of the caisson along depth and displacement field of soil ahead the foundation were measured under the constant load. Comparing the short-term deformation and long-term displacement of root-caisson, soil creep effect and time-dependent displacement of foundation are revealed. The results show that deformation of this root-caisson presents characteristic of the flexible foundation. From the entire process the lateral displacement of caisson vs time curve, the horizontal displacement at the top of caisson increases gradually with the continuous time of the constant load, the displacement rising rate declines gradually, and the horizontal displacement finally tends to a stable value. The deformation of the caisson in the holding load stage is larger than the one in the loading stage, the final horizontal displacement of the caisson top is 2–3 times the instantaneous displacement. The rebound displacements of the caisson are small with experiencing a long-term horizontal load after unloading, the creep deformation of the soil occurs under long-term horizontal load.

Keywords: Root caisson · Lateral load · Long-term lateral displacement
Field test

1 Background and Field Tests

There exists a composite interaction process between subgrade soil and bridge foundation laterally loaded. The lateral displacement of foundation increases with time due to soil consolidation and creep compression deformation [1, 2]. This is a critical issue that cannot be ignored in the current soft soil geological conditions, which should be considered in the designation of bridge foundation. Through the on-site horizontal test and displacement monitoring, the difference and mechanism of long-term and short-term deformation of the foundation are revealed, and the mechanism of long-term deformation and load variation are studied to provide references for contemporary caisson foundation's designation considering long-term horizontal displacement.

© Springer Nature Switzerland AG 2018
W. Wu and H.-S. Yu (Eds.): *Proceedings of China-Europe Conference
on Geotechnical Engineering*, SSGG, pp. 1672–1676, 2018.
https://doi.org/10.1007/978-3-319-97115-5_168

The long-term horizontal load test of the single root caisson foundation was carried out at Wangdong Yangtze River Highway Bridge. The foundation of Approach Bridge adopts the root caisson by imitating the form of tree roots, increasing the contact area between the foundation and the soil. Then the bearing capacity of the foundation can be greatly improved under the limited foundation size and site. The design depth of the test root caisson is 47 m. The main body of foundation is made of reinforced concrete circular caisson with an outer diameter of 5 m. The caisson has a wall thickness of 0.9 m. the bottom of the caisson has a thickness of 3.5 m, and the upper capping platform has a thickness of 1.5 m. Figure 1 shows that structure of root caisson and geological conditions.

    (a) caisson   (b) root key   (c) geological conditions (d) field tests

Fig. 1. Construction of root caisson and Long-term lateraled loaded tests

Figure 2 shows construction process of root caisson. In this horizontal bearing capacity test, hydraulic jack was adopted as horizontal thrust device, and the horizontal reaction can be provided by the adjacent pile.

The horizontal jack is chosen as loading equipment with spherical hinge seat on both sides to ensure horizontal load overcrossing caisson's center. Long-term horizontal bearing test plan layout of root caisson and field test process is shown in Fig. 3.

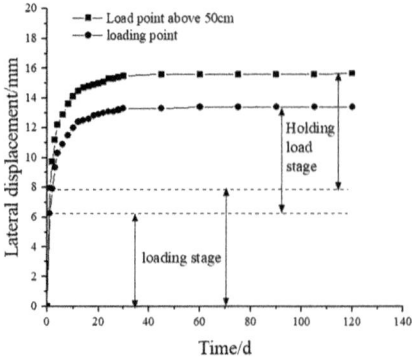
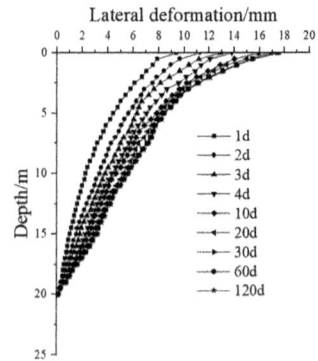

Fig. 2. Time vs displacement curve of roots caisson (left)

Fig. 3. Roots horizontal displacement along the depth profile with time (right)

2 Results and Conclusions

The short-term static load test of rooted caisson was carried out 1 month before long-term horizontal load test. The ultimate level load is 9600kN since the horizontal displacement of root caisson is more than 30 mm (Fig. 4).

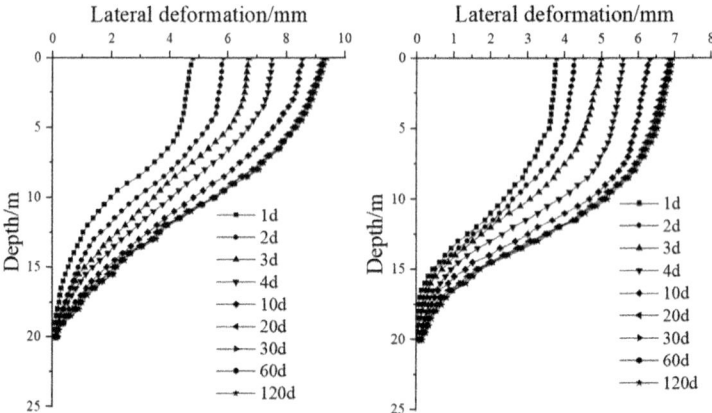

(a) 3 m on the front of the caisson (b) 6 m on the front of the caisson

Fig. 4. Soil horizontal displacement with depth distribution with various time

Considering the safety factor is 2 in the project, the horizontal load in this long-term horizontal load field test is 4800kN. Figure 5 shows the horizontal displacement-time curve of root caisson with a diameter of 5 m in Wangdong Yangtze River Highway Bridge. As shown in the Fig. 5, the horizontal displacement of the root caisson increases with time while the increase rate of the horizontal displacement decreases with the changing time. After about 30 days, the horizontal displacements of the root caisson at the point of 50 mm above the caisson loading position is stable. The horizontal load lasted for about 120d, the long-term horizontal displacement of the load acting point of caisson was 13.4 mm, which is 7.16 mm greater than the final horizontal displacement of 6.24 mm at the initial stage of loading, accounting for 53.43% of the total horizontal displacement. The final horizontal displacement at the point of 50 mm above the caisson is 15.65 mm, which is 7.69 mm larger than the final displacement of 7.96 mm during loading. The long-term horizontal creep deformation of root caisson is 7.69 mm, accounting for 49.14% of the total horizontal displacement. The above test results show that the horizontal displacement of long-term loaded caisson will gradually increase with time in the soft soil area, the long-term horizontal displacement of caisson cannot be ignored in the designation.

Figure 6 shows the horizontal displacement along the depth of the caisson measured at various time by the inclinometer buried in the caisson. As shown in the Fig. 5, the horizontal displacement along the depth of root caisson decreases with the increase of depth, and the horizontal displacement is almost 0 at the depth of 20 m. After about 30 days, the horizontal displacement of the root caisson does not change obviously along the depth. The horizontal displacement of the root caisson along the depth is non-linear distribution, which is in accordance with the deformation law of "flexible long pile".

Figure 7 shows the horizontal displacement distribution of soil along the depth measured by inclinometer pipe in 3 m and 6 m on the front of the caisson along the horizontal load application direction. As shown in the Fig. 7, the horizontal displacement of soil decreases with depth. The horizontal displacement distribution along the depth of soil at various time is slightly different from it along the depth of root caisson. The horizontal displacement of shallow soil is smaller and smaller with the depth decreasing, and even the horizontal displacement first decreases and then increases. The variation of horizontal displacement along depth in 3 m and 6 m on the front of the root caisson is similar to the one of caisson along the depth at different time. The position where soil displacement is zero also appears at a depth of 20 m. At about 30 days later, the horizontal displacement of the soil along the depth direction varies unobvious.

It can also be seen from the Fig. 7 that the initial horizontal displacement of the soil decreases with the distance of the soil from the front edge of the caisson, and the total increase of the displacement will be even greater.

As a result, according to the long-term horizontal load tests of the root caisson in Wangdong Yangtze River Highway Bridge, the following main conclusions are obtained:

(1) The deformation of the caisson of Wangdong Bridge presents characteristic of the flexible foundation.

(2) From the entire process curve of the horizontal displacement of caisson with time, the horizontal displacement at the top of caisson increases gradually with the increasing time of holding load, the increasing rate of displacement decreases gradually, and the horizontal displacement finally tends to a stable value.

(3) The deformation of the caisson in the holding load stage is larger than the deformation in the loading stage, so that the final horizontal displacement of the top of the caisson is 2–3 times the instantaneous horizontal displacement. After unloading, the rebound horizontal displacements of the caisson are small after experiencing a long-term horizontal load, which indicates that the creep deformation of the soil occurs under long-term horizontal load.

Acknowledgements. We thank our collaborators for permission to include data from their simulations and experiment. This work was supported by National Key Research Program of China (2017YFC0703408) and National Natural Science Foundation of China (51678145).

References

1. Nasr, A.M.A., Krishna Rao, S.V.: Behaviour of laterally loaded pile groups embedded in oil-contaminated sand. Géotechnique **66**(1), 58–70 (2015)
2. Sa'don, N.M., Pender, M.J., Karim, A.R.A., et al.: Pile head cyclic lateral loading of single pile. Geotech. Geol. Eng. **32**(4), 1053–1064 (2014)

Geotechnical Solutions for Linear Transport Infrastructure in Mining Areas

Jacek Kawalec[1,2(✉)]

[1] Silesian University of Technology, Akademicka 5, 44-100 Gliwice, Poland
kawalec@tensar.pl
[2] Tensar International s.r.o., Lipova 1965, 737 01 Cesky Tesin, Czech Republic

Abstract. Design and construction of linear structures for transport like road or railway lines in areas with active or historical deep mining operations is always challenging due to various additional influential geotechnical data which needs to be considered. Surface subsidence and its negative impact on road geometry over time needs to be compensated by protection systems designed and implemented prior to time when real influence occurs. Linear or non-linear deformations are one of the most critical parameters to be overlooked at design stage. Also use of mining materials for construction of embankments as alternative to natural soils brings additional challenges and need for good identification of all geotechnical parameters of such material. As Silesia is region with one of the most concentrated deep coal mining activity in Europe problem of constructing safe and environmental friendly infrastructure is always critical. This paper discusses selected aspects of geotechnical practice implemented in the region for protection of transport infrastructure in such challenging geotechnical environment.

Keywords: Mining influence on linear structures
Embankments with man-made soils · Mining waste geotechnical parameters
Geomattress for geosynthetic protection of roads

1 Utilization of Industrial Wastes for Construction

Silesia District of Poland is the most heavy-industrialized part of Poland with the densest network of mines, steel mills and power plants. In consequence it's also the biggest producer of industrial wastes like mining wastes, slags or fly-ashes. Over 1,6 billion tons of coal mining wastes is deposited in heaps with additional yearly increase by tens of thousand tons. Such a big volume of potentially useful for construction materials results in their attractiveness. Utilization of them in geotechnical structures like e.g. embankments are desirable to reduce environmental damage, it's also cost effective and additionally it saves huge volumes of natural soils and aggregates which would need to be used as alternative. Acceptance of industrial wastes for construction is however still poor, mainly due to legal barriers and lack of single standard for them.

© Springer Nature Switzerland AG 2018
W. Wu and H.-S. Yu (Eds.): *Proceedings of China-Europe Conference on Geotechnical Engineering*, SSGG, pp. 1677–1681, 2018.
https://doi.org/10.1007/978-3-319-97115-5_169

1.1   Mining Clay Shales

As mentioned earlier mining wastes are the most popular ounces in Silesia. Majority of deposits wastes are combination of mudstones, claystones, sandstones and clay shales. The last of these materials contains sometimes quite large contents of coal (even above 20%) which makes them not fireproof. Such characteristics makes them very sensitive to semi-fire under surcharge pressure or and wind, loose material has tendencies for auto-ignition and consequently fire within heap. If this happens and clay shales are burned then they significantly change their characteristics and become more stable, tough non-cohesive material. However, if properly build-in and well compacted mining wastes are very good material for construction with many references across Silesia [4] (Fig. 1 and Table 1).

Fig. 1. Example of full-scale test on stability of embankment made from mining wastes [1].

Table 1. Typical physical parameters for mining wastes from Silesia.

Parameter	Typical value range
Organic content (coal) f_C [%]	7,2–23,5
Optimal moisture [%]	8,5–12,4
ρ_{ds} [g/cm³]	1,82–1,97
Frost resistance [%]	55–94

1.2   Slags and Fly Ashes

Slags are second largest by volume after coal mine wastes anthropogenic materials used for constructions of embankments with lot of references [7]. Fly-ashes do occupy the podium then. But instead of these three main materials there are several combinations of materials as mixtures of these three or by additions of cement, lime or chemicals [6].

2 Surface Deformations and Protection Systems for Roads

2.1 Damage to Roads

Mining influences on roads are visible on pavement surface in form of cracks, horizontal and vertical deformations of pavement, local pavement uplifts and in worst cases voids or faults. Mining deformations have always a negative impact on the pavement parameters making almost impossible to maintain longitudinal and transversal evenness. They also heavily affect efficiency of drainage. If deformations are continuous traffic could continue to carry on, with speed limitations, especially on higher-speed roads [3]. Much more complicated situation occurs in the case of discontinuous deformations. In this case if the ground surface continuity is broken and sudden change in the ground profile is created risk becomes too high to continue trafficking and road must be closed for maintenance. Unfortunately, often this is only a short term solution until new deformations from next mining operations pop-up on ground surface (Figs. 2 and 3).

Fig. 2. Example of mining influences on the pavement.

Fig. 3. A surface-type linear discontinuous deformation [2].

2.2 Prevention Measures

Protection of roads and other linear transport structures must be designed in advance to construction of road. At the very first moment of designing road planned and executed mining operations must be studied. It's important to estimate all potential deformations on the surface which could occur on pavement during road live cycle. Modern approach is based on use of FEM models of road subgrade to determine design parameters. But results are as precise as input data so typical lack of precise information about planned mining activity makes this process more complex (Fig. 4).

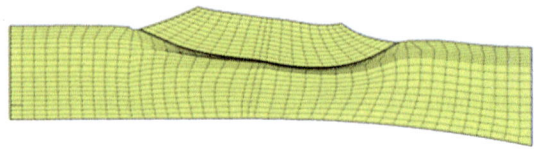

Fig. 4. Example of deformed FEM mesh of road embankment after mining excavation occurs used for designing protection layers to reduce risk of deformation for surface.

Over last two decades in Silesia majority of main roads were designed with protective layers in form of geomattresses made from stabilizing geogrids and crushed aggregate [5]. Such mattresses with typically min. 2 layers of geogrid and min 50 cm thickness in total should be placed at the base of embankment and under pavement structure. Double mattresses prove to be most effective solution (Fig. 5).

Fig. 5. Example of geomattress made with two layers of geogrids and anthropogenic aggregate.

3 Conclusions

There is a lot good case studies from Silesia on both applications of industrial wastes for and for protective geomatresses, including monitoring results over time. This could be extended in a longer article

References

1. Gryczmanski, M., Kawalec, J., Kawalec, B.: Destructive slope stability tests for assessment of mining waste strength parameters. Slovak J. Civ. Eng. **4**, 32–35 (1996)
2. Grygierek, M.: Road pavement damage caused by discontinuous mining deformations. Research Reports of Central Mining Institute. Mining & Environment, no. 4/1, pp. 72–82, Krakow, Poland (2010)
3. Grygierek, M., Kawalec, J.: A4 motorway operation in the area of linear discontinuous surface deformations. In: Lehane, B.M., Acosta-Martinez, H.E., Kelly, R. (eds) 5 International Conference on Geotechnical & Geophysical Site Characterization (ISSMFE TC-102 ISC'5, pp. 1337–1342. Australian Geomechanics Society, Sydney, Australia (2016)

4. Kawalec, J.: Fine fraction mining wastes as material for civil engineering applications. In: The 21st International Conference on Solid Waste Technology and Management Journal of Solid Waste Technology and Management, Philadelphia, USA, pp. 323–332 (2006)
5. Kawalec, J., Koda, E.: Polish experiences in using geogrids in earth structures engineered with anthropogenic materials. In: Proceedings 3rd Pan-American Conference on Geosynthetics GeoAmericas, vol. 1, pp. 287–297, Miami Beach, USA (2016)
6. Zabielska-Adamska, K.: Laboratory compaction of fly ash with cement additions. J. Hazard. Mater. 151(2–3), 481–489 (2008)
7. Zawisza, E.: Odpady hutnicze jako antropogeniczne grunty budowlane. Metody badan i wlasciwosci geotechniczne, Krakow (2012)

Kinematics of Piled Embankments with Defective Piles

Louis King[1,2(✉)] ⓘ, Abdelmalek Bouazza[1] ⓘ,
and Stephen Dubsky[1] ⓘ

[1] Monash University, Clayton, VIC, Australia
louis.king@monash.edu
[2] Golder Associates, Richmond, VIC, Australia

Abstract. Defective piles within piled embankments have the ability to cause significant project delays and require costly remedial works. At present, the ability of the load transfer platform and embankment fill to redistribute loads away from defective piles and reduce differential settlements that develop between defective and non-defective piles is not well understood. Synchrotron X-ray computed tomography was performed on a small scale model piled embankment comprising a defective pile with digital volume correlation analysis implemented on pairs of reconstructed volumes to study the kinematics of soil arching above a defective pile. Results indicate that a plane of equal settlement can still develop above a defective pile.

Keywords: Piled embankments · Computed tomography · Defective pile

1 Introduction

Individual piles that exhibit a relatively softer load-settlement response to other piles within a larger group may be referred to as defective piles. When such piles are structurally connected to other non-defective piles by a pile cap or raft, there is a redistribution of loads away from the defective pile [1]. As such, the consequence of the softer response exhibited by the defective pile within a group is not as severe as if the single defective pile was to act in isolation. However, in earth structures, such as piled embankments, the ability of geomaterials to redistribute loads away from defective piles by the mechanism of soil arching is not well understood.

In piled embankments, the amount of piles that are installed on a typical large footprint infrastructure project frequently leads to the installation and detection of defective piles. Piled embankments often comprise a load transfer platform (LTP) made up of high quality granular material (sometimes with geosynthetic reinforcement) overlying the pile heads. To predict the effects of a defective pile on the performance of a piled embankment, it is necessary to understand the ability of the LTP and embankment fill to redistribute loads away from the defective pile and reduce differential settlements that develop at the pile head level. To understand the kinematics of soil arching above a defective pile within a piled embankment, model tests were performed using synchrotron X-ray computed tomography (CT). Digital volume

© Springer Nature Switzerland AG 2018
W. Wu and H.-S. Yu (Eds.): *Proceedings of China-Europe Conference on Geotechnical Engineering*, SSGG, pp. 1682–1686, 2018.
https://doi.org/10.1007/978-3-319-97115-5_170

correlation (DVC) was applied to pairs of CT volumes to obtain full-field three dimensional displacement vectors.

2 Experimental Setup

A small scale model piled embankment was built to simulate subsoil settlement using a mechanical tray (settlement plate) and the settlement of a single defective pile. The piles in the model were arranged on an equilateral triangular grid, with a center-to-center spacing of 40 mm and a pile head diameter of 12.6 mm. The model comprised six rigid non-defective piles surrounding a single defective pile. The piles and settlement plate were surrounded by a polycarbonate cylinder (142 mm in diameter), which provided confinement to the sand. The piles penetrated through the settlement plate, which was mechanically lowered at a rate of 100 µm/s. Simultaneous to the lowering of the settlement plate, the central defective pile was lowered at approximately 1/3rd the rate of the settlement plate, simulating a softer load-settlement response of a defective pile to the surrounding non-defective piles. The model was designed and built to fit on an imaging rotation stage, such that the model did not have to be removed for the settlement plate and defective column to be lowered.

Silica sand was used in the tests to model the LTP granular material and embankment fill, with properties listed in Table 1 [2]. The sand was prepared at two different relative densities, $D_r = 96\%$ (dense) and $D_r = 61\%$ (medium dense). The sand was air-pluviated into the models, and vibrated in the case of the dense sample, achieving a final sand height of 100 mm. A uniform surcharge of 6.3 kPa was applied to the surface of the sand prior to commencing the test.

Table 1. Sand properties

Property	Values
Specific gravity, G_s	2.67
Average particle size, d_{50}	0.18 mm
Maximum dry density, ρ_{max}	1497 kg/m^3
Minimum dry density, ρ_{min}	1774 kg/m^3
Angle of internal friction, critical state, ϕ'_{cv}	31.6°

Imaging of the sand overlying the pile heads was performed using the Imaging and Medical Beamline (IMBL) at the Australian Synchrotron. A narrow band (monochromatic) of the broad energy spectrum emitted from the source was used for imaging. It was found that an x-ray beam consisting of photons with an energy of between 60 and 70 keV produced radiographs of the sand with the highest contrast, and hence image texture. The radiographs were reconstructed using the software X-TRACT [3], in which a filtered back projection CT algorithm was adopted.

CT volumes were collected as the settlement plate and defective pile were lowered in increments. Given that the model was not removed from the imaging stage between displacement increments, Digital Volume Correlation (DVC) could be applied to pairs of CT volumes. The DVC software [4] was applied to pairs of reconstructed CT volumes. The sand used in the present study was particularly well suited to DVC as it contained trace amounts of a relatively denser mineral, and as such, resulted in images with fine texture. This resulted in the DVC analysis calculating displacements vectors with high precision.

3   Results

The results in the following section are presented at specific increments of settlement plate displacement, δ_{sp}, which has been normalized by an equivalent axisymmetric clear spacing [5], b', equal to 29.4 mm for the pile layout in the model tests. The vertical displacements estimated from DVC analysis are normalized by δ_{sp} and presented in Fig. 1 along with the maximum natural shear strain, γ_{max}. The slices are taken through the middle of the model, with the central defective pile positioned beneath the center of the slice, and non-defective piles positioned beneath the edges of the slice. Due to the symmetry of the model around the central defective pile, slices are divided in half, with the left half of each slice displaying the results from the model comprising dense sand and the right half displaying results from the model comprising medium dense sand.

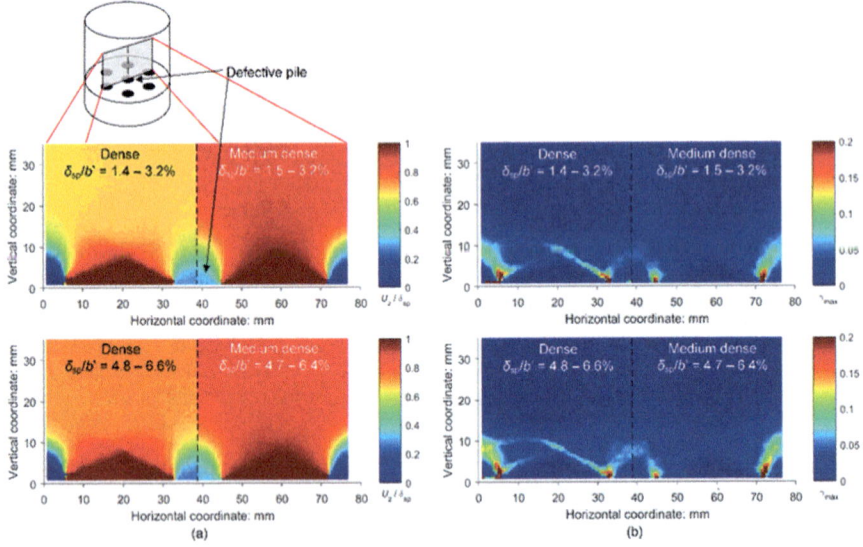

Fig. 1. Slices of (a) incremental normalized vertical displacements; and (b) maximum natural shear strains, with dense sand on the left and medium dense sand on the right of each slice

The incremental vertical displacements in Fig. 1a show that while the defective pile underwent additional settlement relative to the rigid non-defective piles, soil arching still developed. This is evidenced by the reduction in differential settlement with increasing height above the pile heads. Strain localizations developed above both defective and non-defective piles in Fig. 1b for both material densities. These localizations developed as a result of the differential settlement between the settlement plate and the pile head. The amount of shear strain that occurred within the localizations above the defective pile head were relatively less than within the localizations above the non-defective piles. This is due to differential settlement between the plate and the defective pile head being less than for the non-defective piles. The relatively smaller differential settlement resulted in less shear strain being mobilized in the localizations above the defective pile. As such, it is likely that the defective pile would have attracted less load relative to non-defective piles, although load was not measured.

An understanding of the reduction of differential settlements within the LTP and embankment fill is of importance when designing piled embankments to ensure that differential settlements are not experienced at the embankment surface. It has been shown that in a sufficiently high piled embankment without defective piles, a plane of equal settlement develops [6]. The height above the pile heads at which this plane of equal settlement first develops is referred to as the critical height, H_{cr}. It is shown in Fig. 1a that at some height in the sand overlying the pile heads, the displacements above pile heads and amid pile heads approached uniformity, and thus, a plane of equal settlement developed. This plane of equal settlement occurred in both sample densities, and at both settlement plate displacements.

4 Conclusions

By implementing the advanced non-destructive imaging technique of synchrotron X-ray CT, high quality images of small scale piled embankments comprising a defective pile were obtained. DVC analysis on pairs of reconstructed CT volumes allowed the three-dimensional displacements vectors to be estimated, which in turn allowed the kinematics of soil arching above a defective pile to be investigated.

Results indicate that soil arching may still develop above a pile that exhibits a relatively softer load-settlement response compared to the surrounding non-defective piles. The sand used within the models was also able to reduce the differential settlements experienced at the pile head level to a plane of equal settlement at some height above the pile heads.

References

1. Poulos, H.: Behaviour of pile groups with defective piles. In: Proceedings of the 14th International Conference on Soil Mechanics and Foundation Engineering, Hamburg, Germany (1997)
2. Chow, S.H., O'Loughlin, C.D., Gaudin, C., Lieng, J.T.: Drained monotonic and cyclic capacity of a dynamically installed plate anchor in sand. Ocean Eng. **148**, 588–601 (2018)
3. Gureyev, T.E., et al.: Toolbox for advanced X-ray image processing. In: Proceedings of Advances in Computational Methods for X-Ray Optics II. SPIE, San Diego, USA (2011)

4. Dubsky, S., Hooper, S.B., Siu, K.K., Fouras, A.: Synchrotron-based dynamic computed tomography of tissue motion for regional lung function measurement. J. R. Soc. Interface **9**(74), 2213–2224 (2012)

5. King, D.J., Bouazza, A., Gniel, J.R., Rowe, R.K., Bui, H.H.: Serviceability design for geosynthetic reinforced column supported embankments. Geotext. Geomembr. **45**(4), 261–279 (2017)

6. McGuire, M.P.: Critical height and surface deformation of column-supported embankments, Ph.D. thesis., Virginia Polytechnic Institute and State University (2011)

Seismic Behavior of Pile Supported Railway Track

Jinsun Lee[1(✉)], Mintaek Yoo[2], and Yunwook Choo[3]

[1] Wonkwang University, Iksan Jeollabuk-do 54538, Republic of Korea
blueguy@wku.ac.kr
[2] Korea Railroad Research Institute, Uiwang Gyeonggi-do 16105,
Republic of Korea
[3] Kongju National University, Cheonan Chungcheongnam-do 31080,
Republic of Korea

Abstract. Verification of nonlinear dynamic numerical analysis was performed using dynamic centrifuge test results for pile supported track under seismic loadings. The numerical analysis models were constructed in the same dimensions with the centrifuge tests in prototype scale. Verification was conducted by comparing the measurement data obtained from the centrifuge test and numerical analysis output directly. Total 4 different numerical models corresponding centrifuge test are verified with changing thickness of soft soil layer, embankment, input motion intensity and history. Numerical analyses were conducted in time domain by an explicit time integration method. Dynamic hysteretic damping model was adopted for nonlinearity behavior of soil under seismic excitation. The piles were modeled with shell element that can show interface behavior with surrounding soils. The other structural members are modeled in a hexahedron solid element that has an interface. Seismic responses of the pile supported track were quite similar for both numerical analysis and centrifuge test.

Keywords: Dynamic centrifuge test · Numerical analysis · Seismic analysis

1 Introduction

Pile-supported slab track has been widely constructed as a practical solution for track construction on soft soil in Europe and China. The Pile-supported slab track system consists of pile foundation, cross beams and platforms as shown in Fig. 1.

This system has an advantage for soil layers which expected to be settled after construction in forms of either instantaneous or long-term settlement. Thus, this system has been applied to a high-potential collapsible or soft ground embankment system [1]. In Europe, pile-supported slab track system has been applied to soft ground section of submarine tunnel [2]. Pile foundation system is highly effective for restraining settlement of superstructure, however, it is vulnerable to a lateral load, such as earthquake. Therefore, it is necessary to check lateral stability of the system under seismic loading condition.

Today, numerical simulation is the most popular and powerful solutions to estimate seismic response of infra structures. However, reliability of the analysis is still in

© Springer Nature Switzerland AG 2018
W. Wu and H.-S. Yu (Eds.): *Proceedings of China-Europe Conference
on Geotechnical Engineering*, SSGG, pp. 1687–1691, 2018.
https://doi.org/10.1007/978-3-319-97115-5_171

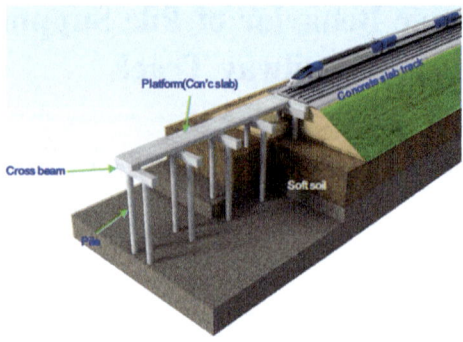

Fig. 1. Concept drawing of pile supported railway track

question with a lack of the verification process. Thus, any numerical analysis model for seismic analysis should be verified and refined using measurement data obtained from historical earthquakes or well-performed physical test. This paper presents results of verification for a dynamic numerical analysis of the pile-supported track system. A dynamic numerical analysis was conducted using finite-difference program FLAC 3D, which solves dynamic equation of motions using an explicit time integration method. The numerical analysis results were verified using dynamic centrifuge test results.

2   Verification

2.1   Dynamic Centrifuge Test

The centrifuge tests were conducted at the Geo-Centrifuge Testing Center at Korea Advanced Institute of Science and Technology (KAIST). The maximum capacity of the KAIST centrifuge, which has a 5-m radius, is 2400 kg for up to 100 gc of centrifugal acceleration. The earthquake simulator is mounted on the centrifuge.

The models were constructed in an equivalent shear beam (ESB) container with an internal length, width, and height of 490 mm × 490 mm × 630 mm, respectively. The centrifugal acceleration used in the experiments was 58 gc. All of the results are presented with respect to prototype units, unless otherwise stated according to the scaling laws. Figure 2 shows standard drawing of the pile supported track system and corresponding centrifuge test setup. Soft soil layer used in the experiment was formed by kaolin slurry and consolidated to the target pre-consolidation pressure by increasing centrifugal acceleration. The other layers, base and embankment were made by weathered soil which is the most popular soil in Korean peninsula. All the structural members used in centrifuge test were made in aluminum alloy. Model dimensions of the structural members are determined to satisfy bending stiffness (EI) of the prototype members.

Plasticity index of the clay is about 20 and shear wave velocity ranges from 75 m/s to 140 m/s. Bender element array was installed to obtain shear wave velocity profile.

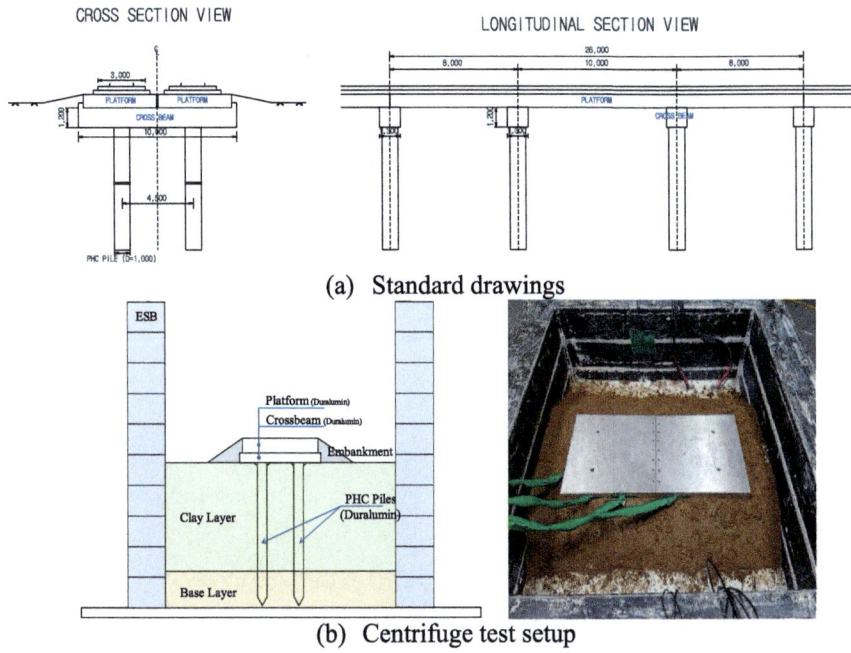

(a) Standard drawings

(b) Centrifuge test setup

Fig. 2. Standard drawing and centrifuge test setup of the pile supported track system

Measurement system was composed by accelerometer, LVDT, strain gauges etc. Total three types of input motion were applied in a staged manner. Ofunato type earthquake for short period motion, Hachinohe type earthquake for long period motion and 1 Hz of Sinusoidal motion were used.

2.2 Numerical Analysis

The maximum size of the hexahedron finite-difference grid (0.5 m) enables to pass waves of less than 20.0 Hz. To eliminate wave reflection at the side boundaries, ESB was modeled as a free-field boundary. The initial shear modulus of the sandy soil layer was updated after the first static equilibrium condition by considering the mean effective stress.

The soil layers were modeled using Mohr-Coulomb model and the cyclic nonlinearity before plastic failure was considered by using a hyperbolic model. The hysteretic model was fitted and adopted to the numerical model by using shear modulus degradation curve for each soil layers. Compliant (elastic half-space) base condition was adopted to minimize the input motion error. The hollow PHC piles were modeled with a liner element in diagonal form of a discrete Kirchhoff triangle-constant strain triangle (DKT-CST) shell and a Coulomb-slider interface. The other structural members, cross beam and platform are modeled in solid element with interface element to consider kinematic interaction with surrounding soils.

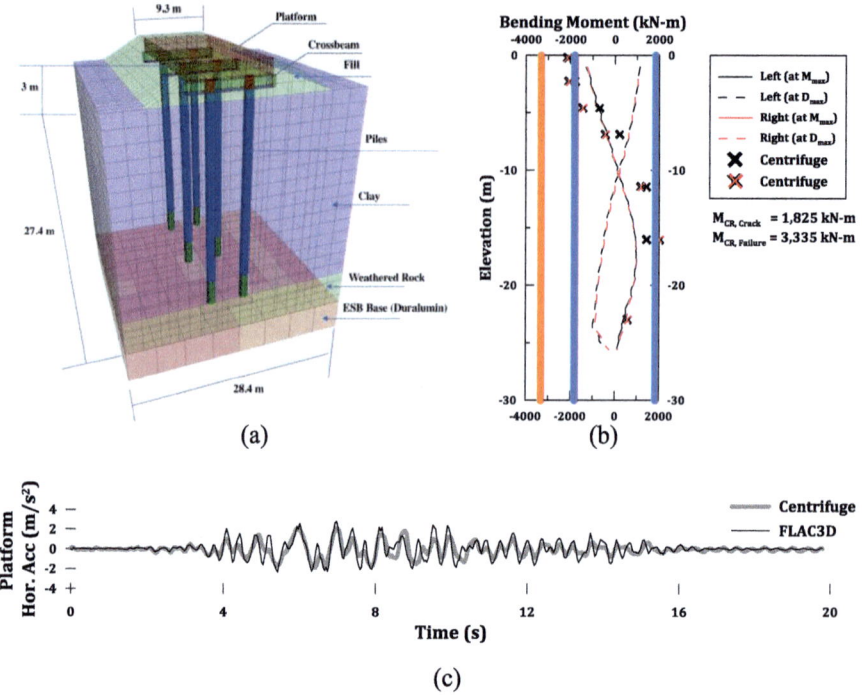

Fig. 3. Numerical model and verification results (a) Numerical model (b) Bending moment profile of piles at the maximum bending moment occurrence time (c) Time histories comparison at the side of concrete platform

3   Conclusions

Figure 3 shows representative numerical model and verification results. Both physical test and numerical analysis are in good agreement in terms for structural member forces as well as horizontal acceleration at the side of concrete platform. The rigorously verified numerical model will be used for a parametric study which can figure out the critical design parameter of the pile-supported track system under seismic loading.

Acknowledgement. Support for this research was provided by Korea Railroad Research Institute and the project entitled "Development of performance-based seismic design technologies for advancement in design codes for port structures," funded by the Ministry of Oceans and Fisheries of Korea.

References

1. Xiao, J.H., Wang, B.L., Wang, Z.D., Yang, L.C., Gong, Q.M.: Differential Settlement of Subgrade and its Control for High Speed Railway. Tongji University Press (2015)
2. Wángfēng: The theory and practice of concrete deck on pile system in a high speed ballastless track. China Railway Publishing House, Beijing (2012)

Initial Sinking Method for Large Open Caisson in a Highway Bridge Project

Peng Li[(⊠)][iD], Erxiang Song, and Tianliang Zheng

Department of Civil Engineering, Tsinghua University, Beijing 100084, China
lipengcivil@mail.tsinghua.edu.cn

Abstract. Open caisson foundation has been widely used as anchor foundation for cable-stayed bridge in recent years. Large scale open caisson displays different mechanical characters compared with small and medium-sized open caisson in both construction process control and spatial mechanical performance. Generally, the drainage sinking of a large open caisson is more dangerous than the non-drainage sinking. At the beginning of caisson subsidence, the blade foots tend to deviate outwards because there is no buoyancy force or lateral constraint at the blade edge, which may lead to larger tensional force at the bottom of the caisson structure. Therefore, proper design of the first sinking scheme is particularly important for large-scale open caisson. A three dimensional soil-structure interaction finite element modeling, which is based on a large-scale open caisson for anchor foundation of a highway bridge project, is carried out to analyze the influence of the caisson structure diameter on the internal force in condition of first sinking. Numerical results indicate for large-scale open caisson the initial sinking method has a great influence on the tensional force generated in the caisson structure. It is also revealed the model of structure on Winkler foundation may lead to significant error in calculating the deformation pattern and internal force of the caisson structure. The numerical results and conclusion will provide reference for properly design of initial sinking method of similar large-scale open caisson projects.

Keywords: Open caisson foundation · Initial sinking · Numerical modeling
Non-drainage sinking · Soil-structure interaction

1 Introduction

Open caisson-sinking techniques is to let the structure gradually be sunk either under its own weight or by excavation of the soil inside it. The initial sinking method of a large scale open caisson is an essential consideration in their design because almost all the cases of crack occur at the very beginning of sinking. Different excavation method may result in variable stress condition in the structure, which gives rise to the importance of proper choice of sinking method through numerical analysis.

The engineering background of this paper is a famous open caisson foundation project which has been used as anchor foundation for cable-stayed bridge crossing Yangtze River. In real construction process of this caisson project, the initial sinking method has been changed from global groove to precise cross groove excavation. This paper presents the

© Springer Nature Switzerland AG 2018
W. Wu and H.-S. Yu (Eds.): Proceedings of China-Europe Conference
on Geotechnical Engineering, SSGG, pp. 1692–1696, 2018.
https://doi.org/10.1007/978-3-319-97115-5_172

results of three-dimensional finite element analysis of the initial sinking method, in which the traditional global groove excavation method and the newly proposed cross groove excavation method are compared. The stress responses in the caisson structure under different conditions are investigated in order to facilitate the design of initial sinking method for this important project.

2   Project Layout and Problem Description

The tested open caisson foundation project is of 100.7 m long and a width of 72.1 m, with 48 compartments in total, as shown in Fig. 1. The design height of the caisson is 56 m with 10 sections totally. The structural form of the first section is plain concrete filled steel shell, with a height of 8 m. The second to 10th sections adopted reinforced concrete structure form, with the second section 6 m, 3rd to 8th sections 5 m each, 9th section 4 m and 10th section 8 m in height.

(a) Real open caisson project

(b) Top view

Fig. 1. Layout of the test open caisson project

At the very beginning, the first three sections are cast in place, after harden of which the structure will be sunk by excavation. The original excavation method is to form a global groove by gradually pump out the soil from the central compartments to surrounding compartments. Later the designers decide to change this method to cross groove excavation inside each compartment with a shape of an inverted parabola considering that the former one may lead to tension crack produced at the bottom of the caisson structure. This paper discusses and compares these two methods by numerical modeling, and the approximated excavation method is shown in Fig. 2, in which the structure supporting area is colored.

3   Numerical Model and Discussion

3.1   Numerical Model and Material Parameters

The test open caisson is simulated by 3D FEA in the platform of PLAXIS 3D. Due to symmetry, only a quarter of the caisson structure is analyzed. The soil and structures

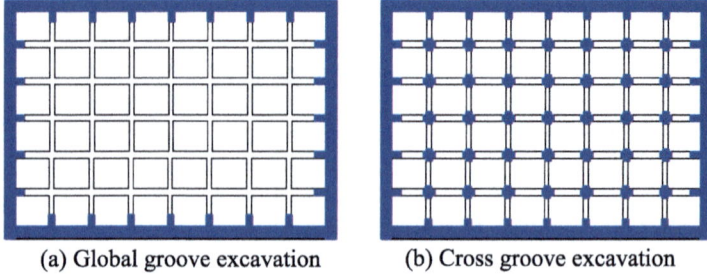

(a) Global groove excavation   (b) Cross groove excavation

Fig. 2. Schematic diagram of supporting boundary in the numerical analysis

are modelled by continuous solid elements. Compared with a structure on winkler foundation, the 3-D model can fully consider the soil-structure interaction effects, and give reasonable results of settlements and stress conditions of the caisson structure. The constitutive models of each material are as follows: The caisson structure is modelled by elastic material properties while the foundation soil layers are all using hardening soil model. The finite element model is shown in Fig. 3.

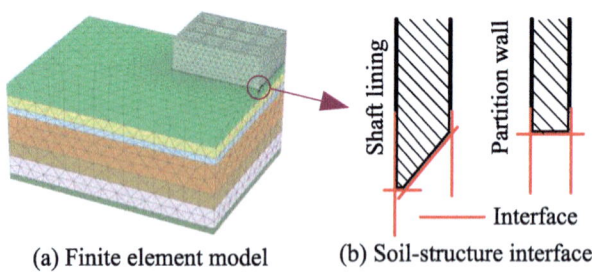

(a) Finite element model   (b) Soil-structure interface

Fig. 3. Numerical model for initial sinking analysis of caisson structure

3.2   Modeling of Construction Steps

In order to analyze the impact of the initial sinking method on the displacement and stress in the caisson structure, several simulation construction steps are followed in the numerical model: (1) Initial ground equilibrium; (2) Construction of the first three sections in which the bottom of the blade foots are at a depth of 3 m; (3) Excavation of a global groove or cross groove in each compartment. PLAXIS can output results of each step, benefiting the study of structure response at different construction stage.

4 Results and Discussion

The numerical results are obtained by using the soil parameters those are provided by the construction unit. Numerical results show that after construction of the first three sections, tension stress produced at the bottom of the caisson along the long side is 1.8 MPa compared with 1.6 MPa of short side. The largest settlement occurs at the middle of the caisson, but one should notice the middle of long side shaft lining also produce settlement. Therefore, boundary condition of the caisson structure is different from two-way slab with simply supported at four sides. This has demonstrated the importance of 3-D soil-structure interaction analysis, from which it is deduced the tension stress along the long side is larger.

For global groove excavation method large tension stress will be produced at the bottom of the caisson structure. The largest tension stresses along the long side and short side are 6.8 MPa and 6.2 MPa, respectively, which are all beyond the tensile limit of C30 concrete. Meanwhile, during sinking of caisson structure passive earth pressure will apply on the inclined cutting-edge, resulting in tensile stress concentration near the blade foots, the maximum of which is up to 7.8 MPa from the calculated results. This also exceeds the concrete tensile limit.

If cross groove excavation is adopted as sinking method, the tensile stress increments in the middle of caisson structure for long side and short side are 1.15 MPa and 1.3 MPa in maximum, respectively. The largest tensile stress increment near the blade foot is 0.8 MPa. Therefore, by using cross groove excavation method the internal force can be effectively controlled, which can guarantee the safety of the caisson structure.

5 Conclusion

In this paper, a 3-D finite element analysis of soil- structure interaction model has been carried out to investigate the influence of initial sinking method of an open caisson structure on its stress and settlement responses. Cross groove excavation method has been suggested and comparison is made with the originally designed global groove approach. This study provides an important reference for future similar projects. Following conclusions can be drawn from this study: (1) 3-D soil-structure interaction model is more reliable than structure on winkler foundation model in precisely analyzing the displacement pattern and stress distribution in caisson structure; (2) The caisson structure cannot be simply treated as a two-way slab with simply supported at four sides due to its complicated boundary conditions; (3) For large open caisson project the proper choice of initial sinking method is of great importance, as the traditional global groove excavation method may lead to large tensile stress and increase the risk of concrete crack; (4) Cross groove excavation method is comparable a good choice for initial sinking method in which the internal force can be best controlled.

Acknowledgement. This work is financially supported by the National Key Fundamental Research and Development Program of China (Project no. 2014CB047003), the National Natural Science Foundation of China (Project No. 51408331) and the Science and Technology Research and Development Program of China Railway Corporation (Project No. 2014G004-C).

Mechanism of Isolating Piles in Reducing Tunnel Settlement of Hong Kong-Zhuhai-Macao Bridge Project

Peng Li[1(✉)] [iD], Erxiang Song[1], Abbas Haider[1] [iD], and Xiaodong Liu[2]

[1] Department of Civil Engineering, Tsinghua University, Beijing 100084, China
lipengcivil@mail.tsinghua.edu.cn
[2] CCCC Hong Kong-Zhuhai-Macao Bridge Island and Tunnel Project
Department, Zhuhai 519080, Guangdong, China

Abstract. Artificial islands are adopted in Hong Kong-Zhuhai-Macao Bridge (HZMB) project to form the transition between the bridges and tunnels. A transition foundation solution along the longitudinal direction is introduced, including cast in place piles and PHC piles. Large area of backfill at both sides of the tunnel structure can result in unacceptable settlement as well as differential settlements along the transverse direction. Several measures have been suggested in the real project to deal with this problem, including increasing the length of piles underneath the tunnel structure, and adding two rows of isolating piles at both sides. These treatments are systematically introduced in this paper and a 3-D soil-structure interaction finite element analysis is also carried out which thoroughly models the construction steps. Mechanism of isolating piles in reducing tunnel settlement and uneven settlement is investigated. The study outcome provides reference for similar immersed tunnel projects.

Keywords: Hong Kong-Zhuhai-Macao bridge · Isolating pile
Settlement

1 Introduction

Hong Kong-Zhuhai-Macao Bridge project is the world's largest water crossing connecting three metropolises in southern China. The project comprises of a 6 km long immersed tunnel which has been constructed between two artificially created islands. Soil conditions between the transition of these islands and immersed tunnel are relatively poor requiring effective ground treatment measures to avoid large settlement as well as differential settlement. The top layer of subsoil beneath the HZMB tunnel is silty soft soil with thickness ranging from 10–24 m having weak engineering properties. This layer is followed by a silty-clay cohesive soil which has a continuous distribution with an uneven base. It is a high compressive cohesive soil layer in the middle of the stratum with its thickness ranging from 10–20 m. The third layer comprises of a coarse gravel based sandy soil which is the thickest of all with a range of 35–45 m. This layer can be used as the tunnel foundation. The fourth layer is a hard

© Springer Nature Switzerland AG 2018
W. Wu and H.-S. Yu (Eds.): *Proceedings of China-Europe Conference
on Geotechnical Engineering*, SSGG, pp. 1697–1701, 2018.
https://doi.org/10.1007/978-3-319-97115-5_173

foundation of mixed schist or mixed granite and can be considered as rigid during the analysis.

Dredging operations can be carried out to remove the first two silty layers but, there still remain highly compressible silt deposits in the third layer which are of poor engineering properties. Foundation treatment measures will directly affect the subsidence and internal force of the immersed tunnel.

2 Construction Stage and Problem Description

Considering the changes of loading conditions as well as the thickness of the soft ground along the longitudinal direction near the west artificial island, different ground treatment should be carried out to control differential settlement. A transition foundation comprising of Prestressed High Strength Concrete Piles (PHC) placed below the artificial island and high-pressure jet grouting placed below the first two immersed tunnel elements E1-S1 and E1-S2 which are part of the artificial island. Meanwhile, because the subsoil is of poor quality, surcharge preloading combined with dewatering measures is adopted to improve the character of soft soil near the head of island. Soil layer distribution and ground improvement techniques near the west artificial island are introduced by Figs. 1 and 2.

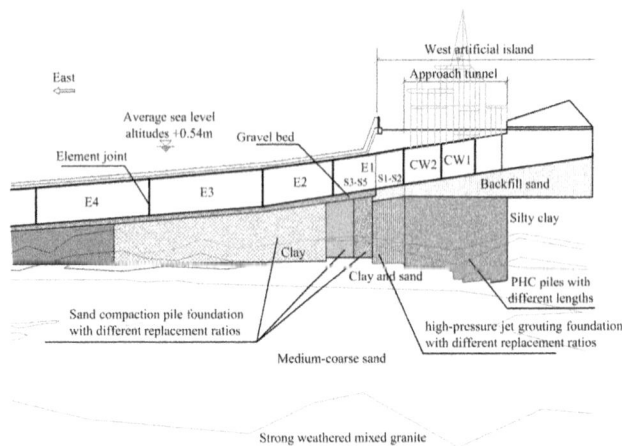

Fig. 1. Layout of the ground and treatment near the west artificial island

Backfilling of the two sides is carried out after the tunnel has been docked. To reduce the impact of the backfill, sand pile replacement is carried out on both sides of the tunnel. Following arrangements of isolating piles outside the tunnel are chosen for comparative analysis: (1) A-1, Single row of isolating piles; (2) A-2, Single row of isolating piles with pile spacing reduced to half; (3) A-3, The same as A-2 and located close to the tunnel; (4) A-4, Two rows of isolating piles in staggered layout.

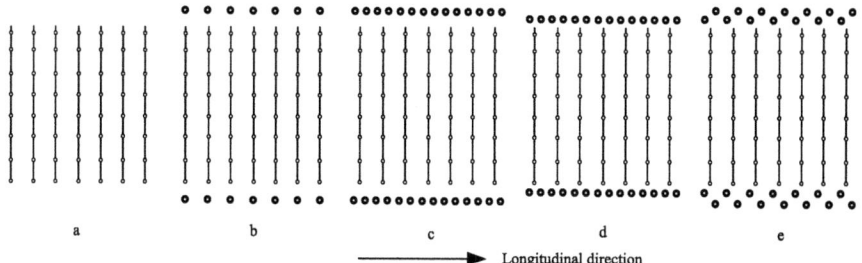

Fig. 2. Plan view of isolating piles: (a) No isolating piles (b) A-1 (c) A-2 (d) A-3 (e) A-4

3   Numerical Model and Discussion

3.1   Numerical Model and Material Parameters

The control section chosen here is K12+327 (E3) for calculation of the stress and the settlement of the tunnel and piles. In order to accurately consider the mechanism of composite foundation as well as local three-dimensional effect of the contact between pile and soil, this study still takes a local 3-D model as the research object, of which the thickness is chosen as one pile space along the length of the tunnel. Friction between pile and soil can be accurately simulated, making this model reasonable.

The constitutive models of each material are as follows: PHC piles and rock armor are modelled as linear elastic material; Tunnel top and bottom as well as side walls are modelled using shell elements, of which material behaviors are of linear elastic; locking backfill and gravel cushion adopted Mohr Coulomb model; remaining coarse sand, replaced sand and original sand are modelled using hardening soil model.

3.2   Modeling of Construction Steps

In order to analyze the impact of the side backfilling, several simulation construction steps are followed in the numerical model: (1) Initial equilibrium; (2) Excavation, dug out silty soil; (3) Backfill, backfill the medium coarse sand to the tunnel bottom including 0.5 m of gravel bed; (4) Piling, place the piles below and on the sides of the tunnel and carryout sand replacement; (5) Place gravel bed, sand pile replacement is completed; (6) Immersed tunnel is placed and rock armor is backfilled; (7) Backfill all the gravel on both sides of the tunnel which is the main part of the large backfill.

4   Results and Discussion

The first example did not include the isolating piles on both sides of the tunnel. The total settlement of the tunnel increases by 12–15 cm as a result of backfill load. Before backfilling, the settlement in the middle of tunnel section is larger. However, tunnel settles gradually from middle to the outer side as backfill load is applied. Two reasons contribute to this result: (1) the lateral soil has pulling effect on the side walls of the tunnel due to down-drag frictional force; (2) the outer piles become negative skin frictional piles, which further increases the settlements at both tunnel sides.

Under different arrangements of isolating piles, the maximum settlement of the tunnel and the maximum horizontal displacement of the pile in the composite foundation are shown in Table 1. It is evident that arrangement A-4 performs best in reducing the settlements. Without isolating piles, the backfill load on both sides diffuses into the composite foundation to increase its settlement. The principle of the isolating pile is its role in converting the load transferring path from both laterally and downwards to mostly downwards, resulting in the decrease of vertical compression underneath the tunnel structure. This effect becomes greater as the length of the isolating pile or its diameter increases. Numerical orthogonal experiments have also been conducted on the effect of isolating piles. It is found that isolating pile length has the greatest impact, followed by pile diameter, and followed by pile density. Although the validation of these orthogonal experiment results depends on proper choice of numerical parameters, this conclusion is still valuable for engineering reference.

Table 1. Numerical calculation results for different types of isolating piles

Arrangement type	Tunnel subsidence (mm)	Maximum horizontal displacement of pile in composite foundation (mm)
Without isolating piles	183	32
A-1	160	26
A-2	156	22
A-3	140	26
A-4	120	22

5   Conclusion

A local 3-D finite element analysis of a cross section of west artificial island of HZMB immersed tunnel has been carried out to determine the impact of large backfill on the settlement of the tunnel. Installation of isolating piles on both sides of the tunnel has been suggested and comparisons are made between four different arrangement schemes. Following conclusions can be drawn from this study: (1) Backfilling at both sides will have a great impact on the settlement of the tunnel as it will both increase the negative friction of the pile in composite foundation and apply a down-drag frictional force at the outer walls of tunnel structure; (2) Construction of isolating piles on both sides of the tunnel can reduce the impact of backfilling to a certain extent; (3) The location and arrangement of the isolating piles are analyzed comparatively. The nearer they are located to the tunnel, the more effective they are in reducing tunnel settlement. The further they are located; the better control they give on the horizontal pile displacement.

Acknowledgement. This work is financially supported by the National Key Fundamental Research and Development Program of China (Project no. 2014CB047003), the National Natural Science Foundation of China (Project No. 51408331) and the Science and Technology Research and Development Program of China Railway Corporation (Project No. 2014G004-C).

Lateral Decompression Behaviors of a Hard Claystone in Excavation-Damaged Zone of Galleries

Zaobao Liu[1,2($\boxtimes$)] (iD), Jianfu Shao[2], Shouyi Xie[2], and Nathalie Conil[3]

[1] Key Laboratory of Ministry of Education on Safe
Mining of Deep Metal Mines, Shenyang, China
zaobao.liu@yahoo.com
[2] University of Lille, FRE 2016 - Lamcube, 59000 Lille, France
[3] RD, Andra, Laboratoire de Recherche Souterrain de Meuse/Haute-Marne,
Bure, France

Abstract. We propose a new type of mechanical tests, named the lateral decompression tests, to evaluate the mechanical properties of the surrounding rocks in the shifts based on elastic stress solutions of circular openings. We carried out lateral decompression test with a developed autonomous device to evaluate the mechanical properties of the Callovo-Oxfordian (COx) claystone, the hosted rock and geological barriers of the repositories for radioactive waste disposal in France. In the tests, the samples were firstly subjected to a hydrostatic stress that corresponds to the representative in situ mean stress (p = 12 MPa). Then, the samples were loaded to a possible failure at the given constant mean stress (Δp = 0). The results show that the COx claystone failed in the lateral decompression tests at the in situ mean stress of 12 MPa. A shear band controls the failure pattern in lateral decompression tests, which is similar to that in conventional triaxial tests. However, the failure prior to rupture is more brittle than the failure in conventional triaxial tests. The failure of the COx claystone under the tested mean stress suggests that the excavated galleries need external supports to keep its stability after tunneling in the COx formation.

Keywords: Lateral decompression test · Rock deformation · Rock strength
Clayey rocks · Radioactive waste disposal

1 Context of Study

The safety requirement of underground repositories for radioactive waste disposal demands a full investigation and examination of the mechanical behaviors of the surrounding rocks in the excavation-damaged zones (EDZ) of the tunnels. In the context of the French concept of radioactive waste disposal, the Callovo-Oxfordian claystone (COx) has been confirmed to be the hosted rock and geological barriers of the repositories.

The mechanical behaviors of the COx claystone have been widely studied by conventional triaxial tests in various conditions in laboratory [1, 2]. Despite of those achievements obtained, the mechanical behaviors of the COx claystone are still in need

© Springer Nature Switzerland AG 2018
W. Wu and H.-S. Yu (Eds.): *Proceedings of China-Europe Conference on Geotechnical Engineering*, SSGG, pp. 1702–1706, 2018.
https://doi.org/10.1007/978-3-319-97115-5_174

for further investigation since new types of cracks, e.g. the extension ones, have been observed in the EDZ around the galleries at the underground research laboratory at Bure of France. The observed fissures do not correspond to the failure cracks observed in conventional triaxial tests. In consequence, the experimental results are not sufficient enough to give a full investigation of the mechanical behaviors of the COx claystone. Therefore, further new types of tests are in need to give a full characterization of the COx claystone. In this way, the French National Agency for Radioactive Waste Management (Andra) has launched a new series of investigation of the mechanical behaviors of the COx claystone which involves in new types of experimental methods. This paper presents the premier results in this context that proposes a new type of experimental method, named the lateral decompression test, to characterize the mechanical properties of the COx claystone based on analysis of the elastic stress solutions of underground circular openings.

2   Method Development and Materials

The mechanical model shown in Fig. 1 can simplify the stresses in openings of tunnels with the assumptions of homogeneity, elasticity and isotropy of the rocks. The stress solution before opening shown in Fig. 1(a) can be given by Eq. (1)

$$\sigma_r = \sigma_0, \sigma_\theta = \sigma_0, \tau_{\theta r} = \tau_{r\theta} = 0 \tag{1}$$

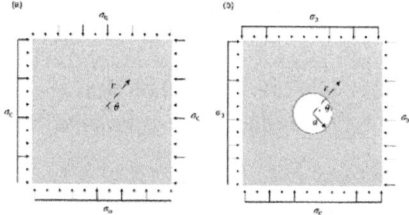

Fig. 1. Mechanical model of underground circular openings: (a) Initial state, (b) after opening.

After tunnel opening as shown in Fig. 1(b), the stress solution can be given by Eq. (2)

$$\sigma_r = \sigma_0 \cdot \left(1 - \frac{a^2}{r^2}\right), \quad \sigma_\theta = \sigma_0 \cdot \left(1 + \frac{a^2}{r^2}\right), \quad \tau_{\theta r} = \tau_{r\theta} = 0 \tag{2}$$

The stresses in both radius and tangential directions around the tunnels have changed due to the opening indicated in Eqs. (2) and (1). The stress paths do not agree with those of conventional triaxial tests in which the rock specimens are loaded axially to failure under constant radius stress, i.e. confining pressure. However, it is interesting and inspiring if one observes the mean stress before and after openings.

The initial mean stress p_0 before opening indicated in Eq. (1) is $p_0 = 1/3\sigma_{ii} = \sigma_0$. Similarly, the mean stress after opening as indicated in Eq. (2) is $p = 1/3\sigma_{ii}$, which also turns out to be σ_0. Therefore, the mean stress change $\Delta p = 0$ before and after opening. Based on this finding, we developed the lateral decompression tests to evaluate the mechanical behaviors of the surrounding rocks in underground rock engineering. In the lateral decompression tests, we increase the axial stress σ_1 and decrease the lateral stress σ_3, which is compromised by $\Delta p = 0$ to realize rock failure.

Thus, we developed the autonomous experimental device in [3] by incorporating an internal force sensor and modifying the loading components to conduct the lateral decompression tests on the COx claystone. The axial strains are the global solution calculated from values of LVDTs. Radius strains are measured by circumferential rings. The samples are firstly applied to a hydrostatic stress of 12 MPa at a rate of 0.5 MPa/min to reduce the drainage effects in the very low permeability claystone. After stabilization of strains, the samples are then subjected to the lateral decompression phase in which the mean stress $\Delta p = \Delta(\sigma_1 + 2\sigma_3) = 0$, i.e., $\Delta\sigma_1 = -2\sigma_3$, is controlled by decreasing at a constant gradient of σ_3.

The tested COx claystone is cored in the URL at Bure of France. At micrometer scale, the COx claystone is constituted of clay minerals, calcite and quartz. The average pore size is about some ten nanometers [4] and the majority of pores are inside the clay minerals. The average distribution of mineral groups in the claystones throughout the whole formation is clay fraction (phyllosilicates) $\sim 42 \pm 11\%$ of the rock, carbonates $\sim 30 \pm 12\%$ of minerals, ectosilicates $\sim 25 \pm 8\%$ of minerals; and ancillary minerals constitute less than 4% [5]. Mineralogical analyses show the mineral content exhibits low variability in the bed plane but significant perpendicular to the bedding [6]. The COx clay-rich rock porosity lies between 14% and 20% at the URL site and is close to 18% at the URL main level, and natural water content ranges between 5% and 8%. The COx claystone sample tested was drilled from the rock core EST30447 with saturation degree estimated 84% by the method presented in [1].

3   Experimental Results

The stress and strain outputs are shown in Fig. 2 for the lateral decompression test. The evolutions of material strains are shown in Fig. 2(a) with the change of axial stress, lateral stress and mean stress. Besides, the evolutions of material strains with respect to the differential stress are shown in Fig. 2(b). In Fig. 2, σ denotes stress, q is differential stress, p is mean stress, σ_1 and σ_3 are respectively axial and lateral stress, and ε_1, ε_3 and ε_V are respectively the axial, radius and volumetric strains.

It is shown in Fig. 2(a) that the mean stress keeps the same while the axial stress increases and lateral stress deceases with increasing axial strains during the test. The behaviors indicated by the differential stress versus axial strain in Fig. 2(b) shows the tested COx claystone in lateral decompression tests exhibits an elastic-plastic feature. Plastic strains accumulate to generate micro cracks that further accumulates to form a localized shear band which induces the volumetric dilatancy of the material as indicated by the volumetric strains shown in Fig. 2.

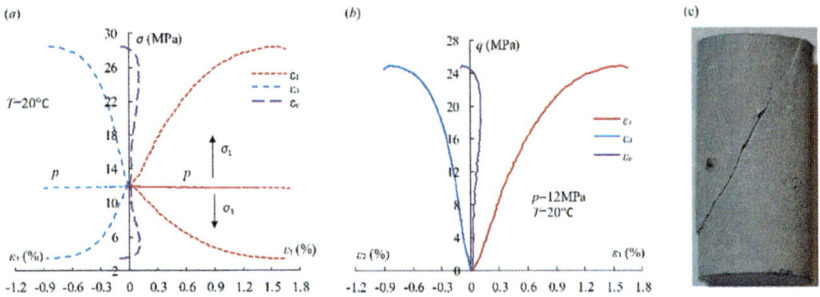

Fig. 2. Stress - strain outputs during lateral decompression tests and sample photo after failure

The peak axial, lateral and differential stresses obtained from the lateral decompression tests are respectively 28.6, 3.7 and 25.0 MPa under mean stress of 12 MPa at the tested saturation. The lateral stress at failure indicates that the COx claystone needs at least 3.7 MPa of support to avoid failure during stress redistribution in lateral decompression path from the initial stress state after opening.

4 Remarks

The tested COx claystone can fail at the means stress of 12 MPa under lateral decompression path. The failure is featured by accumulation of plastic strains that creates micro cracks to form finally a localized shear band, which induces volumetric dilatancy. The COx gallery needs external supports after opening during stress redistribution in lateral decompression path. Anisotropic and thermo effects will be further studied.

Acknowledgements. Supports from Andra and Key Lab of Chinese Ministry of Education on Safe Mining of Deep Metal Mines (Z016013) are acknowledged.

References

1. Liu, Z., Shao, J., Xie, S., Conil, N., Zha, W.: Effects of relative humidity and mineral compositions on creep deformation and failure of a claystone under compression. Int. J. Rock Mech. Min. Sci. **103C**, 68–76 (2018)
2. Liu, Z.B., Xie, S.Y., Shao, J.F., Conil, N.: Multi-step triaxial compressive creep behaviour and induced gas permeability change of clay-rich rock. Géotechnique **16**, 117 (2017)
3. Liu, Z.B., Shao, J.F., Liu, T.G., Xie, S.Y., Conil, N.: Gas permeability evolution mechanism during creep of a low permeable claystone. Appl. Clay Sci. **129**, 47–53 (2016)
4. Robinet, J.C.: Minéralogie, porosité et diffusion des solutés dans l'argilite du Callovo-Oxfordien de Bure (Meuse, Haute-Marne, France) de l'échelle centimétrique à micrométrique. Université de Poitiers, Poitiers (2008)

5. Armand, G., Leveau, F., Nussbaum, C., et al.: Geometry and properties of the excavation-induced fractures at the Meuse/Haute-Marne URL drifts. Rock Mech. Rock Eng. **47**(1), 21–41 (2014)

6. Armand, G., Noiret, A., Zghondi, J., Seyedi, D.M.: Short - and long-term behaviors of drifts in the Callovo-Oxfordian claystone at the Meuse/Haute-Marne underground research laboratory. J. Rock Mech. Geotech. Eng. **5**(3), 221–230 (2013)

Dynamic Shakedown of Cohesive-Frictional Materials Under Moving Traffic Load

Yuchen Dai, Jiangu Qian$^{(\boxtimes)}$, Xiaoqiang Gu, and Maosong Huang

Department of Geotechnical Engineering, Tongji University,
Shanghai 200092, China
qianjiangu@tongji.edu.cn

Abstract. Shakedown theory provides a rational tool for prediction of the long-term plastic behavior of pavement subjected to variable or repeated loads. A dynamic lower-bound shakedown solution has been proposed to estimate the critical shakedown limit load, over which plastic collapse or excessive permanent deformation of the pavement takes place. However, dynamic effects on the shakedown limit remains unexplored, particularly when rolling and sliding contact between vehicle and pavement are involved. In this paper, a finite-infinite (FE-IF) dynamic numerical method is presented to calculate the dynamic elastic stresses resulting from rolling and sliding contact at different moving speed for computing the shakedown limit. It is found that the shakedown limit decreases with the increasing moving speed initially and then turns to increase when the moving speed exceeds the Rayleigh wave speed of the pavement system. This dynamic effect is more profound as the horizontal force component reduces. The influence of frictional coefficient on shakedown limit is also discussed.

Keywords: Dynamic · Shakedown · Moving load · Rolling and sliding contact
FE-IE

1 Introduction

In pavement engineering, shakedown limit can be considered as a criterion to predict the failure of pavement structure subjected to repeated traffic load considering the long-term plastic behavior of pavement structure [1]. If the load level is relatively low (within the shakedown limit), the structure will adapt itself to a stable state and no further plastic deformation accumulates after a certain number of load cycles. However, if the load is higher than the shakedown limit, the plastic deformation will accumulate at each load application and excessive rutting will eventually occur [2].

Based on Melan's static shakedown theorem or Koiter's kinematic shakedown theorem, some lower-bound or upper-bound shakedown solutions were proposed in last decades to estimate the critical shakedown limit [3–5]. Considering rolling and sliding contact, the influence of frictional coefficient on static shakedown limit was studied in terms of Hertzian and trapezoidal contacts [6, 7]. However, since traffic load is one type of moving loads, the induced dynamic responses to moving speed failed to be taken account into static shakedown analyses previously. Hence, this work is

© Springer Nature Switzerland AG 2018
W. Wu and H.-S. Yu (Eds.): *Proceedings of China-Europe Conference on Geotechnical Engineering*, SSGG, pp. 1707–1710, 2018.
https://doi.org/10.1007/978-3-319-97115-5_175

intended to explore a more realistic shakedown limit for the pavement, including the influence of various vehicle speeds.

2   Problem Definition in Rolling and Sliding Contact

A three-dimensional flexible pavement system subjected to a moving rolling and sliding point contact load is considered here. The contact area is modeled by a circular-shaped patch and a 3D Hertz load is representative of a wheel load distribution, with vertical and horizontal components as follows:

$$p = \frac{3P}{2\pi a^3}(a^2 - x^2 - y^2)^{1/2} \tag{1a}$$

$$q = \frac{3Q}{2\pi a^3}(a^2 - x^2 - y^2)^{1/2}, \quad (x + vt)^2 + y^2 < a^2 \tag{1b}$$

where a is the radius of the contact area and $a = 0.125$ m is chosen here. x, y, z represent Cartesian coordinates and the x-axis is the moving direction. P and Q are total normal and tangential forces, respectively. $p_0 = 3P/2\pi a^2$ is the peak pressure. The normal and tangential loads are linked by the frictional coefficient μ as the following:

$$\mu = Q/P \tag{2}$$

3   Numerical Approach

In this paper, a dynamic shakedown analysis method based on the lower-bound shakedown theorem [8] is proposed to determine the shakedown limit of pavement subjected to moving load with various speeds. Similar to previous static shakedown methods, rolling and sliding contact between wheel and pavement surface were considered. In order to determine the dynamic stresses due to traffic speeds, dynamic finite element (FE) modeling, coupled by an infinite element (IE) artificial boundary as reported by Qian et al. [8], has been established in ABAQUS. Specifically, a thin-layer is built to represent the moving contact interface between wheel and pavement surface. Both vertical and horizontal moving forces can be simultaneously applied on the thin layer via VDLOAD, an ABAQUS subroutine.

4   Shakedown Limit Analysis

The dynamic shakedown solution is now applied to analyze the shakedown limit for a uniform cohesive-frictional half space. The Rayleigh wave speed of uniform materials is considered to be 60 m/s about. The normalized shakedown limits are represented as $\lambda' p_0/c$ as shown in Fig. 1. As a validation, the shakedown limits at an extremely low

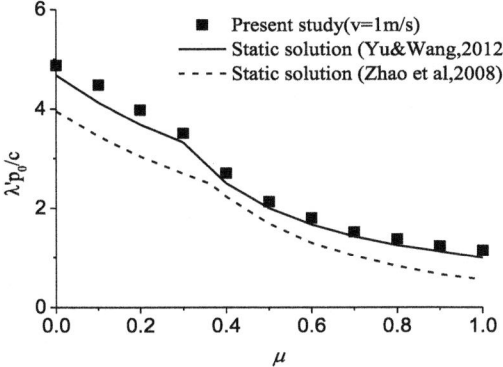

Fig. 1. Comparison between the present study and static solution ($\phi = 0°$).

speed state ($v = 1$ m/s) is compared with previous static shakedown solutions [6, 7]. Clearly, the present dynamic solutions can be reduced to conventional static ones.

Figure 2 presents the change of the normalized shakedown limit with the moving speed. It is found that the shakedown limit almost decreases as the moving speed grows, but eventually turns to increase once the speed exceeds some critical level, representing Rayleigh wave speed of the pavement. Specifically, the dynamic effect tends to become more significant with reducing friction coefficient (μ) or growing internal friction angle (ϕ).

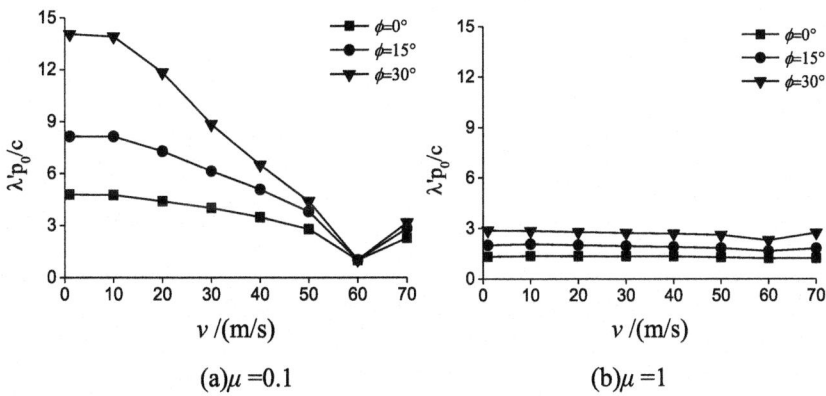

Fig. 2. Shakedown limits versus moving speeds for various internal friction angles.

Figure 3 presents the influence of frictional coefficient on the shakedown limits at various moving speeds. Clearly, the shakedown limit decreases with increasing frictional coefficient, in agreement with previous work [6, 7]. However, as the moving speed approaches Rayleigh wave speed, the shakedown limits is gradually reducing to the same level, representing a minimum limit, with increasing frictional coefficient.

Besides, as the friction coefficient increases, the failure mode changes from subsurface failure to surface failure when the load moving speed is relatively small. Meanwhile, it should be noticed that the increase of moving speed will make the failure more likely to occur below the surface. Particularly, the critical point always lies below the surface (i.e. subsurface failure occurs) when the load moving speed is relatively high (beyond Rayleigh wave speed).

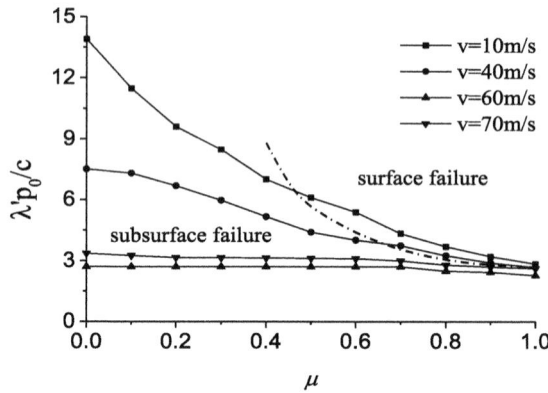

Fig. 3. The influence of frictional coefficient on the shakedown limit for various values of moving speed ($\phi = 30°$).

References

1. Sharp, R.W., Booker, J.R.: Shakedown of pavements under moving surface loads. J. Transp. Eng. ASCE **110**, 1–14 (1984)
2. Zhao, J.D.: A revisit to the shakedown of pavements under moving surface loads. In: Pan-American Conference on Soil Mechanics and Geotechnical Engineering in Conjunction with, Canadian Geotechnical Conference (2011)
3. Wang, J., Yu, H.S.: Shakedown analysis for design of flexible pavements under moving loads. Road Mater. Pavement Des. **14**(3), 703–722 (2013)
4. Collins, I.F., Boulbibane, M.: Geomechanical analysis of unbound pavements based on shakedown theory. J. Geotech. Geoenviron. Eng. **126**(1), 50–59 (2000)
5. Yu, H.S.: Three-dimensional analytical solutions for shakedown of cohesive-frictional materials under moving surface loads. In: Proceedings of the Royal Society A: Mathematical, Physical and Engineering Science, vol. 461 (2059), pp. 1951–1964 (2005)
6. Zhao, J.D., Sloan, S.W., Lyamin, A.V., Krabbenhøft, K.: Bounds for shakedown of cohesive-frictional materials under moving surface loads. Int. J. Solids Struct. **45**(11–12), 3290–3312 (2008)
7. Yu, H.S., Wang, J.: Three-dimensional shakedown solutions for cohesive-frictional materials under moving surface loads. Int. J. Solids Struct. **49**(26), 3797–3807 (2012)
8. Qian, J.G., Wang, Y.G., Wang, J., Huang, M.S.: The influence of traffic moving speed on shakedown limits of flexible pavements. Int. J. Pavement Eng. http://dx.doi.org/10.1080/10298436.2017.1293259

Influence of Rock Joint Orientation on the Natural Frequency of Dam-Foundation System

Ajay Rampal, Prasun Halder[(⊠)], Bappaditya Manna,
and K. G. Sharma

Indian Institute of Technology, Delhi 110016, New Delhi, India
prasun.siliguri@gmail.com

Abstract. The natural frequency of dam-foundation system is a function of its stiffness and the participating mass. In the present study, an attempt is made to understand the effect of rock joint orientation on the natural frequency of the dam-foundation system. Discrete element simulations are carried out using UDEC (1993) software for the intact rock foundation and rock foundations having joint inclinations of 0°, 30°, 60°, 90°, 120° and 150° with 20 m, 10 m, 5 m, 2.5 m and 1 m joint spacing in the rock mass. Results, confirm that the joint orientation affects the natural frequency of the system significantly. The natural frequency of the dam-foundation system is found higher for the intact rock foundation as compared to other jointed rock foundations. Results also reveal that the magnitude of the natural frequency of the system increases as the joint spacing increases in the rock mass.

Keywords: Dam foundation · Natural frequency · Stiffness · Joint orientation

1 Introduction

The natural frequency of a system can be defined as the rate at which it oscillates without the application of any external driving force. Mathematically,

$$\omega_n = (K/M)^{0.5} \tag{1}$$

where, ω_n = natural frequency of the system, K = stiffness of the system and M = participating mass. Increase in the system's stiffness will lead to increased natural frequency when its mass is constant. Natural frequency plays a key role in resonance.

Calculation of the natural frequency is an important aspect regarding the dynamic response of dam foundations during earthquakes. In this regard, Gazetas and Dakoulas [1] reported the effect of cohesive soil composting on the natural frequency of earthen dam with the help of shear slice procedure. Watanabe et al. [2] calculated the natural frequencies and their corresponding modes of a rockfill dam from Fourier Spectra. Tsai et al. [3] investigated the influence of clay core on the natural frequency of earthen dam. Parish et al. [4] determined the natural frequencies of a dam-foundation with

© Springer Nature Switzerland AG 2018
W. Wu and H.-S. Yu (Eds.): *Proceedings of China-Europe Conference on Geotechnical Engineering*, SSGG, pp. 1711–1715, 2018.
https://doi.org/10.1007/978-3-319-97115-5_176

Fourier analysis of free vibration response of the dam. Hasani et al. [5] considered the dam-foundation flexibility for calculating dam's natural frequency.

In the present study, an attempt is made to understand the effect of rock joint orientation on the natural frequency of the dam-foundation system.

2   Details of the Study

2.1   Dam Geometry and Material Properties

The concrete gravity dam is considered 100 m high having base width of 80 m and crest width of 12 m. The base width of foundation is 500 m and depth 250 m as shown in Fig. 1. The analysis is carried out for the intact rock and jointed rock with joint inclinations of 0°, 30°, 60°, 90°, 120°, and 150° for different joint spacings i.e. 20 m, 10 m, 5 m, 2.5 m, and 1 m.

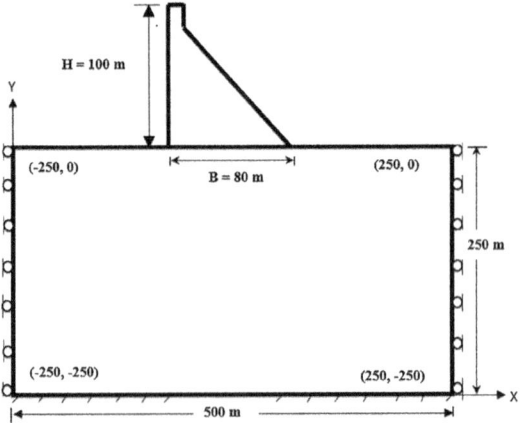

Fig. 1. Model and geometry of the dam foundation.

Properties of both concrete and rock are summarized in Table 1. The normal stiffness (k_n) and shear stiffness (k_s) of rock joints are considered 13000 and 5000 MPa/m respectively. The properties of the dam foundation are taken from the Detailed Project Report of the hydro-electric project in Himalayan region of India. The rock material is coarse gneiss which is a foliated metamorphic rock. Grade of concrete used is M15.

2.2   UDEC Analysis and Fast Fourier Transformation (FFT)

A modal analysis is carried out using Universal Distinct Element Code [6]. In the analysis, Mohr-Coulomb model is used for rock and Coulomb slip model is used for joints. Model of the dam-foundation system is created by assembling the discrete blocks and interfaces between them. The foundation block is assumed to be restrained at the base. The lateral boundaries of the block are restrained in the X-direction but kept

Table 1. Properties of rock and concrete used in the analysis.

Parameters	Rock	Concrete
Young's modulus, E (MPa)	18000	19364.9
Bulk modulus, K (MPa)	15000	12909.9
Shear modulus, G (MPa)	6923	7745.9
Poisson's ratio, v	0.3	0.2
Mass density, ρ (g/cm^3)	2.69	2.40
Cohesion, c (MPa)	5	3
Angle of internal friction (ϕ)	40°	40°

free along the Y-direction. Gravity load is applied on the dam-foundation system, which leads to the vibration of the whole dam-foundation system. The Y-displacement of the crest of the dam is recorded against the time for the undamped condition. FFT of this history is then carried out in order to find the natural frequency of the dam-foundation system for different joint inclinations and spacings.

3   Results and Discussion

The variation of natural frequency of the dam-foundation system with the change in joint spacing for horizontal joint set (0°) is presented in Fig. 2. It can be clearly seen that natural frequency of the dam-foundation system is decreasing with the decrease in spacing of joint set. The magnitude of natural frequency is found maximum (1.79 Hz) for intact rock and minimum (1.23 Hz) for jointed rock with 1 m joint spacing.

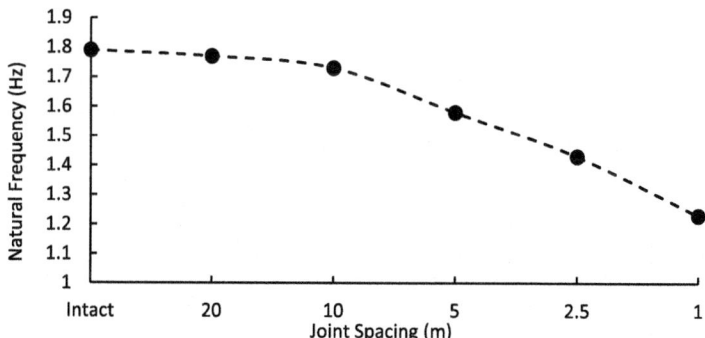

Fig. 2. Variation of natural frequency with joint spacing for horizontal (0°) joints.

This can be attributed to the fact that the decrease in the joint spacing i.e. higher number of joints makes the rock weaker resulting a reduction in the deformation modulus of the rock. The stiffness of the dam-foundation system is related to the deformation modulus of the rock mass foundation. So, a decrease in the deformation modulus will result in the reduction of the stiffness of the dam-foundation system. It can be seen that increase in the joint spacing makes rock mass foundation stiffer and for higher spacing, rock mass starts behaving like an intact rock. For example, rock foundations with 20 m and 10 m spacing possess natural frequency closer to the intact rock.

Figure 3 shows the variation of the natural frequency of the dam-foundation system with the change in the joint inclination for 5 m, 2.5 m and 1 m joint spacings. The natural frequency of the jointed rock foundation having vertical joint (90°) set for all the joint spacings is found higher than all other joint inclinations. Foundation with 30° joint shows the lowest value of natural frequency i.e. 1.13 Hz for 1 m joint spacing.

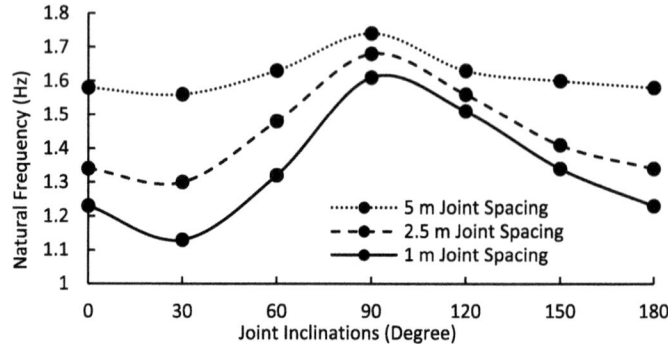

Fig. 3. Variation of natural frequency with joint inclination.

4 Conclusions

From the work presented in this paper, the following conclusions can be made:

- Natural frequency of the intact rock dam-foundation system is found higher as compared to other jointed rock foundations considered in this study.
- Stiffness of the dam-foundation system decreases with reducing joint spacings resulting a reduction in the natural frequency of the system.
- Rock mass with higher spacing i.e. 10 m and higher starts behaving like an intact rock.
- Natural frequency of the rock with 90° joint inclination for all joint spacings is found higher than all other joint inclinations considered in this study.
- Foundation with 30° joint inclination and joint spacing of 1 m possesses the lowest natural frequency value of 1.13 Hz.

References

1. Gazetas, G., Dakoulas, P.: Seismic analysis and design of rockfill dams: State of the art. Soi Dyn. Earthq. Eng. **11**(1), 27–61 (1992)
2. Watanabe, H., Kikuchi, K., Cao, Z.: Vibration modes of rockfill dam based on the observations of microtremors and an earthquake. Research report, 2, Thammasat Universit (1996)

3. Tsai, P., Hsu, S., Lai, J.: Effects of core on dynamic responses of earth dam. In: Slope Stability, Retaining Walls and Foundations: GeoHuman International Conference, pp. 8–13. ASCE (2009)

4. Parish, Y., Sadek, M., Shahrour, I.: Numerical analysis of the seismic behavior of earth dam. Nat. Hazards Earth Syst. Sci. **9**, 451–458 (2009)

5. Hasani, M., Ghanbari, A., Hosseini, S.: Analytical estimation of natural frequency with respect to foundation effects. Numer. Meth. Civ. Eng. **1**(1), 7–13 (2014)

6. UDEC, Universal Distinct Element Code user's manual, ver. 4.0, Itasca Consulting group, Inc., Minneapolis (1993)

Sheet Pile Bridge Abutments: Faster, Economic and Viable Solution for Urban Transportation Needs

Abhishek Jain[(⊠)] and Anirban Sen

Atkins member of the SNC-Lavalin Group,
Global Design Centre, Gurgaon, India
abhishek.jain2@atkinsglobal.com

Abstract. Although the concept of sheet pile bridge abutments is not new, they do not enjoy as much widespread usage as reinforced concrete bridge abutments. This paper provides a design case-history involving replacement of two existing road bridges with proposed new bridges supported by a combination of steel sheet piles and concrete bearing piles as abutments. The steel sheet piles have been designed to provide lateral earth support while concrete piles, vertical deck loads. Sheet piles of different types: anchored, tied-back and cantilever, have also been proposed for wingwalls. The arrangement for abutments was selected to minimize traffic interruption over existing road bridge, to enable faster construction in a restricted space environment and reduce 'possession' of railway track, and to provide an economic and durable solution. The two road bridges are a part of preparatory works for electrification program of Denmark state railway network between Aarhus and Lindholm.

Keywords: Abutments · Sheet pile walls · Ground anchors · Concrete piles

1 Introduction

As part of preparatory works for electrification of 145 km of double-track railway line between Aarhus and Lindholm in Denmark, as many as 50 overbridges need to be reconfigured to have required clearance for the power lines underneath. Viborgvej and Parkboulevarden in Randers are the first two bridges to be rebuilt by Banedanmark under this scheme between January 2018 and June 2019.

Viborgvej bridge carries a state road crossing the railway line over a single span in western Randers. The bridge will be shifted six meters to the north and raised by one and a half meters from its existing configuration. The existing reinforced concrete arch bridge crosses the railway line at a skew angle of 22° with its north and south abutments extending over 100 m. Parboulevarden bridge, also an arch bridge built in 1938 approximately half kilometers to the northwest of Viborgvej location, carries a municipality road crossing the railway line over a single span perpendicular to the railway tracks. Its abutments measure approximately 18 m. The aerial satellite images of existing Viborgvej and Parkboulevarden bridges are shown in Fig. 1.

© Springer Nature Switzerland AG 2018
W. Wu and H.-S. Yu (Eds.): *Proceedings of China-Europe Conference on Geotechnical Engineering*, SSGG, pp. 1716–1720, 2018.
https://doi.org/10.1007/978-3-319-97115-5_177

Fig. 1. Existing Viborgvej (left) and Parkboulevarden (right) Bridges – Aerial Images (Source: Google)

Both bridges were proposed to be replaced by prestressed concrete bridges with their decks supported on sheet pile abutments and raked square reinforced concrete driven piles. This was viewed as the fastest method of construction minimizing the traffic interruptions and providing an economic and durable solution.

2 Construction Considerations

Viborgvej road supports vehicular traffic of 15,000 cars a day. Hence, minimal interruption to both road traffic and railways was desired at Viborgvej location, requiring a temporary diversion bridge from the western side of the existing bridge. The diversion bridge (see Fig. 2) would route the traffic utilizing an existing side road connecting the state road from its north (see Fig. 1). Temporary retaining walls to protect existing residential structures south of the bridge near railway tracks were also proposed.

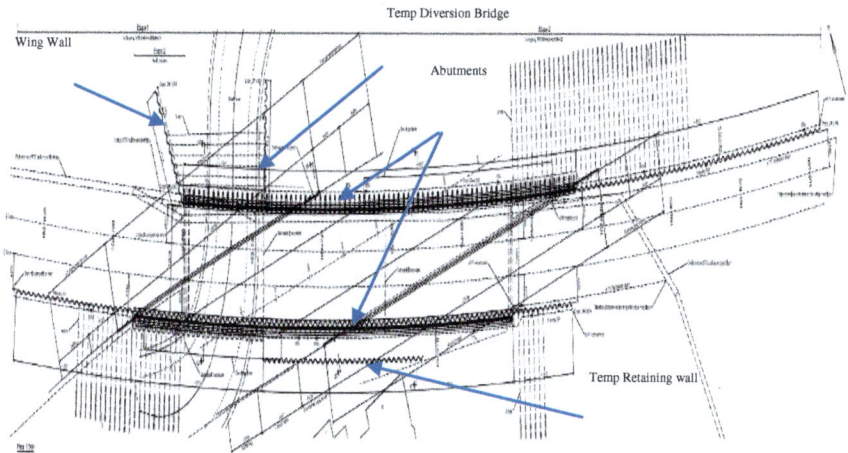

Fig. 2. Proposed Viborgvej Bridge (Plan View Showing Abutments, Wingwalls, and Temporary Diversion Bridge and Retaining Walls)

As Parkboulevarden bridge is much smaller in comparison, and carries relatively light traffic of 3,500 vehicles, diversion of traffic was not planned during demolition of existing bridge and construction of new bridge.

3   Ground Conditions

The two bridge project sites comprise fill/topsoil near ground surface underlain by glacial clay till and sand till, with local embedded layers of meltwater silt and clay, and meltwater sand and gravel. As several different sheet pile type profiles (with or without anchors or tied-back) were proposed in different zones of abutments and retaining walls depending on retained height, surcharge and sensitivity to deflections, their embedment lengths varied between different zones, resulting in their termination in clay till or sand gravel layers. The characteristic geotechnical parameters are shown in Table 1.

Table 1. Characteristic Geotechnical Parameters

Soil type	Density γ [kN/m^3]	Undrained shear strength, C_u [kPa]	Friction angle, $\varphi'[°]$	Effective shear strength, c'[kPa]	Drained Poisson ratio ϑ	E_{oed} (MPa)
Fill and Top soil	18.0	50	30	0	0.25	10
Clay/Till	21.0	200	32	20	0.25	60
Sand/Gravel	19.0	–	40	0	0.2	60
Clay/Silt	20.0	150	30	15	0.25	50
Compacted Granular fill material	18.0	–	38	0	0.2	–

[a] where needed to proposed final configuration and levels

High groundwater was encountered approximately 5–6 m below existing bridge levels. Artesian water level was also encountered at Viborgvej location within 3 m.

4   Design Methodology and Key Considerations

The geotechnical design was carried out in accordance with Eurocode 7 and Danish National Annex (DS/EN 1997-1 DK NA 2015, A3) using limit state design philosophy

The sheet pile design was carried out using WALLAP software program that utilizes limit equilibrium method for ultimate limit state factor of safety, and subgrade reaction method for shear force (SF), bending moment (BM) and deflections. The vertical loads for raked concrete piles were obtained from LUSAS finite element software method.

The pile lengths were determined using static analytical method for bearing resistance as presented in Danish National Annex.

The structural design checks for sheet pile cross-sectional capacities were performed in accordance with Eurocodes 3 and 2 (DS/EN 1993-5:2007, DS/EN 1992) and relevant Danish National Annexes based on elastic theory. The structural design also involved design of bridge deck connection with abutments to ensure that sheet pile abutments take all lateral loads and concrete piles take most of the vertical loads. Hence, the connection design between bridge deck and abutment was a key element of design in terms of load distribution. This connection could be divided into two parts: connection of sheet pile with cill beam and that of cill beam with bridge deck using dowel bars.

The following key design considerations were made for design:

1. Different groundwater levels were assumed for ULS, ALS and SLS limit states on active side of the wall. Groundwater was considered at terrain level on passive side.
2. Excavation up to 1.34 m for any future utilities in front of the walls was considered.
3. Surcharges of 20 kPa for abutments and 5 kPa for other walls were considered for temporary construction loads. Loads due to braking and acceleration were applied.
4. The cross-sectional structural checks were considered at three levels: anchor level, max. BM above terrain level and max. BM below terrain level.
5. Long-term reduction in effective area of sheet piles was considered over 120-year design life due to atmospheric and soil corrosion based on Eurocode 3 - Part 5.
6. Shaft resistance for concrete piles were considered on three out of four sides till abutment sheet pile tip and for all four sides below, to account for soil disturbance.
7. Tip resistance of concrete piles in non-cohesive soils was capped to 400 kN per pile.
8. The temperature loads due to expansion and contraction of bridge deck was considered in sheet pile design.

5   Results and Conclusions

The two replacement bridges at Viborgvej and Parboulevarden were successfully designed using sheet pile and raked pile abutments meeting various design requirements and construction constraints. Different Arcelor Mittal AZ and CAZ profile types, with or without anchors, were proposed in different zones for abutments, wing walls and other temporary walls based on ground type, retained height and other deflection and structural requirements. The abutments were designed such that lateral deflections were within 25 mm. The deflections at other retaining wall locations were limited to 80 mm that were within the acceptable limits.

References

1. Atkins: Preparation of Electrification Aarhus-Lindholm, Geotechnical Design Report, BRO 21004 – Parkboulevarden, Doc. Ref. Ar-Lih-6066-027-Bridge 21004. GDR, September 2017
2. Atkins: Preparatory Works for Electrification Aarhus-Lindholm, Geotechnical Design Report, BRO 20998 – Viborgvej, Southwest, Northeast and Southeast Wingwalls, July 2017
3. Atkins: Preparatory Works for Electrification Aarhus-Lindholm, Geotechnical Design Report, BRO 20998 – Viborgvej, Northwest Wingwalls, October 2017

Robust Geotechnical Design of Transition Zone Material Parameters Based on Fuzzy Set Theory

Yao Shan, Li Su$^{(\boxtimes)}$, and Yi Lu

Key Laboratory of Road and Traffic Engineering of the Ministry of Education,
Tongji University, Shanghai 201804, China
1632411@tongji.edu.cn

Abstract. A safe and economical design of high-speed railway subgrade-bridge transition zone nowadays relies heavily on the accuracy of the geotechnical parameter measurements, while inherent variability, testing error and transformation error often stand in the way. This paper presents a fuzzy set-based robust geotechnical design (RGD) methodology for the transition zone, freeing the design from the requirement of precise geotechnical parameters. Based on the theory of vehicle-track coupling dynamics, a plane stress finite-infinite element model of the transition zone is proposed to investigate the influence of 3 simultaneously varied subgrade material parameters on the system dynamic response. Using the vertex method, the subgrade system's robustness is evaluated by the signal-to-noise ratio. The results indicate that the influence of the dynamic elastic modulus of the graded broken stone on the robustness of the vehicle system is considerably larger than that of the dynamic elastic modulus of the subgrade bed surface layer. The system is more robust when the dynamic elastic modulus of the graded broken stone is between 625 Mpa and 750 Mpa.

Keywords: Transition zone · Vertex method · Robust geotechnical design
Finite-infinite element model

1 Introduction

A reasonable stiffness configuration providing a smooth transition between different sub-rail structures is crucial for the design of transition zone. Many scholars have conducted researches to analyze the influence of subgrade stiffness matching on the dynamic response of the vehicle-track system [1, 2]. However, the uncertainty of geotechnical parameters is often neglected in existing literature.

In this paper, a plane stress model is developed to calculate the dynamic response of a vehicle-track-subgrade system with high computational efficiency. A fuzzy set-based robust geotechnical design (RGD) methodology is proposed for the optimizing design of transition zone material parameters. The robustness of the system dynamic response against the fluctuating stiffness of subgrade filling within 1-m distance behind the bridge abutment is considered as a criterion for the evaluation of the stiffness configuration of the transition zone.

© Springer Nature Switzerland AG 2018
W. Wu and H.-S. Yu (Eds.): *Proceedings of China-Europe Conference on Geotechnical Engineering*, SSGG, pp. 1721–1725, 2018.
https://doi.org/10.1007/978-3-319-97115-5_178

2 Fuzzy Set-Based Robust Geotechnical Design Methodology

Existing RGD methodologies mainly requires the probability density function of the uncertain variables, which is not easy to find, sometimes even doesn't exist. Based on fuzzy set theory, an uncertain geotechnical parameter can be modeled with only the knowledge of its highest conceivable value (HCV) and lowest conceivable value (LCV). A fuzzy set is a set of ordered pairs, $[x, \mu(x)]$, where a member belongs to the set with a certain level of confidence, called membership grade, $\mu(x)$ [3]. A fuzzy set with a membership function that is convex in shape, and with its highest membership grade equal to 1, is a special fuzzy set called fuzzy number.

The mean and standard deviation of the resulting factor of system response can be readily calculated by following equations [3]:

$$p_i = \frac{(\alpha_i)^n}{2 \sum_{i=1}^{i=m} (\alpha_i)^n + (\alpha_{m+1})^n} \tag{1}$$

$$E[X] = \sum_{i=1}^{i=m} p_i (X_{\alpha_i}^- + X_{\alpha_i}^+) + p_{m+1} X_{\alpha_{m+1}} \tag{2}$$

$$\sigma^2[X] = \sum_{i=1}^{i=m} p_i [(X_{\alpha_i}^- - E[X])^2 + (X_{\alpha_i}^+ - E[X])^2] + p_{m+1} (X_{\alpha_{m+1}} - E[X])^2 \tag{3}$$

where p_i is the probability for noise factor $X = X_{\alpha_i}^- (X_{\alpha_i}^+)$, α_i is the membership degree of the system response factor, n is the number of the noise factors, m is the number of intervals between the LCV to the HCV, $E[X]$ and $\sigma^2[X]$ are the mean and standard deviation of the resulting factor of system response.

3 Vehicle-Track-Subgrade Model and Optimal Design Case Study

Based on the theories of vehicle-track coupling dynamics, a coupled vehicle-track-subgrade model is established by the finite-infinite element method. In this model, the vehicle is simplified as a multi-body system, where the vehicle body, bogie and wheelset form the vehicle system are considered rigid bodies. The vertical and pitch motion of vehicle body and bogie as well as the vertical motion of the wheelset are considered. The suspensions are simulated as springs and dampers. In this case, the vehicle system has 10 degrees of freedom (DOF). The infinite element method (I-FEM) is employed to prevent wave reflection on the boundaries of the transition zone model Nodes on the bottom of the bridge abutment are fixed. A free boundary is utilized on the left side of the bridge abutment. The infinite elements are employed on the bottom and the right-hand side of the subgrade. The coupling of the vehicle and the track is the wheel-rail contact force.

Using the FEM-IFEM coupled vehicle-track-subgrade model described above, the system dynamic response under the influence of 3 simultaneously changing parameters is calculated. The three parameters are the dynamic elastic modulus of the surface layer of the subgrade bed, the graded broken stone within 1-m distance behind the bridge abutment, and the graded broken stone more far away from the abutment. The near-abutment area is considered as the noise factor in the system, and the other two are considered signal factors.

The total length of the bridge-to-subgrade transition zone model is 120 m, in which the bridge is 32.7 m long, the abutment is 1.8 m long, the transition zone itself is 24.3 m long, and the subgrade section is 61.2 m long. The weight of the rail is 60 kg/m. The speed of the vehicle is 350 km/h, and the moving distance is 120 m. The total degree of freedom of the system is 5690. The simulation program is written in MATLAB. A single run of this model takes 1321.8 s. The model is shown in Fig. 1.

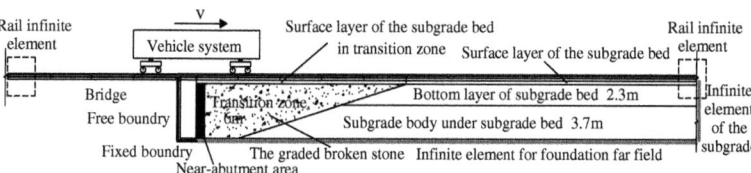

Fig. 1. Vehicle-track-subgrade model of the inverted trapezoid transition zone.

Five levels are chosen for each of the signal factor parameters, 1000 Mpa, 1250 Mpa, 1500 Mpa, 1750 Mpa, 2000 Mpa for the surface layer and 500 Mpa, 625 Mpa, 750 Mpa, 875 Mpa, 1000 Mpa for the graded broken stone. For each stiffness combination, seven sets of fuzzy number are chosen to describe the noise factor. The fuzzy-sets are (500, 0), (583.33, 0.33), (666.67, 0.67), (750, 1), (833.33, 0.67), (1000, 0). Using Eqs. (2) to (4) in Sect. 3, 25 sets of response under the influence of two signal factors are calculated. The amplitude of wheel-rail contact force and rail dynamic displacement are chosen as the system dynamic response criteria. The dynamic elastic modulus of the subgrade bed surface layer is referred to as E_1, and that of the graded broken stone is referred to as E_2 in following figures.

As shown in Fig. 2, the amplitude of the rail vertical displacement decreases when the subgrade stiffness increases (both the surface layer and the graded broken stone), while the amplitude of the wheel-rail contact force increase along with the subgrade stiffness. For both criteria the influence of the graded broken stone is much greater than of the surface layer. The results are in accordance with those of Shan [2].

The system's dynamic response robustness is represented by the signal-to-noise ratio (SNR), which quantifies the influence of the signal factors relative to the noise factors on the system. The bigger the SNR value is, the more robust can the system be considered. SNR can be calculated according to the following equation [3]:

$$SNR = 10 \log_{10}(\frac{E^2[X]}{\sigma^2[X]}) \qquad (4)$$

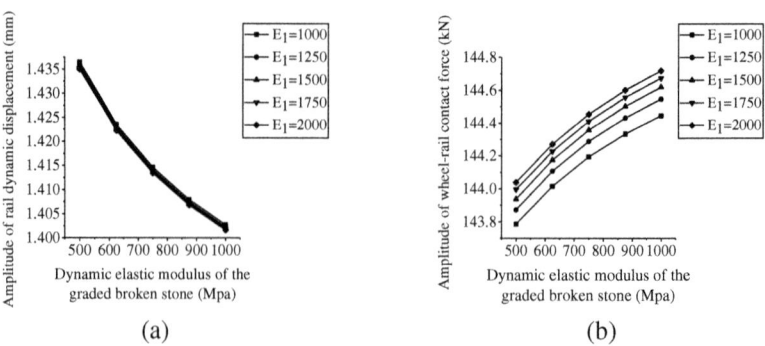

(a)   (b)

Fig. 2. (a) Rail vertical displacement; (b) Wheel-rail contact force under influence of the dynamic elastic modulus of the graded broken stone and the subgrade bed surface.

in which $E^2[X]$, $\sigma^2[X]$ can be calculated by Eqs. (3) and (4). In the following figures, SNR_1 and SNR_2 refers to the signal-to-noise ratio of the amplitude of the rail vertical displacement and the wheel-rail contact force separately (Fig. 3).

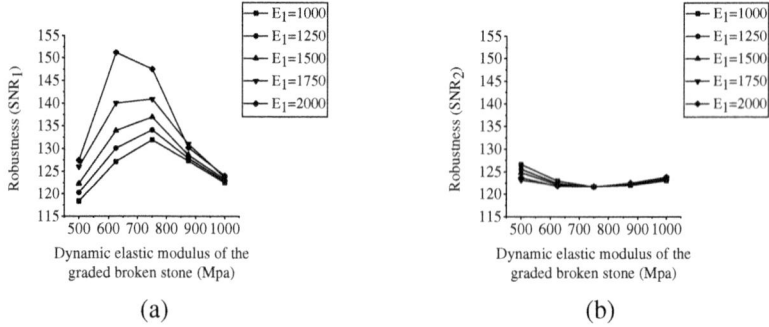

(a)   (b)

Fig. 3. System robustness (a) SNR_1 (b) SNR_2 under influence of the dynamic elastic modulus of the graded broken stone and the subgrade bed surface.

It can be concluded that for the subgrade stiffness matching influences SNR_2 much more than SNR_1. SNR_1 mostly increases with the increase of E_1, and firstly increase then decrease with the increase of E_2. The relations for SNR_2 is the exact opposite on a much smaller scale. For an optimal design, E_1 should be relatively large, E_2 should be between 625 Mpa to 750 Mpa.

References

1. Lei, X.Y.: Influences of track transition on track vibration due to the abrupt change of track rigidity. China Railway Sci. **27**(5), 42–45 (2006)
2. Shan, Y., Shu, Y., Zhou, S.: Finite-infinite element coupled analysis on the influence of material parameters on the dynamic properties of transition zones. Constr. Build. Mater. **148**, 548–558 (2017)
3. Gong, W., Wang, L., Juang, C.H., et al.: Robust geotechnical design of shield-driven tunnels. Comput. Geotech. **56**(1), 191–201 (2014)

Tentative Investigation of Structure Size Effect of High-Filled Geotechnical Structures

Erxiang Song[(⊠)] ⓘ, Tianliang Zheng, and Yufei Kong

Department of Civil Engineering, Tsinghua University, Beijing 100084, China
songex@mail.tsinghua.edu.cn

Abstract. The structure size effect with regard to the deformation behavior of high filled geotechnical structures is investigated. Theoretical analyses and numerical Monte Carlo simulations reveal that the relatively large variability of the mechanical properties, especially those related with plastic deformations, can be the main cause of this kind of size effect. Although variability of material properties implies both degradation and upgradation of the material parameters by certain probability, computations show that the parameter degradation dominates the final results and gives rise to the structure size effect.

Keywords: Structure size effect · High-Filled structure
Monte Carlo simulation

1 Introduction

In recent years many high filled engineering structures are constructed in China, such as high filled dams and high filled ground for airports in the mountainous regions. The high filled ground for airports are often of a height of 50–100 m, and the high filled dams are often of a height of 200–300 m [1]. Promoted by the engineering practice much research has been conducted on the engineering properties of the rockfills and the deformation and stability analysis of such high filled structures [1, 2].

Much knowledge regarding the engineering properties of the rockfills has been gained through laboratory tests, field observations and theoretical studies. Some numerical models have been further developed to simulate the complex properties of such filling materials, including compression hardening, shear dilatancy, nonlinear strength due to particle breakage, wetting and time dependent deformation (creep), etc. [1, 2]. At the same time attention has been paid to the difference between the laboratory tests on samples of reduced-size particles and the on-site filling material of much larger particle size [6, 7].

Based on all the studies above, numerical analysis of the post construction deformation of high filled structures were conducted. However, it is noted by many researchers that the calculated deformations are often larger than the measured values for relatively low structures, whereas they are too small for much higher structures. This is especially true for high filled dams [3]. This phenomenon is summarized as "the calculated deformation too large for low dams and too small for high dams" [4].

© Springer Nature Switzerland AG 2018
W. Wu and H.-S. Yu (Eds.): *Proceedings of China-Europe Conference on Geotechnical Engineering*, SSGG, pp. 1726–1729, 2018.
https://doi.org/10.1007/978-3-319-97115-5_179

The above phenomenon is directly related with the height of the structure. It is categorized here as a type of structure size effect. What is the reason behind this structure size effect? In this paper attempt is made to give an explanation, through both theoretical analyses and numerical calculations.

2 Theoretical Thinking on the Mechanism of Structure Size Effect

As mentioned previously, when talking about size effect in relation with high filled structures, one thinks first the particle size effect, on which much research has been done in China and in the international world [5–7]. It is commonly realized that filling material of relatively larger particles may undergo larger deformation and the deformation developing period is also relatively long [6]. However, the structure size effect described above is obviously not the same as the particle size effect, since both the lower dams and the higher ones were constructed by using basically the same type of filling materials. Distinction must be made between these two types of size effect.

The most evident structure size effect should be the strength behavior of columns under compression. The strength of a column is dependent not only on the material strength but also on its slenderness ratio. But, with normally designed high filled structures there should be no bulking problem as columns under compression. The difference between a low dam and a higher one, as we can think out, should be that the stresses in a higher dam are larger, which may in turn influence the shear dilatancy and reduce the friction angle of the filling material due to particle breaking under high compressive stresses. But all these properties have been included, at least in some advanced calculation models.

With regard to structure size effect much research has been done in recent decades, but it is mainly concentrated on quasi-brittle structures [5], such as concrete structure members, ice plates, which are obviously different from high filled structures. Among the available theories, the Weibull theory [8] is well known. The physics behind this theory is that the larger the structure is, the more flaws exist in the structure.

Hunt et al. [9] once analyzed the influence of permeability uncertainty on the seepage through a finite size medium. For a regular flow net consisting of $N \times N$ flow channels, the average percolation will remain constant for any N values if all the flow channels are of the same permeability. If each channel has a certain probability to be blocked, then the average percolation will decrease with increasing N. The average percolation ability will be dramatically reduced with the increase of N when the failure probability of the flow channels reaches a critical value. This is actually a kind of structure size effect for seepage problem.

Enlightened by the above perceptions, we have made the inference that the structure size effect for high filled dams and slopes may be attributed to the variability of the mechanical properties of the filling material [10]. This inference will be verified through numerical analyses of some structures in the following.

3 Case Verification Through Numerical Calculations

Several types of structures have been calculated by using FEM to verify the above inference, and the material variability is considered through Monte Carlo method. Here only the results of a type of high filled slopes (Fig. 1), similar in deformation behavior to high filled dams, are presented to limit the length of the paper.

The slope ratio is 1:1.5. The well-known software PLAXIS is used for the deformation calculation. The material model used is the widely accepted hardening soil model. The model parameters for a certain proportion of elements are assumed to reduce and/or enhance from the normal values according to various given probability. The deformation of two different height slopes, $H = 50$ m and $H = 100$ m, is calculated, and then the *softening or hardening ratios* are obtained by dividing the calculated averaged displacements with those of the same slope calculated without any model parameter variation. The vertical displacement at the slope shoulder, point A (see Fig. 1), is taken for the comparison (Fig. 2).

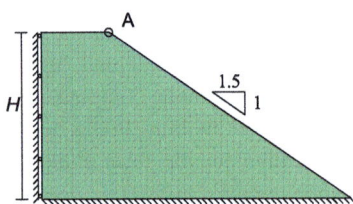

Fig. 1. The high fill slope

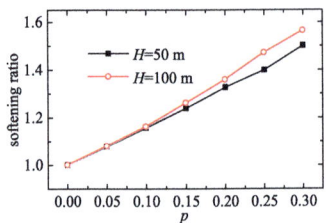

Fig. 2. Softening ratio via the material variation probability.

About the variation of the material parameters, four different schemes have been considered: (1) only reducing the material moduli, (2) reducing both the moduli and the strength parameters; (3) enhancing both the moduli and the strength parameters; (4) both reducing and enhancing of the material moduli and strength parameters by the same probability. Obviously, only the last scheme is close to reality, and the first three schemes are just for examining the influence of different variations of the material.

Table 1. Model parameters for the slopes.

Type	γ kN/m^3	E_{50}^{ref} MPa	E_{oed}^{ref} MPa	E_{ur}^{ref} MPa	m	ν	c kPa	φ	ψ
Normal values	18	30	30	90	0.5	0.2	30	33	0
Reduced values	18	10	10	30	0.5	0.2	20	26	0
Enhanced values	18	50	50	150	0.5	0.2	40	39	0

The material parameters are listed in Table 1. The results show that both scheme 1 and 3 give hardly any structure size effect. Namely, there is little difference in the calculated softening or hardening ratios for the low slope and high slope. In contrast, scheme 2 and 4, both involving strength parameter deduction, give obvious structure size effect, that is the softening ratio for high slopes is significantly larger for relatively greater variation probability of the material properties. Hence, it is the variation in material parameters closely related with plastic deformation which induces the size effect.

4 Concluding Remarks

The problem that the calculated deformation is often too large for lower dams and too small for higher dam, as noted in recent years' engineering practice in China is discussed. It is categorized as a kind of structure size effect and is believed to be a different concept from the particle size effect of the filling material.

Numerical studies are conducted which reveal that the variability of material properties can be the main cause of the structure size effect, especially the variation in parameters closely related with plastic deformation of the structure.

Further study along this line will consider the probability distribution of the material parameters, including their spatial correlations.

Acknowledgement. This research is finically supported by the National Basic Research Program of China (973 Program, 2014CB047003).

References

1. Ma, H., Chi, F.: Major technologies for safe construction of high earth-rockfill dams. Engineering **2**, 498–509 (2016)
2. Yang, Z., et al.: Research summary on safety and key technologies of 300 m-level face rockfill dam. Water Power **47**(9), 41–45 (2016). (in Chinese)
3. Cheng, Z., Pan, J.: Analysis of monitoring data of stress and deformation for Shuibuya concrete face rockfill dam. Chin. J. Geotech. Eng. **34**(12), 2299–2306 (2012). (in Chinese)
4. Chen, S.: Experimental techniques for earth and rockfill dams and their applications. Chin. J. Geotech. Eng. **37**(1), 1–26 (2015). (in Chinese)
5. Bazant, Z.: Size effect. Inter. J. Solids Struct. **37**, 69–80 (2000)
6. Alonso, E., Tapias, M., Gili, J.: Scale effects in rockfill behavior. Geotech. Lett. **2**(3), 155–160 (2012)
7. Kong, X., Liu, J., Zou, D.: Scale effect of rockfill and multiple-scale triaxial test platform. Chin. J. Geotech. Eng. **38**(11), 1941–1947 (2016). (in Chinese)
8. Weibull, W.: A statistical distribution function of wide applicability. J. Appl. Mech. **18**(3), 293–297 (1951)
9. Hunt, A., Ewing, R., Ghanbarian, B.: Percolation theory for flow in porous media, 3rd edn. Springer, Heidelberg (2014)
10. Kong, K.: Research on the long-term deformation of soil-rock mixtures in mountainous airports. Ph.D. thesis, Dept. of Civil Engineering, Tsinghua University, Beijing (2017)

Numerical Analysis of Stone Columns for Road Embankment Construction

Jakub Stacho[✉], Jana Frankovska, and Peter Mušec

University of Technology in Bratislava,
Radlinskeho 11, 810 05 Bratislava, Slovakia
jakub.stacho@stuba.sk

Abstract. The highway, R2, between cities Trencin and Prievidza in Slovakia was constructed under adverse geological conditions. Some sections of the motorway were designed on embankments and require the use of soil improvement because of the unsuitable subsoil. Original subsoil was improved using vertical drains, stone columns, dynamic compaction, preloading and stone drainage ribs. This paper deals with the numerical analysis of selected profiles where the subsoil for road embankment can be improved using stone columns. The subsoil consisted of compressible soft soils with low permeability. Stone columns reduced the settlement of a road embankment and reduce the time required for consolidation. Stone columns with a diameter of 600 mm and length of 5 m were designed in a square mesh with an axial distance of 2 m. Numerical modelling was executed with Plaxis software using FEM. Stone columns are a displacement technique which significantly changes properties of the original subsoil. The geotechnical monitoring included results of road embankment settlement using horizontal inclinometers, as well as geodetical measurements. The analysis included the comparison of road embankment settlement determined using measurements with results of numerical models, which assumed different ways of stone column modelling. The results showed that numerical modelling leads to the acceptable determination of road embankment settlement. Advantages and disadvantages of different ways of stone columns modelling are discussed.

Keywords: Stone column · Road embankment · Soil improvement

1 Introduction

The highway, R2, in the Ruskovce – Pravotice section between the cities of Trencin and Prievidza, was designed in some areas on an embankment and required the use of soil improvement because of a very soft subsoil. These areas were improved using vertical drains, stone columns, dynamic compaction, preloading and stone drainage ribs. This article is focused on numerical analysis of road embankment in cross-section KM 1.475, which is based on subsoil improved by stone columns. Geological conditions of the original subsoil are described in Fig. 1. The settlement of the embankment was measured using horizontal inclinometers, which were integrated immediately after installation of the stone columns and allowed analysis of the settlement in time. The design of the stone columns is a very extensive topic. Numerical modeling of the stone

© Springer Nature Switzerland AG 2018
W. Wu and H.-S. Yu (Eds.): *Proceedings of China-Europe Conference
on Geotechnical Engineering*, SSGG, pp. 1730–1733, 2018.
https://doi.org/10.1007/978-3-319-97115-5_180

columns was analyzed by many authors [1, 2]. Castro and Sagaseta [1] presented different ways of numerical modeling of stone columns, such as unit cell, transformation to walls for 2D plane strain models and homogenization of the subsoil. Wood et al. [3] presented numerical analyses which essentially confirmed results of many previous researchers, and focused on the interaction and failure mechanism of a single stone column, as well as a group of columns.

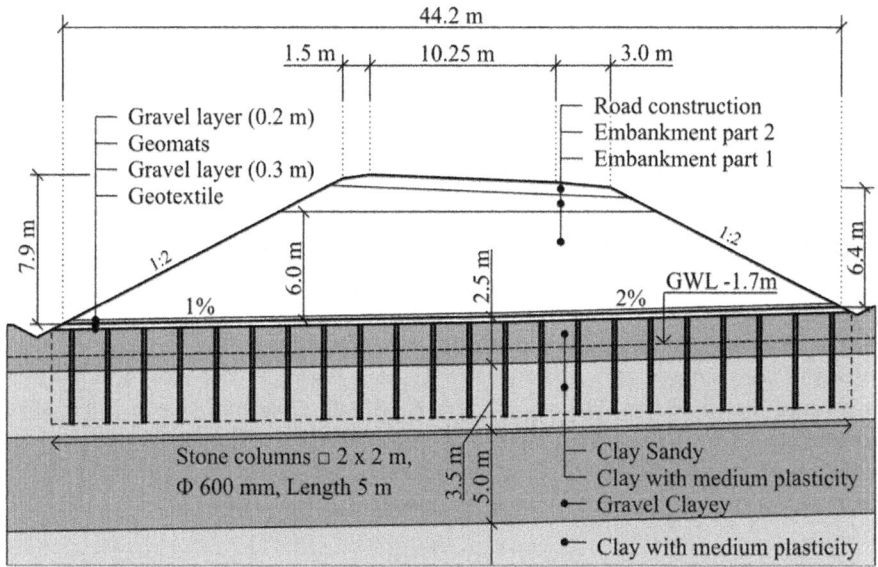

Fig. 1. Cross section of road embankment based on improved subsoil

2 Numerical Model of Road Embankment with Stone Columns

The numerical model of a road embankment constructed on improved subsoil was created using Plaxis software. A schematic of the numerical model is shown in Fig. 1. The model was created as a plane strain model, with a length of 130 m, height of 30 m and 6 node triangular elements. Soils were modeled using a hardening soil material model. The calculation was divided into 11 phases. All of the phases were defined as a consolidation of phases with respect to corresponding time. A function of updated mesh and updated pore pressures was used in all cases. The initial phase of the calculation represented the original geological conditions. The second one included installation of the stone columns. The following phases modeled lying of the geotextile, bearing layer of gravel, geomats and a second layer of gravel. The construction of an embankment body was modeled in two parts (height of 6 m and full height), according to records from building site.

The improved subsoil was analyzed using different numerical models, which took into account different ways of modeling stone columns. The following methods of modeling of stone columns were included:

- *Model 1:* Stone columns were transferred to 2D elements, compaction of subsoil between stone columns was considered, activation of drains and change of coefficient of filtration for 2D task were done according to [3];
- *Model 2:* All the same, but without reduction of coefficient of filtration for 2D task;
- *Model 3:* All the same, but without reduction of coefficient of filtration for 2D task and modeling of stone columns transferred to 2D elements;
- *Model 4:* Only soil improvement modeled as a "homogenized subsoil" in improved area;
- *Model 5:* Only modeling of homogenized subsoil with drains for reduction of consolidation time.

3   Results and Discussion

The analysis was primarily focused on comparison of time – settlement curves of the base of the road embankment. The results of analysis are presented in Fig. 2. Exact results of the analyses are summarized in Table 1. Measurement was recorded using horizontal inclinometers at specific times, which represent construction of an embankment to a height 6 m, consolidation of this part of embankment and construction of the second part of the embankment to the full height. The results of the analyses include results from 5 numerical models.

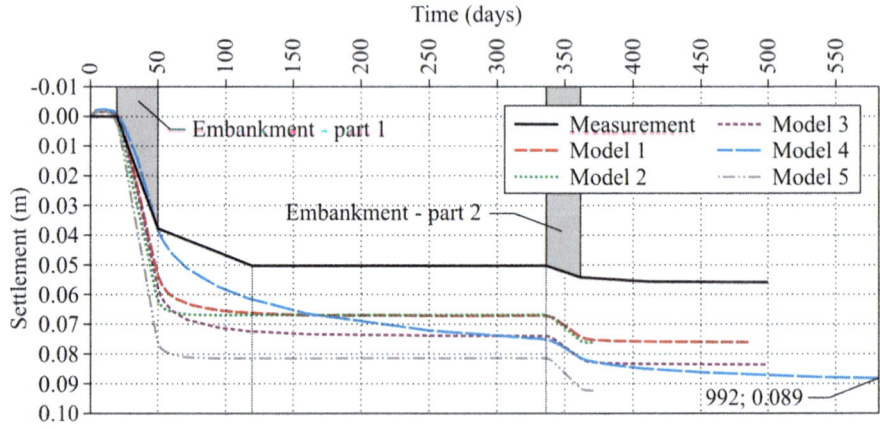

Fig. 2. Results of time – settlement curves determined by measurement and calculations

The deformation of the first part was steady in about 3 months, and deformation of second part was steady in about 50 days. The most optimal results of time - settlement curve were obtained using Model 1. Model 2 didn't consider reduction of the filtration

Table 1. Settlement of the road embankment determined in selected times

Measurement	Time (day)	50	120	337	362	412
	Settlement (mm)	37.7	50.2	50.2	54.2	55.8
Model 1	Time (day)	50	120	337	362	486
	Settlement (mm)	53.8	65.7	67.2	74.3	76.0
Model 2	Time (day)	50	120	337	362	370
	Settlement (mm)	62.1	67.0	70.0	75.5	76.0
Model 3	Time (day)	50	120	337	362	498
	Settlement (mm)	57.9	72.2	74.0	81.7	83.5
Model 4	Time (day)	50	120	337	362	992
	Settlement (mm)	38.4	61.2	75.2	81.5	89.4
Model 5	Time (day)	50	120	337	362	371
	Settlement (mm)	76.2	81.5	81.5	91.6	92.3

coefficient in the compaction zone, and the Model 3 didn't contain stone columns as 2D wall elements. Both models predicted less time required for consolidation and a bigger settlement. Similar results were obtained in the case of the Model 5. A bigger settlement was obtained because of disregarding improved soil properties. Model 4 included only modeling of the homogenized subsoil and the time required for settlement was very inaccurate.

4 Conclusion

The numerical modeling of soil improvement using stone columns in a plane strain model is a difficult task. Stone columns, among other things, cause improvement of soil properties of the original subsoil and reduce time required for consolidation. This article presents results of a settlement analysis of a road embankment along the highway R4 in Slovakia. The road embankment was based on stone columns. Geotechnical monitoring included measurements of the settlement of the embankment, and allowed its consolidation analysis. The most optimal results were obtained using a numerical model which included modeling of stone columns transferred to 2D walls, activation of drains. It also included changing soil properties between stone columns, which were affected by installation of the stone columns using the vibro replacement technique, as well as reduction of the filtration coefficient in the smear zone.

References

1. Castro, J., Sagaseta, C.: Consolidation and deformation around stone columns: numerical evaluation of analytical solutions. Comput. Geotech. **38**(3), 354–362 (2011)
2. Wood, D.M., Hu, W., Nash, D.F.T.: Group effects in stone column foundations: model tests. Geotechnique **50**(6), 689–698 (2000)
3. Tran, T.A., Mitachi, T.: Equivalent plane strain modeling of vertical drains in soft ground under embankment combined with vacuum preloading. Comput. Geotech. **35**, 655–672 (2008)

Using the FE-Method for Lock Repair Measures at the Main-Danube-Canal

Oliver Stelzer[(✉)] and Annette Richter

Federal Waterways Engineering and Research Institute (BAW),
Karlsruhe, Germany
{oliver.stelzer,annette.richter}@baw.de

Abstract. This paper focuses on the experiences and benefits of using the Finite Element Method (FEM) in order to describe soil-structure interaction behavior taking the example of two successfully completed lock repair projects. It turned out that the influences and interactions of the repair works on the existing lock, in particularly the deformation of the lock, earth pressures acting on the lock and structural forces in different lock sections, could only be determined in a realistic way by calculations with the FEM. Based on the calculation results, limit values for the monitoring system, which is necessary to check the system behavior and to verify the calculation assumptions during the repair works, were defined. An additional focus is on the relevance of small-strain stiffness in the analyses. For this purpose, an extension of the Hardening Soil model (HS-model) that accounts for a higher stiffness at small strains was used.

Keywords: Navigable lock · Finite element calculation
Soil-structure interaction · Small-strain stiffness
Monitoring system · Deep pit

1 Introduction

The connection between the North Sea and the Black Sea, the Main-Danube-Canal (MDC), is an important waterway in Europe with 16 modern locks. A structural inspection of Bamberg Lock, carried out by the German Waterways and Shipping Authority in 2004, revealed severe damage to the eastern chamber wall. After discovery of this damage, intensified inspections and structural investigations of the other MDC-locks were conducted. They showed that further repairs to other locks like Eibach Lock are necessary and that two locks (Kriegenbrunn and Erlangen) even have to be completely reconstructed adjacent to the existing lock.

The two projects discussed here are Bamberg and Eibach Lock. They have in common that the lock has to be in operation while repairing and that the stresses on the damaged lock may not exceed beyond the existing level during the repair. Soil-structure interaction related issues and the use of the FEM played an important role during the planning phase.

Meanwhile, the lock projects Bamberg and Eibach are successfully completed while the projects Kriegenbrunn and Erlangen are still in the execution planning.

© Springer Nature Switzerland AG 2018
W. Wu and H.-S. Yu (Eds.): *Proceedings of China-Europe Conference
on Geotechnical Engineering*, SSGG, pp. 1734–1737, 2018.
https://doi.org/10.1007/978-3-319-97115-5_181

2 Bamberg Lock

The damage detected at Bamberg Lock comprised pronounced cracking near the drainage channel of the eastern chamber wall extending downwards to the longitudinal channel (Fig. 1, left). The earth-side reinforcement close to the drainage channel was fractured.

As short-term emergency measures, the top of the chamber wall was rear-anchored to a sheet pile wall with tendons, the fractured reinforcement was reactivated by means of sleeves and a deformation monitoring system was installed.

For long-term stability the construction of an earth-side concrete section to strengthen the chamber wall was needed to repair the lock. A pit excavated to a depth of around 17 m was required to enable the additional concrete section to be constructed (see Fig. 1, right). When the lock is full (load case: upper water level, UWL) the chamber wall leans towards the wall of the construction pit and is supported by bracing struts. When the lock is empty a total of five rows of pre-stressed ground anchors ensured that the stresses, acting on the chamber wall - due to the deformation of the retaining wall during excavation of the pit, remain at tolerable levels.

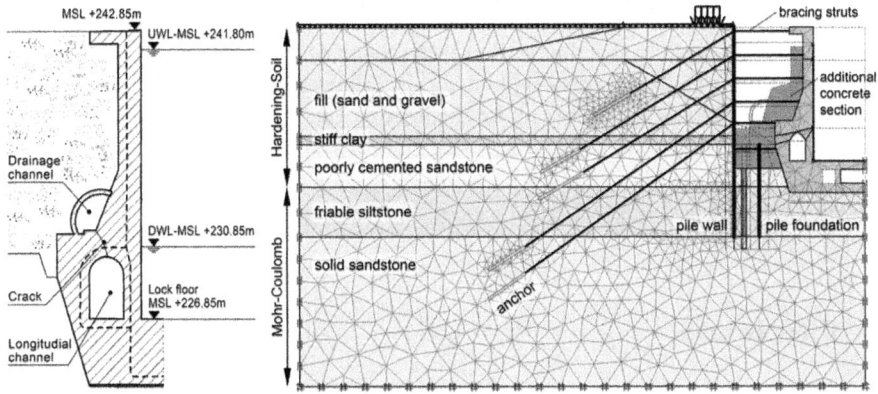

Fig. 1. Part of the Cross-Section of Bamberg Lock (left) and calculation model with construction pit and soil strata [1]

A calculation model, which takes the mechanical connection between the in-situ soil, the wall of the construction pit and the lock into account, was established in the course of drawing up the concept for the construction pit. Owing to the complex interactions involved, it was necessary to apply the FEM to achieve as realistic an estimate of the earth pressures, of the strut forces and of the chamber wall deformations as possible. The calculations were performed with a 2D-model using the Plaxis FE-software. The HS-model was applied as the constitutive law for the fill, clay and poorly cemented sandstone. A characteristic feature of the model is that it considers the dependence of the soil stiffness on initial and un-/reloading and includes a stress-dependent definition of the stiffness. The linear-elastic ideal-plastic constitutive law according to Mohr-Coulomb was applied to the rock strata beneath the base of the lock. In order to take into account, the complete loading history and the fluctuations in the loads between the lower and upper water levels in the lock (initial and repeated loading) totally 133 phases were calculated in the FE-model.

For automated structural verifications of the existing lock during the repair a second calculation model of the lock without the surrounding soil was used. The earth pressure distribution and strut forces acting on the lock wall in the different construction phases were taken as input from the results of the Plaxis-model.

The deformation measurements, taken at the structure prior to the commencement of the repair work, have proved particularly valuable for verifying the soil parameters. They enabled the deformations of the top of the chamber wall, measured in-situ at different water levels in the lock, to be compared with the deformations calculated with the finite element model. It was therefore possible to check how well the calculation model reflects the actual conditions.

Owing to the difficulties involved in executing the construction work while the lock is in operation, a monitoring system was installed during the repair works to check the behavior of the system predicted in the calculations. It was therefore possible to respond promptly to any deviations from the forecast behavior (observational method). Emergency plans for a variety of scenarios were developed accordingly. The principal components of the monitoring system were the measurement of the horizontal deformations at the top of the chamber wall in each section of the chamber with an automatic tachymeter and of all strut forces with load cells. The measurements were continuously evaluated as the work progressed to ensure that the limit values specified in the structural design were not exceeded. The comparison showed sufficient agreement between the calculations and the measurements. Details on the repair solution, the construction process, the numerical model, the monitoring system and the measurements are presented in [1].

3   Eibach Lock

The second project dealt with is Eibach Lock. First preliminary structural calculations showed a safety deficit in the eastern chamber wall. To simulate the deformation behavior of the lock in a more realistic way a 2D FE-model was created (Fig. 2, left). The soil stiffness parameters for the soil and rock layers were determined from laboratory (oedometer) and field tests (borehole widening) and were used as input for the model. It was obvious that the wall backfill plays an important role for the eastern wall deformation.

Like at Bamberg Lock the horizontal deformations of the wall tip were measured for different water levels in the lock chamber, which represents a well-defined load situation. When the lock is filled, the horizontal deformation of the eastern wall tip is about 4 mm, the western wall tip deforms about 12 mm. The results of the calculations using the HS-model exhibited deformations of the eastern chamber wall which were 3 times higher than the measured values. For the western wall, however, there was a good agreement between measurement and calculation, as this wall is not backfilled and thus less affected by soil behavior.

A possible reason for the deviation for the eastern wall deformation was neglecting the stiffer soil behavior at small strains, in particular in the wall backfill. To account for small-strain stiffness the HSsmall-model was used for the backfill material, which is an extension of the HS-model. The small-strain stiffness parameter E_0 necessary therefore was taken from surface wave seismic field tests (Fig. 2, right). Details on soil investigations and the numerical model used are given in [2].

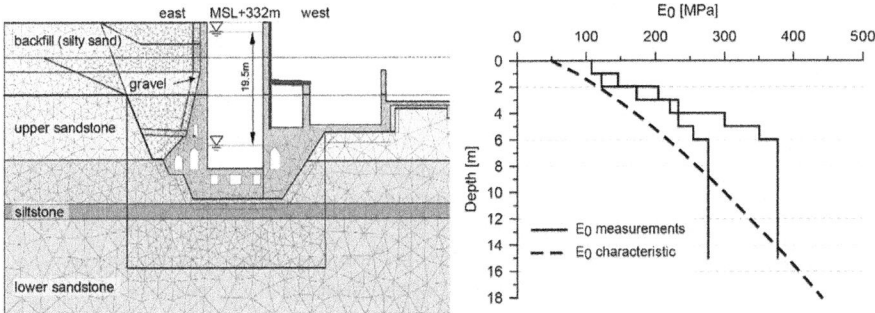

Fig. 2. FE-model of Eibach Lock (left) and small-strain stiffness from seismic field tests [2]

The comparison of wall deformation revealed that the HSsmall-model showed a stiffer deformation behavior and therefore a better agreement with the measured wall-tip deformation. The use of the HSsmall-model improved both the quality and the trustworthiness of the numerical calculations, with the consequence that the cyclic load safety deficits were no longer as high as the initial calculations had predicted. The measures to recondition the lock structure could therefore be optimized. Essentially a part of the backfill behind the wall was replaced by lightweight concrete to reduce the load acting on the wall.

4 Summary

The FEM was successfully used to model the complex interaction between the structure, the construction pit and the ground when planning the repairs to Bamberg Lock. It was subsequently possible to draw up a numerical prediction of the system behavior to serve as a reference when monitoring the construction work by the observational method. For Eibach Lock it was possible to optimize measures to recondition the lock on the basis of FEM results using advanced soil models.

References

1. Stelzer, O., Espert, M., Schum, S.: Instandsetzung der Schleuse Bamberg unter laufendem Schifffahrtsbetrieb. Geotechnik **3**(4), 241–249 (2010)
2. Stelzer, O., Kauther, R.: Relevance of small-strain stiffness in the deformation analysis of navigable locks. In: Proceedings of the 5th International Symposium on Deformation Characteristics of Geomaterials, pp. 1231–1238. IOS Press, Seoul (2011)

Performance Evaluation of Ballast-Subballast Interface Stabilized with Geogrids

Kumari Sweta[(⊠)] and Syed Khaja Karimullah Hussaini

Department of Civil and Environmental Engineering,
Indian Institute of Technology Patna, Patna, India
{sweta.pce13,hussaini}@iitp.ac.in

Abstract. Large-scale direct shear tests were carried out to explore the shear behavior of the ballast-subballast interface with and without the inclusion of geogrids. Fresh granite ballast and subballast with an average particle sizes (D_{50}) of 42 mm and 3.5 mm respectively, and triaxial geogrids with different aperture sizes were used in this study. Tests were performed at different normal stresses (σ_n) ranging from 20 to 100 kPa at a constant shearing rate (S_r) of 2.5 mm/min. The experimental results reveal that the shear strength of the ballast-subballast interface improved significantly with the inclusion of geogrids. The interface efficiency factor (α), defined as the ratio of the shear strength of ballast-geogrid-subballast interface to the shear strength of ballast-subballast varies from 1.15 to 1.17 and the ballast-geogrid-subballast friction angle (φ) lies in the range of 51.1° to 67°. These test results emphasize the role of triaxial geogrids in stabilizing the ballasted rail tracks and thus reducing the maintenance cost.

Keywords: Ballast-Subballast interface · Triaxial geogrid · Friction angle

1 Introduction

The ballasted railway track comprises of ballast, subballast and the subgrade. While the ballast distributes the applied train load uniformly from the sleepers to the subgrade, the subballast acts as a separation layer between the ballast and the subgrade. However, both ballast and subballast are the granular layers and they undergo deformation and deterioration upon continuous train loading that subsequently leads to track irregularities and instability. Therefore, it is needed to strengthen the ballast-subballast interface in order to maintain track stability.

Several researchers have highlighted the role of geogrids in improving the interface shear strength of ballast using direct shear apparatus [1–3]. Of the various shape geogrids available in the market, Dong et al. [4] established the triaxial geogrid to be more efficient than that of biaxial geogrids. However, the studies conducted by Indraratna et al. [1] established the triangular aperture geogrid (36 × 36 mm) to be less effective than that with rectangular apertures, primarily due to its smaller aperture size. Hussaini et al. [2] further envisaged that triaxial geogrids with bigger aperture size might work well for stabilizing rail ballast. Subsequently, the recent investigation done by Sweta and Hussaini [3] established that triangular geogrids with bigger aperture size (69 × 69 mm) is more efficient on ballast than the biaxial geogrids. While th

© Springer Nature Switzerland AG 2018
W. Wu and H.-S. Yu (Eds.): *Proceedings of China-Europe Conference on Geotechnical Engineering*, SSGG, pp. 1738–1741, 2018.
https://doi.org/10.1007/978-3-319-97115-5_182

performance of triangular aperture geogrids as in case of ballast has already been studied, its effectiveness when used at the ballast-subballast interface has not been explored. Therefore, the current study is conducted to evaluate the effectiveness of triangular aperture geogrids on ballast-subballast interface.

2 Materials and Methods

Laboratory tests were conducted using large-scale direct shear apparatus having a size of 450 mm × 450 mm and the overall depth of 300 mm. The material used for ballast was granite and the subballast was a mixture of crushed granite and sand. The PSD of both ballast and subballast adapted in the current study (Fig. 1) are as per the guidelines provided by the Indian Railways [5, 6].

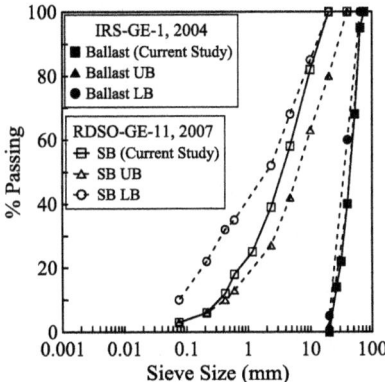

Fig. 1. Particle size distribution of ballast and subballast used in the current study

Table 1. Physical and technical specifications of geogrids used in the current study.

Geogrid type	Aperture type	t_r (mm)	Aperture size (mm)	T_{ult} (kN/m)
G2	Triangular	1.5	46 × 46	19
G4		2.0	69 × 69	21

The specifications of triangular aperture geogrids (labeled G2 and G4) are listed in Table 1. The maximum and average particle sizes (D_{max} and D_{50}) of ballast are 65 and 42 mm and that of subballast are 20 and 3.5 mm respectively. The sample was prepared by the thorough mixing of the sieved ballast and subballast separately as per the gradation curve (Fig. 1). The mixed subballast was placed in the lower box and compacted in two layers to achieve a required density of 2000 kg/m³. Thereafter, a layer of geogrid was placed at the interface of upper and lower boxes of shear box. Subsequently, the mixed ballast was placed in the upper box followed by compaction in two layers to achieve a desired density of 1470 kg/m³. The tests were conducted up to a shear displacement of 67.5 mm at different normal stresses (σ_n) of 20, 35, 70 and 100 kPa and at constant shearing rate (S_r) of 2.5 mm/min.

3 Results and Discussion

3.1 Shear Characteristics of Ballast-Subballast Interface With and Without Geogrids

The effect of geogrid on the shear behavior of the ballast-subballast interface is shown in Fig. 2a. The inclusion of both the geogrids $G2$ and $G4$ increases the stress ratio (τ/σ_n) when compared to the unreinforced ballast-subballast interface (Fig. 2a). This may due to the interlocking of ballast particles with the aperture of the geogrids. Further, it is observed that the inclusion of geogrids reduces the extent of dilation. A similar kind of behavior was observed for unreinforced and reinforced ballast in past [1–3].

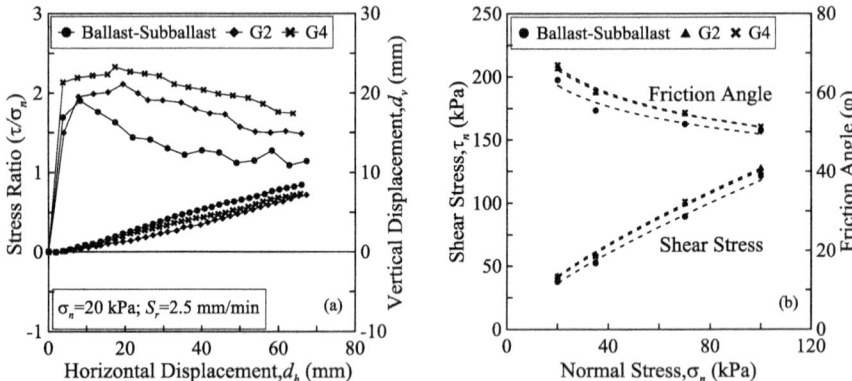

Fig. 2. Variation of (a) stress ratio (τ/σ_n) and vertical displacement (d_v) with horizontal displacement (d_h) for unreinforced and reinforced ballast-subballast interface, (b) shear stress (τ_n) and friction angle (φ) with normal stress (σ_n) for unreinforced and reinforced ballast-subballast interface

Effect of Applied Normal Stress (σ_n) on the Shear Strength of Ballast-Subballast
The effect of applied normal stress (σ_n) on the shear strength of ballast-subballast interface is shown in Fig. 2b. The shear strength in case of both unreinforced and geogrid-reinforced ballast-subballast interface follows a non-linear trend with the increase in σ_n. However, the shear strength of the reinforced interface is found to be higher than unreinforced conditions with $G4$ performing better than that of $G2$. This implies that triangular geogrid with bigger aperture sizes is more beneficial in stabilizing ballast-subballast interface. It is further observed that friction angle (φ) of unreinforced ballast-subballast decreases from $63.2°$ to $50.3°$ as σ_n increases from 20 to 100 kPa. The apparent friction angle (δ) of the said interface when reinforced with geogrid $G2$ and $G4$ decreases from $66.2°$ to $51.1°$ and $67°$ to $51.3°$ respectively for same values of σ_n.

3.2 Interface Efficiency Factor (α)

The effectiveness of geogrid in strengthening the ballast-subballast interface is evaluated in terms of interface efficiency factor (α) (Table 2). The value of α for both the geogrids $G2$ and $G4$ is more than unity indicating that they both are effective in stabilizing the ballast-subballast interface. However, $G4$ exhibits a slightly higher value of α of 1.17 in comparison to 1.15 in case of $G2$. This is primarily because the geogrid $G4$ having bigger aperture size allows better interlocking with the particles and thus, provides higher strength to the ballast-subballast interface.

Table 2. Interface efficiency factors for ballast-subballast interfaces.

Geogrid type	Interface efficiency factor (α)
$G2$	1.15
$G4$	1.17

4 Conclusions

The current study presented the behavior of ballast-subballast interface reinforced with triangular aperture geogrids. The internal friction angle (φ) of unreinforced interface decreased non-linearly from 63.2° to 50.3° as σ_n was increased from 20 to 100 kPa. A similar trend was seen in case of geogrid-reinforced ballast-subballast interface. However, the shear strength of ballast-subballast reinforced with geogrids was found to be greater than that of unreinforced ballast-subballast. The value of α was found to be 1.15 in case of ballast-subballast interface with $G2$ and 1.17 with geogrid $G4$ indicating that $G4$ with bigger aperture size is more suitable as reinforcement at ballast-subballast interface.

References

1. Indraratna, B., Hussaini, S.K.K., Vinod, J.S.: On the shear behavior of ballast- geosynthetic interfaces. Geotech. Test. J. **35**(2), 305–312 (2012)
2. Hussaini, S.K.K., Indraratna, B., Vinod, J.S.: Performance of geosynthetically reinforced rail ballast in direct shear conditions. In: Narsilo, G.A., Arulrajah, A., Kodikara, J. (eds.) 11th Australia New Zealand Conference on Geomechanics: Ground Engineering in a Changing World, Engineers Australia, Australia, pp. 1268–1273 (2012)
3. Sweta, K., Hussaini, S.K.K.: Effect of shearing rate on the behavior of geogrid-reinforced railroad ballast under direct shear conditions. Geotext. Geomembr. **46**, 251–256 (2018)
4. Dong, Y.L., Han, J., Bai, X.H.: Numerical analysis of tensile behavior of geogrids with rectangular and triangular apertures. Geotext. Geomembr. **29**, 83–91 (2011)
5. IRS-GE-1: Specifications for Track Ballast. Research Design and Standard Organisation (RDSO), Ministry of Railways, India (2004)
6. RDSO-11-GE: Guidelines for blanket layer provision on track formation. Design and Standard Organisation (RDSO), Ministry of Railways, India (2007)

The Relation Between Static and Dynamic Shakedown Limits of Slab Track Substructures Under Moving Train Loads

Juan Wang and Shu Liu[(⊠)]

Ningbo Nottingham New Materials Institute,
University of Nottingham Ningbo China, Ningbo 315100, China
shu.liu@nottingham.edu.cn

Abstract. Shakedown limits can be used to predict different long-term responses of geo-structures under repeated moving loads. Based on lower-bound shakedown theorems, static and dynamic shakedown limits were obtained respectively for slab substructures under train loads of various speeds. The determination of the dynamic shakedown limit is relatively complicated due to the calculation of stable dynamic elastic stresses in the substructures. It was found the change of the dynamic shakedown limit from the corresponding static one is highly dependent on soil friction angle and a velocity factor (train speed over critical speed of the substructure). A simple relation between the dynamic shakedown limits at various speeds and the static one is then proposed which can serve as an efficient approach to estimate the influence of the train speed on the long-term performance of substructure.

Keywords: Shakedown · Slab track substructures · Critical speed

1 Introduction

Slab tracks have been widely used for high speed railways. The good performance of a slab track requires very limited post-construction settlement or differential settlement of track substructure which is a consequence of permanent deformation of granular materials and soils. Train speed has a significant influence on the long-term response of the substructure. Typically, the influence of the train speed is considered in design standards by multiplying the static axle load by an amplification factor. These amplification factors were suggested mainly considering the increase of the elastic vertical stress at a critical point due to the raised train speed. However, the permanent deformation of the material is related to the plastic properties of the materials.

Shakedown limit is a load limit, below which a stable long-term response of a geo structure to a repeated moving load is expected [1]. The shakedown limit can be predicted using shakedown analysis [2–4]. For the problem of slab track substructures, shakedown limits can be calculated considering quasi-static or dynamic situations. This study will present static and dynamic shakedown limits for typical substructures as well as their relations.

© Springer Nature Switzerland AG 2018
W. Wu and H.-S. Yu (Eds.): *Proceedings of China-Europe Conference on Geotechnical Engineering*, SSGG, pp. 1742–1745, 2018.
https://doi.org/10.1007/978-3-319-97115-5_183

2   Typical Substructure and Train Loads

A typical substructure composed of an anti-frozen layer, a prepared subgrade layer and subsoil is used in this study. Four axle loads belonging to two adjacent bogies on two carriages are considered. Each axle load is denoted by λP, where P is a unit axle load and λ is a load factor. The train loads are applied on the superstructure which acts as a single beam on a Winker foundation. Using the Beam on Elastic Foundation Analysis, the train loads are converted into a distributed load on the surface of the substructure [5]. It is assumed the distributed load moves at a constant speed along x-direction. The transverse and vertical directions are denoted as y-direction and z-direction respectively (Table 1).

Table 1. Material properties.

Layer name	h (m)	E_d (MPa)	v	ρ (kg/m^3)	ϕ ($^\circ$)	c (kPa)
Anti-frozen layer	0.4	290	0.3	2000	50	1
Prepared subgrade	2.3	190	0.3	1850	40	2
Subsoil	∞	160	0.3	1800	0,15,30	2

3   Shakedown Solutions and Discussions

Static and dynamic shakedown limit of the substructure can be obtained by solving a unified mathematical optimization problem [5–7]:

$$\max \lambda$$
$$s.t. \begin{cases} f\left(\sigma^r_{xx}(\lambda\sigma^e), \lambda\sigma^e\right) \leq 0 \text{ forallpoints} \\ \sigma^r_{xx}(\lambda\sigma^e) = \sigma^r_{xx-l} \text{ or } \sigma^r_{xx-u} \end{cases} \qquad (1)$$

where σ^e is elastic stress fields induced by a unit train load, σ^r_{xx} is a residual stress field; and f is the yield criterion of the material. The maximum load factor satisfying this equation then gives the shakedown limit of the problem $\lambda_{sd}P$. For the static shakedown analysis, the elastic stress fields can be calculated relatively easily by using a static train load assuming low travelling speed; while the dynamic shakedown analysis requires dynamic elastic stress fields fulfilling the dynamic equilibrium conditions. In the latter case, a large finite element model with an artificial boundary to absorb body waves is needed [5, 8]. It has been found that the elastic stress fields will change with the train speed. However, at any specific speed, elastic stresses only change with the source (i.e. load) location and therefore a stabilized dynamic elastic stress field can be obtained.

3.1   Effect of Train Speed and Elastic Modulus

Shakedown limits for this typical slab track substructure are shown as the solid line in Fig. 1(a). The static shakedown limit is 452 kN/axle, and the dynamic shakedown limits are reduced from it with increasing train speed. When the train speed exceeds the shear wave velocity of the subsoil, the dynamic shakedown limit is recovered slightly. This means this shakedown limit plot is able to capture the critical speed of the substructure V_{cr}. Figure 1(a) also shows the shakedown limits of a modified case where the elastic modulus of each layer is reduced to 44% of the original case. It is found that when the shakedown limits are plotted against a velocity factor (Eq. 2) instead, the shakedown limits of the two cases are identical. As a result, the shakedown limit is dependent on the velocity factor and the layer elastic modulus ratio.

$$\alpha = V/V_{cr} \tag{2}$$

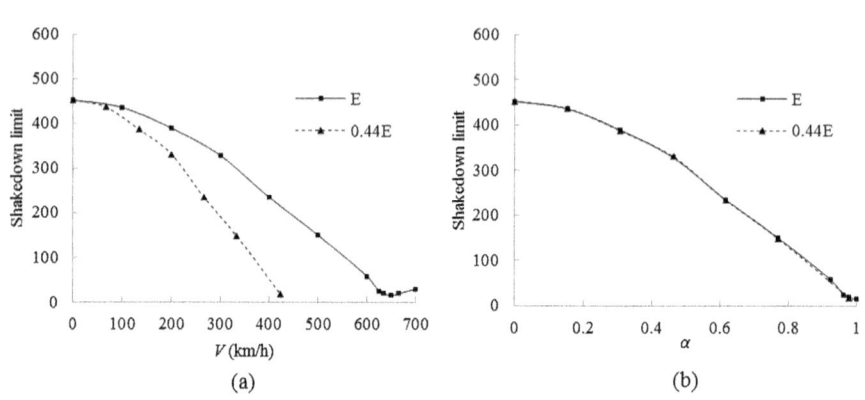

Fig. 1. Shakedown limit against train speed.

3.2 Effect of Friction Angle and Attenuation Factor

Figure 2(a) demonstrates that decreasing friction angle of the subsoil leads to reducing static and dynamic shakedown limits. The deviation of the dynamic shakedown limit from the static value for cases with different friction angles and velocity factors are exhibited in Fig. 2(b) in which η is an attenuation factor, defined as the shakedown limit of the current velocity factor over that of the static case. A higher friction angle leads to a lower attenuation factor. Therefore, though the high friction angle has a positive effect on the long-term performance of the track substructure, one should be very careful when trying to increase the train speed at those cases.

The attenuation factor barely changes when the velocity factor is smaller than 0.1; otherwise it can be approximated using the equation on Fig. 2(b) where n is a coefficient depending on the friction angle of the subsoil. Therefore, Eq. 2 can be used to predict the shakedown limit at any velocity factor once the static shakedown limit $\lambda_{sd-s}P$ is known. The value of n varies between 1.2 and 1.8 for $0 \leq \phi \leq 30°$. Note this equation as well as the n-values also apply to many other cases including the case of a homogenous half-space. This equation allows an efficient estimation of the dynamic shakedown limit. As a result, the influence of train speed on the long-term response of the slab track substructure can be quantified in a relatively rational manner

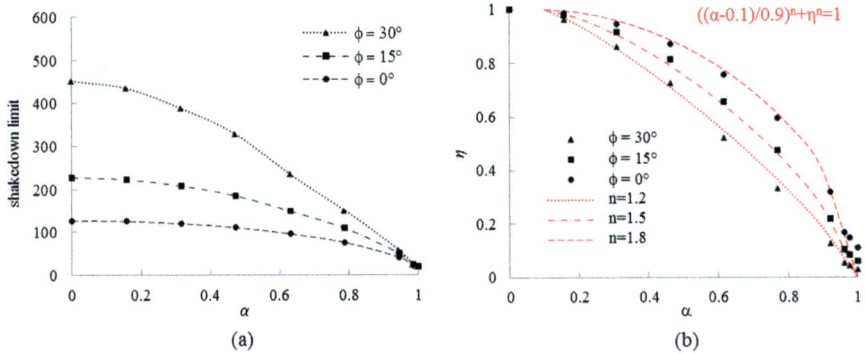

Fig. 2. Shakedown limit and attenuation factor against velocity factor.

$$\lambda_{sd}P = \eta\lambda_{sd-s}P \tag{3}$$

$$\text{where } \eta = \begin{cases} = 1 & \text{for } \alpha \leq 0.1 \\ = \sqrt[n]{1 - \left(\frac{\alpha-0.1}{0.9}\right)^n} & \text{for } 0.1 < \alpha < 1 \end{cases} \tag{3}$$

References

1. Yu, H.S.: Plasticity and Geotechnics. Springer, UK (2006)
2. Boulbibane, M., Weichert, D.: Application of shakedown theory to soils with non associated flow rules. Mech. Res. Commun. **24**(5), 513–519 (1997)
3. Collins, I.F., Wang, A.P., Saunders, L.R.: Shakedown in layered pavements under moving surface loads. Int. J. Num. Anal. Methods Geomech. **17**(3), 165–174 (1993)
4. Ponter, A.R.S., Chen, H.F., Ciavarella, M., Specchia, G.: Shakedown analyses for rolling and sliding contact problems. Int. J. Solids Struct. **43**(14–15), 4201–4219 (2006)
5. Wang, J., Liu, S., Yang, W.: Dynamics shakedown analysis of slab track substructures with reference to critical speed. Soil Dyn. Earthquake Eng. **106**, 1–13 (2018)
6. Yu, H.S., Wang, J.: Three-dimensional shakedown solutions for cohesive-frictional materials under moving surface loads. Int. J. Solids Struct. **49**(26), 3797–3807 (2012)
7. Liu, S., Wang, J., Yu, H.S., Wanatowski, D.: Shakedown for slab track substructures with stiffness variation. Geotechnical research (2018). https://doi.org/10.1680/jgere.17.00018
8. Kouroussis, G., Parys, L.V., Conti, C., Verlinden, O.: Using three-dimensional finite element analysis in time domain to model railway-induced ground vibrations. Adv. Eng. Softw. **70**(2), 63–76 (2014)

Modelling Stress Distribution in Subgrade Due to Construction of Enlarged Embankment

Heng Wang, Cong Mou, Jianwen Ding[(✉)], and Xing Wan

Institute of Geotechnical Engineering, School of Transportation,
Southeast University, Nanjing 210096, People's Republic of China
jwding@seu.edu.cn

Abstract. Many exiting embankments along river have been enhanced by enlarged construction. This paper proposed a practical method of determining enlarged embankment loads induced vertical superimposed stress in subsoil by considering the effect of existing embankment and variations in foundation properties underneath existing and enlarged embankment. The Stochastic stress diffusion theory is used to establish an equation of calculating contact stress for overcoming the assumptions of traditional trapezoidal stress distribution. The elastic solution is used to calculate the vertical superimposed stress. An empirical coefficient is employed for incorporating the effect of variations in properties of subsoil. The method was applied to analyze one case in Korean. Comparisons of the calculated values with field data indicate that the proposed method is useful for designing the enlarged embankment on soft subsoil.

Keywords: Enlarged embankment · Contact stress · Properties of subsoil

1 Introduction

Embankments constructed on soft soils are often reinforced by widening and elevating the height and often encountered with differential settlement problems [1, 2]. The differential settlements depend on contact stress (σ_b) and vertical superimposed stress in subsoil (σ_s).

Note that the values of σ_b were determined by equivalent trapezoidal stress distribution, this approach was established based on the assumptions of homogeneity and flexible loaded area along the base of embankments [1]. Importantly, the loaded area cannot be considered as perfectly flexible, and the time and properties difference between the existing and enlargement embankments must be taken into account [2].

The σ_s were calculated by the Osterberg method, which was based on linear elastic theory, but the real soil is elastoplastic [1, 3]. The stress distribution in subsoil are sensitive to subsoil properties [4]. In practice, considering the difficulty in situ test under existing embankment, only the subsoil properties of the enlarged area are used for overall design without considering the variations in subsoil properties due to consolidation under old embankment loads [2].

The method considering the effect of existing embankment and variations in foundation properties underneath existing and enlarged embankment is proposed

© Springer Nature Switzerland AG 2018
W. Wu and H.-S. Yu (Eds.): *Proceedings of China-Europe Conference on Geotechnical Engineering*, SSGG, pp. 1746–1749, 2018.
https://doi.org/10.1007/978-3-319-97115-5_184

A simple equation was suggested for calculating the contact stress. Then, the vertical superimposed stress was calculated using the elastic solution. Application of the proposed equation to case history in Korean is described.

2 Methodology

Instead of assuming homogeneity and continuity of material, Harr [5] proposed an equation of calculating vertical stress as:

$$\sigma(x,z) = P^* \left\{ \psi \left[\frac{x+a}{z\sqrt{v}} \right] - \psi \left[\frac{x-a}{z\sqrt{v}} \right] \right\} \tag{1}$$

Figure 1 shows the cross-section view of the typical enlargement embankment.

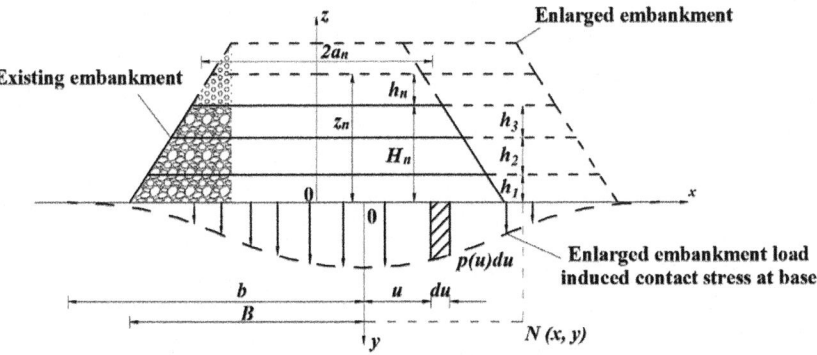

Fig. 1. The cross-section view of the typical enlargement embankment.

Moreover, considering the effect of existing embankment, the proposed equation by Kandaurov was introduced:

$$\overline{h}_{n-1} = h_1 \sqrt{\frac{v_1}{v_n}} + h_2 \sqrt{\frac{v_2}{v_n}} + \ldots + h_{n-1} \sqrt{\frac{v_{n-1}}{v_n}} \tag{2}$$

Thus, the proposed equation for calculating σ_b was given as:

$$\sigma_b(x,z) = \sum_1^n \gamma h_n \cdot \left[\psi \left(\frac{x+a_n}{\left(\overline{h}_{n-1} + H_n \right) \cdot v_n} \right) - \psi \left(\frac{x-a_n}{\left(\overline{h}_{n-1} + H_n \right) \cdot v_n} \right) \right] \tag{3}$$

Then elastic method and empirical coefficient R^* are introduced for determining σ_s

caused by σ_b:

$$\sigma_s = R^* \cdot \int_{-b}^{b} \frac{2y^3 \sigma_b}{\pi((u-x)^2 + y^2)^2} du \tag{4}$$

Note that R^* was related to mechanical property of soil in considering the varied properties of subsoil.

3   Analyses of Case History

3.1   Description of Centrifuge Model

The case of centrifuge model of enlarged embankment in Korean was used to validate the proposed method [2]. The detail of the enlarge embankment is shown in Fig. 2. The thickness of foundation is 28.4 cm. The mechanical behavior of the foundation after the centrifuge test were listed in Table 1.

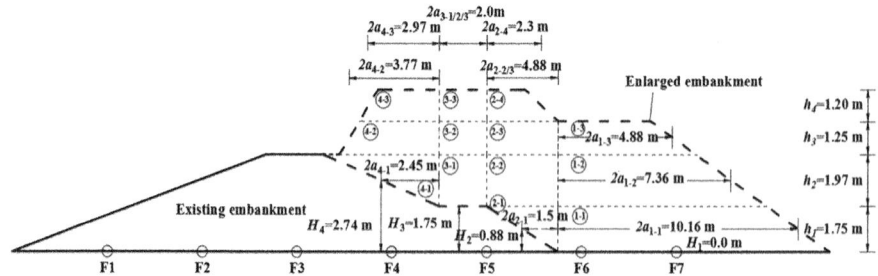

Fig. 2. Cross-sectional profiles of the centrifuge model enlargement embankment.

Table 1. Mechanical properties of foundation materials after consolidation tests

Section	F2	F3	F4/F5	F6/F7
Cc	0.46	0.43	0.41	0.48

3.2   Settlement Result

The measurements of settlement are used to verify the Eq. (4) of calculating σ_s. For normally consolidated clays, the consolidation settlement can be calculated using following Eq. (5), in which the σ_s was determined by Eq. (4).

$$S = \frac{H^*}{1 + e_0} C_c \log_{10} \frac{\sigma'_{v0} + \sigma_s}{\sigma'_{v0}} \tag{5}$$

It can be seen that obvious difference between the proposed and Osterberg method in calculating vertical superimposed stresses in Fig. 3(a). Figure 3(b) shows the comparisons between calculated settlements and field measurements. The predicted settlements using the proposed equation approximate the field observations, indicating the validity of the proposed methods of determining σ_s and σ_b.

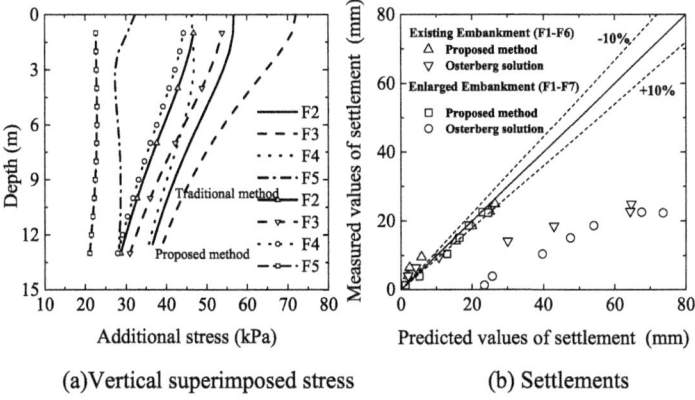

(a)Vertical superimposed stress   (b) Settlements

Fig. 3. Calculated result

4   Conclusions

A practical equation for estimating the enlarge embankment induced vertical super-impose stress is proposed. The method considers the effect of existing embankment and variations in foundation properties underneath existing and enlarged embankment. The method was applied to analyze the case in Korean. The predicted settlements consistent with the field measurements indicate that the proposed method can be used to design the enlarge embankment on soft clay.

Acknowledgement. This study is supported by the National Key R&D Program of China (No. 2016YFC0800200) and the National Key Basic Research Program of China (973 Program) (No. 2015CB057803).

References

1. Leroueil, S., Magnan, J.P., Tavenas, F.: Embankments on Soft Clays. Ellis Horwood, Chichester (1990)
2. Jin, S.W., Choo, Y.W., Kim, Y.M., Kim, D.S.: Centrifuge modeling of differential settlement and levee stability due to staged construction of enlarge embankment. KSCE J. Civil Eng. **18**(4), 1036–1046 (2014)
3. Zeng, L.L., Hong, Z.S., Han, J.: Experimental investigations on discrepancy in consolidation degrees with deformation and pore pressure variations of natural clays. Appl. Clay Sci. **152**, 38–43 (2018)
4. Lamande, M., Schjonning, P.: Transmission of vertical stress in real soil profile. Part III: Effect of soil water content. Soil Tillage Res. **114**(2), 78–85 (2011)
5. Harr, M.E.: Mechanics of Particulate Media: A Probabilistic Approach. McGrawHill, New York (1977)

Geosynthetic Strain During Filling Stage of Geosynthetic-Reinforced Pile-Supported (GRPS) Embankments

Haiyun Yan[1,2(✉)], Binglong Wang[1,2], and Chengyu Liu[1,2]

[1] College of Transportation Engineering, Tongji University,
Shanghai 201804, China
vanitas0@tongji.edu.cn
[2] Key Laboratory of Road and Traffic Engineering of the Ministry of Education,
Tongji University, Shanghai 201804, China

Abstract. This paper discusses designing of geosynthetic tension strain in Geosynthetic-Reinforced Pile-Supported (GRPS) embankments. Traditionally, filling period of the embankment is not considered among design codes, and the arching model is applied to the final state. The paper attempts to adapt existing designing methods to increasing filling height and static compaction load. Adapted designing approaches include codes distinguishing partial-arching and full arching. The variations of stress concentration ratio and tension in the geosynthetic layer are studied with increasing filling height and different overcharge load. Results show that, for certain cases, the maximum geosynthetic strain is acquired while filling, during partial arching with compaction load. This phenomenon is analyzed and validated with two reported cases of field-test data and full-scale experiment, and relative design suggestions are proposed.

Keywords: Geosynthetic reinforcement · Piled embankments · Soil arching
Geosynthetic reinforcement strain · Design approach

1 Introduction

Soil arching effect is one of the load transfer mechanisms of Geosynthetic-Reinforced Pile-Supported (GRPS) embankments. To calculate the load distribution and geosynthetic strain, different countries have proposed design codes. Existing design codes do take partial arching into consideration by either definition of the full arch height or different shape of the rigid arches. However, these existing models are applied to the final state of the embankment, and the geosynthetic strain is determined using the calculated pressure on it by the tension-membrane effect. Due to different load transfer mechanism for partial arching and full arching, the ignoring of filling period and compaction load may lead to not conservative design of the geosynthetic reinforcement.

The paper attempts to adapt existing designing methods to increasing filling height and static compaction load. The Concentric Arches (CA) model proposed by Van Eekelen [1, 2] has been validated to provide reliable calculation for different embankment height and surcharge loads, and the EBGEO [3] is also adapted. Two different reported field-test data of GRPS embankments are investigated in this paper to illustrate the phenomeno

© Springer Nature Switzerland AG 2018
W. Wu and H.-S. Yu (Eds.): *Proceedings of China-Europe Conference on Geotechnical Engineering*, SSGG, pp. 1750–1753, 2018.
https://doi.org/10.1007/978-3-319-97115-5_185

regarding the change of strain during filling stage for certain cases. Influence of several critical parameters are studied to reveal the trigger of the phenomenon illustrated in this paper. Finally, conclusions are given and designing advice are proposed.

2 Case Study

Some field-test of GRPS embankments shows that for certain cases the maximum GR strain is achieved during partial arching instead of showing up in the final state. Figures 1, 2 respectively drawn by Briançon [4] and Cai [5] depicts how the stresses on soft soil and the tensile strain is changed during filling. Inspired by that, two full-scale cases [6, 7] are then studied. The surcharge load p (Fig. 3) is considered as a load applied above the fill during filling to simulate the compaction load produced by trucks and other construction equipment. The given results validate the phenomenon using CA model observed from the field-test measurements.

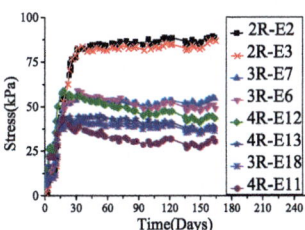

Fig. 1. Stresses on soft soil during filling

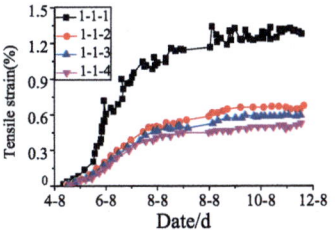

Fig. 2. Tensile strain during filling

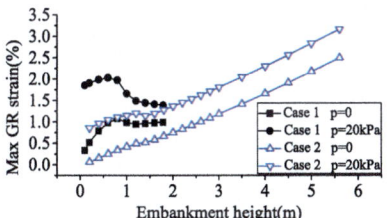

Fig. 3. Maximum GR strain of two cases

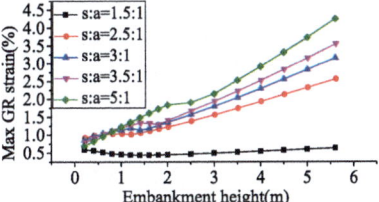

Fig. 4. Influence of s/a.

3 Parametric Study

Since most existing models are applied to the final state of the arch, which may lead to unsafe design in reality, several parameters influencing the change rules of the GR strain are studied. The parametric study is based on Case 2 in the previous case study, each time one single parameter is in variance. with the detail shown in Table 1.

Table 1. Parameters used in the calculation

	Ctc distance of piles $s_x = s_y$	Width square pile caps a	Embankment height H	Unit weight of the fill γ	Friction angle φ	Subgrade reaction K	GR stiffness $J_x = J_y$	Surcharge load P
Unit	m	m	m	kN/m³	deg	kN/m³	kN/m	kPa
Influence of φ	3	1	5.6	18.5	20° ~45°	300	2250	20
Influence of k	3	1	5.6	18.5	30°	0.1 ~ 1000	2250	20
Influence of p	3	1	5.6	18.5	30°	300	2250	0 ~ 50
Influence of J	3	1	5.6	18.5	30°	300	0 ~ 1000	20
Influence of s/a	1.5 ~ 5	1	5.6	18.5	30°	300	2250	20

Figures 4, 5, 6, 7 and 8 illustrate the calculation results. From the figures, it might be interesting to note that: (1) The increase of surcharge load, especially during initial filling, significantly increases the strain in geosynthetic reinforcement. The GR strain at 5 m height without surcharge is smaller than 40 kPa surcharge at 1 m height, thus the influence of compaction load and filling period should be considered in the reinforcement design. (2) For a constant surcharge, P = 20 kPa, the increase of friction angle changes the increasing trend of reinforcement strain, the strain isn't monotonically increasing with filling height for friction angle larger than 25°. As is shown in Table 2, the increase of friction angle reduces the filling height at the maximum reinforcement strain. (3) The influence of subgrade reaction K, and GR stiffness J is remarkable on the GR strain, but not that obvious in the pattern of the increasing of reinforcement strain. (4) The ratio of the distance between the pile s to the size of pile cap has a strong impact on the appearance of maximum GR strain during partial arching, by varying the height of full-arching.

Table 2. Filling height when the GR strain reaches maximum (full arch height is 2.12 m)

φ	20°	25°	30°	35°	40°	45°
H (m)	Full arching	1.3	1.2	1.15	1.05	1

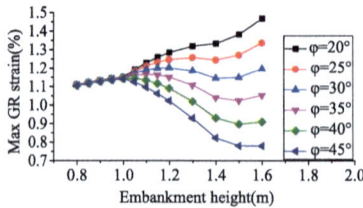

Fig. 5. Influence of friction angle φ

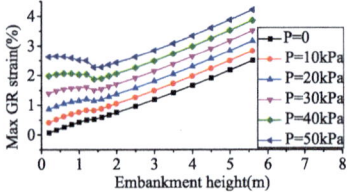

Fig. 6. Influence of surcharge load P

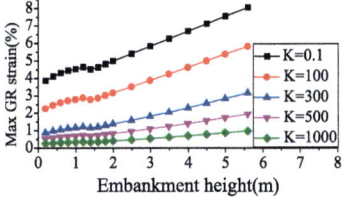

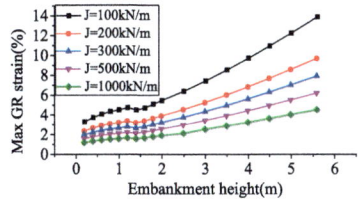

Fig. 7. Influence of subgrade reaction K

Fig. 8. Influence of GR stiffness J

4 Conclusions

Reported field-test data of GRPS embankments show that, in certain cases, the maximum geosynthetic strain may reach its maximum while filling, during partial arching with compaction load. Combined with the calculation results, the conclusions of this paper are listed as follows: (1) The increase of surcharge load, especially during initial filling, significantly increases the strain in geosynthetic reinforcement. The influence of compaction load and filling period should be considered in the reinforcement design. (2) The geosynthetic strain isn't monotonically increasing with filling height, the increase of friction angle reduces the filling height at the maximum reinforcement strain. The influence of subgrade reaction K, and GR stiffness J is not that obvious in the pattern of the increasing of reinforcement strain. (3) Advice is given that for all height of subgrades, both the potential maximum point of strain during filling and the final state should be calculated for safer design.

References

1. Eekelen, S.J.M.V., Bezuijen, A., Tol, A.F.V.: An analytical model for arching in piled embankments. Geotext. Geomembr. **39**(39), 78–102 (2013)
2. Eekelen, S.J.M.V., Bezuijen, A., Lodder, H.J., et al.: Model experiments on piled embankments. Part I. Geotext. Geomembr. **32**(1), 82–94 (2012)
3. EBGEO: Empfehlung für den Entwurf und die Berechnung von Erdkörpermit Bewehrungenals Geokunststoffen. In: Bewehrte Erdkörper auf punkt-oder linienförmigen Traggliedern. Deutsche Gesellschaft für Geotechnik e.V., (German Geotechnical Society), Ernst & Sohn (2010). ISBN 978-3-433-02950-3
4. Briançon, L., Simon, B.: Performance of pile-supported embankment over soft soil: full-scale experiment. J. Geotech. Geoenviron. Eng. **138**(4), 551–561 (2012)
5. Cai, D.G., Ye, Y.S., Zhang, Q.L., et al.: Field test study on the mechanical behaviors of the geosynthetic-reinforced pile-supported embankment and the deformation of the reinforced beddin. China Railway Sci. **30**(5), 1–8 (2009)
6. Eekelen, S.J.M.V., Bezuijen, A., Tol, A.F.V.: Validation of analytical models for the design of basal reinforced piled embankments. Geotext. Geomembr. **43**(1), 56–81 (2015)
7. Fei, K., Liu, H.L.: Field test study and numerical analysis of a geogrid reinforced and pile-supported embankment. Rock Soil Mech. **30**(4), 1004–1012 (2009)

Design Improvement of Sandy Soil Levees in Hydro-engineering Projects, China

Chunbao Yang[✉], Feng Zhu, An Zhang, and Shuli Jiang

General Institute of Water Resources and Hydropower Planning and Design,
The Ministry of Water Resources, Beijing 100120, China
yangcbchina@163.com

Abstract. Soil levees are widely constructed in thousands of rivers in China. While due to the limitation of on-site materials, some projects have to use sandy soil as the filling material. The South-North Water Transport Project, flood control projects of the Yangtze River and Yellow River, water diversion irrigation projects in the northwest of China, and some coastal seawall projects all contain sandy soil levees. Considering the slope stability, many sandy soil levees are designed with slope ratio 1:4 or 1:5, which need a lot of land cover and soil filler. So the section design and construction technology of sandy soil levees are improved. To increase the stability and narrow the section, methods such as the geotextile reinforcement, sand geotube construction, "sandwich" filling and soil modifying methods are common ways. Among them, the geotextile reinforcement and sand geotube construction are quite effective and widely applied. And engineering applications are listed. So, the sandy soil levee can be widely promoted.

Keywords: Sandy soil levee · Section design · Geotextile reinforcement
Geotube construction

1 Introduction

Many cities and more than 40 million hectares of land are located along the rivers in China. In 1998, a catastrophic flood disaster happened in the Yangtze River, and caused great economic losses. So, levee construction is very necessary. The fill material is the key point for the levee. Soil levee is the main form. Sandy soil is not appropriate to be used as the levee fill material. However, due to the limitation of on-site materials, some projects have to use sandy soil as the fill material. Thus the sandy soil levees are applied. The South-North Water Transport Project, flood control projects of the Yangtze River and Yellow River, water diversion irrigation projects in the northwest of China, and some coastal seawall projects all contain sandy soil levees. For example, there are 600 km sandy soil levees in the Songhua River. This paper summarizes the application and development of sandy soil levee construction technology in China's hydro-engineering projects.

© Springer Nature Switzerland AG 2018
W. Wu and H.-S. Yu (Eds.): *Proceedings of China-Europe Conference
on Geotechnical Engineering*, SSGG, pp. 1754–1758, 2018.
https://doi.org/10.1007/978-3-319-97115-5_186

2 The Code for Levee Design

The Code for design of levee project [1] stipulates that the anti-sliding stability safety factor of soil levee slopes should not be less than the value specified in Table 1 in accordance with the Sweden's Arc Method or simplified Bishop Method.

Table 1. Anti-sliding stability safety factor of levee slopes

Scale			1	2	3	4	5
Safety factor	Swedish Arc Method	Normal condition	1.30	1.25	1.20	1.15	1.10
		Non-normal condition I	1.20	1.15	1.10	1.05	1.05
		Non-normal condition II	1.10	1.05	1.05	1.00	1.00
	Simplified Bishop Method	Normal condition	1.50	1.35	1.30	1.25	1.20
		Non-normal condition I	1.30	1.25	1.20	1.15	1.10
		Non-normal condition II	1.20	1.15	1.15	1.10	1.05

Note: Non-normal condition I means construction period, Non-normal condition II means the average water level and an earthquake occurs.

Researches show that the anti-sliding stability safety factor of sandy soil levees can basically meets the requirement when the levee slopes being designed as ratio 1:4 or 1:5, but the margin of safety factor is little. Furthermore, if the sandy soil levees are designed with slope ratio 1:4 or 1:5, that need a lot of land cover and soil filler. Meanwhile, the cost for land acquisition and construction is large. Therefore, to narrow the levee section and reduce the land acquisition, the design and construction technology of sandy soil levee needs to be improved.

3 Design Improving for Sandy Soil Levees

The common design improving methods for sandy soil levees are summarized as the followings.

(1) Geotextile reinforcement method. The geotextiles are layered with intervals along the height of the soil levee, as the same as the reinforcing steel bars in the concrete. And gravel or clay is wrapped at the end of each layer. The detailed design of the geotextile reinforcement method can be found in the Technical code for application of geosynthetics [2].

(2) Sand geotube levee construction method. The sand is rushed by the use of high-pressure water, then the sand and water mixture are pumped into the bags. Due to the elastic restoring force of the bag material and the gravity effect of the sand

itself, and with the permeability of the bag, water can quickly filtered through the bag, so the sand will consolidate in the bag and have some shear strength. Then bags are stacked into the levee core, and a protective layer is built outside to form the levee.

(3) "Sandwich" filling method. The sand layer and the other material layer are stratified, which the layers are interlaced like a sandwich. Each layer has different quality control standard.

(4) Soil modifying method. The physical properties and engineering characteristics of the filling sand can be improved by two ways. The physical way is to mix the sand with other soils which are better for levee construction. The chemical way is to mix the sand with cement.

For the methods, geotextile reinforcement method can enhance the slope stability and also prevent the slope from the erosion of wind, rain and river wave through the wrapped gravel (or clay). The sand geotube levee can be constructed rapidly, so it' widely used in coastal reclamation projects. "Sandwich" filling method is mainly used in the highway subgrade. Soil modifying method is a reliable method. However, the complex construction process, the difficulty of uniform mixing, and the high construction cost make it not easy to be applied.

So the geotextile reinforcement and sand geotube method are the most effective ways for design improving of the sandy soil levees construction.

4   Engineering Applications

Geotextile Reinforcement Method in Sandy Soil Levees. Some sandy soil levees are constructed in the South-North Water Transport Project. For example, there is planned 7.32 km length levee near Shijiazhuang city. The original designed clay levee need to expropriate clay land and have a long distance transport. While the local material is mainly sand. In addition, 280000 m^3 sand had been excavated in the channel and have no place to stack. So the sand levee design is decided. The levee slope ratios are designed to be 1:3 inside the river and 1:2.5 outside.

While the slope stability safety factor is calculated to be 1.34 in accordance with the Simplified Bishop Method [3], failing to meets the code requirement 1.5. So the geotextile reinforcement is applied. The total tension required for reinforcement T_s [2] is calculated as

$$T_s = (F_{sr} - F_{su}) \frac{M_D}{D}$$ (1

where: F_{sr} is the required safety factor, F_{su} is the calculated safety factor, M_D is the bending moment, D-Force arm of T_s. Also, some levees in the Songhua River are constructed by sand. To narrow the levee section and reduce the land cover. Geotextile reinforcement method is applied. So the slope ratio is decreased from 1:4 to 1:3. High strength woven cloth is used for the reinforcement material. The thickness of each soil layer of is 40 cm. The length of the bottom layer is 8 m. The top layer is gradually reduced to 4 m. The length of the wrapped geotextile is 2 m. The safety factor is calculated to be 1.42, which is greater than the required 1.25.

Li and Jie [4] analyzed the reinforcement effect relating to the strength, quantity, distance, and length of the geotextile. And the slope form is also discussed. In addition, the reduction coefficient of the geotextile strength under long operating period is taken into account.

Sand Geotube Levees. There are a lot of fine sand resources in the estuary of the Yangtze River, while the clay material is quite scarce. So the sand geotube levees are widely applied. Due to the non-cohesion and poor anti-erosion ability of sand, it is easy to lose body soil under the hydrodynamic force in front of the levee. While the reliability of geotube levee is to prevent the sand loss of the levee body essentially. With the sand filled into bags, the levee slope ratio can also be design as 1:3 inside the river and 1:2.5 outside. Cai et al. [5] developed a method for the optimal design of the structural section of sand geotube levee. The project cost can be reduced by more than 20%. Peng et al. [6] summarized the field testing results of many practical engineering geotextiles, concluded that the durability of sand bags can be more than 30 years, and proposed some anti-aging measures.

5 Conclusions and Suggestions

The background and situation of the sandy soil levee are summarized. Then the design improving methods and their applications are studied. The following conclusions are got. (1) Sandy soil levee are widely applied in China's hydro-engineering projects. The levees with slope ratio 1:4 or 1:5 need a lot of land cover and soil filler. So improving the design and construction technology is necessary; (2) The common design improving methods for sandy soil levee are geotextile reinforcement, sand geotube construction, "Sandwich" filling, soil modifying and so on. While geotextile reinforcement and geotube construction are the most effective ways; (3) The engineering applications of the sandy soil levees with section design improved are studied. The improved design can enhance the slope stability and narrow the section. Also some major factors are comprehensively considered; (4) Based on the engineering experiences, Sandy soil levees can be widely promoted.

References

1. The Ministry of Water Resources: GB50286-2013. Code for design of levee project. China Planning Press, Beijing (2013)
2. The Ministry of Water Resources: GB/T50290-2014. Technical code for application of geosynthetics. China Planning Press, Beijing (2014)
3. Dong, X.Y.: Design of sand levee in Jingshi section of middle route of south-north water transport project. Haihe Water Resour. **2**, 35–38 (2014)
4. Li, G.X., Jie, Y.X.: Design method on geosynthetics reinforced soil slopes. In: 4th National Symposium on Geosynthetics Reinforced Earth, Wuhan, pp. 61–84 (2013)

5. Cai, X., Yan, W., Zhu, J., Guo, X.W., Jiang, Q.: Optimal design of structural section of geotextile tube dam. J. Hohai Univ. (Nat. Sci.) **43**(1), 1–5 (2015)
6. Peng, L.Q., Li, L.C., Huang, J.He.: Discussion on the technology of geotube levee in Yangtze River estuary. Yangtze River **46**(22), 70–74 (2015)

Deflection Mechanism and Stress Analysis of Large Steel Structure Derrick in Mine

Zhi-shu Yao[1,3(✉)], Ming-kai Liu[2], Xiao-jian Wang[1], Bin Tang[1], and Wei-pei Xue[1]

[1] School of Civil Engineering and Architecture, Anhui University of Science and Technology, Huainan, Anhui, China
yao.zs@163.com
[2] Huaihu Coal Power Co., Ltd., Dingji Coal Mine, Huainan, Anhui, China
[3] 168 Taifeng Street, Huainan, Anhui, China

Abstract. The issue of large steel structure derrick in mine of freeze sinking on deep alluvium suffering from deflection and security enhancement is studied using the auxiliary shaft derrick of Dingji Coal Mine as a model system. First, the study analyses the mechanisms of uneven settling of the foundations in frost-thawed soil, the experimental results show that the bearing capacity and compression modulus of the soil mass decrease after freeze thawing. Since the soil texture and artificial freeze temperature field are uneven, resulting in uneven settlement of the foundation soil of the derrick footing and causing the deflection of the derrick. Secondly, the analysis of finite-element numerical values indicates that in the event of uneven settling, the greatest tensile stress in the derrick structure of Dingji auxiliary shaft increased by 39.83% and the largest pressure stress increased by 33.33%. These stresses have seriously affected the security enhancement of the derrick and ground stabilisation and deviation rectification of the derrick is urgently required.

Keywords: Mine · Large steel structure derrick · Deflection mechanism Uneven settlement · Stress analysis

1 Introduction

Some studies have reported on differential settlement of permanent headframe foundation and reinforcement technology in shaft constructed by surface freezing technique [1–3]. But no systematic study on the differential settlement mechanism of permanent headframe foundations and stress analysis in shafts constructed by the surface freezing technique have been carried out. Particularly, the subject lacks research on the declination mechanism of permanent headframes in ultra-thick soil to depths of 500 m and corresponding control technologies. Due to this lack of research, a systematic study on the declination mechanism of permanent headframes in ultra-thick soil freezing shaft and corresponding control technologies was conducted on the permanent headframe deviation control in the auxiliary shaft in Dingji Coal Mine.

© Springer Nature Switzerland AG 2018
W. Wu and H.-S. Yu (Eds.): *Proceedings of China-Europe Conference on Geotechnical Engineering*, SSGG, pp. 1759–1762, 2018.
https://doi.org/10.1007/978-3-319-97115-5_187

2 Engineering Background

Dingji Coal Mine is a newly constructed large mine. The main shaft, auxiliary shaft, and air shaft were designed in the surface plant. The headframe used an independent reinforced concrete foundation. The strength grade of concrete was designed to C30. The total height of the two principal inclined leg bases (JC1 and JC4) was 7.0 m, including 1.0 m above ground and 6.0 m underground. The size of the foundation base was 9.5 m × 8.5 m. The total height of two auxiliary inclined leg bases (JC3 and JC4) was 5.0 m, including 1.0 m above ground and 4.0 m underground. The size of the foundation base was 6 m × 5 m. The four foundations are numbered as is shown in Fig. 1. The bases of all four foundations rested on the hardcore bed. The bearing capacity eigenvalue of the hardcore bed was 270 kPa and the bearing capacity eigenvalue of the silty clay below the hardcore was 200 kPa.

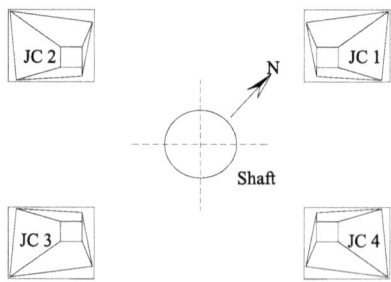

Fig. 1. Numbering of auxiliary shaft headframe foundation

Dingji Coal Mine was completed and entered operation on 1 October 2007. However, differential settling of the headframe foundation was soon recognized [4]. The differential settling of the headframe foundation and the declination of the hoisting sheave, threatening the lifting safety of the auxiliary shaft. The mechanisms behind the headframe foundation settling process is analysed below.

3 Settlement Mechanism Analysis of Headframe Foundation

Coring at the project site was conducted before and after freezing and thawing at different moisture content of the compressive modulus test. As shown by these tests, the bearing capacity and compression modulus of soils decreased after thawing compared to those before freezing. Differential settling of the headframe foundation occurred because of the non-uniformity of the soil texture, which further caused deflection of the headframe and affected the lifting safety.

The measured settlement curve from 3 June 2007 to 21 September 2011 is shown in Fig. 2. The average settlement rate was 1–2 mm/month, without any sign of attenuation. The accumulative settlement volumes of the four foundations were 120.5 mm, 88.8 mm, 93.5 mm and 113.2 mm, showing evident differences. Such differential settlement adversely affected the stress bearing of the headframe, these stresses on the headframe are analysed below.

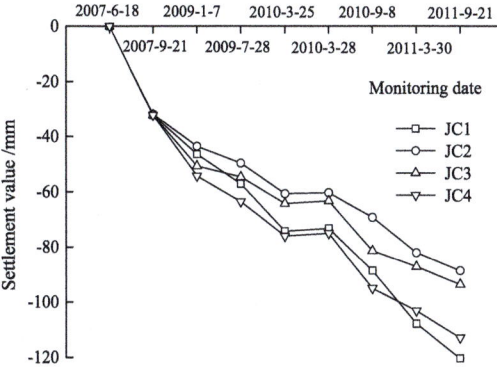

Fig. 2. Settlement curve of auxiliary shaft headframe foundation

4 Stress Analysis on the Headframe Structure and Foundation

The headframe is a large hyperstatic structure. Stresses on headframe were analysed by ANSYS and the results reported in the following text. A calculation model was constructed according to the actual construction map. The numbering of the inclined legs of the headframe was the same as the foundations. Four-hundred 3D BEAM188 beam elements were meshed, and the connections between headframe inclined legs and the foundations were simplified into hinge joints, while other nodes used a rigid connection. Axial forces of inclined legs were calculated (Fig. 3).

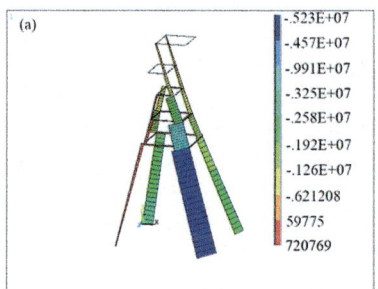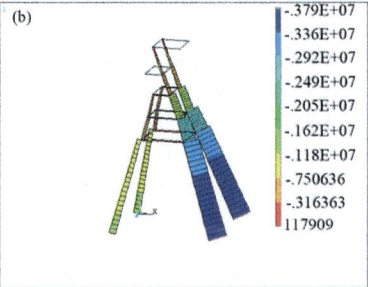

Fig. 3. The oblique leg axis of the derrick: (a) the early period of freezing and thawing, (b) the period of freezing and thawing

In Fig. 3, the differential settlement due to freezing and thawing caused the axial forces of the four inclined legs changed dramatically, specifically, the axial forces of JC1 and JC3 dropped sharply, while the axial forces of JC2 and JC4 rose sharply. The headframe is a hyperstatic rigid structure, and the original four foundations were at the same level and their stresses tended to be uniform. After the differential settling, the inclined legs with relatively high settlement volumes stretched, which was beneficial in releasing the axial compressive force. Opposed to this, the inclined legs with relatively small amounts of settlement were compressed and the axial compressive force increased accordingly. They required key reinforcement in late construction.

At the same time, the relevant principle stress in the headframe structure was calculated. It was found from the analysis of the stress strength that the stress concentration at the intersection of the transverse beam and inclined leg was intensified. The maximum tension stress of the headframe structure was originally designed to be 23.6 MPa, but it increased by 39.83% to 33 MPa after differential settlement. The maximum compressive stress in the structure was designed 39 MPa and it increased by 33.33% to 52 MPa after differential settlement. They will lead to material fatigue and compromising the lifting safety of the headframe structure. Thus, it was proposed that the foundation be reinforced urgently, and the deviation of the headframe rectified.

5   Conclusions

The following conclusions can be drawn.

(1) This experimental study showed that the bearing capacity and compression modulus of soil mass decline after thawing compared to those before freezing. Due to the non-uniformity of soil textures and artificial freezing temperature field, there is a differential settlement of the headframe foundation, which further causes headframe deviation. According to the field investigations of the auxiliary shaft in Dingji Coal Mine, the headframe foundation still subsided at a rate of 1.5–2 mm five years post construction, which led to serious impacts on the lifting safety in the shaft.

(2) According to finite element analysis of the stresses on the headframe structure of auxiliary shaft in Dingji Coal Mine, the maximum tension stress in the headframe structure after differential settlement increased by 39.83% and the maximum pressure stress increased by 33.33% when the loads increased normally. This intensified structural damage and material fatigue strength, thus influencing lifting safety in the shaft. Therefore, it was proposed that there was an urgent need for foundation reinforcement and headframe deviation rectification.

References

1. Liu, B., Wang, W.M., Heng, Z.D.: Analysis on permanent settlement of auxiliary shaft of Liangbaoji coal mine. Mine Constr. Technol. 26(3), 36–38 (2005)
2. Xiao, Q.F., Li, M., Chen, B.J., et al.: Dealling with foundation uneven settlement of froze vertical headframe. Mine Constr. Technol. 30(4), 37–39 (2009)
3. Wang, S.H., You, C.A., Ma, Z.Z., et al.: Analysis on unequal settlement of auxiliary sha tower of Xinli. Geotech. Invest. Surv. 40(2), 33–37 (2012)
4. Wang, S.: Analysis on settlement and deformation of auxiliary shaft headframe foundation i Dingji mine. Coal Technol. 27(5), 15–16 (2008)

Numerical Simulation on Pore Pressure in Electro-Osmosis Combined with Vacuum Preloading

Tianjiao Zhang[1,2], Fanglei Zhan[1,2], Jian Zhou[1,2(✉)], Cunyi Li[1,2], and Xiaonan Gong[1,2]

[1] Coastal and Urban Geotechnical Engineering Research Center, Zhejiang University, Hangzhou 310058, China
zjelim@zju.edu.cn
[2] Zhejiang Urban Underground Space Development Engineering Research Center, Zhejiang University, Hangzhou 310058, China

Abstract. Electro-osmosis combined with vacuum preloading is numerically studied. The influence of vacuum degree attenuation, simultaneously vacuuming at both electrodes and permeability of sand interlayer on pore water distribution were analyzed. The results of comparison show that excess pore water pressure is consistent with the analytical solution. In the center area and vicinity of anode, the absolute value of excess pore pressure is smaller than that of no-attenuation firstly, and then it becomes larger. Vacuuming at both electrodes can greatly shorten the reinforce period by 70%. The impact of permeability coefficient was investigated through a sand interlayer with different permeability. Permeability ratio between the sand interlayer and the clay will affect pore pressure dissipation time, and the higher the ratio, the shorter the consolidation time. The analyses of pore water pressure in electro-osmosis combined with vacuum preloading can help to predict settlement and optimize design.

Keywords: Electro-osmosis · Vacuum preloading · Numerical analysis Pore water pressure

1 Introduction

In order to improve ground treatment effect, researchers tried to combine electro-osmosis with vacuum preloading [1–3]. However, the corresponding consolidation theory remains uncertainty. Xu (2011) deduced the two-dimensional consolidation equation for electro-osmosis combined with vacuum preloading [4], in which the attenuation of vacuum degree was not considered. Wu (2012) set up an axial symmetrical model of electro-osmosis combined with vacuum preloading [5]. Based on their studies, this paper tends to simulate excess pore pressure distribution by COMSOL to consider the effect of vacuum degree attenuation, simultaneously vacuuming at both electrodes and permeability of sand interlayer . The scenario in this paper is more in line with the actual project, therefore the results are more reasonable and useful.

Numerical analysis was conducted through a two-dimensional model. Three different loading systems, i.e. vacuum preloading, electro-osmosis, and electro-osmosis

© Springer Nature Switzerland AG 2018
W. Wu and H.-S. Yu (Eds.): *Proceedings of China-Europe Conference on Geotechnical Engineering*, SSGG, pp. 1763–1766, 2018.
https://doi.org/10.1007/978-3-319-97115-5_188

combined with vacuum preloading, were studied with the same model, in which vacuum preloading or electro-osmosis can be simulated by setting electric potential or vacuum degree to zero. In the analyses the vacuum degree of -80 kPa, the electro-osmotic permeability coefficient k_e of 5×10^{-9} m²/(s · V) and the volumetric compressibility coefficient m_v of 2.5 MPa^{-1} are used. The vertical and horizontal consolidation coefficients, C_v and C_h, are 8×10^{-8} m²/s and 4×10^{-8} m²/s separately. The electrodes are arranged with a spacing L of 1 m, an inserted depth H of 2 m, and the applied effective electric potential U_0' is 8 V. The water density ρ_w is 1.0×10^3 kg/m³. All these parameters are exactly the same as used in Xu (2011) to validate the numerical model. It needs to be addressed that the permeability coefficient in Xu (2011) is set to 2×10^{-9} m/s, both vertically and horizontally. So, permeability anisotropy was not included in his research. To simulate water transfer from anode to cathode, the anode is set as an impermeable boundary, and the cathode, permeable boundary. The results of numerical analysis show that excess pore pressure in electro-osmosis combined with vacuum preloading method gradually decreases from anode to cathode, and it decreases sharply under the combined effect. From day 10th to 12th, dramatic decrease in excess pore pressure occurred, i.e. from -177 kPa to -256 kPa. However small variation observed in the late period, for example, from day 40th to 60th, pore pressure only dropped from -296 kPa to -300 kPa. These results demonstrate that nearly 90% consolidation was accomplished within the first half period, which shows the importance of initial stage of electro-osmosis combined with vacuum preloading. These results are consistent with the analytical solution proposed by Xu (2011).

2   Vacuum Degree Attenuation

Liu (2015) studied the vacuum degree attenuation along the depth with a rate of minimum 2.0 kPa/m to maximum 6.3 kPa/m [6]. In this section, these two extreme rates are both studied. Results reveal that vacuum attenuation has little influence on the pattern of excess pore pressure distribution. In order to study the effect more clearly, the excess pore pressures at three different places between anode and cathode are compared under maximum attenuation 6.3 kPa/m and no-attenuation situation. The absolute value of excess pore pressure is slightly increased with the attenuation of vacuum degree near the cathode, and the increment increased along the depth. While in the center area and vicinity of anode, the absolute value of excess pore pressure is smaller than that of no-attenuation firstly, and then it becomes larger. That is because of the low vacuum degree in deep cathode area, and the electro-osmosis has more effect on excess pore pressure, which makes the absolute value of excess pore pressure in deep area increased. This influence is benefit for deep reinforcement.

3   Simultaneously Vacuuming

In the combined construction field, simultaneously vacuuming at anode and cathode is far more practical than only vacuuming at cathode. So, the vacuuming influence at both anode and cathode are investigated. Both the anode and the cathode are set as free draining boundaries. And the ground surface is regarded as a free draining boundary. The effective electric potential is 20 V. The vacuum degree is −80 kPa. Isotropic permeability is assumed, so the horizontal and vertical permeability coefficients are 2×10^{-9} m/s. If only vacuuming at cathode, total consolidation time is about 50 days, and the maximum absolute value of excess pore pressure is 300 kPa. Vacuuming on both electrodes, it only costs 15 days to fully consolidate, with a 70% reduction in consolidation time, as seen in Fig. 1. In addition, the absolute value of excess pore pressure is about 325 kPa, increased by 8.3% and it gets stable very soon after the sharp decrease in the first ten days. The results show that the consolidation time can be shortened significantly and the improvement effect, particularly the initial result, is much better, which provides a sensible method to save the construction period.

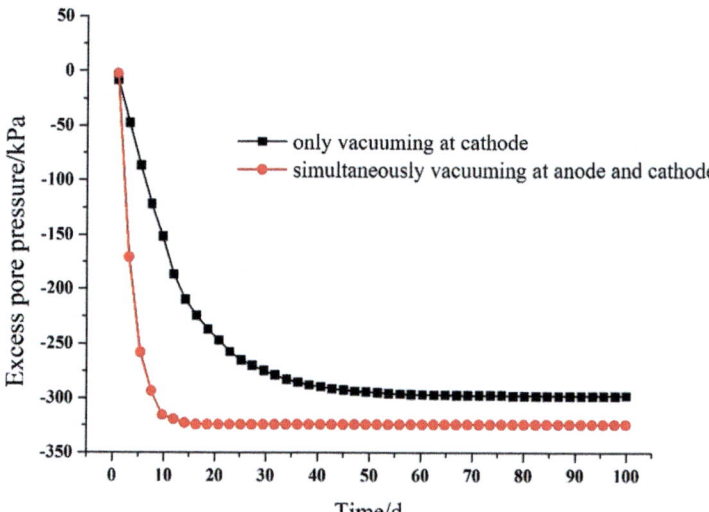

Fig. 1. The variation of excess pore pressure under the conditions of vacuuming at both electrodes simultaneously and only at cathode

4   Permeability of Sand Interlayer

When the soil contains a sand interlayer, the drainage capacity will be greatly improved. This paper studied the influence of permeability coefficient ratio between the sand interlayer and the clay (k_s/k) on excess pore pressure. A 0.2 m sand interlayer is set in the middle of the clay with the distance up to the ground of 2.0 m, and down to the bottom of 0.8 m. The anode is set as a non-draining boundary and the cathode is set as a free draining boundary. The surface of the model is set as a free draining boundary. These boundary conditions and the initial value of the model are the same as used in Xu 2011). The permeability coefficient of clay layer (K) is 2×10^{-9} m/s. And the permeability coefficient of sand interlayer (k_s) is set as 1, 2, 4, 8, 20 and 100 times of the value of clay layer. Besides, the vacuum degree is −80 kPa, and the electric potential between anode and cathode is 10 V. The results show that the absolute value of excess pore pressure nonlinearly decreases with the permeability ratio k_s/k increased. Large permeability coefficient of the sand interlayer (k_s) results in rapid pore water pressure dissipation. For example, pore water pressure in case of $k_s/k = 2$ is 71% of that of $k_s/k = 1$. However, when k_s/k is 20, the pore pressure is less than half (only 45%) of the value of $k_s/k = 1.0$. Large permeability of the sand interlayer is helpful for drainage and consolidation.

The variation of pore water pressure is the key evaluation aspect for improvement. The analyses in this paper can help to predict settlement and optimize design.

References

1. Gao, Z.Y., Zhang, M.Y., Zhang, J.: Laboratory model test of vacuum preloading in combination with electro-osmotic consolidation. China Harbour Eng. **05**, 58–61 (2000)
2. Wang, J., Zhang, L., Liu, F.Y., Fu, H.T.: Experimental study of vacuum preloading combined reinforcement with electro-osmosis in soft clay ground. Chin. J. Rock Mech. Eng. **33**(S2), 4181–4192 (2014)
3. Qiu, C.C., Shen, Y., Li, Y.D., You, Y.F., Rui, X.X.: Laboratory tests on soft clay using electro-osmosis in combination with vacuum preloading. Chin. J. Geotech. Eng. **39**(S1), 251–255 (2017)
4. Xu, W., Liu, S.H., Wang, L.J.: Analytical theory of soft ground consolidation under vacuum preloading combined with electro-osmosis. J. Hohai Univ. (Nat. Sci.) **39**(2), 169–175 (2011)
5. Wu, H., Hu, L.M.: Analytical models of the coupling of vacuum preloading and electro-osmosis consolidation for ground stabilization. J. Tsinghua Univ. (Sci. Technol.) **52**(2), 182–185 (2012)
6. Liu, Z.Z., Deng, D.S., Ding, J.W.: Analysis of vacuum damping law based on extended analytical solution of pore pressure. J. China Three Gorges Univ. (Nat. Sci.) **37**(04), 51–54 (2015)

Author Index

A
Abd, A., 1539
Abood, Ahmed S., 893
Adair, Desmond, 1022
Agrawal, Silky, 1136
Aktas, Tugce, 885
Aleixo, Vanessa, 889
Alonso, Eduardo E., 1117
Al-Suhaily, Ahmed S., 893
An, Peng, 1412
Appel, Silke, 1032
Assadollahi, Hossein, 1251

B
Baraković, Alen, 1561
Bhuiyan, Mohammad Zahidul Islam, 898
Bi, Yuzhang, 1256
Biswas, Sanjit, 915
Bouazza, Abdelmalek, 1316, 1591, 1682
Burak Ekmen, A., 973

C
Cai, Guoqing, 1609
Cai, Yuanqiang, 1651, 1656
Cai, Zhengyin, 1636
Calisan, Oguz, 885
Cao, Jie, 1156
Cao, Weiwei, 1660
Cao, Xue-shan, 902
Cao, Yongyong, 927
Cao, Zhigang, 1651, 1656
Capobianco, Vittoria, 1260, 1312
Carvalho, Cláudia, 906
Cascini, L., 1260
Chan, Kit, 1453

Chandrasekaran, S. S., 1516
Chang, Dan, 1335
Chang, Xiaoli, 1359
Chen, Dun, 1340, 1376, 1394
Chen, Fei, 1128
Chen, Fuqiang, 1064
Chen, Guoliang, 1399
Chen, Jiangang, 1495
Chen, Jingyu, 1656
Chen, Jinjian, 1660
Chen, Longzhu, 1081
Chen, Renren, 1132, 1201
Chen, Rui, 1278
Chen, Tao, 1399
Chen, Xiaoqing, 1495
Chen, Z. L., 1274
Cheng, Ke, 987
Cheng, Wen-Chieh, 910
Choo, Yunwook, 1687
Choudhary, Shiva Shankar, 915
Cichy, Barbara, 1664
Cokca, Erdal, 885
Conil, Nathalie, 1702
Coo, J. L., 1312
Crosta, Giovanni B., 1574
Cudmani, Roberto, 1345
Cui, Feilong, 1160

D
Dai, Ching, 1668
Dai, Feng, 1521
Dai, Guangyu, 1216
Dai, Guo-liang, 1111
Dai, Guoliang, 1192, 1672
Dai, Yuchen, 1707

© Springer Nature Switzerland AG 2018
W. Wu and H.-S. Yu (Eds.): Proceedings of China-Europe Conference
on Geotechnical Engineering, SSGG, pp. 1767–1772, 2018.
https://doi.org/10.1007/978-3-319-97115-5

Dang, Boxiang, 1433
Dang, Faning, 1264
Dang, Jinqian, 1328
Dano, Christophe, 1005
Datta, M., 1017
Dave, Kulin, 1136
Deng, Jianhui, 1128
Deng, Yahong, 1197
Deng, Yongfeng, 1282
di Prisco, Claudio, 1140
Ding, Jianwen, 1746
Ding, Jiulong, 1264
Ding, Xuanming, 987
Dominijanni, Andrea, 1618
Dong, Zibo, 1152
Du, G. Y., 1274
Du, Yanjun, 1256, 1269
Dubsky, Stephen, 1682
Dziadziuszko, Piotr, 919

E
E, Li-su, 902
Egiazarova, Diana, 1458
Eremin, Mikhail, 1462
Eslami, Abolfazl, 1036

F
Faizal, Mohammed, 1591
Fan, Chengwen, 983, 987
Fan, Henghui, 1328
Fan, Xinping, 1483
Fang, Jianhong, 1350
Fareniuk, Gennadiy, 1466
Fareniuk, Iegor, 1466
Fattah, Mohammed Y., 893
Fatty, Abdoulie, 1470
Feng, Decheng, 1385, 1442
Feng, Yan, 1308
Feng, Yasong, 1269
Flessati, Luca, 1140
Foresta, V., 1260
Frankovska, Jana, 1730
Fu, Longlong, 1041, 1181

G
Galkin, A. F., 1144
Gang, Li, 1447
Gao, Wenhua, 956
Gao, Yufeng, 1216
Gaudio, Domenico, 1474
Gondar, João, 923

Gong, Quanmei, 1234
Gong, Weiming, 1192, 1583, 1672
Gong, Xiao-Nan, 1077
Gong, Xiaonan, 1106, 1239, 1244, 1763
Grass, Bertram, 1032
Großwig, Stephan, 1149
Gruchot, A., 1530
Gu, Chuan, 1656
Gu, Desheng, 1504
Gu, Xiaoqiang, 1707
Guan, Yunfei, 927
Guarena, Nicolò, 1618
Guo, Jiaqi, 932
Guo, Xiaoxia, 1644
Guo, Xueluan, 1355
Guo, Yan, 1376

H
Haider, Abbas, 1697
Halder, Prasun, 1711
Han, Congcong, 937
Han, Tongchun, 1479
Hashimoto, Tadashi, 1152
He, Ben, 942
He, Rui, 947
He, Ruixia, 1359
Hong, Yi, 942
Hu, Wei, 956, 1085
Hu, Xiaoying, 1364
Hu, Yifu, 1504
Hu, Zhi, 1160
Huang, Linchong, 952, 1164
Huang, Long, 1372
Huang, Maosong, 1483, 1707
Huang, Shuai, 952
Huang, Wei, 1595
Huang, Xubin, 1368
Hussaini, Syed Khaja Karimullah, 1738

I
Ibrayev, Askar, 1022
Ivanik, Olena, 1487

J
Jain, Abhishek, 1716
Jardine, R. J., 961
Jia, Kai, 1064
Jiang, Chong, 1504
Jiang, Haixi, 1156
Jiang, Shuli, 1754
Jiang, Yan, 1064

Jin, Dehai, 1055
Jin, Huijun, 1359
Jing, Hongyuan, 1380

K
Kaczmarczyk, R., 1526
Kaczmarek, Ł., 1530
Kaliukh, Iurii, 1466
Kalpakcı, Volkan, 973
Kawalec, Jacek, 1677
Khosravi, Mohammad Hossein, 1172
Kim, Jong, 1022, 1512, 1565
King, Louis, 1682
Kong, Yufei, 1726
Kraszewski, Cezary, 1664
Krokoszyński, P., 1526
Ku, Taeseo, 1512
Kumari, Nisha, 978
Kumari, W. G. P., 1600
Kung, Lianfei, 1447
Kurta, I. V., 1144

L
Lacasse, Suzanne, 1551
Lai, Jianying, 987
Lai, Xi-yang, 902
Lai, Yuanming, 1335, 1438
Lavasan, Arash A., 1230
Lee, Deuckhang, 1512
Lee, Jinsun, 1687
Lei, G. H., 1216
Lei, Guoping, 1491
Lei, Lele, 1340, 1376, 1394
Leung, A. K., 1287
Li, A. J., 1470
Li, Cunyi, 1763
Li, Dongqing, 1403
Li, Guoyu, 1380, 1424, 1508
Li, Haili, 1226
Li, Jian, 1609
Li, Jun-Yuan, 1077
Li, Lihua, 1160
Li, Longjin, 1156
Li, Miao-Kun, 1059
Li, Mingguang, 1660
Li, Ning, 1420
Li, Peng, 1692, 1697
Li, Pengfei, 1094
Li, Ping, 983, 987
Li, Shuai, 1495
Li, Tao, 1442
Li, Weihua, 1609
Li, Xiaoyue, 1604
Li, Xu, 1350

Li, Zhiyun, 1064
Liang, Yu, 952, 1164
Lin, Bo, 1385, 1442
Liu, Baoguo, 1168
Liu, Chengyu, 1750
Liu, Chenyinan, 1350
Liu, Fengyin, 1412
Liu, Jian, 1278
Liu, Jianguo, 1041
Liu, Jiankun, 1350
Liu, Jun, 937
Liu, Li, 1282
Liu, Ming-kai, 1759
Liu, Qianwen, 1282
Liu, Qingbing, 1324
Liu, S. Y., 1274
Liu, Shu, 1742
Liu, Sihong, 1308, 1390
Liu, Songyu, 1068
Liu, Wei, 992
Liu, Xiaodong, 1697
Liu, Xin, 1197
Liu, Yujian, 1152
Liu, Z. B., 1274
Liu, Zaobao, 1702
Liu, Zhi, 1206
Liu, Zhongqiang, 1551
Livingston, James, 1668
Long, Chengbi, 956
Long, Jianbing, 1206
Lu, Dechun, 1094
Lu, Hong-qian, 1111
Lu, Hongqian, 1583
Lu, L. L., 1274
Lu, Xilin, 1483
Lu, Yang, 1390
Lu, Yi, 1181, 1721
Luo, Hongyu, 1500
Luo, Jin, 1595
Łydżba, Dariusz, 1613

M
Ma, Bo, 947
Ma, Li, 1504
Ma, Shiguo, 1479
Ma, Wei, 1340, 1376, 1380, 1394, 1424, 1508
Makarov, Pavel, 1462
Manassero, Mario, 1618
Manna, B., 1017
Manna, Bappaditya, 915, 1711
Mao, Yuncheng, 1424, 1508
Mašín, David, 1632
Masini, Luca, 1474
Mehra, Sagar, 997

Mei, Guoxiong, 1001
Mešić, Majda, 1561
Ming, Liang Kee, 898
Moon, Sung-Woo, 1512, 1565
Mou, Cong, 1746
Moussaei, Nader, 1172
Mu, Yanhu, 1340, 1394, 1399, 1424, 1508
Murat Algın, H., 973
Mušec, Peter, 1730
Mussabayeva, Aigul, 1512

N
Najser, Jan, 1632
Nakai, Teruo, 1152
Navarrete, Miguel Benz, 1005
Navarta, Gustavo, 1556
Ng, C. W. W., 1278, 1287, 1312
Ni, J. J., 1287
Ni, James C., 910
Ni, Jin, 1256
Ni, Pengpeng, 1001, 1068
Ni, Wankui, 1399
Niklasch, Christoph, 1640
Niu, Fujun, 1399
Nowamooz, Hossein, 1251

O
Oldecop, Luciano A., 1556
Öztürk, Şevki, 973

P
Pei, Wansheng, 1438
Peng, Erxing, 1364
Peng, Jianbing, 1197
Pinto, Alexandre, 889, 906, 923
Popielski, P., 1530
Proprenter, Michael, 1177
Pu, Hefu, 1304

Q
Qi, Jilin, 1355, 1407, 1412, 1420, 1433
Qi, Yi, 1168
Qian, Jiangu, 1707
Qiu, Jinwei, 1304
Qu, Yonglong, 1399

R
Rafalski, Leszek, 1664
Ralli, Rohit, 1017
Ramana, G. V., 1017
Rampal, Ajay, 1711
Rampello, Sebastiano, 1474

Ranjith, P. G., 1600
Reddy, Krishna R., 1269
Richter, Annette, 1734
Rowe, R. Kerry, 1316
Różański, Adrian, 1613, 1627

S
Sahriar, Kourosh, 1172
Salama, Imane, 1005
Schaller, Maria-Barbara, 1149
Schanz, Tom, 1230
Schneider, Nikolaus, 1009
Sen, Anirban, 1716
Senthilkumar, V., 1516
Shahin, Hossain Md., 1152
Shan, Yao, 1181, 1721
Shang, Xiangyu, 1447
Shao, Jianfu, 1702
Shao, Longtan, 1644
Sharifzedeh, Mostafa, 1172
Sharma, K. G., 1711
Shen, Huawei, 1239, 1244
Shen, Jack Shuilong, 910
Shen, Kanmin, 942
Shen, Weigang, 1521
Sheng, Daichao, 898, 1051
Sheng, Yu, 1364, 1368, 1372
Shi, Peixin, 992
Shi, Xiangyang, 1403
Shi, Xiaomeng, 1168
Shi, Yujin, 1660
Silva, Matias, 1005
Sinkinson, Mike, 1027
Sloan, Scott William, 898
Sloan, Scott, 1051
Sobala, Dariusz, 919
Sołowski, Wojciech T., 1328
Song, Chunyu, 1081
Song, Dongri, 1495
Song, Erxiang, 1055, 1692, 1697, 1726
Song, Zhaoyang, 1609
Soranzo, Enrico, 1177
Stacho, Jakub, 1013, 1730
Stanisz, J., 1526, 1530
Stefaniuk, Damian, 1613
Stelzer, Oliver, 1734
Su, Aijun, 1546
Su, Li, 1181, 1721
Sulovska, Monika, 1013
Sun, Haiquan, 1632
Sun, Jian, 1345
Sun, Si, 1651

Sun, Xiaoyu, 1407
Sun, Zhizhong, 1429
Sweta, Kumari, 1738

T
Tandon, Kavita, 1017
Tang, Bin, 1759
Tang, Jianhui, 983
Tang, Qiang, 992
Tang, Yang, 1535
Tang, Yi, 1636
Tao, Wenyan, 947
Tasnim, R., 1312
Tazabekova, Alima, 1022
Tchkonia, Zurab, 1458
Teparaksa, Jirat, 1027
Teparaksa, Wanchai, 1027
Tian, Yanzhe, 1420
Tincopa, Mayu, 1316
Tomásio, Rui, 889
Tributsch, Alexander, 1032
Trivedi, Ashutosh, 978, 997

U
Uljarević, Mato, 1561
Utili, S., 1539

V
Valikhah, Fatemeh, 1036
Veenhof, Rick, 1320
Veiskarami, Mehdi, 1036
Volkov, Nikolay, 1380
Vu, Ba Thao, 1186

W
Wan, Xing, 1746
Wan, Zhi-hui, 1111
Wan, Zhihui, 1192, 1583
Wang, Binglong, 1206, 1750
Wang, Bo, 1424
Wang, Dayan, 1340, 1376, 1394
Wang, Di, 1098
Wang, Fei, 992, 1380, 1424, 1508
Wang, Haoran, 1483
Wang, Heng, 1746
Wang, Hu, 1216
Wang, Jianhua, 1660
Wang, Jiliang, 1416
Wang, Jinge, 1546
Wang, Jingyu, 1041
Wang, Juan, 1742
Wang, Liping, 1420

Wang, Lizhong, 942
Wang, Qinze, 1412
Wang, Rui, 1132, 1201
Wang, Shanyong, 898, 1051
Wang, Songhe, 1264, 1412
Wang, Tao, 1308
Wang, Xiao, 1046, 1239, 1244
Wang, Xiao-jian, 1759
Wang, Xuetao, 1640
Wang, Y., 1274
Wang, Yan, 1168
Wang, Yansong, 1355, 1407
Wang, Yongtao, 1376
Wang, Zhenhua, 1324
Wei, Hua, 1211
Wei, Xinjiang, 1046
Wejrzanowski, T., 1530
Wen, Tiande, 1644
Wen, Zhi, 1429
Wenhua, Zha, 1090
Wu, Ming, 1197
Wu, Nan, 932
Wu, Wei, 1320, 1491, 1535
Wu, Xueting, 1324
Wu, Yuedong, 1278

X
Xia, Jie, 1234
Xia, Weiyi, 1269
Xiang, Wei, 1324, 1546, 1595
Xiao, Henglin, 1160
Xiao, Junhua, 932
Xiao, Mian, 1201
Xiao, Te, 1453
Xiao, Xiong, 1051
Xie, Kanghe, 1098
Xie, Shouyi, 1702
Xiong, Chengren, 1546
Xiong, Congcong, 1206
Xu, Chuanbao, 1064
Xu, Ding-ye, 1072
Xu, Guangwei, 1041
Xu, Meijuan, 1001
Xu, Ming, 1055
Xu, Minghui, 1579
Xu, Nuwen, 1521, 1574
Xu, Qianwei, 1186
Xu, Ri-Qing, 1077
Xu, Riqing, 1479
Xu, Shuanhai, 1420
Xu, Yongfu, 1604

Xue, Wei-pei, 1759
Xue, Xiu-Li, 1059, 1085

Y
Yan, Haiyun, 1750
Yan, Wei, 1345
Yan, Zhongrui, 1438
Yang, Beibei, 1551
Yang, Chunbao, 1754
Yang, Guanghua, 1064
Yang, Meng, 1094, 1390
Yang, Xiujuan, 1328
Yang, Z. X., 961
Yao, Xiaoliang, 1433
Yao, Zhi-shu, 1759
Ye, Weitao, 1041, 1206
Yi, Yaolin, 1068
Yin, KunLong, 1535
Yin, Kunlong, 1551
Ying, Hong-wei, 1072
Ying, Hongwei, 1098, 1211, 1239, 1244
Yoo, Mintaek, 1687
You, Xingyuan, 1256
Yu, Guangming, 1672
Yu, Haitao, 1156, 1221
Yu, Jian-Lin, 1077
Yu, Qihao, 1438
Yu, Xingfu, 1046
Yuan, Kaixuan, 1256
Yuan, Xiaoyi, 1081
Yuan, Yong, 1221
Yuan, Zhi-Cheng, 1085
Yuan, Zonghao, 1651

Z
Zabala, Francisco, 1556
Zaobao, Liu, 1090
Zekan, Sabid, 1561
Zeng, Chao-Feng, 1059, 1085
Zeng, Chaofeng, 956
Zhamanbay, Akzhunis, 1565
Zhan, Fanglei, 1763
Zhang, An, 1754
Zhang, Chenrong, 1226
Zhang, Chenxi, 1416
Zhang, Fei, 1216
Zhang, Feng, 1385, 1442
Zhang, Ga, 1570
Zhang, JianMin, 1132

Zhang, Jianmin, 1201
Zhang, Jinghua, 1221
Zhang, Jinhong, 1211
Zhang, Limin, 1453, 1500
Zhang, Lisha, 1098
Zhang, Mingju, 1094
Zhang, Mingyi, 1438
Zhang, Rihong, 1106
Zhang, Rongjun, 1304
Zhang, Runlai, 1234
Zhang, Shujuan, 1429
Zhang, Tianjiao, 1763
Zhang, Xiao, 932
Zhang, Yonggan, 1390
Zhang, Ze, 1403
Zhao, Changjun, 1579
Zhao, Chenggang, 1609
Zhao, Chenyang, 1230
Zhao, Jiapeng, 1102
Zhao, Junlin, 1355, 1407
Zhao, Tao, 1521, 1574
Zhao, Wei, 937, 1206
Zhao, Yanlin, 1001
Zhao, Yiying, 1570
Zhao, Yu, 1234, 1282
Zheng, Gang, 1102
Zheng, Junjie, 1304
Zheng, Tianliang, 1692, 1726
Zheng, X. J., 1604
Zhong, Jia-Nan, 1077
Zhou, Gongdan, 1495
Zhou, Haizuo, 1102
Zhou, Jiajin, 1106
Zhou, Jian, 1763
Zhou, Jiawen, 1521
Zhou, Lianying, 1046
Zhou, Zhiwei, 1340, 1394
Zhu, Cheng-wei, 1072, 1098
Zhu, Chengwei, 1211, 1239, 1244
Zhu, Feng, 1754
Zhu, Hehua, 1186
Zhu, Hong, 1453
Zhu, Ming-xing, 1111
Zhu, Mingxing, 1192, 1583
Zhu, Qiyin, 1447
Zhu, Tao, 947
Zhu, Yaohong, 1152
Zhuang, Xiaoying, 1186
Zydroń, T., 1530

Printed by Printforce, the Netherlands